Toxicology Cases for the Clinical and Forensic Laboratory

Toxicology Cases for the Clinical and Forensic Laboratory

Edited by

Hema Ketha
Technical Director, Toxicology, Mass Spectrometry and Metals,
Discipline Director of Clinical Toxicology,
Center for Esoteric Testing, Laboratory Corporation of America Holdings,
Burlington, North Carolina, USA
kethah@labcorp.com

Uttam Garg
Director of Division of Laboratory Medicine,
Director of Clinical Chemistry and Toxicology Laboratories,
Children's Mercy Hospital;
Professor of Pediatrics Pathology,
University of Missouri School of Medicine,
Kansas City, Missouri, USA
ugarg@cmh.edu

Academic Press is an imprint of Elsevier
125 London Wall, London EC2Y 5AS, United Kingdom
525 B Street, Suite 1650, San Diego, CA 92101, United States
50 Hampshire Street, 5th Floor, Cambridge, MA 02139, United States
The Boulevard, Langford Lane, Kidlington, Oxford OX5 1GB, United Kingdom

Copyright © 2020 Elsevier Inc. All rights reserved.

No part of this publication may be reproduced or transmitted in any form or by any means, electronic or mechanical, including photocopying, recording, or any information storage and retrieval system, without permission in writing from the publisher. Details on how to seek permission, further information about the Publisher's permissions policies and our arrangements with organizations such as the Copyright Clearance Center and the Copyright Licensing Agency, can be found at our website: www.elsevier.com/permissions.

This book and the individual contributions contained in it are protected under copyright by the Publisher (other than as may be noted herein).

Notices
Knowledge and best practice in this field are constantly changing. As new research and experience broaden our understanding, changes in research methods, professional practices, or medical treatment may become necessary.

Practitioners and researchers must always rely on their own experience and knowledge in evaluating and using any information, methods, compounds, or experiments described herein. In using such information or methods they should be mindful of their own safety and the safety of others, including parties for whom they have a professional responsibility.

To the fullest extent of the law, neither the Publisher nor the authors, contributors, or editors, assume any liability for any injury and/or damage to persons or property as a matter of products liability, negligence or otherwise, or from any use or operation of any methods, products, instructions, or ideas contained in the material herein.

British Library Cataloguing-in-Publication Data
A catalogue record for this book is available from the British Library

Library of Congress Cataloging-in-Publication Data
A catalog record for this book is available from the Library of Congress

ISBN: 978-0-12-815846-3

For Information on all Academic Press publications
visit our website at https://www.elsevier.com/books-and-journals

Publisher: Stacy Masucci
Acquisitions Editor: Ana Claudia Garcia
Editorial Project Manager: Mona Zahir
Production Project Manager: Maria Bernard
Cover Designer: Christian J. Bilbow

Typeset by MPS Limited, Chennai, India

Contents

List of contributors ... xv
Preface ... xxi

Part I
Basic principles ... 1

1. **An introduction to clinical and forensic toxicology** ... 3

   Hema Ketha and Uttam Garg

   Introduction ... 3
   Scope of clinical and forensic toxicology practice ... 4
   Analytical workflows in a toxicology laboratory ... 4
   Conclusion ... 6
   References ... 6

2. **Pharmacokinetics and pharmacodynamics** ... 7

   Susan M. Abdel-Rahman

   Introduction ... 7
   Pharmacokinetics ... 7
   Toxicokinetics (and nonlinear pharmacokinetics) ... 12
   Pharmacodynamics ... 14
   Toxicodynamics ... 15
   Conclusion ... 17
   References ... 17

3. **Laboratory methods in toxicology** ... 19

   Patrick B. Kyle

   Introduction ... 19
   Methods for general screening ... 19
   Methods for targeted analysis ... 24
   Other considerations ... 24

   References ... 25
   Further reading ... 26

4. **Management of an overdose patient** ... 27

   Stephen Thornton

   Introduction ... 27
   Evaluation of overdosed patients ... 27
   Treatment ... 29
   References ... 31

Part II
Overview and case studies ... 35

5. **Alcohols: volatiles and glycols** ... 37

   Uttam Garg and Hema Ketha

   Introduction ... 37
   Ethanol ... 37
   Specimen collection and the analysis of ethanol ... 39
   Methanol ... 44
   Isopropanol ... 45
   Ethylene glycol ... 45
   Other glycols ... 46
   Conclusion ... 48
   References ... 48

5.1 **Ethylene glycol toxicity** ... 51

   Jaswinder Kaur and Patrick B. Kyle

   Case description ... 51
   Conclusion ... 52
   Acknowledgment ... 53
   References ... 53

5.2 Falsely elevated ethylene glycol results in a patient with diabetic ketoacidosis ... 55

Mushal Noor, Chelsea Milito and Y. Victoria Zhang

- Case description ... 55
- Discussion ... 56
- References ... 57

5.3 Recurrent inhalational methanol toxicity during pregnancy ... 59

Zoë Piggott, Aaron Guinn, Wesley Palatnick and Milton Tenenbein

- Case description ... 59
- Acknowledgment ... 61
- Disclosure statement ... 61
- Funding ... 61
- References ... 61

5.4 A patient with high anion gap metabolic acidosis and increased serum osmolal gap ... 63

Zengliu Su, Roger W. Stone and Yusheng Zhu

- Case description ... 63
- Discussion ... 64
- References ... 65

6. Analgesics and anti-inflammatory drugs ... 67

Christopher McCudden

- Introduction ... 67
- Mechanism of action ... 70
- Pharmacokinetics, toxicokinetics, and treatment ... 71
- Analytical methods ... 72
- Summary ... 73
- References ... 73

6.1 Acetaminophen toxicity ... 75

D. Adam Algren

- Case history ... 75
- Case discussion ... 75
- References ... 77

6.2 Massive ibuprofen ingestion in an adolescent treated with plasma exchange ... 79

Umar Salimi, Marita Thompson, Darcy Weidemann, Judith Sebestyen VanSickle, John D. Nolen, Uttam Garg and Michael R. Christian

- Case description ... 79
- Discussion ... 80
- References ... 82

7. Antibiotics ... 83

Deborah French

- Introduction ... 83
- Mechanism of action ... 86
- Pharmacokinetics, toxicokinetics, and treatment ... 89
- Analytical methods and clinical management implications ... 93
- References ... 95

8. Antipsychotics and antidepressants ... 99

Michael M. Mbughuni, Caren J. Blacker, Jesse Seegmiller, Hemamalini Ketha and Amy B. Karger

- Introduction ... 99
- Antidepressants ... 99
- Antipsychotics ... 100
- Pharmacokinetics and toxicokinetics and treatment ... 102
- Analytical methods for antipsychotics and antidepressants ... 105
- Conclusion ... 107
- References ... 107

8.1 Olanzapine toxicity in an infant ... 109

Theresa Swift and Hemamalini Ketha

- Case description ... 109
- References ... 111

8.2 Genetic polymorphism leading to drug overdose fatality while in custody ... 113

Milad Webb and Jeffrey Jentzen

- Case description ... 113
- References ... 116

8.3 A death involving flubromazepam and methadone — 117

Uttam Garg, Robert Krumsick, C. Clinton Frazee III, Robert Pietak and Diane C. Peterson

Case description	117
Discussion	117
References	119

9. Anticonvulsants — 121

Angela M. Ferguson

Introduction	121
Mechanism of action	122
Pharmacokinetics and toxicokinetics	123
Clinical management	125
Analytical considerations	127
References	128

9.1 Free versus total phenytoin measurements—a case study of phenytoin toxicity — 131

Heather A. Paul, Jason L. Robinson and S.M. Hossein Sadrzadeh

Case description	131
Discussion	131
References	133

9.2 A fatality involving massive overdose of gabapentin — 135

Bheemraj Ramoo, Marius C. Tarau, Mary Dudley, C. Clinton Frazee III and Uttam Garg

Case history	135
Discussion	135
References	137

9.3 An oxcarbazepine overdose in a 23-month-old child — 139

Stephen Thornton and Uttam Garg

Case description	139
Case discussion	139
References	140

10. Antineoplastic drugs — 141

Alejandro R. Molinelli and Kristine R. Crews

Introduction	141
Childhood cancer	141
Therapeutic drug monitoring	141
Cytotoxic agents	142
Pathway-targeted therapies	144
Hormones and hormone regulators	145
Discussion of methotrexate and the thiopurine analogs	145
Mechanisms of action	145
Pharmacokinetics	146
Treatment	147
Analytical methods and clinical management implications	148
References	149

10.1 Case study—methotrexate toxicity, treatment, and measurement — 151

Valkal Bhatt, Michael Scordo and Dean C. Carlow

Case description	151
Discussion	152
References	154
Further reading	156

10.2 Methotrexate toxicity—case study — 157

Alejandro R. Molinelli and Kristine R. Crews

Case description	157
Discussion	157
References	160

10.3 The importance of selecting an appropriate method for measuring methotrexate concentration after glucarpidase rescue: immunoassay or LC—MS/MS? — 161

Fang Wu, Andrew W. Lyon and Martha E. Lyon

Introduction	161
Case report	161
Discussion	163
References	163

10.4 **Different cross-reactivity profiles of methotrexate immunoassays and the clinical management of methotrexate treatment** 165

Yifei Yang and Kiang-Teck J. Yeo

Case description 165
Discussion 166
References 167

11. **Cannabinoids** 169

Hema Ketha and Uttam Garg

Introduction 169
Cannabinoid compounds 170
Pharmacokinetics and toxicokinetics 170
Analytical methods for cannabinoids 172
Conclusion 173
References 174

11.1 **A review of impaired drivers under the influence of 5F-ADB** 177

Amanda Chandler, Elizabeth Wehner, Brianna Peterson, Brian Capron and Fiona Couper

Case description 177
Discussion 178
References 181

11.2 **Can cannabidiol use cause a false-positive tetrahydrocannabinol urine drug screen?** 183

Mushal Noor and Y. Victoria Zhang

Case description 183
Discussion 183
References 184

11.3 **Clobazam intoxication due to cannabidiol consumption** 187

Mikail Kraft, Ara Hall and Uttam Garg

Case description 187
Discussion 187
References 189

11.4 **Phencyclidine and marijuana exposure in utero** 191

John O. Ogunbileje, Paul E. Young and Anthony O. Okorodudu

Case description 191
Discussion 191
References 192

12. **Cardiac drugs** 195

Hari Nair

Introduction 195
Lifestyle and genetic triggers 196
Drug-induced cardiotoxicity 196
Drug–drug interaction 197
Chemotherapy associated myocardial toxicity 197
Cardiotoxicity associated with noncancer therapeutic drugs 199
Cardiotoxicity associated with drugs of abuse 199
Endocrine therapy 201
Therapeutic drug monitoring of cardiac drugs 201
Summary 201
Acknowledgment 201
References 201

12.1 **Use of therapeutic plasma exchange to enhance propafenone elimination following intentional drug overdose** 203

Yong Y. Han, Cindy George, Uttam Garg and Gabor Oroszi

Case description 203
Case discussion 204
References 206

13. **CNS depressants: benzodiazepines and barbiturates** 209

Christine L.H. Snozek

Introduction 209
Mechanism of action 210
Pharmacokinetics and toxicokinetics 211
Treatment 212
Analytical methods and clinical-management implications 213
References 215

13.1 False-negative results in urine benzodiazepine immunoassay screening — 219

Sheng Feng, Paul R. Hess, Ping Wang, Michael C. Malone and Leslie M. Shaw

- Case description — 219
- Discussion and follow-up — 219
- References — 221

13.2 A suicide involving zolpidem — 223

Bheemraj Ramoo, C. Clinton Frazee III, Melissa Beals, Diane C. Peterson and Uttam Garg

- Case history — 223
- Discussion — 223
- References — 224

14. Central nervous system stimulants — 227

Erin Kaleta

- Introduction — 227
- References — 235

14.1 Caffeine: massive accidental caffeine overdose treated with continuous veno-venous hemodiafiltration — 239

Ruben Thanacoody

- Case description — 239
- Discussion and follow-up — 239
- References — 241

14.2 Delayed presentation and conservative management of an intraarterial injection of crushed amphetamine/dextroamphetamine salts — 243

Cornelia McDonald and Justin Arnold

- Case description — 243
- Discussion — 244
- References — 245

14.3 Now you see it, now you don't: ecstasy or not? — 247

Jayson V. Pagaduan, Marianne Benyon and Sridevi Devaraj

- Case description — 247
- Discussion — 247
- Key points — 248
- References — 248

14.4 What is in the cocaine? The vessels never lie — cocaine-induced vasculitis — 251

Alexandra Rapp and Anthony O. Okorodudu

- Case description — 251
- Discussion — 251
- References — 252

14.5 Polysubstance abuse and rhabdomyolysis — 253

Paul E. Young, John O. Ogunbileje and Anthony O. Okorodudu

- Case description — 253
- Patient follow-up — 253
- Discussion — 254
- Rhabdomyolysis — 254
- References — 255

14.6 Unexpected finding of amphetamine and methamphetamine in a patient on monoamine oxidase inhibitor — 257

Geza S. Bodor

- Case description — 257
- Discussion — 257
- Conclusion of our case — 258
- References — 259

14.7 A death due to acute nicotine intoxication — 261

Tiffany Hollenbeck, Marius Tarau, Lindsey Haldiman and Uttam Garg

- Case description — 261
- Discussion — 261
- References — 263

14.8 Cocaethylene—ethanol adding fuel to cocaine's fire — 265
Alexandra Rapp and Anthony Okorodudu

Case description — 265
Discussion — 265
References — 266

14.9 A fatality caused by acute methamphetamine intoxication due to intravaginal absorption — 267
Melissa Beals, Megan Weitzel, Diane C. Peterson, C. Clinton Frazee III and Uttam Garg

Case history — 267
Discussion — 267
References — 268

15. Designer drugs — 269
Gregory Janis

Introduction — 269
Designer opiates — 272
Cathinones — 273
Other designer drugs — 274
Conclusion — 275
References — 276

15.1 A death involving extremely high levels of 3,4-methylene dioxymethamphetamine — 277
Leo Johnson, C. Clinton Frazee, III, Diane C. Peterson and Uttam Garg

Case description — 277
Discussion — 277
References — 279

15.2 Rhabdomyolysis associated with laboratory-confirmed FUB-AMB use — 281
Stephen L. Thornton and Roy Gerona

Case description — 281
Case discussion — 281
References — 282

15.3 Clinical and pathological findings in fatal cases involving the ingestion of methylone — 283
Diane M. Boland

Case descriptions — 283
Discussion — 284
References — 285

15.4 A death involving a "bath salt" methylenedioxypyrovalerone and tramadol — 287
Uttam Garg, Clinton Frazee and Diane Peterson

Case description — 287
Discussion — 287
References — 289

15.5 Clonazolam abuse: a report of two cases — 291
Heath A. Jolliff

Case descriptions — 291
Discussion — 291
References — 293

16. Hallucinogens—psychedelics and dissociative drugs — 295
Mark Petersen, Uttam Garg and Hemamalini Ketha

Introduction — 295
Toxicokinetics and clinical management of overdose — 299
Analysis of psychedelics, dissociatives, and other hallucinogens — 300
Conclusion — 301
References — 301

16.1 Ketamine—a review of published cases — 305
Manoj Tyagi, Amit Bansal and Alina G. Sofronescu

Clinical cases description — 305
Discussion — 306
References — 308

16.2 **Tryptamine trauma: *N,N*-dipropyltryptamine-associated trauma and rhabdomyolysis**   311

Stephen L. Thornton and Roy G. Gerona

Case description	311
Discussion	311
References	312

16.3 **Case reports involving the use of lysergic acid diethylamide**   313

Diane M. Boland

Case descriptions	313
Discussion	314
References	315

17. **Therapeutic drug monitoring of immunosuppressants**   317

Sami Albeiroti, Vincent Buggs, Bjoern Schniedewind, Kimia Sobhani, Uwe Christians and Kathleen A. Kelly

Case	317
Introduction	317
Mammalian target of rapamycin (mTOR) inhibitors	324
Concluding remarks	326
References	326
Further reading	331

18. **Opioids**   333

Jessica A. Hvozdovich, Meagan L. Wisniewski and Bruce A. Goldberger

Introduction	333
Mechanism of action	333
Pharmacokinetics and pharmacodynamics	334
Opioid use disorder and treatment	335
Analytical methods	337
Commonly encountered opioids	339
Conclusion	341
References	341

18.1 **Heroin or not: a case for timing of specimen collection**   343

Chelsea Milito and Y. Victoria Zhang

Case description	343
Discussion	343
References	345

18.2 **A death involving fentanyl-laced pills**   347

Bheemraj Ramoo, Robert B. Pietak, C. Clinton Frazee III and Uttam Garg

Case history	347
Discussion	347
References	349

18.3 **Opioid metabolism: impact of concomitant medications and genetic variations on the interpretation of drug tests**   351

Hila Shaim, Paul E. Young and Anthony O. Okorodudu

Case 1 description	351
Case 2 description	351
Discussion	352
References	353

18.4 **Emergence and sudden disappearance of pink: U-47700 in southeast Michigan**   355

Milad Webb and Jeffrey Jentzen

Case description	355
Discussion	355
References	358

19. **Toxic herbals and plants in the United States**   359

Amitava Dasgupta

Introduction	359
Popularity and safety of herbal supplements	359
Toxic herbal supplements: an overview	360
Herbs toxic to the liver	360
Cardiotoxic herbs	363
Herbs with nephrotoxicity	365
Other toxic herbs	365
Poisonous plants	365
Conclusion	366
References	366
Further reading	368

19.1 Toxicity due to kava tea consumption in conjunction with alcohol and multiple antidepressants — 369

Theresa Swift, Brian Wright and Hema Ketha

Case presentation	369
Discussion	369
References	372

19.2 Kratom, a novel herbal opioid in a patient with benzodiazepine use disorder — 373

Heather M. Stieglitz and Steven W. Cotten

Case presentation	373
Discussion	373
Pharmacology	374
Toxicity	374
Kratom analysis in a toxicology laboratory	375
Legality	375
Conclusion	375
References	376

19.3 Accidental death involving psilocin from ingesting "magic mushroom" — 379

Uttam Garg, Jeff Knoblauch, C. Clinton Frazee III, Adrian Baron and Mary Dudley

Case history	379
Discussion	379
References	382

19.4 A US Army Captain faced discipline after cocaine positive test from coca tea consumption — 383

Cecilia M. Rosales and Uttam Garg

Case description	383
Discussion	383
References	385

20. Toxic gases — 387

Saswati Das

Introduction	387
Carbon monoxide	387
Chlorine	390
Hydrogen cyanide	391
Hydrogen sulfide	393
Phosgene	394
References	395
Further reading	396

20.1 Evaluation of toxicity following ammonia exposure: a case report — 397

Andrew W. Lyon, Viktor A. Zherebitskiy and Fang Wu

Case description	397
Discussion	397
References	399

20.2 Two fatalities involving 1,1-difluoroethane — 401

C. Clinton Frazee, Lindsey J. Haldiman, Diane Peterson and Uttam Garg

Case history 1	401
Case history 2	401
Discussion	402
References	403

20.3 A case of difluoroethane toxicity—sudden sniffing death syndrome — 405

Milad Webb and Jeffrey Jentzen

Case description	405
References	407

20.4 Carbon monoxide poisoning — 409

Julia E. Esswein and D. Adam Algren

Case history	409
Case discussion	409
References	411

21. Toxic metals — 413

Frederick G. Strathmann and Riley Murphy

Introduction	413
Mechanisms of action	413
Pharmacokinetics and toxicokinetics	414
Treatment	415
Analytical methods and clinical management implications	416

Forensic considerations	418	
References	418	

21.1 Toxicity of heavy metals—case study 421

Frederick G. Strathmann

Case description	421
Discussion and follow-up	421
Findings from therapeutic exposures	422
Specific interpretation	423
Further reading	423

21.2 Lithium toxicity—case study 425

Prashant Nasa and Deven Juneja

Case description	425
Discussion and follow-up	425
References	427

21.3 Mercury poisoning from a high seafood diet: a case report 429

Vijayalakshmi Nandakumar, Sarah Delaney and Paul J. Jannetto

Case report	429
Discussion	430
Conclusion	431
References	431

21.4 Trust your gut: a case of persistent gastrointestinal disturbances 433

Sarah R. Delaney, Vijayalakshmi Nandakumar and Paul J. Jannetto

Case report	433
Discussion	433
References	435

22. Venoms 437

Jennifer A. Lowry

Introduction	437
Terrestrial snakes	437
Arthropods	440
Marine species	445
Mammals	446
References	446

23. Case Studies on Other Drugs 449

23.1 Naloxone-responsive respiratory depression in a patient with a negative urine drug screen 451

Nicola J. Rutherford-Parker and Jennifer M. Colby

Case description	451
Discussion and follow-up	451
References	453

23.2 Assessing medication compliance in palliative care: what methodology should be utilized? 455

Stacy E.F. Melanson

Case description	455
Discussion	455
References	457

23.3 A case of suicide involving diphenhydramine 459

Lindsey J. Haldiman, Andrea Ho, Diane C. Peterson, C. Clinton Frazee III and Uttam Garg

Case history	459
Discussion	459
References	463

23.4 Tizanidine intoxication in a postmortem case 465

Ross J. Miller, Brehon Davis, C. Clinton Frazee III, Marius C. Tarau, Mary H. Dudley and Uttam Garg

Case description	465
Discussion	466
References	467

23.5 Methemoglobinemia due to dietary nitrate 469

Devin L. Shrock and Matthew D. Krasowski

Case histories	469
Discussion	469
Case follow-up	471
References	471

23.6 Cyanide toxicity—a case study 473

Kamisha L. Johnson-Davis

Case description	473
Introduction	473
Signs and symptoms of toxicity	476
Laboratory evaluation	477
References	478

23.7 A case of unknown pill ingestion—bupropion toxicity 481

Heather A. Paul and S.M. Hossein Sadrzadeh

Case description	481
Discussion	481
References	483

23.8 Laboratory confirmed massive donepezil ingestion 485

Stephen Thornton and Todd Crane

Case description	485
Discussion	485
References	486
Further reading	486

23.9 Missing oxycodone metabolites confirm suspected drug diversion 489

Geza S. Bodor

Case presentation	489
Discussion	489
Conclusion	491
References	491

23.10 Cocaine, is it really there? Differing sensitivities of immunoassay drug screen and mass spectrometry 493

Hana Vakili, Khushbu Patel and Patricia M Jones

Case description	493
Discussion	493
References	495

23.11 False-positive levorphanol (opioid) due to dextromethorphan on urine drug screen by time-of-flight MS 497

Patricia M Jones and Khushbu Patel

Case description	497
Discussion and follow-up	497
Conclusion	499
Further reading	499

23.12 Comparison of methamphetamine detection in urine and oral fluid 501

Sarah Smiley and Amadeo Pesce

Case description	501
Discussion	501
References	502

Index 505

List of contributors

Susan M. Abdel-Rahman Division of Clinical Pharmacology, Toxicology, and Therapeutic Innovation, Children's Mercy Hospital, Kansas City, MO, United States; Department of Pediatrics, University of Missouri-Kansas City School of Medicine, Kansas City, MO, United States

Sami Albeiroti Sutter Health Shared Laboratory, Livermore, CA, United States

D. Adam Algren Division of Clinical Pharmacology, Toxicology, Therapeutic Innovation, Children's Mercy Hospital, Kansas City, MO, United States; Department of Emergency Medicine, Truman Medical Center, University of Missouri Kansas City, Kansas City, MO, United States; University of Kansas Hospital Poison Control Center, Kansas City, MO, United States

Justin Arnold Department of Internal Medicine, Division of Emergency Medicine, University of South Florida, Tampa, FL, United States

Amit Bansal Department of Medicine, Rochester Regional Health, Rochester, NY, United States

Adrian Baron Office of the Jackson County Medical Examiner, Kansas City, MO, United States

Melissa Beals Department of Pathology and Laboratory Medicine, Children's Mercy Hospital, Kansas City, MO, United States; University of Missouri-Kansas City School of Medicine, Kansas City, MO, United States

Marianne Benyon Department of Pathology and Immunology, Baylor College of Medicine, Houston, TX, United States; Section of Clinical Chemistry, Department of Pathology, Texas Children's Hospital, Houston, TX, United States

Valkal Bhatt Incyte Corporation, Wilmington, DE, United States

Caren J. Blacker Department of Psychiatry and Psychology, Mayo Clinic, Rochester, MN, United States

Geza S. Bodor Department of Pathology, University of Colorado School of Medicine, UC Health Laboratories, Aurora, CO, United States

Diane M. Boland Miami-Dade Medical Examiner Department, Toxicology Laboratory, Miami, FL, United States

Vincent Buggs UCLA Medical Center, Geffen School of Medicine, Los Angeles, CA, United States

Brian Capron Toxicology Laboratory Division, Washington State Patrol, Seattle, WA, United States

Dean C. Carlow Department of Laboratory Medicine, Memorial Sloan Kettering Cancer Center, New York, NY, United States

Amanda Chandler Toxicology Laboratory Division, Washington State Patrol, Seattle, WA, United States

Michael R. Christian Medical Toxicology, Children's Mercy Hospital, Kansas City, MO, United States

Uwe Christians iC42 Clinical Research and Development, Department of Anesthesiology, University of Colorado Anschutz Medical Campus, Aurora, CO, United States

Jennifer M. Colby Department of Pathology, Microbiology and Immunology, Vanderbilt University Medical Center, Nashville, TN, United States

Steven W. Cotten Department of Pathology and Laboratory Medicine, University of North Carolina at Chapel Hill, Chapel Hill, NC, United States

Fiona Couper Toxicology Laboratory Division, Washington State Patrol, Seattle, WA, United States

Todd Crane Department of Emergency Medicine, University of Kansas Health System, Kansas City, KS, United States

Kristine R. Crews Department of Pharmaceutical Sciences, St. Jude Children's Research Hospital, Memphis, TN, United States

Saswati Das Ram Manohar Lohia Hospital, New Delhi, India

Amitava Dasgupta Department of Pathology and Laboratory Medicine, University of Texas McGovern Medical School at Houston, Houston, TX, United States

Brehon Davis Department of Pathology and Laboratory Medicine, Children's Mercy Hospital, Kansas City, MO, United States; University of Missouri School of Medicine, Kansas City, MO, United States

Sarah R. Delaney Department of Laboratory Medicine & Pathology, Mayo Clinic, Rochester, MN, United States

Sridevi Devaraj Department of Pathology and Immunology, Baylor College of Medicine, Houston, TX, United States; Section of Clinical Chemistry, Department of Pathology, Texas Children's Hospital, Houston, TX, United States

Mary H. Dudley (retired) Office of the Jackson County Medical Examiner, Kansas City, MO, United States

Julia E. Esswein University of Missouri-Kansas City School of Medicine, Kansas City, MO, United States

Sheng Feng Department of Pathology and Laboratory Medicine, Perelman School of Medicine at the University of Pennsylvania, Philadelphia, PA, United States

Angela M. Ferguson Department of Pathology and Laboratory Medicine, Children's Mercy Hospital, Kansas City, MO, United States; University of Missouri School of Medicine, Kansas City, MO, United States

C. Clinton Frazee III Department of Pathology and Laboratory Medicine, Children's Mercy Hospitals and Clinics, Kansas City, MO, United States; University of Missouri School of Medicine, Kansas City, MO, United States

Deborah French Department of Laboratory Medicine, University of California San Francisco, San Francisco, CA, United States

Uttam Garg Department of Pathology and Laboratory Medicine, Children's Mercy Hospital, Kansas City, MO, United States; University of Missouri School of Medicine, Kansas City, MO, United States

Cindy George Apheresis Program, Children's Mercy Hospital, Kansas City, MO, United States

Roy G. Gerona Department of Laboratory Medicine, University of California, San Francisco, San Francisco, CA, United States

Bruce A. Goldberger Department of Pathology, Immunology and Laboratory Medicine, University of Florida College of Medicine, Gainesville, FL, United States

Aaron Guinn Department of Emergency Medicine, University of Manitoba, Winnipeg, MB, Canada

Lindsey J. Haldiman Office of the Jackson County Medical Examiner, Kansas City, MO, United States

Ara Hall Epilepsy Section, Division of Neurology, Children's Mercy Hospital, Kansas City, MO, United States

Yong Y. Han Department of Pediatrics, Division of Critical Care Medicine, Children's Mercy Hospital, Kansas City, MO, United States

Paul R. Hess Department of Pathology and Laboratory Medicine, Perelman School of Medicine at the University of Pennsylvania, Philadelphia, PA, United States

Andrea Ho Department of Pathology and Laboratory Medicine, Truman Medical Center, Kansas City, MO, United States

Tiffany Hollenbeck Office of the Jackson County Medical Examiner, Kansas City, MO, United States

Jessica A. Hvozdovich Department of Pathology, Immunology and Laboratory Medicine, University of Florida College of Medicine, Gainesville, FL, United States

Gregory Janis MedTox Laboratories, Laboratory Corporation of America Holdings, St. Paul, MN, United States

Paul J. Jannetto Department of Laboratory Medicine & Pathology, Mayo Clinic, Rochester, MN, United States

Jeffrey Jentzen Wayne County Medical Examiner's Office, Detroit, MI, United States; Department of Pathology, University of Michigan Health System, Ann Arbor, MI, United States; Washtenaw County Medical Examiner's Office, Ann Arbor, MI, United States

Leo Johnson Department of Pathology and Laboratory Medicine, Children's Mercy Hospital, Kansas City, MO, United States; University of Missouri School of Medicine, Kansas City, MO, United States

Kamisha L. Johnson-Davis Department of Pathology, University of Utah, Salt Lake City, UT, United States; Clinical Toxicology, ARUP Laboratories, Salt Lake City, UT, United States

Heath A. Jolliff Department of Emergency Medicine, Doctors Hospital, OhioHealth, Columbus, OH, United States

Patricia M. Jones Department of Pathology, University of Texas Southwestern Medical Center, Dallas, TX, United States; Children's Medical Center, Dallas, TX, United States

Deven Juneja Institute of Critical Care Medicine, Max Superspeciality Hospital, Saket, India

Erin Kaleta Department of Laboratory Medicine and Pathology, Mayo Clinic, Scottsdale, AZ, United States

Amy B. Karger Department of Laboratory Medicine and Pathology, University of Minnesota, Minneapolis, MN, United States

Jaswinder Kaur Department of Pathology, University of Mississippi Medical Center, Jackson, MS, United States

Kathleen A. Kelly UCLA Medical Center, Geffen School of Medicine, Los Angeles, CA, United States; Pathology and Laboratory Medicine, Clinical Chemistry and Toxicology Geffen School of Medicine, University of California at Los Angeles, Los Angeles, CA, United States

Hema Ketha Toxicology and Mass Spectrometry, Center for Esoteric Testing, Laboratory Corporation of America Holdings, Burlington, NC, United States

Jeff Knoblauch Department of Pathology and Laboratory Medicine, Children's Mercy Hospital, Kansas City, MO, United States; University of Missouri School of Medicine, Kansas City, MO, United States

Mikail Kraft Pediatric Resident, Children's Mercy Hospital, Kansas City, MO, United States

Matthew D. Krasowski Department of Pathology, University of Iowa Hospitals and Clinics, Iowa City, IA, United States

Robert Krumsick Department of Pathology and Laboratory Medicine, Children's Mercy Hospitals and Clinics, Kansas City, MO, United States; University of Missouri School of Medicine, Kansas City, MO, United States

Patrick B. Kyle Department of Pathology, University of Mississippi Medical Center, Jackson, MS, United States

Jennifer A. Lowry Department of Pediatrics, Children's Mercy Hospital, Kansas City, MO, United States; University of Missouri School of Medicine, Kansas City, MO, United States

Andrew W. Lyon Division of Clinical Biochemistry, Department of Pathology & Laboratory Medicine, Saskatchewan Health Authority, Saskatoon, SK, Canada

Martha E. Lyon Division of Clinical Biochemistry, Department of Pathology & Laboratory Medicine, Saskatchewan Health Authority, Saskatoon, SK, Canada

Michael C. Malone Department of Pathology and Laboratory Medicine, Perelman School of Medicine at the University of Pennsylvania, Philadelphia, PA, United States

Michael M. Mbughuni Department of Laboratory Medicine and Pathology, Minneapolis VA Healthcare System, Minneapolis, MN, United States; Department of Laboratory Medicine and Pathology, University of Minnesota, Minneapolis, MN, United States

Christopher McCudden Division of Biochemistry, The Ottawa Hospital, Ottawa, ON, Canada; Department of Pathology and Laboratory Medicine, University of Ottawa, Ottawa, ON, Canada

Cornelia McDonald Department of Emergency Medicine, University of Alabama at Birmingham, Birmingham, AL, United States

Stacy E.F. Melanson Department of Pathology, Division of Clinical Laboratories, Brigham and Women's Hospital, Harvard Medical School, Boston, MA, United States

Chelsea Milito Department of Pathology and Lab Medicine, University of Rochester Medical Center, Rochester, NY, United States

Ross J. Miller Office of the Chief Medical Examiner, Tulsa, OK, United States

Alejandro R. Molinelli Department of Pharmaceutical Sciences, St. Jude Children's Research Hospital, Memphis, TN, United States

Riley Murphy NMS Labs, Horsham, PA, United States

Hari Nair Boston Heart Diagnostics, Framingham, MA, United States

Vijayalakshmi Nandakumar Department of Laboratory Medicine & Pathology, Mayo Clinic, Rochester, MN, United States

Prashant Nasa Critical Care Medicine, NMC Specialty Hospital, Dubai, UAE

John D. Nolen Laboratory Medicine, Children's Mercy Hospital, Kansas City, MO, United States

Mushal Noor Department of Pathology and Lab Medicine, University of Rochester Medical Center, Rochester, NY, United States

John O. Ogunbileje Department of Pathology, University of Texas Medical Branch, Galveston, TX, United States

Anthony O. Okorodudu Department of Pathology, University of Texas Medical Branch, Galveston, TX, United States

Gabor Oroszi Apheresis Program, Children's Mercy Hospital, Kansas City, MO, United States; Department of Pathology and Laboratory Medicine, Children's Mercy Hospital, Kansas City, MO, United States

Jayson V. Pagaduan Department of Pathology and Immunology, Baylor College of Medicine, Houston, TX, United States; Section of Clinical Chemistry, Department of Pathology, Texas Children's Hospital, Houston, TX, United States

Wesley Palatnick Department of Emergency Medicine, University of Manitoba, Winnipeg, MB, Canada

Khushbu Patel Department of Pathology, University of Texas Southwestern Medical Center, Dallas, TX, United States; Children's Medical Center, Dallas, TX, United States

Heather A. Paul Alberta Precision Laboratories, Calgary, AB, Canada; Department of Pathology and Laboratory Medicine, Cumming School of Medicine, University of Calgary, Calgary, AB, Canada

Amadeo Pesce Precision Diagnostics, LLC, San Diego, CA, United States

Mark Petersen Department of Toxicology, Clinical Mass Spectrometry and Metals Testing, Center for Esoteric Testing, Laboratory Corporation of America Holdings, Burlington, NC, United States

Brianna Peterson Toxicology Laboratory Division, Washington State Patrol, Seattle, WA, United States

Diane C. Peterson Office of the Jackson County Medical Examiner, Kansas City, MO, United States; Johnson County Department of Health and Environment/Medical Examiner's Office, Olathe, KS, United States

Robert B. Pietak Office of the Jackson County Medical Examiner, Kansas City, MO, United States

Zoë Piggott Department of Emergency Medicine, University of Manitoba, Winnipeg, MB, Canada

Bheemraj Ramoo Department of Pathology and Laboratory Medicine, Children's Mercy Hospital, Kansas City, MO, United States; University of Missouri-Kansas City School of Medicine, Kansas City, MO, United States

Alexandra Rapp Department of Pathology, University of Texas Medical Branch, Galveston, TX, United States

Jason L. Robinson Alberta Precision Laboratories, Calgary, AB, Canada; Department of Pathology and Laboratory Medicine, Cumming School of Medicine, University of Calgary, Calgary, AB, Canada

Cecilia M. Rosales Souteastern Pathology Associates, Brunswick, GA, United States

Nicola J. Rutherford-Parker Department of Pathology, Microbiology and Immunology, Vanderbilt University Medical Center, Nashville, TN, United States

S.M. Hossein Sadrzadeh Alberta Precision Laboratories, Calgary, AB, Canada; Department of Pathology and Laboratory Medicine, Cumming School of Medicine, University of Calgary, Calgary, AB, Canada

Umar Salimi Pediatric Critical Care Medicine, Children's Mercy Hospital, Kansas City, MO, United States

Bjoern Schniedewind iC42 Clinical Research and Development, Department of Anesthesiology, University of Colorado Anschutz Medical Campus, Aurora, CO, United States

Michael Scordo Adult Bone Marrow Transplant Service, Department of Medicine, Memorial Sloan Kettering Cancer Center, New York, NY, United States; Department of Medicine, Weill Cornell Medical College, New York, NY, United States

Jesse Seegmiller Department of Laboratory Medicine and Pathology, University of Minnesota, Minneapolis, MN, United States

Hila Shaim Department of Pathology, University of Texas Medical Branch, Galveston, TX, United States

Leslie M. Shaw Department of Pathology and Laboratory Medicine, Perelman School of Medicine at the University of Pennsylvania, Philadelphia, PA, United States

Devin L. Shrock Department of Pathology, University of Iowa Hospitals and Clinics, Iowa City, IA, United States

Sarah Smiley Precision Diagnostics, LLC, San Diego, CA, United States

Christine L.H. Snozek Department of Laboratory Medicine and Pathology, Mayo Clinic in Arizona, Scottsdale, AZ, United States

Kimia Sobhani Cedars-Sinai Medical Center, Los Angeles, CA, United States

Alina G. Sofronescu Department of Pathology and Microbiology, University of Nebraska Medical Center, Omaha, NE, United States

Heather M. Stieglitz Department of Pathology, The Ohio State University, Columbus, OH, United States

Roger W. Stone Department of Pathology and Laboratory Medicine, Medical University of South Carolina, Charleston, SC, United States

Frederick G. Strathmann NMS Labs, Horsham, PA, United States

Zengliu Su Department of Pathology and Laboratory Medicine, Medical University of South Carolina, Charleston, SC, United States

Theresa Swift Department of Pathology, University of Michigan Health System, Ann Arbor, MI, United States

Marius C. Tarau Office of the Jackson County Medical Examiner, Kansas City, MO, United States

Milton Tenenbein Department of Pediatrics and Child Health, University of Manitoba, Winnipeg, MB, Canada

Ruben Thanacoody UK National Poisons Information Service, Newcastle Hospitals NHS Foundation Trust, Newcastle upon Tyne, United Kingdom; Translational and Clinical Research Institute, Newcastle University, Newcastle upon Tyne, United Kingdom

Marita Thompson Pediatric Critical Care Medicine, Children's Mercy Hospital, Kansas City, MO, United States

Stephen L. Thornton Department of Emergency Medicine, University of Kansas Health System, Kansas City, KS, United States; University of Kansas Health System Poison Control Center, Kansas City, KS, United States; Division of Toxicology, Children's Mercy Hospital, Kansas City, MO, United States

Manoj Tyagi Align Laboratories, Clinical Chemistry Consultant, Charlotte, NC, United States

Hana Vakili Department of Pathology, University of Texas Southwestern Medical Center, Dallas, TX, United States

Judith Sebestyen VanSickle Nephrology, Children's Mercy Hospital, Kansas City, MO, United States

Ping Wang Department of Pathology and Laboratory Medicine, Perelman School of Medicine at the University of Pennsylvania, Philadelphia, PA, United States

Milad Webb Wayne County Medical Examiner's Office, Detroit, MI, United States; Department of Pathology, University of Michigan Health System, Ann Arbor, MI, United States

Elizabeth Wehner Toxicology Laboratory Division, Washington State Patrol, Seattle, WA, United States

Darcy Weidemann Nephrology, Children's Mercy Hospital, Kansas City, MO, United States

Megan Weitzel Department of Pathology and Laboratory Medicine, Children's Mercy Hospital, Kansas City, MO, United States; University of Missouri School of Medicine, Kansas City, MO, United States

Meagan L. Wisniewski Department of Pathology, Immunology and Laboratory Medicine, University of Florida College of Medicine, Gainesville, FL, United States

Brian Wright Department of Pathology, University of Michigan Health System, Ann Arbor, MI, United States

Fang Wu Division of Clinical Biochemistry, Department of Pathology & Laboratory Medicine, Saskatchewan Health Authority, Saskatoon, SK, Canada

Yifei Yang Pritzker School of Medicine, Section of Clinical Chemistry, Department of Pathology, The University of Chicago, Chicago, IL, United States

Kiang-Teck J. Yeo Pritzker School of Medicine, Section of Clinical Chemistry, Department of Pathology, The University of Chicago, Chicago, IL, United States

Paul E. Young Department of Pathology, University of Texas Medical Branch, Galveston, TX, United States

Y. Victoria Zhang Department of Pathology and Lab Medicine, University of Rochester Medical Center, Rochester, NY, United States

Viktor A. Zherebitskiy Department of Pathology & Laboratory Medicine, Saskatchewan Health Authority, Saskatoon, SK, Canada

Yusheng Zhu Department of Pathology and Laboratory Medicine, Medical University of South Carolina, Charleston, SC, United States; Department of Pathology & Laboratory Medicine, Pharmacology, Penn State University Hershey Medical Center, Hershey, PA, United States

Preface

Clinical and forensic toxicology laboratories are involved in detection, identification, and measurement of drugs and toxins in various biological and nonbiological specimens. Clinical laboratories mostly analyze samples for patient diagnosis, treatment, and counseling. Forensic laboratories are mostly involved in analyzing drugs and toxins to determine their contribution in impairment or cause of death. Despite these differences, both laboratories use similar approaches and analytical techniques in the determination of xenobiotic toxicity. In fact, many laboratories are involved in both clinical toxicology and forensic toxicology. This book aims to serve as a practical guide to approach, teach, and learn basic and advanced concepts in clinical and forensic toxicology using a case study as an example for a variety of drug classes and clinical scenarios encountered by the clinical and forensic toxicologists. We also hope the book will serve as a teaching guide to clinical laboratory scientists, medical students, and pathology residents.

The chapters are presented in a consolidated format that make concepts in toxicology understandable to approach analytical, clinical, and forensic toxicology concepts. To provide comprehensive understanding approach, the chapters are organized in two sections. The first section provides an overview of toxicokinetic, clinical, forensic, and analytical aspects followed by cases encountered in clinical and forensic settings. This case-based text is designed to answer questions like: how does a typical drug overdose exposure present in a clinical setting? We hope that our vision of combining general information with case reports provides comprehensive approach in understanding analytical, clinical, and forensic toxicology for practicing toxicologists.

We want to thank our authors for their valuable contributions. We would also like to thank our publisher, Elsevier, and our Editorial Project Managers Sandra Harron, Anna Dunbow, Mona Zahir and Maria Bernard for their support and encouragement. And finally, we want to thank our families as their love and support made this challenging journey very meaningful and enjoyable.

Part I

Basic principles

Chapter 1

An introduction to clinical and forensic toxicology

Hema Ketha[1] and Uttam Garg[2,3]

[1]Toxicology, Mass Spectrometry and Metals, Laboratory Corporation of America Holdings, Burlington, NC, United States, [2]Division of Laboratory Medicine, Pathology and Laboratory Medicine, Children's Mercy Hospital, Kansas City, MO, United States, [3]Pathology and Laboratory Medicine, University of Missouri School of Medicine, Kansas City, MO, United States

Introduction

Poisonous or toxicological manifestations in humans have been recognized since antiquity. Naturally occurring plant extracts, animal venoms, and refined mixtures of minerals have been used as toxins and medicines. The venoms of snakes have been used to produce lethal weapons since the ancient historical times—a practice that continues today in some tribal cultures [1]. Arsenic was nicknamed "the inheritance powder" as it was commonly used to poison family members for a fortune in the Renaissance era. Paracelsus, a Swiss physician and alchemist, established the fundamental concept of toxicology "*sola dosis facit venenum*—the dose makes the poison" in the 16th century. During the same era the practice of rational, fact-based crime scene investigation methods led to the emergence of various forensic sciences and the practice of determining the cause and manner of death. Overlap of fundamentals of chemistry, physiology, medicine, and the law began to play an important role in the society. Pioneering work by the 16th and 17th century physicians and chemists who worked at the interface of science and the law laid the foundation of principles of forensic toxicology. In the narrative, it was highlighted that the use of analytical toxicology for forensic purposes using Marsh test for establishing arsenic poisoning first came into existence for finding the cause of death of Charles LaFarge, a wealthy business owner in France who was murdered by his wife with arsenic poisoning in 1840. The forensic toxicologist demonstrated the presence of arsenic using the Marsh test in the tissues of LaFarge's body and in the food material used by his wife as the cornerstone evidence [2–6].

An appreciation of fundamental principles of chemistry, pharmacology, laboratory sciences, physiology, and medicine is indispensable for a rewarding clinical toxicology practice. Clinical toxicology as a discipline is centered on the management of accidental or intentional toxic exposures in humans. The most common settings where clinical toxicologists practice are poison control centers and hospitals in which active management of overdosed patients is necessary. Other areas of practice include occupational and environmental safety. In a hospital setting, facilitating prompt and appropriate management of an overdosed patient is one of the most valuable roles a clinical toxicologist plays. Often, successful management of a poisoned patient is dependent on collaborative efforts from multidisciplinary teams including emergency physicians or medical and clinical toxicologists, pathologist, and the clinical laboratory professionals. With an increasing emphasis and a growing need for prescription drug monitoring, the scope of clinical toxicology practice has grown to encompass compliance monitoring and even overlaps with pharmacogenomics. Another aspect of toxicology relevant to clinical practice is therapeutic drug monitoring.

The term "forensic" finds its root in the Latin word *forensis* or forum referring to an open communal arbitration of crimes in open court in the Roman era. In contemporary context, forensic science encompasses the intersection of several technical and scientific fields with the legal practice including postmortem pathology, ballistics, anthropology, DNA analysis, bloodstain–pattern analysis, engineering, and digital sciences. Forensic toxicology pertains to the practice of toxicology in the medicolegal realm including investigation of causes of death, driving under the influence of toxicants, workplace testing, occupational exposures, child protection services, drug-facilitated crime, doping control, bioterrorism, and environmental exposures [7].

Scope of clinical and forensic toxicology practice

The practice of toxicology can be classified into the preclinical, clinical, and forensic settings. Preclinical or nonclinical toxicology pertains to studying the toxicological manifestations of early drug development candidates. Preclinical toxicology studies involve pharmacological assessment including in vivo preclinical toxicology studies. These studies are used to assess the onset, severity, and duration of toxic effects. These studies also explore and determine dose−response relationships and the extent of reversibility of the pharmacological effects of the drug.

Clinical toxicology is the study of physiological and/or adverse effects of pharmacotherapeutic agents, illicit drugs, and other drugs. It also involves an active role in supporting the medical teams for facilitating appropriate treatment and management of patients who have been exposed to toxic levels of a drug or a chemical. A clinical toxicologist plays an important role in multiple areas of clinical medicine. By providing consultative services, clinical toxicologists facilitate the best treatment plan for patients present at the emergency department for the treatment of an acute toxic exposure of accidental or intentional nature. In many cases the causative agent is unknown. A clinical toxicologist often interacts with the clinical team to assess the best analytical methods to identify the causative agents. Other areas include prescription drug monitoring and therapeutic drug monitoring. In cases of prescription drug monitoring where polypharmacy is involved, a broad understanding of the metabolic pathways is needed. Clinical toxicologists commonly get involved in providing technical consultative services in complex prescription drug monitoring cases.

For investigating unusual deaths from a medicolegal standpoint, there are particular systems in place in the United States. The unusual deaths include suspicious, sudden and unexplained, violent, nonnatural, and those that may be considered a threat to the public health. There are two types of medicolegal systems in place—a coroner system and a medical examiner (ME) system. Coroner and ME systems might be structured at the country, regional, or state level. The organizational structure of the medicolegal system depends on the area of jurisdiction. A unique role that a forensic toxicologist plays is that of an expert witness in cases that require judicial enquiry. Forensic toxicologists provide analytical basis for the results and provide a technical feedback in their role as an expert witness in the judicial process [8,9].

Forensic toxicological analysis typically consists of an initial analysis and, in presumptive positive cases, confirmatory analysis with definitive methods such as mass spectrometry (MS). The samples analyzed depend on the question to be answered. If being under the influence is to be ascertained, a blood sample is obligatory. A positive urine laboratory result reveals the use or exposure to a foreign substance. Commonly used alternative samples include hair and oral fluid. Hair provides a large window for detecting drug use, whereas oral fluid is noninvasive and easy to collect without any concern from hesitation from a witnessed collection. Other alternate samples are sweat, meconium, umbilical cord, and nail. The results from forensic toxicological investigations can be collated and stored in for later research, can also be used to explore and study changing trends in illicit drug abuse, and to improve drug public health safety in general. In forensic toxicological investigations the results can be used in a legal context. Therefore judicial processes that govern sampling, sample handling, and the related civil rights must be strictly followed. These rules also apply to the chain of custody regarding the handling of samples.

Analytical workflows in a toxicology laboratory

The analytical methods used in clinical and forensic toxicology laboratories are essentially the same. Forensic toxicology workflows have to follow an established chain of custody process to support any legal proceedings related to the sample in question. Analytical methods used in toxicology laboratories are evolving rapidly. In the 1980s untargeted screening was performed on thin-layer chromatography plates. The assay required approximately 35−40 minutes of analytical time per sample. It was laborious, subject to interferences and did not separate commonly used prescription drugs causing false-positive and -negative results.

In contemporary toxicology workflows, screening in the majority of laboratories is performed by immunoassays. Common formats for immunoassays include Cloned Enzyme Donor Immunoassay, CEDIA; Enzyme-Multiplied Immunoassay Technique, EMIT; Fluorescent Polarization ImmunoAssay, FPIA; Kinetic Interaction of Microparticles in Solution, KIMS; and Immunochromatographic Techniques. Point of care dip-stick immunoassays are also very popular methods for in-service urine toxicology screening. Gas chromatography (GC)−MS was and is still used for qualitative and quantitative confirmation of screening results. Liquid chromatography−tandem MS (LC−MS/MS) is becoming a mainstay in large and small toxicology laboratories as the analytical tool of choice for quantitative confirmation. High-pressure liquid chromatography (HPLC) is also used for quantitative confirmation, although most laboratories are transitioning the HPLC assays used for drug analysis to LC−MS/MS platforms. In some laboratories, even screening is being performed by LC−MS/MS assays designed to detect a few hundred compounds. In addition, LC coupled with high-resolution MS (LC−HRMS) has allowed laboratories to demonstrate the feasibility of nontargeted MS−based screening. LC−HRMS workflows are still very uncommon and require advanced operators to interpret the data. The rapid turnaround time and ease of operation offered by immunoassays for toxicology screening are unparalleled. Therefore for the near future, we can envision that majority of toxicology laboratories will use a combination of qualitative or semiquantitative immunoassays and qualitative GC−MS and/or LC−MS/MS workflows for screening and confirmation. A common workflow used in toxicology laboratories for comprehensive screening and confirmation of multiple drug classes is shown in Fig. 1.1.

The choice of which immunoassay screens a toxicology laboratory utilizes should be guided by the practice needs [10]. For example, a laboratory serving a pain management clinic should consider employing screening methods for a variety of opioids and benzodiazepines. However, a laboratory serving an opioid cessation program needs to implement assays for methadone and buprenorphine. Drug classes or metabolites included in a toxicology screen commonly have drug classes, including opioids, benzoylecognine (cocaine metabolite), amphetamine, 6-monoacetylmorphine (heroin metabolite), benzodiazepines, barbiturates, phencyclidine, and 11-Nor-9-carboxy-Δ^9-tetrahydrocannabinol (THC-COOH, cannabinoid metabolite). Each of these screening assays has a cross-reactivity profile that may include several drug metabolites. For example, most opioid assays have a defined cross reactivity toward morphine, but most opioid immunoassays can also detect other semisynthetic opioids, including codeine, hydrocodone, and hydromorphone. Of note, most opioid screens lack any cross reactivity toward fentanyl and can detect oxycodone only at very high concentration. A common perception about immunoassay screens is that these are more prone to false positives than false negatives. However, this general concept should not be used when developing toxicology-screening workflows. While several common drugs are known to cause a false-positive amphetamine immunoassay screen, benzodiazepine screens are prone to false negatives. Similarly, most general opioid screens cannot detect commonly abused opioids such as fentanyl, buprenorphine, methadone, and oxycodone. Therefore laboratories may need to implement separate screens for the compounds relevant to their practice when designing comprehensive toxicology workflows. Prevalence of a drug's use in the general population also dictates the need to implement a certain assay in a toxicology laboratory.

A key difference between the two types of toxicology laboratories is the matrix types. While urine, serum, and blood are predominantly used in the clinical settings, the forensic laboratory witnesses a wide variety of specimen types for toxicological analysis, including tissue, hair, nails, oral fluid, vitreous fluid, and central or peripheral blood. In addition, forensic toxicology laboratory may receive samples that are in a wide spectrum of biological decay. This poses an additional challenge to the forensic toxicology laboratory to develop methods in which the matrix effects from a variety of specimen types have been evaluated. Forensic toxicology laboratory may also be involved in the identification of unknown substances. Method Validation Guidelines (CLSI C62A) for LC−MS/MS assays have been published and are now being used in clinical and forensic laboratory settings [11,12].

Compliance with pharmacotherapy in some patients can be challenging to assess without supporting analytical information. Clinical toxicology and forensic laboratories offer tests for a variety of drug classes. The drug classes that are most monitored are the ones with high abuse or diversion potential. These drug classes include opioids, benzodiazepines, barbiturates, and amphetamines. Illicit drugs including heroin (its metabolite 6-monoacetylmorphine is commonly monitored for assessing

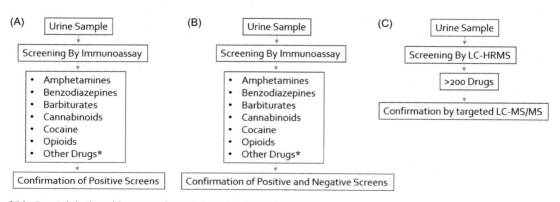

Other Drugs include: Phencyclidine, propoxyphene, LSD, fentanyl, methadone, oxycodone etc.

FIGURE 1.1 A schematic for urine drug screening workflows. (A) Urine sample is subject to immunoassay screening and only positive screens are confirmed by confirmation methods. (B) Urine sample is subject to immunoassay screening and both negative and positive screens are confirmed by confirmation methods. (C) Urine sample is subject to LC−HRMS-based screen that can screen up to 200 drugs; positives are confirmed by targeted LC−MS/MS methods. *LC−HRMS*, Liquid chromatography coupled with high-resolution mass spectrometry; *LC−MS/MS*, liquid chromatography−tandem mass spectrometry.

heroin use), cocaine, and cannabinoids are also monitored as a part of prescription drug monitoring programs. Evaluating and interpreting a comprehensive multidrug prescription drug monitoring report can be a daunting task. Equally daunting can be the task for the physician to pick the appropriate test for the drug of choice. Compliance testing aims to assess the presence of the expected drugs but also the absence of unexpected drugs [13].

Most drug panels offer a screening test for the common prescription and illicit drugs. The testing cascade starts with screening for broad drug classes, and then the confirmation tests for the drug classes that screen positive are performed (Fig. 1.1). However, some commonly abused drugs can be missed by this approach. For example, in case of a drug testing cascade driven by immunoassay positivity, certain benzodiazepines can be missed. In these cases, ordering tests for specific drugs in question can be of value. However, in these cases, any unexpected drugs such as illicit drugs may be missed. The general scheme for the analysis of a urine sample for prescription drug monitoring is as follows. First, the urine sample is subject to a screening test. In most contemporary clinical and forensic toxicology laboratories, urine drug screening is performed by immunoassays. The common drug classes analyzed are amphetamines, opioids, cocaine, benzodiazepines, barbiturates, and cannabinoids. Some laboratories also include screening test for methadone, oxycodone, phencyclidine, LSD, and other drugs that are not detected by common immunoassay screens. Positive results are confirmed by mass-spectrometric methods. This scheme for screening and confirmation, particularly of prescription drugs, is prone to many issues. Drugs that are not detected by the screening methods can be missed by drug testing schemes that rely on initial screening for running confirmatory testing. On the other hand, drug-testing panels that analyze all classes of drugs by confirmatory methods irrespective of the screening test results or sometimes termed as comprehensive drug panels can be cost prohibitive.

Conclusion

The practice of toxicology has been evolving for last many decades. Routine practice of toxicology involves a multilayered approach. While clinical toxicology professionals interact with medical and clinical laboratory professionals who most commonly practice in a hospital-based setting, forensic toxicology professionals interact with MEs, coroner's law enforcement officers, and the judicial system. The analytical aspects of both clinical and forensic toxicology practice are convergent and involve similar workflow approaches. Screening and confirmatory methods used in toxicology laboratories have evolved tremendously in the past decade. Mass-spectrometric methods are a mainstay in analytical toxicology workflows. This book aims to be a tool for a toxicologist that provides a cohesive approach to the practice of clinical and forensic toxicology by combining the use of clinical, forensic, and analytical knowledge for interpretation of data to facilitate the best management of the patient.

References

[1] Utkin YN. Animal venom studies: current benefits and future developments. World J Biol Chem 2015;6(2):28–33. Available from: https://doi.org/10.4331/wjbc.v6.i2.28.
[2] Smith M. Poisons and poisoners: arsenic. Hospital (Lond 1886) 1891;10(236):4.
[3] Kapaj S, Peterson H, Liber K, Bhattacharya P. Human health effects from chronic arsenic poisoning—a review. J Environ Sci Health A: Tox Hazard Subst Environ Eng 2006;41(10):2399–428. Available from: https://doi.org/10.1080/10934520600873571.
[4] Mead MN. Arsenic: in search of an antidote to a global poison. Environ Health Perspect 2005;113(6):A378–86. Available from: https://doi.org/10.1289/ehp.113-a378.
[5] Moffat AC, Osselton DM, Widdop BWJ. Clarke's analysis of drugs and poisons. 4th ed. pharmaceutical press; 2011.
[6] Baselt RC. Disposition of toxic drugs and chemicals in man. 11th ed. Seal Beach, CA: Biomedical Publications; 2017.
[7] Santucci KA, Hsiao AL. Advances in clinical forensic medicine. Curr Opin Pediatr 2003;15(3):304–8. Available from: https://doi.org/10.1097/00008480-200306000-00014.
[8] Pichini S, Busardo FP. Analytical advances in clinical and forensic toxicology. Second part. Curr Pharm Biotechnol 2018;19(2):89–90. Available from: https://doi.org/10.2174/138920101902180628100648.
[9] Busardo FP, Pichini S. Editorial: Analytical advances in clinical and forensic toxicology. Curr Pharm Biotechnol 2017;18(10):784–5. Available from: https://doi.org/10.2174/138920101810180124150401.
[10] Reisfield GM, Goldberger BA, Bertholf RL. Choosing the right laboratory: a review of clinical and forensic toxicology services for urine drug testing in pain management. J Opioid Manage 2015;11(1):37–44. Available from: https://doi.org/10.5055/jom.2015.0250.
[11] Peters FT, Wissenbach DK, Busardo FP, Marchei E, Pichini S. Method development in forensic toxicology. Curr Pharm Des 2017;23(36):5455–67. Available from: https://doi.org/10.2174/1381612823666170622113331.
[12] Stickle DGU. Validation, quality control, and compliance practice for mass spectrometry assays in the clinical laboratory. In: Clarke W, Nair, editors. Mass spectrometry for the clinical laboratory. 1st ed. Academic Press; 2017.
[13] Drummer OH. Good practices in forensic toxicology. Curr Pharm Des 2017;23(36):5437–41. Available from: https://doi.org/10.2174/1381612823666170704123836.

Chapter 2

Pharmacokinetics and pharmacodynamics

Susan M. Abdel-Rahman[1,2]

[1]Division of Clinical Pharmacology, Toxicology, and Therapeutic Innovation, Children's Mercy Hospital, Kansas City, MO, United States,
[2]Department of Pediatrics, University of Missouri-Kansas City School of Medicine, Kansas City, MO, United States

Introduction

The adage "the dose makes the poison," adapted from the writings of Renaissance physician Paracelsus (Text Box 2.1), highlights a recognition of the relationship between dose and response as early as the 1500s [1]. However, it was advances in analytical chemistry, enabling the detection and measurement of drugs in biological fluids, which allowed pharmacokinetics (PK) and pharmacodynamics (PD) to evolve as an independent field of study. Regrettably, the strings of differential and integral equations employed by practitioners of the field give many the impression that PK/PD is a rather esoteric discipline which could not be further from the truth.

Individuals associated with clinical or medical fields are uniquely suited to understanding PK/PD because it is simply a mathematical representation of the physiologic processes that determine how drugs get into the body (absorption), where they go (distribution), how they are modified (metabolism), and eventually removed (excretion). A related set of mathematical equations is used to describe how the drugs interact with a biological system ultimately eliciting a response (whether desired or unintended). Thus a working knowledge of anatomy, physiology, cellular- and molecular biology enhances one's understanding of PK/PD in much the same way that knowledge of patient-specific factors such as age, genetics, diet, and environmental exposures influences our understanding of disease risk. Importantly, the utility of this knowledge is bidirectional such that PK/PD studies can broaden our understanding of biology when those data are unavailable or incomplete.

This chapter is designed to introduce the reader to fundamental pharmacokinetic and pharmacodynamic concepts and perhaps, in the process, reduce the perception that PK/PD is an obscure discipline.

Pharmacokinetics

PK encompasses the processes involved with the absorption, distribution, metabolism, and elimination of drugs, toxicants, and other xenobiotics. Essential to an understanding of PK is a review of some basic terms that are referenced repeatedly when discussing the movement of drugs through the body:

Half-life ($t_{1/2}$): As implied by the name, the $t_{1/2}$ of a drug is the duration of time required for the concentration of a drug in the body to decrease by one-half its value (Fig. 2.1). This PK parameter allows us to determine the fraction of drug that has been removed from the body at any given time after administration and, by extension, the fraction of drug that remains in the body at that time. Since the $t_{1/2}$ reflects the time it takes concentrations to drop by half, 50% of the drug will remain after one half-life has elapsed, 25% will remain after two half-lives, 12.5% after three half-lives, and so on such that ~97% of the drug will have left the body after five half-lives. At therapeutic doses, where the elimination for many drugs follows an exponential or first-order process, the half-life is determined from the slope of the log-transformed plasma concentration versus time curve (a.k.a. the elimination rate constant, kel) according to Eqs. (2.1.0) and (2.2.0).

Half-life also informs how long it will take to achieve steady state and the extent of drug accumulation that can be expected when steady state is attained. A drug administered with a consistent dosing frequency or interval (τ) will accumulate in the body until equilibrium is reached and the amount administered in a given time period is equal to the amount eliminated during that same period of time. At this point, with no perturbations to the system, the drug has

> **TEXT BOX 2.1**
>
> *What is there that is not poison?*
>
> *All things are poison and nothing is without poison.*
>
> *Solely the dose determines that a thing is not a poison.*
>
> Paracelsus (1493–1541).

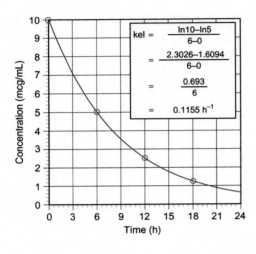

FIGURE 2.1 Plasma concentration versus time curves depicting the concept of half-life that reflects the time required for the total amount of drug in the body to decrease by one-half. The drug pictured here has a $t_{1/2}$ of 6 h. Note the initial concentration is 10 mcg/mL and the amount of time required to drop to 5 mcg/mL is 6 h. As depicted by the red circles, the concentration drops by an additional 50% of what remains for every successive half-life. The slope of the natural log-adjusted concentration versus time curve provides the elimination rate constant which can also be used to calculate $t_{1/2}$ (see Eq. 2.1.0).

achieved steady state. As depicted in Fig. 2.2, $t_{1/2}$ and τ together determine the extent of accumulation (i.e., how high the drug concentrations will get, Eq. 2.3.0) and how long it will take after the last dose for the drug to clear the system. With a rearrangement of these equations, $t_{1/2}$ and desired concentrations can be used to drive decisions about dosing frequency (Text Box 2.2).

It is important to recognize that perturbations to the system (i.e., changes in the rate into or out of the body) can occur for any number of reasons; thus assumptions that a patient is at steady state and that their PK still follow first-order kinetics may not always be valid. The rate into the body, for example, can be disrupted in the face of under- or overutilization (e.g., poor adherence or overdose). It can also change if the drug is given at uneven intervals throughout the day or, for drugs susceptible to food effects, if the size and composition of coadministered foods vary from dose to dose. Similarly, the rate out of the body can be expected to change with physiologic changes that contribute to alterations in drug clearance (Cl), including changes in cardiac output, renal function, and liver function. The impact of these and other variables on changing $t_{1/2}$ will become clearer in subsequent sections on volume of distribution (V_d) and Cl.

We would be remiss not to mention that more than one $t_{1/2}$ can exist for any given drug. The PK profile depicted in Fig. 2.1 suggests that the drug is rapidly and homogenously distributed throughout the body (also referred to as fitting a one-compartment model). This rapid distribution could well be taking place after administration of the drug; however, this conclusion could also be drawn erroneously if an insufficient number of plasma samples have been collected to accurately describe the movement of this drug in the body. When the drug enters different body compartments at measurably different rates (the key here is "measurably"), we refer to the drugs as following a multicompartment model and this is evidenced by a log-transformed plasma concentration versus time curve that takes on a polyexponential decay profile (Fig. 2.3). Every rate constant (k) that we can describe mathematically is associated with its own $t_{1/2}$ such that we might be able to characterize an absorption $t_{1/2}$, a distribution $t_{1/2}$, an elimination $t_{1/2}$, etc. Importantly, the mathematical half-lives describing the movement of drug in the body may not necessarily reflect the $t_{1/2}$ associated with the biological effect (touched on below under *PD*).

Apparent V_d: As with any volume encountered in our everyday lives, V_d is intended to represent a quantitative measure of size or space. In theory, V_d reflects the combined volume of the various compartments and tissues into which a drug enters, or with which a drug associates, after it enters the body. Physicochemical factors that influence the extent of distribution include the drug's octanol/water partition coefficient, ionization constant, and affinity for plasma/tissue

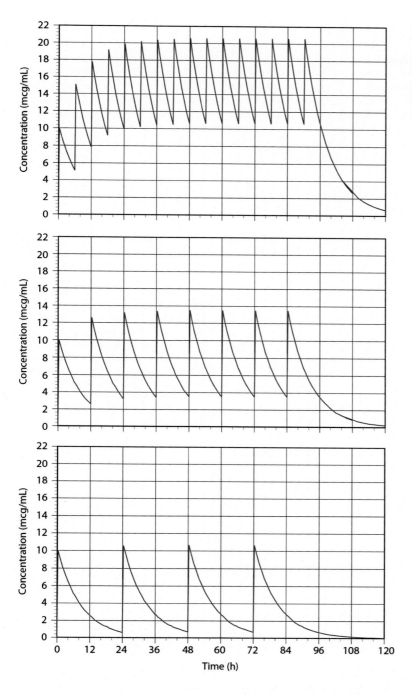

FIGURE 2.2 Plasma concentration versus time curves depicting the steady-state accumulation of a drug possessing a 6 h $t_{1/2}$ dosed every 6 h (upper), 12 h (center), and 24 h (lower). Note that steady-state drug levels are achieved after ∼5 half-lives irrespective of the number of doses they received. Thus the time to achieve steady state is dependent on the $t_{1/2}$ of the drug and independent of the dosing frequency.

proteins. In practice, V_d represents a mathematical proportionality constant that relates the amount of drug administered (i.e., the dose) to the resulting plasma concentration (Eq. 2.4.0). In essence, it reflects the size of a compartment necessary to account for the total amount of drug that was administered at the measured concentration in the blood.

For some drugs (e.g., polar, highly ionized) the V_d is relatively small and approximates a "true" biological space (e.g., the intravascular compartment, extracellular fluid stores, or total body water space). For other drugs that associate widely with proteins and other cellular components, the calculated V_d can be quite large. A V_d that exceeds the physical volume of the body may seem nonsensical in a biological context; however, we simply need to recognize that V_d estimates are based on the measurement of drug concentrations in the blood. When tissue concentrations increase and corresponding plasma concentrations decrease, the denominator in Eq. (2.4.1) becomes smaller and the value of V_d increases.

TEXT BOX 2.2

Adenosine, a class V antiarrhythmic used to treat supraventricular tachycardia, has a $t_{1/2}$ of less than 10 s requiring dosing to be repeated every few minutes until the desired effect is achieved and the tachycardia has been terminated. This ultrashort half-life makes the drug very easy to titrate which significantly limits the likelihood of protracted toxicity but also precludes any real utility for use in long-term or chronic management of patients. By comparison, propranolol, a nonselective beta blocker, also used to treat supraventricular arrhythmias, has a $t_{1/2}$ of 3–6 h. Sustaining the desired effect on heart rate requires administration of the immediate release formulation two to four times per day. In contrast, the class III antiarrhythmic amiodarone has a $t_{1/2}$ that spans months. Though conveniently administered once-daily, amiodarone can be expected to accumulate extensively in the body after repeated administration and pose a much greater challenge for toxicological management in the event the patient experiences untoward side effects.

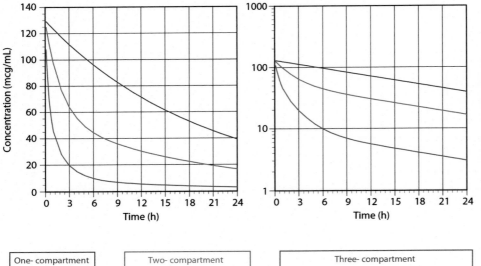

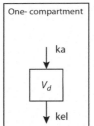
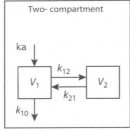
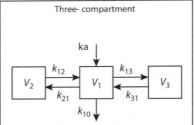

FIGURE 2.3 Plasma concentration versus time curves illustrating the profile observed for a drug following a one-compartment (blue), two-compartment (red), or three-compartment (green) model. When plotted on a semilog graph, the one-compartment model appears linear and the multicompartment models clearly demonstrate more than one phase. The graphics in the lower region of this figure illustrate the rate constants and volumes whose values inform the shape of the pharmacokinetic profiles.

In a pharmacokinetic context the inherent value of V_d lies in its ability to help us estimate the drug dose to recommend for our patient (Text Box 2.3). If we know the volume of the compartment into which we are delivering the drug and we know the concentration that is associated with the response we are hoping to achieve, then we can calculate the desired dose (Eq. 2.4.2). In a toxicokinetic context (i.e., an overdose), these same equations can allow us to estimate a dose that might have been ingested to explain the concentrations which we observe clinically. Of course this relies on the validity of assumptions that drive the other elements of the equation.

When drugs fit a multicompartment model, we can describe several distribution volumes in the same way that we can describe several half-lives. Every compartment described in the models will be associated with a different V_d (Fig. 2.3). Consequently, we can encounter terms such as V_c and V_p which reflect the volumes of the central and peripheral compartments, respectively. We might also see V_1, V_2, V_3, etc. which represent the volumes of compartment 1, 2, and 3, respectively. Just as was discussed with V_d, these various other volume terms allow us to relate the amount of drug administered to the concentrations observed (or anticipated) in various tissue compartments.

TEXT BOX 2.3

Gentamicin is an aminoglycoside antibiotic used for the treatment of infections caused by Gram-negative bacteria. Owing to its hydrophilic nature, this antibiotic distributes into a V_d approximating the extracellular fluid space. The highest extracellular fluid volumes are experienced immediately after birth, dropping steadily through infancy and childhood. As a result, when individuals of different age groups are administered the same weight-based dose of gentamicin, infants experience the lowest peak concentrations, while children experience peak concentrations that are nearly 33% greater and adults experience concentrations almost 50% greater. These findings are reflected in the prescribing recommendations for gentamicin which are 2.5 mg/kg for the newborn, 2 mg/kg for the child, and 1 mg/kg for the adult. Since these youngest patients also demonstrate longer half-lives owing to reduced renal clearance, one could ostensibly deliver the same dose of gentamicin to neonates and manipulate the dosing interval so that these patients eventually accumulate concentrations comparable to those observed in adults. However, the delay in attaining therapeutic peak concentrations can have deleterious clinical consequences for this drug whose optimal activity depends on attaining high peak plasma concentrations early in therapy to promote rapid bacterial killing and limit the risk of adaptive resistance [2–5].

TEXT BOX 2.4

A biologic analog of drug clearance (Cl) familiar to most clinically oriented individuals is creatinine Cl which estimates the rate of glomerular filtration, the first step in process of making urine. The glomerular filtration rate in a normal healthy adult is 120 mL/min/1.73 m². This suggests that ∼120 mL of blood is cleared of creatinine per minute for every 1.73 m² of body surface area. Extending this example to drugs, an infant receiving acetaminophen, with a Cl value of ∼0.29 L/h/kg, completely removes all acetaminophen molecules from ∼0.29 L of blood every hour for every kilogram of body weight.

Cl: Instinctively, we are inclined to think about drug removal from the body as an amount of drug eliminated over a unit of time. However, in a pharmacokinetic context, Cl actually reflects the fraction of the total V_d from which the drug is completely removed in any given unit of time (e.g., L/h) (Text Box 2.4). Mathematically, Cl reflects the proportionality constant that relates the V_d to the rate of elimination (Eq. 2.5.0). Biologically, Cl reflects the sum total of all physiologic processes, typically metabolism and elimination, that are working together to remove the drug from the body. Apart from the innate ability of the kidney and liver to remove the drug, there are other biological factors that influence the access of a drug to the organs of elimination, including the extent of plasma protein binding, tissue extraction ratios, and cardiac output. With respect to its utility in the practice of medicine, Cl helps to determine the frequency with which a drug needs to be delivered to a patient in order to maintain the concentrations that are associated with the response we aim to achieve. Given that both Cl and V_d are mathematically related to kel, a change in either of these parameters will influence the estimated $t_{1/2}$ (Eq. 2.5.1).

Bioavailability (F): For drugs administered by any route other than direct intravenous instillation, properties of both the drug and the patient influence the extent to which the drug finds its way into the systemic circulation. These factors can include the rate and extent to which the dosage forms disintegrates and dissolves, the stability of the drug under conditions present at the site of absorption, and the degree to which the drug serves as a substrate for drug metabolizing enzymes (DMEs) or transporters that may be encountered before the drug reaches the systemic circulation (i.e., first pass effect). The fraction of the administered dose that makes it into the blood as "intact" drug reflects its bioavailability. For intravenously administered drugs, 100% of the administered dose is placed into the systemic circulation thus $F = 1$. For drugs given by any other route (e.g., sublingual, oral, rectal, and percutaneous), F will be less than or equal to 1 depending on the drug factors alluded to above along with factors unique to the patient (Text Box 2.5).

Understanding the fluid nature of bioavailability is important because it allows us to recognize that acute and chronic comorbidities (e.g., diarrheal disease, Zollinger–Ellison syndrome), dietary changes, concurrent foods and medicines, etc. can each alter the dose–exposure relationship. Because of the variable nature of this parameter, F represents an "unknown" modifier that influences the estimates of V_d and Cl. Recall that our calculation of V_d relies on the concentrations we observe in the blood. If concentrations in the blood are low after the administration of an oral drug, we do not know whether it is because the drug has rapidly moved out of the circulation and into the tissues or because the

> **TEXT BOX 2.5**
>
> Drugs administered by the oral route encounter a number of environments and processes that can influence the extent to which the drug is available for absorption. Among the first environments encountered is the lumen of the stomach and the corresponding gastric fluid which can range in pH over several orders of magnitude depending on the age of the patient, the composition of recently ingested foods, and the action of concurrently administered antacid medications. The differences in gastric pH that can be observed are relevant to medications that are acid-labile and undergo chemical degradation at low pH. An example can be found in the acid-labile beta-lactam antibiotic penicillin that suffers from degradation at low pH. Because gastric pH is higher in the neonate than in older children, measured plasma concentrations of this drug are ~5 times higher in the neonate where the higher pH offers protection from degradation during the time the drug spends in the stomach [6,7].
>
> Another example can be found at the next environments encountered by the drug, the proximal small intestine. For the young infant in whom bile acid transport out of the liver and into the intestine has yet to fully mature, rate-limited absorption can be observed for some fat soluble vitamins and drugs such as chloramphenicol palmitate and pleconaril. As we continue to trace the path of an orally administered drug through the gastrointestinal tract, we can identify other process that influences oral bioavailability, including gastric emptying rate, drug metabolizing enzyme activity/expression, intestinal transporter expression, and intestinal microbiome composition [8–11].

drug never made it past the intestine into the circulation in the first place. Thus a more accurate representation of Eqs. (2.4.0) and (2.4.1) is represented by Eqs. (2.4.0.1) and (2.4.1.1). Similarly, the term Cl is more accurately depicted as Cl/F (Eq. 2.5.0.1).

Toxicokinetics (and nonlinear pharmacokinetics)

The term "toxicokinetics" can take on several different meanings. In the context of drug development, the term is used to describe the concentrations measured in the body during preclinical toxicology studies with the goal of linking systemic concentrations to toxicologic findings. In a clinical context, it represents the mathematical characterization of drug when concentrations exceed the therapeutic range (contrasted with PK that largely deals with the therapeutic range). In reality, however, a drug's therapeutic range is defined for an "average" population. This means that while many individuals will experience a therapeutic effect, some will experience no benefit and others will experience toxicity. With this in mind, it is easy to consider toxicokinetics on a continuum with PK.

Where the mathematics can deviate between toxicokinetics and PK is in situations where the dose–exposure relationship becomes nonlinear (though this nonlinearity occurs for some drugs within the therapeutic range, Text Box 2.6). This happens because one or more processes mediating drug movement through the body (i.e., absorption, distribution, metabolism, and elimination) become saturated (Fig. 2.4). When intestinal efflux transporters, DMEs, and active tubular secretion become saturated, an increase in dose or dosing frequency results in a disproportionate increase in concentration. Conversely, exceeding a drug's dissolution limit, saturating intestinal influx transporters, or saturation of active tubular reabsorption processes each result in a disproportionate decrease in concentration as a function of dose. The extent to which saturation will affect the PK profile depends largely on the relative importance of the pathway in question and role of other compensatory pathways.

Mathematically, nonlinear processes are usually represented by zero-order or mixed zero-order/first-order equations. The Michaelis–Menten equation (Eq. 2.6.0) describes the expected reaction rate for saturable processes which can be estimated when the biological concentration of the drug (C), the proteins maximal velocity (V_{max}) for that drug, and the drug concentration at half-maximal velocity (K_m) are all known. In this equation, when the drug concentration is well below the K_m (i.e., $C \ll K_m$) the equation approaches ($V_{max} \times C/K_m$) and the reaction rate increases in direct proportion to concentration. When the concentration markedly exceeds the K_m (i.e., $C \gg K_m$), the equation approaches ($V_{max} \times C/C$) where C in the numerator and denominator cancels and the equation approaches V_{max}. Under this condition the enzyme is saturated and the maximum reaction rate has been attained resulting in no further increase in reaction rate with increasing concentration. The V_{max} is one of the factors that informs the zero-order rate constant (k_0) in the blood and knowledge of this parameter allows one to predict the concentration of drug remaining at any time after the initial measurement according to Eq. (2.7.0).

As was observed in the case of linear or first-order PK, variables such as age, disease, and concurrent medications can alter V_{max} and K_m resulting in a change to the PK profile. Concurrently, administered enzyme inducers (including

TEXT BOX 2.6

Phenytoin, an anticonvulsant used to treat seizures, has an established therapeutic range of 10–20 mcg/mL. However, the metabolism of phenytoin is saturable at concentrations within this range. The patient depicted below with a K_m of 5 mg/L and a V_{max} of 450 mg/day requires ~220 mg to achieve a steady-state concentration of 5 mcg/L. Less than 50% more drug (~300 mg) is required to double the concentration so that the low end of the therapeutic range (10 mcg/mL) is attained and ~20% more drug (~360 mg) will again double the concentration reaching the upper limit of the therapeutic range (20 mcg/mL). Of note a mere 10% more drug (~400 mg) at this stage will place the patient well outside of the therapeutic window at a supratherapeutic steady-state concentration approaching 40 mcg/mL.

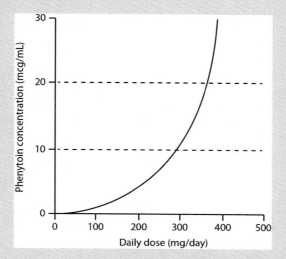

It should be recognized that clearance is decreasing and half-life is increasing in this patient as the concentrations increase. As a result, it takes longer for the patients to get to their new steady state which means that a clinician who does not allow enough time to pass before measuring drug levels after a dosage change may fail to appreciate the magnitude of the change in concentration that can be expected for this patient.

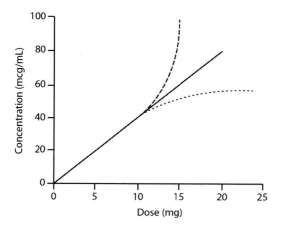

FIGURE 2.4 Dose–exposure relationships for drugs that display linear pharmacokinetics (*solid line*) and nonlinear pharmacokinetics (*dashed lines*) where concentrations begin to change disproportionately with increasing dose.

drugs that are susceptible to autoinduction with repeated exposure) can alter the reaction rate either by enhancing the activity of the protein or by enhancing the expression of that protein at the level of gene transcription. Enzyme inhibitors, by contrast, can change either K_m or V_{max} via a number of mechanisms. Competitive inhibitors, as the name implies, "competes" with the drug to occupy the enzyme's active site. The drug can eventually reach V_{max} but it takes higher concentrations to do so, effectively increasing the K_m. Noncompetitive inhibitors do not affect the ability of a drug to bind to the active site. Instead, they impair the ability of the enzyme to catalyze its reaction, effectively reducing the number of functional proteins and lowering the V_{max}. Like noncompetitive inhibitors, uncompetitive inhibitors also bind to sites other than the active site, but they only bind when the drug of interest is also bound to the catalytic site. As a consequence, these inhibitors can alter both K_m and V_{max}. Nondrug-related changes in V_{max} and/or K_m can occur as a function of growth and maturation, exemplified by the effect of age on the expression of some catalytic proteins, and comorbidities as can be seen with cirrhotic damage to the liver.

Pharmacodynamics

PD describes the relationship between the amount of drug in the body and the resultant biological response. To produce a response the drug molecule is typically required to associate with a receptor and generate a signal or series of signals that trigger the intracellular events which ultimately lead to drug action. Thus an interaction at the molecular level stimulates a response at the cellular level and ultimately actions at the level of organ or organism. Just how well a drug does this is determined by the ability of the drug to bind to the receptor (affinity), the amount of drug required to elicit a response at that receptor (potency), and the maximal effect that the drug can elicit (intrinsic activity).

The shape of the concentration–effect curve determines the pharmacodynamic model that is selected in much the same way that the concentration–time curve determines the pharmacokinetic model that is used. The mathematical equation defining traditional drug–receptor interactions is the E_{max} model (Eq. 2.8.0), an equation similar in structure to the Michaelis–Menten equation introduced earlier. With this model, E_{max} is the maximum effect that the drug can elicit, EC_{50} is the concentration at which half the maximal effect is achieved, and C is the drug concentration for drugs that antagonize or inhibit activity at the receptor we may see E replaced with I (e.g., IC_{50}, I_{max}). When comparing two

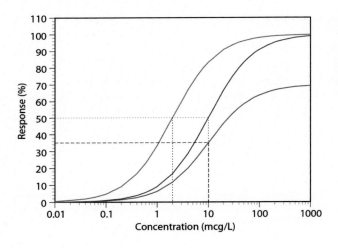

FIGURE 2.5 Exposure–response curves for drugs with varied efficacy and potency. The drugs depicted in red and blue both achieve an E_{max} of 100% making them equally effective and more effective than the drug in green which achieves an E_{max} of 70%. The drugs depicted in blue and green are equally potent with an EC_{50} of 10 mcg/mL but less potent than the drug depicted in red which has an EC_{50} of 2 mcg/mL.

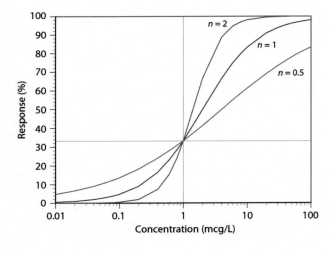

FIGURE 2.6 Exposure–response curves for drugs fit to a Sigmoid E_{max} model with varied exponents.

drugs that act at the same receptor and elicit the same response, the drug with the larger E_{max} will be the more effective of the two and the drug with the smaller EC_{50} will be the more potent of the two though neither of these parameters reveal the duration of effect (Fig. 2.5). The sigmoid E_{max} model (Eq. 2.8.1) is a variation of the E_{max} model with the addition of an exponent (n) that permits the shape of the concentration−effect curve to change (Fig. 2.6).

Irrespective of curve shape, the reader should appreciate that there is a range of concentrations where incremental increases in exposure have a relatively small impact on effect, and a segment of the curve where similar increases in exposure are accompanied by much larger changes in effect. Familiarity with this concept is especially important in the context of medications where clinically relevant concentrations span the totality of exposures that define the concentration−effect curve. However, there are medications for which clinically relevant concentrations span only a fraction of the total curve such that the exposure−response relationship appears relatively linear (i.e., an increase in concentration results in proportional increase in effect). In this setting, effect is predicted according to Eq. (2.9.0). Unfortunately, a maximum response (when the slope of the relationship is positive), or a minimum response (when the slope of the relationship is negative) cannot be defined in a linear model which is not plausible in a biological system. With its y-intercept, however, this equation does account for the fact that there may be a baseline effect when the concentration is negligible and there is no drug in the system (E_0).

For many physiologic measures of interest to the clinician (e.g., heart rate and blood pressure), equations need to integrate a baseline effect because the response of interest exists before we measure a change. The baseline effect can be addressed in the E_{max} and sigmoid E_{max} models as well depending on whether effect is increasing (Eqs. 2.8.0.1 and 2.8.1.1) or decreasing (Eqs. 2.8.0.2 and 2.8.1.2) as a consequence of drug administration. An alternative mathematical approach can use the original equations defining E as a relative (i.e., fractional or percentage) change from baseline, rather than an absolute change, by rearranging the equations to estimate $(E - E_0)$ rather than E.

The previous equations are employed with observations of a direct effect where the time course of both concentration and effect is superimposable. These equations do not account for scenarios where a lag exists between the measured concentration and the effect such that the concentration−effect curve forms a hysteresis loop. In these scenarios the magnitude of effect can differ at the same concentration when occurring at different times (Text Box 2.7).

Clockwise hysteresis loops, where effects are larger when the concentrations are increasing than when they are decreasing, can be observed in scenarios such as those where tolerance to the drug is developing (e.g., there is receptor downregulation) or a metabolite that antagonizes the parent drug at the site of action is starting to accumulate. Counterclockwise hysteresis, wherein effects are smaller when the concentrations are increasing than when they are decreasing, can be observed with the accumulation of an active metabolite, a delay in distribution of the drug to the site of action, or when sensitization (i.e., receptor upregulation) occurs in response to drug exposure [13].

These indirect effects help to explain scenarios where there is a mismatch between the pharmacokinetic half-life and the effect half-life. Examples include the organophosphate insecticide methyl parathion where the concentrations in the blood decay over hours but the effect persists for days or the short-acting barbiturate sedative-hypnotic, pentobarbital which can be measured in the blood for days but the effects only persist for minutes when administration of the drug ceases [14]. Modeling these PD relationships requires the incorporation of an effect compartment or the use of indirect PK/PD models whose discussion is beyond the scope of this chapter.

As was discussed with dose−exposure relationships, patient-specific factors can also influence exposure−response relationships. For example, genetic mutations in the voltage-gated sodium channel SCN1A can reduce a patient's sensitivity to antiepileptic medications, age-dependent changes in the concentration of circulating prothrombin fragments and protein C can make children more susceptible to the effects of warfarin, underlying diseases, such as metabolic syndrome, with the corresponding increase in free fatty acids, can block the ability of insulin to activate the signaling pathways that ultimately lead to glucose transport, and estrogen-mediated reduced repolarization reserves make females more susceptible to drug-induced prolongation of the QT interval [15−18].

Toxicodynamics

In contrast to toxicokinetics which can take on several meanings, the term "toxicodynamics" is typically used to refer to the effects observed after exposure to a toxicant. In this context, one still considers the same molecular and cellular mechanisms as one does with a medication administered for therapeutic intent and uses many of the same mathematical approaches described previously for PD. However, the exposure−response relationship can be more difficult to describe for selected toxicological responses such as mutagenicity and carcinogenicity, and the exposure−response profiles themselves can be more complex. For example, nonmonotonic profiles can be found in the toxicology literature. These

TEXT BOX 2.7

Owing to either a lag in the distribution from plasma to the site of action or a lag in molecular signaling, the time to maximal effect after an oral dose of the hallucinogenic lysergic acid diethylamide (LSD) occurs well after the time that maximal concentrations of LSD are observed. This can be seen for both the subjective psychosensory effects and the autonomic effects (e.g., blood pressure, heart rate, and body temperature) which peak and persist after the drug concentration has peaked in the blood. The following graphs illustrate the counterclockwise hysteresis curves that result when plotting the mean concentration—response relationship of 16 adults after administration of a 200 mcg dose of LSD.

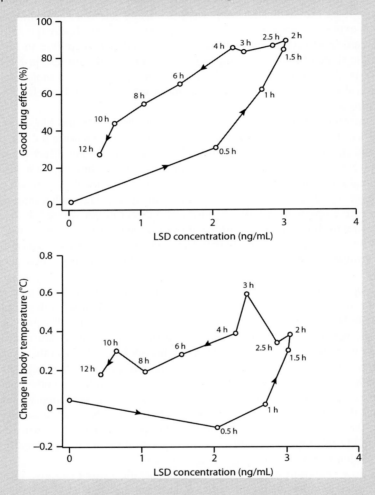

Source: Adapted from Dolder PC, Schmid Y, Steuer AE, Kraemer T, Rentsch KM, Hammann F, et al. Pharmacokinetics and pharmacodynamics of lysergic acid diethylamide in healthy subjects. Clin Pharmacokinet 2017;56(10):1219-30. [12].

curves change direction, most often in a U-shape or inverted U-shape, such that a continual increase in exposure is accompanied by a reversal of response. They are most often reported for toxicants that trigger upstream events (e.g., endocrine disruptors), after short-term exposure, and can be observed in vitro at elevated concentrations (though their relevance in vivo is less clear). Another example is the hermetic, or J-shaped curve, a variant of the nonmonotonic curve where response shifts from beneficial to detrimental as a change in exposure progresses (e.g., an antineoplastic agent that inhibits tumor growth at high concentrations but stimulates proliferation at low exposures) [19]. Toxicodynamic modeling is also applied to groups of individuals and ecosystems where differences in susceptibility and the capacity for detoxification or damage repair differ between different members of the population.

Conclusion

PK represents a valuable tool that allows us to describe the movement of drugs, toxicants, and xenobiotics in the body. Our understanding of their absorption, distribution, metabolism, and excretion can be coupled with knowledge of patient-specific factors to characterize, and in some instances predict, how drug exposures are influenced by alterations in body composition and organ function whether abnormal (e.g., cardiac failure, kidney diseases, and liver damage) or normally occurring (growth and development, pregnancy, and senescence). As a natural extension of PK, PD unites exposure and response enabling clinicians to rationally select the dose and dosing regimen that has the greatest likelihood of maximizing benefit and minimizing harm in the patients for whom they care.

We would be remiss not to accept that there remain gaps in our knowledge with respect to the drivers of disposition and action for many drugs. In these settings, data generated from carefully constructed pharmacokinetic and pharmacodynamic investigations may serve to expand our knowledge of relevant biological, pathological, and environmental factors that influence dose–exposure–response relationships and, in the process, inform the most judicious strategies by which to intervene in patients where pharmacologic or toxicologic management is required.

Equation	Parameter	Abbreviation	Calculation
2.1.0	Elimination rate constant	$k_{el} =$	$\Delta y/\Delta x = (\ln C_1 - \ln C_2)/(t_2 - t_1)$
2.2.0	Half-life	$t_{1/2} =$	$\ln 2 / k_{el}$
2.3.0	Accumulation ratio	$R =$	$1/[1 - \exp(-k_{el} \times \tau)]$
2.4.0	Concentration[a]	$C =$	D/V_d
2.4.0.1	Concentration	$C =$	$(D \times F)/V_d$
2.4.1	Volume of distribution	$V_d =$	D/C
2.4.1.1	Apparent V_d	$V_d/F =$	D/C
2.4.2	Dose	$D =$	$C \times V_d$
2.5.0	Clearance	$Cl =$	$V_d \times k_{el}$
2.5.0.1	Apparent oral Cl	$Cl/F =$	$V_d \times k_{el}$
2.5.1	Half-life	$t_{1/2} =$	$\ln 2 \times V_d/Cl$
2.6.0	Rate	$v =$	$(V_{max} \times C)/(K_m + C)$
2.7.0	Concentration[b]	$C =$	$C_0 - (k_0 \times t)$
2.8.0	E_{max} Effect model	$E =$	$(E_{max} \times C)/(EC_{50} + C)$
2.8.0.1		$=$	$E_0 + [(E_{max} \times C)/(EC_{50} + C)]$
2.8.0.2		$=$	$E_0 - [(E_{max} \times C)/(EC_{50} + C)]$
2.8.1	Sigmoidal E_{max} Effect model	$E =$	$(E_{max} \times C^n)/(EC_{50}^n + C^n)$
2.8.1.1		$=$	$E_0 + [(E_{max} \times C^n)/(EC_{50}^n + C^n)]$
2.8.1.2		$=$	$E_0 - [(E_{max} \times C^n)/(EC_{50}^n + C^n)]$
2.9.0	Linear Effect model	$E =$	$Slope \times C + E_0$
2.9.1		$=$	$Slope \times C - E_0$

[a]After an intravenous bolus.
[b]For zero-order elimination.

References

[1] Grandjean P. Paracelsus revisited: the dose concept in a complex world. Basic Clin Pharmacol Toxicol 2016;119(2):126–32.
[2] Siber GR, Echeverria P, Smith AL, Paisley JW, Smith DH. Pharmacokinetics of gentamicin in children and adults. J Infect Dis 1975;132:637–51.
[3] Noone P, Parsons TMS, Pattison JR, Slack RC, Garfield-Davies D, Hughes K. Experience in monitoring gentamicin therapy during treatment of serious gram negative sepsis. BMJ 1974;1:477–81.
[4] Gerber AU, Craig WA. Aminoglycoside-selected subpopulations of *Pseudomonas aeruginosa*. J Lab Clin Med 1982;100:671–81.
[5] Blaser J, Stone BB, Groner MC, Zinner SH. Comparative study with enoxacin and netilmicin in a pharmacodynamic model to determine importance of ratio of antibiotic peak concentration to MIC for bactericidal activity and emergence of resistance. Antimicrob Agents Chemother 1987;31:1054–60.
[6] Agunod M, Yamaguchi N, Lopez R, Luhby AL, Glass GB. Correlative study of hydrochloric acid, pepsin, and intrinsic factor secretion in newborns and infants. Am J Dig Dis 1969;14:400–14.
[7] Huang NN, High RH. Comparison of serum levels following the administration of oral and parenteral preparations of penicillin to infants and children of various age groups. J Pediatr 1953;42:657–8.
[8] Poley JR, Dower JC, Owen CA, Stickler GB. Bile acids in infants and children. J Lab Clin Med 1964;63:838–46.
[9] Suchy FJ, Balistreri WF, Heubi JE, Searcy JE, Levin RS. Physiologic cholestasis: elevation of the primary serum bile acid concentrations in normal infants. Gastroenterology 1981;80:1037–41.

[10] Kearns GL, Bradley JS, Jacobs RF, Capparelli E, James LP, Johnson KM, et al. Single-dose pharmacokinetics of a pleconaril in neonates. Pediatr Infect Dis J 2000;19:833–9.

[11] Shankaran S, Kauffman RE. Use of chloramphenicol palmitate in neonates. J Pediatr 1984;105:113–16.

[12] Dolder PC, Schmid Y, Steuer AE, Kraemer T, Rentsch KM, Hammann F, et al. Pharmacokinetics and pharmacodynamics of lysergic acid diethylamide in healthy subjects. Clin Pharmacokinet 2017;56(10):1219–30.

[13] Louizos C, Yáñez JA, Forrest ML, Davies NM. Understanding the hysteresis loop conundrum in pharmacokinetic/pharmacodynamic relationships. J Pharm Pharm Sci 2014;17(1):34–91.

[14] Kramer RE, Ho IK. Pharmacokinetics and pharmacodynamics of methyl parathion. Zhonghua Yi Xue Za Zhi (Taipei) 2002;65(5):187–99.

[15] Angelopoulou C, Veletza S, Heliopoulos I, et al. Association of SCN1A gene polymorphism with antiepileptic drug responsiveness in the population of Thrace, Greece. Arch Med Sci 2016;13(1):138–47.

[16] Takahashi H, Ishikawa S, Nomoto S, Nishigaki Y, Ando F, Kashima T, et al. Developmental changes in pharmacokinetics and pharmacodynamics of warfarin enantiomers in Japanese children. Clin Pharmacol Ther 2000;68(5):541–5.

[17] Morino K, Petersen KF, Shulman GI. Molecular mechanisms of insulin resistance in humans and their potential links with mitochondrial dysfunction. Diabetes 2006;55(Suppl. 2):S9–15.

[18] Li G, Cheng G, Wu J, Zhou X, Liu P, Sun C. Drug-induced long QT syndrome in women. Adv Ther 2013;30(9):793–802.

[19] Lushchak VI. Dissection of the hormetic curve: analysis of components and mechanisms. Dose Response 2014;12(3):466–79. Available from: https://doi.org/10.2203/dose-response.13-051 Published 2014 Apr 11.

Chapter 3

Laboratory methods in toxicology

Patrick B. Kyle
Department of Pathology, University of Mississippi Medical Center, Jackson, MS, United States

Introduction

Toxicology laboratories typically utilize a variety of highly specialized techniques and instrumentation. While differences exist between the processes used by clinical versus forensic laboratories, the goal of each involves the detection and/or quantification of toxic substances. Primary differences between clinical and forensic laboratories involve the timeliness in which results are needed, the requirement or lack thereof for chain of custody documentation, and the types of specimen matrices typically submitted for analysis. Due to these differences, clinical and forensic laboratories often employ different methods of specimen preparation, but very similar methods of analysis. Timeliness is generally of lesser importance in forensic laboratories associated with medical examiners because their specimens are usually submitted as a result of a crime that has already taken place. This allows these forensic laboratories to take full advantage of batch-processing methods in the quest to achieve high efficiencies. Conversely, clinical laboratories are expected to provide results quickly for the purposes of clinical decision-making in patient care. As a result, turn-around times in clinical toxicology are often measured in minutes instead of days. Although clinical laboratories must attend to STAT (high-priority, urgent) specimens as they appear, they may still obtain efficiencies by processing batches of routine specimens for drug class confirmation.

In instances of suspected ingestion or unknown exposure, laboratories should seek to employ screening methods that are sensitive to a wide variety of substances [1]. Given the broad array of compounds available to the public as illicit drugs, prescription drugs, over-the-counter medications, and herbal remedies, initial screening methods should be as comprehensive as reasonably possible [1,2]. Laboratories often begin with some forms of immunoassay screening and follow with a broadly sensitive chromatographic analysis. The positive results obtained from these may require confirmation and/or quantification via specific chromatographic methods such as gas chromatography–mass spectrometry (GC–MS) or liquid chromatography–tandem mass spectrometry (LC–MS/MS) methods that employ isotopically labeled internal standards. These are complex instruments that require significant financial investment, technical expertise, and time required to develop and maintain analytical methods. As a result drugs of abuse immunoassays are often the only drug screening tools available in small hospitals.

Methods for general screening

Immunoassay screening

Immunoassay screens are readily available for several classes of commonly used drugs, including amphetamines, barbiturates, benzodiazepines, cannabinoids, cocaine, lysergic acid diethylamide, methadone, methaqualone, opiates, phencyclidine, and propoxyphene. As such, an immunoassay panel that includes 7–9 drug classes may be sensitive to over 100 distinct and commonly abused compounds. Immunoassays engineered for the analysis of urine typically exhibit rapid turn-around times of 15–20 minutes. These are available on many automated clinical chemistry analyzers, which promotes their availability in nearly every hospital laboratory. These immunoassay screens are calibrated to specific cutoff concentrations for negative/positive results, although some are engineered to produce semi quantitative results. Manufacturers often provide multiple calibrators so that laboratories may choose from several cutoff concentrations. For example, laboratories performing preemployment screening are required to use the cutoff concentrations published by Substance Abuse and Mental Health Services Administration (SAMHSA) [3], whereas clinical laboratories are wise to choose the lowest cutoff concentrations available. The specimen types accepted by the lab and the expectations in turn-around time help lab staff choose specific methods for specimen preparation. Forensic laboratories may adopt immunoassay techniques, such as ELISA in a 96-well plate format due to the batch-processing capability and its ability to analyze multiple matrices such as serum, blood, urine, and vitreous humor (reference). Some heterogeneous immunoassays require up to 2 hours for results due the multiple incubation steps and washes required by some methods [4], but this is typically inconsequential in the forensic arena.

Drugs of abuse immunoassays are engineered to be sensitive to common drugs per respective drug class. However, their sensitivity to individual drugs varies from manufacturer to manufacturer. For example, after calibrating two assays to 300 ng/mL morphine, one manufacturer's opiate immunoassay may require over 30,000 ng/mL oxycodone for positivity, whereas a second manufacturer's immunoassay may only require 1500 ng/mL oxycodone [5]. As a result, manufacturers have recently developed new assays to address some of these deficiencies. Recent developments include immunoassay targeting 6-acetylmorphine, buprenorphine/norbuprenorphine, methylenedioxymethamphetamine, oxycodone, and fentanyl. Recent developments also include immunoassays directed to designer drugs such as amphetamine derivatives, synthetic cannabinoids, and fentanyl derivatives [6,7]. These options and differences highlight the need for clinical staff to be aware of the capabilities and limitations of the testing performed at their respective institution. Laboratorians should work to educate clinical staff on the potential for false-negative and false-positive results with immunoassays. Given that immunoassays are designed to detect multiple compounds within a drug class, they may also exhibit cross-reactivity to structurally similar but unrelated compounds. Due to this lack of specificity, positive immunoassay results should not be relied upon in courts of law. Positive immunoassay results should be followed up by confirmatory procedures with definitive results.

Chromatographic methods

Chromatographic methods may also be performed for broad spectrum screening and detection. Prior to the 1990s, thin layer chromatography was widely used for screening due to its inexpensive design and sensitivity to a wide variety of compounds. Lab-developed and commercially available designs were widely utilized, both of which required drug extraction prior to analysis. The commercially available Toxi-Lab [8] system included liquid–liquid extraction materials, chromatographic plates, and chemical reagents for drug detection. After chromatography the plate would be viewed under ultraviolet light in order to visualize fluorescing compounds. Chemical reagents sprayed onto the chromatographic plates produced color changes with specific classes of compounds, based on the molecular functional groups they contained. The thin layer chromatographic procedures were relatively rapid with analytical runs completed in approximately 2 hours. Unfortunately, these results were not definitive, and many reports were issued with statements that included, "A compound with chemical and chromatographic properties consistent with... was detected." Although this could be helpful to a clinician treating a child with new onset of seizures or altered mental status, definitive compound identification was greatly preferred. The advent of widespread computerization in the 1990s decreased instrument prices and ushered in the era of computerized instrumentation.

Gas chromatography

Gas chromatography is a powerful chromatographic technique used for the analysis of small molecules such as drugs and volatile alcohols. A gas chromatograph (GC) may be outfitted with one of a number of different types of detectors, including flame ionization, nitrogen–phosphorus, and mass spectrometry detectors. During the chromatographic procedure, analytes in a mixture are separated and ideally delivered to the detector one at a time. Initially, an analyte mixture is injected into the injection port, which is heated to approximately 250°C. Drugs of low molecular weight are immediately volatilized and positive gas pressure moves them into the analytical column. Each drug interacts with the analytical column based on its boiling point, molecular weight, and polarity, all of which determine its retention time in the column at a given temperature. Packed chromatographic columns were originally used, but these have almost entirely been replaced by fused silica (aka capillary) columns. A wide variety of drugs can be separated using the capillary columns currently in production. Fused silica columns are of the open tube type in which the interior wall is coated with stationary phase. Columns are chosen for each application based on their internal diameter, length, stationary phase, and thickness of the stationary phase. Thicker stationary phase coatings and longer column lengths not only tend to provide higher resolution but also increase retention times. For the analysis of drugs the GC oven is often programed with a temperature gradient to decrease the analytical time. During sample injection a relatively low oven temperature of

50°C−75°C is often used to promote condensation of the analytes onto the head of the column, which helps prevent band broadening. The column temperature is then increased to 250°C or more over a set amount of time in order to elute the analytes. Labs may choose to hold the column at a high temperature for several minutes at the end of the analytical run in order to "boil off" late eluting compounds.

Gas chromatography with nitrogen−phosphorus detection

Gas chromatography with nitrogen−phosphorus detection is a long-standing screening tool that has been used since the 1970s [9]. Nitrogen−phosphorus detectors (NPD) are sensitive to virtually any compound that contains nitrogen and/or phosphorus. This includes many drugs such as the amphetamines, antidepressants, antihistamines, barbiturates, benzodiazepines, cannabinoids, cocaine metabolites, opiates, opioids, phenothiazines, muscle relaxants, and pesticides [10]. The NPD is composed of a heated rubidium bead positioned in the path of the column effluent. The bead may be heated via electrical current or a flame, and the early detectors were described as modified flame ionization detectors. As compounds exit the chromatographic column they interact with the rubidium bead and produce ions that are detected as changes in electrical current between the anode and cathode. These detectors are sensitive to gas flow rates, so the GC should be set to maintain a constant flow instead of constant pressure, throughout the temperature gradient. The rubidium bead is robust but is consumed over time. Therefore many operators turn off the detector when not in use.

Gas chromatography with nitrogen−phosphorus detection is sensitive to drug concentrations of 100 ng/mL in urine/serum/blood [10], and to 10 ng/mg in tissue [11]. Specimen preparation for analysis with GC varies with the target analytes and specimen matrix. Solid phase extraction (SPE) and liquid−liquid extraction techniques are often used to isolate drugs from urine and serum, whereas liquid−liquid extraction techniques are typically used to isolate drugs from tissue and putrefied blood. Some choose to derivatize drugs to improve their sensitivity or resolution, and some employ methods to hydrolyze glucuronidated drugs to improve their chromatography and/or sensitivity. These methods will be addressed in detail later in this chapter. The advantages of GC−NPD detection include sensitivity to a wide variety of compounds, ability to separate complex mixtures, and the potential for quantitation due to linear detector response. Disadvantages of GC−NPD include the requirement for complex instrumentation, labor-intensive sample preparation, the lack of definitive identification, and it is limited to small molecules that can be volatilized but are stable in high temperatures.

Gas chromatography−mass spectrometry

GC−MS is sensitive to a broad number of compounds and has long been regarded as the gold standard method in drug identification [12,13]. The combination of chromatographic separation and generation of reproducible mass spectra make this one of the most revered methods of drug identification. The GC is set up as described earlier except that the last few inches of the chromatographic column is routed through a heated transfer tube, and into the ion source of the detector. As compounds exit the chromatographic column, they are fragmented into ions that are characteristic for the compound. Most laboratories rely on electron-impact (EI) ionization that provides the most robust fragmentation. Long ago it was determined that the use of a given fragmentation energy from injection to injection would result in reproducible ion fragments from a given compound. Even the abundance of the individual fragment ions was found to be reproducible. At some point the fragmentation energy used in EI ionization was standardized to 70 eV, which allowed the development of commercial mass spectral libraries. As early as 1985, the Pfleger/Maurer/Weber library included EI mass spectra for 1500 drugs, metabolites, and poisons. Current mass spectral libraries are quite comprehensive with the NIST/EPA/NIH and Wiley Registry of Mass Spectral Data libraries containing EI mass spectra for over 250,000 compounds [14]. The positive drug identification produced by reproducible fragmentation and the availability of comprehensive commercial drug libraries combine to make GC−MS an unsurpassed tool for comprehensive toxicology screening.

The most common GC−MS detector employed by toxicology labs is the single quadrupole detector, which is considered a workhorse in every laboratory. A quadrupole detector is composed of four parallel rods in a square array connected to an electrical circuit. Voltage is applied to the rods in an alternating fashion, which also generates an RF field. The combination of voltage and RF select the m/z of ions that will travel to the electron multiplier to generate a signal. Nearly every clinical and forensic toxicology laboratory employs one or more single quadrupole GC−MS. The quadrupole detector offers adequate sensitivity, mass resolution to 0.1 m/z, and linear response. A fully configured GC−MS with quadrupole detector currently costs approximately US$90,000 making it an attractive and affordable option for a lab seeking to bring in gold standard methodology. A single quadrupole GC−MS functions well for drug screening as well as drug quantitation and confirmation. For general unknown screening a detector may be set to continuously scan from 50 to 500 m/z, with a single analysis being completed in approximately 30 minutes. Although sample preparation is required, a laboratory should be able to report results of a comprehensive GC−MS screen within 1.5 hours of sample receipt. This offers the clinical laboratory enormous potential for clinical involvement when a toxic ingestion is suspected. Clinicians are often impressed by the comprehensive results of GC−MS, which often include caffeine, nicotine, over-the-counter (OTC) medications, prescription medications, and illicit drugs. It is not uncommon to detect 8−12 drugs and pharmaceuticals in a specimen obtained after poly drug overdose. In addition to comprehensive screening, single quadrupole GC−MS is often employed for drug quantitation due to their linear responses. Analyte sensitivity can be increased by setting the detector to selective ion monitoring (SIM) mode. In SIM mode the detector focuses on a few target ions instead of a large mass range. The instrument dwell time for the target ions may be increased 100-fold in SIM mode. Other types of mass spectral detectors can be obtained for GC−MS, including ion trap, triple quadrupole, Fourier transform (Orbitrap), time-of-flight (TOF), and hybrid variants such as quadrupole-TOF (QTOF), and quadrupole-Orbitrap.

Ion trap detectors are also frequently used in toxicology laboratories and are touted as having higher sensitivity than single quadrupole detectors. However, ion trap detectors do not exhibit linear analyte response due to their ion concentrating characteristics, which generally precludes their use for accurate quantitative analyses. This can be overcome by manually setting the ion accumulation time, but this decreases sensitivity of the detector unless a narrow range of concentrations is used. It has been found that analyte sensitivity can be related to the internal design of a given detector and its associated electronics. Higher sensitivity was noted in a newer single quadrupole instrument compared to an ion trap model only a few years old [15]. This was due to faster scan rates and the incorporation of an S-shaped lens used to route ions to the quadrupole. It was found that negative and uncharged species could not navigate the S-shaped lens and were not routed to the quadrupole, which greatly reduced background ions and noise. Laboratories often employ other types of GC−MS detectors for select analyses. However, many of these are more costly and often not be used for day-to-day operation.

While GC−MS offers laboratories state-of-the-art analytical capability, it is not without its limitations. Analytes must be extracted out of biological matrices into an organic solvent for injection, because the injection of water would ruin a polysiloxane analytical column. In addition, target analytes must exhibit low polarity, heat stability, and relatively low boiling points. Factors that influence an analyte's boiling point include molecular weight, cyclic ring saturation, and its functional side groups. Therefore analytes for GC−MS are generally limited to compounds equal to or less than 450 amu. This can eliminate the analysis of some oral hypoglycemics (glyburide, 494 amu), antihypertensives (verapamil, 455 amu), and anticoagulants (brodifacoum, 523 amu). While GC−MS instruments are not perfect, they are relatively affordable and allow laboratories to utilize gold standard methodies for the analysis of drugs, metabolites, chemicals, and pesticides.

A method that works well for comprehensive screening involves the injection of 1 μL sample extract onto a GC−MS injection port at 250°C fitted with a Restek RTx-5MS, 30 m × 0.25 mm, 25 μm column (Restek, Bellefonte, PA). The baseline oven temperature is held at 60°C for 1 minute before ramping to 280°C at 20°C/min, and then held at 280°C for 20 minutes. This system is set to 1:10 split injection using 10 mL/min split flow, a sweep gas of 5 mL/min, and injector flow rate of 40 mL/min. A filament start delay of 3.5 minutes prolongs filament life. The detector is set to positive ionization and operated in full-scan mode from 40 to 500 m/z for a total analytical time of 32 minutes [16].

Liquid chromatography−mass spectrometry

The popularity of liquid chromatography−mass spectrometry (LC−MS) is increasing in the field of toxicology [17]. Several advantages exist in using LC−MS versus GC−MS, including the ability to analyze larger compounds (up to 1000−2000 amu), analytes are not required to be volatile, and analytes do not have to be extracted out of an aqueous matrix [18]. While LC−MS has been employed for targeted drug confirmation for over 10 years, more recent high-resolution designs are now being used for general unknown screening. High-resolution mass spectrometry (HRMS) detectors include TOF, QTOF, and Fourier transform (Orbitrap) designs, which exhibit accurate mass resolution (100,000) to approximately 1 ppm of the parent and fragment ions. The mass resolution of these analyzers typically allows one to screen and "rule in" a matching mass for only 8−10 compounds at the exclusion of all others [19]. The HRMS instruments can determine the number of carbon, hydrogen, nitrogen, and oxygen atoms in an unknown compound relatively accurately, and then compare its isotopic pattern to that of known species. These capabilities in combination with LC retention time matching confer great analytical capability to the laboratory. Current LC−MS/MS systems are quite expensive with purchase prices ranging from US$200,000 to 400,000 depending upon the detector, LC, autosampler, software, and compound libraries purchased.

Mass spectral libraries for LC−MS still seem to be in their infancy when compared to commercial GC−MS libraries. This is a result of the lack of standardization of fragmentation energies in LC−MS, which is due to the number of fragmentation modalities and their inherent complexities [14]. The AB Sciex (Foster City, CA) iMethod Forensic Toxicology Library has been available for several years and contains more than 1250 compounds. HighChem LLC (Bratislava, Slovakia) recently began offering spectral matching from their online database termed mzCloud [20] (https://www.mzcloud.org/). The database contains MS/MS spectra from more than 8100 compounds that were acquired with multiple fragmentation energies from each compound. Current HRMS Thermo Fisher Scientific (Waltham, MA) instruments can perform online spectral searching via mzCloud. The mzCloud database also offers substructure searching that can link fragments of an unknown compound to potential drug classes, which can be useful in a time when new designer drugs are rapidly being developed. The mzCloud database is also useful during analytical method development as users can scroll through the MS/MS spectra to predict the best fragmentation energy to use with a given compound.

Regardless of the mass spectral library used, one should also include chromatographic matching for definitive drug identification. Laboratories should inject standard formulations of known drugs in order to set and match expected retention times in their chromatographic system. Reference standards should also be obtained for injection, whenever spectra of new compounds are detected. Isobaric drugs are not uncommon and, while many can be differentiated by their MS/MS spectra, some must be differentiated chromatographically.

Although sample preparation for LC−MS analysis can be less rigorous than that for GC−MS, the injection of whole blood, serum, or unfiltered urine would quickly degrade the chromatographic system and reduce detector sensitivity. Some choose rudimentary protein precipitation steps, but most advocate drug extraction from blood/sera via solid phase or liquid−liquid methods, which not only removes interferences but also concentrates the analytes [12]. Many users find that urine can be injected without extraction after centrifugation or filtration and subsequent dilution. A comprehensive method for drugs in urine that has worked well at the University of Mississippi involves a 1:10 dilution of centrifuged urine injected onto a Restek Ultra Biphenyl 100×2.1 mm, $3\,\mu$m column. A mobile phase consisting of (1) 0.1% formic acid and (2) 0.1% formic acid in acetonitrile is ramped from 90% of (1) to 90% of (2) over 9 minutes before being held at 90% of (2) for 3 minutes to flush the column. The instrument was set to positive ionization mode using electrospray ionization with a 3.5 kV spray voltage and capillary set to 250°C. The detector was set to continuously scan from 100 to 500 m/z and fragment the largest peaks in each scan. The collision energy was set to a constant 35%. These conditions were not optimum for all analytes, but the long elution time accommodates a range of analytes from the small amphetamines to the larger oral hypoglycemic and synthetic cannabinoids.

Specimen preparation for chromatographic analysis

Chromatographic methods often require preanalytical sample preparation in order to eliminate endogenous interferences, remove analytes from aqueous matrices, concentrate analytes, remove conjugated glucuronides, or to make analytes more amenable for chromatography. As always, appropriate internal standards should be mixed with each specimen prior to extraction and analysis. It is generally recommended that isotopic-labeled compounds be used for mass spectrometry analyses. Use of isotopic-labeled drug analogs can accommodate for ion suppression and maintain quantitative accuracy when appropriately included. As previously discussed, specimens are often extracted prior to GC−MS analysis via liquid−liquid or SPE. One method that provides good recovery of acid, neutral, and basic drugs involves SPE extraction via a Bond Elut Certify (Agilent, Santa Clara, CA) column. A 100 mg column is prepared via sequential aspiration of 2 mL methanol and 2 mL 100 mM phosphate buffer (pH 6.0). Either 2 mL serum or 5 mL urine, respectively, is combined with 2 mL or 5 mL 100 mM phosphate buffer (pH 6.0) and then aspirated through the column. The column is washed via sequential aspiration of 1 mL water and 0.5 mL 0.01 M acetic acid then allowed to dry using a minimum of 15 mm Hg vacuum for 5 minutes. Acidic and neutral drugs are eluted into a 10 mL conical glass tube via aspiration of 4 mL acetone/chloroform (1:1), whereas 3 mL ethyl acetate/ammonium hydroxide (98:2). Compressed nitrogen is used to evaporate the eluate to 50 μL prior to injection onto GC−MS [15]. For LC−MS the eluate should be evaporated to dryness prior to reconstitution with methanol and mixed with mobile phase for injection.

It is often desirable to cleave analytes from their conjugated glucuronide forms prior to analysis of urine in order to increase their detectable concentrations. Laboratories may employ acid hydrolysis or enzymatic hydrolysis, however, acid hydrolysis has been shown to convert oxycodone to oxymorphone, hydrocodone to hydromorphone, and to degrade 6-monoacetylmorphine [21]. Enzymatic hydrolysis is often performed using beta-glucuronidase obtained from (1) *Helix pomatia* (snail), (2) abalone (mollusk), (3) *Escherichia coli*, (4) *Patella vulgata*, or other sources. This is easily performed prior to LC–MS analysis as follows. Add 2500 IU of beta-glucuronidase solution (volume varies by manufacturer) to 0.5 mL urine, 50 μL 2.0 M sodium acetate buffer (pH 4.5), and internal standards in a 1.7 mL plastic microfuge tube. Vortex mix for 5 seconds and incubate at 60°C for 30–60 minutes. Centrifuge at 10,000 × *g* for 5 minutes. Mix 100 μL of the supernatant with 900 μL mobile phase and inject 10 μL onto the analytical column.

Methods for targeted analysis

Targeted analyses are often used to (1) detect analytes not covered by general screening methods and (2) to confirm and/or quantify analytes potentially detected during general screening. Laboratories often use targeted analyses for specific drug classes, heavy metals, and volatile alcohols/ketones. For example, a laboratory may establish a targeted method to confirm the results of positive opiate immunoassays, or a targeted method to detect synthetic opioids and fentanyl derivatives that are not covered during initial screening tests.

Most toxicology laboratories maintain a targeted method for the analysis of volatile alcohols, ketones, and ethylene glycol. This may involve detection of acetone, ethanol, isopropanol, methanol, 1,2-propanediol, acetaldehyde, diethylene glycol monomethyl ether, methyl ethyl ketone, and others. Gas chromatography with flame ionization detection is typically used after direct or headspace injection. One method utilizing headspace injection involves incubating an aliquot of the aqueous specimen in a sealed 10 mL headspace vial at 80°C for 5 minutes with constant shaking. One microliter of the headspace is then injected into the injection port heated to 200°C fitted with a Restek RTX-BAC 1 or RTX-BAC 2 fused silica column held to a constant 40°C. With the helium carrier gas flow set to 24 mL/min, acetaldehyde, acetone, ethanol, isopropanol, methanol, and *N*-propanol (internal standard), all elute within 5 minutes. The analysis of ethylene glycol may be performed by direct injection into a large-bore glass column such as a Supelco #2-0624 packed with GP 60/80 Carbopack B and 5% Carbowax, but an oven temperature of 150°C should be used. More recent methods involve the derivatization of ethylene glycol and glycolic acid with isobutyl chloroformate for analysis by GC–MS [22].

The confirmation of select drug classes is often performed via GC–MS, although laboratories tend to prefer LC–MS, if available. The confirmation of opiates and benzodiazepines in urine by GC–MS requires hydrolysis, derivatization, and analyte extraction, whereas only hydrolysis is required for these for LC–MS analysis. The confirmation of δ-9-tetrahydrocannabinol (THC) and metabolites is typically performed by GC–MS after derivatization of the carboxy-THC metabolite to improve its chromatography. Without derivatization the carboxy-THC elutes from the GC column so slowly that no "peak" is observed. One should not overlook applications of LC with ultraviolet/visible detection. The LC–UV/Vis systems are relatively inexpensive and easily placed into service because they do not require the technical expertise needed for MS systems. The laboratory should recognize that the development of methods for new drugs/classes can be an ongoing process. These may include methods for the synthetic cannabinoids, synthetic opioids, and other designer drugs as warranted. The use and abuse select drugs are regionally specific and tends to change over time [23]. For these reasons the toxicology laboratory must maintain a number of different analytical tools and the flexibility to employ them quickly.

Other considerations

Alternate specimen types

Laboratories may receive requests for the analysis of alternate (atypical) specimen types. Specimens such as plant material, crystalline material/rocks, vaporizer additives/oils, syringe contents, tablets, and powders are commonly submitted to forensic laboratories but can be out of the ordinary for a clinical laboratory. If possible, laboratories should seek to maintain a variety of analytical techniques as the analysis of such specimens can be of great service to submitting clinicians. Many of these specimens are obtained from patients' pockets or personal belongings by emergency department personnel, which seems to occur most often at approximately 4:30 p.m. on a Friday afternoon. Such alternate specimen types are often viewed as problem cases because they can require long hours for analysis. In one recent case, two twin boys were admitted to the pediatric emergency department with altered mental status. A family member related that the boys drank an unknown liquid from a grandparent's garage and immediately "went crazy." Blood and urine specimens from both boys as well as a clear liquid in an unmarked glass container were submitted for analysis. Analysis of the specimens by immunoassay and comprehensive GC–MS only revealed caffeine and theobromine in the biological specimens. SPE of the clear liquid was performed, but no drugs or toxic compounds were detected by GC–MS. A full

UV/Vis spectrophotometry scan also revealed no useful information. A liquid—liquid extraction of the liquid was performed and subjected to immunoassay analysis with negative results. As a final effort, the pH of the liquid was measured and was determined to be >12.0, a very basic solution. The emergency clinicians were advised of the pH and that the solution was likely corrosive, information that was valuable for the clinical management of the two boys. It was later determined that the solution was a commercial drain cleaner a grandparent had decanted into a glass jar for storage. The boys apparently "went crazy" because of the burning sensations in their oropharynx. While laboratorians have a responsibility to help clinicians identify offending toxic agent/s, we should recognize the limitations as well as the opportunities in our laboratories.

Use of clinical information

The clinical toxicologist should evaluate a patient's clinical history and/or presentation whenever necessary as this information can be valuable in determining suspected toxicants and analyses that may be required. Cardiac disturbances, respiratory rate/pattern, pupil size, reflexes, seizures, Glasgow coma scale, naloxone response, and other indicators may suggest a pattern of toxicity that excludes many drugs from a long list of potential agents. Basic core laboratory assays for glucose, anion gap, osmolality, c-peptide, acetylcholinesterase, ferritin, and prothrombin time (PT) are always useful to help rule out suspected agents. Obviously therapeutic drug levels for acetaminophen, carbamazepine, cardiotropic, and anticonvulsants should be considered. Those who staff the toxicology laboratory should be taught to consider the clinical aspects of emergency cases as this valuable information can be used to help direct the analyses.

References

[1] Thoren KL, Colby JM, Shugarts SB, Wu AHB, Lynch KL. Comparison of information-dependent acquisition on a tandem quadrupole TOF vs a triple quadrupole linear ion trap mass spectrometer for broad-spectrum drug screening. Clin Chem 2016;62(1):170—8.
[2] Colby JM, Thoren KL, Lynch KL. Suspect screening using LC-QqTOF is a useful tool for detecting drugs in biological samples. J Anal Toxicol 2018;42:207—13.
[3] US Department of Health and Human Services. Mandatory guidelines for federal workplace drug testing programs. Federal Register 2017;82 (13):7920—70.
[4] Benzodiazepine Group. ELISA kit instructions. Lexington, KY: Neogen® Corporation; 2014. Product #130119 & 130115.
[5] Melanson SE, Baskin L, Magnani B, Kwong TC, Dizon A, Wu AHB. Interpretation and utility of drug of abuse immunoassays lessons from laboratory drug testing surveys. Arch Pathol Lab Med 2010;134:735—9.
[6] Franz F, Angerer V, Jechle H, Pegoro M, Ertl H, Weinfurtner G, et al. Immunoassay screening in urine for synthetic cannabinoids — an evaluation of the diagnostic efficiency. Clin Chem Lab Med 2017;55(9):1375—84.
[7] Helander A, Stojanovic K, Villen T, Beck O. Detectability of fentanyl and designer fentanyls in urine by 3 commercial fentanyl immunoassays. Drug Test Anal 2018;10:1297—304.
[8] Jarvie DR, Simpson D. Drug screening: evaluation of the Toxi-Lab TLC system. Ann Clin Biochem 1986;23(1):76—84.
[9] Burgett CA, Smith DH, Bente HB. The nitrogen-phosphorus detector and its applications in gas chromatography. J Chromatogr A 1977;134 (1):57—64.
[10] Lillsunde P, Michelson L, Forsstrom T, Korte T, Schultz E, Ariniemi K, et al. Comprehensive drug screening in blood for detecting abused drugs or drugs potentially hazardous for traffic safety. Forensic Sci Int 1996;77(3):191—210.
[11] Terada M. Determination of methamphetamine and its metabolites in rat tissues by gas chromatography with a nitrogen-phosphorus detector. J Chromatogr 1985;318(2):307—18.
[12] Lee J, Park J, Go A, Moon H, Kim S, Jung S, et al. Urine multi-drug screening with GC-MS or LC-MS-MS using SALLE-hybrid PPT/SPE. J Anal Toxicol 2018;42:617—24.
[13] Maurer HH. Systematic toxicological analysis of drugs and their metabolites by gas chromatography-mass spectrometry. J Chromatogr 1992;580:3—41.
[14] Zedda M, Zwiener C. Is nontargeted screening of emerging contaminants by LC-HRMS successful? A plea for compound libraries and computer tools. Anal Bioanal Chem 2012;403:2493—502.
[15] Kyle PB, Bhaijee F, Magee L, Booth D. A quadrupole GC/MS with dual-role capability in forensic toxicology. In: Annual meeting of the society of forensic toxicologists; 2013. p. 52.
[16] Kyle PB, Spencer JL, Purser CM, Hume AS. Drugs detected in suspected pediatric ingestions: a three-year review. J Miss State Med Assoc 2004;45:35—40.
[17] Liu L, Wheeler SE, Venkataramanan R, Rymer JA, Pizon AF, Lynch MJ, et al. Newly emerging drugs of abuse and their detection methods—an ACLPS critical review. Am J Clin Pathol 2018;149:105—16.
[18] Wu AHB, Gerona R, Armenian P, French D, Petrie M, Lynch KL. Role of liquid chromatography-high resolution mass spectrometry (LC-HR/MS) in clinical toxicology. Clin Toxicol 2012;50(8):733—42.

[19] Wu AHB, Colby J. High-resolution mass spectrometry for untargeted drug screening. In: Garg U, editor. Clinical applications of mass spectrometry in drug analysis. Methods in molecular biology, vol. 1383. New York: Humana Press; 2016.
[20] MZCLOUD. Advanced mass spectral database, <https://www.mzcloud.org/> [accessed 14.11.18].
[21] Sitasuwan P, Melendez C, Marinova M, Mastrianni KR, Darragh A, Ryan E, et al. Degradation of opioids and opiates during acid hydrolysis leads to reduced recovery compared to enzymatic hydrolysis. J Anal Toxicol 2016;40(8):601–7.
[22] Hlozek T, Bursova M, Cabala R. Simultaneous and cost-effective determination of ethylene glycol and glycolic acid in human serum and urine for emergency toxicology by GC-MS. Clin Biochem 2015;48(3):189–91.
[23] Sloboda Z. Changing patterns of "drug abuse" in the United States: connecting findings from macro- and microepidemiologic studies. Subst Use Misuse 2002;37(8–10):1229–51.

Further reading

Moss MJ, Warrick BJ, Nelson LS, McKay CA, Dube PA, Gosselin S, et al. ACMT and AACT position statement: preventing occupational fentanyl and fentanyl analog exposure to emergency responders. Clin Toxicol (Phila) 2018;56(4):297–300.

Chapter 4

Management of an overdose patient

Stephen Thornton
University of Kansas Health System Poison Control Center, Kansas City, KS, United States

Introduction

The misuse of drugs and associated overdoses are a major health concern worldwide. The United Nations Office on Drugs and Crime estimates that almost 200,000 people die from drug overdoses per year, with a disproportionate amount of them in North America [1]. Emergency department visits and hospitalizations due to substance abuse and overdoses have risen dramatically over the last 10 years, with estimates of over 2 million visits per year [2]. Much of this rise in morbidity and mortality is due to the opioid epidemic which is estimated to cause over 40,000 deaths per year in the United States alone [3]. Substance abuse and overdoses disproportionately affect younger populations with highest rates in 20–30 age group [3] worldwide. This leads to estimates of 28 million years of "healthy" life lost worldwide because of premature death and disability caused by drug use [1]. Overdoses can result from recreational use of drugs or intentional suicide attempts. Illicit drugs, over the counter medications, and prescription drugs can all be abused and overdosed on. A majority of overdoses are the result of an oral ingestion, though exposure via intravenous and other parenteral routes can be associated with severe toxicity [4]. The overdose patient can be a complex and dynamic patient. Early consultation with your local poison control center or local toxicology expert is recommended for any overdose patient to assist in evaluation and treatment.

Evaluation of overdosed patients

Overdose patients can present for health care either via emergency medical services or by private vehicle. As with all emergent medical conditions, immediate attention to the overdose patient's airway status, breathing and circulation is critical. Impairment in any one of these areas can lead to rapid clinical deterioration and possibly death. Many overdoses, such as opioids or ethanol, can lead to loss of airway control or respiratory depression and result in aspiration or frank apnea [5]. Similarly, some overdoses can impair circulation and lead to hypotension and shock. Toxin-induced cardiogenic shock can be very difficult to treat. Early identification and appropriate treatment of substance-induced shock can be lifesaving. In particular, overdoses of calcium channel blockers and beta-antagonists can result in severe cardiogenic shock refractory to typical treatments [6]. Conversely, cocaine and methamphetamine can cause profound tachycardia and potentially life-threatening hypertension [7]. While evaluating the patient's airway, breathing and circulation, vital signs should be obtained, including blood pressure, heart rate, respiratory rate, temperature, and pulse oximetry. Capnography or end tidal CO_2 may also be useful as it allows for a true evaluation of ventilation which is frequently impaired in overdoses. However, recent literature suggests that it has limited utility in predicting severe outcomes in the overdose patient [8].

Determination of blood glucose levels should also be rapidly performed. Hypoglycemia is a potentially easily reversible condition that can mimic many signs and symptoms of overdoses [9]. In addition, many overdoses, such as sulfonylureas, venlafaxine, and tramadol, can cause hypoglycemia [10].

After addressing and stabilizing the patient's airway, breathing and circulation status, obtaining pertinent history, if possible, should be attempted. Information on what substance or substances were taken, route of exposure (oral vs intravenous, etc.), and how much was involved in the exposure can be critical in the management of the overdose patient. For instance, identifying if a modified release substance was involved can impact duration of required monitoring. Determining when the exposure took place and what symptoms have been observed can be crucial to further management, as well. Unfortunately, this information is frequently only partially available, if at all. In general, single substance ingestions are most common but polysubstance overdose is associated with more severe outcomes [4]. Patients typically present 2—3 hours after their exposure [11]. Some maybe asymptomatic on arrival, especially those who took acetaminophen that has no reliable clinical signs of acute intoxication. However, many others will be manifesting signs of intoxication which can be detected on a thorough physical exam.

Physical exam of the overdose patient should be systematic and focus on findings that can help determine what the patient may have been exposed to. There are constellations of physical exam findings, known as toxidromes. Presence of a particular toxidrome can help identify what substance a patient may have been exposed to. There are five common toxidromes which should be evaluated for the following:

1. The opioid toxidrome consists of central nervous system (CNS) depression, respiratory depression, pin-point pupils or miosis, and decreased bowel motility. Typically, heart rate and blood pressure are only minimally affected. Not all opioids cause miosis. It is important to note that the most clinically significant component of the opioid toxidrome is the respiratory depression that can lead to apnea and death. Classically, this toxidrome was associated with heroin or prescription opioid use. However, now an increasing percentage of cases are associated with illicit fentanyl use which is used to either adulterate heroin or completely substituted for heroin [12].
2. The sedative—hypnotic toxidrome is typified by profound CNS depression with relatively preserved heart rate, blood pressure, and respiratory rate. While apnea from a sedative—hypnotic is rare, the significant CNS depression can lead to loss of airway reflexes resulting in aspiration and other complications. Ethanol is by far the number one cause of sedative—hypnotic toxidrome, though benzodiazepine toxicity makes up a significant and rising proportion [13]. Synergistic toxicity can be seen in cases where ethanol and benzodiazepines are present.
3. The sympathomimetic toxidrome consists of psychomotor agitation, tachycardia, hypertension, mydriasis, diaphoresis, and, in severe cases, hyperthermia. Drugs such as cocaine and methamphetamine are common causes of sympathomimetic toxidrome, though in recent years the substituted cathinones (often called "bath salts") also have to be considered as possible causes [14]. The sympathomimetic toxidrome is one of the leading causes of excited delirium, which is associated with sudden death [15]. The hypertension associated with sympathomimetic toxidrome can be severe and lead to complications such as myocardial infarctions and strokes [16]. Hyperthermia, if present, carries with it a poor prognosis and should be treated aggressively [17].
4. The anticholinergic or antimuscarinic toxidrome manifests very similarly to the sympathomimetic toxidrome, though via a completely different mechanism. Whereas the sympathomimetic toxidrome is caused by the excessive effects of monoamine neurotransmitters such as norepinephrine and dopamine, the anticholinergic or, more aptly named, antimuscarinic toxidrome is the result of inhibition of parasympathetic and sympathetic muscarinic receptors. The result is a constellation of delirium, hallucinations, tachycardia, anhidrosis, mydriasis, urinary retention, and dry mouth or xerostomia. Hypertension, though usually not severe, and hyperthermia can be present due to agitation and lack of ability to dissipate heat via sweating. Multiple prescription drugs are potent antimuscarinic agents, including tricyclic antidepressants, cyclobenzaprine, and older antipsychotics. Diphenhydramine, an over the counter first-generation antihistamine, is the most common antimuscarinic drug involved in overdoses [4].
5. Cholinergic toxidrome is the least commonly encountered toxidrome. It is caused by excessive stimulation of cholinergic receptors resulting in salivation, lacrimation, emesis, diarrhea, diaphoresis, bronchorrhea, bronchoconstriction, and miosis. In more severe cases delirium, seizures, fasciculations, and paralysis can occur. Most cases of cholinergic toxidrome are attributed to organophosphate and carbamate insecticide toxicity. Nerve agents, such as sarin, cause severe cholinergic toxidrome. While cholinergic toxicity is rare in North America, it is commonly seen in many Southeast Asian countries due to suicide attempts with organophosphate pesticides [18].

It should be noted that in cases where multiple drugs are involved, the result will be a mixing of toxidromes and a resultant nonspecific physical exam. For instance, patients who inject cocaine and heroin simultaneously may present not only with hypertension and tachycardia but also CNS depression and respiratory depression.

Along with looking for signs consistent with toxidromes, there are several other physical exam findings that can help narrow down what substance or substances may be involved in an overdose. Nystagmus or repetitive, uncontrolled movements of the eyes are classically associated with phencyclidine (angel dust) but can also be seen in overdoses involving multiple other substances, including phenytoin and other anticonvulsants [19,20]. Bruxism or grinding of the teeth is associated with amphetamine toxicity, especially methylenedioxymethamphetamine (ecstasy) [21]. Clonus, which is involuntary, rhythmic, muscular contractions, and relaxation, particular in the lower extremities, is considered a pathognomonic sign of proserotonergic drug toxicity such as the monoamine oxidase inhibitor antidepressants and selective serotonin reuptake inhibitor antidepressants [22]. These drugs can cause a particular type of toxicity called serotonin syndrome or serotonin toxicity. Serotonin toxicity is a hyperkinetic state typified by CNS dysfunction, autonomic dysfunction (classically hyperthermia) and neuromuscular dysfunction (classically spontaneous clonus). Serotonin toxicity is usually associated with acute exposures to two or more serotonergic drugs [22]. Choreoathetosis that is an involuntary movement disorder characterized by writhing and twisting can be seen with cocaine (crack dancing) or methamphetamine abuse [23]. The presence of seizures in an overdose patient may indicate an exposure to a proconvulsant substance such as cocaine, bupropion, citalopram, tricyclic antidepressants, or isoniazid [24]. Conversely, many drugs can mimic brain death in overdoses. Tricyclics antidepressants, baclofen, and valproate acid are among CNS depressing drugs which have been associated with mimicking brain death [25]. Compartment syndrome, which is muscle ischemia and necrosis caused by increased tissue pressure within a closed fascial space, should be suspected and evaluated for in any overdose with altered mental status or who was immobilized for any period of time [26]. Gluteal compartment syndrome seems especially common in overdoses [27]. Missed compartment syndrome can lead to renal failure and significant permanent morbidity [28].

The electrocardiogram (EKG) can be a valuable test in the overdose patient that should be obtained early in the evaluation process. Bradycardia, including complete heart block, may be seen in overdoses of calcium channel blockers, beta blockers, digoxin, and clonidine [29]. Multiple drugs such as cocaine, tricyclic antidepressants, lamotrigine, and diphenhydramine can cause sodium channel blockade that manifests as a widening of the QRS complex which is the duration of the Q wave, R wave, and S wave on the EKG and an R wave in lead aVR [30]. Identification of these abnormalities on the EKG can allow for rapid intervention with sodium bicarbonate which may help prevent life-threatening dysrhythmias [31]. QT prolongation should also be looked for on the EKG as many drugs, such as methadone and antipsychotics, may prolong the QT interval putting patients at risk of torsade de pointes [32].

Laboratory tests are frequently obtained during the evaluation of the overdose patient. In cases where history and physical exam are not revealing, laboratory tests can be helpful but should always be correlated with the patient's clinical picture. The most important laboratory test to obtain in any overdose patient is a serum acetaminophen level [33]. Detection of a toxic acetaminophen level allows for the effective early treatment of what otherwise could be a fatal ingestion. Serum levels of other substances, such as aspirin, iron, lithium or digoxin, may also be useful in the appropriate clinical situation. While the complete blood count is rarely useful, basic metabolic labs can identify an anion gap metabolic acidosis which could be caused by ingestions of substances such as iron, ethylene glycol, methanol, or aspirin [34]. Routine evaluation of serum creatinine kinase levels may have benefit as rhabdomyolysis, the breakdown of muscle tissue, is a common complication of drug overdoses [35]. Calculation of the osmolar gap may help determine the presence of one of the so-called toxic alcohols: isopropyl alcohol, ethylene glycol, methanol, or diethylene glycol [36]. The osmolar gap is calculated by subtracting the calculated serum osmolarity from the lab-derived serum osmolality. A large (>30) osmolar gap can help confirm the ingestion of a toxic alcohol but it is important to remember that a normal osmolar gap does not rule out such an ingestion and overall the osmolar gap is a test of limited utility [37]. The urine immunoassay drug screen (UDS) is commonly obtained when evaluating the overdose patient but unfortunately is a qualitative test plagued by false-positives that has no utility in determining if a patient's current symptoms are due to an overdose. Multiple studies have demonstrated that the UDS rarely alters acute clinical management [38].

Radiographs are typically of limited utility in evaluating the overdose patient. In cases where the history is unclear and the patient has altered mental status or focal neurological deficits, computerize tomography of the head may be needed to rule out intracranial processes such as hemorrhage or stroke. Plain radiographs of the chest may be indicated if there is concern for aspiration. Abdominal radiographs are sometimes obtained to evaluate for the presence of ingested substances. These plain radiographs have poor sensitivity but can detect certain ingestions such as iron and calcium containing pills along with phenothiazine antipsychotics, tricyclic antidepressants, and chloral hydrate [39]. Computerized tomography of the abdomen and pelvis can be used to evaluate for the presence of swallowed packets of drugs that are frequently missed on plain radiographs [40].

Treatment

Treatment of the overdose patient is based on excellent symptomatic and supportive care. This begins with addressing airway, breathing and circulation. Patients may require supplemental oxygen or in more severe cause, endotracheal intubation and mechanical ventilation. Hypotension associated with overdoses should first be treated with aggressive intravenous fluids. If refractory, then vasopressors along with ingestion-specific treatments should be initiated. For beta antagonist induced hypotension glucagon can be an effective treatment and for calcium channel blocker overdoses, high-dose insulin euglycaemia therapy can be lifesaving [6]. In cases of hypertensive emergencies due to cocaine or methamphetamine toxicity, chemical sedation with benzodiazepines should be first line and use of an alpha antagonist such as phentolamine can be considered in refractory cases. Use of beta blockers in this population remains controversial but is likely safer than is often stated in older literature [41]. Psychomotor agitation caused by any overdose should be controlled with aggressive chemical sedation first and foremost. Benzodiazepines, antipsychotics, and ketamine appear to all be effective to some degree and frequently combination therapy may be needed [42,43]. Physical restraints may be required for both patient and staff safety but should always be initiated with concurrent aggressive chemical sedation.

The decontamination of the overdose patient should be considered as a part of early management but routine implementation is not recommended due to multiple limitations, contraindications, and lack of evidence demonstrating clinical benefit. The theoretical goal of decontamination is to reduce the absorption of whatever substance the patient may have ingested. Options for decontamination include gastric lavage, activated charcoal, and whole bowel irrigation. Gastric lavage involves the placement of a large bore naso- or orogastric tube and subsequent lavage with normal saline to remove gastric contents. There is no evidence that routine use of gastric lavage has any clinical benefit and it should only be reserved for the rare patients who may have ingested a potentially life-threatening amount of a substance, who present within 60 minutes of ingestion and their airway is intact or protected [44]. Activated charcoal is much more commonly used in the treatment of the overdose patient but even it should not be used routinely. Due to its massive surface area, activated charcoal readily absorbs most substances, though it has limited effectiveness for metals and other polar molecules, such as alcohol [45]. It should be considered for patients who present within 60 minutes of a potentially toxic ingestion who are not at risk of airway compromise or seizures and have not ingested a caustic, hydrocarbon, alcohol, or metal [45]. The typical dose of activated charcoal is 1 g/kg. Multidose-activated charcoal is a variation of charcoal decontamination which may be of benefit in selected few ingestions. It can be considered for patients who have ingested a life-threatening amount of carbamazepine, dapsone, phenobarbital, quinine, or theophylline and have a secure airway [46]. Multidose-activated charcoal is typically given as an initial dose of 1 g/kg then 0.5 g/kg every 4—6 hours for 24 hours. Whole bowel irrigation is a decontamination technique where a polyethylene glycol (PEG) solution is used to enhance bowel motility and results in the rapid expulsion of ingested substances. Doses of up to 2 L of PEG per hour may be needed and is given till rectal effluent is clear. Whole bowel irrigation can be considered for potentially toxic ingestions of sustained-release or enteric-coated drugs and is widely used for acute iron and lithium overdoses and for those who ingest packets of drugs (body stuffers) [47]. It should only be used if the patient has a functional, intact gastrointestinal tract, and a secure airway.

While the true clinical benefit of decontamination is unclear, the adverse effects are unfortunately well documented. Gastric lavage is associated with esophageal perforations and activated charcoal can cause significant lung injury if aspirated [48]. These complications highlight why the decision to use decontamination techniques should not be routine but rather made after careful risk/benefit analysis.

While symptomatic and supportive care remains the gold-standard treatment for a vast majority of overdoses, there are some specific antidotes that can be potentially lifesaving. Naloxone, a competitive opioid receptor antagonist, is one of the most commonly used antidotes today due to the ongoing opioid epidemic. When given intravenously or intranasally it rapidly reverses the effects of opioids, including the life-threatening respiratory depression [49]. The dose of naloxone is variable and practitioner dependent but typically doses of 0.4 mg are effective for most opioid overdoses [50]. Care should be taken to avoid precipitating acute opioid withdrawal as iatrogenic opioid withdrawal is associated with rare but significant morbidity [51]. Flumazenil, a benzodiazepine receptor antagonist, can be used to rapidly reverse the effects of benzodiazepines but there remains significant controversy around its use [52]. It should not be used routinely but rather in small (0.2 mg), titrated doses for select cases of benzodiazepine ingestions where intubation is being considered. It should never be used in polysubstance ingestions, especially those involving seizure causing drugs such as tricyclic antidepressants [53]. N-Acetylcysteine (NAC) is the antidotal treatment for acetaminophen toxicity. Whether given intravenous or orally, NAC serves to prevent acetaminophen-induced liver injury by replenishing glutathione that detoxifies the hepatotoxic metabolite of acetaminophen. So long it is administered within 8 hours of acetaminophen ingestion it is highly effective [54]. Other antidotes such as digoxin immune FAB for digoxin overdoses, octreotide for sulfonylurea overdoses, pyridoxine for isoniazid toxicity, and deferoxamine for iron overdoses can all help significantly reduce morbidity and mortality [55—58].

While not a true antidote, intravenous lipid emulsion (ILE) is a unique treatment that can be considered for specific overdoses that result in cardiovascular instability or even in cardiac arrest [59]. Intralipid 20% is the most commonly used form of lipid emulsion. Though the exact mechanism of action is unclear, administration of ILE is thought to create a "third space" that sequesters lipophilic drugs [60]. The best evidence for ILE use is for the treatment of local anesthetic overdoses, particularly bupivacaine [61]. There are also reports of using ILE for severe overdoses of several drugs, including verapamil, propranolol, and amitriptyline [62—64]. Use of ILE is associated with severe adverse events, including interference with certain laboratory testing and pancreatitis [65].

The removal of drugs via hemodialysis and other renal replacement therapy options can be lifesaving interventions for several overdoses [66]. Typically for hemodialysis to be effective in removing a substance is must be small (<500 Da), have a small volume of distribution (<1 kg/L), and not be highly protein bond [66]. Some overdoses for which hemodialysis may be indicated include aspirin, methanol, ethylene glycol, lithium, valproic acid, theophylline, and metformin [66]. A less commonly used form of extracorporeal drug removal is plasmapheresis or therapeutic plasma exchange that removes and replaces the patient's plasma portion of the blood. The literature supporting plasmapheresis in the management of overdose is limited but this technique has been used for rare overdose of antibiotics and immunosuppressants [67].

In the last decade the use of extracorporeal membrane oxygenation (ECMO) for severe overdoses has become more common treatment option at larger health centers [68]. ECMO can be a potentially lifesaving intervention for those patients who have ingested calcium channel blockers, beta blockers, and several other toxins known to cause cardiovascular collapse [69]. It supports the patient's basic circulatory function, allowing them the necessary time to metabolize and eliminate the offending substances.

The disposition of the overdose patient will vary greatly, depending on what signs and symptoms are present and their severity. The overdose patients who present with symptoms will need to be treated and observed till they have returned to their premorbid baseline. While occasionally these patients can be treated and recover in the emergency department, more often than not they will need to be admitted to a health-care facility. This may require admission to a critical care unit if they are extremely agitated, acidotic, sedated, or hypotensive [70]. Duration of symptoms may be hours or even days depending on the exposure. Overdoses of modified release substances are classically associated with prolonged toxicity. Other drugs may have prolonged toxicity due to altered metabolism, such as phenytoin [71]. In massive ingestions of any substance, toxicity can be prolonged due to erratic absorption caused by pharmacobezoar formation [72].

Suspected overdose patients who present without symptoms should be observed for a minimum of 6 hours for any signs of toxicity [11]. With rare exceptions, if the patients remain asymptomatic during this time frame, they can be medically cleared. This "6 hour rule" only holds true if the exposure was to an immediate release substance. Any overdose of a modified release substance warrants a 24-hour observation period due to risk of delayed effects [73,74]. On the extreme end of the spectrum, some rare overdoses, such as thyroid hormone and vitamin K antagonist, can result in toxicity that does not manifest for several days [75,76].

Regardless of whether they are admitted to the hospital or discharged from the emergency department, all overdose patients should be evaluated for suicidality prior to final disposition and referred to appropriate psychiatric or drug rehabilitation resources.

References

[1] United Nations Office on Drugs and Crime. World drug report. United Nations Publication, Sales No. E. 17. XI.6; 2017, ISBN: 978-92-1-148291-1. eISBN: 978-92-1-060623-3.

[2] Moore BJ, Stocks C, Owens PL. Trends in emergency department visits, 2006–2014. HCUP statistical brief #227. Rockville, MD: Agency for Healthcare Research and Quality. September 2017.

[3] Hedegaard H, Warner M, Miniño AM. Drug overdose deaths in the United States, 1999–2016. NCHS data brief, no 294. Hyattsville, MD: US Department of Health and Human Services, CDC, National Center for Health Statistics; 2017.

[4] Gummin DD, Mowry JB, Spyker DA, Brooks DE, Fraser MO, Banner W. 2016 Annual Report of the American Association of Poison Control Centers' National Poison Data System (NPDS): 34th Annual Report. Clin Toxicol 2017;55(10):1072–254.

[5] Fox LM, Hoffman RS, Vlahov D, Manini AF. Risk factors for severe respiratory depression from prescription opioid overdose. Addiction 2018;113(1):59–66.

[6] Graudins A, Lee HM, Druda D. Calcium channel antagonist and beta-blocker overdose: antidotes and adjunct therapies. Br J Clin Pharmacol 2016;81(3):453–61.

[7] Papadopoulos DP, Sanidas EA, Viniou NA, Gennimata V, Chantziara V, Barbetseas I, et al. Cardiovascular hypertensive emergencies. Curr Hypertens Rep 2015;17(2):5.

[8] Viglino D, Bourez D, Collomb-Muret R, Schwebel C, Tazarourte K, Dumanoir P, et al. Noninvasive end tidal CO_2 is unhelpful in the prediction of complications in deliberate drug poisoning. Ann Emerg Med 2016;68(1):62–70.e1.

[9] Brady Jr WJ, Duncan CW. Hypoglycemia masquerading as acute psychosis and acute cocaine intoxication. Am J Emerg Med 1999;17(3):318–19.

[10] Soderstrom J, Murray L, Daly FF, Little M. Toxicology case of the month: oral hypoglycaemic overdose. Emerg Med J 2006;23(7):565–7.

[11] Hollander JE, McCracken G, Johnson S, Valentine SM, Shih RD. Emergency department observation of poisoned patients: how long is necessary? Acad Emerg Med 1999;6(9):887–94.

[12] O'Donnell JK, Gladden RM, Seth P. Trends in deaths involving heroin and synthetic opioids excluding methadone, and law enforcement drug product reports, by census region — United States, 2006–2015. MMWR Morb Mortal Wkly Rep 2017;66(34):897–903.

[13] Bachhuber MA, Hennessy S, Cunningham CO, Starrels JL. Increasing benzodiazepine prescriptions and overdose mortality in the United States, 1996–2013. Am J Public Health 2016;106(4):686–8.

[14] Warner M, Trinidad JP, Bastian BA, Minino AM, Hedegaard H. Drugs most frequently involved in drug overdose deaths: United States, 2010–2014. Natl Vital Stat Rep 2016;65(10):1–15.

[15] Vilke GM, Bozeman WP, Dawes DM, Demers G, Wilson MP. Excited delirium syndrome (ExDS): treatment options and considerations. J Forensic Leg Med 2012;19(3):117–21.

[16] Richards JR, Garber D, Laurin EG, Albertson TE, Derlet RW, Amsterdam EA, et al. Treatment of cocaine cardiovascular toxicity: a systematic review. Clin Toxicol (Phila) 2016;54(5):345–64.

[17] Callaway CW, Clark RF. Hyperthermia in psychostimulant overdose. Ann Emerg Med 1994;24(1):68–76.

[18] Mew EJ, Padmanathan P, Konradsen F, Eddleston M, Chang SS, Phillips MR, et al. The global burden of fatal self-poisoning with pesticides 2006–15: systematic review. J Affect Disord 2017;219:93–104.

[19] Dominici P, Kopec K, Manur R, Khalid A, Damiron K, Rowden A. Phencyclidine intoxication case series study. J Med Toxicol 2015;11(3):321–5.

[20] Praveen-kumar S, Desai M. Ocular motor abnormalities in a patient with phenytoin toxicity—case report and minireview. Clin Neurol Neurosurg 2014;127:116–17.

[21] Dinis-Oliveira RJ, Caldas I, Carvalho F, Magalhães T. Bruxism after 3,4-methylenedioxymethamphetamine (ecstasy) abuse. Clin Toxicol (Phila) 2010;48(8):863–4.

[22] Dunkley EJ, Isbister GK, Sibbritt D, Dawson AH, Whyte IM. The hunter serotonin toxicity criteria: simple and accurate diagnostic decision rules for serotonin toxicity. QJM 2003;96(9):635–42.

[23] Narula N, Siddiqui F, Katyal N, Krishnan N, Chalhoub M. Cracking the crack dance: a case report on cocaine-induced choreoathetosis. Cureus 2017;9(12):e1981.

[24] Reichert C, Reichert P, Monnet-Tschudi F, Kupferschmidt H, Ceschi A, Rauber-Lüthy C. Seizures after single-agent overdose with pharmaceutical drugs: analysis of cases reported to a poison center. Clin Toxicol (Phila) 2014;52(6):629–34.

[25] Neavyn MJ, Stolbach A, Greer DM, Nelson LS, Smith SW, Brent J, et al. American College of Medical Toxicology. ACMT position statement: determining brain death in adults after drug overdose. J Med Toxicol 2017;13(3):271–3.

[26] Torrens C, Marin M, Mestre C, Alier A, Nogues X. Compartment syndrome and drug abuse. Acta Orthop Belg 1993;59(2):143–6.

[27] Benns M, Miller K, Harbrecht B, Bozeman M, Nash N. Heroin-related compartment syndrome: an increasing problem for acute care surgeons. Am Surg 2017;83(9):962–5.

[28] Kumar R, West DM, Jingree M, Laurence AS. Unusual consequences of heroin overdose: rhabdomyolysis, acute renal failure, paraplegia and hypercalcaemia. Br J Anaesth 1999;83(3):496–8.

[29] Jang DH, Spyres MB, Fox L, Manini AF. Toxin-induced cardiovascular failure. Emerg Med Clin North Am 2014;32(1):79–102.

[30] Lester L, McLaughlin S. SALT: a case for the sodium channel blockade toxidrome and the mnemonic SALT. Ann Emerg Med 2008;51(2):214.

[31] Bruccoleri RE, Burns MM. A literature review of the use of sodium bicarbonate for the treatment of QRS widening. J Med Toxicol 2016;12(1):121–9.

[32] Miura N, Saito T, Taira T, Umebachi R, Inokuchi S. Risk factors for QT prolongation associated with acute psychotropic drug overdose. Am J Emerg Med 2015;33(2):142–9.

[33] Lucanie R, Chiang W, Reilly R. Utility of acetaminophen screening in unsuspected suicidal ingestions. Vet Hum Toxicol 2002;44(3):171–3.

[34] Casaletto JJ. Differential diagnosis of metabolic acidosis. Emerg Med Clin North Am 2005;23(3):771–87.

[35] Köppel C. Clinical features, pathogenesis and management of drug-induced rhabdomyolysis. Med Toxicol Adverse Drug Exp 1989;4(2):108–26.

[36] Liamis G, Filippatos TD, Liontos A, Elisaf MS. Serum osmolal gap in clinical practice: usefulness and limitations. Postgrad Med 2017;129(4):456–9.

[37] Glaser DS. Utility of the serum osmol gap in the diagnosis of methanol or ethylene glycol ingestion. Ann Emerg Med 1996;27(3):343–6.

[38] Montague RE, Grace RF, Lewis JH, Shenfield GM. Urine drug screens in overdose patients do not contribute to immediate clinical management. Ther Drug Monit 2001;23(1):47–50.

[39] Chan YC, Lau FL, Chan JCS, Hon TWY. A study of drug radiopacity by plain radiography. Hong Kong J Emerg Med 2004;11:205–10.

[40] Schulz B, Grossbach A, Gruber-Rouh T, Zangos S, Vogl TJ, Eichler K. Body packers on your examination table: how helpful are plain x-ray images? A definitive low-dose CT protocol as a diagnosis tool for body packers. Clin Radiol 2014;69(12):e525–30.

[41] Richards JR, Hollander JE, Ramoska EA, Fareed FN, Sand IC, Izquierdo Gómez MM, et al. β-Blockers, cocaine, and the unopposed α-stimulation phenomenon. J Cardiovasc Pharmacol Ther 2017;22(3):239–49.

[42] Zeller SL, Rhoades RW. Systematic reviews of assessment measures and pharmacologic treatments for agitation. Clin Ther 2010;32(3):403–25.

[43] Riddell J, Tran A, Bengiamin R, Hendey GW, Armenian P. Ketamine as a first-line treatment for severely agitated emergency department patients. Am J Emerg Med 2017;35(7):1000–4.

[44] Vale JA. Position statement: gastric lavage. American Academy of Clinical Toxicology; European Association of Poisons Centres and Clinical Toxicologists. J Toxicol Clin Toxicol 1997;35(7):711–19.

[45] Chyka PA, Seger D. Position statement: single-dose activated charcoal. American Academy of Clinical Toxicology; European Association of Poisons Centres and Clinical Toxicologists. J Toxicol Clin Toxicol 1997;35(7):721–41.

[46] [No authors listed]. Position statement and practice guidelines on the use of multi-dose activated charcoal in the treatment of acute poisoning. American Academy of Clinical Toxicology; European Association of Poisons Centres and Clinical Toxicologists. J Toxicol Clin Toxicol 1999;37(6):731–51.

[47] [No authors listed]. Position paper: whole bowel irrigation. J Toxicol Clin Toxicol 2004;42(6):843–54.

[48] Askenasi R, Abramowicz M, Jeanmart J, Ansay J, Degaute JP. Esophageal perforation: an unusual complication of gastric lavage. Ann Emerg Med 1984;13(2):146; Elliott CG, Colby TV, Kelly TM, Hicks HG. Charcoal lung. Bronchiolitis obliterans after aspiration of activated charcoal. Chest 1989;96(3):672–4.

[49] O'Brien CP, Greenstein R, Ternes J, Woody GE. Clinical pharmacology of narcotic antagonists. Ann NY Acad Sci 1978;311:232–40.

[50] Rzasa Lynn R, Galinkin JL. Naloxone dosage for opioid reversal: current evidence and clinical implications. Ther Adv Drug Saf 2018;9(1):63–88.

[51] Buajordet I, Naess AC, Jacobsen D, Brørs O. Adverse events after naloxone treatment of episodes of suspected acute opioid overdose. Eur J Emerg Med 2004;11(1):19–23.

[52] Goldfrank LR. Flumazenil: a pharmacologic antidote with limited medical toxicology utility, or . . . an antidote in search of an overdose. Acad Emerg Med 1997;4(10):935–6.

[53] Haverkos GP, DiSalvo RP, Imhoff TE. Fatal seizures after flumazenil administration in a patient with mixed overdose. Ann Pharmacother 1994;28(12):1347–9.

[54] Heard K, Green J. Acetylcysteine therapy for acetaminophen poisoning. Curr Pharm Biotechnol 2012;13(10):1917–23.

[55] Chan BS, Buckley NA. Digoxin-specific antibody fragments in the treatment of digoxin toxicity. Clin Toxicol (Phila) 2014;52(8):824–36.

[56] Glatstein M, Scolnik D, Bentur Y. Octreotide for the treatment of sulfonylurea poisoning. Clin Toxicol (Phila) 2012;50(9):795–804.

[57] Tajender V, Saluja J. INH induced status epilepticus: response to pyridoxine. Indian J Chest Dis Allied Sci 2006;48(3):205–6.

[58] Levison LH. Specific chelation therapy with desferrioxamine in massive iron overdose. Appl Ther 1969;11(3):153.

[59] Ozcan MS, Weinberg G. Intravenous lipid emulsion for the treatment of drug toxicity. J Intensive Care Med 2014;29(2):59–70.

[60] Kryshtal DO, Dawling S, Seger D, Knollmann BC. In vitro studies indicate intravenous lipid emulsion acts as lipid sink in verapamil poisoning. J Med Toxicol 2016;12(2):165–71.

[61] Corman SL, Skledar SJ. Use of lipid emulsion to reverse local anesthetic-induced toxicity. Ann Pharmacother 2007;41(11):1873–7.

[62] Young AC, Velez LI, Kleinschmidt KC. Intravenous fat emulsion therapy for intentional sustained-release verapamil overdose. Resuscitation 2009;80(5):591–3.

[63] Dean P, Ruddy JP, Marshall S. Intravenous lipid emulsion in propranolol [corrected] overdose. Anaesthesia 2010;65(11):1148–50.

[64] Agarwala R, Ahmed SZ, Wiegand TJ. Prolonged use of intravenous lipid emulsion in a severe tricyclic antidepressant overdose. J Med Toxicol 2014;10(2):210–14.

[65] Bucklin MH, Gorodetsky RM, Wiegand TJ. Prolonged lipemia and pancreatitis due to extended infusion of lipid emulsion in bupropion overdose. Clin Toxicol (Phila) 2013;51(9):896–8.

[66] Mirrakhimov AE, Barbaryan A, Gray A, Ayach T. The role of renal replacement therapy in the management of pharmacologic poisonings. Int J Nephrol 2016;2016:3047329. Available from: https://doi.org/10.1155/2016/3047329. Epub 2016 Nov 30.

[67] Ibrahim RB, Liu C, Cronin SM, Murphy BC, Cha R, Swerdlow P, et al. Drug removal by plasmapheresis: an evidence-based review. Pharmacotherapy 2007;27(11):1529–49.

[68] Wang GS, Levitan R, Wiegand TJ, Lowry J, Schult RF, Yin S. Toxicology investigators consortium. Extracorporeal membrane oxygenation (ECMO) for severe toxicological exposures: review of the toxicology investigators consortium (ToxIC). J Med Toxicol 2016;12(1):95–9.

[69] de Lange DW, Sikma MA, Meulenbelt J. Extracorporeal membrane oxygenation in the treatment of poisoned patients. Clin Toxicol (Phila) 2013;51(5):385–93.

[70] van den Oever HLA, van Dam M, van 't Riet E, Jansman FGA. Clinical parameters that predict the need for medium or intensive care admission in intentional drug overdose patients: A retrospective cohort study. J Crit Care 2017;37:156–61.

[71] Craig S. Phenytoin poisoning. Neurocrit Care 2005;3(2):161–70.

[72] Magdalan J, Zawadzki M, Słoka T, Sozański T. Suicidal overdose with relapsing clomipramine concentrations due to a large gastric pharmacobezoar. Forensic Sci Int 2013;229(1–3):e19–22.

[73] Starr P, Klein-Schwartz W, Spiller H, Kern P, Ekleberry SE, Kunkel S. Incidence and onset of delayed seizures after overdoses of extended-release bupropion. Am J Emerg Med 2009;27(8):911–15.

[74] Brubacher JR, Dahghani P, McKnight D. Delayed toxicity following ingestion of enteric-coated divalproex sodium (Epival). J Emerg Med 1999;17(3):463–7.

[75] Hays HL, Jolliff HA, Casavant MJ. Thyrotoxicosis after a massive levothyroxine ingestion in a 3-year-old patient. Pediatr Emerg Care 2013;29(11):1217–19.

[76] Isbister GK, Hackett LP, Whyte IM. Intentional warfarin overdose. Ther Drug Monit 2003;25(6):715–22.

Part II

Overview and case studies

Chapter 5

Alcohols: volatiles and glycols

Uttam Garg[1,2] and Hema Ketha[3]

[1]Department of Pathology and Laboratory Medicine, Children's Mercy Hospital, Kansas City, MO, United States, [2]Pathology and Laboratory Medicine, University of Missouri School of Medicine, Kansas City, MO, United States, [3]Toxicology and Mass Spectrometry, Center for Esoteric Testing, Laboratory Corporation of America Holdings, Burlington, NC, United States

Introduction

An alcohol is an organic compound that has a hydroxyl functional group on a straight or branched chain aliphatic hydrocarbon. Clinically relevant alcohols can be roughly divided into volatiles and glycols. Volatile alcohols include ethanol, methanol, and isopropanol. Ethanol is the most commonly encountered volatile in clinical and forensic toxicology. Glycols have two hydroxyl groups (diol) and are not considered volatiles due to their high boiling point. In clinical settings, ethylene glycol is the most commonly encountered glycol. Both enzymatic and chromatographic methods are available for the determination of ethanol and ethylene glycol. Other volatiles and glycols are primarily measured by gas chromatography (GC) linked to flame ionization detectors (FIDs) or by mass spectrometry (MS). The toxicity of alcohols varies from relatively nontoxic to highly toxic. Specific treatments are available for certain volatiles and glycols. A number of preanalytical, analytical, and postanalytical factors can impact results and interpretation [1–3]. Preanalytical factors are more variable for forensic samples; thus results should be interpreted in light of these factors.

Ethanol

Ethanol is a two-carbon aliphatic alcohol that is weakly polar and is both water and fat soluble. It is a fermentation product of sugars present in a variety of foods such as grains and fruits. During fermentation, one molecule of glucose produces two molecules of ethanol and two molecules of carbon dioxide. Yeast fermentation is the most common way to produce ethanol with concentrations up to approximately 15%. Since yeast cannot grow at concentrations above 15%, only ciders, beer, and wine can be produced by direct fermentation. Liquors with higher ethanol concentrations are created by ethanol supplementation produced by distillation. The typical concentration of alcoholic beverages is as follows: beers 3%–8%, wines 10%–16%, and liqueurs and spirits 35%–60%. Other common sources of ethanol are mouthwashes, cough and cold medications, hand sanitizers, antiseptic solutions, perfumes, and many household products. The term alcohol by volume (alc/vol) is a measure of the amount of ethanol in an alcoholic beverage also defined as volume percent. An alternative terminology used to is alcohol proof which is twice the alc/vol in the United States compared to 1.75 times the alc/vol in the United Kingdom. For example, an 80 proof alcoholic drink in the United States is 40% ethanol by volume.

Ethanol is the most widely used social drug in the world, and its abuse has been associated with high morbidity and mortality [4]. According to the 2015 National Survey on Drug Use and Health, 86.4% of people ages 18 or older reported that they drank alcohol at some point in their lifetime; 70.1% reported that they drank in the past year; and 56.0% reported that they drank in the past month. It is estimated that 88,000 people (approximately 62,000 men and 26,000 women) die annually from alcohol-related causes [4]. It is the leading cause of motor vehicle accidents, drowning, occupational injury, and criminal violence. It is estimated that driving under the influence of ethanol results in more than 10,000 fatalities every year in the United States alone.

Pharmacology of ethanol

Pharmacokinetics

Although ethanol can reach the blood stream by inhalation, dermal exposure, and intravenous administration, oral consumption is the most common mode of ethanol exposure. When consumed orally, bioavailability of ethanol is almost 100% with peak concentration reaching at 20–60 minutes [5–7]. Ethanol is absorbed passively, primarily in the small intestine (>75%). Significant amounts of ethanol can also be absorbed through the stomach, particularly in the presence of food and slower gastric emptying; however, the presence of food, overall, delays the absorption of ethanol. Ethanol absorption is also influenced by other factors. For example, carbonated beverages and gastrointestinal motility increase and fatty foods decrease ethanol absorption. Beverages with ethanol concentrations ranging from 10% to 30% are absorbed more rapidly than very dilute or concentrated beverages [5,6,8].

Ethanol is mostly (>90%) metabolized in the liver. Approximately 2% is excreted, unchanged, through urine, and a small amount is eliminated through exhaled air and sweat [9]. Primarily, at concentrations above 0.01 g/dL, ethanol metabolism follows zero-order kinetics (i.e., a certain amount of ethanol is eliminated per unit time, independent of blood concentration). A significant amount of research has been done on the metabolism of ethanol, including the pioneering work of Widmark [7,10]. Many factors such as alcohol dehydrogenase (ADH) isoenzymes, enzyme induction, coadministration of other drugs, liver disease, and genetic factors can significantly affect ethanol metabolism and clearance. It can vary from 0.01 to over 0.03 g/dL/h. The average elimination rate of ethanol is 0.015–0.018 g/dL/h or 7–10 g/h, while the average volume of distribution (V_d) of ethanol is 0.7 L/kg in males and 0.6 L/kg in females [7,10]. V_d decreases with age and adiposity. It is important to keep in mind that ethanol elimination can be prolonged if absorption from the gut is not complete.

The major pathway of ethanol metabolism is shown in Fig. 5.1. It involves ADH and acetaldehyde dehydrogenase. Ethanol metabolism can also happen by another enzyme system called CYP2E1, originally designated as microsomal ethanol oxidizing system. CYP2E1 is inducible and can play a significant role in ethanol metabolism in chronic alcoholics. It can also play a larger role in metabolism at high ethanol concentrations.

Pharmacodynamics

Ethanol effects can be broadly divided into two categories: acute and chronic. The central nervous system (CNS) is most markedly affected by acute ethanol intake. Ethanol causes CNS depression leading to sedation, slurred speech, impaired judgment, uninhibited behavior, euphoria, and impaired sensory and motor skills. Continued increase in ethanol concentration and CNS depression leads to confusion, stupor, coma and, finally, death. Ethanol's CNS-depressant effects are potentiated by other CNS depressants such as barbiturates, benzodiazepines, antihistamines, and opioids. Other acute effects include depression in myocardial contractibility and the inhibition of an antidiuretic hormone leading to increased urine output, dehydration, and irritation of gastric mucosa. Chronic effects of

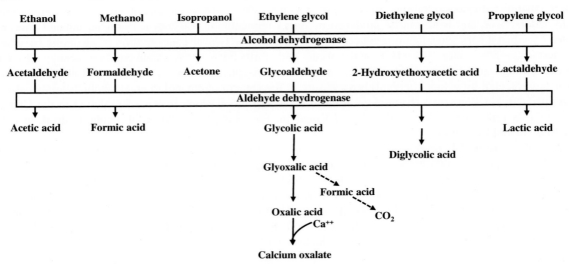

FIGURE 5.1 Metabolism of ethanol and other alcohols. Alcohol dehydrogenase and aldehyde dehydrogenase sequentially metabolize alcohols to generate toxic metabolites. Alcohol dehydrogenase is the target for enzyme inhibition therapy.

ethanol include liver injury, ethanol dependence, impaired memory, anemia resulting from folate and iron deficiency, hyperlipidemia, and cardiomyopathy. Ethanol use during pregnancy and intrauterine exposure to ethanol may result in fetal alcohol syndrome (FAS) and fetal alcohol spectrum disorders (FASD). FAS and FASD are the most common non-heritable causes of intellectual disability and are characterized by facial abnormalities, stunted growth, microcephaly, low IQ, poor coordination, and behavior problems [11].

Treatment of ethanol abuse

Alcohol addiction and dependency is associated with many clinical, financial, and social consequences for the patient and the immediate family members. The cornerstone of ethanol abuse treatment is based on counseling by a healthcare provider. The complete treatment process involves acute detoxification, long-term counseling, and pharmacological treatment to prevent relapse [12,13]. Benzodiazepines are commonly used to treat withdrawal symptoms associated with ethanol detoxification. Disulfiram, an inhibitor of acetaldehyde dehydrogenase, is used in early intervention. It results in the accumulation of acetaldehyde, thus an unpleasant taste which discourages drinking; however, due to an unpleasant feeling, patient compliance with disulfiram is poor. Naltrexone, an opioid receptor antagonist, has shown good efficacy in long-term prevention of alcoholism. Craving and anxiety have been treated with bromocriptine, a dopamine antagonist. Antiepileptic drugs that inhibit CNS excitation have also been used to treat alcohol dependency.

Specimen collection and the analysis of ethanol

Specimen types

Clinical specimens generally include plasma, serum, and urine. For driving under the influence of alcohol, most jurisdictions consider whole blood and breath alcohol as equivalent. Postmortem specimens include plasma, serum, whole blood, urine, vitreous fluid, bile, and tissues. If an ethanol analysis cannot be performed within a few hours, the samples should be collected in potassium oxalate/sodium fluoride containing tubes. Breath is a commonly used specimen in evidential breath alcohol testing. In recent years, oral fluid (saliva) is gaining popularity due to noninvasive sample collection and its use in point-of-care testing.

Specimen collection

Sufficient blood (2–5 mL) should be collected in an appropriate tube type. The collection site should not be cleaned with ethanol or isopropanol, and the sample should not be taken from veins in which intravenous fluids are being administered, as these factors may cause erroneous results. For legal testing and when testing cannot be performed within a few hours, blood should be collected in potassium oxalate/sodium fluoride containing tubes. Potassium oxalate is an anticoagulant and sodium fluoride inhibits activities of various enzymes and the growth of microorganisms. For postmortem testing, vitreous fluid and urine can also be collected in sodium fluoride/potassium oxalate containing tubes. Oral fluid, which is used for screening ethanol, is collected on a swab or other collection devices.

Information such as the subject's name, collector's name, time of collection, and other relevant information should be well documented. For forensic testing the samples should be sent to the laboratory under chain of custody.

Blood ethanol analysis

Early methods of blood ethanol estimation include microdiffusion and spectrophotometric methods [6,7]. The microdiffusion method uses a Conway cell containing two concentric wells. The specimen is added to the outer well, and a chemical reagent, generally potassium dichromate and sulfuric acid, is added to the inner well. Ethanol and any other volatile reducing agent will diffuse from the outer well and react with the coloring reagent. The color of the coloring reagent changes from yellow-orange to green as dichromate is converted to chromic ions. This is the chemical basis of a widely used breathalyzer developed by Borkenstein in 1954. This method can be made semiquantitative/quantitative by using ethanol standards and measuring the color spectrophotometrically at 450 nm [7].

Enzymatic methods have been in use for several decades. In clinical practice, they are the most commonly used methods for measuring ethanol in serum, plasma, and urine [6]. The most common enzymatic method utilizes ADH. As in vivo, ADH converts ethanol and NAD to acetaldehyde and NADH. Spectrophotometrically, NADH is measured at 340 nm. The concentration of NADH generated is proportional to the ethanol concentration in a specimen. Abbott's assay is based on Radiative Energy Attenuation (REA) technology in which NADH reacts with thiozyl blue dye to form a chromagen that attenuates the intensity of the fluorophore [7]. In general, enzymatic assays are suitable for clinical use but have limitations. Some assays have cross-reactivity with isopropanol. Increased lactate and lactate dehydrogenase (LDH) may cause spurious results [14,15]. This is due to the fact that lactate is converted to pyruvate by LDH, and this reaction generates NADH. Increased lactate and LDH are seen in postmortem samples. Due to these reasons and hemolyzed samples, enzymatic methods are not preferred for postmortem ethanol testing.

In forensic laboratories, GC-FID is the most commonly used method for the measurement of ethanol [5–7]. It is highly specific with good sensitivity, accuracy and precision. Also, the method has the ability to separate other volatiles such as methanol, isopropanol, acetone, and acetaldehyde. Two commonly used GC-FID procedures are direct injection and headspace analysis. In a typical direct injection analysis a specimen is diluted with an aqueous solution containing an internal standard such as 1-propanol and 2-butanone. The method is easy and reproducible. The major disadvantage of this method is the build-up of proteins and other nonvolatiles in the injection port and front end of the analytical column, leading to increased instrument maintenance. On the other hand, headspace analysis is relatively clean and requires less instrument maintenance. In headspace analysis the sample is diluted with aqueous solution containing internal standard in an air tight vial. The vial is heated to 65°C and gas phase "headspace" is injected onto the column. Calibrators and controls are treated the same way, and ethanol concentrations are calculated based on the relative peak areas. A representative chromatogram from headspace GC-FID is shown in Fig. 5.2.

Breath ethanol

Despite some controversies, breath is a commonly used specimen in evidential or screening ethanol use [16,17]. Ethanol distributes between blood and alveolar air according to Henry's law which states that, at a given temperature, there is a direct concentration correlation of a volatile substance between the liquid and the gas phase. A blood/breath ethanol ratio of 2100:1 is commonly used in breath analyzers, but recent data suggests a blood/breath ethanol ratio of 2300:1 [17–19]. For breath ethanol testing, breath is generally collected 15–20 minutes after the direct observation of a suspect to make sure that nothing is ingested prior to the sample collection. This also avoids problems with any ethanol in the mouth. Two samples are generally collected and should agree within 0.02 g/210 L of breath.

Infrared (IR) spectrometry is a commonly used technology in evidential breath-testing devices. Certain IR wavelengths, between 800 and 4000 nm, are absorbed by various chemical bonds such as C–H, O–H, C–C, and C–O and are characteristic of a particular compound. Since these bonds are common in various compounds, the specificity is

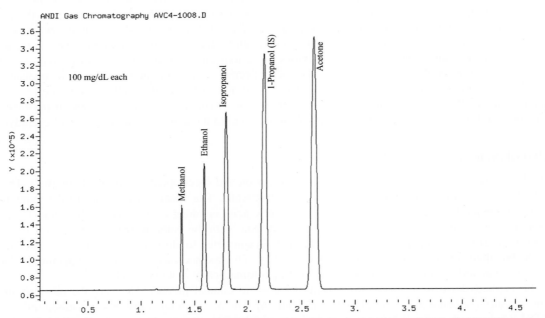

FIGURE 5.2 A representative chromatogram for volatiles quantification from the author's (UG) laboratory. The method involves headspace GC-FID. The chromatogram is from a quality control sample containing 100 mg/dL of each volatile. Internal standard is 1-propanol. Alcohol concentrations are calculated from the relative peak areas of calibrators and unknown samples. *GC-FID*, Gas-chromatography with a flame ionization detector.

increased by measurements at multiple wavelengths. The concentration of ethanol in a breath sample is calculated from the relative absorbance of IR energy by ethanol in the exhaled air to that of air without ethanol.

Electrochemical fuel-cell sensors involving the oxidation of ethanol are commonly used in handheld screening devices for breath analysis. In the fuel cell, ethanol is oxidized to acetic acid, producing a current that is measured. The electric current is directly proportional to the ethanol concentration. Acetone, which is the most abundant endogenous volatile, does not undergo oxidation and, thus, does not interfere; however, methanol and isopropanol are oxidized and can interfere at high concentrations. Prolonged fasting and diabetic ketoacidosis can produce high amounts of acetone and isopropanol. Electrochemical and IR detectors can be used in combination for breath ethanol testing. Use of this combination increases specificity and decreases potential interferences.

Oral fluid (saliva) ethanol

Since oral fluid is a convenient sample to collect without the need for a specially trained individual, interest in oral fluid testing is growing [20]. Due to higher water content, the concentration of ethanol in oral fluid is 10% higher as compared to whole blood. There are several CLIA-waived and Department of Transportation approved oral fluid ethanol screening point-of-care devices available. An oral fluid specimen is collected on a swab or pad and inserted into the point-of-care device. In many devices, ethanol detection is based on an ADH-diaphorase color-generating reaction. Some patients taking anticholinergic drugs, such as tricyclic antidepressants, may not produce oral fluid.

Vitreous fluid ethanol

Vitreous fluid is an excellent sample for postmortem ethanol testing. Vitreous fluid remains sterile for 2–4 days after death. Since vitreous fluid and blood establish equilibrium fairly quickly, vitreous fluid ethanol concentrations can reliably provide estimation of antemortem blood ethanol concentrations. On average, vitreous ethanol concentrations are 10%–20% higher than corresponding blood concentrations, due to the higher water content of vitreous fluid. Much higher variation has been reported [21,22].

Biomarkers of ethanol use

Since ethanol is cleared fairly quickly, biomarkers which can provide information on long-term ethanol use/abuse are desirable. Various biomarkers for ethanol use include as γ-glutamyltransferase (GGT), mean corpuscular volume (MCV), carbohydrate deficient transferrin (CDT), fatty acid ethyl esters (FAEEs), ethyl glucuronide (EtG), ethyl sulfate (EtS), and acetaldehyde-protein adducts [23–27]. Historically, GGT and MCV have been used as markers of alcohol abuse. Both GGT and MCV increase with days to weeks of ethanol use. Although fairly sensitive, these markers are not very specific and increase in many other conditions.

Due to different degrees of sialylation with sialic acid residues, transferrin exists in many isoforms from asialo- to hexasialotransferrin. The main isoform is tetrasialotransferrin. Chronic ethanol use decreases the degree of sialylation, resulting in an increase in asialo-and disialotransferrin (CDTs). The pattern of abnormal transferrin reverts to normal after a few weeks of abstinence of ethanol use. It is important to keep in mind that certain conditions, such as genetic disorders of glycosylation and liver disease, can result in CDT. Testing of CDT should not be used for general screening but with other laboratory and clinical information. Various isoforms of transferrin can be measured by electrophoresis or chromatography, such as liquid chromatography tandem MS.

FAEEs are synthesized by an esterification reaction between fatty acids and ethanol. Sample types include serum, hair, meconium, and tissues. In blood, FAEEs can be detected up to 48 hours after ethanol consumption. In hair and meconium, detection of FAEEs can provide a long-term window of ethanol use. GC/MS is the method of choice for the measurement of FAEEs.

EtG and EtS are formed in the body by glucuronidation and sulfation of ethanol, respectively. EtG and EtS are detected in various sample types including urine, blood, meconium and hair, with urine as the specimen of choice. Using GC–MS or LC–MS/MS, EtG and EtS can be detected in urine for several days. LC–MS is a commonly used and preferred method for the detection of EtG and EtS. The presence of EtG and EtS does not confirm the use of ethanol. Ethanol production during putrefaction and fermentation can cause false-positive results. Bacterial degradation can cause false-negative EtG results.

Acetaldehyde-protein adducts are formed from acetaldehyde and various proteins after ethanol consumption. Most of the work has been done on acetaldehyde-hemoglobin adducts. They can be detected for several weeks after ethanol consumption. Patients with decreased acetaldehyde dehydrogenase activity who are also being treated with disulfiram show increased concentrations of acetaldehyde-hemoglobin adducts. Phosphatidylethanol is another biomarker of ethanol use. It can be detected for several weeks after ethanol consumption. Various markers of ethanol abuse and their detection window are shown in Table 5.1.

Interpretation issues

A number of preanalytical, analytical, and postanalytical factors should be considered for proper interpretation [6,7,24].

Preanalytical considerations

Due to variable water content, ethanol concentration varies significantly among different specimen types. Ethanol distribution is directly proportional to the water content of the specimen. Water content of serum and plasma is 92% and whole blood is 80%, therefore, ethanol concentration in serum or plasma is 15% (range 4%–25%) higher as compared to the whole blood. In a legal setting, it is important to keep track of specimen types as these differences may lead to legal ramifications. The average ratio of ethanol in different specimen types is given in Table 5.2. A number of preanalytical factors that may affect results and, thus, interpretation are listed as follows:

1. Evaporation of ethanol happens in the air space of a collection tube. Ethanol can dissipate when the container is opened. To minimize this artifact the tube should be filled with blood.
2. Ethanol concentration can decrease due to microbial consumption.

TABLE 5.1 Ethanol biomarkers and their detection windows.

Ethanol Biomarker	Detection Window
Ethanol	1–15 h (varies with amount of ethanol consumed)
GGT	2–3 weeks
MCV	2–4 months
CDT	2–3 weeks
FAEEs	24 h in blood; months in hair and meconium
EtG and EtS	1–5 days in blood/urine
Acetaldehyde-hemoglobin adducts	1–6 weeks
PEth	1–4 weeks

CDT, Carbohydrate deficient transferrin; *EtG*, ethyl glucuronide; *EtS*, ethyl sulfate; *FAEEs*, fatty acid ethyl esters; *GGT*, γ-Glutamyltransferase; *MCV*, mean corpuscular volume; *PEth*, phosphatidylethanol.

TABLE 5.2 Ethanol Ratios in Various Specimen Types as Compared to Whole Blood.

Specimen	Ratio
Serum/plasma	1.12
Vitreous fluid	1.15
Saliva	1.10
Urine	1.30
Bile	1.00

3. Ethanol can increase due to microbial fermentation of carbohydrates in the sample. This can be avoided by collecting the sample in fluoride containing collection tubes. Ethanol formation due to microbial fermentation can be a challenge in postmortem toxicology particularly when the sample collection in a deceased person is significantly delayed. Several factors can assist with the proper interpretation of ethanol levels in postmortem samples. Information on the history of ethanol use and postmortem conditions of the body may help to indicate ethanol consumption or postmortem ethanol formation. An analysis on multiple samples is useful. Ethanol analysis on vitreous fluid and urine can provide important information. These specimens are relatively resistant to the putrefaction process. For example, positive blood but negative vitreous fluid and/or urine results would suggest postmortem ethanol formation, whereas positive vitreous fluid and/or urine results would suggest ethanol consumption. An analysis of other volatiles such as acetaldehyde, acetone, and n-propanol can also assist to determine if changes were due to putrefaction. In addition to ethanol, these volatiles are produced in small quantities during microbial activity.
4. Ethanol concentration will be spuriously high if ethanol containing skin cleansers are used.
5. A blood draw from a vein which is being used for intravenous fluid administration may falsely decrease ethanol concentration.
6. Urine specimens from diabetics may contain high glucose and yeast or bacteria. This may result in falsely high ethanol.

Analytical considerations

Once in the lab, the samples and paper work should be inspected closely for sample integrity and the completion of paper work, including chain of custody, if applicable. Blood samples should not be clotted and cells and plasma should be adequately mixed before analysis. The analysis should be performed with properly validated methods with acceptable accuracy, precision, linearity, and other analytical characteristics. The laboratory should have strict quality control and a quality assurance program that can withstand legal challenges in a court of law. The laboratory should also participate in an external proficiency testing program.

Postanalytical considerations

Postanalytical factors include proper data analysis with well-established rules for acceptability of analytical results. The report should contain relevant information such as the subject's name, date of birth, collection time, name of the analyst (if a legal sample), and the analytical method. The laboratory should be prepared to provide preanalytical, analytical, and postanalytical information that is defensible in the court of law.

Ethanol levels should be interpreted in context with clinical or legal information. In a clinical and human performance setting, there is a wide interindividual variation. Generally, most drinkers are affected at an ethanol concentration of 0.04–0.05 g/dL. Chronic drinkers and alcoholics may develop significant tolerance and can perform well at an ethanol concentration of >0.2 g/dL [28]. Irrespective of human performance, in the United States and many other countries, a blood ethanol concentration of 0.08 has been established as the per se limit for driving under the influence and violation of the law. The legal drinking age in the United States is 21 years. Any detectable ethanol (>0.02 g/dL) in a person under the age of 21 years is considered illegal and results in revocation of their driving license; the same applies to the workplace service standard. For commercial drivers under Federal Highway Standards, the legal ethanol limit is 0.04 g/dL.

Common scenarios toxicologists are asked to participate in include the calculation of blood ethanol concentration at the time of an accident and the amount of ethanol consumed. In motor vehicle accidents, it is desirable to know the ethanol concentration at the time of an accident by extrapolating the result from a sample collected after the accident. Many factors affect the extrapolation of results. There is a wide interindividual variation in ethanol clearance (0.01–0.03 g/dL/h). Multiple samplings and measurements, which do happen in legal settings, can provide ethanol clearance for a particular person, and facilitate the extrapolation of results. These estimates are only valid in a postabsorptive state. In the absorptive phase, measured ethanol concentrations in a sample collected after the accident will overestimate the blood ethanol concentration at the time of the accident. The more time that passes in the sample collection after the accident, the more unreliable the calculations become.

Estimation of dose, with limitations, can be estimated using the following equation:

$$\text{Dose (g)} = \text{Body weight (kg)} \times \text{ethanol concentration (g/L)} \times \text{volume of distribution}, V_d \text{ (L/kg)}$$

As described previously, V_d may vary significantly among individuals. Despite limitations, large discrepancies (such as the consumption of two vs six alcoholic drinks) can be resolved with a certain degree of confidence.

Methanol

Methanol (methyl alcohol and wood alcohol) is a clear, colorless, volatile, and highly flammable liquid. It is miscible with water and a number of organic solvents such as ethanol, acetone, chloroform, ethylene glycol, and acetone. It is a widely used commercial solvent and is commonly used as antifreeze, wind-shield wiper fluid, paint remover, ethanol denaturant, and varnish solvent. Exposure to methanol can occur though ingestion, dermal absorption, or inhalation. Oral ingestion is the most common and toxic exposure to methanol. Chronic alcoholics, desperate to consume anything mimicking ethanol, are frequently victims of methanol poisoning. Drinking windshield wiper fluid is a common cause of methanol poisoning. Adulteration of ethanol with methanol is a common cause of methanol poisoning in poor countries. Continuous exposure through inhalation can result in irritation and serious eye injury.

Pharmacology of methanol

Like ethanol, when consumed orally, methanol is rapidly and completely absorbed through the gastrointestinal tract [29]. Methanol is oxidized to formaldehyde by ADH. Formaldehyde is further metabolized by aldehyde dehydrogenase to formic acid and carbon dioxide (Fig. 5.1). Methanol is metabolized by zero-order kinetics with a half-life of 2–24 hours. It has V_d of 0.4–0.6 L/kg [30].

The majority of toxic effects of methanol are attributed to formic acid. Toxic effects include metabolic acidosis, visual disturbances, vomiting, and encephalopathy. Acute ingestion of as little as 10 mL has been reported to cause blindness, and 100–200 mL is fatal in most cases [30]. Metabolic acidosis leads to an elevated anion gap. The anion gap (mmol/L) is calculated as follows:

$$\text{Sodium} + \text{potassium} - \text{bicarbonate} - \text{chloride (reference range: } 8-16).$$

An increased anion gap is mostly due to the consumption of bicarbonate because of metabolic acidosis. In the beginning, only formic acid leads to metabolic acidosis. Later on, due to disturbed cellular metabolism, increased lactic acid also contributes to metabolic acidosis. If the baseline anion gap is low, it may not rise above the upper limit of the reference range. Visual disturbances and blindness are caused by formic acid by inhibiting the respiratory chain in the photoreceptors, Muller glial cells, and optic nerve. Methanol toxicity correlates better with metabolic acidosis and formic acid concentration than methanol levels. The measurement of formic acid can be very valuable; however, testing formic acid is not readily available.

Methanol and formic acid assay

GC is the most commonly used method for the quantitation of methanol. This method can also be used to assay formic acid. The enzymatic assay for methanol and formic acid has been described. Unfortunately, these methods have limitations and are not readily available in most hospital-based laboratories [31,32]. Healthcare providers have to rely on history, patient clinical symptoms, and laboratory values such as anion gap and osmolal gap. When methanol assay is not available, osmolal gap may be used to roughly estimate methanol concentration as follows [33]:

$$\text{Osmolal gap} = \text{Measured osmolality} - \text{Calculated osmolality}$$

$$\text{Calculated osmolality} = [(2 \times \text{sodium}) + (\text{glucose}/18) + (\text{BUN}/2.8) + (\text{ethanol}/4.6)]$$

$$\text{Methanol (mg/dL) estimation} = \text{Osmolal gap} \times 2.6$$

The unit of measurement for sodium is mmol/L, and the units for glucose, BUN, and ethanol are mg/dL. Glucose, BUN, and ethanol are divided by 18, 2.8, and 4.6, respectively, to convert their concentrations from mg/dL to mmol/L. The reference range for osmolal gap is −10 to +10. Most often, osmolality is measured by freezing point depression. Significantly elevated osmolal gap (>25) is fairly specific for toxic alcohol ingestion [18].

Treatment and management of methanol toxicity

In addition to supportive therapy, a specific therapy of methanol poisoning is treatment with 4-methylpyrazole (4-MP) [29,30]. Supportive treatment includes maintaining ventilation and oxygenation. Metabolic acidosis is treated with sodium bicarbonate. 4-MP (Fomepizole or Antizol) is a specific ADH inhibitor. It inhibits the formation of toxic metabolites. The alternate is intravenous administration of 5%–10% ethanol to keep the ethanol levels of 100–150 mg/dL. At this ethanol level, ADH is almost completely saturated, inhibiting the formation of formic acid. Due to unpredictable ethanol kinetics, CNS depression, and hypoglycemia, ethanol treatment is not preferred. Though expensive, 4-MP treatment is safe and predictable. Hemodialysis is preferred in severe poisoning indicated by severe metabolic acidosis and plasma methanol levels of >50 mg/dL. Folic acid and folinic acid may also be added to intravenous fluid. This promotes the elimination of formic acid to carbon dioxide and water. Despite these treatments, methanol poisoning causes high morbidity and mortality; a mortality rate of 20% has been reported. Among survivors, 20%−25% suffer from permanent visual defects, and ∼10% from neurological sequelae [34].

Isopropanol

Isopropanol is a colorless and flammable liquid with a strong odor. It is miscible in water and many organic solvents such as ethanol, methanol, chloroform, and acetone. It is readily available to the general public as rubbing alcohol which is 70%−90% isopropanol. Most cases of isopropanol poisoning are due to oral ingestion, though poisonings due to topical application through sponge baths have been reported [34]. Prolonged vapor inhalation may result in headache, dizziness, flushing, and hypotension. Isopropanol is 2−3 times more CNS depressant compared to ethanol. It is metabolized to acetone by ADH. Half-lives of isopropanol and acetone are 1−6 and 17−27 hours, respectively. Isopropanol ingestion's CNS depressant effects are prolonged by a very long half-life of acetone. A very small amount of acetone is metabolized to acetate and generally does not cause metabolic acidosis.

Like ethanol, isopropanol intoxication results in ataxia, slurred speech, dysarthria, and confusion. Hypotension is commonly seen in isopropanol toxicity due to vasodilation. Severe intoxication can result in respiratory depression, coma, and death. Isopropanol can cause gastrointestinal irritation and gastritis [35]. A toxic dose is between ∼0.5 and 1.0 g/kg. Isopropanol levels of 50−100 mg/dL lead to toxicity. Levels more than 150−200 mg/dL can cause coma and fatality, but patients with isopropanol levels >500 mg/dL have survived with supportive care and dialysis [36]. High levels of isopropanol, >200 mg/dL, may require hemodialysis. The decision for hemodialysis should be taken in context with the patient's clinical status.

Patients with diabetes type I can show significant levels (>30 mg/dL) of acetone and isopropanol without consumption of isopropanol. These findings may have clinical and forensic implications.

A commonly used method for isopropanol and acetone testing is GC with direct or headspace injection. Acetone can be estimated using Acetest tablets. As levels of isopropanol declines, levels of acetone rise. When the isopropanol assay is not available, osmolal gap may be used to roughly estimate the isopropanol concentration.

$$\text{Isopropanol (mg/dL) estimation} = \text{Osmolal gap} \times 6$$

The calculation of osmolal gap is discussed previously.

Ethylene glycol

Ethylene glycol is a colorless, odorless, and relatively nonvolatile liquid. Due to its low freezing and high boiling point, it is a commonly used ingredient in antifreeze and deicing solutions. In a clinical setting, it is the most commonly encountered glycol. It is ingested intentionally to inflict self-harm or as a substitute for ethanol. Due to its sweet taste, it may be accidently ingested by children. Untreated ethylene glycol ingestion can cause significant morbidity and mortality.

Pharmacology of ethylene glycol

When ingested orally, ethylene glycol is rapidly and completely absorbed. After oral ingestion, peak concentrations are reached after 1−2 hours. Ethylene glycol distributes throughout the body water and has a volume of distribution (V_d) of 0.5−0.8 L/kg [37,38]. The half-life of ethylene glycol is 3−5 hours. Approximately 20% of ethylene glycol is excreted unchanged through the kidneys. Other than causing mild gastritis, ethylene glycol is virtually nontoxic; however, its metabolic products are toxic and lead to morbidity and mortality. Like ethanol and other alcohols, ethylene glycol is metabolized by ADH and aldehyde dehydrogenase to produce glycolic, glyoxalic, and oxalic acids [39]. These metabolites, particular glycolic acid, lead to severe metabolic acidosis. Oxalic acid binds to calcium to form calcium oxalate crystals which cause widespread tissue injury and precipitates in the kidneys, causing renal failure. Further toxicity may lead to CNS depression, cardiopulmonary failure, convulsions, and coma. Ethylene glycol toxicity generally happens in distinct phases [3,37].

Phase 1 (the first few hours): During this phase the symptoms are due to the presence of ethylene glycol and are similar to that of ethanol intoxication. Symptoms include ataxia, nausea, vomiting, and slurred speech. Since ethylene glycol is an osmotically active substance, the osmolal gap may be elevated and can be used to roughly estimate ethylene glycol concentration.

$$\text{Ethylene glycol concentration (mg/dL)} = \text{osmolal gap multiplied by } 6.2.$$

Phase 2 (4–24 hours): During this phase the clinical effects are due to ethylene glycol metabolites. The patient develops metabolic acidosis, hypocalcemia, and crystalluria. Clinical manifestations may include renal failure, hyperventilation, arrhythmias, convulsions, and coma. Metabolic acidosis leads to an elevated anion gap. Calculation of anion gap is discussed previously.

Phase 3 (2–4 days): In this phase the untreated patient may show renal insufficiency with hematuria, proteinuria, and oliguria. Ethylene glycol and its metabolites are generally not detectable.

Treatment and management of ethylene glycol toxicity

Like methanol, the toxicity of ethylene glycol can be prevented by inhibiting its metabolism. Ethanol and 4-MP are competitive inhibitors of ADH and prevent the metabolism of ethylene glycol to its toxic metabolites [3,37,40,41]. Ethylene glycol levels of 20 mg/dL are considered toxic and may need treatment with ethanol or 4-MP. The levels ≥ 50 mg/dL are considered critical and indicate a potential need for hemodialysis. Treatment with 4-MP is preferred as it is more predictable. When treated with ethanol or 4-MP, the half-life of ethylene glycol increases to 17–20 hours, and its elimination is entirely renal. Metabolic acidosis is treated with sodium bicarbonate and hypocalcemia is treated with calcium gluconate or chloride [37].

In an untreated individual the approximate lethal dose of ethylene glycol is 1.0–1.5 g/kg body weight; however, survivals with very large amounts of ethylene glycol intake have been reported. In one report, in a suicide attempt, a 36-year-old man with a history of depression consumed approximately 3 L of ethylene glycol containing antifreeze. The patient developed nausea, emesis, lethargy, metabolic acidosis, and acute renal failure. The patient was treated with ethanol and hemodialysis and did not develop any chronic problems. The patient's blood ethylene glycol level was 1889 mg/dL [42]. In another report a 34-year-old patient consumed two 1-L bottles of ethylene containing antifreeze. The patient's ethylene glycol level was 1234 mg/dL [43]. The patient developed severe metabolic acidosis with an arterial blood pH of 6.79, pCO_2 of 37 mmHg, pO_2 of 115 mmHg, and bicarbonate of 5.5 mEq/L. Urine showed calcium oxalate crystals. Despite treatment with 4-MP and dialysis, the patient expired 22 hours after admission.

In addition to ethylene glycol levels, certain routine laboratory tests can be helpful in the management of a patient with ethylene glycol and other glycols toxicity [44]. These tests include renal and hepatic function tests, lactate, and glucose. Glucose is helpful to rule out hypoglycemia as the cause of any alteration in mental status. To avoid misinterpretation, it is important to rule out interference from ethylene glycol metabolites in lactate assays; glycolate is known to interfere in some lactate assays [45]. Urinary ketones and serum-hydroxybutyrate levels may help in distinguishing ethylene glycol poisoning from alcoholic or diabetic ketoacidosis.

Other glycols

Ethylene glycol is the most commonly encountered glycol. Other less frequently encountered glycols include diethylene glycol, propylene glycol, polypropylene glycol, ethylene glycol monobutyl ether and ethylene glycol monomethyl ether [1,37,46]. Diethylene glycol is used as antifreeze, lubricant, brake fluid, and industry solvent. Similar to ethylene glycol, it tastes sweet. It is more toxic than ethylene glycol, with an estimated human lethal dose of 0.014–0.170 g/kg [47]. Renal failure, coma, metabolic acidosis, and death have been reported after ingestion [48]. Like ethylene glycol and many other alcohols, it is metabolized by ADH and aldehyde dehydrogenase with a half-life of ~3 hours.

Propylene glycol is used as antifreeze and vehicle solvent for medications. It has relatively low toxicity and primarily only produces an increased osmolal gap. It occurs in various forms such as 1,2-propanediol and 1,3-propanediol. 1,2-Propanediol form is the most studied form and metabolizes to lactic acid, pyruvic acid, and acetic acid. Its toxicity includes metabolic acidosis, hypoglycemia, CNS depression, seizures, and coma [35]. Polypropylene glycols are a group of compounds with molecular weights ranging from 200 to more than 4000. They are relatively nontoxic and are commonly used as a liquid vehicle in cosmetics and topical medications. Polypropylene glycols with a molecular weight of >500 are poorly absorbed and rapidly excreted by the kidneys. Low molecular weight compounds (200–400) may result in metabolic acidosis, renal failure, and hypercalcemia after massive oral ingestions or repeated dermal applications in patients with extensive burn injuries.

Ethylene glycol monobutyl ether's toxic effects include lethargy, coma, anion gap metabolic acidosis, hyperchloremia, hypotension, respiratory depression, hemolysis, renal and hepatic dysfunction. Serum levels in poisoning cases have ranged from 0.005 to 432 mg/L [37]. Ethylene glycol monomethyl ether may lead to cerebral edema, hemorrhagic gastritis, and the degeneration of the liver and kidneys. Toxicity happens after 8–18 hours.

Management of toxicity from these glycols is supportive or involves the treatment with ethanol or 4-MP.

Analysis of glycols and other related analytes

For ethylene glycol, both enzymatic and chromatographic methods have been reported. Enzymatic methods are rapid and available on routine chemistry analyzers [49–52]. One screening method is based on the interference of ethylene glycol in the quantitative triglyceride method. In this method, triglycerides are measured by two methods and the "triglyceride gap" is used to estimate the ethylene glycol concentration [50]. Another enzymatic method uses glycerol dehydrogenase purified from *Enterobacter aerogenes*. This enzyme catalyzes the oxidation of ethylene glycol and produces NADH in the presence of NAD [51]. Juenke et al. [50] developed an improved method for the determination of ethylene glycol. In contrast to the original 2-point method, they used a multipoint assay. With this strategy, interferences from propylene glycol, various butanediols, and other related compounds were minimal.

Despite the convenience of being available on an automated analyzer, enzymatic methods lack specificity. Chromatographic methods are considered the gold standard methods for the assay of glycols. GC with FID or MS detector (GC/MS) have been described for the analysis of ethylene glycol and other glycols [53–57]. Due to better specificity, GC/MS methods are preferred over GC/FID methods. Most GC methods involve protein precipitation by organic solvents, such as acetonitrile followed by derivatization of glycols. A commonly used derivatizing agent is phenylboronic acid. Glycols and phenylboronic acid form cyclic phenylboronate esters which have a higher volatility and are amenable to GC. A representative GC/FID chromatogram for ethylene glycol analysis is shown in Fig. 5.3. Although methods for underivatized ethylene glycol have been described, they may pose problems due to low volatility and poor chromatographic properties. In a capillary column GC method, simultaneous detection of diethylene glycol, ethylene glycol, methanol, isopropanol, acetone, and ethanol has been described [57]. Since measurement of glycolic acid is also desired, GC/MS methods for the simultaneous estimation of ethylene glycol and glycolic acid have been described [56,58].

When specific testing is not available, a diagnosis of ethylene glycol poisoning can be challenging [44]. Other tests have been attempted to diagnose ethylene glycol exposure. Some ethylene glycol preparations contain fluorescein dye to detect radiator leak. Urine from a patient with suspected glycol exposure may show fluorescence under ultraviolet light; however, this test lacks both specificity and sensitivity. This problem is particularly seen in children. In one study, urine samples were collected from 150 healthy children and examined under Wood's lamp; 81% of them showed fluorescence [59].

Calcium oxalate crystals can be seen in urine from patients with ethylene glycol exposure. It is important to keep in mind that the detection of fluorescein or calcium oxalate crystals is not very specific or sensitive. The absence of oxalate crystals in the presence of high concentrations of ethylene glycol have been reported [60]. Garg et al. [61] reported a fatal case with a blood ethylene glycol level of 2340 mg/dL. Oxalate crystals were not seen, despite urine ethylene glycol level of 2261 mg/dL. Also, a urine organic acid analysis did not show elevated levels of glycolic and oxalic

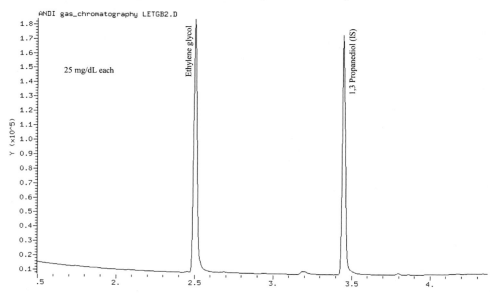

FIGURE 5.3 A representative chromatogram for ethylene glycol quantification from the author's (UG) laboratory. The method involves GC-FID. Internal standard, 1,3 propanediol, and phenylboronic acid are added to the sample. Ethylene glycol and the internal standard react, in the injection port, with the phenylboronic acid to form a cyclic phenylboronate esters. The esters are then detected by a FID. Ethylene glycol concentration is calculated from the relative peak areas of calibrators and unknown samples. *GC-FID*, Gas-chromatography with a flame ionization detector.

acids. Furthermore, calcium oxalate crystals can be seen in other conditions. In primary hyperoxaluria, calcium oxalate crystals are seen in the patient's urine. Certain foods such as tomatoes, garlic, and spinach are very rich in oxalate and may produce calcium oxalate crystalluria.

Conclusion

Ethanol is the most commonly encountered alcohol in clinical and forensic settings. A number of preanalytical, analytical, and postanalytical factors can influence results interpretation with legal ramifications. Ethylene glycol is the most commonly encountered glycol. Although enzymatic methods are available for the estimation of ethanol and ethylene glycol, these methods have limitations. Chromatographic methods, in particular GC, are the method of choice for the determination of alcohols and are essential in forensic setting.

References

[1] Garg U, Lowry J, Algren A. Ethylene glycol and other glycols: testing and interpretation issues. In: Dasgupta A, editor. Alcohol and drugs of abuse testing. Washington, DC: AACC Press; 2009. p. 29–39.
[2] Kraut JA. Diagnosis of toxic alcohols: limitations of present methods. Clin Toxicol (Phila) 2015;53:589–95.
[3] Kraut JA, Mullins ME. Toxic alcohols. N Engl J Med 2018;378:270–80.
[4] Alcohol Facts and Statistics. National institute on alcohol abuse and alcoholism. <https://pubs.niaaa.nih.gov/publications/AlcoholFacts&Stats/AlcoholFacts&Stats.pdf>. [Updated December 2019].
[5] Baselt RC. Ethanol. In: Baselt RC, editor. Disposition of toxic drugs and chemicals in man. Seal Beach, CA: Biomedical Publications; 2017. p. 818–22.
[6] Jones GR. Ethanol. In: Magnani B, Bissell MG, Kwong TC, Wu AHB, editors. Clinical toxicology testing: a guide for laboratory professionals. Northland, IL: CAP Press; 2012. p. 97–103.
[7] Levine B, Caplan YH. Alcohol. In: Levine B, editor. Principles of forensic toxicology. Washington, DC: AACC Press; 2003. p. 157–72.
[8] Norberg A, Jones AW, Hahn RG, Gabrielsson JL. Role of variability in explaining ethanol pharmacokinetics: research and forensic applications. Clin Pharmacokinet 2003;42:1–31.
[9] Dubowski KM. Absorption, distribution and elimination of alcohol: highway safety aspects. J Stud Alcohol Suppl 1985;10:98–108.
[10] Andreasson R, Jones AW. The life and work of Erik M. P. Widmark. Am J Forensic Med Pathol 1996;17:177–90.
[11] Denny L, Coles S, Blitz R. Fetal alcohol syndrome and fetal alcohol spectrum disorders. Am Fam Physician 2017;96:515–22.
[12] McMillin GA, Melis R, Bornhorst J. Alcohol abuse and dependency: genetics of susceptibility. In: Dasgupta A, editor. Alcohol and drugs of abuse testing. Washington, DC: AACC Press; 2009. p. 13–27.
[13] Reus VI, Fochtmann LJ, Bukstein O, Eyler AE, Hilty DM, Horvitz-Lennon M, et al. The American Psychiatric Association Practice Guideline for the pharmacological treatment of patients with alcohol use disorder. Am J Psychiatry 2018;175:86–90.
[14] Nine JS, Moraca M, Virji MA, Rao KN. Serum-ethanol determination: comparison of lactate and lactate dehydrogenase interference in three enzymatic assays. J Anal Toxicol 1995;19:192–6.
[15] Powers RH, Dean DE. Evaluation of potential lactate/lactate dehydrogenase interference with an enzymatic alcohol analysis. J Anal Toxicol 2009;33:561–3.
[16] Hartung B, Schwender H, Pawlik E, Ritz-Timme S, Mindiashvili N, Daldrup T. Comparison of venous blood alcohol concentrations and breath alcohol concentrations measured with Dräger Alcotest 9510 DE Evidential. Forensic Sci Int 2016;258:64–7.
[17] Jones AW. Evidential breath alcohol analysis and the venous blood-to-breath ratio. Forensic Sci Int 2016;262:e37–9.
[18] Emerson VJ, Holleyhead R, Isaacs MD, Fuller NA, Hunt DJ. The measurement of breath alcohol. The laboratory evaluation of substantive breath test equipment and the report of an operational police trial. J Forensic Sci Soc 1980;20:3–70.
[19] Jaffe DH, Siman-Tov M, Gopher A, Peleg K. Variability in the blood/breath alcohol ratio and implications for evidentiary purposes. J Forensic Sci 2013;58:1233–7.
[20] Bueno LHP, da Silva RHA, Azenha AV, de Souza Dias MC, De Martinis BS. Oral fluid as an alternative matrix to determine ethanol for forensic purposes. Forensic Sci Int 2014;242:117–22.
[21] Chao TC, Lo DS. Relationship between postmortem blood and vitreous humor ethanol levels. Am J Forensic Med Pathol 1993;14:303–8.
[22] Jones AW, Holmgren P. Uncertainty in estimating blood ethanol concentrations by analysis of vitreous humour. J Clin Pathol 2001;54:699–702.
[23] Andresen-Streichert H, Muller A, Glahn A, Skopp G, Sterneck M. Alcohol biomarkers in clinical and forensic contexts. Dtsch Arztebl Int 2018;115:309–15.
[24] Hammett-Stabler C. Alcohol use and abuse. In: Dasgupta A, editor. Alcohol and drugs of abuse testing. Washington, DC: AACC Press; 2009. p. 1–12.
[25] Jastrzebska I, Zwolak A, Szczyrek M, Wawryniuk A, Skrzydlo-Radomanska B, Daniluk J. Biomarkers of alcohol misuse: recent advances and future prospects. Prz Gastroenterol 2016;11:78–89.
[26] Nanau RM, Neuman MG. Biomolecules and biomarkers used in diagnosis of alcohol drinking and in monitoring therapeutic interventions. Biomolecules 2015;5:1339–85.

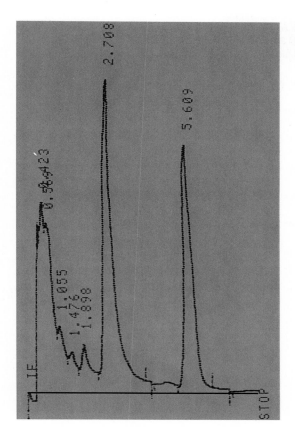

FIGURE 5.1.2 A chromatogram of the patient's serum illustrating 383 mg/dL ethylene glycol and the internal standard (3-Bromo-1-propanol) at 2.7 and 5.6 minutes, respectively. A chromatographic baseline was manually added for emphasis.

determined by its freezing point depression. The measured value is then compared with the calculated osmolality to determine if there is a significant gap in the two values. The presence of a large osmolal gap supports a diagnosis of EG poisoning. A metabolic acidosis may also suggest EG poisoning although, metabolic acidosis has a large number of differential diagnoses, including poisoning from methanol, salicylates, iron, isoniazid, or from conditions such as uremia or diabetic and alcoholic ketoacidosis. Microscopy of the urine can reveal needle or envelope-shaped calcium oxalate crystals in the urine in late stages of poisoning. The most reliable method of diagnosis is the direct measurement EG, which is usually performed via gas chromatography (GC) or GC–mass spectrometry (GC–MS). These specialized methods are often only available at large academic medical centers and regional toxicology laboratories.

In the laboratory an aliquot of serum was mixed 1:1 with the internal standard 3-bromo-1-propanol (25 mg/dL) prior to analysis by GC with flame ionization detection. One microliter of the sample/internal standard mixture was injected onto a 2–0624 column (Supelco/Sigma-Aldrich, Bellefonte, PA) packed with GP 60/80 Carbopack B 5% Carbowax 20 M held at a constant 160°C. With a carrier gas flow of 13 mL/min helium, EG elutes at approximately 2.7 minutes in this system (Fig. 5.1.2). More recent methods involve the derivatization of EG and glycolic acid prior to analysis by GC–MS [7]. Methods that include the glycolic acid metabolite are more preferable as it is the primary metabolite responsible for metabolic acidosis and cardiopulmonary toxicity [8,9].

Acknowledgment

The authors gratefully acknowledge Debbie Rigney M.D. for obtaining photographs of oxalate crystals in a recent patient.

References

[1] Baselt RC. Ethylene glycol. In: Baselt Randall C, editor. Disposition of toxic drugs and chemicals in man. Eleventh ed. Seal Beach, CA: Biomedical Publications; 2017. p. 840.

[2] Gummin DD, Mowry JB, Spyker DA, Brooks DE, Fraser MO, Banner W. 2016 Annual Report of the American Association of Poison Control Centers' National Poison Data System (NPDS): 34th Annual Report. Clin Toxicol 2017;55(10):1072−252.

[3] Agency for Toxic Substances and Disease Registry (ASTDR). Toxicological profile for ethylene glycol. Atlanta, GA: US Department of Health and Human Services Public Health Service; 2010.

[4] Ford MGL. Alcohols and glycols. In: Rippe JM, Irwin RS, Alpert JS, Fink MP, editors. Intensive care medicine. 2nd ed. Boston, MA: Little, Brown and Co; 1991. p. 1160−73.

[5] Viinamaki J, Sajantila A, Ojanpera I. Ethylene glycol and metabolite concentrations in fatal ethylene glycol poisonings. J Anal Toxicol 2015;39:481−5.

[6] Levey AS, Stevens LA, Schmid CH, Zhang YL, Castro 3rd AF, Feldman HI, et al. A new equation to estimate glomerular filtration rate. Ann Intern Med 2009;150(9):604−12.

[7] Perala AW, Filary MJ, Bartels MJ, McMartin KE. Quantitation of diethylene glycol and its metabolites by gas chromatography mass spectrometry or ion chromatography mass spectrometry in rat and human biological samples. J Anal Toxicol 2014;38:184−93.

[8] Moreau CL, Kerns 2nd W, Tomaszewski CA, McMartin KE, Rose SR, Ford MD, et al. Glycolate kinetics and hemodialysis clearance in ethylene glycol poisoning. META Study Group 1998;36(7):659−66.

[9] Porter W, Rutter P, Bush B, Pappas A, Dunnington J. Ethylene glycol toxicity: the role of serum glycolic acid in hemodialysis. J Toxicol − Clin Toxicol 2001;39:607−15.

Chapter 5.2

Falsely elevated ethylene glycol results in a patient with diabetic ketoacidosis

Mushal Noor, Chelsea Milito and Y. Victoria Zhang
Department of Pathology and Lab Medicine, University of Rochester Medical Center, Rochester, NY, United States

Case description

A 41-year old male with a past history of type 2 diabetes, hypertension, and schizophrenia was brought to the emergency department. He presented with altered mental status, hypotension, and hypothermia, with presenting vitals: BP 63/33, HR 68, and temperature 87.3°F. Naloxone was administered for suspected heroin or opioid overdose. EKG showed J waves which were attributed to hypothermia and initial and 3 hour troponin levels were within normal range. Head CT was negative. The initial lab results (Table 5.2.1) indicated severe hyperglycemia, low sodium, elevated potassium, and creatinine. There was also severe metabolic acidosis with markedly low bicarbonate, an elevated anion gap, normal lactate, elevated acetone and β-hydroxybutyrate, and only slightly elevated urinary ketones. Based on these results, the patient was administered IV fluids, insulin, bicarbonate, and antibiotics. Repeat venous blood gases 5 hours later demonstrated an improved pH of 7.1, but persistently low bicarbonate (10 mmol/L), and elevated anion gap of 24. The blood glucose had lowered to 874 mg/dL with treatment.

TABLE 5.2.1 Initial and Posthemodialysis Tests Results for the Patient

Analyte	Results (initial)	Results (posthemodialysis)	Reference Intervals
Glucose	1550 mg/dL	197 mg/dL	60–99 mg/dL
Sodium	116 mmol/L	144 mmol/L	133–145 mmol/L
Potassium	6.7 mmol/L	2.9 mmol/L	3.4–4.7 mmol/L
Creatinine	4.54 mg/dL	1.65 mg/dL	0.67–1.17 mg/dL
pH	6.9	7.4	7.32–7.42
Bicarbonate	6 mmol/L	23 mmol/L	21–28 mmol/L
Base excess	−26 mmol/L	−1 mmol/L	−3 to 1 mmol/L
Anion gap	21	19	7–16
Lactate	1.4 mmol/L	1.6 mmol/L	0.5–2.2 mmol/L
Acetone (blood)	38 mg/dL	n.a.	0–9
β-hydroxybutyrate	7.08 mmol/L	5.28 mmol/L[a]	0.02–0.27 mmol/L
Urinary ketones	1+	n.a.	NEG
Ethylene glycol	44 mg/L	13 mg/L	0–9 mg/L

[a]The second β-hydroxy butyrate reading is from 2 days later. No orders were placed for β-hydroxy butyrate, acetone, or urinary ketones right after hemodialysis.

Urine toxicology screen was negative, and only showed evidence of the patient's known prescription medications. Diabetic ketoacidosis and hyperosmolar hyperglycemic state were in the differential. At this point, ethylene glycol (EG) level was obtained and found to be elevated at 44 mg/L. The patient underwent urgent hemodialysis and fomepizole was administered. Further testing was done soon after hemodialysis, and the results were listed in the table along with the initial results (Table 5.2.1). EG concentration after dialysis was 13 mg/L. β-Hydroxybutyrate was not ordered along with other tests right after hemodialysis. Result obtained 2 days later indicated sustained elevated level at 5.28 mmol/L. Further analysis of urine samples did not reveal any urine crystals associated with EG poisoning. The patient's family strongly believed EG poisoning was unlikely, and the patient's vital signs did not indicate EG poisoning at this point. The question is what caused the continuously elevated EG levels in this patient.

Discussion

EG is a colorless, odorless, sweet-tasting synthetic liquid that is found in antifreeze, deicing solutions, and windshield wiper fluids. EG is highly toxic, with an oral LD_{LO} of 786 mg/kg for humans (about 50 mL or 1.7 oz for a 70-kg person) [1]. EG is an important cause of intentional and unintentional poisonings. In the United States, from January 2006 to December 2013, there were 45,097 cases of EG exposures resulting in 154 deaths [2]. EG itself is relatively nontoxic, and its initial central nervous system effects resemble those of ethanol. The metabolism of EG by alcohol dehydrogenase results in the formation of toxic metabolites glycolic acid and oxalic acid, which are responsible for the clinical manifestations: confusion, nausea, vomiting, central nervous system dysfunction, cardiovascular compromise, elevated anion gap metabolic acidosis, and acute kidney injury [2].

EG can be detected in the lab by enzymatic assay and by gas chromatography. Gas chromatography is the gold standard and the more accurate diagnostic method; however, it is not widely available and frequently must be performed at a reference laboratory. Hence, such results are rarely instrumental in aiding clinical decision-making [3,4]. Enzyme-based screening methods are readily automated and more easily available. They are the more common methods available in laboratories which perform EG testing. All EG enzymatic assays are based on the reaction between nicotinamide adenine dinucleotide (NAD) and EG in the presence of glycerol dehydrogenase [5]. We use the enzymatic assay in our lab for routine testing, with the option to send out tests for confirmation with gas chromatography.

The disadvantage of enzymatic assays is they can lead to false-positive results, for example, from interference by propylene glycol, 2,3-butanediol, glycolate, or severe acetaminophen toxicity. Elevated lactate dehydrogenase, lactate, or ketoacids may also cause a positive interference with enzymatic methods [5,6]. Similarities in chemical structure between EG and lactate can cause spurious detection of EG by spectrophotometry [7]. The enzymatic assay for EG uses glycerol dehydrogenase to oxidize EG. During the reaction, NAD is produced and it is then detected by spectrophotometry. In critically ill patients with increased serum lactate, oxidation of lactate to pyruvate increases the production of NAD, which may cross-react with the EG assay, causing false elevation of EG [7,8]. Ketone bodies could also interfere with the spectrophotometric assay because β-hydroxybutyrate and EG share a similar molecular structure [7].

Many efforts have been made to reduce interferences in the EG enzymatic assay. Many laboratories, including ours, have modified the enzymatic assay parameters for more accurate results for routine functions. In the standard enzymatic assay the difference between the absorbance readings at two time points is used to determine the EG concentration. In a modified assay the slope of the line is determined by measuring absorbance differences at several points, starting at a later time point than the original 2-point design [8]. For samples containing EG the difference between the 2 determinations is minimal. For samples containing compounds that mimic EG (e.g., propylene glycol), the difference between the two methods is substantial because the slope of the line flattens after an initial increase in absorbance [8]. However, despite these modifications, the interference concerns still exist.

In situations where EG testing is not available, or there is a delay in results, indirect tests can help make a presumptive diagnosis or give an indication of the likelihood of EG toxicity. Urine examination for the presence of calcium oxalate crystals, together with an elevated anion gap or osmolal gap, strongly suggests EG poisoning. However, care should be taken not to over interpret positive or negative results. Oxalate crystals can be seen in normal individuals with high dietary intake of foods containing either oxalates or vitamin C, giving rise to potential false-positive results [4]. Detection of oxalate crystals is also not a sensitive finding. In one series, only 4 of 12 patients with confirmed EG toxicity had calcium oxalate crystals detected in their urine [9]. Hence, while detection of crystals is potentially important, it is not pathognomonic [4].

The challenges in accurately detecting EG have created concerns in treatment of toxicity. EG toxicity requires immediate medical attention. Patients with suspicion of EG poisoning will be treated with 4-methylpyrazole (4-MP; fomepizole) or ethanol to saturate the enzyme alcohol dehydrogenase and prevent conversion of EG to its toxic metabolites, while continuous monitoring takes place. In addition, supportive care and hemodialysis in cases of severe acidemia or end organ injury are also recommended.

In this case the patient had a positive EG result and was given supportive treatment with urgent hemodialysis to remove any EG and its metabolites. Fomepizole was administered, as well as bicarbonate, until acidosis had resolved. The blood glucose had decreased to <200 mg/dL with insulin and IV fluids. The patient's clinical condition was greatly improved, but when EG level was repeated following dialysis, it was found to be 13 mg/L. It was unusual for the EG to remain elevated after treatment and did not correlate with the patient's improved clinical condition. In the meantime the toxicologist performed microscopic examination of urine for crystals and found no evidence of oxalate crystals. All the results prompted the clinician to consider a false-positive result and triggered the decision to send the samples out to a reference lab for confirmation by gas chromatography. The gas chromatography results were negative for EG, thus confirming that the enzymatic method yielded a false-positive EG result.

There are many possible causes for the false-positive results for EG in this case, as indicated previously. Lactate was within the reference range; however, ketone bodies remained high even after dialysis, despite a reduction in concentration, which could still cause false-positive reading for EG. Hence, in the case of our patient, it seems most probable that elevated ketoacids may have contributed to the false-positive result. This has been reported in other patients as well, for the same method [7].

Overall, this is a case which indicates a false-positive EG level by enzymatic assay. When in doubt, a confirmatory gas chromatography assay should be ordered to determine true levels of EG. Close collaboration with physicians and clinical toxicologists is important in managing these patients to ensure the best possible care.

References

[1] Singh R, Arain E, Buth A, Kado J, Soubani A, Imran N. Ethylene glycol poisoning: an unusual cause of altered mental status and the lessons learned from management of the disease in the acute setting. Case Rep Crit Care 2016;2016:9157393.
[2] Jobson MA, Hogan SL, Maxwell CS, Hu Y, Hladik GA, Falk RJ, et al. Clinical features of reported ethylene glycol exposures in the United States. PLoS One. 2015;10(11):e0143044-e.
[3] Rooney SL, Ehlers A, Morris C, Drees D, Davis SR, Kulhavy J, et al. Use of a rapid ethylene glycol assay: a 4-year retrospective study at an Academic Medical Center. J Med Toxicol 2016;12(2):172–9.
[4] McQuade DJ, Dargan PI, Wood DM. Challenges in the diagnosis of ethylene glycol poisoning. Ann Clin Biochem 2014;51(2):167–78.
[5] Fraser AD. Clinical toxicologic implications of ethylene glycol and glycolic acid poisoning. Ther Drug Monit 2002;24(2):232–8.
[6] Martinez C, Lubbos H, Rose LI, Swartz C, Kayne F. False-positive ethylene glycol levels in patients with diabetic ketoacidosis. Endocrine Pract 1998;4(5):272–3.
[7] Boissier F, Weiss N, Faisy C. False positive ethylene glycol determination by spectrophotometry in the presence of severe lactic acidosis and ketosis. Ann Emerg Med 2010;56(1):75–6.
[8] Juenke JM, Hardy L, McMillin GA, Horowitz GL. Rapid and specific quantification of ethylene glycol levels adaptation of a commercial enzymatic assay to automated chemistry analyzers. Am J Clin Pathol 2011;136(2):318–24.
[9] Moriarty RW, McDonald RH. The spectrum of ethylene glycol poisoning. Clin Toxicol 1974;7(6):583–96.

Chapter 5.3

Recurrent inhalational methanol toxicity during pregnancy

Zoë Piggott[1,*], Aaron Guinn[1,*], Wesley Palatnick[1] and Milton Tenenbein[2]
[1]Department of Emergency Medicine, University of Manitoba, Winnipeg, MB, Canada, [2]Department of Pediatrics and Child Health, University of Manitoba, Winnipeg, MB, Canada

Case description

A 31-year-old woman, gravida 4 para 3, presented to hospital four times during her third trimester of pregnancy with complaints of abdominal pain, dyspnea, and/or intoxication. She admitted to frequent inhalational use of a methanol-containing lacquer-thinner but denied other ingestions of illicit drugs or toxic substances. Metabolic acidosis with elevated anion gap was documented during each visit and initial methanol serum concentrations ranged from 8.5 to 11.9 mmol/L (27.2–38.1 mg/dL). During her fourth presentation, she was in active labor and had a serum methanol concentration of 8.5 mmol/L (27.2 mg/dL). Less than 12 hours later, she delivered a term infant at 37 weeks 5 days gestational age (GA), weighing 3338 g. Apgar scores were 8 and 9 at 1 and 5 minutes, respectively. No neonatal resuscitation was required.

Serum ethanol during all presentations was undetectable. At 34 weeks GA, she was treated with a 15 mg/kg dose of intravenous fomepizole and a session of hemodialysis (HD) resulting in an undetectable methanol concentration and resolution of acidosis. At 34^4 and 37^5 GA a dose of fomepizole 15 mg/kg IV was administered. In both instances, she was asymptomatic and left hospital against medical advice with methanol measurement in the range of 5–5.6 mmol/L (16–17.9 mg/dL). Her longest admission began at 35^3 GA, when she was treated with fomepizole alone (15 mg/kg loading dose followed by 10 mg/kg every 12 hours for 7 doses). Her methanol concentration was undetectable upon discharge.

Initial laboratory values and treatments administered during each visit are shown in Table 5.3.1.

Discussion

Methanol poisoning after inhalational solvent abuse in this pregnant patient was temporally associated with a compositional change of a toluene-based, lacquer-thinner product from 5%–9% to 20%–25% methanol. This product was

TABLE 5.3.1 Initial Laboratory Findings and Treatment Received at Each Presentation

Gestational Age	pH	HCO$_3$	Anion Gap	Methanol (mmol/L)	Treatment
34^0	7.27	9	27	9.5	Fomepizole + dialysis
34^4	7.43	19	18	10	Fomepizole × 1
35^3	7.33	11	20	11.9	Fomepizole × 8
37^5	7.35	13	16	8.5	Fomepizole × 1

*Contributed equally to this work.

widely available in her community. While there are a number of cases of toxicity secondary to methanol inhalation reported in the literature [1–5], there is only one prior documented case during pregnancy [6]. There are only three previous cases of methanol toxicity during pregnancy via any route in the medical literature [6–8].

Methanol is a simple alcohol with a single-carbon side chain. It is a volatile, colorless liquid at room temperature, with high water solubility, and is absorbed through the respiratory tract with an absorption fraction of 65%–75% [9]. It is rapidly distributed throughout the body, with a distribution half-life of 8 minutes and a volume of distribution of 0.77 L/kg [10]. Onset of symptoms occurs within 1–2 hours when methanol is ingested orally, owing to rapid and complete absorption from the upper gastrointestinal tract [11]. The pharmacokinetics of inhaled methanol likely differ in pregnant patients. During pregnancy, increased tidal volume and lower functional residual capacity may lead to comparatively increased absorption via this route, as well as alterations in distribution.

The parent alcohol molecule is relatively nontoxic, with clinical effects similar to those produced by ethanol intoxication. Central nervous system effects including sedation, ataxia, nystagmus, depression of respiratory drive, and loss of airway-protective reflexes may be observed with significant ingestions [12].

Methanol is metabolized by hepatic enzymes alcohol dehydrogenase (ADH) and aldehyde dehydrogenase. This two-step metabolic pathway results in the production of formaldehyde and subsequently formic acid (formate) [11]. Formate is a clinically important toxic metabolite and is responsible for the toxicity and end-organ damage observed in patients with methanol ingestions. Without treatment, the elimination of methanol (and production of formic acid) follows zero-order kinetics [13].

Formic acid damages the retina, resulting in optic disk swelling and visual loss, which may be permanent [14]. Pancreatitis, upper gastrointestinal bleeding, and basal ganglia ischemia or hemorrhage may also occur [11,12,14]. Production of formic acid results in an anion gap metabolic acidosis. When severe, immediately life-threatening hemodynamic instability with vasodilation, decreased myocardial contractility, hypotension and hypoxemia may result [15,16].

Competitively blocking the hepatic metabolism of methanol with fomepizole significantly slows production of formic acid. Elimination of the parent alcohol becomes first-order (half-life 48–54 hours) via pulmonary and renal routes [9]. Fomepizole remains the cornerstone of pharmacologic treatment, in addition to supportive care and HD. Concurrent administration of folate may accelerate the elimination of formic acid [11].

In order for methanol to have toxic effects on a fetus, it must either be metabolized to formic acid by the fetus, or formic acid produced by maternal metabolism must cross the placenta. Methanol would be expected to cross the placenta given its molecular similarity to ethanol, which crosses the placenta freely [17]. Indeed, data from pregnant rats support this [18] as does data from two human case reports [7,8]. However, ADH expression in the fetal liver is only 10% of that of an adult during the first half of pregnancy, though levels likely increase rapidly during the second half of pregnancy [17]. The placenta itself also expresses small amounts of ADH; however, based on its effect on ethanol metabolism, it is unlikely to play a significant role in the formation of formic acid [19].

Whether formic acid produced by maternal metabolism can cross the placenta to enter the fetus is less clear. Given that the pKa of formic acid is 3.8, most of this weak acid would be in the form of its conjugate base, formate, at physiological pH, and not expected to cross the placenta due to its negative charge [17]. However, ex vivo experiments with human placentas showed that formic acid rapidly crosses the placenta at physiologic pH, though the exact mechanism for this is unclear [20]. The single published case to measure a fetal formic acid level found it to be undetectable; however, this measurement was performed 6 days after the initial maternal ingestion of methanol [8]. It is possible that small amounts of undissociated formic acid present at normal maternal pH could cross the placenta, given its neutral charge, and accumulate in the fetus; however, this has not been documented. In addition to the potential effects of formic acid on the developing fetus, maternal acidosis itself may lead to fetal distress. While the fetus is protected from short periods of maternal acidosis by the buffering capacity of the placenta, sustained maternal acidosis will eventually overwhelm this, leading to fetal acidosis and distress [17].

In adult patients, serum methanol levels exceeding 6 mmol/L have been shown to result in end-organ damage and clinically significant toxicity [16,21]. Laboratory detection of methanol in the serum is usually accomplished with gas chromatography [22,23]. For acutely ill patients, this testing may not be available at all laboratories within a timeframe appropriate for clinical decision-making. Clinicians may also request serum lactate levels and serum pH from blood gases, which may identify the anion gap metabolic acidosis caused by formic acid production. Other helpful laboratory testing includes measured serum osmoles, which allow the calculation of an osmole gap. An unexplained elevated osmole gap, while nonspecific, may suggest the presence of unmetabolized methanol in the serum in a suspicious clinical presentation.

The patient in this case was treated on four occasions with fomepizole, and once with dialysis. Fomepizole is an ADH inhibitor with a pregnancy risk rating of C. It has been shown to cross the placenta and accumulate to sufficient levels to provide fetal protection from toxic alcohols in rats [24]. However, very little information is available regarding teratogenicity or other developmental toxicity [25]. While the developmental toxicity of ethanol is well described in gestation, particularly during organogenesis, fomepizole would appear to be a reasonable method of ADH-inhibition during pregnancy. The side-effect profile of fomepizole in adults is benign, and other cases in the literature describe its successful use in pregnant patients [6,7]. Given that the correlation between maternal methanol levels (or pH) and fetal toxicity is incompletely understood, the role of maternal treatment for the purpose of fetal protection remains unclear. We suggest that the indication for fomepizole use in the pregnant patient should not differ from those in her nonpregnant counterpart.

Though methanol concentrations after inhalation are generally lower than those after oral ingestion, serious toxicity is still possible. This case is the third to report the use of HD after methanol inhalation [1,2] and the first to report this after methanol inhalation during pregnancy. Despite our incomplete understanding of the teratogenic profile of fomepizole, monotherapy with HD alone cannot be recommended due to concerns over increased maternal morbidity with this method [17].

Unfortunately, follow-up information on the infant beyond the immediate perinatal period was unavailable. This is an important limitation. While the normal APGAR scores and lack of requirement of neonatal resuscitation are reassuring, we are unable to comment on the longer term effects of exposure to methanol or fomepizole therapy.

Acknowledgment
None.

Disclosure statement
None.

Funding
None.

References

[1] Wallace EA, Green AS. Methanol toxicity secondary to inhalant abuse in adult men. Clin Toxicol (Phila) 2009;47(3):239–42.
[2] Frenia ML, Schauben JL. Methanol inhalation toxicity. Ann Emerg Med 1993;22(12):1919–23.
[3] Aufderheide TP, White SM, Brady WJ, Stueven HA. Inhalational and percutaneous methanol toxicity in two firefighters. Ann Emerg Med 1993;22(12):1916–18.
[4] LoVecchio F, Sawyers B, Thole D, Beuler MC, Winchell J, Curry SC. Outcomes following abuse of methanol-containing carburetor cleaners. Hum Exp Toxicol 2004;23(10):473–5.
[5] Bebarta VS, Heard K, Dart RC. Inhalational abuse of methanol products: elevated methanol and formate levels without vision loss. Am J Emerg Med 2006;24(6):725–8.
[6] Velez LI, Kulstad E, Shepherd G, Roth B. Inhalational methanol toxicity in pregnancy treated twice with fomepizole. Vet Hum Toxicol 2003;45(1):28–30.
[7] Belson M, Morgan BW. Methanol toxicity in a newborn. J Toxicol Clin Toxicol 2004;42(5):673–7.
[8] Hantson P, Lambermont JY, Mahieu P. Methanol poisoning during late pregnancy. J Toxicol Clin Toxicol 1997;35(2):187–91.
[9] Barceloux DG, Bond GR, Krenzelok EP, Cooper H, Vale JA. American Academy of Clinical Toxicology Ad Hoc Committee on the Treatment Guidelines for Methanol P. American Academy of Clinical Toxicology practice guidelines on the treatment of methanol poisoning. J Toxicol Clin Toxicol 2002;40(4):415–46.
[10] Graw M, Haffner HT, Althaus L, Besserer K, Voges S. Invasion and distribution of methanol. Arch Toxicol 2000;74(6):313–21.
[11] Sivilotti ML. Methanol and ethylene glycol poisoning. Waltham, MA: UpToDate Inc. Available from: <http://www.uptodate.com/>.
[12] Wiener SW. Toxic alcohols ed In: Hoffman RS, Howland MA, Lewin NA, Nelson L, Goldfrank LR, Flomenbaum N, editors. Goldfrank's toxicologic emergencies. 10th ed. New York: McGraw-Hill Education; 2015. p. xxii, 1882 pages.
[13] Hojer J. Severe metabolic acidosis in the alcoholic: differential diagnosis and management. Hum Exp Toxicol 1996;15(6):482–8.
[14] Sivilotti ML, Burns MJ, Aaron CK, McMartin KE, Brent J. Reversal of severe methanol-induced visual impairment: no evidence of retinal toxicity due to fomepizole. J Toxicol Clin Toxicol 2001;39(6):627–31.

[15] McMartin K, Jacobsen D, Hovda KE. Antidotes for poisoning by alcohols that form toxic metabolites. Br J Clin Pharmacol 2016;81(3):505−15.

[16] Kerns II W, Tomaszewski C, McMartin K, Ford M, Brent J. Alcohols MSGMfT. Formate kinetics in methanol poisoning. J Toxicol Clin Toxicol 2002;40(2):137−43.

[17] Tenenbein M. Methanol poisoning during pregnancy—prediction of risk and suggestions for management. J Toxicol Clin Toxicol 1997;35(2):193−4.

[18] Pollack GM, Brouwer KL. Maternal-fetal pharmacokinetics of methanol. Res Rep Health Eff Inst 1996;(74):1−48 discussion 9-53.

[19] Pares X, Farres J, Vallee BL. Organ specific alcohol metabolism: placental chi-ADH. Biochem Biophys Res Commun 1984;119(3):1047−55.

[20] Hutson JR, Lubetsky A, Eichhorst J, Hackmon R, Koren G, Kapur BM. Adverse placental effect of formic acid on hCG secretion is mitigated by folic acid. Alcohol Alcohol 2013;48(3):283−7.

[21] Liesivuori J, Savolainen H. Methanol and formic acid toxicity: biochemical mechanisms. Pharmacol Toxicol 1991;69(3):157−63.

[22] Church AS, Witting MD. Laboratory testing in ethanol, methanol, ethylene glycol, and isopropanol toxicities. J Emerg Med 1997;15(5):687−92.

[23] Fraser AD. Clinical toxicologic implications of ethylene glycol and glycolic acid poisoning. Ther Drug Monit 2002;24(2):232−8.

[24] Gracia R, Latimer B, McMartin KE. Kinetics of fomepizole in pregnant rats. Clin Toxicol (Phila) 2012;50(8):743−8.

[25] Bailey B. Are there teratogenic risks associated with antidotes used in the acute management of poisoned pregnant women? Birth Defects Res A Clin Mol Teratol 2003;67(2):133−40.

Chapter 5.4

A patient with high anion gap metabolic acidosis and increased serum osmolal gap

Zengliu Su[1], Roger W. Stone[1] and Yusheng Zhu[1,2]

[1]Department of Pathology and Laboratory Medicine, Medical University of South Carolina, Charleston, SC, United States, [2]Department of Pathology & Laboratory Medicine, Pharmacology, Penn State University Hershey Medical Center, Hershey, PA, United States

Case description

A 61-year-old African-American man being found unconscious with an overdose of unknown etiology was sent to the emergency department. Upon arrival, the patient was in respiratory failure with an increased anion gap metabolic acidosis, increased serum osmolal gap, and negative volatiles (Table 5.4.1). Gas chromatography (GC) analysis for glycols revealed significantly elevated ethylene glycol (EG) at 22 mg/dL and propylene glycol (PG) was negative (Fig. 5.4.1A) [1]. Then the patient was treated with emergent hemodialysis followed by continuous veno-venous hemofiltration and

TABLE 5.4.1 Laboratory Findings for the Patient During Hospitalization.

Tests	At Admission	After 13 hours	Reference Ranges
BUN (mg/dL)	13.0	3.0	8.0–20.0
Creatinine (mg/dL)	1.2	1.2	0.4–1.0
Sodium (mmol/L)	143.0	139.0	135.0–145.0
Chloride (mmol/L)	119.0	103.0	98.0–107.0
Potassium (mmol/L)	6.7	4.9	3.50–5.00
Calcium (mg/dL)	9.4	7.9	8.4–10.2
Phosphorus (mg/dL)	N/A	5.1	2.4–4.7
Anion gap (mmol/L)	19.0	11.0	2–11
pH	7.03	7.39	7.35–7.45
pCO_2 (mmHg)	20.0	43.0	35–45
pO_2 (mmHg)	179.0	132	80–100
sO_2 (%)	99.0	99.6	96–97
HCO_3 (mmol/L)	6.0	25.7	22–26
Base excess	−24.0	1.2	−2.0 to 3.0
Osmolality (mOsm/L)	326.0	312.0	280–300
EG (mg/dL)	22.0	<5	Negative
PG (mg/dL)	<5	27	Negative

EG, Ethylene glycol; *PG*, propylene glycol.

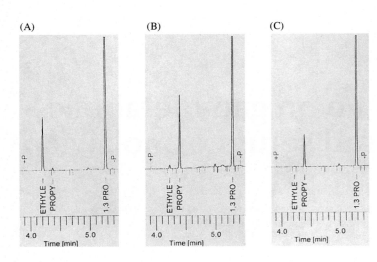

FIGURE 5.4.1 Chromatograph of EG, PG and 1,3-propanediol (internal standard). (A) On admission, EG is detected at 22 mg/dL, PG is undetectable; (B) 13 hours later, EG is undetectable, but a PG peak (27 mg/dL) is observed; and (C) 38 hours later, EG is undetectable and PG decreased to 13 mg/dL. *EG*, Ethylene glycol; *PG*, propylene glycol; *1,3 PRO*, 1,3-propanediol; +*P*, enablement of peak detection; −*P*, disablement of peak detection [1].

fomepizole. At the same time, the patient also received phenytoin and a high dose of lorazepam drip for a witnessed seizure. After 13 hours, the EG level was undetectable. However, a new peak identified as PG was at level of 27 mg/dL (Fig. 5.4.1B) [1] and 38 hours later, PG decreased to 13 mg/dL (Fig. 5.4.1C).

Discussion

EG and PG have similar physical properties and chemical structures. However, EG is much more toxic than PG. The former is a predominant constituent of antifreeze, coolant, household cleaners, and deicers. EG poisoning cases are not uncommon in hospital emergency rooms. According to the American Association of Poison Control Center, 7114 cases of EG poisoning were reported in 2010, and 87% were due to the consumption of automotive products, such as antifreeze, and 20 of them died. Usually, EG is ingested by accident or consumed as alcohol substitute or as a suicide agent [2,3]. Compared to EG, PG is less toxic and considered as a relatively safe vehicle for drugs or vitamins given intravenously. However, poisoning cases associated with the overdose of medications using PG as diluent have been described [4–6].

EG is highly water-soluble and easily absorbed in the gastrointestinal tract. Approximate 80% of absorbed EG is metabolized in the liver. The other 20% is excreted without changes by the kidney [7]. The serum concentration reaches a peak 1–4 hours after the ingestion of EG. At the early stage (first phase), EG can directly cause ethanol-like central nervous system effects such as euphoria, intoxication, coma, or seizures. With the metabolism proceeds, serum EG concentration decreases due to its rapid clearance. At this stage (second phase, usually after 12–36 hours of ingestion), the osmolal gap may become normal, but organ damages may occur due to the accumulation of numerous metabolized organic acids, such as glycolic acid, glyoxylic acid, and oxalic acid. Accumulated oxalate can chelate calcium and deposit in the kidney, brain, or other organs. Thus finding of calcium oxalate crystal in urine or deposition in the kidney or other organs is a strong indication of EG poisoning. Common clinical symptoms in this stage include tachycardia, hypertension, and metabolic acidosis. Unsuccessful treatment will result in acute renal injury, which usually happens after 24–72 hours of the ingestion [7].

Unlike EG, approximately 55% of PG is metabolized to lactic acid and then converted to pyruvic acid joining the Krebs cycle in the liver, with the remainder excreted without any changes by the kidney [8]. It has been suggested that renal clearance may reach to saturation at higher level of PG doses. Elevated osmolal gap and high anion gap metabolic acidosis are the most common laboratory findings for patients with PG accumulation. Significant clinical deterioration has been less likely observed for those patients [9,10]. However, a case of acute tubular necrosis was reported after 9 days of continuous infusion of lorazepam and 3 days of intravenous trimethoprim-sulfamethoxazole [10]. A study of a group of patients with clinical deterioration showed that PG levels were 104–144 mg/dL. However, patients with serum PG levels up to 127 mg/dL could only have metabolic abnormalities [4]. The overlapping of PG concentrations between mild and severe toxicity indicates that the PG serum concentration is not closely correlated with the severity of the toxicity. Also, a safe dose of PG has not been established yet [11].

A basic metabolic panel is recommended for all patients with suspected glycols poisoning. Blood gas test also provides valuable information regarding the patient's acid-base status. Increased anion gap accompanied by a metabolic acidosis support the presumptive diagnosis of glycol poisoning. A large osmolal gap implies the presence of unmeasured osmotically active substance such as volatiles or glycols. In our case, all tested volatiles (ethanol, methanol, isopropanol, and acetone) were negative. Therefore EG or PG could be the main cause of the elevated osmolal gap. On the other hand, a normal osmolal gap does not rule out glycol poisoning due to their rapid metabolisms. The quantitation of glycols in patient serum is the most definitive method for the diagnosis of the poisoning. In most clinical laboratories, only EG is analyzed, but EG and PG can be simultaneously measured using GC (Fig. 5.4.1). It is worth to note that serum concentration of EG does not closely correlate with the severity of the toxicity, because its metabolites, not EG itself, cause the severe toxicity. EG poisoning is a medical emergency and an immediate and appropriate treatment is required. If analysis of EG is not available or a quick response is needed, criteria for the presumptive diagnosis of EG poisoning are suggested: suspected ingestion of EG within the last 1 hour plus osmolal gap >10 mOsm/kg or suspected ingestion of EG plus any two of the following factors: (1) arterial pH <7.3, (2) serum bicarbonate <20 mEq/L, (3) osmolal gap >10 mOsm/kg, and (4) presence of urinary oxalate crystals [7]. The treatments of EG poisoning includes supportive care, inhibition of the synthesis of toxic metabolites using antidotes or ethanol, elimination of serum EG with hemodialysis, and correction of metabolic abnormalities.

In our case, the treatment of EG poisoning with hemodialysis followed by continuous veno-venous hemofiltration effectively cleared EG from the patient's blood. Then the patient was treated with phenytoin followed by a high dose of lorazepam drip for a witnessed seizure. The two medications contained PG at 40% (v/v) and 80% (v/v), respectively, which led to the new peak of PG 13 hours later while the EG was cleared. Although PG poisoning is rare, its toxicity due to excessive and rapid infusion of medications containing PG, such as lorazepam or diazepam, is being increasingly reported [9,10]. In general, the acidosis of these patients can be resolved after discontinuing the infusion of medications containing PG without a special treatment. In our case, the PG concentration decreased to 13 mg/mL after discontinuing lorazepam drip 38 hours later (Fig. 5.4.1C). However, patients with high risk factors, such as active ethanol abuse, kidney disease, liver disease, pregnant women or children under the age of 4, are prone to PG accumulation that may cause severe toxicity. In these cases, hemodialysis and treatments to correct the metabolite abnormalities are necessary.

References

1. Su Z, Stone RW, Zhu Y. Positive propylene glycol result in a patient with ethylene glycol poisoning. Clin Chem 2014;60:697—8.
2. Bronstein AC, Spyker DA, Cantilena Jr. LR, Green JL, Rumack BH, Dart RC. 2010 Annual Report of the American Association of Poison Control Centers' National Poison Data System (NPDS): 28th annual report. Clin Toxicol (Phila) 2011;49:910—41.
3. Litovitz TL, Smilkstein M, Felberg L, Klein-Schwartz W, Berlin R, Morgan JL. 1996 annual report of the American Association of Poison Control Centers Toxic Exposure Surveillance System. Am J Emerg Med 1997;15:447—500.
4. Wilson KC, Reardon C, Theodore AC, Farber HW. Propylene glycol toxicity: a severe iatrogenic illness in ICU patients receiving IV benzodiazepines: a case series and prospective, observational pilot study. Chest 2005;128:1674—81.
5. Cawley MJ. Short-term lorazepam infusion and concern for propylene glycol toxicity: case report and review. Pharmacotherapy 2001;21:1140—4.
6. Yaucher NE, Fish JT, Smith HW, Wells JA. Propylene glycol-associated renal toxicity from lorazepam infusion. Pharmacotherapy 2003;23:1094—9.
7. Brent J. Current management of ethylene glycol poisoning. Drugs 2001;61:979—88.
8. Wilson KC, Reardon C, Farber HW. Propylene glycol toxicity in a patient receiving intravenous diazepam. N Engl J Med 2000;343:815.
9. Arroliga AC, Shehab N, McCarthy K, Gonzales JP. Relationship of continuous infusion lorazepam to serum propylene glycol concentration in critically ill adults. Crit Care Med 2004;32:1709—14.
10. Hayman M, Seidl EC, Ali M, Malik K. Acute tubular necrosis associated with propylene glycol from concomitant administration of intravenous lorazepam and trimethoprim-sulfamethoxazole. Pharmacotherapy 2003;23:1190—4.
11. Zar T, Graeber C, Perazella MA. Recognition, treatment, and prevention of propylene glycol toxicity. Semin Dialysis 2007;20:217—19.

Chapter 6

Analgesics and anti-inflammatory drugs

Christopher McCudden[1,2]

[1]*Division of Biochemistry, The Ottawa Hospital, Ottawa, ON, Canada,* [2]*Department of Pathology and Laboratory Medicine, University of Ottawa, Ottawa, ON, Canada*

Introduction

Analgesics and anti-inflammatory drugs

Analgesic and anti-inflammatory drugs are a broad category of compounds that are used to treat swelling, redness, and pain. They are the most widely used drugs in the world. It is estimated that some 7 billion grams of acetaminophen alone is consumed in the United States each year, with over 60 million individuals using the drug on a weekly basis [1,2]. Analgesic and anti-inflammatory drugs are available both over-the-counter (OTC) at low doses and by prescription at higher doses. They are available as monotherapy and in combination with numerous other compounds for the treatment of conditions such as flu, aches, pains, and fever.

Analgesic and anti-inflammatory drugs are classified by their therapeutic effects, mechanism of action, and structure. They broadly include acetaminophen and nonsteroidal anti-inflammatory drugs (NSAIDs). These structurally diverse drugs act through a series of different pathways inhibiting pain and inflammation. Acetaminophen is essentially a single compound, whereas NSAIDs have more than a half-dozen mechanistically related compounds, including salicylates (e.g., acetylsalicylic acid—aspirin), propionic acid derivatives (e.g., ibuprofen), acetic acid derivatives (e.g., indomethacin), and selective COX-2 inhibitors (e.g., celecoxib), among many others.

It should be noted that there are numerous other drugs that *individually* act as either analgesics (opioids, cannabinoids, steroids, and other psychotropic agents) *or* anti-inflammatory drugs (antileukotrienes, immune selective anti-inflammatory derivatives, and corticosteroids), which are collectively out of scope for this chapter. Included in this chapter is the history, mechanisms of action, toxicology, and analytical considerations for acetaminophen and NSAIDs.

Discovery and history

Salicylates

The discovery of the analgesic properties of salicylates dates back thousands of years, where white willow (*Salix alba*) bark was used for its medicinal properties some 2500 years ago. These origins correlate with the beginning of modern medicine, where Hippocrates himself is recorded to have prescribed willow leaves and bark for pain relief. Willow contains salicin, the source of the analgesic effects (Fig. 6.1). Shortly thereafter, salicylic acid (Fig. 6.2) was purified from

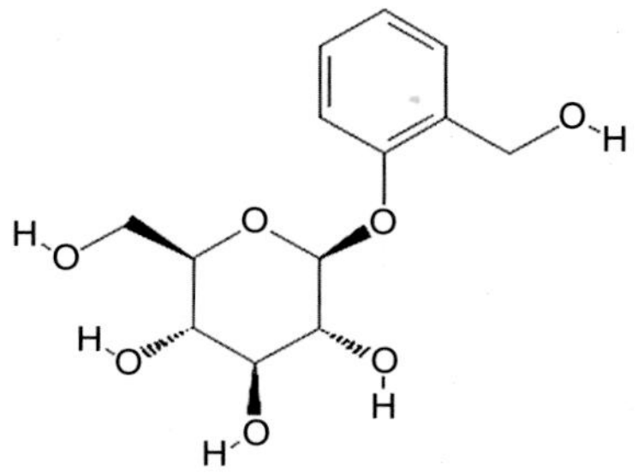

FIGURE 6.1 Structure of salicin.

the bark and became commonly employed to treat fever. However, salicylic acid was found to be irritating to the mouth and stomach, and alternatives were sought. The key to this was the addition of an acetyl group (Fig. 6.3). Chemist Felix Hoffmann is credited with synthesizing acetylsalicylic acid for Bayer. The commercial name aspirin was derived from the chemical combination of (a) from the acetyl group and "spir" from the salicin group, which at that time was sourced from meadow (*Spirea* sp.).

Many years after the success of aspirin, the mechanism of action was determined (see next), which led to the development of other compounds, such as selective cyclooxygenase-2 (COX-2) inhibitors. Selective COX-2 inhibitors were designed to specifically target the enzyme responsible for aspirin's analgesic bioactivity, but avoiding some of the side effects of gastrointestinal (GI) discomfort and peptic ulcer. COX-2 inhibitors have had varying degrees of commercial success largely clouded by increased cardiovascular risk [3]. While the analgesic effects are well-substantiated and there appear to be lower incidence of GI bleeding risk, there remain concerns for cardiovascular safety.

Acetaminophen

Similar to salicylates, the discovery of acetaminophen can be tied back to natural sources. In the 1800s, cinchona bark was used as the source for derivation of quinine. Quinine was not only widely used as an antimalarial but also was noted to function as an antipyretic. As the natural sources became scarce, chemically derived alternatives were sought. In a remarkable twist of fate, two Austrian doctors accidentally prescribed acetanilide (Fig. 6.4) due to a pharmacy error in an effort to treat a parasitic infection [4]. While acetanilide was ineffective at treating the parasite, its antipyretic properties were noted. Shortly thereafter, they sought to market acetanilide for its antifebrile effects. While it is an effective antipyretic, acetanilide was observed to be highly toxic (causes methemoglobinemia), which leads to the search for an alternative. In this search chemist Harmon Morse synthesized *N*-acetyl-*p*-aminophenol (acetaminophen) and phenacetin in the late 1800s (Fig. 6.5). Unfortunately, flawed early clinical trials resulted in widespread use of phenacetin until it was recognized to be nephrotoxic. Finally, Brodie and Axelrod determined that acetaminophen was the active metabolite of phenacetin [5]. Following this key discovery, acetaminophen was first marketed in the 1950s by McNeil laboratories for pain relief and fever in children. It became popular in the 1960s and 1970s with a perception that it had a lower toxicity profile that salicylates. Use increased further as aspirin was tied to Reye syndrome [6] and analgesics in children shifted to acetaminophen.

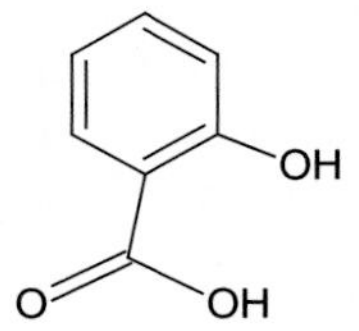

FIGURE 6.2 Structure of salicylic acid.

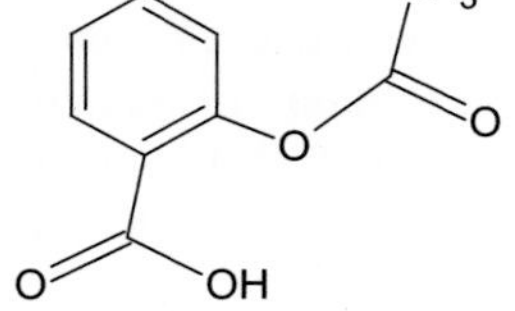

FIGURE 6.3 Structure of acetylsalicylic acid.

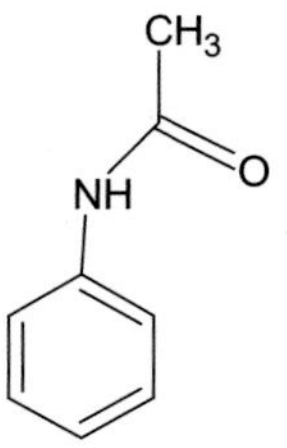

FIGURE 6.4 Structure of acetanilide.

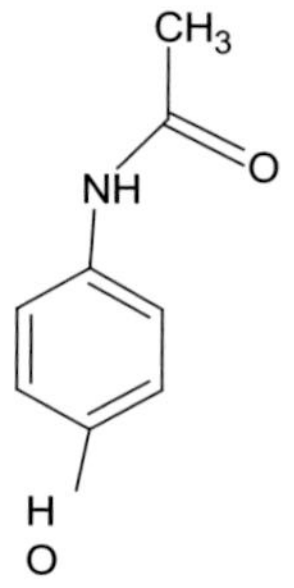

FIGURE 6.5 Structure of acetaminophen.

Toxicity

Nonsteroidal anti-inflammatory drugs

Although NSAIDs are effective treatments for pain and inflammation, they do have the risk of both acute and chronic toxicity. With some 10%–20% of adults >60 years old in the United States using aspirin daily, toxicity is a common occurrence. The American Poison Control Center has estimated more than 10,000 salicylate poisonings annually and more than 50 deaths [7]; these deaths are now more common in adults as use in children has diminished after association with Reye syndrome.

There are numerous different prescription and nonprescription concoctions containing salicylates from pills to powders, creams, and oils [8]. The common topical application is pain relief, but it is also used as a keratolytic (skin thinning and softening) agent. Here applications include warts, acne, psoriasis, and actinic and seborrheic keratosis. OTC topical formulations contain ~5%–10% salicylic acid, but prescription doses may contain up to 40% salicylic acid. Topical formulations include mineral oil, petroleum jelly, polyethylene glycol, and methyl glycol preparations. Percutaneous absorption varies with the additives [9] but is estimated to be ~60% of the dose. Many consumers do not appreciate how many treatments and compounds contain salicylate and that the accumulation is additive. Further, the use of herbal remedies also contributes to risk where products contain salicylates that are not known to the patient; some of these contain large quantities of salicylate [10]. Topical products [11] exhibit additional risk if applied improperly [12], such as covering skin surface area or application of occlusive dressings [13].

Recognition of salicylate toxicity can be a challenge. This is caused by a variety of factors. Medical trainees who treat patients may encounter local products or jargon they may not recognize as potential causative agents of poisoning. In general, patients who have overdosed may be unreliable at providing history useful to make the diagnosis. Younger patients are typically associated with intentional or acute overdose, whereas the elderly are more likely to present with chronic toxicity. Recognition of toxicity is compounded by the broad differential diagnosis and clinical presentation of salicylate poisoning; the differential includes sepsis, psychosis, alcohol withdrawal, and any high anion gap metabolic acidosis. Finally, initial diagnostic testing may fail to detect salicylate early on, requiring serial monitoring if clinical suspicion is high. Rarely, there are analytical interference from salicylate, which cause falsely elevated chloride effectively masking high anion gap that is commonly used to help make the diagnosis [14,15].

Salicylate toxicity typically occurs at doses >150 mg/kg. While useful in guiding clinical decisions, patients cannot be managed using blood salicylate levels in isolation. In terms of blood monitoring of acute toxicity, generally there is only mild toxicity at concentrations of 2.0–3.5 mmol/L. With mild toxicity, symptoms include dizziness, nausea, vomiting, and tinnitus. At blood (serum or plasma) concentrations >3.5 mmol/L severe toxicity can occur, presenting as diaphoresis, tachypnea, and dehydration. Extremely high concentrations (>5 mmol/L) can cause coma, seizures, hallucinations, cardiac failure, coagulopathy, renal failure, and death [16].

In cases of chronic toxicity, lower concentrations may cause symptoms as there is systemic distribution of salicylate in the CNS [17]. CNS toxicity is aggravated in the case of metabolic acidosis as the pH shift enhances passing of salicylate into the CNS. Acute toxicity is more commonly associated with GI problems, whereas chronic toxicity is more commonly associated with CNS symptoms [18]. Multidrug therapy adds risk for toxicity, where clearance may be affected by the additional onboard compounds. While toxicity in children is often avoided as parent's awareness of Reye syndrome rises, the wide availability and perceived safety contribute to the common toxicity. Mechanisms of toxicity are discussed later.

In addition to acute toxicity, aspirin is also associated with additional risk from cerebrovascular and GI bleeding with chronic use [19]. Aspirin is widely used to prevent stroke and heart attack in the general population and even more so in patients with history of these diseases. Indeed, the American Heart Association and European Society of Cardiology recommend low-dose aspirin in all patients with coronary artery disease based on evidence of efficacy [20]. In addition, chronic low-dose aspirin is associated with upper GI complaints, such as burning, pain, ulcer, and reflux. While these cases are not likely to present to the laboratory for acute toxicity management, the mechanisms are noteworthy for a complete understanding. Aspirin and other NSAIDs have both direct and indirect toxicity mechanisms in the GI tract [21]. Direct effects of aspirin are caused by passage of the drug into gastric epithelial cells. Drug enters gastric cells as a weak acid at the low pH of the stomach. Once drug passes into cells, it dissociates at the relatively neutral cellular pH and becomes trapped; trapped drug molecular and metabolite can cause damage to biliary structures and reflux (of metabolites) [21]. During the dissociation a hydrogen ion is released and it is the effect that causes cellular damage [22].

Acetaminophen

Acetaminophen was first linked to liver toxicity in the 1960s. At that time, histology identified liver necrosis in sections from patients with fulminant liver failure. Subsequent reports determined that hepatocyte damage occurred in the first 4 hours, but liver failure did not manifest until 24–96 hours later [23]. Reflecting the damage during this time frame, there are dramatic increases in serum ALT and AST levels. As liver failure progresses, bilirubin and INR also increase. Until the fentanyl crisis, acetaminophen was the single most common agent associated with overdose [24]. It remains a common agent of both accidental and intentional overdose. It is estimated that staggering half of liver transplants are related to acetaminophen toxicity [25]. Acetaminophen overdose is estimated to cost the United States >$80 million annually from poison control centers to emergency departments.

Mechanism of action

Despite the centuries of salicylate-related compound use, it was only in the early 1970s that the chemical mechanisms of action have been elucidated [26]. The primary mechanism of action for salicylates is to inhibit the formation of prostaglandins; prostaglandins are known to enhance the effects of bradykinins on pain receptors. Acetylsalicylates block conversion of arachidonic acid to prostaglandin-G2 by inhibiting the enzyme cyclooxygenase. Cyclooxygenase is blocked by direct acetylation of a serine residue within the active site. Active site acetylation prevents binding to the substrate arachidonic acid and therefore inhibits production of the prostaglandins and other downstream compounds. Cyclooxygenase acetylation is irreversible, which means that enzyme activity will remain reduced until new cyclooxygenase is synthesized. While most tissues synthesize new cyclooxygenase, regeneration does not occur in platelets. Thus platelet cyclooxygenase activity is blocked for the life of the cell.

By blocking cyclooxygenase, salicylic acid (aspirin) also inhibits production of other compounds. Specifically, decreased conversion of arachidonic acid to prostaglandin-G2 affects subsequent formation of other related endoperoxides, such as prostaglandins D2-, E2-, and F2-alpha, and thromboxane A2 [26]. By decreasing thromboxane A2, aspirin indirectly inhibits platelet aggregation. This feature has resulted in the use of low-dose aspirin as a prophylactic to stroke and heart attack. While effective at reducing cardiovascular events, recent evidence indicates that low-dose aspirin carries a substantial risk of bleeding, which may exceed the benefit [20].

Other salicylate-related compounds, such as indomethacin and COX-2 inhibitors, bind directly to cyclooxygenase-2, blocking its catalytic activity. As the name suggests, COX-2 is specific for cyclooxygenase-2, which was created by design to avoid GI side effects attributed to cyclooxygenase-1. An unintended consequence of this specificity was a lack of effect on thromboxane and therefore no effect on platelet aggregation. Without the apparent protective anticlotting activity of aspirin, COX-2 inhibitors were recognized to increase the risk of heart attack and strokes [27]. As a result, most of the COX-2 inhibitors were taken off the US market [3]. The only remaining one (celecoxib) is an effective analgesic with a black box warning about the risk.

Similar to salicylates, the mechanism of action of acetaminophen was determined long after it's widespread commercial success. Indeed, mechanistic actions of acetaminophen are complicated still actively being investigated. Fundamentally, acetaminophen functions similar to aspirin in that it blocks cyclooxygenase activity. However, unlike aspirin, acetaminophen does not function as an anti-inflammatory and does not have GI side effects. The reason for these differences was long of debate with some confusion about mechanisms.

Years of study have elucidated that the cyclooxygenase inhibition occurs under specific conditions in vivo. These conditions involve the intracellular concentrations of peroxide and arachidonic acid. When arachidonic acid concentration is low, cyclooxygenase-2 becomes the dominant pathway for prostaglandin-G2 production. Conversely, when there are high concentrations of arachidonic acid, cyclooxygenase-2 is the dominant pathway. It is hypothesized that arachidonic acid concentration is directly correlated with prostaglandin-G2 levels. Prostaglandin-G2 itself is a peroxide, which is known to inhibit acetaminophen. Thus the inhibitory action of prostaglandin-G2 is the cause of apparent preference of acetaminophen for cyclooxygenase-2. That mechanism also explains the tissue specificity of acetaminophen and inability to function as an anti-inflammatory. Consider that many inflammatory conditions have high prostaglandin-G2, which would inhibit acetaminophen. Likewise, there is no apparent effect on thromboxane A2 and therefore no antiplatelet aggregation activity because of the high concentrations of prostaglandin-G2 (and arachidonic acid) in platelets. Overall, acetaminophen is effectively cyclooxygenase-2 preferential, not through structural specificity, but the chemical milieu.

Interestingly, acetaminophen inhibits cyclooxygenases by serving as a chemical substrate. In particular, acetaminophen serves as a peroxidase substrate for cyclooxygenases; cyclooxygenases have both peroxidase and cyclooxygenase activity.

Acetaminophen acts as an antipyrogen by blocking production of prostaglandin E2 in the CNS. Prostaglandin E2 acts as a pyrogen by acting on the preoptic nucleus of the hypothalamus to reduce the signaling rate of the heat sensitive neurons, effectively increasing the body's thermal set point [28].

More recent work has linked acetaminophen and NSAIDs to the opioid, serotonin, and endocannabinoid pathways. Mechanisms are complex, but the implications are that these long known analgesics have additional connections to other important pain pathways that may explain some of the activity of the drugs [29].

Pharmacokinetics, toxicokinetics, and treatment

Salicylate

Aspirin is absorbed in the stomach with a small volume of distribution of 0.1–0.3 L/kg. Once absorbed, 90% of salicylate is protein-bound (primary to albumin). It is largely metabolized through the liver by conjugation with glycine and glucuronides, which are cleared by the kidney; urinary excretion includes free salicylic acid, salicyluric acid, glucuronides, acylglucuronides, and gentistic acid.

During overdose, metabolic pathways become saturated and reliance on renal clearance increases. When saturated, removal of drug changes from first-order kinetics (exponential removal with time) to zero-order one (constant removal with time). The pH of urinary clearance is important to understand for treatment. At acidic pH, only 2% of the drug is eliminated. In contrast, with alkaline urine up to 30% may be eliminated in urine. During overdose the aspirin half-life of 2–4 hours increases up to 24 hours in overdoses. Conversely, protein binding also decreases with toxic doses to 75%, with a concomitant increased drug availability.

Toxicity is caused by several mechanisms. First, salicylate directly stimulates the respiratory control center in the brain. In doing so, patients begin to hyperventilate, which leads to respiratory alkalosis caused by increased excretion of CO_2 (effectively loss of acid). Second, salicylate itself is an acid, which can directly bind circulating HCO_3^- contributing to metabolic acidosis. More importantly, salicylate is effectively a mitochondrial oxidative phosphorylation poison, which increases the production of other organic acids, such as lactate and beta-hydroxybutyrate. Collectively, these disturbances result in high anion gap metabolic acidosis.

The most dangerous aspect of salicylate overdose is CNS toxicity. CNS abnormalities include central hyperventilation, confusion, agitation, and lethargy. Progression to seizures and coma portends a poor outcome. At high salicylate concentrations, CNS effects are caused by impaired energy metabolism induced by uncoupling of oxidative phosphorylation. Consequently, patient's body temperature is increased and brain cells can become starved for ATP. Other toxic effects include acute lung injury, hemorrhagic gastritis, platelet dysfunction, and rhabdomyolysis [17].

Laboratory manifestations of salicylate poisoning include decreased bicarbonate (high anion gap metabolic acidosis), low pCO_2 (respiratory alkalosis), increased lactate (uncoupled oxidative phosphorylation), prolonged thrombin time (platelet inhibition), and apparent low CSF glucose as compared with serum glucose (uncoupled oxidative phosphorylation combined with higher demand in the CNS). The presence of a mixed acid-base disorder may prompt clinical suspicion for salicylate poisoning. Mixed disorders are more difficult to identify in children but are classic in adults.

Measurement of salicylate is typically helpful for guiding therapy (as mentioned previously) but requires close monitoring of pH in combination (e.g., with a blood gas sample). If blood pH decreases, measured blood salicylate concentrations may also decrease, but this is because of uptake by tissues. At neutral pH, such a decrease would be inferred as a true decrease in the burden of salicylate. Other interpretative limitations of salicylate include incorrect units or confusion about toxic concentration. One mmol/L of salicylate translates to 13.8 mg/dL (multiply mg/dL by

0.0724 to yield mmol/L). To help interpret salicylate measurements, in 1960 Alan Done published a nomogram [30]. The Done nomogram combines the concentration of salicylate with the time since the dose to predict toxicity [31]. However, the conditions used to create the nomogram were limited in that it only included children within 6 hours of the dose (noncoated pills), where the pH was within reference limits. As a result the nomogram has poor predictive value [31], particularly for more severe overdoses. Collectively, clinical manifestations are essentially for patient management.

Acetylsalicylic acid overdoses are treated based on the presentation severity using any of several options, including dialysis, activated charcoal, and IV saline. Unlike acetaminophen, there is no antidote for salicylate poisoning. Primarily, the objective is to limit absorption, increase excretion, and manage toxicity [18]. A single dose of activated charcoal is common practice provided the patient can consume it; activated charcoal is provided as an unpleasant liquid slurry the patient must drink. IV fluid is used to manage hypovolemia that is caused by the hypermetabolic state, high temperature, and vomiting. Glucose is provided if the patient has neurological deficits. As described earlier, urine alkalinization is effective to promote excretion. This is achieved by providing IV sodium HCO_3^-. In severe cases, hemodialysis is the most effective therapy to eliminate salicylate.

Acetaminophen

Acetaminophen is typically taken orally, where it is absorbed rapidly by the intestines. It is a water soluble weak acid with a large volume of distribution (50 L). Peak blood concentration is <200 μmol/L and the half-life is 1–4 hours. The majority of acetaminophen is cleared in urine within 1–4 hours. The drug provides analgesic relief within 30–45 minutes. Acetaminophen is largely removed via glucuronidation in the liver (and kidney to a lesser extent) with the major drug metabolites being conjugates with glucuronide, sulfate, glutathione, and cysteine. In the course of normal metabolism, small amounts of the toxic metabolite N-acetyl-p-benzoquinone imine (NAPQI) are produced. NAPQI is generated by the conversion of acetaminophen via cytochrome p450 enzymes. Low amounts of NAPQI are quickly neutralized by glutathione to nontoxic mercaptate and cysteine compounds, which are then renally excreted [1].

In the case of overdose, glutathione is depleted and NAPQI cannot be cleared fast enough by the liver, resulting in accumulation of the toxin. In these conditions, NAPQI reacts with sulfhydryl groups of liver proteins causing hepatic and renal damage [32]. However, the overall mechanisms of acetaminophen toxicity appear more complex. Toxicity appears to additionally involve production of acetaminophen-protein adducts, oxidative stress, mitochondrial permeability, and ATP depletion [33].

Risk factors for toxicity include the dose taken (toxic effects can occur with doses 150–200 mg/kg or >10 g/day), existing renal or liver disease, alcoholism, anorexia, and multidrug therapy [34,35]. Consumption of several compounds containing acetaminophen often contributes to accidental poisoning cases [36].

Understanding of acetaminophen toxicology led to the development of therapies started in the 1970s. Early versions included methionine and cysteine. Antidotes include methionine, cysteine, and acetylcysteine. Antidotes for acetaminophen poisoning largely focus on replacing the glutathione depleted by NAPQI; N-acetylcysteine (Mucomyst) is thought to detoxify NAPQI by either direct conjugation inactivating the toxin or through increased glutathione synthesis. If given within 10 hours of acetaminophen overdose, acetylcysteine is effective at reducing acute liver injury. Liver injury is reduced from >50% to <10% with rapid administration of antidote [32].

Analytical methods

Salicylates are measured in clinical laboratories typically using two different methods. The first is known as the Trinder reaction, where ferric nitric complexes with salicylate, which is measurable by spectrophotometry at 510 nm. The second method uses salicylate hydroxylase to convert the substrate to catechol in the presence of NADH. Subsequent measurement of NAD+ is calibrated to salicylate concentration; other methods rely on secondary products for measurement, but fundamentally the assays rely on the same concept.

Acetaminophen is routinely measured in clinical laboratories to investigate potential poisoning. Methods include homogenous immunoassay and colorimetric assays. Colorimetric assays rely on enzymes (amidase) to convert acetaminophen to p-aminophenol, which is then bound to dye (e.g., o-cresol and ammoniacal copper sulfate) and measured spectrophotometrically (OD 600 nm). Other methods may use different coupling dyes, but principles are similar.

Irrespective of the method used, acetaminophen measurement can be done on serum or plasma. Measurements may be interpreted using the Rumack–Matthew nomogram [31], though, as always, clinical judgment and presentation are important in making treatment decisions.

Summary

Analgesics and anti-inflammatory drugs are some of the oldest and the most widely used medications in the world. They are effective pain relievers, antifever therapies, and antiswelling treatments with a wide therapeutic window. However, despite the relative safety of these medications, overdoses are common. This occurs largely due to their ubiquity and availability. Both NSAID and acetaminophen overdoses are treatable providing health-care providers recognize and manage the toxicity.

References

[1] Agrawal S, Khazaeni B. Acetaminophen Toxicity. In: StatPearls [Internet]. Treasure Island, FL: StatPearls Publishing; 2019 [cited 2019 Apr 27]. Available from: <http://www.ncbi.nlm.nih.gov/books/NBK441917/>.

[2] Diener H-C, Schneider R, Aicher B. Per-capita consumption of analgesics: a nine-country survey over 20 years. J Headache Pain 2008;9(4):225–31.

[3] Krumholz HM, Ross JS, Presler AH, Egilman DS. What have we learnt from Vioxx? BMJ 2007;334(7585):120–3.

[4] Brune K, Renner B, Tiegs G. Acetaminophen/paracetamol: a history of errors, failures and false decisions. Eur J Pain Lond Engl 2015;19(7):953–65.

[5] Brodie BB, Axelrod J. The fate of acetanilide in man. J Pharmacol Exp Ther 1948;94(1):29–38.

[6] Ap. Aspirin labels to warn about Reye syndrome. The New York Times [Internet]. 1986 Mar 8 [cited 2019 May 7]; Available from: <https://www.nytimes.com/1986/03/08/us/aspirin-labels-to-warn-about-reye-syndrome.html>.

[7] Gummin DD, Mowry JB, Spyker DA, Brooks DE, Osterthaler KM, Banner W. Annual Report of the American Association of Poison Control Centers' National Poison Data System (NPDS): 35th Annual Report. Clin Toxicol 2017;56(12):1213–415 2018.

[8] Wiffen PJ, Kalso EA, Bell RF, Aldington D, Phillips T, Gaskell H, et al. Topical analgesics for acute and chronic pain in adults – an overview of Cochrane Reviews. Cochrane Database Syst Rev [Internet]. 2017 May 12 [cited 2019 May 20];2017(5). Available from: <https://www.ncbi.nlm.nih.gov/pmc/articles/PMC6481750/>.

[9] Schwarb FP, Gabard B, Rufli T, Surber C. Percutaneous absorption of salicylic acid in man after topical administration of three different formulations. Dermatology 1999;198(1):44–51.

[10] Baxter AJ, Mrvos R, Krenzelok EP. Salicylism and herbal medicine. Am J Emerg Med 2003;21(5):448–9.

[11] Cross SE, Anderson C, Thompson MJ, Roberts MS. Is there tissue penetration after application of topical salicylate formulations? Lancet 1997;350(9078):636.

[12] Madan RK, Levitt J. A review of toxicity from topical salicylic acid preparations. J Am Acad Dermatol 2014;70(4):788–92.

[13] Bell AJ, Duggin G. Acute methyl salicylate toxicity complicating herbal skin treatment for psoriasis. Emerg Med Fremantle WA 2002;14(2):188–90.

[14] Jacob J, Lavonas EJ. Falsely normal anion gap in severe salicylate poisoning caused by laboratory interference. Ann Emerg Med 2011;58(3):280–1.

[15] Riethmuller S, Luft FC, Mohebbi N. Where is the gap? Clin Kidney J 2012;5(1):63–4.

[16] Pearlman BL, Gambhir R. Salicylate intoxication: a clinical review. Postgrad Med 2009;121(4):162–8.

[17] Flomenbaum NE, Goldfrank LP, Hoffman RS, Howland MA, Lewlin NA, Nelson LS. Salicylates. In: Goldfrank's toxicologic emergencies. 10th ed. New York, NY: McGraw-Hill.

[18] O'Malley GF. Emergency department management of the salicylate-poisoned patient. Emerg Med Clin North Am 2007;25(2):333–46.

[19] Lavie CJ, Howden CW, Scheiman J, Tursi J. Upper gastrointestinal toxicity associated with long-term aspirin therapy: consequences and prevention. Curr Probl Cardiol 2017;42(5):146–64.

[20] Zheng SL, Roddick AJ. Association of aspirin use for primary prevention with cardiovascular events and bleeding events: a systematic review and meta-analysis. JAMA 2019;321(3):277–87.

[21] Wolfe MM, Lichtenstein DR, Singh G. Gastrointestinal toxicity of nonsteroidal antiinflammatory drugs. N Engl J Med 1999;340(24):1888–99.

[22] Schoen R. Mechanisms nonsteroidal anti-inflammatory drug-induced gastric damage. Am J Med 1989;86(4):449–58.

[23] Boyer TD, Rouff SL. Acetaminophen-induced hepatic necrosis and renal failure. JAMA 1971;218(3):440–1.

[24] Prescott K, Stratton R, Freyer A, Hall I, Le Jeune I. Detailed analyses of self-poisoning episodes presenting to a large regional teaching hospital in the UK. Br J Clin Pharmacol 2009;68(2):260–8.

[25] Reuben A, Tillman H, Fontana RJ, Davern T, McGuire B, Stravitz RT, et al. Outcomes in adults with acute liver failure between 1998 and 2013: an observational cohort study. Ann Intern Med 2016;164(11):724–32.

[26] Botting RM. Vane's discovery of the mechanism of action of aspirin changed our understanding of its clinical pharmacology. Pharmacol Rep 2010;62(3):518–25.

[27] Katz JA. COX-2 inhibition: what we learned—a controversial update on safety data. Pain Med Malden Mass 2013;14(Suppl. 1):S29–34.

[28] Walter EJ, Hanna-Jumma S, Carraretto M, Forni L. The pathophysiological basis and consequences of fever. Crit Care [Internet]. 2016 [cited 2019 May 10];20. Available from: <https://www.ncbi.nlm.nih.gov/pmc/articles/PMC4944485/>.

[29] Graham GG, Davies MJ, Day RO, Mohamudally A, Scott KF. The modern pharmacology of paracetamol: therapeutic actions, mechanism of action, metabolism, toxicity and recent pharmacological findings. Inflammopharmacology 2013;21(3):201–32.

[30] Done AK. Salicylate intoxication: significance of measurements of salicylate in blood in cases of acute ingestion. Pediatrics 1960;26(5):800–7.
[31] Dugandzic RM, Tierney MG, Dickinson GE, Dolan MC, McKnight DR. Evaluation of the validity of the Done nomogram in the management of acute salicylate intoxication. Ann Emerg Med 1989;18(11):1186–90.
[32] Chiew AL, Gluud C, Brok J, Buckley NA. Interventions for paracetamol (acetaminophen) overdose. Cochrane Database Syst Rev 2018;23(2): CD003328.
[33] Hinson JA, Roberts DW, James LP. Mechanisms of acetaminophen-induced liver necrosis. Handb Exp Pharmacol 2010;196:369–405.
[34] Buckley N, Eddleston M. Paracetamol (acetaminophen) poisoning. Clin Evid 2002;(8):1447–53.
[35] Park BK, Dear JW, Antoine DJ. Paracetamol (acetaminophen) poisoning. BMJ Clin Evid 2015; Oct 19. pii: 2101. https://www.ncbi.nlm.nih.gov/pmc/articles/PMC4610347/.
[36] Hodgman MJ, Garrard AR. A review of acetaminophen poisoning. Crit Care Clin 2012;28(4):499–516.

Chapter 6.1

Acetaminophen toxicity

D. Adam Algren[1,2,3]

[1]Division of Clinical Pharmacology, Toxicology, Therapeutic Innovation, Children's Mercy Hospital, Kansas City, MO, United States, [2]Department of Emergency Medicine, Truman Medical Center, University of Missouri Kansas City, Kansas City, MO, United States, [3]University of Kansas Hospital Poison Control Center, Kansas City, MO, United States

Case history

A 37-year-old female presented to an emergency department with abdominal pain, vomiting, and generalized fatigue for 2 days. She denied any fevers, diarrhea, hematemesis, hematochezia, or melena. She disclosed that she had been ingested approximately 12 g of acetaminophen daily for the past 8 days to treat recurrent headaches. She denied any other medication use, herbal/supplement use, or mushroom ingestion. She denied any past medical history, alcohol, or drug use. Physical exam was remarkable for tachycardia, (RUQ) tenderness, and jaundice. Laboratory studies demonstrated acute liver failure with an elevated acetaminophen level (Table 6.1.1). The Regional Poison Center and hepatology service were contacted and agreed that the most likely etiology of the liver failure was acetaminophen. The patient was started in IV *N*-acetylcysteine (NAC) and admitted to the ICU. Other etiologies of liver failure were excluded. Laboratory studies demonstrated slow improvement in liver functions over the next week, and NAC was discontinued on hospital day 11. The patient was discharged home on hospital at day 12.

Case discussion

Acetaminophen remains the leading cause of drug-induced liver injury in the United States. In 2017 the American Association of Poison Control Center reported over 65,000 acetaminophen exposures that were associated with at least 115 deaths [1]. Poisonings can be grouped into acute or chronic. Acute toxicity occurs following a single ingestion greater than 200 mg/kg or 10 g in adults (whichever is less). Chronic toxicity can occur after ingesting 6 g/day for >48 hours in adults or 100 mg/kg/day in children [2].

Acetaminophen is rapidly absorbed. It demonstrated first-order kinetics with a 4 hour elimination half-life; however, this can be prolonged in cases of toxicity. Therapeutic use generally results in serum acetaminophen levels between 10 and 30 mcg/mL [3]. A large portion of acetaminophen is glucoronidated with a smaller portion undergoing sulfation or direct renal elimination. Approximately 5%–15% is metabolized by CYP450 2E1 to produce the toxic metabolite NAPQI [4]. Generally, NAPQI is conjugated with glutathione to produce nontoxic metabolites. However, in cases of overdose, glutathione stores become depleted resulting in hepatotoxicity predominately affecting Zone 3 of the liver (centrilobular).

Patients with acute toxicity present most commonly with nausea, vomiting, and abdominal pain. Massive ingestion can result in central nervous system (CNS) depression and lactic acidosis. Chronic toxicity presents with similar symptoms. Generally, abdominal pain and vomiting are the initial findings; however, other signs of liver injury may be present, including jaundice, encephalopathy, or renal failure. Following an acute overdose, peak hepatotoxicity occurs at 72–96 hours [5].

The laboratory evaluation and management of acetaminophen ingestions/toxicity depend on the circumstances. In cases of unintentional pediatric exploratory ingestion, only an acetaminophen level is required. This level should be obtained at 4 hours (or on presentation if >4 hours) postingestion. The Rumack–Matthew nomogram (Fig. 6.1.1) has been used for decades and is useful in determining which patients require antidotal therapy following an acute ingestion [6]. The nomogram requires assessment of an APAP level between 4 and 24 hours postingestion to identify patients that need treatment. It cannot be used in assessment of chronic toxicity. There has been some investigation into the

TABLE 6.1.1 Laboratory Studies.

	Hospital Day 1	Hospital Day 3	Hospital Day 5	Hospital Day 12
Total bilirubin (mg/dL)	6.1	5.4	3.5	1.7
AST (U/L)	3580	3275	2478	980
ALT (U/L)	3265	2980	2190	720
INR	3.8	3.2	2.6	1.4
Creatinine (mg/dL)	1.8	1.6	1.4	1.3
Serum APAP (mcg/mL)	34	0	0	0

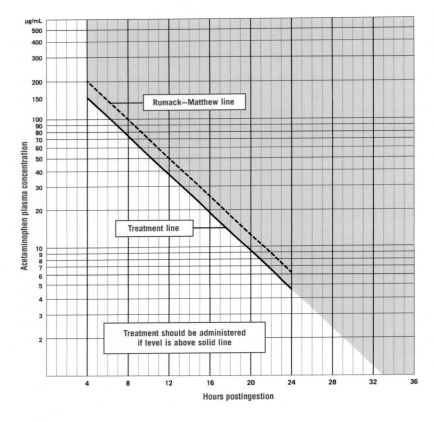

FIGURE 6.1.1 The Rumack–Matthew nomogram. From: *https://en.wikipedia.org/wiki/File:Rumack_Matthew_nomogram_with_treatment_(study)_line.pdf*.

ability of pre-4 hour concentrations to predict the need for antidotal therapy; however, no definitive conclusions can be drawn about pre-4 hours APAP levels [7,8]. Therefore the gold standard remains to obtain a 4-hour APAP level to determine who requires treatment following acute overdoses. In cases of chronic toxicity or in those that are ill-appearing electrolytes, renal function, liver function, and coagulation studies should be assessed. Enzymatic methods are most commonly used in clinical laboratories to quantify acetaminophen levels. Although generally accurate, there are reports of significantly elevated serum bilirubin concentrations, hemolysis, and lipemia resulting in false-positive or false-negative acetaminophen concentrations [9]. Other biomarkers (urine protein adducts, microRNA-122, HMGB-1) are being investigated to help determine etiology in cases of unknown liver failure or identify increase risk of hepatotoxicity prior to Aspartate aminotransferase/Alanine aminotransferase (AST/ALT) elevation [10].

Treatment of acetaminophen toxicity focuses on providing supportive care, including decontamination with activated charcoal when indicated. NAC is the antidote for acetaminophen toxicity and is used both in acute and chronic toxicity [11]. NAC is thought to serve as a glutathione precursor. In addition, it can directly bind NAPQI while serving as a free radical scavenger, antioxidant, and improving microcirculatory blood flow [11]. NAC is available in an oral and IV formulation and both are equally efficacious [12]. In cases of acute toxicity, NAC is essentially universally effective in preventing liver failure and death if initiated within 8 hours of ingestion. However, delayed administration has been shown to also be beneficial. Oral NAC is dosed every 4 hours and has a rotten egg odor and is associated with vomiting. These downsides make IV NAC more attractive. IV NAC is generally well tolerated; however, in very small children (<40 kg), excessive fluid administration may result in hyponatremia and seizures. Therefore close attention should be given to fluid volumes administered to children. IV NAC use is associated with mild spurious elevations in the INR as it has been shown that NAC reacts in vitro with the coagulation assays. Generally, this elevation is minor (INR < 2) [13]. Anaphylactoid reactions are the most significant adverse effect with IV NAC administration. These reactions occur in approximately 5%—10% of cases and are non-IgE-mediated reactions that result in mast cell degranulation [14,15]. A majority of the reactions are mild and consistent of rash, pruritus, flushing, and vomiting. More severe reactions are less common but can cause bronchospasm or hypotension. Risk factors for anaphylactoid reactions are low serum acetaminophen concentrations at this time of NAC initiation (high APAP concentrations stabilize mast cells) and possibly asthma [16]. Treatment involves holding the NAC infusion and administering IV diphenhydramine. In most mild cases the NAC infusion can generally be restarted in 20—30 minutes. NAC is continued for at least 21 hours (presuming no increase in AST/ALT) or until AST and ALT are downtrending with improving liver synthetic function (bilirubin and INR).

Liver transplantation is an option in those patients that have worsening liver failure despite NAC and maximal supportive care. Fortunately, most patients recover. In patients that do recover chronic liver disease or other sequelae is not expected.

References

[1] Gummin DG, Mowry JB, Spyker DA, et al. Annual Report of the American Association of Poison Control Centers' National Poison Data System (NPDS): 35th annual report. Clin Toxicol 2017;56(12):1213—415 2018.

[2] Dart RC, Erdman AR, Olson KR, et al. Acetaminophen poisoning: an evidence-based consensus guideline for out-of-hospital management. Clin Toxicol 2006;44:1—18.

[3] Forrest JAH, Clements JA, Prescott LF. Clinical pharmacokinetics of paracetamol. Clin Pharm 1982;7:93—107.

[4] Manyike PT, Kharasch ED, Kalhorn TF, et al. Contribution of CYP 2E1 and CYP 3Ato acetaminophen reactive metabolite formation. Clin Pharmacol Ther 2000;67:275—82.

[5] Singer AJ, Carracio TR, Mofenson HC. The temporal profile of increased transaminase levels in patients with acetaminophen-induced liver dysfunction. Ann Emerg Med 1995;26:49—53.

[6] Rumack BH. Acetaminophen hepatotoxicity: the first 35 years. J Toxicol Clin Toxicol 2002;40:3—20.

[7] Froberg BA, King KJ, Kurera TD, et al. Negative predictive value of acetaminophen concentrations within 4 hours of ingestion. Acad Emerg Med 2013;20(10):1072—5.

[8] Yarema MC, Green JP, Sivilotti ML, et al. Can a serum acetaminophen concentration obtained less than 4 hours post-ingestion determine which patients require treatment with acetylcysteine? Clin Toxicol 2017;55(2):102—8.

[9] Zhang YV, Fan SL, Kellogg MD. Effect of hemolysis, icterus, and lipemia on three acetaminophen assays: potential medical consequences of false positive results. Clin Chim Acta 2018;487:287—92.

[10] Yoon E, Babar A, Choudhary M, et al. Acetaminophen-induced hepatotoxicity: a comprehensive update. J Clin Transl Hepatol 2016;4 (2):131—42.

[11] Heard KJ. Acetylcysteine for acetaminophen poisoning. N Engl J Med 2008;359:285—92.

[12] Yarema MC, Johnson DW, Berlin RJ, et al. Comparison of the 20 hour intravenous and 72 hour oral acetylcysteine protocols for the treatment of acute acetaminophen poisoning. Ann Emerg Med 2009;54:606—14.

[13] Pizon AF, Jang DH, Wang HE. The in vitro effect of N-acetylcysteine on prothrombin time in plasma samples from healthy subjects. Acad Emerg Med 2011;18:351—4.

[14] Kao LW, Kirk MA, Furbee RB. What is the rate of adverse events after oral N-acetylcysteine administered by the intravenous route to patients with suspected acetaminophen poisoning? Ann Emerg Med 2003;42:741—50.

[15] Sandilands EA, Bateman DN. Adverse reactions associated with acetylcysteine. Clin Toxicol 2009;47:81—8.

[16] Pakravan N, Waring WS, Sharma S, et al. Risk factors and mechanisms of anaphylactoid reactions to acetylcysteine in acetaminophen overdose. Clin Toxicol 2008;46:697—702.

Chapter 6.2

Massive ibuprofen ingestion in an adolescent treated with plasma exchange

Umar Salimi[1], Marita Thompson[1], Darcy Weidemann[2], Judith Sebestyen VanSickle[2], John D. Nolen[3], Uttam Garg[3] and Michael R. Christian[4]

[1]Pediatric Critical Care Medicine, Children's Mercy Hospital, Kansas City, MO, United States, [2]Nephrology, Children's Mercy Hospital, Kansas City, MO, United States, [3]Laboratory Medicine, Children's Mercy Hospital, Kansas City, MO, United States, [4]Medical Toxicology, Children's Mercy Hospital, Kansas City, MO, United States

Case description

A 17-year-old girl was found unresponsive at home approximately 8 hours after she was last seen. A 500-count 200 mg tablet ibuprofen bottle, presumed half-full by her parents, was found with two tablets remaining. The patient had a history of major depressive disorder, posttraumatic stress disorder, substance abuse, and suicide attempt requiring recent inpatient psychiatric treatment. She had access to her prescribed medications: escitalopram, oxcarbazepine, and quetiapine. After the arrival of a medic team, she was breathing spontaneously and appeared to be hyperventilating. Initial vital signs in the emergency department were heart rate of 135 beats per minute, respiratory rate of 12 breaths per minute, oxygen saturation of 91% on room air, and blood pressure of 84/30 mmHg. She was found to be unresponsive to noxious stimuli with no cough or gag reflex and was subsequently intubated for airway protection. She was poorly perfused on exam and remained hypotensive despite several intravenous normal saline boluses. An epinephrine infusion was started and escalated to 0.12 μg/kg/min.

The patient's initial blood chemistries revealed a severely elevated anion gap acidosis and elevated lactic acid (Table 6.2.1). Lab studies showed acute kidney injury with BUN 9 mg/dL, creatinine 1.6 mg/dL, creatine phosphokinase 10,978 IU/L, blood alcohol level of 32 mg/dL, and an undetectable acetaminophen level. The results of a qualitative urine drug screen for nine drugs of abuse were negative. Her electrocardiogram (EKG) was notable for QRS widening. The patient was transported to a pediatric ICU and en route required ongoing support of hypotension with epinephrine and IV fluid boluses. She received sodium bicarbonate for metabolic acidosis.

On arrival to the PICU, she was poorly perfused with cool extremities and prolonged capillary refill time. Her cardiac rhythm displayed progressively widening QRS and PR intervals with a prolonged QTc of 633 ms and peaked T waves. Pupils were described as reactive, although there was no cough or gag reflex. There was no evidence of trauma and a head CT was normal. Significant laboratory derangements were observed with an arterial gas of pH 6.92, pCO_2 20 mmHg, pO_2 120 mmHg, HCO_3 <5 mEq/L, anion gap 29, and lactic acid 14 mmol/L. The BUN was 9 mg/dL, creatinine rose to 2 mg/dL, AST 628 U/L, ALT 88 U/L, potassium 6.5 mEq/L, and there was profound hypocalcemia with total calcium of 6.6 mg/dL and ionized calcium 0.86 mmol/L. The patient was anuric since presentation despite Foley insertion. A broad-spectrum comprehensive urine drug screen by gas chromatography–mass spectrometry was positive for ibuprofen, oxcarbazepine, citalopram, nicotine, carbamazepine, and caffeine. Acetaminophen was undetected by immunoassay. Volatile screen by gas chromatography was negative for ethanol, methanol, isopropanol, and acetone (<10 mg/dL). Plasma ibuprofen concentrations, sent to NMS labs (Horsham, PA) and performed by high-performance liquid chromatography/tandem mass spectrometry, were obtained in the initial hospital course, though not immediately available in clinical management. The first level was 330 μg/mL, 4 hours after patient discovery and 12 hours after the last time known well. Subsequent levels are shown in Table 6.2.1.

TABLE 6.2.1 Laboratory Values at Indicated Time After Patient's Initial Presentation.

	1 h	4 h	7 h	13 h	17 h	23 h	27 h	36 h
Arterial pH	6.91	6.92	7.07	7.07	7.14	7.22	7.29	7.41
pCO_2 (mmHg)	42.7	20	45	45.4	49.1	51.2	50.5	37.9
pO_2 (mmHg)	259	120	55	78	92	145	98	76
Base excess	−24	−29	−17.5	−17.5	−12.9	−7.1	−2.5	−0.2
Sodium (mEq/L)	151	150	137	141	144		142	140
Potassium (mEq/L)	5.5	6.5	4.2	4.2	4.3		3.9	3.4
Bicarbonate (mEq/L)	8	5	13	10	14	20	22	22
Creatinine (mg/dL)	1.6	2.01	1.7	1.72	1.55		2.0	1.61
Lactic acid (mmol/L)	14	14.8	12.4	8.6	6.5	6.3	3.4	3.2
Ibuprofen (μg/mL)		330	230	120		60		

The patient was treated with sodium bicarbonate, calcium chloride, insulin/glucose, and albuterol for her cardiac rhythm disturbance and hyperkalemia. Due to severe metabolic acidosis resistant to therapies and progressive cardiac dysrhythmias, a temporary femoral hemodialysis catheter was placed at her bedside for an emergent hemodialysis session and she was immediately transitioned to continuous renal replacement therapy (CRRT) via continuous venovenous hemodiafiltration using a clearance of 2100 mL/1.73 m^2/h. With ongoing hemodynamic instability and acidosis despite CRRT and vasopressor support, plasma exchange was then performed in tandem with CRRT to remove ibuprofen. A 1.5 × volume exchange was done using albumin as the replacement fluid. The patient's hemodynamics and acidosis improved rapidly over the following 24 hours and her vasopressor support was significantly reduced. She was more wakeful within hours of exchange and was extubated on hospital day 4.

The hospital course was complicated by severe rhabdomyolysis secondary to compartment syndrome of the lower extremities which developed on hospital day 1. The etiology was thought to be multifactorial, as she was initially found in a squatting position causing calf muscle ischemia, may have experienced tetany due to hypocalcemia, and had severe hypotension. Rhabdomyolysis combined with ibuprofen toxicity contributed to prolonged anuric renal failure for which she was supported with CRRT and transitioned to intermittent hemodialysis for a total of 19 days, although her kidney function recovered eventually to normal with serum creatinine of 0.7 mg/dL. Additional three rounds of plasma exchange were performed in those days following the initial exchange to reduce the serum myoglobin burden which peaked at 185,108 ng/mL on hospital day 2. Due to the lower leg injury, she underwent multiple fasciotomies with several subsequent wound debridements in an attempt to preserve her lower extremities. Although amputation was considered, she was ultimately discharged home with braces to allow adequate ambulation. Her hospital course was also complicated by a catheter-associated thrombus of the right internal jugular vein with ensuing pulmonary embolus and lactobacillus bacteremia. Despite these setbacks, she achieved full recovery of her major organ systems and was discharged to an inpatient psychiatric facility on hospital day 60.

Discussion

While ibuprofen overdoses have historically increased since its approval as an over-the-counter medication in 1984 [1], massive ingestions similar to what was observed in this patient are increasingly becoming more common. The drug's wide availability as a commonly used antiinflammatory, analgesic, and antipyretic medication, larger pill bottle sizes, and rising rates of intentional overdose all contribute to the increasing occurrence of massive ingestions. Ibuprofen mediates its therapeutic effects by reversible binding of cyclooxygenase (COX) receptors (COX1 and COX2 isoforms) on prostaglandin synthase, preventing the conversion of arachidonic acid to various prostaglandins. It is rapidly absorbed from the GI tract and reaches peak plasma concentrations within 1−2 hours. As much as 99% of the drug is albumin-bound and it has a low volume of distribution (V_d), ranging from 0.12 to 0.2 L/kg [1]. Ibuprofen is metabolized with a half-life of 1.8−2 hours and completely excreted within 24 hours, although this time can be prolonged in overdose and renal failure [1]. While limited data exists, traditional teaching implies that ibuprofen is not a dialyzable drug due to the extensive protein binding, with drug extraction of less than 4% [2]. Pharmacokinetics do not differ appreciably between children and adults [1]. The short elimination time is important, as the drug's effects persist only as long as it remains in circulation.

Clinical sequelae of massive overdose observed in the patient included coma, metabolic acidosis, shock, cardiac dysrhythmia, and acute kidney injury, which in ibuprofen overdose have only been described in limited case reports. The vast majority of ibuprofen ingestions remain asymptomatic. A retrospective review of 126 single-drug exposures to ibuprofen showed only 19% of all cases and 7% of pediatric patients developed any symptoms [3]. Severe Central Nervous System (CNS) depression and metabolic acidosis were described in two pediatric patients as early as 1987 [4]. The combination of metabolic acidosis and CNS depression has been reported in adolescent presentations since then [5–8]. An elevated anion gap metabolic acidosis often results from the acidic metabolites 2-carboxyibuprofen and 2-hydroxyibuprofen, as well as elevated lactic acid levels caused by tissue hypoperfusion [9]. CNS depression can progress to coma, as reported in a 15-year-old female who ingested 100 g of ibuprofen and required endotracheal intubation due to loss of airway protective reflexes [8]. Profound hypotension requiring aggressive vasopressor support has also been reported [5,7,10]. Marciniak et al. described complete cardiovascular collapse necessitating venoarterial extracorporeal membranous oxygenation in a 14-year-old male who ingested 50 g of ibuprofen [10]. Cardiac rhythm and EKG abnormalities associated with ibuprofen have included atrial fibrillation and ST-segment changes [11,12]. The arrhythmogenic mechanism of ibuprofen was investigated in an animal model and dose-dependent effects on shortening the cardiac action potential, as well as QRS widening, were seen [13]. While our patient's early rhythm disturbance was striking, it is at least partially attributable to hyperkalemia, given concurrently elevated serum potassium and peaked T waves on her EKG. The development of acute kidney injury due to ibuprofen results from the loss of prostaglandin-mediated glomerular afferent arteriolar vasodilation, compromising renal blood flow. Prior experience with this form of kidney injury shows an often-reversible process variable in severity, and support has ranged from IV hydration to renal replacement therapies [5,7,14].

Prediction of severe clinical sequelae from a given ibuprofen overdose remains difficult. Hall et al. devoted attention to ingested ibuprofen doses and toxic effects in their review of 126 cases, noting that a threshold ingestion level of 440 mg/kg could predict severe toxicity, while ingestions less than 99 mg/kg were typically asymptomatic [3]. The same group developed a nomogram to predict the probability of toxicity at a given serum ibuprofen measurement and time postingestion [3]; however, the clinical application of this tool by Mcelwee et al. showed poor correlation between drug levels and symptoms in their studied cohort [15]. Our patient's time of ingestion was not known, but if hypothetically considered at 4 or 12 hours postingestion, her initial level of 330 μg/mL falls well within the "probable toxicity" range of the Hall nomogram.

The use of therapeutic plasma exchange (TPE) to remove ibuprofen in our patient is the first reported therapeutic strategy of extracorporeal blood purification for ibuprofen intoxication in a pediatric patient. Theoretically, ibuprofen is an attractive target for removal by TPE based on its high protein binding and low volume of distribution. The drug adheres to a one-compartment model of distribution, so a rebound in plasma concentration would not be expected after exchange [1]. Our institution has had one prior experience with TPE for severe acidosis and cardiovascular collapse in an adolescent due to massive ibuprofen ingestion, who subsequently survived and made a full recovery (unpublished data). Geith et al. described the only published case of massive ibuprofen ingestion treated with TPE in 2017: a 48-year-old man who ingested 72 g and underwent five daily exchanges due to refractory circulatory failure, with rapid clinical improvement noted [16]. Experience in dogs was described in a review of 11 cases of large NSAID ingestion (six ibuprofen) treated with a single round of TPE, with outcomes of good survival and reduced incidence of kidney injury [17]. In the case we report here, the decision to trial TPE approximately 9 hours after admission to the ICU was based on a persistent, severe acidosis, and poor perfusion despite a high-dose epinephrine infusion. On TPE initiation, our patient's arterial pH was the first clinical parameter to improve, climbing from 7.08 pre-TPE to 7.14 2 hours post-TPE and to 7.30 10 hours after treatment. The lactic acid went from 8.8 pre- to 6.5 postexchange. Her hemodynamics were slower to improve than the case reported by Geith et al., in which there appeared to be substantial reductions in vasopressor support during—and immediately after—TPE. Our patient's hypotension predictably worsened somewhat during TPE initiation, likely due to fluid shifts, and there was a need to add a norepinephrine infusion to the epinephrine infusion already in place. Her hemodynamics then improved dramatically over the ensuing 24 hours and the epinephrine infusion was reduced from 0.17 to 0.05 μg/kg/min, while norepinephrine was discontinued.

Our patient's plasma ibuprofen profile improved following TPE. There was also a spontaneous reduction in the ibuprofen level before TPE; however, from the initial 330 to 230 μg/mL over 3 hours, then to 120 μg/mL 30 minutes into TPE. Eight hours after TPE, the level fell to 60 μg/mL; therefore the otherwise spontaneous decrease in plasma ibuprofen level was only slightly augmented by TPE. However, the fall in ibuprofen preexchange did not correlate to clinical improvement, similar to observations by Geith et al. in their case, which again speaks to the poor correlation between plasma ibuprofen concentration and clinical toxicity. There may be a several-day delay in reporting of levels from clinical laboratories, making it difficult for the drug levels to impact the daily clinical management of these patients. Numerous factors may also confound the assessment of plasma ibuprofen content in critically ill patients, making plasma levels a difficult endpoint by which to judge the efficacy of TPE. There can be delayed drug absorption due to GI ischemia during systemic hypoperfusion. In addition, the protein-bound fraction of ibuprofen which TPE removes may make a relatively smaller contribution to the total measured amount due to the influence of pH on albumin's ability to bind drugs. With diclofenac, for instance, there is reduced drug–protein binding in acidosis [18]. In a massive overdose, both acidosis and a saturation in protein binding would lead to a larger amount of unbound, free ibuprofen [16]. As the pH normalizes in a stabilizing patient, the excess drug may be mobilized from the periphery to bind plasma proteins, leaving a potential role for serial plasma exchanges. The measurement of protein-bound and free ibuprofen concentrations, currently unavailable, would be highly useful in this setting to prospectively plan a course of TPE for severe toxicity. While hemodialysis and CRRT have not traditionally been used to remove ibuprofen in massive ingestions due to its high protein binding, CRRT in tandem with TPE may have worked to remove the unbound ibuprofen, enhancing our patient's recovery from hemodynamic instability and acidosis. Currently, TPE is an American Society for Apheresis category III treatment for overdoses and poisonings, meaning decision-making on its use should be individualized [19], and we advocate it be considered specifically for ibuprofen-induced circulatory failure.

We report the first use of TPE to reduce plasma ibuprofen levels in a pediatric case of massive overdose. Our patient displayed a quick recovery from severe acidosis and hemodynamic compromise, as well as an excellent outcome. This case demonstrates anticipation for—and early initiation of—plasma exchange in tandem with CRRT by a well-coordinated and trained pediatric intensive care team. TPE should be considered an acute therapeutic option for patients in extremis from the ingestion of massive amounts of ibuprofen.

References

[1] Davies NM. Clinical pharmacokinetics of ibuprofen: the first 30 years. Clin Pharmacokinetic 1998;34(2):101–54.
[2] Senekjian HO, Lee CS, Kuo TH, Au DS, Krothapalli R. Absorption and disposition of ibuprofen in hemodialyzed uremic patients. Eur J Rheumatol Inflamm 1983;6(2):155–62.
[3] Hall AH, Smolinske SC, Conrad FL, Wruk KM, Kulig KW, Dwelle TL, et al. Ibuprofen overdose: 126 cases. Ann Emerg Med 1986;15(11):1308–13.
[4] Linden CH, Townsend PL. Metabolic acidosis after acute ibuprofen overdosage. J Pediatr 1987;111(6 Pt 1) 922-5. Review.
[5] Holubek W, Stolbach A, Nurok S, Lopez O, Wetter A, Nelson L. A report of two deaths from massive ibuprofen ingestion. J Med Toxicol 2007;3(2):52–5.
[6] Lamkin S, Fraiz A. An acid/base disturbance from ibuprofen toxicity. J Emerg Nurs 2009;35(6):584–5.
[7] Levine M, Khurana A, Ruha AM. Polyuria, acidosis, and coma following massive ibuprofen ingestion. J Med Toxicol 2010;6(3):315–17.
[8] Seifert SA, Bronstein AC, McGuire T. Massive ibuprofen ingestion with survival. J Toxicol Clin Toxicol 2000;38(1):55–7.
[9] Arens A, Smollin C. Case Files of the University of California, San Francisco Medical Toxicology Fellowship: Seizures and a Persistent Anion Gap Metabolic Acidosis. J Med Toxicol 2016;12(3):309–14.
[10] Marciniak KE, Thomas IH, Brogan TV, Roberts JS, Czaja A, Mazor SS. Massive ibuprofen overdose requiring extracorporeal membrane oxygenation for cardiovascular support. Pediatr Crit Care Med 2007;8(2):180–2.
[11] Akçay M. Ibuprofen-induced Kounis syndrome with diffuse ST segment depression and atrial fibrillation. Anatol J Cardiol 2017;18(5):380–1.
[12] McCune KH, O'Brien CJ. Atrial fibrillation induced by ibuprofen overdose. Postgrad Med J 1993;69(810):325–6.
[13] Yang ZF, Wang HW, Zheng YQ, Zhang Y, Liu YM, Li CZ. Possible arrhythmiogenic mechanism produced by ibuprofen. Acta Pharmacol Sin 2008;29(4):421–9.
[14] Kim J, Gazarian M, Verjee Z, Johnson D. Acute renal insufficiency in ibuprofen overdose. Pediatr Emerg Care 1995;11(2):107–8.
[15] McElwee NE, Veltri JC, Bradford DC, Rollins DE. A prospective, population-based study of acute ibuprofen overdose: complications are rare and routine serum levels not warranted. Ann Emerg Med 1990;19(6):657–62.
[16] Geith S, Renner B, Rabe C, Stenzel J, Eyer F. Ibuprofen plasma concentration profile in deliberate ibuprofen overdose with circulatory depression treated with therapeutic plasma exchange: a case report. BMC Pharmacol Toxicol 2017;18(1):81.
[17] Rosenthal MG, Labato MA. Use of therapeutic plasma exchange to treat nonsteroidal anti-inflammatory drug overdose in dogs. J Vet Intern Med 2019;33(2):596–602.
[18] Brune K, Renner B, Hinz B. Using pharmacokinetic principles to optimize pain therapy. Nat Rev Rheumatol 2010;6(10):589–98.
[19] Schwartz J, Padmanabhan A, Aqui N, Balogun RA, Connelly-Smith L, Delaney M, et al. Guidelines on the use of therapeutic apheresis in clinical practice-evidence-based approach from the writing committee of the American Society for Apheresis: the seventh special issue. J Clin Apher 2016;31(3):149–62.

Chapter 7

Antibiotics

Deborah French
Department of Laboratory Medicine, University of California San Francisco, San Francisco, CA, United States

Introduction

Introduction to antibiotics

Traditionally, antibiotics are substances produced by a microorganism that inhibit or kill other microorganisms. Penicillin, the first natural antibiotic, was discovered in 1928 by Alexander Fleming, a Scottish physician—scientist [1]. In medical practice the term antibiotic is used to refer to natural, semisynthetic, or synthetic compounds that inhibit or kill microorganisms [2,3]. For these to be useful drugs, they should be targeted toward the microorganism and not the host to decrease the risk of adverse side effects. Therefore the targets of antibiotics are commonly bacteria-specific cellular components such as the cell wall or ribosomes. This becomes more complicated when the microorganisms are eukaryotic like the host (such as fungi) making good drug targets with acceptable toxicity profiles harder to find [2,3].

Antimicrobials are commonly classified into broad-, extended-, or narrow-spectrum drugs. Broad-spectrum drugs are usually effective against a number of pathogens (including both Gram-positive and Gram-negative organisms), whereas narrow-spectrum antibiotics, as the name suggests, are usually only effective against selected types of pathogen [4]. In clinical practice, when a microbial infection is suspected, the causative organism is determined using a bacterial culture and a Gram stain followed by targeted antimicrobial therapy. Alternatively, a physician can elect to treat using a broad-spectrum antibiotic [5].

Gram stain and susceptibility testing

A Gram stain is a staining technique (using crystal violet, iodine, alcohol or alcohol acetone, and safranin) that determines if the organism is Gram-positive, Gram-negative, Gram-variable or does not stain at all [2,5]. Gram-positive organisms appear purple, whereas Gram-negative organisms appear pinkish or red. The Gram-staining characteristics are determined by the structure of the cell wall. Gram-positive bacteria have a cell wall consisting of a thick peptidoglycan layer. Antibiotics that are effective in treating Gram-positive bacterial infections mainly prevent the formation of this layer. Gram-negative bacteria have a cell wall made of two layers, one is a thinner peptidoglycan layer than in Gram-positive bacteria and an outer membrane consisting of proteins, phospholipids, and lipopolysaccharide [2,5]. Mycobacteria have a Gram-positive cell wall structure but also have another layer with a waxy appearance consisting of glycolipids and fatty acids (mainly mycolic acid) creating an acid-fast cell wall. These organisms are difficult to stain with Gram stain, but since they are Gram-positive, they will produce a faint blue color. Mycoplasma does not have a cell wall but has a membrane consisting of sterols. The differences in cell wall or membrane structure play a role in the effectiveness of antibiotics on each of these types of organisms [2,5].

If a clinician suspects that a patient has a fungal infection, stains such as India ink, potassium hydroxide, and calcofluor white can be used on clinical samples such as cerebrospinal fluid to visualize the fungal elements, including morphologic characteristics that can aid in the identification of the organism [2]. Histological stains using tissue are also useful to give more specific information about the fungal infections. Further, cultures can also be performed to distinguish between yeasts and molds and for speciation [2].

Susceptibility testing of the organism can then be carried out to determine which antibiotics will be effective in treatment of the infection [2]. For bacterial infections, susceptibility testing is commonly carried out using disk diffusion or dilution testing that identifies the minimum inhibitory concentration (MIC); the lowest concentration of the antimicrobial agent required to inhibit the growth of the organism. The organism is then interpreted as nonsusceptible, susceptible, intermediate, or resistant to each antibiotic. A similar disk diffusion method can also be used for antifungal susceptibility testing that will determine if the mold or yeast is susceptible, intermediate, or resistant to the antibiotic.

Antibiotics and toxicity

Once the causative organism has been identified, and the susceptibility established, the appropriate antibiotic treatment must be chosen. Antibiotics fall into different classes and some of the most commonly used classes and the mechanism of action are shown in Table 7.1 [2,3,6,7]. The classes include fluoroquinolones, rifamycins, β-lactams, glycopeptides, aminoglycosides, tetracyclines, macrolides, oxazolidinones, azoles, polyenes, and echinocandins, and the mechanism of action of these antibiotics will be discussed in greater detail later. Like all drugs, antibiotics have the potential to be toxic given the right dose. Since antibiotics that are active against bacteria tend to be directed toward prokaryotic structures, the toxicity profiles are mostly safe for humans. When it comes to antibiotics used to treat eukaryotic organisms such as fungi, it is harder to find a drug target that is not also found in humans, so toxicity can be harder to avoid [2,3,6,7].

TABLE 7.1 Overview of Different Antibiotic Classes.

Antibiotic Class	Drug Names	Drug Type	Species Affected
Fluoroquinolones	Nalidixic acid, ciprofloxacin, levofloxacin, gemifloxacin	DNA synthesis inhibitors	Aerobic Gram-positive and Gram-negative; some anaerobic Gram-negative; *Mycobacterium tuberculosis*
Rifamycins	Rifamycins, rifampicin, rifapentine	RNA synthesis inhibitors	Gram-positive and Gram-negative; *M. tuberculosis*
βLactams	Penicillins (penicillin, ampicillin, oxacillin), cephalosporins (cefazolin, cefoxitin, ceftriaxone, cefepime), and carbapenems (imipenem)	Cell wall synthesis inhibitors	Aerobic and anaerobic Gram-positive and Gram-negative
Glycopeptides	Vancomycin	Cell wall synthesis inhibitors	Gram-positive
Aminoglycosides	Gentamicin, tobramycin, amikacin	Protein synthesis inhibitors	Aerobic Gram-positive and Gram-negative; *M. tuberculosis*
Tetracyclines	Tetracycline, doxycycline	Protein synthesis inhibitors	Aerobic Gram-positive and Gram-negative
Macrolides	Erythromycin, azithromycin	Protein synthesis inhibitors	Aerobic and anaerobic Gram-positive and Gram-negative
Oxazolidinones	Linezolid	Protein synthesis inhibitor	Gram-positive; *M. tuberculosis*
Azoles	Voriconazole, posaconazole, fluconazole, itraconazole, isavuconazole, ketoconazole	Cell wall synthesis inhibitors	Most fungi (e.g., *Candida* and *Aspergillus* species) (fluconazole not affective against molds)
Polyenes	Amphotericin B (deoxycholate and lipid-based preparations)	Disruption of cell wall	Most fungi (e.g., *Candida* and *Aspergillus* species)
Echinocandins	Caspofungin, anidulafungin, micafungin	Cell wall synthesis inhibitors	Mainly yeast

Amphotericin B deoxycholate, a macrolide polyene broad-spectrum antifungal drug, is well known for its severe and potentially lethal side effects, including infusion-related adverse effects such as hypotension, fever, rigors and chills, and renal toxicity [8]. The reported rates of these adverse effects are variable and depend on the organism being treated and the patient's underlying disease. Some studies report >50% of patients developing nephrotoxicity and >60% experiencing infusion-related adverse effects when treated with amphotericin B deoxycholate [8—10]. Newer lipid-based formulations of amphotericin B were aimed at reducing the adverse effects seen with administration of amphotericin B deoxycholate; however, implementation has been limited due to high cost. Further, the relative efficacy and safety of these agents are still under evaluation [11], although a recent metaanalysis showed promising results [10].

Oxazolidinone antibiotics are synthetic drugs of which only linezolid is currently used clinically. It is effective against Gram-positive infections, including methicillin-resistant *Staphylococcus aureus* (MRSA), vancomycin-resistant enterococci, *Streptococcus pneumoniae*, and *Mycobacterium tuberculosis* [2,12]. Linezolid has been reported to cause adverse effects such as gastrointestinal disturbance (~10% of patients), reversible myelosuppression, including thrombocytopenia (~7% of patients) and anemia (~4% of patients), and neuropathy in patients on long-term treatment with this drug [13,14]. However, linezolid has also more recently been reported to cause serotonin syndrome that is characterized by increased neuromuscular tone, autonomic dysregulation, and mental status changes that can range in severity from mild to life threatening [15,16]. The prevalence of serotonin syndrome in patients treated with linezolid is unclear, but it is estimated to be approximately 3% when a selective serotonin reuptake inhibitor is concomitantly prescribed [17].

Cardiac toxicity

Certain antibiotics are known to cause toxicity by inducing prolonged repolarization of the heart, known as QT prolongation, which is also associated with torsades de pointes (TDP), a form of polymorphic ventricular tachycardia. Ketoconazole and itraconazole (azole antifungals) have also been shown to cause QT prolongation and TDP [18], although it is unclear how frequently this adverse effect occurs since to date, only a few case reports have been published [19—24].

Other antibiotics that have been implicated in causing adverse cardiovascular effects include the macrolides, erythromycin, and azithromycin [18]. Since multiple reports of QT prolongation and TDP were reported for azithromycin [25—27], the FDA released a drug safety communication and changed the labeling for the drug to highlight that there is a risk of potentially fatal heart rhythms with the use of this antimicrobial [28]. This risk is increased in patients that are already at high risk for cardiovascular death [28]. In one study, patients who were prescribed azithromycin compared to amoxicillin had a hazard ratio of 2.49 in terms of risk of cardiovascular death and a hazard ratio of 2.02 in terms of risk of death from any cause during a 5-day course of antibiotic therapy [29]. In a similar study, patients who were prescribed erythromycin had a hazard ratio of 2.01 in terms of risk of sudden death from cardiac causes compared to patients who did not use any antibiotics [30].

The fluoroquinolone antibiotics are broad spectrum and so are widely used in infections. However, several have been removed from the market due to toxicity, for example, sparfloxacin that causes QT prolongation in a dose-dependent manner [31]. The fluoroquinolones that are still available have also been reported to cause QT prolongation and TDP [32,33]. In one metaanalysis, the use of fluoroquinolones (moxifloxacin, levofloxacin, or ciprofloxacin) was associated with a significant increase in risk of both arrhythmia and risk for cardiovascular death, with the highest risk observed with the use of moxifloxacin [34].

Neurotoxicity

Almost all classes of antibiotics listed in Table 7.1 have been shown to have neurotoxic effects, including the aminoglycosides, β-lactams, tetracyclines, macrolides, fluoroquinolones, and oxazolidinones [35]. The aminoglycosides (e.g., gentamicin, amikacin, and tobramycin) have most commonly been reported to cause ototoxicity, although other neurotoxicity such as peripheral neuropathy, encephalopathy, and neuromuscular and autonomic blockade have also been reported [35]. Neurotoxicity observed in patients prescribed cephalosporins and penicillins has included encephalopathy and nonconvulsive status epilepticus which seem to be exacerbated by preexisting renal failure and/or central nervous system (CNS) disease [35]. Use of the macrolides has been shown to cause ototoxicity due to damage to the cochlea resulting in equilibrium dysfunction [35]. The fluoroquinolones have been associated with seizures, encephalopathy, and altered mental status in patients prescribed these antibiotics [35]. Linezolid-related encephalopathy has been reported as has Bell's palsy, optic neuropathy, and a painful peripheral neuropathy [35]. Vancomycin has also been shown to cause ototoxicity due to damage to the eighth cranial nerve. This ototoxicity is associated with previously diagnosed hearing loss as well as renal dysfunction [36].

Nephrotoxicity and hepatotoxicity

The aminoglycosides, including gentamicin, tobramycin, and amikacin, cause nephrotoxicity in approximately 10%–25% of patients taking these antibiotics [37]. The nephrotoxicity manifests itself in a number of ways, including renal tubular epithelial cell death, reduced glomerular filtration, and renal blood flow [37]. Use of intravenous vancomycin in the treatment of MRSA is also associated with nephrotoxicity in 7%–17% of patients [36].

Amphotericin B deoxycholate has been reported to cause hepatotoxicity [38]. It is thought that this toxicity is caused by induction of oxidative damage, upregulation of inflammatory cytokines, interference with the CYP450 enzymes, and the damaging effects of the sodium deoxycholate carrier [38]. When using amphotericin B deoxycholate, the reports of hepatotoxicity have been reported in around 13%–22% of patients, seen as increases in alkaline phosphatase, alanine aminotransferase, aspartate aminotransferase, and bilirubin [38]. The azole antifungal drugs have also been reported to cause hepatotoxicity [38]. Ketoconazole has been removed from some markets due to the hepatotoxicity and the availability of alternative antifungal drugs with less toxicity [38]. The azoles have been reported to cause both cholestatic and hepatocellular damage. They are metabolized via the CYP450 enzymes in the liver, so dose adjustments are necessary in patients with liver diseases. The number of patients affected by hepatotoxicity while taking these antifungal agents varies with both the individual drug and the study [38].

Introduction to therapeutic drug monitoring

While it can be seen that there are a large number of antibiotics available to treat patients, and some that have troubling toxicity profiles, very few are monitored through therapeutic drug monitoring (TDM). TDM is used to determine that the serum or plasma concentrations of drugs make sense in terms of efficacy, compliance, optimization of dose, drug–drug interactions, and avoidance of toxicity. It is especially important for drugs that have a narrow therapeutic index or in patients whose kidney or liver function has changed significantly. The antibiotics that are commonly monitored through TDM are gentamicin, tobramycin, vancomycin, amikacin, voriconazole, posaconazole, itraconazole, and fluconazole. The TDM of antibiotics will be discussed in more detail later. Concentrations of antibiotics are not routinely measured forensically.

Mechanism of action

Antibiotic drugs are commonly targeted toward bacteria-specific cellular properties or mechanisms (Table 7.1) [2,3,6,7]. Some common mechanisms are inhibition of cell wall synthesis targeted by β-lactams and glycopeptides, protein synthesis targeted by aminoglycosides, tetracyclines, macrolides and oxazolidinones, and nucleic acid inhibition that is targeted by the fluoroquinolones and rifamycins. The azoles and echinocandins act by inhibiting cell wall synthesis and the polyenes act by disrupting the cell wall. These antibiotic mechanisms of action are discussed in more detail later.

Inhibition of bacterial cell wall synthesis

Since bacterial cell wall components are not present in human cells, they present a number of therapeutic targets for antibacterial drugs [2,3,6]. Antibiotics that inhibit bacterial cell wall synthesis are essentially peptidoglycan synthesis inhibitors that act by preventing the cross-linking of the carbohydrate and peptide components of peptidoglycan, necessary to build bacterial cell walls. Without the cell wall, the bacterial cells are prone to lysis by osmosis. Antibiotics that inhibit bacterial cell wall synthesis are most effective against Gram-positive bacteria since they have no outer membrane [2,3,6,7].

The major antibiotics with this mechanism of action are the β-lactam drugs, including the penicillins, carbapenems, and cephalosporins. They act by binding to penicillin-binding proteins, enzymes involved in peptidoglycan cross-linking and ultimately inhibit the transpeptidation reaction [2]. There are a number of penicillin-related compounds of which some are natural (penicillin G and penicillin V) and some are semisynthetic (oxacillin and ampicillin) (Fig. 7.1) [39]. The natural penicillins are produced by fungi, are fairly narrow spectrum, and are subject to degradation [1]. The only part of the semisynthetic penicillins that is synthetic is the side chain as they have the common β-lactam ring.

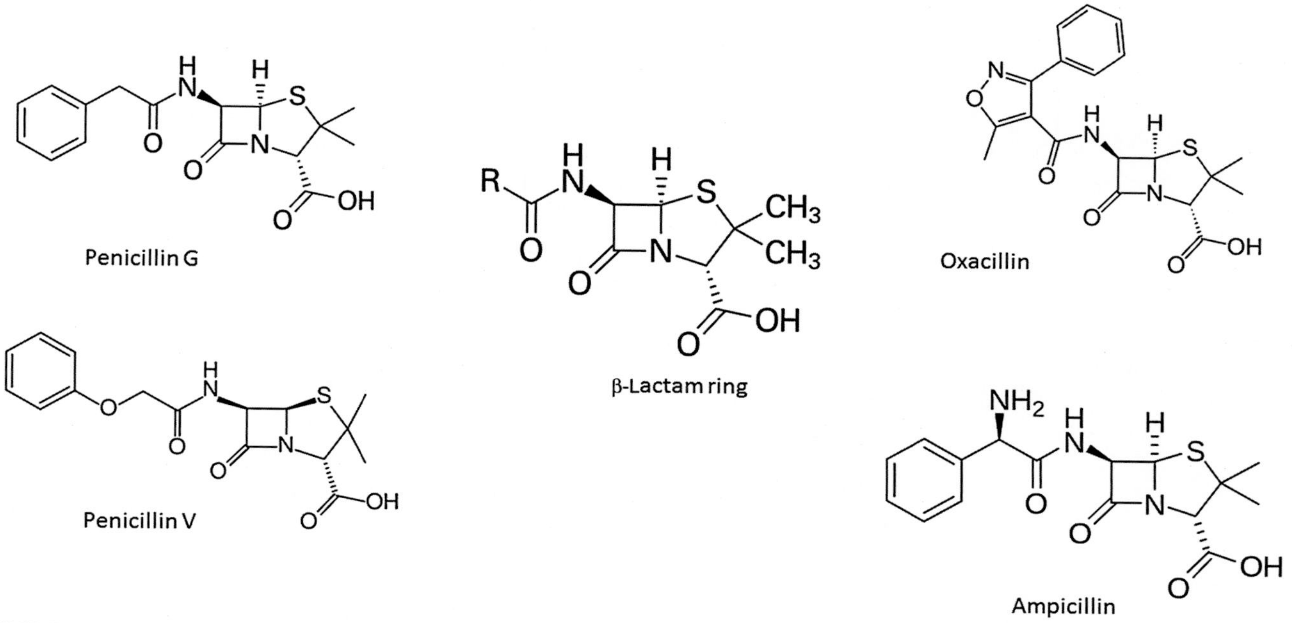

FIGURE 7.1 The β-lactam ring structure and structures of common penicillins.

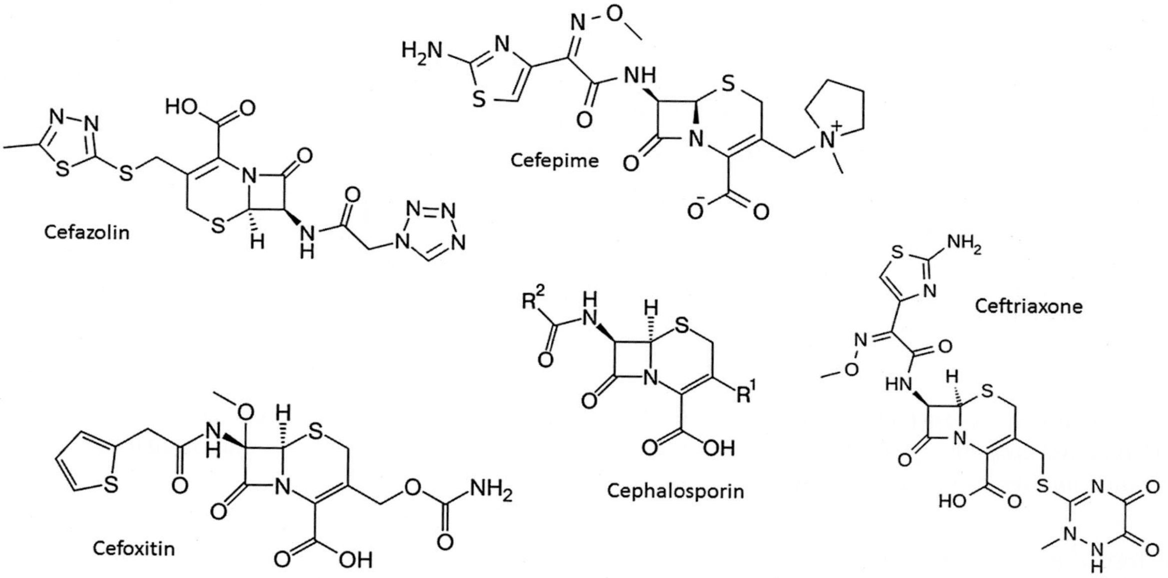

FIGURE 7.2 Structures of common cephalosporins.

They are generally broader spectrum than the natural penicillins, and they are resistant to degradation. Oxacillin is only effective against Gram-positive bacteria, and it is resistant to penicillinase that is produced by bacteria to avoid the effects of penicillin by breaking the β-lactam ring. Ampicillin, on the other hand, is also effective against many Gram-negative bacteria as well as being resistant to penicillinase [2,39,40].

Cephalosporin and its derivatives also inhibit bacterial cell wall synthesis by inhibiting peptidoglycan synthesis and are susceptible to penicillinases (Fig. 7.2) [2,40]. First-generation cephalosporins such as cefazolin are most active against Gram-positive organisms, methicillin-susceptible staphylococci, and nonenterococcal streptococci. Each generation has exhibited less activity against Gram-positive organisms but increased activity against Gram-negative organisms [40]. The fourth-generation drugs such as cefepime are extended-spectrum cephalosporins with activity against *Enterobacter* spp., *Citrobacter freundii*, and *Serratia marcescens*. They are also commonly used with aminoglycosides or fluoroquinolones for the treatment of severe *Pseudomonas aeruginosa* infections [40].

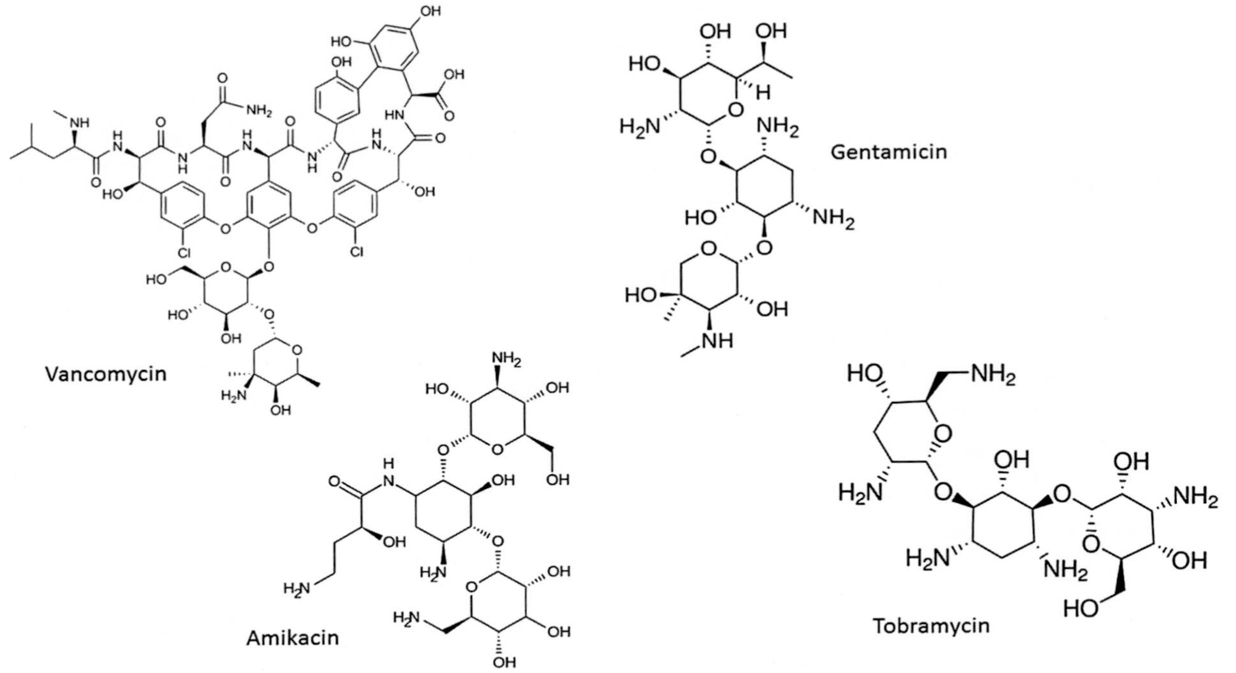

FIGURE 7.3 Structures of vancomycin (glycopeptide antibiotic) and some aminoglycosides (gentamicin, amikacin and tobramycin).

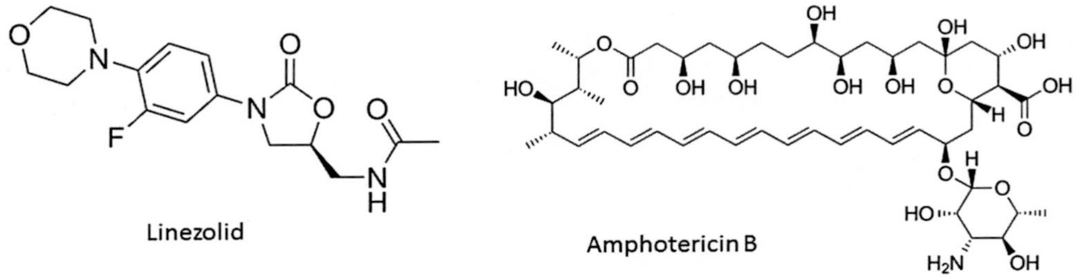

FIGURE 7.4 Structures of linezolid (oxazolidinone) and amphotericin B (polyene).

Vancomycin is a glycopeptide antibiotic that is active against Gram-positive organisms, including MRSA (Fig. 7.3) [2]. It also inhibits peptidoglycan synthesis, but it does it by binding to the peptidoglycan intermediates preventing their incorporation into the peptidoglycan chains [2].

Inhibition of protein synthesis

Bacterial ribosomes have two subunits (30S and 50S) that make up the 70S ribosome. Ribosomes are composed of ribonucleic acid (RNA) and proteins. They bind messenger RNA (mRNA) and transfer RNA (tRNA) and are the sites of protein and polypeptide synthesis in cells [41]. Bacterial ribosomes are slightly smaller than eukaryotic ribosomes and have enough structural differences to make them good therapeutic targets for antibiotics. The tetracycline antibiotics (including tetracycline and doxycycline) bind to 16S ribosomal RNA that inhibits rotation of the tRNA during translation resulting in an early release of the tRNA and termination of peptide bond formation [2,41]. The aminoglycosides (e.g., gentamicin, tobramycin, and amikacin; Fig. 7.3) disrupt the "reading" of the mRNA by changing the shape of the 30S portion of the ribosome and blocking the tRNA leading to mistranslation and therefore production of abnormal proteins [2,42]. Macrolide antibiotics (e.g., erythromycin and azithromycin) inhibit protein synthesis by binding to the 50S ribosome subunit and blocking the exit of the nascent proteins, thereby halting protein synthesis [2,43]. Linezolid (an oxazolidinone antibiotic; Fig. 7.4) also binds to the 50S ribosomal subunit and also the 30S ribosomal subunit preventing the formation of the preinitiation complex and therefore the synthesis of bacterial proteins [2,44].

Inhibition of nucleic acid synthesis

The rifamycins (e.g., rifamycins, rifampicin, and rifapentine) inhibit bacterial transcription via inhibition of mRNA synthesis. They accomplish this by binding to RNA polymerase, the action of which blocks RNA transcription at the initiation step [2,45]. Fluoroquinolones (such as ciprofloxacin and levofloxacin) inhibit bacterial DNA gyrase and/or DNA topoisomerases VI which are necessary for DNA replication [2,46].

Inhibition of fungal cell wall synthesis or cell wall disruption

Fungi are eukaryotic pathogens and therefore resemble human cells in terms of ribosomes and metabolic processes. The similarities between human cellular proteins and fungal pathogens lead to few biochemical pathways that can be exploited for potential drug targets. Fungal cells differ from host cells as their cell walls are made of chitin and fungal membranes have ergosterol instead of cholesterol [47]. The echinocandins (e.g., caspofungin) are lipopeptides and they inhibit 1,3-β-D-glucan synthase that is necessary for synthesis of the chitin cell walls [47]. The azole antifungal drugs (e.g., voriconazole, posaconazole, ketoconazole, fluconazole, and itraconazole; Fig. 7.5) inactivate the enzyme lanosterol demethylase that is involved in production of ergosterol for the fungal membranes [47]. The polyenes (e.g., amphotericin B; Fig. 7.4) act by disrupting the cell wall. They do this by extracting ergosterol from the phospholipid in the cell membrane thereby depleting ergosterol from the cell [47].

Pharmacokinetics, toxicokinetics, and treatment

When a patient presents with an infection, it is important to start antibiotics as soon as possible, especially if it might be sepsis. Treatment is usually started before the causative organism is known and is therefore somewhat empirical [48]. One test carried out in the microbiology laboratory is determination of the MIC that is the minimum concentration of antibiotic required to inhibit visible growth of the organism after 18–24 hours [49]. Antibiotics can be classified into two different groups based on the MIC [50]. β-Lactams, vancomycin, and the macrolides belong to the group that exhibits somewhat concentration-independent efficacy; the time that the microorganism is in contact with the antibiotic is more important than the concentration, although up to the maximum kill rate concentration, they will exhibit concentration-dependent organism death [50]. Therefore the maximum efficacy is normally achieved at concentrations four- to fivefold higher than the MIC [49]. A parameter that is monitored to assess efficacy of this group of antibiotics

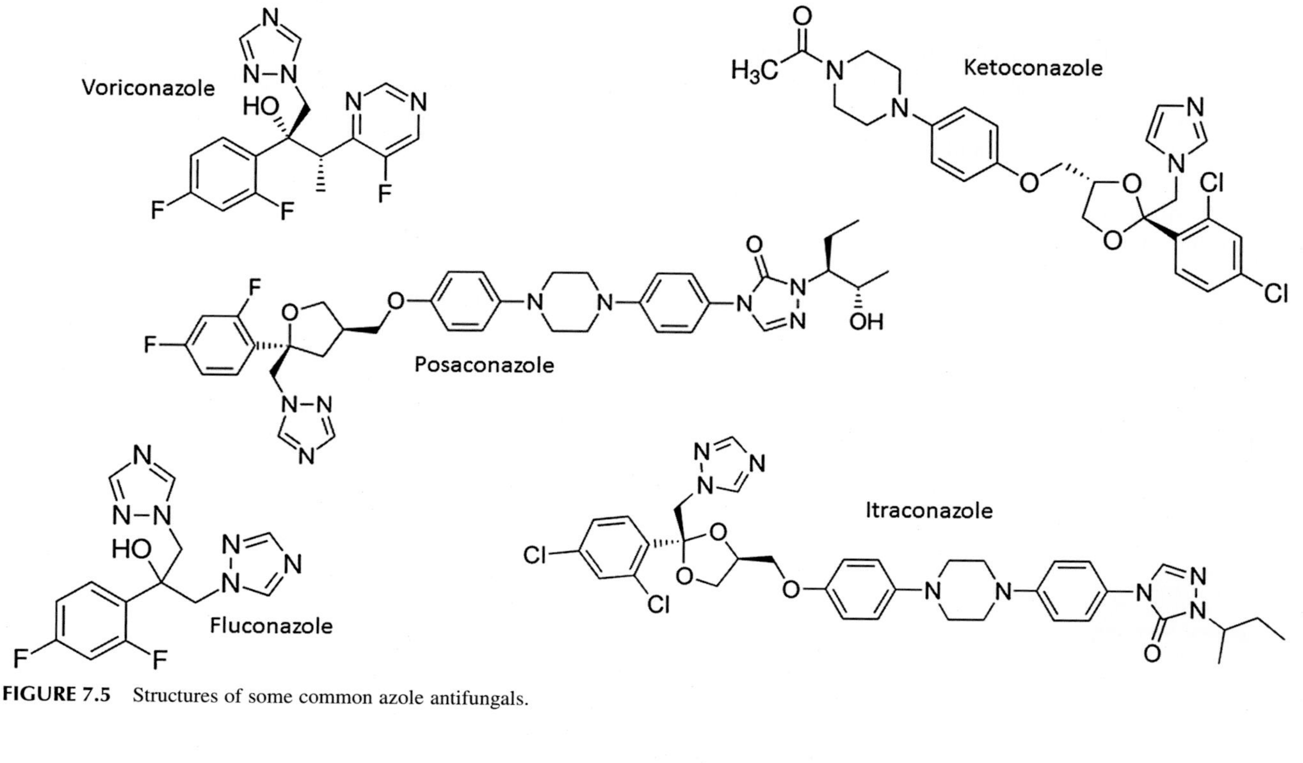

FIGURE 7.5 Structures of some common azole antifungals.

is the time above MIC ($t > $ MIC) which is the time that the serum–drug concentration is above the MIC of the organism [48]. The aminoglycosides and fluoroquinolones demonstrate a concentration-dependent efficacy so it is important to maximize the drug concentration without causing toxicity to the patient [50]. The parameters that are assessed for efficacy for this group of organisms are the Cmax/MIC that is the ratio between the peak serum concentration and the MIC, and the area under the curve should be greater than the MIC (Fig. 7.6) [48,51].

However, using the MIC to dictate dosing decisions has shortcomings as it does not take into account important pharmacokinetic parameters such as the protein binding of the drug, and also the tissue distribution. Only unbound drug can have clinical effects, and since most infections take place outside of the serum, the ability of the drug to enter tissues is an important consideration [48,52].

Once the causative organism is known, the antibiotic treatment becomes more individualized based on the organism and the patient. For example, if the patient has renal failure, they may be at higher risk of toxicity, so a less toxic alternative antibiotic could be used, assuming that the microorganism is susceptible to that antibiotic. Antibiotics commonly cause some side effects such as diarrhea, nausea and vomiting, rash, upset stomach, IV site irritation and with certain antibiotics, or with extended use, fungal infections of the mouth, and digestive tract and vagina. Some patients may also develop an allergic reaction to antibiotics, especially penicillins, where symptoms include a rash, swelling of the tongue and face, and difficulty in breathing [53]. Treatment includes corticosteroids and antihistamines, and for anaphylactic shock, adrenaline would be administered. Overdoses of antibiotics tend to be rare. The rest of this section will focus on the pharmacokinetics of antibiotics that have been shown to have the most frequently observed severe toxicity, and those for which TDM is routinely carried out. Further, the treatment of overdoses of common antibiotics that tend to have the most toxicity is also discussed.

Amphotericin B is administered intravenously (IV), and the pharmacokinetic properties differ among the different preparations (deoxycholate vs lipid preparations). Doses for amphotericin B also differ depending on the preparation with the lipid preparations typically dosed at 5 mg/kg and the deoxycholate preparation dosed at 0.6 mg/kg. The therapeutic drug concentration of amphotericin B deoxycholate is 0.03–1 µg/mL and less than 1 µg/mL for the lipid preparations. It is greater than 90% bound to proteins and distributes well into tissues and fluids except the cerebrospinal fluid. The volume of distribution is 0.23–1.91 L/kg in pediatric patients. Amphotericin B is excreted via the kidneys with 2%–5% of the drug excreted unchanged. The elimination half-life of amphotericin B is approximately 15 days [54,55].

Amphotericin B overdose is usually caused by a dosing error when administering this antifungal drug and is therefore uncommon. There is no antidote to amphotericin B overdose. Toxicity can manifest itself with nausea and vomiting, hypokalemia, hyperkalemia, and liver and renal damage. Symptoms of severe toxicity can include disseminated intravascular coagulation, hypotension, dysrhythmias, renal failure, respiratory failure, and cardiac arrest. Amphotericin B acts by binding to ergosterol in the fungal membrane, but it can also bind cholesterol in the human cell that thought to be responsible for the observed toxicity. Treatment for mild-to-moderate toxicity includes supportive care such as IV fluids and antiemetics, replacement of electrolytes, and salt loading, and use of mannitol to try and prevent renal injury. Treatment of severe toxicity focuses on the patient's airway, breathing, and circulation. Vital signs should be monitored and an electrocardiogram should be obtained to determine cardiac output. Obtaining serum–drug concentrations is not usually possible due to the absence of analytical methods in clinical laboratories and although concentrations would help to confirm an overdose, they are not clinically helpful. However, amphotericin B plasma concentrations of 5–10 µg/mL have been shown to cause toxicity [54–56].

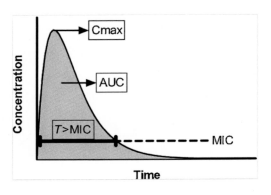

FIGURE 7.6 Graphical representation of pharmacokinetic and pharmacodynamics parameters that pertain to antibiotic treatment. *Used with permission de Velde F, Mouton JW, de Winter BCM, van Gelder T, Koch BCP. Clinical applications of population pharmacokinetic models of antibiotics: challenges and perspectives. Pharmacol Res 2018;134:280–8.*

Linezolid is administered both orally and via IV, and the time to peak drug concentrations is similar for both routes at around 30–60 minutes. The common dose of linezolid is 600 mg every 12 hours except for skin infections where the dose is 400 mg every 12 hours. The minimum desirable therapeutic drug concentration is >2 mg/L during treatment [57]. Linezolid has approximately 100% bioavailability when dosed orally, so it is extensively absorbed into tissues and fluids. The volume of distribution is 40–50 L in adults and linezolid is approximately 31% protein bound. Although the metabolism of linezolid is not completely elucidated, there is minimum metabolism in the liver and there are some known metabolites (aminoethoxyacetic acid, hydroxyethyl glycine, and two metabolites of hydroxyethyl glycine; PNU-142618 and PNU-173558). Linezolid is mainly excreted via the kidneys with approximately 30% excreted unchanged and 50% excreted as metabolites. It is also excreted via the feces as metabolites. The elimination half-life of linezolid is approximately 5 hours in adults, with a shorter half-life observed in pediatric patients [58,59].

Linezolid overdose is rare and information regarding the adverse effects is limited. Common adverse effects include nausea, vomiting, and diarrhea. Symptoms of severe toxicity include rash, optic neuropathy and peripheral neuropathy, liver damage, seizures, lactic acidosis, myelosuppression, and serotonin syndrome. The toxicity is thought to be due to inhibition of mitochondrial protein synthesis. Treatment of overdose is supportive care, including monitoring complete blood counts (CBC), to detect any myelosuppression and transfusions of platelets or packed cells may be required. For seizures, benzodiazepines may be used, followed by barbiturates or propofol if seizures continue. Intubation may also be required in the event of seizures. If the patient presents with serotonin syndrome, cooling measures should be undertaken and benzodiazepines may be administered. Further, treatment with cyproheptadine may be used for mild cases of serotonin syndrome, and chlorpromazine may be used in severe cases. The patient should be monitored via vital signs and liver function tests as well as CBC and platelet counts. Hemodialysis may be considered in patients that are not responding to supportive care and are showing signs of severe toxicity. Serum concentrations of linezolid are not commonly available in clinical laboratories [58,59].

Itraconazole is administered orally and via IV, and the time to peak concentration is approximately 2–5 hours. A typical dose is 200 mg every 12 hours, although loading doses of 200 mg three times daily may be given in the treatment of severe infections. The MIC of common fungal pathogens is <1 μg/mL, and therefore the desired therapeutic trough drug concentration for systemic infection is >1 μg/mL. For localized infections the trough drug concentration should be >0.5 μg/mL. Itraconazole has approximately 55% bioavailability when administered orally, and it is extensively distributed into the tissues, with limited CSF infiltration. Itraconazole has a volume of distribution of approximately 11 L/kg and is 99% protein bound. It is extensively metabolized by the liver, predominantly by CYP3A4. It has many metabolites, with the active metabolite being hydroxyitraconazole. It is mainly excreted via the kidney and into the bile, with an elimination half-life of both the parent compound and the major metabolite of approximately 35–64 hours and 27–56 hours, respectively, when the concentrations are at steady state. Steady state is achieved after approximately 4 days [55,60].

Fluconazole is administered orally and via IV, and the time to peak concentration is approximately 1–2 hours following oral administration. The dose of fluconazole depends on the indication but can range from 100 to 800 mg/day. The therapeutic drug concentrations also depend on the indication but can range from 1 to 40 μg/mL. It takes approximately 5–10 days of treatment with fluconazole to reach steady-state concentrations. Fluconazole has over 90% bioavailability when administered orally, and it is extensively distributed into tissues and CSF. It is approximately 11%–12% protein bound and has a volume of distribution of 0.56–0.82 L/kg. Fluconazole is metabolized minimally by the liver, and it has a 1,2,3-triazole and two N-dealkylated metabolites. It is excreted via the kidney with approximately 80% of the drug being excreted unchanged and approximately 11% excreted as the metabolites. The elimination half-life is approximately 30 hours [55,61].

Voriconazole is administered orally and via IV, and the time to peak concentration is approximately 1–2 hours. A loading dose of 6 mg/kg is commonly administered every 12 hours for 2 doses followed by 3–4 mg/kg every 12 hours. The therapeutic drug concentration is around 1–5 μg/mL. It has an oral bioavailability of approximately 96% and distributes into CSF and other fluids. It is approximately 58% protein bound and has a volume of distribution of 4.6 L/kg. Voriconazole is extensively metabolized by the liver by CYP2C19, CYP2C9, and CYP3A4. The major metabolite of voriconazole is the N-oxide metabolite that is inactive. It is excreted via the kidney and bile as metabolites. The elimination half-life is approximately 6–8.3 hours [55,62].

Posaconazole is administered via IV or orally and the time to peak concentration is 1.4 hours and approximately 4 hours, respectively. A loading dose of 300 mg every 12 hours for 1 day is used, followed by 300 mg daily by IV and for oral administration, the dose is 200 mg every 8 hours. The desired therapeutic drug concentration is a trough concentration of >0.7 μg/mL. Posaconazole has an oral bioavailability of 54%, greater than 98% protein binding and infiltrates the tissues. It has a volume of distribution that varies from 331 to 1341 L depending on the method of administration. Posaconazole is metabolized by the liver mainly by glucuronidation and is excreted predominantly through the feces (71%−77%) and also through the kidney (13%−14%). The elimination half-life also varies depending on the route of administration, with IV, delayed-release tablets, and oral suspension displaying half-lives of 27, 26−31, and 20−66 hours, respectively. Steady-state concentrations are achieved in approximately 7−10 days [55,63].

The azole antifungal drugs are either metabolized via the CYP450 enzymes and are affected by other CYP450 enzymes that either enhance or inhibit the concentrations or they affect the concentrations of other drugs. These interactions may be significant enough to require dose adjustments of all of the drugs being prescribed, or the drugs may be contraindicated for treatment at the same time. Table 7.2 gives an overview of some drugs that are contraindicated when prescribing selected antifungals, and those drugs that cause decreases in antifungal drug−serum concentrations [55,60−63].

Azole antifungal drug overdose is uncommon and although adverse effects have been reported, serious toxicity from an acute ingestion has not been reported. Mild adverse effects include nausea and vomiting, diarrhea, and changes in vision. More serious overdoses could cause symptoms such as liver toxicity, congestive heart failure, thrombocytopenia, neutropenia, and in some cases, seizures. Further, epidermal necrolysis has been reported along with some less serious rashes. As mentioned earlier, QT interval prolongation and TDP have also been reported for itraconazole. Treatment of overdoses involves supportive care, including maintaining electrolyte and fluid balance. In severe toxicity, continuous cardiac monitoring should be implemented, TDP can be treated with magnesium sulfate and if necessary, overdrive pacing may be carried out. If seizures occur, benzodiazepines can be administered followed by barbiturates or propofol if they do not cease. Patients should be monitored using serum electrolytes, CBC, and liver function tests. Fluconazole could be removed by hemodialysis, but the rest of the antifungals listed here are not likely to be removed by hemodialysis. Serum fluconazole, itraconazole (and hydroxyitraconazole), voriconazole, and posaconazole concentrations may be available in some clinical laboratories and may be used to confirm toxic concentrations. Voriconazole serum concentrations of >6 μg/mL and fluconazole concentrations of >20 μg/mL have been associated with toxicity [55,56,60−63].

Vancomycin is administered orally and via IV with a time to peak concentration of the end of the infusion when administered via IV. Doses vary depending upon the indication for treatment but are commonly around 30−45 mg/kg/day divided into 2 or 3 doses for IV administration, and 125 mg every 6 hours for oral administration. The therapeutic trough drug concentration is generally 10−20 or 15−20 mg/L for serious infections. It has an approximately 60% bioavailability following intraperitoneal administration and has limited absorption after oral administration. Vancomycin is

TABLE 7.2 Selected Drug Interactions of the Azole Antifungal Drugs.

Antifungal Drug	Antifungal Interaction With CYPs	Selected Drugs That Cause Antifungal Concentration Decreases	Selected Drugs That Are Contraindicated When Prescribing Azole Antifungals
Itraconazole	CYP3A4, CYP3A5, and CYP3A7 inhibitor and substrate	Antacids, phenobarbital, carbamazepine, H₂ agonists, didanosine, efavirenz, nevirpine, proton pump inhibitors, phenytoin, rifampin, rifabutin	Lovastatin, irinotecan, methadone, simvastatin, ergotamine, dihyroergotamine, quinidine, halofantrine, levomethadyl, pimozide, cisapride
Fluconazole	CYP2C9, CYP2C19, CYP3A4, CYP3A5, and CYP3A7 inhibitor	Antacids, carbamazepine, proton pump inhibitors, phenytoin, rifampin	Cisapride, astemizole, erythromycin, pimozide, quinidine, terfenadine
Voriconazole	CYP2C9, CYP3A4, CYP3A5, CYP3A7, and CYP2C19 inhibitor; CYP2C9 and CYP2C19 substrate	Barbiturates, carbamazepine, efavirenz, nevirapine, proton pump inhibitors, phenytoin, rifampin, rifabutin, ritonavir	Carbamazepine, terfenadine, astemizole, cisapride, pimozide, quinidine, high dose ritonavir, rifampin, sirolimus, St John's wort
Posaconazole	CYP3A4 inhibitor	Antacids, carbamazepine, H₂ antagonists, proton pump inhibitors, phenytoin, rifampin, rifabutin	Methadone, ergotamine, dihydroergotamine, atorvastatin, lovastatin, simvastatin, haloperidol, pimozide, quinidine, risperidone, sunitinib, tacrolimus, halofantrine

approximately 18%–55% protein bound, distributes into the tissues and fluids, and has a volume of distribution ranging from 0.25 to 1.25 L/kg. Vancomycin is not significantly metabolized and is excreted via the kidney, feces, and bile. It has an elimination half-life of 4–6 hours [59,64].

Vancomycin overdose is rare and severe toxicity is uncommon. As already described, vancomycin can cause nephrotoxicity, as well as hypotension, bradycardia, ototoxicity, and cardiac arrest. It can also cause "red man syndrome" that is a collection of signs and symptoms, including sudden hypotension, trouble breathing, wheezing, rashes and flushing, and painful muscle spasms in the chest and back. This toxicity is most commonly associated with rapid intravenous infusion and is due to vancomycin causing mast cells to release histamine. TDM for vancomycin is commonly available in clinical laboratories, and sustained serum concentrations of 80–100 μg/mL have been reported to cause toxicity. Treatment for vancomycin overdose is typically supportive care including treatment for hypotension that may include dopamine or norepinephrine if fluids do not normalize blood pressure. If "red man syndrome" is diagnosed, antihistamines can be administered. If the overdose patient has severe kidney damage, hemodialysis may be performed to reduce the serum vancomycin concentration. Multiple-dose activated charcoal may decrease the half-life of vancomycin administered via IV but is not routinely recommended unless the overdose is large and therefore the clearance may be protracted. The patient should be monitored via vital signs, kidney function tests, electrolytes, urine output, and serum vancomycin concentration. If ototoxicity is observed, auditory function should also be monitored [59,64].

Gentamicin, tobramycin, and amikacin are administered by a number of routes, including IM, IV, ophthalmic, or topical with peak concentrations occurring after approximately 30–90 minutes. Gentamicin IV or IM doses are typically around 3–5 mg/kg/day divided into 3 or 4 doses. Tobramycin IV doses are typically around 4–7 mg/kg once per day, and amikacin IV or IM doses are typically 15 mg/kg/day divided into 2 or 3 doses. Therapeutic trough concentrations are 1–2 μg/mL for gentamicin and tobramycin and <10 μg/mL for amikacin. The bioavailability varies by means of administration and is generally low but can be as high as 76% for gentamicin when administered intraperitoneally. They have low protein binding of 0%–30%, have a volume of distribution of 0.29–0.37 L/kg, and have limited diffusion into tissues and fluids; although high concentrations of gentamicin, tobramycin, and amikacin are found in the renal cortex, gentamicin is found in high concentrations in the inner ear, high concentrations of amikacin are found in extracellular fluids, and tobramycin is found in the aqueous humor and bronchial secretions. Gentamicin, tobramycin, and amikacin are not metabolized to a great degree and are excreted mainly via kidneys with an elimination half-life of approximately 2–4.4 hours [65–68].

Aminoglycoside overdoses are most commonly caused by dosing errors, although they are rare and typically only need to involve supportive care as they are not serious. Toxicity is related to trough concentrations of the aminoglycosides, so TDM is commonly available for these drugs in clinical laboratories. Trough concentrations of >2 μg/mL of gentamicin and tobramycin and >8 μg/mL for amikacin have been associated with toxicity, so serum concentrations can be obtained to confirm the overdose if the assays are clinically available. Symptoms of toxicity include renal damage and ototoxicity and hypersensitivity reactions that are increased with increased total exposure to these drugs. Some of the toxicity may be attributed to the inhibition of acetylcholine release from presynaptic nerve terminals. Treatment for toxicity is supportive care with a focus on a urine output of 3–6 mL/kg/h with use of IV fluids. If an allergic reaction occurs, antihistamines can be administered but if the reaction is severe, airway management is a priority as well as electrocardiogram monitoring and epinephrine use if required. There is no antidote to gentamicin overdose, and hemodialysis can be used but is usually only considered in cases of kidney failure. The patient should be monitored using serum electrolytes and kidney function tests [65–68].

Analytical methods and clinical management implications

TDM is the measurement of drug concentrations in biological fluids to try and ensure that the patient is receiving an optimized dose of the drug so that it reaches concentrations sufficient to be efficacious but prevents high concentrations that may cause toxicity (Fig. 7.7) [70,71]. TDM is most commonly performed on blood, serum, or plasma. TDM can also be used to ensure that the patient is compliant with taking the medication or to determine if there are any drug–drug interactions affecting the serum or plasma concentration when new medications are given to a patient [70,71]. It is based on the premise that there is a relationship between the dose of the drug administered and the concentration of the drug in the body and between the concentration of drug in the body and the therapeutic effect of the drug [70,71]. TDM is not carried out for all drugs, but normally for those that have a narrow therapeutic index, large variability in pharmacokinetics, large interindividual variation, and for those where the toxicity matches the clinical picture of the symptoms that the drug is trying to prevent [70,71]. An example of a drug for which TDM is used is the antiepileptic drug carbamazepine that decreases the action potential firing in the central nervous system by sodium channel blockade. It is used

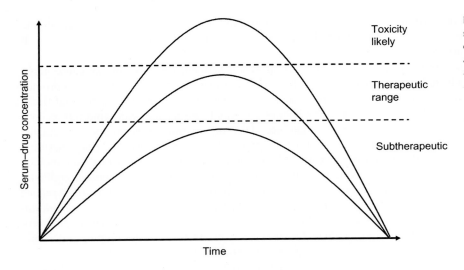

FIGURE 7.7 Serum–drug concentration versus time plot indicating utility of therapeutic drug monitoring. *Adapted from <https://www.slideshare.net/rajat2903/therapeutic-drug-monitoring-31857498>. Slide 27. [accessed on 19.11.18] [69].*

in the treatment of complex partial and tonic–clonic generalized seizures as well as bipolar disorder. The trough therapeutic range is 4–12 μg/mL, and toxicity has been observed at trough concentrations >15 μg/mL. The symptoms of both under dosing and toxicity include seizures and central nervous system depression which are major indications for being prescribed the drug [72,73].

TDM is performed when the concentration of the drug has reached steady state in the body; when the amount of drug being given is equal to the amount of drug being excreted. Steady state is achieved after approximately 5–7 half-lives; after 5 half-lives, over 95% of the drug will have accumulated in the body [70]. For drugs that have a short elimination half-life, steady state is achieved much sooner than those that have a long elimination half-life. It is therefore important to know the time that the drug was administered and the time of collection of the blood samples [70]. Trough drug concentrations are generally the most useful since they are less influenced by the absorption and distribution phases of drug pharmacokinetics, and they tend to be drawn immediately before the next dose of the drug is given. Peak and random concentrations can also be measured, although they are harder to interpret. Clinically, very few antibiotics are routinely monitored by TDM [70]. These include vancomycin, gentamicin, tobramycin, amikacin, and the azole antifungal drugs. For all of these drugs, it is the parent drug concentration that is measured, except for itraconazole where the active metabolite hydroxyitraconazole can also be measured. In the case of the other antibiotics (e.g., amphotericin B and linezolid), if toxicity is suspected due to the observed clinical symptoms, treatment is usually supportive care and removing the antibiotic treatment if appropriate as mentioned previously since serum–drug concentrations are not available [54,58]. Replacement of one antibiotic for another can also be implemented.

Chromatography was historically the method of choice for the TDM of antibiotics. Subsequently, immunoassay became more popular due to both the ease of use, lower sample volume requirements, and the relatively fast analysis time since turn-around time is important in order for TDM results to be clinically meaningful [70,71]. However, immunoassays can prove inaccurate due to the detection of metabolites as well as the parent compound giving falsely increased concentrations [74].

Different types of immunoassay exist for the TDM of antibiotics including enzyme multiplied immunoassay technique (EMIT), cloned enzyme donor immunoassay (CEDIA), fluorescence polarization immunoassay (FPIA), and particle-enhanced turbidimetric inhibition immunoassay. The specific type of assay used in each clinical laboratory depends on the instrumentation manufacturer implemented in that specific laboratory. The EMIT immunoassay is based upon competition between the drug in the patient sample and an enzyme-labeled version of the drug for a specific antibody. In this assay the absorbance increases proportionally with increasing drug concentration [75]. With the CEDIA immunoassay the drug in the patient sample competes with a labeled enzyme donor fragment for the antibody. If there is no drug in the patient sample, the enzyme donor fragment binds to the antibody and there is no absorbance produced. If there is drug in the patient sample, the enzyme donor fragment binds to the enzyme acceptor fragment, which produces an active enzyme that causes an increase in absorbance. Therefore the absorbance produced is directly proportional to the concentration of the drug in the patient sample [75]. With the FPIA immunoassay, there is competition between the drug in the patient sample and a labeled drug analog. If the labeled drug analog binds to the antibody, it causes emission of polarized fluorescence. If the antibody is instead bound to the drug from the patient sample, the emitted fluorescence is not polarized, and so in this case, the drug concentration in the patient sample is inversely proportional to the polarized fluorescence signal [75]. In the PETIA immunoassay, there is competition between latex particles attached to the drug and the drug in the patient sample for antibody binding. If there is no drug in the patient sample, the latex particles bind to the antibody causing insoluble aggregates and therefore increased turbidity in the sample. If there is drug in the patient sample, it binds to the antibodies causing less turbidity due to inhibition of the formation of insoluble aggregates. Therefore the drug concentration in the patient sample is inversely proportional to the rate of formation and the amount of insoluble aggregates [76].

High-performance liquid chromatography (HPLC) coupled to a UV diode array detector is commonly used for TDM of antibiotics [77,78]. While this methodology allows a number of different compounds to be analyzed, it is not specific and many compounds do not absorb UV light. Liquid chromatography−tandem mass spectrometry (LC−MS/MS) is becoming more widely used than HPLC for TDM [79]. This is probably due to the increased specificity afforded by this methodology since the analyte is detected using its molecular weight. There is also a potential increase in sensitivity of this technique, which can lead to the requirement for less sample volume. HPLC and LC−MS/MS both tend to require a sample preparation step that is not required for immunoassay analysis; this can add significantly to the turnaround time [79]. Further, while immunoassay analyzers have the capability to run more than one patient sample at a time, HPLC and LC−MS/MS systems traditionally can only run one patient sample at a time. However, the assay time for HPLC and LC−MS/MS tends to be shorter than for immunoassay, and multiplexing is also an option to increase the throughput.

However, one issue with the availability of all of the methods available to measure these antibiotic drug concentrations is that the calibration of the assays is not standardized. This means that a drug concentration measured on one assay may produce a different result if the same sample was analyzed on a different assay [80,81]. This can be problematic when drugs have defined therapeutic concentration ranges (e.g., for vancomycin that has a trough therapeutic range of 10−20 mg/L). However, since the therapeutic ranges of these antibiotics are relatively wide, it may not be a serious issue in the clinical interpretation of drug concentrations, although one study showed discordant clinical interpretation with one vancomycin immunoassay compared to an LC−MS/MS assay [81].

Clinically, immunoassays are predominantly used for TDM of vancomycin, gentamicin, tobramycin, and amikacin. The concentrations of azole antifungal drugs are measured by either HPLC or LC−MS/MS, although an immunoassay is available for voriconazole [82]. In the future the use of LC−MS/MS methods to detect multiple antibiotics from different classes may overtake the need for multiple immunoassays [83]. New advances in TDM may also include the use of high-resolution mass spectrometry that allows for the analysis of all compounds in a patient sample without prior knowledge of what is present. This technique would also be useful for the elucidation of drug metabolite patterns if relevant [71]. The use of TDM for other antibiotics that are known to cause toxicity such as amphotericin B and linezolid may become routine; an immunoassay for linezolid is available now, but only outside of the United States [84]. The desire to reduce the required sample volume even further may lead to the use of dried blood spots or oral fluid for the purposes of TDM [71,85]. The collection of these samples is also much less invasive than a traditional venipuncture, and the sample is stable enough to allow a patient to draw their own sample and mail it into the clinical laboratory making TDM much more accessible.

References

[1] Tan SY, Tatsumura Y. Alexander Fleming (1881-1955): discoverer of penicillin. Singap Med J 2015;56(7):366−7.
[2] Mahon CR, Lehman DC, Manuselis G, editors. Textbook of diagnostic microbiology. 5th ed. Saunders, an imprint of Elsevier, Inc. Publishers; 2015.
[3] Feingold DS. Antimicrobial chemotherapeutic agents: the nature of their action and selective toxicity. N Engl J Med 1963;269:900−7.
[4] Ory EM, Yow EM. The use and abuse of the broad spectrum antibiotics. JAMA 1963;185(4):273−9.
[5] Bartlett MA. Diagnostic bacteriology: a study guide. F.A Davis Company Publishers; 2000.
[6] Feingold DS. Antimicrobial chemotherapeutic agents: the nature of their action and selective toxicity. N Engl J Med 1963;269:957−64.
[7] Medoff G, Kobayashi GS. Strategies in the treatment of systemic fungal infections. N Engl J Med 1980;302:145−55.
[8] Wingard JR, Kubilis P, Lee L, Yee G, White M, Walshe L, et al. Clinical significance of nephrotoxicity in patients treated with amphotericin B for suspected or proven aspergillosis. Clin Infect Dis 1999;29:1402−7.
[9] Anaissis EJ, Vartivarian SE, Abi-Said D, Uzun O, Pinczowski H, Kontoyiannis DP, et al. Fluconazole versus amphotericin B in the treatment of hematogenous candidiasis: a matched cohort study. Am J Med 1996;101:170−6.
[10] Tonin FS, Steimbach LM, Borba HH, Sanches AC, Wiens A, Pontarolo R, et al. Efficacy and safety of amphotericin B formulations: a network meta-analysis and a multicriteria decision analysis. J Pharm Pharmacol 2017;69(12):1672−83.
[11] Hamill RJ. Amphotericin B formulations: a comparative review of efficacy and toxicity. Drugs 2013;73:919−34.

[12] Bozdogan B, Appelbaum PC. Oxazolidinones: activity, mode of action and mechanism of resistance. Int J Antimicrob Agents 2004;23(2):113–19.
[13] Birmingham MC, Rayner CR, Meagher AK, Flavin SM, Batts DH, Schentag JJ. Linezolid for the treatment of multidrug-resistant, gram-positive infections: experience from a compassionate-use program. Clin Infect Dis 2003;36(2):159–68.
[14] Corallo CE, Paull AE. Linezolid-induced neuropathy. Med J Aust 2002;177(6):332.
[15] Miller DG, Lovell EO. Antibiotic-induced serotonin syndrome. J Emerg Med 2011;40(1):25–7.
[16] Gupta V, Karnik ND, Deshpande R, Patil MA. Linezolid-induced serotonin syndrome. BMJ Case Rep 2013. https://www.ncbi.nlm.nih.gov/pmc/articles/PMC3618736/pdf/bcr-2012-008199.pdf.
[17] Quinn DK, Stern TA. Linezolid and serotonin syndrome. Prim Care Companion J Clin Psychiatry 2009;11(6):353–6.
[18] Mason JW. Antimicrobials and QT prolongation. J Antimicrob Chemother 2017;72:1272–4.
[19] Mok NS, Lo YK, Tsui PT, Lam CW. Ketoconazole induced torsades de pointes without concomitant use of QT interval-prolonging drug. J Cardiovasc Electrophysiol 2005;16(12):1375–7.
[20] Zimmermann M, Duruz H, Guinand O, Broccard O, Levy P, Lacatis D, et al. Torsades de pointes after treatment with terfenadine and ketoconazole. Eur Heart J 1992;13(7):1002–3.
[21] Tsai WC, Tsai LM, Chen JH. Combined use of astemizole and ketoconazole resulting in torsade de pointes. J Formos Med Assoc 1997;96(2):144–6.
[22] Hoover CA, Carmichael JK, Nolan Jr PE, Marcus FI. Cardiac arrest associated with combination cisapride and itraconazole therapy. J Cardiovasc Pharmacol Ther 1996;1(3):255–8.
[23] Pohjola-Sintonen S, Viitasalo M, Toivonen L, Neuvonen P. Itraconazole prevents terfenadine metabolism and increases risk of torsades de pointes ventricular tachycardia. Eur J Clin Pharmacol 1993;45(2):191–3.
[24] NoorZurani MH, Vicknasingam B, Narayanan S. Itraconazole-induced torsade de pointes in a patient receiving methadone substitution therapy. Drug Alcohol Rev 2009;28(6):688–90.
[25] Samarendra PJ, Kumari S, Evans SJ, Sacchi TJ, Navarro V. QT prolongation associated with azithromycin/amiodarone combination. Pacing Clin Electrophysiol 2001;24(10):1572–4.
[26] Russo V, Puzio G, Siniscalchi N. Azithromycin-induced QT prolongation in elderly patient. Acta Biomed 2006;77:30–2.
[27] Matsunaga N, Oki Y, Prigollini A. A case of QT-interval prolongation precipitated by azithromycin. N Z Med J 2003;166:U666.
[28] Food and Drug Administration (FDA) Drug Safety Communication: Azithromycin (Zithromax or Zmax) and the risk of potentially fatal heart rhythms <https://www.fda.gov/Drugs/DrugSafety/ucm341822.htm>; 2013 [accessed 08.11.18].
[29] Ray WA, Murray KT, Hall K, Arbogast PG, Stein CM. Azithromycin and the risk of cardiovascular death. N Engl J Med 2012;366:1881–90.
[30] Ray WA, Murray KT, Meredith S, Narasimhulu SS, Hall K, Stein CM. Oral erythromycin and the risk of sudden death from cardiac causes. N Engl J Med 2004;351:1089–96.
[31] Yap YG, Camm AJ. Drug induced QT prolongation and torsades de pointes. Heart 2003;89:1363–72.
[32] Falagas ME, Rafailidis PI, Rosmarakis ES. Arrhythmias associated with fluoroquinolone therapy. Int J Antimicrob Agents 2007;29(4):374–9.
[33] Mehrzad R, Barza M. Weighing the adverse cardiac effects of fluoroquinolones: a risk perspective. J Clin Pharmacol 2015;55(11):1198–206.
[34] Gorelik E, Masarwa R, Perlman A, Rotschild V, Abbasi M, Muszkat M, et al. Fluoroquinolones and cardiovascular risk: a systematic review, meta-analysis and network meta-analysis. Drug Saf 2018;. Available from: https://doi.org/10.1007/s40264-018-0751-2 Epub ahead of print.
[35] Grill MF, Maganti RK. Neurotoxic effects associated with antibiotic use: management considerations. Br J Clin Pharmacol 2011;72(3):381–93.
[36] Bruniera FR, Berreira FM, Saviolli LR, Bacci MR, Feder D, da Luz Gonçalves Pedreira M, et al. The use of vancomycin with its therapeutic and adverse effects: a review. Eur Rev Med Pharmacol Sci 2015;19(4):694–700.
[37] Lopez-Novoa JM, Quiros Y, Vicente L, Morales AI, Lopez-Hernandez FJ. New insights into the mechanism of aminoglycoside nephrotoxicity: an integrative point of view. Kidney Int 2011;79(1):33–45.
[38] Kyriakidis I, Tragiannidis A, Munchen S, Groll AH. Clinical hepatotoxicity associated with antifungal agents. Expert Opin Drug Saf 2017;16(2):149–65.
[39] Miller EL. The penicillins: a review and update. J Midwifery Womens Health 2002;47(6):426–34.
[40] Zaffiri L, Gardner J, Toledo-Pereyra LH. History of antibiotics. From salvarsan to cephalosporins. J Invest Surg 2012;25(2):67–77.
[41] Chukwudi CU. rRNA binding sites and the molecular mechanism of action of the tetracyclines. Antimicrob Agents Chemother 2016;60(8):4433–41.
[42] Kotra LP, Haddad J, Mobashery S. Aminoglycosides: perspectives on mechanisms of action and resistance and strategies to counter resistance. Antimicrob Agents Chemother 2000;44(12):3249–56.
[43] Vázquez-Laslop N, Mankin AS. How macrolide antibiotics work. Trends Biochem Sci 2018;43(9):668–84.
[44] Hashemian SMR, Farhadi T, Ganiparvar M. Linezolid: a review of its properties, function, and use in critical care. Drug Des Devel Ther 2018;12:1759–67.
[45] Campbell EA, Korzheva N, Mustaev A, Murakami K, Nair S, Goldfarb A, et al. Structural mechanism for rifampicin inhibition of bacterial RNA polymerase. Cell 2001;104(6):901–12.
[46] Hooper DC. Mechanisms of action of antimicrobials: focus on fluoroquinolones. Clin Infect Dis 2001;32(Suppl. 1):S9–15.
[47] Parente-Rocha JA, Bailão AM, Amaral AC, Taborda CP, Paccez JD, Borges CL, et al. Antifungal resistance, metabolic routes as drug targets, and new antifungal agents: an overview about endemic dimorphic fungi. Mediators Inflamm 2017;2017:9870679.
[48] Mueller M, de la Peña A, Derendorf H. Issues in pharmacokinetics and pharmacodynamics of anti-infective agents: kill curves vs MIC. Antimicrob Agents Chemother 2004;48(2):369–77.

[49] Clinical and Laboratory Standards Institute (CLSI). Methods for dilution antimicrobial susceptibility tests for bacteria that grow aerobically. Approved Standard — 10th ed. CLSI document M07-A10. Wayne, PA: Clinical and Laboratory Standards Institute; 2015.

[50] Craig WA. Choosing an antibiotic on the basis of pharmacodynamics. Ear Nose Throat J 1998;77:7—11.

[51] de Velde F, Mouton JW, de Winter BCM, van Gelder T, Koch BCP. Clinical applications of population pharmacokinetic models of antibiotics: challenges and perspectives. Pharmacol Res 2018;134:280—8.

[52] Derendorf H. Pharmacokinetic evaluation of beta-lactam antibiotics. J Antimicrob Chemother 1989;24(3):407—13.

[53] Romano A, Mondino C, Viola M, Montuschi P. Immediate allergic reactions to beta-lactams: diagnosis and therapy. Int J Immunopathol Pharmacol 2003;16(1):19—23.

[54] Amphotericin B. Micromedex® (electronic version). IBM Watson Health, Greenwood Village, CO: Available from < https://www.micromedexsolutions.com/ > [accessed 13.11.18].

[55] MacDougall C. Antifungal agents Accessed November 13th In: Brunton LL, Hilal-Dandan R, Knollmann BC, editors. Goodman & Gilman's: The pharmacological basis of therapeutics, 13e. New York, NY: McGraw-Hill; 2018.

[56] Schulz M, Iwersen-Bergmann S, Andresen H, Schmoldt A. Therapeutic and toxic blood concentrations of nearly 1000 drugs and other xenobiotics. Crit Care 2012;16(4):R136.

[57] Pea F, Furlanut M, Cojutti P, Cristini F, Zamparini E, Franceschi L, et al. Therapeutic drug monitoring of linezolid: a retrospective monocentric analysis. Antimicrob Agents Chemother 2010;54(11):4605—10.

[58] Linezolid. Micromedex® (electronic version). IBM Watson Health, Greenwood Village, CO: Available from < https://www.micromedexsolutions.com/ > [accessed 13.11.18].

[59] MacDougall C. Protein synthesis inhibitors and miscellaneous antibacterial agents Accessed November 13th In: Brunton LL, Hilal-Dandan R, Knollmann BC, editors. Goodman & Gilman's: The Pharmacological Basis of Therapeutics, 13e. New York, NY: McGraw-Hill; 2018.

[60] Itraconazole. Micromedex® (electronic version). IBM Watson Health, Greenwood Village, CO: Available from < https://www.micromedexsolutions.com/ > [accessed 19.11.18].

[61] Fluconazole. Micromedex® (electronic version). IBM Watson Health, Greenwood Village, CO: Available from < https://www.micromedexsolutions.com/ > [accessed 19.11.18].

[62] Voriconazole. Micromedex® (electronic version). IBM Watson Health, Greenwood Village, CO: Available from < https://www.micromedexsolutions.com/ > [accessed 19.11.18].

[63] Posaconazole. Micromedex® (electronic version). IBM Watson Health, Greenwood Village, CO: Available from < https://www.micromedexsolutions.com/ > [accessed 19.11.18].

[64] Vancomycin. Micromedex® (electronic version). IBM Watson Health, Greenwood Village, CO: Available from < https://www.micromedexsolutions.com/ > [accessed 19.11.18].

[65] Gentamicin. Micromedex® (electronic version). IBM Watson Health, Greenwood Village, CO: Available from < https://www.micromedexsolutions.com/ > [accessed 19.11.18].

[66] Tobramycin. Micromedex® (electronic version). IBM Watson Health, Greenwood Village, CO: Available from < https://www.micromedexsolutions.com/ > [accessed 19.11.18].

[67] Amikacin. Micromedex® (electronic version). IBM Watson Health, Greenwood Village, CO: Available from < https://www.micromedexsolutions.com/ > [accessed 19.11.18].

[68] MacDougall C. Aminoglycosides Accessed November 19th In: Brunton LL, Hilal-Dandan R, Knollmann BC, editors. Goodman & Gilman's: The pharmacological basis of therapeutics, 13e. New York, NY: McGraw-Hill; 2018.

[69] < https://www.slideshare.net/rajat2903/therapeutic-drug-monitoring-31857498 >. Slide 27. [accessed on 19.11.18].

[70] Kang JS, Lee MH. Overview of therapeutic drug monitoring. Korean J Intern Med 2009;24(1):1—10.

[71] Eliasson E, Lindh JD, Malstrm RE, Beck O, Dahl MJ. Therapeutic drug monitoring for tomorrow. Eur J Clin Pharmacol 2013;69(Suppl. 1):25—32.

[72] Smith MD, Metcalf CS, Wilcox KS. Pharmacotherapy of the epilepsies Accessed December 3rd In: Brunton LL, Hilal-Dandan R, Knollmann BC, editors. Goodman & Gilman's: The pharmacological basis of therapeutics, 13e. New York, NY: McGraw-Hill; 2018.

[73] Carbamazepine. Micromedex® (electronic version). IBM Watson Health, Greenwood Village, CO: Available from < https://www.micromedexsolutions.com/ > [accessed 03.12.18].

[74] Hu MW, Anne L, Forni T, Gottwald K. Measurement of vancomycin in renally impaired patient samples using a new high-performance liquid chromatography method with vitamin B12 internal standard: comparison of high-performance liquid chromatography, EMIT, and fluorescence polarization immunoassay methods. Ther Drug Monit 1990;12(6):562—9.

[75] Sanavio B, Krol S. On the slow diffusion of point-of-care systems in therapeutic drug monitoring. Front Bioeng Biotechnol 2015;3:20.

[76] Opheim KE, Glick MR, Ou CN, Ryder KW, Hood LC, Ainardi V, et al. Particle-enhanced turbidimetric inhibition immunoassay for theophylline evaluated with the Du Pont aca. Clin Chem 1984;30(11):1870—4.

[77] Lima TM, Seba KS, Gonçalves JCS, Cardoso FLL, Estrela RCE. A rapid and simple HPLC method for therapeutic drug monitoring of vancomycin. J Chromatogr Sci 2018;56(2):115—21.

[78] Zhang M, Moore GA, Barclay ML, Begg EJ. A simple high-performance liquid chromatography method for simultaneous determination of three triazole antifungals in human plasma. Antimicrob Agents Chemother 2013;57(1):484—9.

[79] French D. Advances in clinical mass spectrometry. Adv Clin Chem 2017;79:153—98.

[80] Wilson JF, Davis AC, Tobin CM. Evaluation of commercial assays for vancomycin and aminoglycosides in serum: a comparison of accuracy and precision based on external quality assessment. J Antimicrob Chemother 2003;52(1):78—82.

[81] Oyaert M, Peersman N, Kieffer D, Deiteren K, Smits A, Allegaert K, et al. Novel LC-MS/MS method for plasma vancomycin: comparison with immunoassays and clinical impact. Clin Chim Acta 2015;441:63−70.

[82] Cattoir L, Fauvarque G, Degandt S, Ghys T, Verstraete AG, Stove V. Therapeutic drug monitoring of voriconazole: validation of a novel ARK™ immunoassay and comparison with ultra-high performance liquid chromatography. Clin Chem Lab Med 2015;53(5):e135−9.

[83] Neugebauer S, Wichmann C, Bremer-Streck S, Hagel S, Kiehntopf M. Simultaneous quantification of nine antimicrobials by LC-MS/MS for therapeutic drug monitoring in critically ill patients. Ther Drug Monit 2018;. Available from: https://doi.org/10.1097/FTD.0000000000000570 Epub ahead of print.

[84] Castoldi S, Cozzi V, Baldelli S, Fucile S, Clementi E, Cattaneo D. Comparison of the ARK™ immunoassay with HPLC-UV for therapeutic drug monitoring of linezolid. Ther Drug Monit 2017;. Available from: https://doi.org/10.1097/FTD.0000000000000473 Epub ahead of print.

[85] Martial LC, van den Hombergh E, Tump C, Halmingh O, Burger DM, van Maarseveen EM, et al. Manual punch versus automated flow-through sample desorption for dried blood spot LC-MS/MS analysis of voriconazole. J Chromatogr B Anal Technol Biomed Life Sci 2018;1089:16−23.

Chapter 8

Antipsychotics and antidepressants

Michael M. Mbughuni[1,2], Caren J. Blacker[3], Jesse Seegmiller[2], Hemamalini Ketha[4] and Amy B. Karger[2]

[1]Department of Laboratory Medicine and Pathology, Minneapolis VA Healthcare System, Minneapolis, MN, United States, [2]Department of Laboratory Medicine and Pathology, University of Minnesota, Minneapolis, MN, United States, [3]Department of Psychiatry and Psychology, Mayo Clinic, Rochester, MN, United States, [4]Laboratory Corporation of America, North Carolina, United States

Introduction

Antidepressants and antipsychotics are neuropsychiatric medications used to treat depressive disorders and psychosis, respectively. Clinical depression is etiologically different than the depressive state that can be caused by tragic life-events. Clinical depression is most commonly diagnosed using Diagnostic and Statistical Manual of Mental Disorders (DSM-5) symptom criteria. Some symptoms of clinical depression include prolonged sadness, irritability, loss of interest in normal activities, insomnia, and unexplained physical conditions such as chronic pain and suicidal ideation. On the other hand, psychosis is a symptom that presents commonly as a manifestation of an underlying mental illness such as schizophrenia or bipolar disorder. During a psychotic episode a patient loses connection with tangible reality and can experience anxiety, insomnia, and social withdrawal. Symptoms of psychosis include delusions and hallucinations.

The use of drugs to manage mental health disorders is increasing in the United States. Due to the overlap of clinical symptomology with pain-related syndromes, antidepressant and antipsychotic drugs are also used for off-label management of chronic pain, anxiety, and insomnia. Both classes of drugs act on various neurotransmitter receptors, altering brain chemistry and ameliorating the symptoms of these disorders. It is important to be familiar with the receptor specificity and affinity profiles of each drug to understand the associated side effects and toxicities in cases of overdose. Each class of drugs is associated with side effects, including drowsiness, dizziness, weight gain, nausea, vomiting, and constipation. Response to treatment with antidepressants and antipsychotics varies wherein some patients respond better to these drugs than others. Therefore effectiveness of these drug classes for symptom management and clinical outcomes of antidepressants and antipsychotic therapy has been debated.

Toxic exposures of antidepressants and antipsychotics leading to a clinical overdose and/or fatalities are also increasing. This is particularly true in populations with comorbid substance abuse disorders. Data from epidemiological studies from National Institute of Drug Abuse show high rates of comorbidity of substance use disorders with mental health disorders such as bipolar disorder, attention-deficit hyperactivity disorder, anxiety disorder, and posttraumatic-stress disorder. A renewed interest in laboratory assessment of these drug classes in clinical and forensic setting is emerging. In this chapter, we describe the most commonly used antidepressants and antipsychotic drugs from the perspective of toxicologist's routine practice. A guide for the management of overdose of these drug classes is provided. Finally, analytical methods for clinical laboratory assessment are discussed.

Antidepressants

Depressive disorders are very common, impacting more than 15% of the general population at some time during their life [1]. Given the high prevalence of depression, antidepressants are widely prescribed and, therefore, highly accessible for potential misuse. Antidepressants represent the fifth most common substance involved in human poisoning exposures, responsible for 4.4% of all exposures [2]. Antidepressants also account for 2.7% of emergency department (ED) visits for adverse drug events (ADEs) in the United States [3], with a 12.4% hospitalization rate [4].

Antidepressant drugs belong to several drug classes. The most commonly used antidepressant drug classes include tricyclic antidepressants (TCAs), selective serotonin reuptake inhibitors (SSRIs), serotonin—norepinephrine reuptake inhibitors (SNRIs), monoamine oxidase inhibitors (MAOIs), serotonin modulators (SRMs) serotonin antagonists and reuptake inhibitors, NMDA receptor antagonists, and others (Table 8.1).

Prior to the 1950s, opioids and antihistamines were used as antidepressants. During the 1950s and the 1960s, several research groups and pharmaceutical companies explored the use of compounds, including isoniazid, chlorpromazine, and imipramine. Further work expanded the number of first generation of antidepressants to two classes—TCAs and MAOIs. First-generation TCA and MAOI classes of compounds are effective treatments for depression discovered in the 1950s but have significant, dose-related side effects and toxicities [5,6] (Table 8.1). Because of this, they are no longer recommended as an initial treatment but may be rarely used in patients with treatment refractory or resistant depression. Newer antidepressants include second-generation antidepressants, such as SSRIs, SNRIs, SRMs, atypical antidepressants, and tetracyclics, all of which have been developed over time with superior side effect profiles and lower risks for toxicity and are most commonly used in contemporary medicine [5,7–10].

Therefore second-generation antidepressants are currently associated with more ED visits for adverse events than first-generation antidepressants (Table 8.2) [4]. By contrast, less than 10% of ED visits related to antidepressant adverse events are due to first-generation TCAs or MAOIs [4].

It is important to note that antidepressant-related ED visits for ADEs are more common in children, with antidepressants accounting for 3.8% of all ADE-associated ED visits in ages 6–19, but only 0.8% of ADE-associated ED visits in ages 65–79 [3]. Despite being used for more than five decades in clinical settings, the efficacy of antidepressant drugs has been debated.

Antipsychotics

Psychosis is associated with several neuropsychiatric disorders, including schizophrenia, major depressive disorder with psychotic features, bipolar disorder with psychotic features, drug-induced psychoses, and psychoses secondary to medical illnesses. Schizophrenia is the classic disorder associated with psychosis for which antipsychotic medications are

TABLE 8.1 Examples of Antipsychotics and Antidepressant Medications.

Antipsychotics	
Class	Examples
First-generation antipsychotics (typical)	Fluphenazine, haloperidol, loxapine, perphenazine, pimozide, thiothixene, trifluoperazine, chlorpromazine, and thioridazine
Second-generation antipsychotics (atypical)	Aripiprazole, asenapine, brexpiprazole, cariprazine, clozapine, iloperidone, lurasidone, olanzapine, paliperidone, pimavanserin, quetiapine, risperidone, and ziprasidone

Antidepressants	
First-Generation Antidepressants	Examples
TCAs	Amitriptyline, clomipramine, imipramine, doxepin, trimipramine, nortriptyline, protriptyline, amoxapine, and desipramine
MAOIs	Iproniazid, tranylcypromine, phenelzine, and selegiline
Newer Generation Antidepressants	Examples
SSRIs	Citalopram, escitalopram, fluoxetine, fluvoxamine, paroxetine, and sertraline
Selective norepinephrine–serotonin reuptake inhibitors (SNRIs)	Desvenlafaxine, duloxetine, levomilnacipran, milnacipran, and venlafaxine
Serotonin modulators (SRM and SARIs)	Nefazodone, trazodone, vilazodone, and vortioxetine
Atypical antidepressants	Bupropion, mirtazapine, and agomelatine[a]
Tetracyclics	Maprotiline

MAOIs, Monoamine oxidase inhibitors; *SARIs*, serotonin antagonists and reuptake inhibitors; *SSRIs*, selective serotonin reuptake inhibitors; *TCAs*, tricyclic antidepressants.
[a]Not available in the United States.

used and occurs in 1% of the population [11]. Similar to antidepressants, the first generation of antipsychotics was effective but was found to have significant side effects, with subsequent development of second-generation antipsychotics with improved side effect profiles and lower morbidity risk (Table 8.1).

Antipsychotics account for 2.7% of all US Emergency Department visits for ADEs [3] with a 15.3% hospitalization rate [4], with the majority of antipsychotic-related ED visits secondary to the more commonly prescribed second-generation or atypical antipsychotics (Table 8.3) [4].

In a study that looked at adult ED visits for psychiatric medication adverse events from 2009 to 2011, antipsychotics accounted for 3 of the top 10 most common drugs responsible for ADEs [4]. In addition, haloperidol, a typical antipsychotic, had by far the highest rate of annual ED visits relative to its outpatient use (Table 8.4) [4]. Even the atypical antipsychotics, with an improved side effect profile compared to typical antipsychotics, had two times as many ED visits relative to outpatient prescription use when compared to antidepressants, sedatives, anxiolytics, or stimulants [4].

Of note, antipsychotic ADEs are most common in children and young adults, accounting for 4.5% and 5.2% of ADE-associated ED visits in the United States for the age groups of 6–19 years and 20–34 years, respectively. With increasing age the prevalence of antipsychotic ADEs falls to 0.5% of all ADE-associated ED visits for ages 65–79 [4].

TABLE 8.2 The Top Four Most Common Antidepressant Drugs/Drug Classes Associated With ED Visits Due to Adverse Events in Adults [4].

Drug or Drug Class	Antidepressant-Related ED Visits for Adverse Events (%)	Estimated Annual ED Visits per 10,000 Outpatient Prescription Visits (n)
SSRIs	47.4	1.9
SNRIs	15.3	1.8
Trazodone	13.6	3.7
Bupropion	10.1	2.1

ED, Emergency department; SSRIs, selective serotonin reuptake inhibitors.

TABLE 8.3 Data on Adverse Drug Events—Associated Emergency Department (ED) Visits for Typical Versus Atypical Antipsychotics.

Drug or Drug Class	Antipsychotic-Related ED Visits for Adverse Events (%)
First-generation or typical antipsychotics	26.9
Second-generation or atypical antipsychotics	70.8

TABLE 8.4 Three Most Commonly Implicated Drugs Associated With Emergency Department (ED) Visits Due to Antipsychotic Adverse Events in Adults [4].

Antipsychotic Drug	Percentage of ED Visits for Psychiatric Medication ADEs	Estimated Annual ED Visits per 10,000 Outpatient Prescription Visits (n)
Quetiapine fumarate	7.7	10.8
Haloperidol	5.5	43.3
Risperidone	4.1	11.3

ADEs, Adverse drug events.

In addition, an analysis looking at ADEs secondary to antipsychotics, stimulants, and antidepressants in children and adolescents found that antipsychotics had a significantly higher rate of annual ED visits relative to their outpatient prescription rate when compared to either stimulants or antidepressants [12].

Pharmacokinetics and toxicokinetics and treatment

Management of SSRI, SNRI, and atypical antidepressant overdose

SSRIs are typically not dangerous in overdose, which has been one of their major advantages over older antidepressants. Most SSRI overdoses respond well to supportive treatment. However, massive overdose and/or coingestion are more likely to lead to significant symptoms. Hence, establishing what has been ingested, and how much, is important. Serotonin syndrome, QT prolongation/arrhythmias, and seizures are rare. Citalopram has cardiotoxic metabolites. Dialysis is ineffective as most SSRIs are highly protein bound.

In addition to the SNRIs (e.g., venlafaxine, desvenlafaxine, duloxetine, and milnacipran), commonly used atypical antidepressants include the monocyclic-aminoketone bupropion, the alpha-adrenergic antagonist mirtazapine, and the serotonin modulator/stimulator antidepressants vilazodone and vortioxetine. Like the SSRIs, SNRIs and atypical antidepressants are generally well-tolerated in overdose. However, venlafaxine overdose has been associated with greater cardiotoxicity and lethality than SSRIs, and both venlafaxine and bupropion are associated with higher seizure rates, with potential for recurrent seizures. These two antidepressants also have active metabolites; venlafaxine is metabolized to desvenlafaxine via CYP2D6 (thus poor metabolizers are at risk of elevated drug levels) and bupropion is metabolized to active compounds that have significantly longer half-lives than the parent drug. Both drugs can be associated with delayed toxicity and may require 24 hours of monitoring. Dialysis is not effective in SNRI and atypical antidepressant overdose due to the large volume of distribution of these drugs. The effects of SSRI, SNRI, and atypical antidepressant overdose include neurological: sedation (except bupropion is associated with agitation), tremor, serotonergic symptoms (such as hyperreflexia and clonus), and rarely seizures and coma; cardiopulmonary: tachycardia and hypertension; gastrointestinal: abdominal pain, vomiting, and diarrhea. No specific antidotes are available for SSRI and atypical antidepressant overdose, and treatment is supportive with management of complications as they arise. It is not recommended to obtain drug serum concentrations as it does not influence the treatment plan. (See Table 8.5 for management.)

Management of tricyclic antidepressant overdose

The dangers of TCAs in overdose have led to a decrease in their clinical use for depression. However, they are still frequently used in refractory depression and in many neurological and geriatric conditions (e.g., headache and insomnia). TCAs have a narrow therapeutic range and can be potent central nervous system and cardiovascular toxins at moderate doses. Approximately 30% of TCA metabolites are eliminated via the biliary system, while 70% are renally excreted as inactive metabolites. TCA half-lives very widely, and many of the metabolites are pharmacologically active.

TCA overdose patients have unexpected clinical courses, and even initially well, they can deteriorate rapidly due to variable absorption kinetics. Dialysis is ineffective in TCA overdose due to the large volume of distribution of these lipophilic drugs. Serum TCA concentrations are not recommended as they do not influence treatment. (See Table 8.6 for management.)

The effects of TCA overdose include the following:

- *Neurological*: sedation, confusion, delirium/psychosis, seizures, and coma;
- *Cardiopulmonary*: hypotension, cardiac conduction delays (QRS > 100 ms), and arrhythmias; and
- *Anticholinergic signs*: hyperthermia, flushing, dilated pupils, dry mouth, absent bowel sounds/intestinal ileus, and urinary retention.

Management of monoamine oxidase inhibitor overdose

MAOIs were the first class of antidepressant drugs to be developed. They irreversibly (other than moclobemide) inhibit the monoamine oxidase enzyme that breaks down norepinephrine, serotonin, and dopamine, thus raising levels of all three neurotransmitters. Monoamine oxidase also breaks down the amino acid tyramine that is associated with hypertension, and as a result, for years it was thought that taking an MAOI and eating high-tyramine foods could cause a life-threatening hypertensive emergency. Partly due to these dietary restrictions, MAOIs fell out of favor when alternative antidepressant treatments became available.

TABLE 8.5 Suggested Management Plan for Selective Serotonin Reuptake Inhibitors and Atypical Antidepressant Overdoses.

History	Establish what has been ingested: medications, formulations (immediate or extended release), dose, quantity, timing of ingestion(s), and coingestants.
Physical	Continually assess circulation, airway, breathing—support these as clinically indicated.
	Typically, physical examination will be normal; however, monitor for drowsiness and vomiting with consideration for airway management.
	Monitor for the rare complication of seizures, which are more common with coingestants.
	Monitor for the rare complication of serotonin syndrome (may include any combination of the following: ocular or other clonus, hyperthermia, diaphoresis, hyperreflexia, tremor, and hypertonia).
Testing	Laboratory testing should include the following: • serum salicylate and acetaminophen levels (for coingestion); • serum ethanol concentration (for coingestion); • urine drug screen (for coingestion); • serum bicarbonate concentration; • blood glucose, to rule out hypoglycemia as cause of seizures/altered mental status; • electrolytes to evaluate risks for cardiac conduction abnormalities; and • ECG (repeat serial ECGs and start cardiac monitoring if QTC or QRS are prolonged).
	It may be appropriate to also consider the following: • Pregnancy test • Specific testing in cases of serotonin syndrome should include: renal function, liver function, clotting (for disseminated intravascular coagulation), creatine kinase and urine myoglobin (for rhabdomyolysis), blood gas (for metabolic acidosis).
Management	Consider a single dose of oral activated charcoal without additives (1 g/kg) if airway can be adequately protected—but do not intubate for the sole purpose of administering charcoal.
	Avoid administering any further serotonergic agents, or agents that prolong QT interval.
	Treat seizures with benzodiazepines as first-line therapy. Second-line treatment includes barbiturates. Refractory seizures may require intubation and propofol. Do *not* administer flumazenil, even if a benzodiazepine coingestant is suspected, to avoid precipitating seizures.
	Treat prolonged QRS interval with sodium bicarbonate until either QRS narrows or serum pH reaches 7.45–7.55.
	Monitor for torsades, and treat with intravenous magnesium sulfate, and consider magnesium sulfate infusion.
	Treat serotonin syndrome with supportive care or serotonin antagonists such as cyproheptadine if indicated. Severe cases may require neuromuscular paralysis, intubation, and external cooling.

ECG, Electrocardiogram.

MAOIs also have a low therapeutic index. Ingestion of as little as 2 mg/kg of a nonselective MAOI (e.g., phenelzine, isocarboxazid, and tranylcypromine) can cause severe toxicity. In addition, severe toxicity can also be induced with administration of any drug that increases monoamine levels, either concurrently or within 14 days after MAOI cessation. These drugs include extremely common pharmacological agents such as dextromethorphan, linezolid, SSRIs, tramadol, and methylene blue. These restrictions on treatment options are another reason MAOIs are infrequently used, but as a result, clinicians are much less familiar with MAOI overdoses. Since almost all the MAOIs irreversibly inhibit the enzyme, monoamine oxidase activity can only be recovered by synthesis of the enzyme, meaning the effects of these medications can persist for weeks.

The main toxic effects are related to increased adrenergic tone. The initial effects of tachycardia and hypertension can be followed by potentially catastrophic cardiovascular collapse with hypotension and bradycardia. Seizures may occur and serotonin syndrome is also more common with MAOIs than many other antidepressants. The time course of symptom emergence can be highly variable. The effects of MAOI overdose include the following:

- *Neurological*: agitation, altered mental status, seizures, coma, myoclonus, hyperthermia, muscle rigidity, and nystagmus (described as watching "a ping pong game");

TABLE 8.6 Suggested Management of Tricyclic Antidepressant (TCA) Antidepressant Overdoses.

History	Establish what has been ingested: medications, formulations (immediate or extended release), dose, quantity, timing of ingestion(s), and coingestants.
Physical	Typically, the patient will be sedated or moribund, but even if initially alert, this can deteriorate rapidly.
	Continually assess circulation, airway, and breathing—intubate the airway and provide supplemental oxygen if clinically indicated.
Testing	Laboratory testing should include the following: • serum salicylate and acetaminophen levels (for coingestion); • serum ethanol concentration (for coingestion); • urine drug screen (for coingestion); • serum bicarbonate concentration—if QRS is widened >100 ms in TCA overdose, strong consideration should be given to administration of sodium bicarbonate; • blood glucose, to rule out hypoglycemia as cause of seizures/altered mental status; • electrolytes; • electrocardiogram (recommend cardiac monitoring in TCA overdose); • arterial blood gas (consider arterial line if frequent pH testing is needed during sodium bicarbonate administration); and • pregnancy test (if clinically indicated).
Management	Given the potential for rapid deterioration in cardiopulmonary status and consciousness, continuous monitoring (including pulse oximetry and cardiac monitoring) is necessary.
	Manage airway (intubation as needed) and breathing (supplemental oxygen as needed).
	Consider a single dose of oral activated charcoal without additives (1 g/kg) if airway can be adequately protected—but do not intubate for the sole purpose of administering charcoal.
	Manage hypotension aggressively with IV isotonic crystalloid fluid boluses as first-line therapy. Second-line treatment can include alpha-adrenergic agonist vasopressors (e.g., norepinephrine and phenylephrine). Third-line treatment of hypertonic saline should only be used if hypotension has failed to respond to adequate alkalinization with sodium bicarbonate, aggressive fluid resuscitation, and vasopressor treatment.
	Manage conduction disturbances of prolonged QRS > 100 ms with IV sodium bicarbonate. Second-line treatment with magnesium sulfate or lidocaine should only be considered if sodium bicarbonate therapy has been ineffective.
	Manage seizures with benzodiazepines as first-line treatment; second-line treatment includes barbiturates, but they may cause hypotension. Do *not* use phenytoin due to its sodium channel antagonism which may worsen QRS interval widening. Do *not* administer flumazenil, even if a benzodiazepine coingestant is suspected, to avoid precipitating seizures.
	Do *not* administer physostigmine for anticholinergic toxicity due to the association with cardiac arrest in TCA overdose.

- *Cardiopulmonary*: tachycardia, tachypnea, diaphoresis hypertension, and arrhythmia;
- *Gastrointestinal*: vomiting;
- *Hematological*: decreased fibrinogen, decreased PTT, and increased fibrin products; and
- *Other*: rhabdomyolysis, myocardial infarction, intracranial hemorrhage, renal failure, and hypertensive emergencies.

There are no specific antidotes for MAOI overdose. Dialysis is not effective for elimination of the drug, although it can be used for supportive management if the patient develops renal failure as a result of the overdose. Drug serum concentrations do not influence treatment. (See Table 8.7 for management.)

Management of antipsychotic overdose

Overdose of antipsychotic agents typically involves an exacerbation of adverse effects seen at therapeutic doses. CNS depression can be seen as a response to the H1 histamine receptor antagonism. Anticholinergic effects are also seen with many typical (e.g., chlorpromazine, loxapine, and thioridazine) and atypical (e.g., clozapine, quetiapine, risperidone, and olanzapine) antipsychotics. The classic anticholinergic effects of dry skin, hyperthermia, flushing, urinary retention, tachycardia, and altered mental status can be expected. Importantly, the classic anticholinergic sign of dilated

TABLE 8.7 Suggested Management of Monoamine Oxidase Inhibitor (MAOI) Antidepressant Overdoses.

History	Establish what has been ingested: medications, formulations (immediate or extended release), dose, quantity, timing of ingestion(s), and coingestants.
Physical	Patients can deteriorate rapidly, and symptom emergence has highly variable time courses.
	Continually assess circulation, airway, and breathing—intubate the airway and provide supplemental oxygen if clinically indicated.
Testing	Laboratory testing should include the following: • serum salicylate and acetaminophen levels (for coingestion); • serum ethanol concentration (for coingestion); • urine drug screen (for coingestion); • blood glucose, to rule out hypoglycemia as cause of seizures/altered mental status; • electrolytes (T-waves have been observed on ECG in the absence of hyperkalemia); • coagulation panel; • CBC (for leukocytosis); • Rrnal function (renal failure has been reported); • creatine kinase and urine myoglobin; • ECG; and • pregnancy test (if clinically indicated).
Management	Continually monitor circulation, airway, and breathing—and treat as clinically appropriate.
	ECG and cardiac monitoring.
	Consider a single dose of oral activated charcoal without additives (1 g/kg) if airway can be adequately protected—but do not intubate for the sole purpose of administering charcoal.
	Manage hypertension only with short-acting agents to prevent catastrophic hypotension.
	Manage hypotension with IV saline, norepinephrine if needed.
	Manage seizures with benzodiazepines.
	Manage hyperthermia with cooling measures, and consider paralytic muscle agents if there is persistent rigidity with risk of rhabdomyolysis.
	Do *not* give any medications that interact with MAOIs and precipitate a hypo hypertensive crisis, these include tramadol, dextromethorphan, cyclobenzaprine, and propoxyphene.

CBC, complete blood count; *ECG*, Electrocardiogram.

pupils is often *missing* due to the competing anti-alpha-adrenergic effect of antipsychotics. Orthostatic hypotension is common due to alpha-adrenergic antagonism, putting the patient a very high fall risk. QT prolongation is common, but torsades de pointes is uncommon. Although antipsychotics can lower the seizure threshold, seizures remain a relatively rare complication of overdose. Extrapyramidal side effects are common, particularly akathisia (an internal sense of restlessness that may be accompanied by external motor agitation), although acute dystonias are important especially if they cause spasm of the airway muscles. The primary treatment of antipsychotic overdose is supportive care (Table 8.8).

Analytical methods for antipsychotics and antidepressants

Laboratory analysis of antipsychotics and antidepressant compounds can be critical for various clinical and forensic cases. This section will provide some background of the laboratory technologies and measurement procedures and the evolution of these systems to provide clinicians and toxicologists with results from a multitude of sample types.

There have been numerous approaches to analyze antipsychotics and antidepressants. Some of the first measurement procedures to explore circulating concentrations in biological samples for some of these compounds utilized technologies such as ultraviolet (UV) spectrophotometry [13], colorimetry [14], and isotope derivatization [15] approaches. These measurement procedures focused on specific properties of the drugs and exploited these properties to perform the analysis in biological fluids. While effective, many of these early approaches were subject to interferences, were limited in scope, and lacked analytical specificity for a particular drug. This leads to employing chromatographic systems, such as gas chromatography (GC), to provide increased analytical specificity for analysis in complex biological fluids

TABLE 8.8 Suggested Management of Antipsychotic Overdoses.

History	Establish what has been ingested: medications, formulations (immediate or extended release), dose, quantity, timing of ingestion(s), and coingestants.
Physical	Presentation may include no symptoms or nonspecific symptoms.
	There are no pathognomonic signs. Presentation may only include CNS depression and minor vital sign perturbations.
	Findings could include signs of anticholinergic toxicity, and hypotension.
	Life-threatening presentations may include NMS.
	Acute EPS such as akathisia and dystonic reactions may occur at any dose.
Testing	Laboratory testing should include the following: • serum salicylate and acetaminophen levels (for coingestion); • serum ethanol concentration (for coingestion); • urine drug screen (for coingestion); • blood glucose, to rule out hypoglycemia as cause of seizures/altered mental status; • electrocardiogram (recommend cardiac monitoring in TCA overdose); and • pregnancy test (if clinically indicated).
	Antipsychotic medications have been associated with hyponatremia, serum sodium should be measured in patients with an altered mental status or seizures.
	Creatinine should be monitored for patients who develop hypotension, rhabdomyolysis, prolonged urinary retention, or NMS due to risk of acute kidney injury.
Management	Manage with supportive care until signs of overdose have resolved.
	Continually assess circulation, airway, and breathing—intubate the airway and provide supplemental oxygen if clinically indicated.
	Continuous cardiac monitoring with reevaluation of mental status is recommended.
	Manage hypotension with IV saline, norepinephrine if needed.
	QRS prolongation is treated with sodium bicarbonate.
	Acute extrapyramidal syndromes, such as dystonic reactions and akathisia, are treated with diphenhydramine or benzodiazepines.

EPS, Extrapyramidal syndrome; *NMS*, malignant neuroleptic syndrome; *TCA*, tricyclic antidepressant.

[16,17]. Immunometric systems were also pursued to analyze some of these drugs, such as radioimmunoassay [18], where the creation of these early immunoassays required a great deal of work for laboratories as they had to construct radioactive conjugates and perform the immunizations in animals and harvest the antibodies. Then a rigorous qualification took place to understand antibody specificity and finally formulate the immunoassay followed by laboratory measurement procedure validation. Later, commercial diagnostic research companies took over these tasks and began to market immunoassay systems [19]. The adoption of high-performance liquid chromatography (HPLC) has been instrumental to the analysis of antipsychotics and antidepressants. This technique allows for compounds to be concentrated and separated in the solution phase. If the drug compound is soluble in solution, it can likely be analyzed by HPLC measurement procedures without the need for derivatization, which is required for many of the nonvolatile when analyzing using GC platforms. While the first few HPLC measurement procedures looked at a small number of drugs at a time, later efforts multiplexed several drug compounds during a single run [20–22]. Multiplexing of several compounds in a single measurement procedure allows laboratories to analyze many specimens using the same methodology, which can be beneficial for precious laboratory space and minimizes laboratory procedures. Multiplexing can also be very beneficial in cases of unknown overdose where many compounds and classes can be assessed during a single run. HPLC is also compatible several detection systems such as UV, electrochemical, and mass spectrometric detectors. The most common technology used currently by laboratories to analyze antipsychotics and antidepressants is liquid chromatography tandem mass spectrometry (LC–MS/MS). While techniques such as HPLC–UV provided excellent measurement procedures, any coeluting peaks would have to be resolved chromatographically. This often times extended measurement procedure run times to achieve such separation. Laboratories typically operate the LC–MS/MS systems

using quantitative experiments such as selective reaction monitoring (SRM) or multiple reaction monitoring. In these mass spectrometric experiments, the mass of a compound (precursor mass) can be isolated and then later fragmented (product ion). These product ions can then be isolated and detected by the mass spectrometer. These techniques offer selectivity such that if the compounds have a different mass and/or a different fragment mass, the mass spectrometer can resolve these compounds. This allows for isotopically labeled internal standards to be used during the analysis also. If compounds cannot be resolved by the mass spectrometer then they will need to be dealt with and separated in the chromatographic profile prior to entering the tandem mass spectrometer. It is because of these many advantages LC−MS/MS measurement procedures are able to perform multiplex analysis on many different drugs in a single sample. Quantitative determinations of 48 antidepressants and antipsychotics from a single human serum sample have been reported [23]. Many times the sample preparations for these LC−MS/MS measurement procedures are very simplistic incorporating protein precipitation, liquid extraction, or solid phase extraction. A review on sample preparation for various methods has been reported and outlines the approaches for atypical antipsychotics [24]. Most recently, immunoassays for the analysis of antipsychotic drugs have been approved by the FDA, with the promise of improving turnaround time for routine monitoring of antipsychotic medications.

Conclusion

Evolution of antidepressant and antipsychotic medications has given rise to diverging subclasses of medications with evolving clinical indications, efficacy, and unique side effect profiles. Since these medications differ in safety, presentation, and treatment in overdose, it is important to understand their mechanisms of action as related to efficacy in treatment, adverse reactions, symptoms of toxicity, drug−drug interactions, pharmacodynamics, and pharmacokinetics. Both classes of medications target neurotransmission in the CNS, altering brain chemistry. With clinical use of antidepressants and antipsychotic medications on the rise, toxic incidents involving these medications have also become common place. Direct drug testing has clinical utility in guiding care by identifying coingestions and when confirmation of overdose diagnosis is indicated. Diagnosis and management of overdose cases often focus on history, clinical evaluation, followed by appropriate laboratory testing in order to monitor and guide treatment strategies.

References

[1] O'Donnell JM, Bies RR, Shelton RC. Drug therapy of depression and anxiety disorders. In: Brunton L, Hilal-Dandan R, Knollmann BC, editors. Goodman & Gilman's: the pharmacological basis of therapeutics. 13th ed. New York: McGraw-Hill Education; 2018.

[2] Erickson MA, Penning TM. Drug toxicity and poisoning. In: Brunton L, Hilal-Dandan R, Knollmann BC, editors. Goodman & Gilman's: the pharmacological basis of therapeutics. 13th ed. New York: McGraw-Hill Education; 2018.

[3] Shehab N, Lovegrove MC, Geller AI, Rose KO, Weidle NJ, Budnitz DS. US Emergency Department visits for outpatient adverse drug events, 2013-2014. JAMA 2016;316(20):2115−25.

[4] Hampton LM, Daubresse M, Chang HY, Alexander GC, Budnitz DS. Emergency department visits by adults for psychiatric medication adverse events. JAMA Psychiatry 2014;71(9):1006−14.

[5] Hirsch M, Birnbaum RJ. Tricyclic and tetracyclic drugs: pharmacology, administration, and side effects. In: Post TW, editor. UpToDate. Waltham, MA: UpToDate; 2018.

[6] Hirsch M, Birnbaum RJ. Monoamine oxidase inhibitors (MAOIs) for treating depressed adults. In: Post TW, editor. UpToDate. Waltham, MA: UpToDate; 2018.

[7] Hirsch M, Birnbaum RJ. Selective serotonin reuptake inhibitors: pharmacology, administration, and side effects. In: Post TW, editor. UpToDate. Waltham, MA: UpToDate; 2018.

[8] Nelson C. Serotonin-norepinephrine reuptake inhibitors: pharmacology, administration, and side effects. In: Post TW, editor. UpToDate. Waltham, MA: UpToDate; 2018.

[9] Hirsch M, Birnbaum RJ. Serotonin modulators: pharmacology, administration, and side effects. In: Post TW, editor. UpToDate. Waltham, MA: UpToDate; 2018.

[10] Hirsch M, Birnbaum RJ. Atypical antidepressants: pharmacology, administration, and side effects. In: Post TW, editor. UpToDate. Waltham, MA: UpToDate; 2018.

[11] Meyer JM. Pharmacotherapy of psychosis and mania. In: Brunton L, Hilal-Dandan R, Knollmann BC, editors. Goodman & Gilman's: the pharmacological basis of therapeutics. 13th ed. New York: McGraw-Hill Education; 2018.

[12] Hampton LM, Daubresse M, Chang HY, Alexander GC, Budnitz DS. Emergency department visits by children and adolescents for antipsychotic drug adverse events. JAMA Psychiatry 2015;72(3):292−4.

[13] Salzman NP, Moran NC, Brodie BB. Identification and pharmacological properties of a major metabolite of chlorpromazine. Nature 1955;176 (4493):1122−3.

[14] Hetzel CA. Method for the estimation of phenothiazine derivatives in urine and blood. Clin Chem 1961;7:130−5.

[15] Hammer WM, Brodiet BB. Application of isotope derivative technique to assay of secondary amines: estimation of desipramine by acetylation with H3-acetic anhydride. J Pharmacol Exp Ther 1967;157(3):503—8.
[16] Braithwaite RA, Goulding R, Theano G, Bailey J, Coppen A. Plasma concentration of amitriptyline and clinical response. Lancet 1972;299(7764):1297—300.
[17] Jorgensen A. A gas chromatographic method for the determination of amitriptyline and nortriptyline in human serum. Acta Pharmacol Toxicol (Copenh) 1975;36(1):79—90.
[18] Brunswick DJ, Needelman B, Mendels J. Specific radioimmunoassay of amitriptyline and nortriptyline. Br J Clin Pharmacol 1979;7(4):343—8.
[19] Pankey S, Collins C, Jaklitsch A, et al. Quantitative homogeneous enzyme immunoassays for amitriptyline, nortriptyline, imipramine, and desipramine. Clin Chem 1986;32(5):768—72.
[20] Duverneuil C, de la Grandmaison GL, de Mazancourt P, Alvarez JC. A high-performance liquid chromatography method with photodiode-array UV detection for therapeutic drug monitoring of the nontricyclic antidepressant drugs. Ther Drug Monit 2003;25(5):565—73.
[21] Mercolini L, Bugamelli F, Kenndler E, Boncompagni G, Franchini L, Raggi MA. Simultaneous determination of the antipsychotic drugs levomepromazine and clozapine and their main metabolites in human plasma by a HPLC-UV method with solid-phase extraction. J Chromatogr B Anal Technol Biomed Life Sci 2007;846(1—2):273—80.
[22] Mercolini L, Grillo M, Bartoletti C, Boncompagni G, Raggi MA. Simultaneous analysis of classical neuroleptics, atypical antipsychotics and their metabolites in human plasma. Anal Bioanal Chem 2007;388(1):235—43.
[23] Kirchherr H, Kuhn-Velten WN. Quantitative determination of forty-eight antidepressants and antipsychotics in human serum by HPLC tandem mass spectrometry: a multi-level, single-sample approach. J Chromatogr B Anal Technol Biomed Life Sci 2006;843(1):100—13.
[24] Fragou D, Dotsika S, Sarafidou P, Samanidou V, Njau S, Kovatsi L. Atypical antipsychotics: trends in analysis and sample preparation of various biological samples. Bioanalysis 2012;4(8):961—80.

Chapter 8.1

Olanzapine toxicity in an infant

Theresa Swift[1] and Hemamalini Ketha[2]

[1]Department of Pathology, University of Michigan Health System, University Hospital, Ann Arbor, MI, United States, [2]Clinical Mass Spectrometry, Toxicology, and Metals Testing, Center for Esoteric Testing, Laboratory Corporation of America Holdings, Burlington, NC, United States

Case description

An 18-month-old baby boy was brought into the emergency department for evaluation of altered mental status. The child had no past significant medical history and was up-to-date on vaccinations. He did not attend day care, had no sick contacts and no recent travel. He lived with his parents and siblings. The patient's mother stated that he had been acting normally in the morning and early afternoon. He was put down for a nap in the middle of the afternoon. When his mother went to wake him before dinner, she had a very difficult time waking the baby. He had episodes of crying, he woke up and then fall immediately back asleep. When the patient was awake, he was not interactive with his mother. Emergency medical service was called and the patient was taken to the emergency room. In the emergency room the patient exhibited the same symptoms; he continued to have episodes of crying without being fully awake, then sleeping. There, he received a 240 mL bolus of normal saline. The initial set of labs included a complete blood count, basic metabolic panel, urinalysis, and an immunoassay urine drug screen and they were all unremarkable. A head CT was negative for any abnormalities. A chest X-ray did not show any abnormalities or evidence of a foreign body. An abdominal ultrasound was negative for intussusception.

The patient appeared well developed and nourished. He was intermittently crying and sleeping. He had no signs of injury. His ears, nose, and throat appeared normal. He had normal conjunctivae but his pupils were 2 mm and sluggish bilaterally. His neck had a normal range of motion, and no adenopathy. He had a normal heart rhythm, normal breath sounds, normal bowel sounds, no abdominal distension nor hepatosplenomegaly. The patient would cry and fall asleep intermittently throughout the physical exam. The patient's mother did not find any open pill bottles in the home, and any medications in the home were stored out of the patient's reach. The only medications reported in the home were acetaminophen, aspirin, ibuprofen, and olanzapine.

Immunoassay-based urine drug screen (Roche Integra) was negative for all the drug classes tested amphetamine, barbiturates, benzodiazepines, cannabinoids, cocaine metabolite, methadone, opiates, oxycodone, phencyclidine, and propoxyphene. Qualitative urine drug confirmation by gas chromatography–mass spectrometry (GC–MS) showed a very large peak for olanzapine. Some other laboratory results are shown in Table 8.1.1.

The patient's mother had a prescription for the medication. She believed that one of the pills may have fallen on the floor when she was taking out her dose and the child consumed it. The dose of olanzapine was not disclosed. Poison control was contacted and stated that no further evaluation was needed and that he only needed supportive care. The patient's vital signs were monitored overnight and the patient slept until the following afternoon. When he woke up, he was able to eat, drink, and play without difficulty. The patient was discharged home with his parents.

Olanzapine (Zyprexa) is an atypical antipsychotic in the thienobenzodiazepine class that is a serotonin and dopamine receptor antagonist. It was first approved by the US Food and Drug Administration in 1996. Other atypical commonly used antipsychotics include aripiprazole, quetiapine, risperidone, and ziprasidone. Olanzapine is approved for use in adults and children over the age of 12 in the treatment of acute psychotic disorders such as schizophrenia and bipolar disorder. The most common side effects of olanzapine use are drowsiness and dizziness [1].

In cases of accidental exposure to olanzapine in children under 6 years old, the most common side effects reported were mental status changes, anticholinergic symptoms, and respiratory distress [2]. There are no specific antidotes for olanzapine overdose or any of the other antipsychotic medications [2]. If administered quickly activated charcoal can

TABLE 8.1.1 Selected Laboratory Results for the Baby at the Time of Presentation With Olanzapine Toxicity.

Salicylate	<3 mg/dL
Acetaminophen	<10 µg/mL
Ethanol	None detected
Sodium	141 mmol/L
Potassium	4.5 mmol/L
CO_2	22 mmol/L
Urea nitrogen	15 mg/dL
Creatinine	0.12 mg/dL
Glucose	89 mg/dL
Calcium	9.8 mg/dL
Venous pH	7.34
pCO_2	40 mmHg
pO_2	69 mmHg
Ionized calcium	1.34 mmol/L
Total hemoglobin	12.0 g/dL

decrease the amount of drug absorbed. Monitoring and providing supportive care is necessary after exposure due to the potential for adverse cardiac, respiratory, and anticholinergic effects. The most common clinical symptoms seen in adults who have taken olanzapine in an overdose are similar to what is seen in children: lethargy, tachycardia, anticholinergic syndrome, confusion, and agitation [3]. Children and adults who are not prescribed olanzapine or other atypical antipsychotic medication are going to be more sensitive to the toxic effects of the medications.

Very few laboratory tests can definitively confirm exposure to olanzapine or other atypical antipsychotics. Atypical antipsychotics are not typically screened for in a clinical toxicology laboratory. They are not detected by immunoassay-based drug screens which typically screen for amphetamines, barbiturates, benzodiazepines, cocaine, opiates, oxycodone, methadone, and phencyclidine. Atypical antipsychotic can be detected by mass spectrometric methods, including GC–MS and time-of-flight-MS, but all hospital laboratories may not have these types of assays for toxicological screening available. In most cases of toxicity the sample is sent out to a referral laboratory and the results may take a few days to be received.

In almost 10% of olanzapine-treated patients, elevated liver enzymes were seen. There is no evidence of olanzapine affecting the bone marrow, compared to the toxicity that can be seen with clozapine. Olanzapine may produce elevated serum prolactin levels, since it antagonizes the 5-HT2 and dopamine D2 receptors. Patients can show elevated prolactin levels within the first few weeks of treatment then the elevation tapered off [1].

In another published case of an 18-month-old male who ingested 30–40 mg of olanzapine similar side effects to what our patient experienced were observed such as drowsiness and agitation. Additional side effects included sinus tachycardia, dry mucous membranes, and hypoactive bowel sounds and he required intubation due to respiratory distress [4]. It was known that the patient was exposed to olanzapine because a prescription bottle was found empty near the patient, and pill fragments were found in the patient's mouth. The patient was given syrup of ipecac in an attempt to induce vomiting. At the hospital the patient was given activated charcoal and a gastric lavage was also positive for pill fragments.

In another case of olanzapine ingestion by a 17-month-old female who was found with an open bottle of olanzapine, she became lethargic and drowsy and was taken to the hospital. Olanzapine was confirmed by measuring serum drug levels. The olanzapine level in the patient was 137 ng/mL which is above the therapeutic range for adults (10–78 ng/mL) [5]. This case estimated the elimination half-life of olanzapine to be 13.7 hours, the reported half-life of olanzapine for toddlers is 11.6 hours [5]. The estimate half-life of olanzapine in adults is about 21–54 hours. Higher clearance of olanzapine is seen in younger children and when combined with any preventative effects seen with activated charcoal might be protective factor in olanzapine exposure.

Exposure or an overdose of olanzapine which is also a sedative hypnotic may exhibit similar symptoms to opiates overdose. Symptoms can include central nervous system depression, pinpoint pupils, and hypotension, but they will not respond to naloxone and screen negative for opiates [6]. In another case a 12-year old who took 210 mg of olanzapine on a dare was presented to an emergency room with opioid-like presentation. He had pinpoint pupils, decreased responsiveness and he was lethargic. The patient was combative at first but then he was very drowsy and difficult to arouse. After 36 hours the patient was fully awake and had normal mentation. At a 3-month follow-up the patient had no long-term effects from the overdose [7].

A postmarket retrospective study surveyed the US National Poison System Data of exposure to atypical antipsychotic of children under 6 years old. From 2005 to 2013, there were 23,653 exposures in children. Of those, 6718 were not followed in a health-care facility. Of the 16,935 exposures who were followed in a health-care facility, 1735 were from olanzapine, 5018 aripiprazole, 3904 quetiapine, 4778 risperidone, and 1500 ziprasidone. The average age of the patient was 2.3 years but ages ranged from 1week old to 5 years old. The majority of these exposures were accidental. Almost half of these accidental exposures experienced side effects. The most common observed side effect was drowsiness across all of the atypical antipsychotics, including olanzapine. Following drowsiness, the other common side effects were agitation and tachycardia. With olanzapine tachycardia was reported more commonly than the other atypical antipsychotics [6].

References

[1] Beasley Jr. CM, Tollefson GD, Tran PV. Safety of olanzapine. J Clin Psychiatry 1997;58(Suppl. 10):13–17.
[2] Theisen FM, et al. Olanzapine overdose in children and adolescents: two case reports and a review of the literature. J Child Adolesc Psychopharmacol 2005;15:986–95. Available from: https://doi.org/10.1089/cap.2005.15.986.
[3] Burns MJ. The pharmacology and toxicology of atypical antipsychotic agents. J Toxicol Clin Toxicol 2001;39:1–14.
[4] Catalano G, Cooper DS, Catalano MC, Butera AS. Olanzapine overdose in an 18-month-old child. J Child Adolesc Psychopharmacol 1999;9:267–71. Available from: https://doi.org/10.1089/cap.1999.9.267.
[5] Tanoshima R, Chandranipapongse W, Colantonio D, Stefan C, Nulman I. Acute olanzapine overdose in a toddler: a case report. Ther Drug Monit 2013;35:557–9. Available from: https://doi.org/10.1097/FTD.0b013e3182953ed8.
[6] Stassinos G, Klein-Schwartz W. Comparison of pediatric atypical antipsychotic exposures reported to U.S. poison centers. Clin Toxicol (Philadelphia, Pa) 2017;55:40–5. Available from: https://doi.org/10.1080/15563650.2016.1233342.
[7] Kochhar S, et al. Olanzapine overdose: a pediatric case report. J Child Adolesc Psychopharmacol 2002;12:351–3. Available from: https://doi.org/10.1089/104454602762599907.

Chapter 8.2

Genetic polymorphism leading to drug overdose fatality while in custody

Milad Webb[1,3] and Jeffrey Jentzen[2,3]
[1]Wayne County Medical Examiner's Office, Detroit, MI, United States, [2]Washtenaw County Medical Examiner's Office, Ann Arbor, MI, United States, [3]Michigan Medicine, Department of Pathology, Ann Arbor, MI, United States

Case description

The decedent was a 57-year-old white male inmate in a psychiatric facility. The decedent had a medical history significant for schizophrenia with auditory command hallucinations, traumatic brain injury, bipolar disorder, depression, hypertension, hyperlipidemia, hypothyroidism, diabetes mellitus, alcoholism, depression, and polydipsia. He was last known to be alive and in his usual state of health approximately 30 minutes prior to being found unresponsive in a secured room (verified by video surveillance). Emergency response personnel at the facility initiated cardiopulmonary resuscitation; however, without success. The decedent was rapidly transported to the local hospital emergency department in asystole. He was pronounced dead shortly after arrival. There was no evidence of foul play.

External examination

On initial examination of the body in the hospital emergency department, it was noted that the decedent had a thin linear—patterned abrasion under the chin. The core body temperature was 96.2°F, and the body was not in rigor. There was posterior lividity present which blanched easily. There were no petechiae of the sclerae or conjunctiva. There were two small abrasions and a laceration present on the posterior left hand. There were small clusters of contusions present on the knees, bilaterally. The decedent pants were wet with urine. There were multiple well-healed self-inflicted cutting scars present on both arms and the abdomen.

Clinical history

Review of the decedent's medical records revealed that he had attempted suicide in the past. In addition, the decedent was hospitalized multiple times due to water intoxication—causing the medical staff to monitor his weight daily. On the day of his death, he was at his base weight of 201 lb. His routine medications included trazodone (50 mg), olanzapine (5 mg), lisinopril (40 mg), levothyroxine (7 μg), docusate sodium (100 mg), simvastatin (20 mg), tramadol HCL (50 mg), and ibuprofen (400 mg). Other medications included clozapine, haloperidol, tramadol, duloxetine, and bupropion.

Postmortem examination

Postmortem examination revealed a recent fracture of the fifth and sixth cervical vertebral bodies with soft tissue hemorrhage in a region of the spine with prominent dystrophic calcification and osteophytic hypertrophy. There were remote cerebral contusions on the right temporal lobe identified and confirmed with microscopy to show hemosiderin laden macrophages and gliosis, consistent with a history of traumatic brain injury. Notably, there was no evidence of ischemic encephalopathy in the hippocampus or cerebellum. The remainder of the internal examination was largely unremarkable (Table 8.2.1). Reinvestigation of the scene revealed that the decedent was found unresponsive in the prone position with legs on the ground, and chin propped on the edge of his bed frame and his neck in a hyperextended position (Fig. 8.2.1), consistent with a possible fall and fracture of the cervical spine.

TABLE 8.2.1 The Postmortem.

Postmortem Examination: Salient Findings
External Findings Summary
Rectal temperature: 96.2°F
Compression abrasion under chin (8 cm × 0.3 cm)
No evidence of significant injury or trauma
Internal Findings Summary
Cervical spine osteoarthritis with C5–C6 fracture
Hepatomegaly
Minimal atherosclerotic and hypertensive changes
Cut pieces of playing cards found in the mouth and stomach

Findings were limited to a cervical spine fracture that may have been a result of a terminal fall. Other findings were unlikely to be contributory to death.

FIGURE 8.2.1 The decedent found unresponsive. This image is a reenactment of the position the decedent was found in, performed by the security staff who discovered him. The scene is otherwise exactly as it was found.

Toxicology

Postmortem toxicology performed on the decedent's iliac blood (Table 8.2.2) revealed significant concentrations of clozapine and norclozapine (a major active metabolite), olanzapine, and tramadol. Of note, the toxicology report indicated that the concentration of olanzapine metabolite, *O*-desmethylolanzapine was much lower than expected (3% of the concentration of olanzapine). Further examination was warranted since the decedent (an inmate) was not in control of his medications, which called into question the manner of death.

Clozapine (Clozaril) is a tetracyclic antipsychotic which is indicated for use in the treatment of schizophrenia in patients who have failed first-line therapies [1]. This drug is metabolized into two major products in the liver, norclozapine and clozapine-*N*-oxide. Both major metabolites are believed to have significant pharmacologic activity. Clozapine is known for having volatile blood concentrations from week to week in the same individual taking a consistent dose [2,3]. In a report, 25 patients on chronic clozapine therapy (200 mg daily), the mean steady-state plasma concentration was 230 ng/mL, and 190 ng/mL for norclozapine [1,4]. In patients given 400 mg for 5 weeks, the mean steady-state plasma concentration of clozapine was 470–650 ng/mL. Clozapine toxicity can cause hypotension, dysrhythmia, respiratory depression, coma, and death. In nine reported fatalities from clozapine, postmortem blood concentrations were reported in the range of 1200–13,000 ng/mL [1].

The decedent had a clozapine concentration of 1100 ng/mL, with a norclozapine concentration of 680 ng/mL. Although the activity coefficient of norclozapine relative to clozapine has not been described, it is believed to have significant contribution to the therapeutic potential and hence, toxicity. The combined effect of clozapine and its metabolite was assumed to be well above the reported fatal threshold of 1200 ng/mL.

TABLE 8.2.2 Toxicology Performed on the Decedent's Peripheral Blood Revealed Elevated Levels of Clozapine, Olanzapine, and Tramadol.

Toxicology Findings			
Compound	Concentration	Units	Method
Caffeine	Positive		LC/TOF-MS
Bupropion	530	ng/mL	LC–MS/MS
Hydroxybupropion	610	ng/mL	LC–MS/MS
Clozapine	1100	ng/mL	LC–MS/MS
Norclozapine	680	ng/mL	LC–MS/MS
Duloxetine	72	ng/mL	LC–MS/MS
Haloperidol	56	ng/mL	LC–MS/MS
Olanzapine	810	ng/mL	GC
Tramadol	3000	ng/mL	LC–MS/MS
O-desmethyltramadol	100	ng/mL	LC–MS/MS
mCPP	61	ng/mL	GC/MS
Trazodone	370	ng/mL	GC

Findings also found limited O-desmethyltramadol, a major metabolic product of tramadol.

Olanzapine (Zyprexa) is a therapeutic used in the treatment of psychotic disorders such as schizophrenia and bipolar manic disorders. It is a tetracyclic compound with structural resemblance to clozapine [1]. Plasma concentrations to monitor therapeutic ranges have not been established [5]. Trough plasma concentrations in patients receiving 10, 15, and 20 mg of olanzapine daily were reported to average 9, 19, and 26 ng/mL, respectively.

Toxicity of olanzapine presents as disturbances in body temperature, cardiovascular disturbances, altered mental status, and tardive dyskinesia [1,5,6]. Fatalities due to olanzapine toxicity have been reported with concentrations as low as 1000 ng/mL [1]. The decedent had an olanzapine concentration of 810 ng/mL, significantly outside the range of expected plasma concentrations for the dosage.

Tramadol

Tramadol (Ultram) is a synthetic opioid receptor agonist which is used for the management of moderate to moderately severe pain [7]. Peak plasma levels of tramadol following a single 100 mg oral dose range from 230 to 380 ng/mL and peak levels of the active metabolite, O-desmethyltramadol, range from 35 to 75 ng/mL. Steady-state plasma levels following an oral dosage regimen of 100 mg of tramadol administered four times a day range from 420 to 770 ng/mL [1]. Toxicity of tramadol may cause agitation, tachycardia, hypertension, and seizures. The mean postmortem femoral blood concentration of tramadol in five reported overdose deaths was 6100 ng/mL [1]. The decedent had a tramadol concentration of 3000 ng/mL and an O-desmethyltramadol concentration of 100 ng/mL; greater than six times the expected plasma concentration.

Genetic testing

Genetic testing was requested and obtained to characterize and document a possible genetic mutation in the drug metabolism pathway, specifically interrogating CYP2D6.

Cytochrome P450 2D6 is an enzyme (coded by CYP2D6 gene) that is primarily expressed in hepatocyte within the liver and in selected regions of the central nervous system [8,9]. As a member of the P450 mixed-function oxidase system, CYP2D6 is one of the primary pathways utilized for metabolism and elimination of pharmaceutical and endogenous compounds via hydroxylation, demethylation, and dealkylation. In fact, although 2D6 accounts for ~3% of the total CYP content of the liver, it metabolizes as much as 25% of all pharmaceuticals, including antidepressant, antiarrhythmics, beta-blockers, opioid analgesics, and many others [7,8].

TABLE 8.2.3 2D6 Variants.

Common Variants of CYP2D6	
Gene	Protein Activity
CYP2D6*1	Most common form (wild type), considered fully functional
CYP2D6*2	Normal function, except for 2XN subtype
CYP2D6*3	Nonfunctional variant
CYP2D6*4	Nonfunctional variant, most common variant
CYP2D6*5	Nonfunctional variant (complete deletion)
CYP2D6*6	Nonfunctional variant
CYP2D6*9	Partially functional variant
CYP2D6*10	Partially functional variant
CYP2D6*17	Partially functional variant

Genome sequencing of the decedent's CYP2D6 found him to be heterozygous with a CYP2D6*4 variant.

Although drug inhibition is of consideration, the predominant factor in the determination of CYP2D6 activity is genetics. There are a considerable number of variants of CYP2D6 (Table 8.2.3) which exist in the population (approximately 100 documented alleles) and these variants have considerable differences in efficacy of the resultant protein, manifesting as ultrarapid metabolizers versus poor metabolizers [8–10]. Therefore dosing of certain medications in individuals may become critically important to balance therapeutic effects and potentially lethal outcomes.

Genetic analysis performed on the decedent's liver tissue identified a CYP2D6*4 allele, indicating a complete nonfunctional allele, and therefore, heterozygous status. In addition, the decedent was also positive for bupropion, duloxetine, and haloperidol, which are all well-known inhibitors of CYP2D6; of which, bupropion is known to be a strong inhibitor. This combination of CYP2D6 heterozygosity and inhibition of the remaining functional protein ultimately had fatal results.

In consideration of the investigation, postmortem examination, postmortem toxicology, and genetic testing, the death was certified as mixed drug toxicity (clozapine, olanzapine, and tramadol) and the manner of death as natural. This toxicity was due to an underlying genetic condition (CYP2D6 heterozygosity) as well as accidental pharmaceutical inhibition of the remaining effective enzyme. This toxicity resulted in sudden cardiac death and terminal collapse causing fracture of the C5–C6 vertebral bodies on the edge of the decedent's bed.

References

[1] Baselt RC, Randall C. In: Baselt Randall C, editor. Disposition of toxic drugs and chemicals in man. 11th ed. Foster City, CA: Biomedical Publications; 2017.
[2] Stark A, Scott J. A review of the use of clozapine levels to guide treatment and determine cause of death. Aust NZ J Psychiatry 2012;46(9):816–25.
[3] Liu HC, Chang WH, Wei FC, Lin SK, Lin SK, Jann MW. Monitoring of plasma clozapine levels and its metabolites in refractory schizophrenic patients. Ther Drug Monit 1996;18(2):200–7.
[4] Volpicelli SA, Centorrino F, Puopolo PR, Kando J, Frankenburg FR, et al. Determination of clozapine, norclozapine, and clozapine-N-oxide in serum by liquid chromatography. Clin Chem 1993;39(8):1656–9.
[5] Aravagiri M, Ames D, Wirshing WC, Marder SR. Plasma level monitoring of olanzapine in patients with schizophrenia: determination by high-performance liquid chromatography with electrochemical detection. Ther Drug Monit 1997;19(3):307–13.
[6] Chue P, Singer P. A review of olanzapine-associated toxicity and fatality in overdose. J Psychiatry Neurosci 2003;28(4):253–61.
[7] Miotto K, Cho AK, Khalil MA, Blanco K, Sasaki JD, Rawson R. Trends in tramadol: pharmacology, metabolism, and misuse. Anesth Analg 2017;124(1):44–51.
[8] Zhou S-F. Polymorphism of human cytochrome P450 2D6 and its clinical significance: Part I. Clin Pharmacokinet 2009;48(11):689–723.
[9] Yang Y, Botton MR, Scott ER, Scott SA. Sequencing the CYP2D6 gene: from variant allele discovery to clinical pharmacogenetic testing. Pharmacogenomics 2017;18(7):673–85.
[10] Gaedigk A. Complexities of CYP2D6 gene analysis and interpretation. Int Rev Psychiatry 2013;25(5):534–53.

Chapter 8.3

A death involving flubromazepam and methadone

Uttam Garg[1,2], Robert Krumsick[1,2], C. Clinton Frazee, III[1,2], Robert Pietak[3] and Diane C. Peterson[3]
[1]Department of Pathology and Laboratory Medicine, Children's Mercy Hospitals and Clinics, Kansas City, MO, United States, [2]University of Missouri School of Medicine, Kansas City, MO, United States, [3]Office of the Jackson County Medical Examiner, Kansas City, MO, United States

Case description

A 62-year-old male with a history of heart disease, hypertension, diabetes, stroke, sinus bradycardia, depression, anxiety, drug abuse, and hepatitis C was found by his wife at the base of his home's stairs. The subject was last seen by his wife prior to retiring for the night and was assumed to be well at that time. When found, the subject was unconscious and taken to a nearby hospital. Shortly after arrival at the hospital, the subject was pronounced dead. External examination revealed bruising of the soft tissue of the right upper orbit, blood coming from the right ear, and a palpable subcutaneous defect on the right side of the head. Radiographs of the body revealed multiple cranial fractures involving the right temporal and frontal bones as well as the bilateral parietal bones.

The subject's body was transported to Jackson County Medical Examiner's Office. Less than 8 hours after death, femoral blood and vitreous samples were collected and sent to the laboratory for testing. Postmortem femoral blood was screened for volatiles (ethanol, methanol, isopropanol, and acetone) by headspace gas chromatography with flame ionization detector. Femoral blood was also used for comprehensive broad spectrum drug-screening that involved enzyme immunoassays (EIAs) for amphetamines, barbiturates, benzodiazepines, cannabinoids, cocaine metabolite, methadone, opiates, phencyclidine (PCP) and propoxyphene, and drug-screening for >200 drugs by gas chromatography mass spectrometry (GC–MS). Drug screening by GC–MS involved liquid–liquid alkaline extraction using bicarbonate buffer (pH 11.0) and butyl acetate, and mass spectrometer operation in full scan mode. Presumptive identification of analytes was made by spectral library match and relative retention time comparison with reference standards.

No volatiles were detected in the sample. EIA screen was positive for benzodiazepines and methadone. GC–MS showed the presumptive presence of methadone, methadone metabolite 2-ethylidene-1,5-dimethyl-3,3-diphenylpyrrolidine (EDDP), and flubromazepam. Methadone and EDDP levels in femoral blood were 380 and 30 ng/mL, respectively (LC–MS/MS). At the time of analysis, we could not locate a reference laboratory which offered quantification of flubromazepam. Due to lack of availability of testing, the presence of flubromazepam was confirmed using LC–QTOF–MS. Semiquantification of flubromazepam was achieved using reference standards spiked at 100, 500, and 1000 ng/mL into drug-free hemolyzed blood. Both GC–MS and LC–QTOF–MS were used for semiquantification of flubromazepam. The flubromazepam concentration was measured in femoral blood and determined to be 500–1000 ng/mL. While doing confirmation for benzodiazepines, 7-amino clonazepam was found at a subtherapeutic level of 20 ng/mL. Subsequent to these findings, designer benzodiazepines, clonazolam, diclazepam, etizolam, flubromazepam, flubromazolam, phenazepam, and pyrazolam were purchased from Cayman Chemicals and analyzed by EIA and GC/MS. It was determined that each of these designer benzodiazepines cross-reacted at elevated or positive levels as low as 50 ng/mL with our Siemens Diagnostics EMIT II Plus Benzodiazepines Assay with cutoff of 200 ng/mL lormetazepam (Tarrytown, NY). GC/MS analysis detected each of these benzodiazepines at relatively higher levels (Table 8.3.1).

TABLE 8.3.1 Cross-Reactivity as Change in Absorbance (dAb) for Various Designer Benzodiazepines in Siemens Diagnostics Benzodiazepine Assay.

Designer Benzodiazepine	50 ng/mL	100 ng/mL	300 ng/mL	500 ng/mL	1000 ng/mL	5000 ng/mL	Cutoff 200 ng/mL	Negative Urine	GC/MS Limit of Detection (ng/mL)
Clonazolam	0.348	0.368	0.380	0.400	0.424	0.492	0.391	0.307	1000
Diclazepam	0.365	0.387	0.479	0.507	0.532	0.579	0.391	0.307	60
Etizolam	0.328	0.344	0.378	0.391	0.409	0.463	0.386	0.302	60
Flubromazepam	0.353	0.381	0.468	0.501	0.536	0.587	0.408	0.307	100
Flubromazolam	0.336	0.363	0.469	0.509	0.543	0.588	0.391	0.307	100
Phenazepam	0.329	0.353	0.420	0.449	0.481	0.535	0.382	0.306	40
Pyrazolam	0.328	0.341	0.373	0.389	0.415	0.476	0.391	0.307	1000

GC/MS limits of detection are also listed.

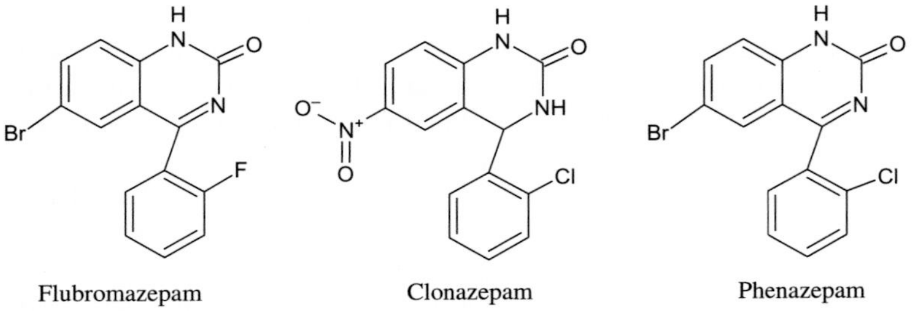

FIGURE 8.3.1 Structures of flubromazepam, clonazepam, and phenazepam.

Discussion

Recreational use of designer benzodiazepines has increased substantially since first hitting the market in 2012 [1,2]. A recent publication [3] reported an increase in websites selling designer benzodiazepines: diclazepam (49 in 2014 to 55 in 2016), pyrazolam (33−35), and flubromazepam (39−45). Unlike most benzodiazepines which usually have a relatively low acute toxicity, many of the designer benzodiazepines could pose a risk of severe life-threatening intoxications [2−4]. The first of these designer benzodiazepines to appear on the market was pyrazolam. Flubromazepam [7-bromo-5-(2-fluorophenyl)-1,3-dihydro-2H-1,4-benzodiazepin-2-one] soon followed and began to appear in online stores in the late 2012 [5,6]. Structurally, it is closely related to clonazepam and phenazepam (Fig. 8.3.1). Currently, flubromazepam has no medicinal use and is exclusively manufactured for illicit use [3]. Like other benzodiazepines, flubromazepam elicits sedation, impaired coordination, respiratory depression, muscle relaxation, and CNS depression [6]. One of the dangerous characteristics of flubromazepam is its extremely long elimination half-life of 106 hours [6]. The long elimination half-life can lead to accumulation of toxic concentrations of the drug after repeated consumption. Currently, there is no well-defined literature on the toxic levels of flubromazepam. However, a study conducted on 24 criminal offenders who were positive for flubromazepam showed the drug levels ranging from 4.7 to 1200 ng/mL [7]. One of the subjects with a flubromazepam level of 600 ng/mL and no other drugs detected was clinically impaired. It is not known if the subjects in this study were naïve users of the drug. However, the combination of designer benzodiazepines with other CNS depressants can potentially produce synergistically fatal results [1,8,9].

The cause of death in our case was traumatic head injury. The manner of death was accident. Based on the semiquantitative analysis performed, a flubromazepam level between 500 and 1000 ng/mL was present in the decedent's femoral blood at the time of death. Although semiquantitative, this level seems substantially higher than reported levels in the literature [6−8]. In one study a researcher who ingested a 4 mg capsule of flubromazepam reached a peak level of 78 ng/mL [6], and in another study involving 24 criminal offenders, a median peak level of 55 ng/mL was determined [7]. Given the aforementioned studies, it seems reasonable to assume that the decedent in our case study experienced flubromazepam toxicity which could have caused him to fall down the stairs. While the levels of methadone and flubromazepam found on their own are not fatal in a nonnaïve subject, the combination of these two CNS depressants can potentially lead to serious, life-threatening toxicity. As a result of its partial agonist effect, methadone can decrease respiration through agonist action at the mu receptors in the medullar respiratory center. Benzodiazepines, including the designer benzodiazepines, act synergistically by facilitating inhibition at the GABA receptors and, thus, can lead to respiratory arrest, cardiac arrest, coma, and/or death when administered in high concentrations. Because potential users of flubromazepam may not have experience with the drug and/or the duration of its effects on CNS, users may incorrectly assume that the drug has been cleared from their system much sooner than is actually true. In our case, flubromazepam and methadone, both CNS depressants, were found in toxic concentrations.

In conclusion, this case highlights the growing abuse potential of synthetic benzodiazepines. As novel designer benzodiazepines continue to be introduced into the illicit drug market, it is imperative that toxicology laboratories consider appropriate designer benzodiazepines testing when positive immunoassay benzodiazepines responses are detected or indicated by the case history.

References

[1] Moosmann B, King LA, Auwarter V. Designer benzodiazepines: a new challenge. World Psychiatry 2015;14:248.
[2] Moosmann B, Auwarter V. Designer benzodiazepines: another class of new psychoactive substances. Handb Exp Pharmacol 2018;252:383–410.
[3] Abouchedid R, Gilks T, Dargan PI, Archer JRH, Wood DM. Assessment of the availability, cost, and motivations for use over time of the new psychoactive substances—benzodiazepines diclazepam, flubromazepam, and pyrazolam—in the UK. J Med Toxicol 2018;14:134–43.
[4] Lukasik-Glebocka M, Sommerfeld K, Tezyk A, Zielinska-Psuja B, Panienski P, Zaba C. Flubromazolam—a new life-threatening designer benzodiazepine. Clin Toxicol (Phila) 2016;54:66–8.
[5] O'Connor LC, Torrance HJ, McKeown DA. ELISA detection of phenazepam, etizolam, pyrazolam, flubromazepam, diclazepam and delorazepam in Blood Using Immunalysis(R) Benzodiazepine Kit. J Anal Toxicol 2016;40:159–61.
[6] Moosmann B, Huppertz LM, Hutter M, Buchwald A, Ferlaino S, Auwarter V. Detection and identification of the designer benzodiazepine flubromazepam and preliminary data on its metabolism and pharmacokinetics. J Mass Spectrom 2013;48:1150–9.
[7] Hoiseth G, Tuv SS, Karinen R. Blood concentrations of new designer benzodiazepines in forensic cases. Forensic Sci Int 2016;268:35–8.
[8] Partridge E, Trobbiani S, Stockham P, Charlwood C, Kostakis C. A case study involving U-47700, diclazepam and flubromazepam-application of retrospective analysis of HRMS data. J Anal Toxicol 2018;42:655–60.
[9] Koch K, Auwarter V, Hermanns-Clausen M, Wilde M, Neukamm MA. Mixed intoxication by the synthetic opioid U-47700 and the benzodiazepine flubromazepam with lethal outcome: pharmacokinetic data. Drug Test Anal 2018;10(8):1336–41.

Chapter 9

Anticonvulsants

Angela M. Ferguson[1,2,3,4]

[1]*Clinical Chemistry, Children's Mercy, Kansas City, MO, United States,* [2]*Allergy and Immunology and Point of Care Testing, Children's Mercy, Kansas City, MO, United States,* [3]*Department of Pathology and Laboratory Medicine, Children's Mercy Hospitals and Clinics, Kansas City, MO, United States,* [4]*University of Missouri School of Medicine, Kansas City, MO, United States*

Introduction

Seizures are the clinical manifestation of epilepsy, a common neurological disorder. Seizures can be divided into two categories, generalized and partial, with generalized involving both cerebral hemispheres at the same time and partial focusing on a specific region of the brain. Anticonvulsants, also termed antiepileptic drugs (AEDs), are primarily used to control and prevent seizures. However, this class of drugs can also be used for treating neuropathic pain, used as sedatives, for treating migraines, or managing addictions [1]. Anticonvulsants can be grouped into three main groups, or generations, based on their development and introduction into clinical use (Table 9.1). First-generation drugs, also termed classical anticonvulsants, including phenytoin, ethosuximide, carbamazepine, primidone, valproic acid, clonazepam, diazepam, and phenobarbital, were developed in the early 1900s through the 1970s and were the first-line therapy until the 1990s. The first-generation AEDs had many disadvantages, including zero-order kinetics for some, pronounced toxic effects, and high protein binding. The second-generation drugs such as gabapentin, felbamate, lamotrigine, oxcarbazepine, tiagabine, topiramate, pregabalin, zonisamide, and levetiracetam were introduced in the 1990s [2,3]. The main goal of developing the second-generation drugs was to move beyond the disadvantages of the first-generation drugs with newer drugs that were both highly effective and well tolerated [3]. The third-generation drugs were approved starting in 2008 and include lacosamide, eslicarbazepine acetate, rufinamide, brivaracetam, perampanel, vigabatrin, and clobazam.

TABLE 9.1 Anticonvulsants by Generation

First Generation	Second Generation	Third Generation
Carbamazepine	Felbamate	Brivaracetam
Clonazepam	Gabapentin	Clobazam
Diazepam	Lamotrigine	Eslicarbazepine acetate
Ethosuximide	Levetiracetam	Lacosamide
Phenobarbital	Oxcarbazepine	Perampanel
Phenytoin	Pregabalin	Vigabatrin
Primidone	Rufinamide	
Valproic acid	Tiagabine	
	Topiramate	
	Zonisamide	

Anticonvulsants were one of the first family of drugs that qualified for therapeutic drug monitoring (TDM). This is due to the fact that both an overdose or an underdose can result in seizure activity, and most of the drugs have therapeutic ranges that are quite narrow [2]. Phenytoin and phenobarbital, two of the earliest anticonvulsants, have clinical and toxic effects that could be correlated to serum concentration in patients [2]. TDM also allowed for monitoring of patient compliance with these medications and the effect of polypharmacy [2,4]. In spite of the fact that TDM is quite common, there is not a clear correlation between the dose given to a patient and the clinical response of the patient for many anticonvulsants due to varied individual response to the medication [4]. It is very likely that a patient could have clinical resolution of their seizure activity on a dose that is below the reference range of a drug or a toxic reaction to a dose that is not outside of the reference range, and most clinicians are advised to treat the patient, not the level of drug. One of the challenges to TDM of anticonvulsants is making an assessment of the clinical impact of a drug dosage, which is difficult due to the irregular nature of seizure activity [4]. In addition, the adverse events that can be seen from these drugs are very similar to the patient's underlying condition. There are also no simple procedures or laboratory tests that can determine the clinical efficacy of the drug and drug dose [4].

The first-generation anticonvulsants are more heavily monitored than later generations of drugs due to their noticeable side effects and the readily available commercial automated immunoassays. Most of the second- and third-generation drugs are monitored via chromatographic methods involving mass spectrometry.

Mechanism of action

With a group of drugs this large and diverse, it is not unexpected that different drugs act in different pathways. Those that have biochemical processes that have been well described will be discussed. The mechanisms of action of anticonvulsants can be grouped into four main families. These include acting on gamma-aminobutyric acid (GABA) and its receptors, affecting ion channels, attenuating excitatory neurotransmission, and a group containing several novel mechanisms [5]. The benzodiazepines clonazepam and diazepam act by reducing neuronal excitation through agonist activity at the GABA receptor via increasing the duration of chloride flow into the synapse [2]. GABA is an inhibitor of pre- and postsynaptic discharges. Tolerance to diazepam develops quickly, so it is only used in emergent situations and not for long term seizure control.

Carbamazepine is structurally similar to tricyclic antidepressants. Carbamazepine interferes with the sodium channel at the synapse, prolonging activation and reducing the ability of the neuron to respond at high frequency [2,6]. This reduces transmission at the central synapse to control abnormal neuronal excitability [2]. Carbamazepine is metabolized to the active metabolite carbamazepine-10,11-epoxide (CBZE), which is responsible for some of the drug's adverse effects [6]. Oxcarbazepine is an analog of carbamazepine and also acts on sodium channels, as well as calcium channels via its main metabolite, 10-hydroxy-10,11-dihydrocarbamazepine, or monohydroxycarbamazepine [2]. Eslicarbazepine acetate is also related to carbamazepine but lacks the toxic metabolites. It also acts on sodium channels by binding to the inactive state to inhibit normal neuronal firing [3].

Phenobarbital has been in use since the early 1900s, and it acts on the GABA receptor by increasing the flow of chloride into the synapse. This potentiates synaptic inhibition, resulting in an increase in seizure threshold. Primidone is metabolized to phenobarbital and thus has a similar mechanism of action to stop seizure activity. Its secondary metabolite, phenylethylmalonamide, also shows antiseizure properties [2,6].

Phenytoin affects sodium channels and prolongs inactivation, which reduces the speed of the neuronal response. This leads to a reduction in synaptic transmission, helping to control abnormal neuronal excitability [2,6].

Valproic acid has two mechanisms of action; it increases the concentration of GABA in the brain through the inhibition of the GABA transaminase enzyme. It also acts on sodium channels to prolong inactivation, which does not allow the neuron to respond at high frequency [2].

The mechanism of action of ethosuximide involves reducing the calcium flow through T-type channels in the synapse of thalamic neurons, slowing the rate of absence seizure−inducing pulses [2,6].

Rufinamide was approved by the Food and Drug Administration (FDA) in 2008 for the treatment of Lennox−Gastaut syndrome, and its mechanism of action is to prolong the inactivated state of voltage-gated sodium channels [3,4].

The tiagabine mechanism of action is to bind the GABA uptake transporter, which results in increased synaptic and extracellular concentrations of the inhibitory transmitter. It is believed that the increased concentration of GABA is responsible for preventing seizure activity [2,4,7].

Gabapentin and pregabalin are closely related and are derivatives of the GABA neurotransmitter and known inhibitors of voltage-dependent calcium channels, where they reduce synaptic vesicle exocytosis [8]. Gabapentin is structurally related to the neurotransmitter GABA, but it does not appear to interact with GABA receptors in the brain or inhibit glutamic acid decarboxylase that controls concentrations of GABA [2,4].

Vigabatrin is an irreversible inhibitor of GABA transaminase, which catalyzes the elimination of GABA [2,4]. The duration of its pharmacological effect is related to the regeneration time of the enzyme and thus is dissociated from its concentration in serum [9].

Levetiracetam has novel mechanism of action and is thought to bind to the synaptic vesicle protein SV2A that is involved in neurotransmitter vesicle exocytosis [2,4]. Brivaracetam has a similar mechanism of action as levetiracetam that binds to the synaptic vesicle protein 2A, but with higher affinity and greater selectivity [3,6].

Zonisamide is a sulfonamide that stabilizes neuronal membranes by acting on voltage-sensitive sodium channels and T-type calcium channels [2,10].

Felbamate also has multiple mechanisms of action. It potentiates GABAergic neurotransmission, blocks sodium channels, and also inhibits glutamate excitation by interacting with the N-methyl-D-aspartate receptor [2,6]. Felbamate is not commonly prescribed as first-line treatment due to the risk of serious side effects, including aplastic anemia and liver failure.

Lamotrigine is another drug that operates on multiple mechanisms to control seizures. It is thought to reduce glutamate release as well as act on both sodium and calcium channels to reduce repetitive nerve firings [2,6].

Topiramate has multiple mechanisms of action that work to stop the spread of seizures. It blocks both sodium and calcium channels, inhibits glutamate release, and potentiates the activity of GABA [2,6].

The lacosamide mechanism of action is to enhance slow inactivation of voltage-gated sodium channels, and it might also mitigate neuronal rearrangement by inhibiting the collapsin response—mediated protein 2 [3].

Perampanel is thought to decrease the occurrence of seizures by reducing neuronal hyperexcitability by acting as an agonist at the alpha-amino-3-hydroxy-5-methyl-4-isoxazolepropionic acid (AMPA) receptor. Action as an agonist will reduce AMPA receptor—mediated synaptic transmission [3,6]. Clobazam is a benzodiazepine that binds to GABA receptors and enhances binding of GABA, which increases chloride conduction and causes neuronal hyperpolarization.

Pharmacokinetics and toxicokinetics

The first-generation AEDs have significant interindividual variation in their pharmacokinetics, which is one of the prime reasons they are good candidates for TDM [4]. This variability in the metabolism of the drugs can be due to impaired function of the kidney or liver or interactions with other drugs that are being taken [4]. The pharmacokinetic properties of most of the AEDs discussed in this chapter are presented in Table 9.2.

As AEDs are often taken in combination with other drugs, frequently other AEDs, pharmacokinetic interactions between the drugs can lead to a change in their distribution or metabolism. An important interaction between AEDs is induction and inhibition of drug metabolism. Most AEDs, with the exception of gabapentin, pregabalin, and vigabatrin, are metabolized by the liver and are susceptible to changes in concentration due to enzyme inhibition or induction [9]. When a drug is metabolized more quickly due to an increase in enzyme activity via enzyme induction, serum concentrations of the drug can fall more quickly out of the therapeutic range. If the drug has an active metabolite, such as carbamazepine and CBZE, this can lead to a toxic concentration [9]. Other drugs that are associated with inducing enzyme activity are phenobarbital, phenytoin, primidone, felbamate, oxcarbazepine, topiramate, and lamotrigine [9]. Drugs that result in enzyme inhibition can be more associated with toxicity, as inhibiting metabolizing enzymes leads to a reduction in drug metabolism and an increase in serum concentrations. Drugs that are enzyme inhibitors include valproic acid, oxcarbazepine, and felbamate [9].

Felbamate is used for the treatment of partial seizures and for Lennox—Gastaut syndrome, a childhood epilepsy syndrome that can be refractory to drug treatment. It has high bioavailability (90%) and is metabolized by the liver [4].

Lamotrigine is used in the treatment of partial seizures and is one of the most common AEDs prescribed during pregnancy, due to a solid safety record. It is metabolized via glucuronidation to an inactive metabolite [4].

Levetiracetam shows linear pharmacokinetics and is not protein bound. It is excreted by the kidneys in a ratio of 2/3 parent drug and 1/3 LO57 metabolite [4].

Oxcarbazepine is structurally related to carbamazepine and is metabolized to 10-hydroxycarbazepine that is the metabolite monitored during TDM [4].

Gabapentin and pregabalin are not metabolized, have low interaction potential with other drugs, and have no protein binding. They are secreted by the kidney, and their dose should be decreased in the face of renal insufficiency [8].

Clobazam is metabolized by CYP3A4 in the liver to N-desmethylclobazam that is also an active compound, but it is unknown which compound is more clinically potent [3].

TABLE 9.2 Pharmacokinetic Parameters of Anticonvulsant Drugs

Drug	Types of Seizures	Therapeutic Range (mg/L)	Toxic Range (mg/L)	Half-Life (h)	Protein Binding (%)	Metabolizing Enzymes
Brivaracetam	Partial and generalized	0.2–2.0	Not established	7–8	35	NA
Carbamazepine	Partial and generalized	4–12	15–20	12–17	70–80	CYP3A4
Clobazam	Partial and generalized	0.03–0.3	0.5	10–30	86	CYP3A4
Clonazepam	Status epilepticus	0.02–0.08	0.1	17–56	85	CYP3A4
Diazepam	Status epilepticus	0.1–2.5	3–5		High	CYP2C19
Eslicarbazepine acetate	Partial and generalized	3–35	Not established	13–20	40	NA
Ethosuximide	Absence	30–100	150–200	40–60	0	CYP3A4
Felbamate	Partial and generalized	30–110	150–200	20–30	25	CYP3A4
Gabapentin	Partial and generalized	2–20	25–105	5–7	0	NA
Lacosamide	Partial and generalized	10–20	20	13	15	CYP2C19
Lamotrigine	Partial and generalized	3–14	20–30	22	55	NA
Levetiracetam	Partial and generalized	12–46	400	6–8	0	NA
Oxcarbazepine	Partial	3–35	45	8–15	40	NA
Perampanel	Partial and generalized	0.05–0.4	Not established	105	95	CYP3A4
Phenobarbital	Partial and generalized	10–40	40–60	70–140	55	CYP2C9, 2C19
Phenytoin	Partial and generalized	10–20	20–25	30–100	90	CYP2C9, 2C19
Pregabalin	Partial and generalized	2–8	10	5–7	0	NA
Primidone	Partial and generalized	5–10	20–50	7–22	20	CYP2C9, 2C19
Rufinamide	Partial and drop	10–30	40	6–10	30	NA
Tiagabine	Partial	0.02–0.2	0.5–3	5–9	20	CYP3A4
Topiramate	Partial and generalized	5–20	16	18–23	15	NA
Valproic acid	Partial and generalized	50–100	150–200	5–20	90	CYP2C9, 2C19 2B6, 2E1, 2A6
Vigabatrin	Partial	2–15	20	7.5	0	NA
Zonisamide	Partial and generalized	10–40	40–70	50–70	50	CYP2C9, 3A4

NA, Not applicable.

Clinical management

Anticonvulsants are not a family of drugs that are often abused, but there has been an increase in cases as all prescription drug abuse is on the rise. In addition, as the number of prescriptions for anticonvulsants is increasing, the opportunity for abuse increases as well. A 2016 study examining the care of poisoned patients in hospital observation units found that the most common reason for a drug exposure in both adults and pediatrics was intentional ingestion, and the second most common agent in pediatrics was anticonvulsants, at 10.7% of exposures [11]. In the majority of cases reported, there is no specific remedy or antidote for each drug; treatment for an overdose is specific to each clinical case and symptomology present. Overdose data for the third-generation AEDs are limited as they have only been recently approved for clinical use, and it will be critical to review the adverse events for each drug and to publish case reports as they occur to gain a better understanding of their toxicity.

As previously mentioned, it can be difficult to determine if a seizure is caused by an organic issue or is due to a medication used to control seizures. This can be seen when the correct drug is not being used for the type of condition the patient presents with or when the correct drug is given at the incorrect dose [12]. Examples of this have been seen in patients that were treated with carbamazepine, whose seizures were controlled when they switched to valproic acid [12]. Similar events have been reported with tiagabine, levetiracetam, and valproic acid [12]. There are no specific treatments when this occurs, besides supportive management of the patient's symptoms. Measuring the concentration of AED in cases of suspected toxicity can be useful, but a lower than expected dose of drug does not necessarily rule out a reaction to that drug, especially in cases where multiple drugs have been ingested [9].

Gabapentin has been increasingly prescribed for nonepileptic patients for off-label treatment of acute and chronic pain, with limited scientific evidence to support this use. Prescriptions for gabapentin have increased recently for use in combination with opioids for pre- and postsurgical pain, but concomitant use of gabapentin and opioids was found to increase the risk of a fatal opioid overdose [13]. A survey of postmortem toxicology results in five geographic areas in the United States found that 22% of all overdoses tested positive for gabapentin, and all overdose deaths that involved gabapentin also involved at least one other drug [13]. In another study looking at abuse of pregabalin and gabapentin, there were reports of 37 case studies of acute overdose [14]. Presenting symptoms included seizures and reduced consciousness, drowsiness, dizziness, ataxia, hypotension, tachycardia, and slurred speech. While most patients experienced only these benign symptoms and had a full recovery, there were five fatal ingestions reported [14]. Most lethal overdoses came when pregabalin or gabapentin was combined with other drugs, especially opioids and sedatives [8]. As gabapentin is not metabolized and is secreted by the kidney, patients with renal failure can have significant toxicity. A case study of a patient who had switched from hemodialysis to peritoneal dialysis without monitoring or adjusting her dose of gabapentin presented with fever and muscle twitching [15]. This was attributed to gabapentin toxicity, and the patient was treated with continuous cycling of peritoneal dialysis for 20 exchanges, which resolved her symptoms. While this was an effective treatment of gabapentin toxicity since the patient was stable, much more rapid removal of gabapentin can be accomplished via hemodialysis [15].

Lamotrigine is one of the most commonly prescribed anticonvulsants of the newer generation of drugs, but there are few case reports of overdoses. Grosso et al. reported an acute intoxication in a 3-year-old child [16] who presented with generalized tonic—clonic status epilepticus followed by hyperkinesia and ataxia. The patient received intravenous midazolam and fluids until seizures ceased. Serum lamotrigine concentrations were measured at 28.4 mg/L, with the therapeutic range of the drug 1.0—14.0 mg/L.

Carbamazepine is a tricyclic drug that can produce a sense of euphoria, which increases its potential for abuse. Carbamazepine overdose can present with altered mental status, coma, respiratory failure, seizures, and cardiac arrhythmias [17]. In addition to supportive care, treatment can also include removal of drugs and metabolites via extracorporeal removals such as hemodialysis, hemoperfusion, continuous venovenus hemofiltration, and continuous renal replacement therapy (CRRT) [18]. One report of an intentional overdose with extended-release capsules resulted in a peak serum concentration of 138 μg/mL, with normal therapeutic concentrations of carbamazepine falling between 4 and 12 μg/mL [19]. The patient was treated with multiple modalities, including activated charcoal, lipid emulsion therapy, plasmapheresis, hemodialysis, continuous venovenous hemodiafiltration, and endoscopic removal of undissolved carbamazepine capsules. The patient was discharged after 14 days in the hospital at her baseline condition. The toxicity of a carbamazepine overdose is increased by the active metabolite CBZE that has equivalent physiologic effects [17]. It is estimated that 40% of carbamazepine is converted into this metabolite, and as the conversion is highly variable, it can result in a delayed peak serum concentration of drug and could explain the late clinical decompensation of some patients [18]. There can be some cross-reactivity with the CBZE metabolite in carbamazepine immunoassays, leading to higher reported peak concentrations than are seen using mass spectrometry methods [18].

Valproic acid intoxication can present with central nervous system depression, hypotension, metabolic acidosis, cerebral edema, hyperammonemia, and elevated lactate levels. In the majority of over 300 cases presented in a review of the literature by Tichelbacker et al., a favorable outcome was reported, with patients having a full recovery after ingestions of valproic acid from 4 to 160 g [20]. The patients were treated with various combinations of intubation, hemodialysis, hemoperfusion, CRRT, and liver support therapy. One of the complications of valproic acid overdose is hyperammonemia, and a small study looked at the effect of L-arginine on ammonia levels during treatment [21]. L-Arginine increases the activity of the *N*-acetylglutamate synthase enzyme that aids in the initiation of the urea cycle and leads to a reduction in serum ammonia concentration. Patients in the study given L-arginine showed a rapid fall in ammonia concentration soon after treatment, but further studies are needed to refine the dose necessary and provide further evidence of treatment utility [21].

Phenobarbital is one of the oldest AEDs in use today, and overdoses result in cortical suppression and coma. In one case report of an overdose with a phenobarbital level of 180 μg/mL, the patient presented not only in a deep coma but also nonconvulsive status epilepticus [22]. After 7 days the patient regained consciousness and left the hospital.

Tiagabine is another example of an anticonvulsant that has proconvulsant properties, especially when supratherapeutic concentrations are ingested [7]. Other symptoms of tiagabine overdose in addition to seizures include lethargy, confusion, and coma. In 2005 the FDA placed a warning on tiagabine warning nonepileptic patients who were taking the drug for off-label conditions that they had a risk of seizure, even when taking tiagabine at therapeutic levels [7].

CRRT is often used as a treatment modality in an anticonvulsant overdose, but there have been no randomized clinical trials to study its efficacy. A study by Mahmoud examined 31 published case reports where CRRT was used to either remove toxic levels of anticonvulsants or was used as a treatment in patients who were on therapeutic doses of anticonvulsants for evidence of anticonvulsant removal by CRRT [23]. They found several factors that influenced the removal of anticonvulsant drugs: the amount of drug that is protein bound, the drug's volume of distribution, and how the drug is metabolized and eliminated. Based on these criteria, anticonvulsant drugs were grouped into three categories: group A, including the drugs most effectively removed by CRRT, group B, including drugs that can be significantly removed by CRRT, and group C, including drugs that CRRT will not have much effect on the concentration [23]. Group A has drugs that exhibit low protein binding and variable hepatic metabolism and renal excretion. These drugs include ethosuximide, brivaracetam, rufinamide, lamotrigine, phenobarbital, oxcarbazepine, zonisamide, lacosamide, primidone, felbamate, and topiramate. Group B has drugs that are renally excreted and have low protein binding, including levetiracetam, vigabatrin, pregabalin, gabapentin, and eslicarbazepine. Group C contains drugs that are mainly hepatically metabolized and highly protein bound, including carbamazepine, clobazam, clonazepam, valproic acid, phenytoin, perampanel, and tiagabine. Group C drugs can be more effectively eliminated by CRRT if the free fraction of the drugs is increased [23].

Overdose data on newer AEDs can be lacking, especially when a combination of drugs is ingested. A case report of a patient who overdosed on both levetiracetam and topiramate twice in a span of 10 days is a rare example of an overdose with no complications [24]. The 21-year-old woman took 20 times the maximum dose of levetiracetam (60 g) and 4 times the maximum dose of topiramate (1.5 g) followed by 10 times the maximum dose of levetiracetam (30 g) and 2 times the maximum dose of topiramate (175 mg) 10 days later. The patient was treated with activated charcoal both times, and neither time suffered from any of the known side effects of either drug, including dizziness, somnolence, acute psychosis, hepatic failure, or ataxia with topiramate, or respiratory distress with levetiracetam [24].

There are fewer reports of overdose of the third-generation AEDs, but mainly because they have not been in use as long. Lacosamide overdoses have resulted in status epilepticus, hypotension, and other effects on the cardiovascular system, including a fatality due to cardiac conduction abnormalities [3].

There have not been many case reports of overdoses of zonisamide, but the majority of them resulted in a full recovery [10,25]. Common adverse events include somnolence, anorexia, dizziness, headache, nausea, and agitation—irritation. Following an overdose, symptoms include coma, hypotension, metabolic acidosis, seizures, and in the one fatality, cardiac arrest [10,25]. Several of these cases were complicated by the ingestion of more than one AED.

Vigabatrin has not caused any deaths after overdose, and serious adverse events after ingestion include coma, psychosis, respiratory compromise, bradycardia, hypotension, vertigo, and status epilepticus [3].

The literature on the side effects of perampanel is contradictory. It carries a black box warning for several potential psychiatric adverse events, including homicidal and suicidal ideation, but data from several studies only showed an increase in irritability and aggression but not suicidal ideation when compared to placebo [3]. There are nine cases of perampanel overdose reported in the literature, but eight of the nine were unintentional. The symptoms included nausea, vomiting, somnolescence, ataxia, and dysarthria, but none were fatal [3].

Common side effects of clobazam include somnolence, aggression, drooling, and ataxia [3]. A series of case reports in both adults and children reported mild cognitive effects and no severe symptoms. Two fatal overdoses have been reported—one due to respiratory depression and the other was undetermined. Levels of clobazam and its metabolite desmethylclobazam in each case were 720 and 36,000 ng/mL (normal concentrations are between 300 and 3000 ng/mL) and 3900 ng/mL, with desmethylclobazam not determined [3]. Flumazenil is a drug that reverses the effects of benzodiazepines, but its use as an antidote to clobazam overdose is to be avoided due to the life-threatening side effects of status epilepticus that can be caused by sudden withdrawal of clobazam [3].

For eslicarbazepine acetate, rufinamide, and brivaracetam, there are no case reports of overdose [3].

A new therapy that has been in the news for several years is cannabidiol (CBD), one of the active constituents of cannabis. Most of the accounts of CBD halting seizure activity were anecdotal or the result of small studies lacking control groups. In 2017 a double-blind, placebo-controlled clinical trial was published examining 120 patients with Dravet syndrome, a rare genetic form of epilepsy associated with drug-resistant seizures [26]. Patients in the study received an oral solution of CBD or placebo in addition to their standard antiepileptic treatment for 14 weeks. While patients in the treatment arm had more adverse events than the placebo arm, the frequency of seizures per month decreased from 12.4 to 5.9, compared with a decrease of 14.9 to 14.1 in the placebo group. In addition, the percentage of patients who had a 50% or more decrease in seizure frequency was 43% in the CBD group compared to 27% in the placebo group [26]. The majority of the adverse events were classified as mild and included vomiting, fatigue, pyrexia, upper respiratory tract infection, convulsion, lethargy, somnolence, and diarrhea. Serious events that were reported included status epilepticus and elevated liver aminotransaminases. No overdoses or deaths were reported [26].

A similar study in patients with Lennox—Gastaut syndrome, another rare genetic form of epileptic encephalopathy, was done in 2018, with similar results [27]. This double-blind, placebo-controlled clinical trial included 171 patients who received CBD or placebo as add on therapy to their existing AEDs for 14 weeks. Patients in the CBD group showed a 43.9% median percentage reduction in drop seizures, compared to a 21.8% reduction in the control group. 86% of the patients in the study arm experienced mild-to-moderate adverse events, including diarrhea, somnolence, pyrexia, decreased appetite, and vomiting [27]. Elevated liver enzymes were the most common serious adverse events that led to withdrawal from the study, and one patient in the treatment arm of the study died due to acute respiratory distress syndrome, but this was not thought to be treatment related. These studies contributed to the evidence that led to CBD being approved for treatment of both disorders in June 2018 [28]. As this drug becomes more commonly used, the literature should be monitored for case reports of overdose, but none have been published at this time.

Analytical considerations

Most of the first-generation anticonvulsants are readily measurable using automated commercial analyzers, as these drugs have been in use since the 1970s, and in vitro diagnostics manufactures have had ample time to develop assays for use on automated platforms. College of American Pathologists (CAP) proficiency testing survey data list multiple commercial platforms for the following AEDs: carbamazepine, phenobarbital, phenytoin and free phenytoin, primidone, and valproic acid and free valproic acid [29]. CAP surveys list at least one immunoassay that can be run on an automated analyzer for the following AEDs: ethosuximide, gabapentin, lamotrigine, levetiracetam, oxcarbazepine metabolite, topiramate, and zonisamide [29]. For some of the newer anticonvulsants and those prescribed less frequently, commercial assays have not been developed, and laboratories must rely on developing other means of detecting the drugs using gas chromatography—mass spectrometry or liquid chromatography—tandem mass spectrometry (LC—MS/MS) or send specimens to reference laboratories for analysis.

Carbamazepine and its metabolite CBZE have shown variable cross-reactivity in commercial immunoassays. The metabolite accumulates in pediatric patients and monitoring the ratio of the parent compound and the metabolite can be useful when assessing for compliance [2]. A study looking at clearance rates of carbamazepine and its metabolite, CBZE, showed a higher serum concentration of carbamazepine when measured by immunoassay compared to time-of-flight mass spectrometry (TOFMS) at all time points tested [18]. In addition, as TOFMS carbamazepine concentrations were decreasing, the TOFMS CBZE concentration concomitantly increased, resulting in a simultaneous rise of carbamazepine concentration detected by immunoassay. This suggests that the carbamazepine concentration detected by the immunoassay is measuring both the parent drug and its metabolite and is not a reflection of continued absorption of drug [18].

A study in 2014 described the development and validation of a simple LC—MS/MS method for the simultaneous detection and quantification of 22 AEDs in serum, plasma, and postmortem blood [30]. This is quite useful as the use of newer AEDs is becoming more common, and this method can be used for TDM of the drugs in patients as well as for routine forensic toxicology in the case of drug abuse and overdose [6]. This method improved upon previous methods that use gas chromatography and high-performance liquid chromatography coupled to ultraviolet detection in several ways, including a quick and simple protein precipitation extraction and a short run time of only 17 minutes [30].

Alternate samples for the analysis of AEDs have been used. Hair analysis has been validated and used to test for environmental compounds, drugs of abuse testing, and doping substances for many years. The use of hair samples in place of urine or blood specimens can remove the issue of specimen adulteration seen with urine testing or provide a new, identical specimen for testing in the instance of a potential specimen mix up or need to retest an identical specimen. It is also very easy to obtain. It has not been used widely for testing of AEDs, but the use of this specimen has been used in measurements of carbamazepine, phenytoin, valproic acid, oxcarbazepine, lamotrigine, and pregabalin concentrations [31]. To aid in detection in forensic cases, a study was done to validate a method for the detection of pregabalin in hair [32]. In addition to seizure control, pregabalin can also be prescribed for opioid, benzodiazepine, and alcohol dependence, which can lead to its being misused or abused. The method described included an acetonitrile extraction followed by ultrahigh-performance LC−MS/MS, and it achieved a limit of detection of 10 pg/mg. The assay was linear from 10 to 2000 pg/mL, and this should allow detection of abuse of pregabalin in the mg range [32].

Testing for AEDs in saliva offers many of the same benefits as using hair as a matrix. Saliva can be sampled repeatedly and is very easy to collect in a noninvasive manner. Concentrations of drugs in saliva generally reflect the free nonprotein bound version of the drug that is biologically active [31]. In addition, standard methods of drug analysis can be easily adapted to this specimen type, as long as the concentration is adequate. AEDs that have been detected in saliva for means of TDM include carbamazepine, clobazam, ethosuximide, lacosamide, lamotrigine, levetiracetam, oxcarbazepine, phenytoin, primidone, rufinamide, and topiramate [31]. The following drugs have either not been detected in saliva or it is not known if saliva is a valid specimen to determine their concentration: clonazepam, eslicarbazepine acetate, felbamate, gabapentin, phenobarbital, pregabalin, tiagabine, valproic acid, vigabatrin, and zonisamide [31].

References

[1] Spina E, Perugi G. Antiepileptic drugs: indications other than epilepsy. Epileptic Disord 2004;6(2):57−75.
[2] Snozek CLH, Mcmillin GA, Moyer TP. Therapeutic drugs and their management. In: Burtis CA, Ashwood ER, Bruns DE, editors. Tietz textbook of clinical chemistry and molecular diagnostics. St. Louis, MO: Elsevier; 2012. p. 1057−108.
[3] LaPenna P, Tormoehlen LM. The pharmacology and toxicology of third-generation anticonvulsant drugs. J Med Toxicol 2017;13(4):329−42.
[4] Krasowski MD. Therapeutic drug monitoring of the newer anti-epilepsy medications. Pharm (Basel) 2010;3(6):1909−35.
[5] Sankaraneni R, Lachhwani D. Antiepileptic drugs—a review. Pediatr Ann 2015;44(2):e36−42.
[6] Abou-Khalil BW. Update on antiepileptic drugs 2019. Continuum (Minneap Minn) 2019;25(2):508−36.
[7] Spiller HA, et al. Review of toxicity and trends in the use of tiagabine as reported to US poison centers from 2000 to 2012. Hum Exp Toxicol 2016;35(2):109−13.
[8] Bonnet U, Scherbaum N. How addictive are gabapentin and pregabalin? A systematic review. Eur Neuropsychopharmacol 2017;27(12):1185−215.
[9] Patsalos PN, et al. Antiepileptic drugs—best practice guidelines for therapeutic drug monitoring: a position paper by the subcommission on therapeutic drug monitoring, ILAE commission on therapeutic strategies. Epilepsia 2008;49(7):1239−76.
[10] McStay C, Pierce R, Riley C. Complete recovery after acute zonisamide overdose in an adolescent female. Pediatr Emerg Care 2018;34(2):e30−1.
[11] Judge BS, et al. Utilization of observation units for the care of poisoned patients: trends from the toxicology investigators consortium case registry. J Med Toxicol 2016;12(1):111−20.
[12] Cock HR. Drug-induced status epilepticus. Epilepsy Behav 2015;49:76−82.
[13] Slavova S, et al. Prevalence of gabapentin in drug overdose postmortem toxicology testing results. Drug Alcohol Depend 2018;186:80−5.
[14] Evoy KE, Morrison MD, Saklad SR. Abuse and misuse of pregabalin and gabapentin. Drugs 2017;77(4):403−26.
[15] Ibrahim H, et al. Treatment of gabapentin toxicity with peritoneal dialysis: assessment of gabapentin clearance. Am J Kidney Dis 2017;70(6):878−80.
[16] Grosso S, et al. Massive lamotrigine poisoning. A case report. Brain Dev 2017;39(4):349−51.
[17] Spiller HA. Management of carbamazepine overdose. Pediatr Emerg Care 2001;17(6):452−6.
[18] Smollin CG, Petrie MS, Kearney T. Carbamazepine and carbamazepine-10,11-epoxide clearance measurements during continuous venovenous hemofiltration in a massive overdose. Clin Toxicol (Phila) 2016;54(5):424−7.
[19] Agulnik A, et al. Combination clearance therapy and barbiturate coma for severe carbamazepine overdose. Pediatrics 2017;139(5):e20161560.
[20] Tichelbacker T, et al. Hemodiafiltration treatment for severe valproic acid intoxication: case report and updated systematic literature review. Front Med (Lausanne) 2018;5:224.
[21] Schrettl V, et al. L-Arginine in the treatment of valproate overdose—five clinical cases. Clin Toxicol (Phila) 2017;55(4):260−6.
[22] Hassanian-Moghaddam H, et al. Phenobarbital overdose presenting with status epilepticus: a case report. Seizure 2016;40:57−8.

[23] Mahmoud SH. Antiepileptic drug removal by continuous renal replacement therapy: a review of the literature. Clin Drug Investig 2017;37(1):7–23.
[24] Sarfaraz M, Syeda RH. Levetiracetam and topiramate poisoning: two overdoses on those drugs with no lasting effects. Drug Discov Ther 2017;11(2):115–17.
[25] Hofer KE, et al. Moderate toxic effects following acute zonisamide overdose. Epilepsy Behav 2011;21(1):91–3.
[26] Devinsky O, et al. Trial of cannabidiol for drug-resistant seizures in the Dravet syndrome. N Engl J Med 2017;376(21):2011–20.
[27] Thiele EA, et al. Cannabidiol in patients with seizures associated with Lennox-Gastaut syndrome (GWPCARE4): a randomised, double-blind, placebo-controlled phase 3 trial. Lancet 2018;391(10125):1085–96.
[28] Sanmartin PE, Detyniecki K. Cannabidiol for epilepsy: new hope on the horizon? Clin Ther 2018;40(9):1438–41.
[29] Pathologists, CoA, College of American Pathology education programs. In: C.T.D.M. C-A, editor; 2019.
[30] Deeb S, et al. Simultaneous analysis of 22 antiepileptic drugs in postmortem blood, serum and plasma using LC-MS-MS with a focus on their role in forensic cases. J Anal Toxicol 2014;38(8):485–94.
[31] Patsalos PN, Berry DJ. Therapeutic drug monitoring of antiepileptic drugs by use of saliva. Ther Drug Monit 2013;35(1):4–29.
[32] Kintz P, Ameline A, Raul JS. Assessment of pregabalin use by hair testing. Subst Use Misuse 2018;53(12):2093–8.

Chapter 9.1

Free versus total phenytoin measurements—a case study of phenytoin toxicity

Heather A. Paul[1,2], Jason L. Robinson[1,2] and S.M. Hossein Sadrzadeh[1,2]
[1]*Alberta Precision Laboratories, Calgary, AB, Canada,* [2]*Department of Pathology and Laboratory Medicine, Cumming School of Medicine, University of Calgary, Calgary, AB, Canada*

Case description

A 57-year-old female with relevant history of hepatorenal syndrome, cirrhosis, recent acute kidney injury, and epilepsy presented to the emergency department with headache, confusion, and seizure activity. Upon admission, seizures were confirmed as partial status epilepticus by electroencephalogram (EEG) that was refractive to increasing the patient's scheduled levetiracetam (Keppra) dose. Fortunately, treatment with the anticonvulsant drug phenytoin (Dilantin) was effective, and the patient remained seizure free for the remainder of her 15-day admission. Prior to discharge, the patient received a final phenytoin dose with instructions to remain on an increased scheduled dose of Keppra. Soon after arriving home, the patient experienced a witnessed episode of altered level of consciousness with associated anxiety, lightheadedness, rapid breathing, rapid eye movements, and stuttering that lasted several minutes. The patient returned to the emergency department, and a similar episode occurred. There was no impaired recollection of the events or seizure-associated postictal phase. Laboratory investigations were unremarkable except for elevated total bilirubin and gamma-glutamyl transferase and a reduced albumin (Table 9.1.1). The patient's total phenytoin was within the target therapeutic range (73 μmol/L, therapeutic range: 40–80 μmol/L). The patient's EEG was normal, and there was no suspicion of hepatic encephalopathy. However, the following morning free phenytoin was requested in addition to routine total phenytoin testing. While the total phenytoin was within the therapeutic range (65 μmol/L), the free phenytoin result was high at 11.2 μmol/L (therapeutic range, 4–8 μmol/L), resulting in a phenytoin ratio of 17%. The patient was subsequently admitted for phenytoin toxicity.

Discussion

Therapeutic drug monitoring (TDM) is used to tailor medication dosing when drugs (1) have a narrow therapeutic index, (2) display variable pharmacokinetics, or (3) their efficacy and/or toxicity is challenging to assess early in the therapeutic course [1]. Phenytoin is a prevalent anticonvulsant medication used to prevent, control, and treat various types of seizures [1,2]. Phenytoin is usually sold as the prodrug fosphenytoin that is metabolized to phenytoin after intramuscular injection, as oral phenytoin, which exhibits slow and incomplete gastrointestinal absorption [1], or in the acute care setting as intravenous phenytoin [2]. Due to phenytoin's narrow therapeutic index and complex pharmacokinetics, it is a prime candidate for TDM using blood levels to ensure safe, effective use [2]. Indeed, phenytoin levels above the therapeutic range can result in significant dose-related toxicity, causing symptoms such as nystagmus, ataxia, confusion, and coma [3,4], and in extreme cases (≥ 140 μmol/L) can lead to seizures [1]. A key aspect of phenytoin TDM is that it circulates highly protein bound ($\approx 90\%$), and only a small fraction of phenytoin dose (the free fraction) is pharmacologically active [5]. Despite this, total phenytoin, and not free phenytoin, is measured and reported by most clinical laboratories. Therefore free phenytoin level is mostly estimated and reported as approximately 10% (8%–12%) of total phenytoin in blood [6]. Further, due to this high degree of protein binding, any condition that affects phenytoin

TABLE 9.1.1 Laboratory Results

	Analyte	Result (Reference Interval)
Chem panel	Albumin	32 (33–48 g/L)
	Calcium	2.23 (2.10–2.60 mmol/L)
	Estimated glomerular filtration rate (eGFR)	103 ($\geq$60 mL/min/1.73 m^2)
	Magnesium	0.57 (0.65–1.05 mmol/L)
	Phosphate	1.37 (0.80–1.50 mmol/L)
	Anion gap	8 (4–16 mmol/L)
	Chloride	105 (98–111 mmol/L)
	Creatinine, serum	52 (40–100 µmol/L)
	Carbon dioxide content	23 (21–31 mmol/L)
	Glucose, random	5.4 (3.3–11.0 mmol/L)
	Potassium	4.7 (3.5–5.0 mmol/L)
	Sodium	136 (133–145 mmol/L)
	Urea	2.8 (2.5–8.5 mmol/L)
Liver panel	Alanine transaminase	6 (1–40 U/L)
	Alkaline phosphatase	120 (30–145 U/L)
	Bilirubin, total	28 (0–24 µmol/L)
	Gamma-glutamyl transferase	60 (8–35 U/L)
	Lactate dehydrogenase	228 (100–235 U/L)
	Lipase	21 (0–80 U/L)

partitioning to protein can drastically affect the ratio of free to bound drug, and therefore the balance between toxicity and efficacy. It is important to note that phenytoin partitioning can be altered by several conditions that are commonly encountered in the acute care setting, including hypoalbuminemia, uremia, coadministration of other antiepileptic medications such as valproate, pregnancy, and critical illness [5,6]. In these cases, free phenytoin can comprise up to 25% of the total phenytoin [5], which has considerable implications due to its narrow therapeutic index. Conversely, the opposing situation is also possible where free phenytoin is disproportionately high relative to total phenytoin, which could lead to erroneous dosing decisions and possible recurrence of seizure activity [7]. Ultimately, the physician must be aware of the physiological factors affecting phenytoin binding and adjust their approach accordingly. Unfortunately, there is no specific antidote to phenytoin, and if phenytoin toxicity does occur, treatment for phenytoin toxicity generally involves discontinuation or decreased dosing of phenytoin and supportive care [4,8].

The patient above had several factors that could have led to greater circulating levels of free phenytoin. For example, phenytoin clearance is dependent on hepatic metabolism that is limited by hepatic enzyme capacity [9]. Therefore any factor that impairs hepatic enzyme activity, such as cirrhosis, may prolong phenytoin half-life and affect its elimination. Further, any factor affecting phenytoin binding to protein, such as low albumin levels, would be expected to alter the level of free drug. In this case, intermittent hypoalbuminemia could have increased the free phenytoin fraction [10]. In addition, bilirubin has been shown to compete with albumin for phenytoin binding sites [4]. However, because indirect bilirubin was not measured in this case, it is difficult to assert that displacement of phenytoin by bilirubin contributed to the observed toxicity. Ultimately, this patient was determined to have phenytoin toxicity as a result of an impaired ability to clear phenytoin due to cirrhosis.

Unfortunately, the measurement of free phenytoin is not available in all clinical laboratories, mainly because total phenytoin measurements are automated and less costly versus free phenytoin testing [11]. Diagnosis of phenytoin toxicity therefore requires careful consideration of factors affecting free drug concentrations. One approach has been to estimate free phenytoin concentrations using equations [11]. Such equations are primarily based on patient albumin levels because its testing is widely available and its levels significantly affect free phenytoin concentrations [11]. For example, the Sheiner−Tozer equation corrects the total phenytoin measurement based on the patient's albumin levels [11]. Free phenytoin levels are then calculated as 10% of this corrected value; this overall estimation of free phenytoin is also known as the Winter−Tozer (WT) equation [12]. The WT equation has also been modified for end-stage renal disease (ESRD) patients [13]. This ESRD WT equation accounts for ESRD-associated changes in phenytoin−albumin binding affinity, although the accuracy of this derivative is under debate [10]. These equations have also been further modified for ESRD patients on home dialysis [10]. However, it is critical that factors besides hypoalbuminemia and ESRD are considered for TDM of phenytoin patients. Indeed, the Sheiner−Tozer equation was biased and imprecise in critical care patients that suffered head trauma, and those with neurological disorders [11]. Importantly, had this equation been used to estimate free phenytoin in our case, estimated free phenytoin would have been 7.9 μmol/L or 12% of total phenytoin.

In summary, phenytoin is an anticonvulsant medication that is highly protein bound. During conditions that disrupt phenytoin partitioning into protein, it is prudent to quantify the levels of free phenytoin in patients to avoid phenytoin toxicity. For example, free phenytoin levels may be disrupted in patients with uremia, chronic liver disease, hypoalbuminemia, pregnancy, or due to drug−drug interactions [6]. While equations exist to estimate the free phenytoin, these may not be applicable in patients with other maladies or conditions, and in some cases, an inaccurate assessment of free phenytoin levels may ultimately harm the patient. Ideally, as is evident from this case, the direct measurement of free phenytoin rather than estimation should be available to all modern laboratories and used whenever total drug levels do not correlate with patient history and symptoms. It is critical that all health-care providers are aware of these factors to prevent phenytoin toxicity.

References

[1] Rifai N. Tietz textbook of clinical chemistry and molecular diagnostics, Sixth edition. St. Louis, Missouri: Elsevier;2018.
[2] Tobler A, Mühlebach S. Intravenous phenytoin: a retrospective analysis of Bayesian forecasting versus conventional dosing in patients. Int J Clin Pharm 2013;35:790−7.
[3] Imam SH, et al. Free phenytoin toxicity. Am J Emerg Med 2014;32:1301.e3−4.
[4] Robertson K, Von Stempel CB, Arnold I. When less is more: a case of phenytoin toxicity. BMJ Case Rep 2013;2013: bcr2012008023.
[5] Burt M, Anderson DC, Kloss J, Apple FS. Evidence-based implementation of free phenytoin therapeutic drug monitoring. Clin Chem 2000;46:1132−5.
[6] Dasgupta A. Usefulness of monitoring free (unbound) concentrations of therapeutic drugs in patient management. Clin Chim Acta 2007;377:1−13.
[7] Selioutski O, et al. Evaluation of phenytoin serum levels following a loading dose in the acute hospital setting. Seizure 2017;52:199−204.
[8] Gupta A, Yek C, Hendler RS. Phenytoin toxicity. JAMA—J Am Med Assoc 2017;317:2445−6.
[9] Buckley MS, Reeves BA, Barletta JF, Bikin DS. Correlation of free and total phenytoin serum concentrations in critically ill patients. Ann Pharmacother 2016;50:276−81.
[10] Soriano VV, Tesoro EP, Kane SP. Characterization of free phenytoin concentrations in end-stage renal disease using the Winter-Tozer equation. Ann Pharmacother 2017;51:669−74.
[11] Kiang TKL, Ensom MHH. A comprehensive review on the predictive performance of the Sheiner-Tozer and derivative equations for the correction of phenytoin concentrations. Ann Pharmacother 2016;50:311−25.
[12] Cheng W, Kiang TKL, Bring P, Ensom MHH. Predictive performance of the Winter-Tozer and derivative equations for estimating free phenytoin concentration. Can J Hosp Pharm 2016;69:269−79.
[13] Mauro LS, Mauro VF, Bachmann KA, Higgins JT. Accuracy of two equations in determining normalized phenytoin concentrations. DICP 1989;23:64−8.

Chapter 9.2

A fatality involving massive overdose of gabapentin

Bheemraj Ramoo[1], Marius C. Tarau[2], Mary Dudley[2], C. Clinton Frazee, III[1] and Uttam Garg[1,3]
[1]Department of Pathology and Laboratory Medicine, Children's Mercy Hospital, Kansas City, MO, United States, [2]Office of the Jackson County Medical Examiner, Kansas City, MO, United States, [3]University of Missouri, School of Medicine, Kansas City, MO, United States

Case history

A 67-year-old white male was found unresponsive by his family in his residence. The subject was last seen alive by his family several days prior to his death. He was found prone to the bedroom floor with his head and shoulders elevated. The subject was in advance stage of decomposition with foul odor, green discoloration, bloating, skin slippage and blistering, and purge. There were no obvious signs of trauma or foul play. The subject had a history of hypertension and insulin-dependent diabetes. Several medications, including citalopram hydrobromide, gabapentin, hydrocodone/acetaminophen, lisinopril, and pregabalin, were found on the scene.

The body was transported to the Jackson County Medical Examiner's Office. As per protocol for advanced age, significant medical history and no trauma, only an external body exam was performed. The whole-body X-rays were unremarkable. Postmortem chest blood and liver tissue were submitted for toxicological analysis. The blood was screened for volatiles (ethanol, methanol, isopropanol, and acetone) by headspace gas chromatography with flame ionization detector. The blood was also used for comprehensive broad-spectrum drug screening that included enzyme immunoassays (EIAs) for amphetamines, barbiturates, benzodiazepines, cannabinoids, cocaine metabolite, methadone, opiates, phencyclidine and propoxyphene, and drug screening for >200 commonly detected drugs by gas chromatography–mass spectrometry (GC–MS). Drug screening by GC–MS involved liquid–liquid alkaline extraction using bicarbonate buffer (pH 11.0) and butyl acetate, and mass spectrometer operation in full scan mode. Presumptive identification of analytes was made by spectral library match and relative retention time comparison with reference standards.

The EIA drug screen was positive for opiates class drugs. However, on opiates confirmation by GC/MS, morphine, codeine, hydrocodone, and oxycodone were not detected. Volatile screen testing for isopropanol, acetone, methanol, and ethanol indicated the presence of ethanol at 22 mg/dL. The GC/MS drug screen detected the presence of gabapentin and acetaminophen. High-performance liquid chromatography/tandem mass spectrometry (LC–MS/MS) testing for gabapentin was performed on postmortem chest blood and liver tissue by a reference laboratory (NMS Labs, Willow Grove, PA). The chest blood gabapentin level was 180 μg/mL and the liver tissue measured 42 μg/g. All results are summarized in Table 9.2.1.

Discussion

Gabapentin (Gralise, Neurontin, Horizant) is a widely prescribed drug for the treatment of epilepsy and neuropathic pain [1,2]. As an anticonvulsant drug, it is used as an adjunctive therapy or monotherapy. Its other uses include pain relief in postherpetic neuralgia, bipolar disorder, movement disorder, migraine prophylaxis, and cocaine dependence [2–4]. Gabapentin is a structural analog of the inhibitory neurotransmitter gamma-aminobutyric acid (GABA) (Fig. 9.2.1). Though gabapentin is structurally related to neurotransmitter GABA, its mechanism of action is not completely understood since it does not bind to $GABA_A$ or $GABA_B$ receptors. Also, the drug has minimal effect on the synthesis or uptake of GABA. The mechanism of action favors the selective inhibitory effect on voltage-gated calcium

TABLE 9.2.1 Postmortem Toxicology Results on Chest Blood

EIA	Results
Amphetamine	Negative
Barbiturates	Negative
Benzodiazepines	Negative
Cannabinoids	Negative
Cocaine metabolite	Negative
Methadone	Negative
Opiates	**Positive**
Phencyclidine	Negative
Propoxyphene	Negative
GC-FID Volatile Screening	
Ethanol	22 mg/dL
Acetone	<5 mg/dL
Methanol	<5 mg/dL
Isopropanol	<5 mg/dL
GC–MS screening	LC–MS/MS
Gabapentin	180 µg/mL (blood)
	42 µg/g (liver)

EIA, Enzyme immunoassay; *GC-FID*, gas chromatography with flame ionization detector; *GC–MS*, gas chromatography–mass spectrometry; *LC–MS/MS*, liquid chromatography/tandem mass spectrometry.
Bold is to highlight positive results.

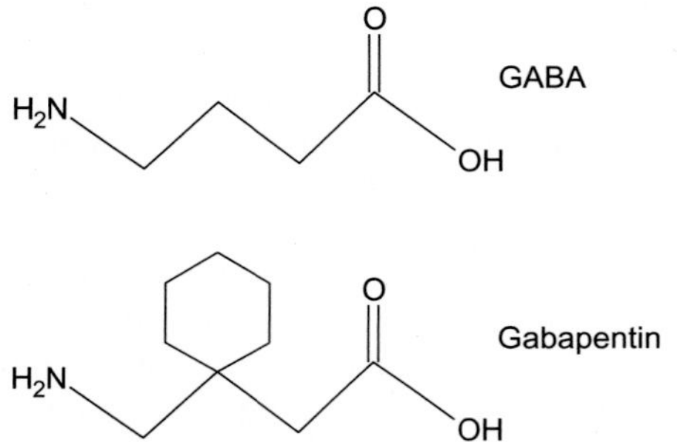

FIGURE 9.2.1 Structure of GABA and gabapentin. *GABA*, Gamma-aminobutyric acid.

channels specifically possessing the alpha-2-delta-1 subunit. Gabapentin is a unique drug that is not metabolized, does not bind to plasma proteins, and is solely eliminated unchanged by renal excretion [4]. Approximately 76%–81% of a single oral dose is eliminated in the urine and 10%–23% in feces [5]. The recommended dose of gabapentin is 900–1800 mg/day in adults and 25–35 mg/kg/day in children. The drug dose is significantly lower in patients with reduced glomerular filtration. Oral bioavailability of gabapentin varies from 27% to 60% and is inversely proportional to the dose [5]. The drug reaches a peak concentration at 1.5–4 hours with a half-life of 5–9 hours with normal renal function [4,5]. Therapeutic levels of the drug are 2–12 µg/mL. The methods of measurement include high performance liquid chromatography (HPLC), LC–MS/MS, and GC–MS.

Gabapentin is a relatively safe drug in patients with normal renal functions. Side effects from an acute overdose include double vision, somnolence, nystagmus, ataxia, slurred speech, drowsiness, lethargy, and diarrhea, but in more serious cases, drug-induced coma, hypotension, and respiratory depression develop [6–8]. Long-term usage causes weight gain, drowsiness, dizziness, and fatigue. The serious side effects that are generally reported in patients with impaired renal functions or intentional overdose include coma, hypotension, and respiratory depression [6−9]. These adverse reactions are reversible with reduction of dosage or discontinuation of therapy with gabapentin.

Because of its safe profile, cases of severe toxicity or death related to gabapentin overdose are relatively uncommon [5,10]. However, many cases of unintentional and intentional overdose and toxicity have been described [7−15]. A 71-year-old patient with renal failure, receiving 1200 mg/day gabapentin, developed confusion and hallucinations. The serum gabapentin level in this patient was 27 μg/mL [14]. In two patients with impaired renal functions, one of the patients became comatose and the other needed intubation. Gabapentin serum concentrations in these cases were 22.6 and 85.0 μg/mL [8,11]. Many cases of intentional overdose have also been reported [10,13,15,16]. In these cases the gabapentin concentrations ranged from 44.5 to 104.5 μg/mL. All subjects reported with unintentional and intentional overdose survived with supportive therapy and dialysis. Suicides involving gabapentin have been rarely reported. In fact, our research revealed only one case of suicide by gabapentin overdose [17]. In this case a 62-year-old woman was found unresponsive in her hotel room with handwritten notes of suicidal attempt and drug overdose. The peripheral blood was positive for gabapentin (88 μg/mL), clonazepam (7.7 ng/mL), and its 7-aminoclonazepam metabolite (56 ng/mL). In our case the chest blood gabapentin concentration quantified at 180 μg/mL. To our current awareness, this is the highest reported gabapentin level. Although peripheral blood was not available for testing in our case due to extreme decomposition, the peripheral blood level was likely comparable to the chest blood level. We make this assumption given that gabapentin does not undergo protein binding and has a low volume of distribution. In fact, one recent study reported that the central-to-peripheral-blood ratio mean and median for gabapentin were 0.90 and 0.97, respectively, with no significant redistribution of the drug [18]. Other drugs found in our case seem to be insignificant.

In view of the significant toxicological findings the cause of death in this case was ruled "acute gabapentin intoxication" and the manner of death accident. In conclusion, prescribing gabapentin for disorders other than epilepsy is increasing. Although deaths due to gabapentin overdose are rare, toxicity is not uncommon. Our case report augments previously reported fatalities in which gabapentin was determined contributory to cause of death and highlights the importance of determining gabapentin levels in all suspected drug-related fatalities.

References

[1] Moore A, Derry S, Wiffen P. Gabapentin for chronic neuropathic pain. JAMA 2018;319:818−19.
[2] Moore RA, Wiffen PJ, Derry S, Toelle T, Rice AS. Gabapentin for chronic neuropathic pain and fibromyalgia in adults. Cochrane Database Syst Rev 2014;CD007938. Available from: https://doi.org/10.1002/14651858.CD007938.pub2.
[3] Magnus L. Nonepileptic uses of gabapentin. Epilepsia 1999;40(Suppl. 6):S66−72 discussion S73-4.
[4] Rose MA, Kam PC. Gabapentin: pharmacology and its use in pain management. Anaesthesia 2002;57:451−62.
[5] Baselt RC. Gabapentin. In: Baselt RC, editor. Disposition of toxic drugs and chemicals in man. Seal Beach, CA: Biomedical Publications; 2017. p. 961−2.
[6] Hung TY, Seow VK, Chong CF, Wang TL, Chen CC. Gabapentin toxicity: an important cause of altered consciousness in patients with uraemia. BMJ Case Rep 2009;2009. Available from: https://doi.org/10.1136/bcr.11.2008.1268.
[7] Bookwalter T, Gitlin M. Gabapentin-induced neurologic toxicities. Pharmacotherapy 2005;25:1817−19.
[8] Jones H, Aguila E, Farber HW. Gabapentin toxicity requiring intubation in a patient receiving long-term hemodialysis. Ann Intern Med 2002;137:74.
[9] Miller A, Price G. Gabapentin toxicity in renal failure: the importance of dose adjustment. Pain Med 2009;10:190−2.
[10] Fischer JH, Barr AN, Rogers SL, Fischer PA, Trudeau VL. Lack of serious toxicity following gabapentin overdose. Neurology 1994;44:982−3.
[11] Verma A, St Clair EW, Radtke RA. A case of sustained massive gabapentin overdose without serious side effects. Ther Drug Monit 1999;21:615−17.
[12] Wahba M, Waln O. Asterixis related to gabapentin intake: a case report and review. Postgrad Med 2013;125:139−41.
[13] Fernandez MC, Walter FG, Petersen LR, Walkotte SM. Gabapentin, valproic acid, and ethanol intoxication: elevated blood levels with mild clinical effects. J Toxicol Clin Toxicol 1996;34:437−9.
[14] Mansfield AS, Qian Q. 71-year-old man with chronic kidney failure and sudden change of mental status. Mayo Clin Proc 2009;84:e5−8.
[15] Stopforth J. Overdose with gabapentin and lamotrigine. S Afr Med J 1997;87:1388.
[16] Spiller HA, Dunaway MD, Cutino L. Massive gabapentin and presumptive quetiapine overdose. Vet Hum Toxicol 2002;44:243−4.
[17] Middleton O. Suicide by gabapentin overdose. J Forensic Sci 2011;56:1373−5.
[18] Hamm CE, Gary RD, McIntyre IM. Gabapentin concentrations and postmortem distribution. Forensic Sci Int 2016;262:201−3.

Chapter 9.3

An oxcarbazepine overdose in a 23-month-old child

Stephen Thornton[1] and Uttam Garg[2]
[1]Division of Toxicology, Children's Mercy Hospital, Kansas City, MO, United States, [2]Department of Pathology and Laboratory Medicine, Children's Mercy Hospital, Kansas City, MO, United States

Case description

A 23-month-old boy with a history of focal epilepsy was brought to an emergency department after family noted he was sedated and vomiting. Per family report he was found with an open bottle of his seizure medication earlier in the day. Though no ingestion was witnessed, the bottle was empty. Initially, it was reported to the emergency department staff that this medication was Tegretol (carbamazepine). The child's initial vital signs were a pulse of 113 bpm, respiratory rate of 20 breaths per minute, a blood pressure of 69/47 mmHg. While in the emergency department, he was noted to be sedated and would arouse only to painful stimuli. He had an episode of right arm shaking that was concerning for a seizure. Laboratory evaluation was normal except for a serum carbamazepine level of 7.9 μg/mL measured by immunoassay on the Vitros 5600 chemistry analyzer (Ortho-Clinical Diagnostics, Rochester, NY). Broad-spectrum urine drug screening by gas chromatography/mass spectrometry (GC/MS) (Agilent, Santa Clara, CA) showed the presence of carbamazepine and oxcarbazepine. No radiology imaging was performed. Due to the continued altered mental status and possible seizure, the child was admitted to the hospital.

Further questioning discovered that the child was not currently nor had ever been prescribed carbamazepine and there was no one in the home who had been prescribed carbamazepine. Rather, the child had been prescribed oxcarbazepine (Trileptal) 300 mg/5 mL, and it was this bottle he had been found with at home. It was estimated the child could have ingested up to 4.5 g of oxcarbazepine. Further testing by high-performance liquid chromatography (HPLC) with UV detector confirmed the presence of oxcarbazepine and its metabolite 10,11-dihydro-10-hydroxycarbamazepine (DiCBZ) at the concentrations of 20.4 and 49.3 μg/mL, respectively. Carbamazepine was not detected by HPLC.

In the hospital an EEG was performed which was normal and the child had no further events concerning for seizure activity. The seizure activity noted in the emergency department was attributed to a dystonic reaction. He remained irritable but gradually improved over the next 2 days and was discharged on hospital day 3. He was continued on his previous oxcarbazepine dose.

Case discussion

Oxcarbazepine is commonly used for seizure disorders and mood disorders. It is a prodrug and structurally is similar to carbamazepine (Fig. 9.3.1). It is rapidly metabolized in the liver to DiCBZ, its active metabolite [1]. DiCBZ blocks voltage-sensitive sodium channels resulting in stabilization of hyperexcited neuronal membranes and thereby inhibiting repetitive firing and decreasing the propagation of synaptic impulses [2].

Overdoses of oxcarbazepine are not commonly reported in the medical literature. The few that are suggest that mild-to-moderate CNS sedation would be expected as was seen in this child [3]. In particular, pediatric exposures to oxcarbazepine seem to be associated with little morbidity [4]. This case is the youngest oxcarbazepine overdose reported in the medical literature.

There is lack of published cases associating severe outcomes or deaths to oxcarbazepine exposures. One poison control center study did find oxcarbazepine exposures associated with more seizures than other "newer" anticonvulsants

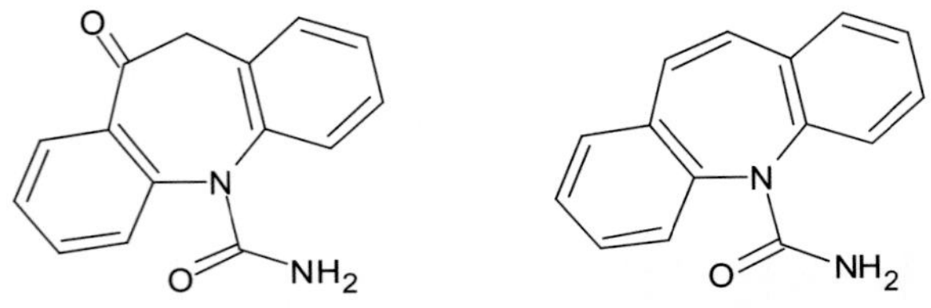

FIGURE 9.3.1 Structures of oxcarbazepine and carbamazepine.

but still was not associated with severe outcomes [5]. Initially, there was concern that this child may have been having seizures but this was not confirmed with EEG monitoring and, rather, was attributed to a dystonic reaction. Dystonic reactions are rarely described with oxcarbazepine, especially with overdoses, as the one case in the medical literature details an 8-year old on chronic therapy who developed dystonia [6]. Hyponatremia is a complication that occurs with chronic, therapeutic use of oxcarbazepine through an unclear mechanism [7]. It has not been described after an acute exposure, and this patient's sodium was normal throughout his hospitalization.

This case was complicated by the false-positive carbamazepine serum level that was initially ordered due to the emergency department believing the child was on carbamazepine. In general, obtaining oxcarbazepine or DiCBZ levels is not clinically useful or routinely recommend in acute exposures. Subsequent history and testing confirmed that this was an oxcarbazepine exposure. Interestingly, both immunoassay and GC/MS produced false-positive carbamazepine results. Further experiments showed that oxcarbazepine but not DiCBZ produced false-positive results by immunoassay [8]. On the other hand, DiCBZ, but not oxcarbazepine, was responsible for false-positive carbamazepine on GC/MS [8].

This case suggests that oxcarbazepine overdoses, even in young pediatric patients, are associated with minimal morbidity but health-care providers should be aware of possible laboratory confusion with carbamazepine.

References

[1] Hooper WD, Dickinson RG, Dunstan PR, Pendlebury SC, Eadie MJ. Oxcarbazepine: preliminary clinical and pharmacokinetic studies on a new anticonvulsant. Clin Exp Neurol 1987;24:105−12.
[2] McLean MJ, Schmutz M, Wamil AW, Olpe HR, Portet C, Feldmann KF. Oxcarbazepine: mechanisms of action. Epilepsia 1994;35(Suppl. 3): S5−9.
[3] Pedrini M, Noguera A, Vinent J, Torra M, Jiménez R. Acute oxcarbazepine overdose in an autistic boy. Br J Clin Pharmacol 2009;67 (5):579−81.
[4] Spiller HA, Strauch J, Essing-Spiller SJ, Burns G. Thirteen years of oxcarbazepine exposures reported to US poison centers: 2000 to 2012. Hum Exp Toxicol. 2015;. Available from: https://doi.org/10.1177/0960327115618246.
[5] Wills B, Reynolds P, Chu E, Murphy C, Cumpston K, Stromberg P, et al. Clinical outcomes in newer anticonvulsant overdose: a poison center observational study. J Med Toxicol 2014;10(3):254−60.
[6] Hergüner MÖ, Incecik F, Altunbaşak S. Oxcarbazepine-induced tardive dyskinesia: a rare adverse reaction. J Pediatr Neurosci 2010;5(1):85−6.
[7] Dong X, Leppik IE, White J, Rarick J. Hyponatremia from oxcarbazepine and carbamazepine. Neurology 2005;65(12):1976−8.
[8] Garg U, Johnson L, Wiebold A, Ferguson A, Frazee C, Thornton S. False-positive carbamazepine results by gas chromatography−mass spectrometry and VITROS 5600 following a massive oxcarbazepine ingestion. J Appl Lab Med 2018;3(1):1−4.

Chapter 10

Antineoplastic drugs

Alejandro R. Molinelli and Kristine R. Crews
Department of Pharmaceutical Sciences, St. Jude Children's Research Hospital, Memphis, TN, United States

Introduction

In its most reductive definition cancer is the name given to a collection of diseases in which cells divide without control and invade nearby tissues [1]. During the last few decades, great strides have been made in understanding the development and complexities of neoplastic disease [2]. As such many new drugs targeting different facets of tumor development have reached clinical practice. Antineoplastic drugs now include hundreds of compounds with dozens of mechanisms of action [3]. Modern cancer treatment seldom includes a single approach but rather includes the use of multiple combinations of drugs with differing mechanisms of action together with radiotherapy, and/or surgery when indicated. The type of tumor being treated will dictate the therapeutic management of the patient and increasingly the molecular characterization of the tumor guides the use of different pathway-targeted therapies. In this chapter, we will describe the breadth of antineoplastic drug categories including traditional cytotoxic drugs, pathway-targeted therapies, and hormone regulators. We will also describe in detail two cytotoxic medications for which laboratory measurements are essential: methotrexate (MTX) and 6-thiopurine analogs. The case study will describe the management of MTX-induced nephrotoxicity in a pediatric oncology patient undergoing treatment for metastatic osteosarcoma.

Childhood cancer

Childhood malignancies are relatively rare with an estimated 10,590 new cases per year in the United States. The most common childhood malignancies are the leukemias comprising 29% of all childhood cancers (ages 0–14). Of the leukemias, acute lymphocytic leukemia (ALL) is the most common comprising 26% of all childhood cancers (ages 0–14). Five-year survival for ALL is 91%. Bone tumors account for 4% of all childhood cancers (ages 0–14) and osteosarcoma is the most common. The 5-year survival rate for osteosarcoma is 71%, although for metastatic osteosarcoma, it is lower [4].

Therapeutic drug monitoring

Therapeutic drug monitoring (TDM) can be used to assess toxicity, optimize drug efficacy, and to monitor compliance with treatment regimes. It is most useful for drugs that have a narrow therapeutic index, have a relatively steep relationship between exposure and response, have substantial pharmacokinetic variability, are administered repeatedly, and there are no clinical parameters that can be used to determine dosage (e.g., biomarkers). It is also important to have a complete understanding of the relationship between drug levels and therapy outcomes. For most antineoplastic agents, current dosing practices focus on body-surface area and standardized dosing instead of pharmacokinetically guided dosing. This is likely because of the paucity of clinical outcome studies correlating specific blood levels with therapy endpoints. This lack of outcomes data could be the result of the long-lag time between antineoplastic therapy and response (e.g., complete remission), the use of combination therapies including multiple drugs, and the complexities of tumor biology. Additional complicating factors include needing to use advanced analytical instrumentation (e.g., chromatography and mass spectrometry), low testing volumes making the cost of test development and maintenance difficult to justify, and the need for properly timed collections. Given these factors, only a few drugs from the hundreds of antineoplastic agents' available benefit from TDM.

Cytotoxic agents

The cytotoxic agents comprise some of the oldest drugs traditionally used in cancer chemotherapy. Cytotoxic agents prevent cell replication or growth and exert their effects systemically. Some common cytotoxic agents are listed in Table 10.1 [5]. The alkylating agents react with electron-rich atoms in macromolecules forming covalent bonds. Most alkylating agents tend to form carbonium ion intermediates that covalently bind DNA. Commonly observed targets of alkylating agents include the N7 position of guanine, the N1 and N3 positions of adenine, the N3 position of cytosine, and the O6 position of guanine. Amino and sulfhydryl groups of proteins are also common targets. Bifunctional alkylating agents such as the nitrogen mustards have two functional groups that can form DNA or protein adducts. Many nitrogen mustards such as cyclophosphamide require metabolic activation to an active moiety. The alkyl sulfonate busulfan alkylates DNA through the release of methyl radicals. Busulfan exhibits selective toxicity toward myeloid cells and is used as a preparative regimen in allogeneic bone marrow transplantation. Pharmacokinetically guided dosing of busulfan is recommended as low exposures can lead to graft-rejection and high exposures increase the risk of sinusoidal obstruction syndrome. Multiple chromatographic and mass spectrometric methods to monitor busulfan have been published in the literature [6]. Although platinum coordination complexes are listed as alkylating agents these compounds form DNA adducts.

Antimetabolites are compounds that interfere with the enzymes or reactions involved in nucleotide synthesis. An important example used in the treatment of several childhood malignancies including ALL and osteosarcoma is MTX which will be discussed in detail in subsequent sections. Other important antimetabolites include the thiopurine analogs and the fluoropyrimidines. The fluoropyrimidine 5-fluorouracil (5-FU) is an important component in the treatment of carcinomas involving the colon, upper gastrointestinal tract, and the breast. 5-FU requires ribosylation and phosphorylation to the nucleotide form to be incorporated into DNA or RNA. The enzyme dihydropyrimidine dehydrogenase (DPD) which is found in the liver, intestinal mucosa, and tumor cells can inactivate 5-FU by the reduction of the pyrimidine ring. The gene encoding DPD (termed DPYD) exhibits inherited variability. Genotyping of the DPYD gene prior to the beginning of therapy with 5-FU can reduce the occurrence of severe 5-FU toxicity [7]. Cytarabine (cytosine arabinoside, Ara-C) is a cytosine analog that prevents DNA base pairing by hindering the rotation of the pyrimidine base around the nucleoside bond.

Other cytotoxic drugs include the microtubule damaging agents which interfere with microtubule functions and can lead to apoptosis. The vinca alkaloids bind to β-tubulin blocking its polymerization with α-tubulin to form microtubules. The interference with microtubule formation results in cell division arrest as the mitotic spindle cannot form and duplicated chromosomes cannot align in the division plate. The camptothecin analogs are DNA topoisomerase inhibitors. Topoisomerases reduce torsional stress in supercoiled DNA allowing replication and transcription by DNA and RNA polymerases. Camptothecins are s-phase specific as DNA synthesis needs to be occurring for the drugs to render their effect. As such prolonged exposure is necessary. Pharmacokinetically guided dosing of topotecan utilizing population-based models produces more accurate and precise estimates of exposure than a fixed-dosing approach and can be used to individualize drug dosing [8]. The antibiotics affect DNA replication and transcription; anthracyclines intercalate DNA and can form complexes with topoisomerase II.

The cytotoxic drug mitotane [1,1-(dichlorodiphenyl)-2,2-dichloroethane] (o,p'-DDD) is chemically similar to the insecticides DDT and DDD. Mitotane is used in the treatment of adrenocortical carcinoma. The mechanism of action and pharmacokinetics of mitotane are not completely understood, but its antineoplastic effect is correlated with drug plasma levels; therefore therapeutic drug monitoring is recommended [9,10].

L-Asparaginase is a standard agent in the treatment of ALL. Malignant lymphoid cells lack sufficient amounts of asparagine synthase to enable protein synthesis and require circulating asparagine. L-Asparaginase hydrolyses asparagine to aspartic acid and ammonia, thus depriving malignant cells of this amino acid. A side effect of L-asparaginase use is hypersensitivity reactions which can occur in 5%−20% of patients [5]. Enzyme inactivation by antibodies can lead to treatment failure. Laboratory based methods to monitor asparaginase activity [11] and asparaginase antibodies [12] can aid in the selection of agents or facilitate clinical decisions on whether to rechallenge patients after prior hypersensitivity reactions [13].

The cytotoxic agents predominantly affect all rapidly replicating cells in the body; thus it is common to observe similar toxicities. However, the occurrence and severity of side effects is highly dependent on the medications used, and the quality of the supportive care offered. Some of the side effects observed during treatment with cytotoxic agents include damage to dividing mucosal cells causing ulceration (e.g., mucositis); damage to hair follicles causing alopecia; and intestinal denudation causing nausea and vomiting, that in severe cases could devolve into neutropenic enterocolitis (typhlitis). Neutropenic enterocolitis is the breakdown of the intestinal mucosa in the presence of neutropenia leading to

TABLE 10.1 Selected Cytotoxic Antineoplastic Agents

Drug Action	Drug Category	Drug Examples	Selected Structures
Alkylating agents	Nitrogen mustards	Cyclophosphamide	Cyclophosphamide
	Ethyleneimines	Thiotepa	
	Alkyl sulfonates	Busulfan	Busulfan
	Nitrosoureas	Carmustine	
	Triazenes	Dacarbazine, temozolomide	
	Methylhydrazines	Procarbazine	
	Platinum coordination complexes	Cisplatin, carboplatin	
Antimetabolites	Folic acid analogs	Methotrexate	Methotrexate
	Pyrimidine analogs	5-Fluorouracil	
	Cytidine analogs	Cytarabine	
	Purine analogs	6-Thioguanine	
Microtubule-damaging agents	Vinca alkaloids	Vincristine	Vincristine
	Taxanes	Paclitaxel, docetaxel	
Camptothecin analogs	DNA topoisomerase inhibitors	Topotecan, irinotecan	Topotecan
Antibiotics	Anthracyclines	Doxorubicin	Doxorubicin
	Epipodophyllotoxins	Etoposide	
Other drugs	L-asparaginase (e.g., pegaspargase), hydroxyurea, mitotane (o,p′-DDD), tretinoin (all-*trans* retinoic acid), arsenic trioxide		Mitotane

bacterial or fungal infections. Additional side effects include dose-limiting toxicity to marrow cells (e.g., myelosuppression); neurotoxicity including altered mental status, generalized seizures, and coma; sinusoidal obstruction syndrome (veno-occlusive disease) which is characterized by nonthrombotic occlusion of the terminal hepatic venules and sinusoids causing portal hypertension and ischemic necrosis of the liver. Some alkylating agents can also induce acute nonlymphocytic leukemia.

Pathway-targeted therapies

As the molecular changes that define neoplastic disease are elucidated, a diverse array of targeted therapies has been developed. Hanahan and Weinberg [2] provide a conceptual framework of the biological characteristics' tumors acquire. These include sustaining proliferative signaling, evading growth suppressors, resisting cell death, enabling replicative immortality, inducing angiogenesis, activating invasion and metastasis, reprogramming of energy metabolism, and evading immune destruction. The authors also highlight the important role genomic instability and inflammation play in the development of neoplastic disease. As increased understanding of the pathophysiology of cancer has been gained, progress in the development of pathway-targeted therapies has yielded hundreds of new compounds.

The two main classes of pathway-targeted drugs are monoclonal antibodies (mAbs) and small molecules. mAbs exert their antitumor action by blocking cell surface receptors such as the epidermal growth factor receptor (EGFR). mAbs also recruit immune system cells to the antigen—antibody complex. Small molecules can enter cells and inhibit specific enzymes such as protein kinases [16].

The first pathway-targeted drug to gain FDA approval in the United States was the small molecule imatinib mesylate which is used in the treatment of chronic myelogenous leukemia (CML). A defining characteristic of CML is the Philadelphia chromosome translocation where the ABL1 gene on chromosome 9 is translocated onto the BCR gene of chromosome 22 leading to the transcription of the BCR-ABL protein kinase which drives cell proliferation, one of the defining features of malignant cells. Imatinib exerts its inhibitory effect by binding to the active site of the tyrosine kinase, thus making the site unavailable for binding by adenosine triphosphate. Several second generation tyrosine kinase inhibitors (TKI's) have been subsequently developed including dasatinib and nilotinib. The second-generation compounds have higher inhibitory activity and are often used in patients that have developed imatinib resistance.

Therapeutic drug monitoring outside of cytotoxic agents is rare. However, drugs such as imatinib and other TKI's fulfill many of the indications for TDM including significant inter-individual variability in plasma concentrations. An additional indication is the monitoring of therapy compliance. Studies have shown that oral therapy over long periods of time can often lead to decreased compliance once symptoms resolve. The lack of studies correlating specific blood drug concentrations to treatment outcomes is probably one of the reasons TDM of TKI's is not performed. However, in the case of imatinib, studies have demonstrated correlations between trough plasma levels and response to therapy, including molecular response and complete cytogenetic response [17]. The European Society for Medical Oncology recommends TDM for imatinib in the cases of treatment failure or adverse effects during the treatment of CML [18].

Similar to the small molecule TKI's, therapeutic drug monitoring of mAbs is not common. The pharmacokinetics of mAbs can be influenced by multiple factors. These include demographic variables such as body—surface area; immunogenicity, that is the formation of antidrug antibodies; blood chemistry variables such as serum albumin and alkaline phosphatase; and treatment variables such as distribution dependent on the drug administration modality. Disease variables can also affect the pharmacokinetics of mAbs. For some mAbs, such as trastuzumab, used in the treatment of HER2 positive metastatic breast cancer, the number of metastatic sites is an important variable for trastuzumab clearance [19]. Bruno et al. [20] report 22% higher trastuzumab clearance in patients with four or more metastatic sites resulting in lower exposure at steady state. When compared to small molecules, mAbs demonstrate enhanced tolerability leading to ambiguity in the selection of upper concentration limits. Additional concerns include the argument that the high pharmacokinetic variability of mAbs is not as important as their pharmacodynamic variability. Thus pharmacodynamic endpoints might be better variables to measure [21].

The family of receptor tyrosine kinases (ErbB) to which EGFR belongs plays important roles in cell proliferation, survival, and differentiation. When ligands bind to the extracellular domain of EGFR, receptor dimerization stimulates the tyrosine kinase activity of the intracellular domain. The downstream effects include the stimulation of multiple signaling pathways including mitogen-activated protein kinases, the phosphatidylinositol-3-kinases/protein kinase B (PI3K/AKT), and STAT (signal transducer and activator of transcription) pathways. Cetuximab is a recombinant antibody that binds the extracellular domain of EGFR and prevents ligand binding. Cetuximab is indicated for the treatment of metastatic colon cancer and head and neck squamous cell carcinomas. Although not routinely monitored several studies report improved outcomes such as progression-free survival and higher trough concentrations when TDM is performed [19,22]. Other mAbs that could benefit from TDM based on existing exposure–response relationships include alemtuzumab, obinutuzumab, rituximab, and trastuzumab [19].

Hormones and hormone regulators

Hormones are another important group of drugs used in neoplastic disease therapy. These molecules exert their effects by interacting with hormone nuclear receptors which act as ligand-dependent transcription factors upon binding by the target hormone. The downstream effects of hormone receptors are varied and dependent on the target tissue. In oncology some of the observed effects include diminished cell proliferation, immunosuppression, and apoptotic responses. Hormone receptors include the glucocorticoid receptor, estrogen and androgen receptors, and the progesterone receptor. Prednisone is a synthetic glucocorticoid used in the treatment of multiple hematological malignancies in combination with other targeted or cytotoxic agents [23]. Tamoxifen is a selective estrogen receptor modulator used in the treatment of estrogen receptor positive breast cancer. Tamoxifen is metabolized by the CYP2D6 enzyme to the active metabolites 4-hydroxytamoxifen and N-desmethyl-4-hydroxytamoxifen (endoxifen). The CYP2D6 enzyme exhibits genetic polymorphisms, which may place certain patients at risk for disease recurrence. In addition, patients taking strong CYP2D6 inhibitors also have an increased risk of recurrence. Knowledge of CYP2D6 status should be used to guide therapy [24]. Although of limited availability, the TDM of tamoxifen could also be performed to assess compliance or to ensure adequate treatment [6].

Discussion of methotrexate and the thiopurine analogs

Methotrexate

MTX is a mainstay in the therapy of childhood ALL where it is used in high-dose regimes (greater than 500 mg/m^2 dosing) during the remission induction and consolidation phases of treatment. During the maintenance phase, it is typically administered orally at low doses. High steady state levels during induction are associated with lower relapse rates. MTX is also utilized in the treatment of childhood osteosarcoma, non-Hodgkin lymphoma, medulloblastoma, and carcinomas of the head, neck, lung, breast, cervix, ovaries, testes, and bladder [14]. MTX is also utilized in the treatment of several nonmalignant conditions such as rheumatoid arthritis, psoriasis, and ectopic pregnancy.

Thiopurine analogs

The thiopurine medications include azathioprine (a prodrug for mercaptopurine), 6-mercaptopurine (6-MP), and 6-thioguanine. In general, azathioprine and 6-MP are used in the treatment of nonmalignant immunologic disorders, 6-MP is used in the treatment of lymphoid malignancies, and 6-thioguanine is used in the treatment of myeloid leukemias [15]. The thiopurine analogs also have indications in nonmalignant conditions such as inflammatory bowel disease, severe rheumatoid arthritis, and autoimmune disease. The thiopurines are involved in a complex metabolic pathway described in detail in a subsequent section.

Mechanisms of action

Methotrexate

MTX is a folic acid analog that inhibits the enzyme dihydrofolate reductase which normally reduces folic acid to tetrahydrofolic acid (FH$_4$). FH$_4$ is a cofactor that provides methyl groups for the synthesis of nucleic acid precursors and amino acids such as thymidylate, purine nucleotides, methionine, glycine, and serine. This leads to inhibition of DNA, RNA, and protein synthesis. Extracellular MTX exists in its monoglutamate form. However, once it is transported into cells, it undergoes a series of sequential glutamic acid residue additions catalyzed by the enzyme folypolyglutamate synthetase. The resulting intracellular MTX polyglutamate reservoir increases the inhibitory potency of the drug.

Thiopurine analogs

The 6-thiopurine analogues undergo a complex metabolic pathway involving multiple enzymes (Fig. 10.1). The cytotoxic effects of the thiopurine drugs is mediated in part through their conversion to thioguanine nucleotides (TGNs) which inhibit de novo purine synthesis and become incorporated into nucleic acids where they induce strand breaks and base mispairing.

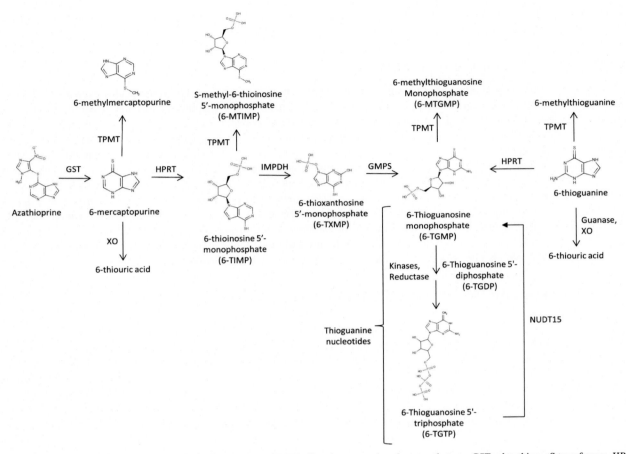

FIGURE 10.1 Metabolic pathway of the thiopurine analogs. *GMPS*, Guanine monophosphate synthetase; *GST*, glutathione *S*-transferase; *HPRT*, hypoxanthine-guanine phosphoribosyltransferase; *IMPDH*, inosine-5′-monophosphate dehydrogenase; *TPMT*, thiopurine methyltransferase; *XO*, xanthine oxidase. Source: *Chemical structures obtained from The Human Metabolome Database (http://www.hmdb.ca).*

Pharmacokinetics

Methotrexate

MTX doses higher than 25 mg/m^2 cannot be completely absorbed orally and need to be administered intravenously. Following IV administration MTX undergoes a rapid distribution phase followed by a second phase characteristic of renal clearance ($t_{1/2}$ of 2–3 hours). A third phase with a $t_{1/2}$ of 8–10 hours can be prolonged by renal failure. MTX is excreted in urine mostly unchanged. However, after high doses metabolites such as 7-hydroxy-MTX are detectable. 7-Hydroxy-MTX is a potentially nephrotoxic metabolite. Glomerular filtration and tubular secretion account for the renal excretion of MTX. The use of drugs that reduce renal blood flow such as NSAID's, that are nephrotoxic, or that are weak organic acids can delay excretion and cause myelosuppression. MTX can be distributed into third spaces (pleural or peritoneal cavities), and in the presence of pleural effusions or ascites, these spaces may act as reservoirs. The slow release of the drug from third spaces can lead to prolonged elevation of serum/plasma concentrations. MTX undergoes moderate protein binding (approximately 50%), and many drugs such as phenytoin, sulfonamides, salicylates, tetracycline, and chloramphenicol can displace it from plasma albumin [5,14].

Thiopurine analogs

Azathioprine is a prodrug that is converted to 6-MP. Absorption of oral 6-MP is incomplete (10%–50%) due to first-pass metabolism in the liver. 6-MP can be inactivated to 6-thiouric acid by xanthine oxidase or to 6-methylmercaptopurine (6-MMP) by thiopurine *S*-methyltransferase (TPMT). Oral doses of 6-MP should be reduced in patients receiving allopurinol, an inhibitor of xanthine oxidase. In the anabolic pathway, 6-MP is converted to

6-thioinosine 5′-monophosphate (6-TIMP) by hypoxanthine-guanine phosphoribosyltransferase (HPRT). 6-TIMP is then converted to 6-thioguanosine monophosphate (6-TGMP) through the action of inosine monophosphate dehydrogenase and guanine monophosphate synthetase. 6-TGMP can then be phosphorylated to its diphosphate and triphosphate moieties. 6-Thioguanosine triphosphate (6-TGTP) is the active compound. 6-Thioguanine (6-TG) is metabolized to 6-TGMP by the action of HPRT and can be inactivated to 6-thiouric acid by guanase and xanthine oxidase or to 6-methylthioguanine by TPMT. 6-TIMP and 6-TGMP can also undergo S-methylation by TPMT. The methylation of 6-TIMP generates S-methylthioinosine-5′-monophosphate (6-MTIMP), an active metabolite that is a potent inhibitor of de novo purine synthesis. Fig. 10.1 summarizes the metabolic pathways for the thiopurine analogs.

As described above TPMT is a cytosolic enzyme that catalyzes S-methylation of MP and thioguanine yielding S-methyl nucleobases that are inactive, but also S-methylated nucleotides which are active. TPMT activity is inherited as a monogenic, autosomal, codominant trait, and three single nucleotide polymorphisms account for over 90% of low activity phenotypes [15]. Individuals who are homozygous or compound heterozygous TPMT deficient are at high risk for life-threatening myelosuppression. Heterozygous patients are at higher risk for toxicity from azathioprine and 6-MP than wild-type patients although some heterozygous patients have high tolerance. This could be due to heterozygous patients having high levels of TGNs but lower concentrations of methyl MP nucleotides [25].

NUDT15 belongs to the Nudix hydrolase superfamily that catalyzes the hydrolysis of nucleoside diphosphates and nucleoside triphosphates. In the thiopurine metabolic pathway NUDT15 catalyzes the conversion of 6-TGTP (active) to 6-TGMP (inactive). Defects in NUDT15 mediated degradation of 6-TGTP result in higher concentrations of this metabolite available for incorporation into DNA. The first NUDT15 variant (rs 116855232; c.415C > T) results in nearly complete loss of enzymatic activity. Patients carrying this allele are at risk for severe myelosuppression [26]. Patients with ALL who are homozygous for this allele tolerated only 8% of the standard 6-MP dose, whereas heterozygous and wild-type patients tolerated 63% and 83.5% of the standard dose [27]. NUDT15 and TPMT polymorphisms differ among ancestral groups. Inherited TPMT deficiency affect patients of European and African descent, whereas NUDT15 explains the majority of thiopurine-related myelosuppression in Asians and has been found in Hispanics [15].

Treatment

Methotrexate

The goal of any antineoplastic therapy is to contain toxicity while maintaining adequate concentrations to ensure proper treatment. With high-dose MTX therapy (greater than 500 mg/m^2 dosing) the goal is to prevent nephrotoxicity by prehydrating the patient with IV sodium bicarbonate. MTX precipitates in acidic urine causing renal tubule damage, thus maintaining a urine pH of 6.5 will increase MTX solubility. Avoiding other nephrotoxic drugs such as amphotericin and NSAID's and drugs that could compete for MTX excretion (e.g., penicillins and proton pump inhibitors) is an important consideration. Leucovorin is a fully reduced folate coenzyme that restores the intracellular pool of FH$_4$ cofactors. Leucovorin is used during high-dose MTX therapy to terminate toxicity, and its dosing should be guided by MTX serum/plasma concentrations. Patients with delayed MTX elimination, MTX-induced nephrotoxicity, or accidental intrathecal overdose who do not respond adequately to leucovorin rescue could be treated with glucarpidase (Voraxaze). Glucarpidase is a recombinant bacterial enzyme (carboxypeptidase G2) that cleaves the carboxyl-terminus glutamate residue from MTX, thereby forming the noncytotoxic metabolites 4-amino-4-deoxy-N10-methylpteroic acid (DAMPA) and glutamic acid. Glucarpidase administration can cause MTX levels to fall by 99% within 5–15 minutes following administration. It is most effective when administered within 48 hours from the start of MTX treatment [5]. Delaying the use of glucarpidase will result in diminished effects as MTX is incorporated into cells and not available in blood for cleavage by glucarpidase.

Thiopurine analogs

6-MP is utilized in the maintenance phase of therapy for ALL. Important laboratory values to monitor are white blood cell and platelet counts. MTX use is synergistic with 6-MP as MTX elevates intracellular concentrations of alpha-phosphoribosylpyrophosphoric acid (PRPP) which is a cofactor in several reactions involving the transfer of phosphate groups. Prior to the beginning of therapy with thiopurine medications it is essential to gain knowledge of the patient's TPMT and NUDT15 status through genetic testing. There is substantial evidence linking TPMT and NUDT15 genotype with phenotypic variability and preemptive dose adjustment has been shown to reduce adverse effects without compromising therapy [15]. Hepatotoxicity and myelosuppression are the most serious toxicities observed during treatment with thiopurine medications. Dose reduction or discontinuation of therapy is the main courses of action in the management of thiopurine-induced toxicities. Adherence to 6-MP therapy plays an important role in lowering the risk of ALL relapse. Patients that do not adhere to their 6-MP regimen (adherence rate <95%) are at a 2.7-fold risk of relapse compared to patients with a 95% or higher adherence rate. Among patients that adhere to their 6-MP treatment regimens high intra-individual variability in TGN levels contributes to an increased risk of relapse. Therefore 6-MP regimen adherence and maintaining a steady thiopurine exposure are important factors in the prevention of ALL relapse [28]. Therapeutic drug monitoring of TGN levels can help clinicians ensure patient adherence to therapy.

Analytical methods and clinical management implications

Methotrexate

The use of therapeutic drug monitoring in antineoplastic therapy has been limited. MTX is possibly the most commonly monitored antineoplastic drug in part due to knowledge of its pharmacokinetics and the relationship between blood drug concentrations and toxicity. Widespread monitoring of MTX is very likely possible due to the existence of commercial immunoassays. During the last several decades, two FDA-approved commercial MTX immunoassays methods dominated the market in the United States. The first one consisted of an enzyme-multiplied immunoassay technique developed by the Syva Corporation (later acquired by Siemens Diagnostics, Tarrytown, NJ) in the late 1970s (reviewed by Ref. [29]). The other technique consisted of a Fluorescence Polarization Immunoassay (FPIA) developed by Abbott Diagnostics (Lake Forest, Illinois) and designed to be used with the TDx line of instruments. This method was approved by the FDA in 1983 and was subsequently improved upon (reviewed by Ref. [30]). Proficiency testing surveys performed by the College of American Pathologists showed that the Abbott TDx FPIA was the method of choice for most laboratories [31]. In July 2017 the Abbott TDx FPIA was retired from the US market. In the year 2011 the FDA approved a new MTX homogeneous enzyme immunoassay manufactured by ARK Diagnostics (Fremont, California) to be used as a user-defined assay in various chemistry analyzers. A chemiluminescent microparticle immunoassay for use with the Abbott Architect i-series instruments is available for use outside of the United States but does not have FDA-approval [32–34]. All MTX immunoassays suffer from cross-reactivity with MTX metabolites (e.g., 7-hydroxy-MTX) especially when blood concentrations are compared to results obtained using chromatographic or mass spectrometry-based techniques. It is also important to note that when compared to the Abbott TDx FPIA assay the ARK Diagnostics assay overestimates MTX concentrations, particularly at the low end of the analytical measurement range [35,36]. It is important for laboratories to consider that due to the wide use of the Abbott TDx FPIA method throughout several decades, many treatment guidelines have been developed using blood concentrations obtained using this assay. Numerous high performance liquid chromatography (HPLC) or liquid chromatography-tandem mass spectrometry (LC-MS/MS) assays to monitor MTX have been published in the literature [6], but their routine use has been limited. An important consideration when utilizing MTX immunoassays is that in patients where glucarpidase is administered, the MTX metabolite DAMPA will cross-react with all the commercial immunoassays yielding unreliable results. In such patients, it is necessary to follow MTX levels utilizing other methods.

Thiopurine analogs

The use of thiopurine medications should be guided by the use of four laboratory tests. Due to the important roles TPMT and NUDT15 metabolism plays in the use of these drugs it is recommended to know the TPMT and NUDT15 genotypes of patients prior to the beginning of therapy [15]. Analyzing TPMT activity is a useful indicator of TPMT phenotype. Monitoring the thiopurine metabolites (6-TG and 6-MMP) is useful to monitor compliance with therapy and as an adjunct in the evaluation of TPMT phenotype. Thiopurine metabolite concentrations should be obtained following standard, uninterrupted dosing over at least 2 weeks. Multiple methods exist for the determination of thiopurine metabolites. All the methods involve the hydrolysis of the glycosidic bond between the ribose and the nitrogen base. Methods based on acid hydrolysis and detection by HPLC have been in use for several decades [37,38]. Enzymatic hydrolysis methods coupled with HPLC have been published but do not enjoy wide utilization [39,40]. Methods based on acid hydrolysis and detection with LC–MS/MS are also common [41,42]. TPMT phenotype can be measured utilizing a radiochemical assay [43], HPLC [44] or LC–MS/MS [45,46].

References

[1] National Cancer Institute. Dictionary of cancer terms, Retrieved from: <https://www.cancer.gov/publications/dictionaries/cancer-terms/def/cancer>; 2018.

[2] Hanahan D, Weinberg RA. Hallmarks of cancer: the next generation. Cell 2011;144(5):646−74.

[3] Wellstein A. General principles in the pharmacotherapy of cancer. In: Brunton LL, et al., editors. Goodman & Gilman's pharmacol basis therapeutics. 13th ed. McGraw-Hill Education; 2018. p. 1161−6.

[4] American Cancer Society. Cancer facts & figures 2018. Atlanta, GA: American Cancer Society; 2018.

[5] Wellstein A, et al. Cytotoxic drugs. In: Brunton LL, et al., editors. Goodman & Gilman's pharmacol basis therapeutics. 13th ed. McGraw-Hill Education; 2018. p. 1167−202.

[6] Dasgupta A. Therapeutic drug monitoring in cancer patients: application of chromatographic techniques. In: Dasgupta A, editor. Advances in chromatographic techniques for therapeutic drug monitoring. CRC Press/Taylor & Francis; 2010. p. 299−322.

[7] Amstutz U, Henricks LM, Offer SM, Barbarino J, Schellens JHM, Swen JJ, et al. Clinical Pharmacogenetics Implementation Consortium (CPIC) guideline for dihydropyrimidine dehydrogenase genotype and fluoropyrimidine dosing: 2017 update. Clin Pharmacol Ther 2018;103(2):210−16.

[8] Schaiquevich P, Panetta JC, Iacono LC, Freeman III BB, Santana VM, Gajjar A, et al. Population pharmacokinetic analysis of topotecan in pediatric cancer patients. Clin Cancer Res 2007;13(22 Pt 1):6703−11.

[9] Feliu C, Cazaubon Y, Guillemin H, Vautier D, Oget O, Millart H, et al. Therapeutic drug monitoring of mitotane: analytical assay and patient follow-up. Biomed Chromatogr 2017;31(11).

[10] Kerkhofs TM, Derijks LJ, Ettaieb H, den Hartigh J, Neef K, Gelderblom H, et al. Development of a pharmacokinetic model of mitotane: toward personalized dosing in adrenocortical carcinoma. Ther Drug Monit 2015;37(1):58−65.

[11] Fernandez CA, Cai X, Elozory A, Liu C, Panetta JC, Jeha S, et al. High-throughput asparaginase activity assay in serum of children with leukemia. Int J Clin Exp Med 2013;6(7):478−87.

[12] Wang B, Hak LJ, Relling MV, Pui CH, Woo MH, Storm MC. ELISA to evaluate plasma anti-asparaginase IgG concentrations in patients with acute lymphoblastic leukemia. J Immunol Methods 2000;239(1−2):75−83.

[13] Bleyer A, Asselin BL, Koontz SE, Hunger SP. Clinical application of asparaginase activity levels following treatment with pegaspargase. Pediatr Blood Cancer 2015;62(6):1102−5.

[14] Dasgupta A, et al. Therapeutic drug monitoring of antineoplastic drugs. In: Hammett-Stabler CA, Dasgupta A, editors. Therapeutic drug monitoring data: a concise guide. 3rd ed. AACC Press; 2007. p. 209−20.

[15] Relling MV, Schwab M, Whirl-Carrillo M, Suarez-Kurtz G, Pui CH, Stein CM, et al. Clinical Pharmacogenetics Implementation Consortium (CPIC) guideline for thiopurine dosing based on TPMT and NUDT15 genotypes: 2018 update. Clin Pharmacol Ther 2018. Available from: https://doi.org/10.1002/cpt.130? [Epub ahead of print].

[16] Wellstein A, et al. Pathway-targeted therapies: monoclonal antibodies, protein kinase inhibitors, and various small molecules. In: Brunton LL, et al., editors. Goodman & Gilman's the pharmacological basis of therapeutics. 13th ed. McGraw-Hill Education; 2018. p. 1203−36.

[17] Herviou P, Thivat E, Richard D, Roche L, Dohou J, Pouget M, et al. Therapeutic drug monitoring and tyrosine kinase inhibitors. Oncol Lett 2016;12(2):1223−32.

[18] Baccarani M, Dreyling M, ESMO Guidelines Working Group. Chronic myelogenous leukemia: ESMO clinical recommendations for diagnosis, treatment and follow-up. Ann Oncol 2009;20(Suppl. 4):105−7.

[19] Oude Munnink TH, Henstra MJ, Segerink LI, Movig KL, Brummelhuis-Visser P. Therapeutic drug monitoring of monoclonal antibodies in inflammatory and malignant disease: translating TNF-α experience to oncology. Clin Pharmacol Ther 2016;99(4):419−31.

[20] Bruno R, Washington CB, Lu JF, Lieberman G, Banken L, Klein P. Population pharmacokinetics of trastuzumab in patients with HER2 + metastatic breast cancer. Cancer Chemother Pharmacol 2005;56(4):361−9.

[21] Stroh M, Lum BL. Should therapeutic drug monitoring for monoclonal antibodies remain the exception or become the norm? Clin Pharmacol Ther 2016;100(3):215−17.

[22] Azzopardi N, Lecomte T, Ternant D, Boisdron-Celle M, Piller F, Morel A, et al. Cetuximab pharmacokinetics influences progression-free survival of metastatic colorectal cancer patients. Clin Cancer Res 2011;17(19):6329−37.

[23] Isaacs C. Hormones and related agents in the therapy of cancer. In: Brunton LL, et al., editors. Goodman & Gilman's the pharmacological basis of therapeutics. 13th ed. McGraw-Hill Education; 2018. p. 1237−48.

[24] Goetz MP, Sangkuhl K, Guchelaar HJ, Schwab M, Province M, Whirl-Carrillo M, et al. Clinical Pharmacogenetics Implementation Consortium (CPIC) guideline for CYP2D6 and tamoxifen therapy. Clin Pharmacol Ther 2018;103(5):770−7.

[25] Higgs JE, Payne K, Roberts C, Newman WG. Are patients with intermediate TPMT activity at increased risk of myelosuppression when taking thiopurine medications? Pharmacogenomics 2010;11(2):177−88.

[26] Moriyama T, Nishii R, Perez-Andreu V, Yang W, Klussmann FA, Zhao X, et al. NUDT15 polymorphisms alter thiopurine metabolism and hematopoietic toxicity. Nat Genet 2016;48(4):367−73.

[27] Yang JJ, Landier W, Yang W, Liu C, Hageman L, Cheng C, et al. Inherited NUDT15 variant is a genetic determinant of mercaptopurine intolerance in children with acute lymphoblastic leukemia. J Clin Oncol 2015;33(11):1235−42.

[28] Bhatia S, Landier W, Hageman L, Chen Y, Kim H, Sun CL, et al. Systemic exposure to thiopurines and risk of relapse in children with acute lymphoblastic leukemia: a Children's Oncology Group Study. JAMA Oncol 2015;1(3):287−95.

[29] Buice RG, Evans WE, Karas J, Nicholas CA, Sidhu P, Straughn AB, et al. Evaluation of enzyme immunoassay, radioassay, and radioimmunoassay of serum methotrexate, as compared with liquid chromatography. Clin Chem 1980;26(13):1902–4.

[30] Pesce MA, Bodourian SH. Evaluation of a fluorescence polarization immunoassay procedure for quantitation of methotrexate. Ther Drug Monit 1986;8(1):115–21.

[31] College of American Pathologists. Chemistry/therapeutic drug monitoring survey C-A 2017. Northfield, IL; 2017.

[32] Florin L, Lemahieu C, Stove V. Evaluation of the new Methotrexate CMIA assay on the Architect i2000SR. Clin Chem Lab Med 2016;54(1):e15–17.

[33] Bouquié R, Grégoire M, Hernando H, Azoulay C, Dailly E, Monteil-Ganière C, et al. Evaluation of a methotrexate chemiluminescent microparticle immunoassay: comparison to fluorescence polarization immunoassay and liquid chromatography-tandem mass spectrometry. Am J Clin Pathol 2016;146(1):119–24.

[34] Aumente MD, López-Santamaría J, Donoso-Rengifo MC, Reyes-Torres I, Montejano Hervás P. Evaluation of the novel methotrexate architect chemiluminescent immunoassay: clinical impact on pharmacokinetic monitoring. Ther Drug Monit 2017;39(5):492–8.

[35] Godefroid MJ, von Meyer A, Parsch H, Streichert T, Verstraete AG, Stove V. Multicenter method evaluation of the ARK™ methotrexate immunoassay. Clin Chem Lab Med 2014;52(2):e13–16.

[36] Günther V, Mueller D, von Eckardstein A, Saleh L. Head to head evaluation of the analytical performance of two commercial methotrexate immunoassays and comparison with liquid chromatography-mass spectrometry and the former fluorescence polarization immunoassay. Clin Chem Lab Med 2016;54(5):823–31.

[37] Lennard L. Assay of 6-thioinosinic acid and 6-thioguanine nucleotides, active metabolites of 6-mercaptopurine, in human red blood cells. J Chromatogr 1987;423:169–78.

[38] Armstrong VW, Shipkova M, von Ahsen N, Oellerich M. Analytic aspects of monitoring therapy with thiopurine medications. Ther Drug Monit 2004;26(2):220–6.

[39] Dervieux T, Chu Y, Su Y, Pui CH, Evans WE, Relling MV. HPLC determination of thiopurine nucleosides and nucleotides in vivo in lymphoblasts following mercaptopurine therapy. Clin Chem 2002;48(1):61–8.

[40] Shipkova M, Armstrong VW, Wieland E, Oellerich M. Differences in nucleotide hydrolysis contribute to the differences between erythrocyte 6-thioguanine nucleotide concentrations determined by two widely used methods. Clin Chem 2003;49(2):260–8.

[41] Dervieux T, Meyer G, Barham R, Matsutani M, Barry M, Boulieu R, et al. Liquid chromatography-tandem mass spectrometry analysis of erythrocyte thiopurine nucleotides and effect of thiopurine methyltransferase gene variants on these metabolites in patients receiving azathioprine/6-mercaptopurine therapy. Clin Chem 2005;51(11):2074–84.

[42] Kirchherr H, Shipkova M, von Ahsen N. Improved method for therapeutic drug monitoring of 6-thioguanine nucleotides and 6-methylmercaptopurine in whole-blood by LC/MSMS using isotope-labeled internal standards. Ther Drug Monit 2013;35(3):313–21.

[43] Weinshilboum RM, Raymond FA, Pazmiño PA. Human erythrocyte thiopurine methyltransferase: radiochemical microassay and biochemical properties. Clin Chim Acta 1978;85(3):323–33.

[44] Indjova D, Shipkova M, Atanasova S, Niedmann PD, Armstrong VW, Svinarov D, et al. Determination of thiopurine methyltransferase phenotype in isolated human erythrocytes using a new simple nonradioactive HPLC method. Ther Drug Monit 2003;25(5):637–44.

[45] Kalsi K, Marinaki AM, Yacoub MH, Smolenski RT. HPLC/tandem ion trap mass detector methods for determination of inosine monophosphate dehydrogenase (IMPDH) and thiopurine methyltransferase (TPMT). Nucleosides Nucleotides Nucleic Acids 2006;25(9–11):1241–4.

[46] Ma J, Sies CW, Pike LS. Analytical and clinical validation of an LC-MS/MS method to measure thiopurine S-methyltransferase activity by quantifying d3-6-MMP. Clin Biochem 2018;54:100–5.

Chapter 10.1

Case study—methotrexate toxicity, treatment, and measurement

Valkal Bhatt[1], Michael Scordo[2,3] and Dean C. Carlow[4]

[1]*Incyte Corporation, Wilmington, DE, United States,* [2]*Adult Bone Marrow Transplant Service, Department of Medicine, Memorial Sloan Kettering Cancer Center, New York, NY, United States,* [3]*Department of Medicine, Weill Cornell Medical College, New York, NY, United States,* [4]*Department of Laboratory Medicine, Memorial Sloan Kettering Cancer Center, New York, NY, United States*

Case description

A 63-year-old man was in his usual state of health until he presented to a local emergency department with complaints of new, intermittent headaches and gait imbalance for two weeks. An MRI of the brain with contrast showed a large enhancing brain lesion involving both cerebral hemispheres. Lumbar puncture with cerebral spinal fluid (CSF) analysis demonstrated an abnormally elevated CSF protein level. He underwent a surgical brain biopsy and the pathology was consistent with primary central nervous system diffuse large B cell lymphoma. A whole-body positron emission tomography (PET) scan showed no evidence of systemic lymphoma, and a bone marrow aspirate and biopsy showed no evidence of lymphoma.

Induction systemic therapy with R-MPV chemotherapy (1 cycle equal to 14 days) was implemented as follows: day 1, Rituximab 500 mg/m^2 IV; day 2, methotrexate (MTX) 3.5 g/m^2 IV, Vincristine 1.4 mg/m^2, and Procarbazine 100 mg/m^2/day given on days 2–8 during odd cycles. Standard hydration with sodium bicarbonate by continuous infusion pre- and post-MTX infusion and leucovorin rescue (25 mg IV every 6 hours) were given per institutional guidelines. A baseline comprehensive metabolic panel (CMP) performed 1 day prior to the start of induction chemotherapy was notable for a normal serum creatinine (SCr) of 0.7 mg/dL (Table 10.1.1) with an estimated glomerular filtration rate (eGFR) of 90 mL/min. Repeat CMP performed 6 hours after completion of high-dose MTX on day 2 showed acute kidney injury (AKI) with a SCr of 2.2 mg/dL (eGFR 35 mL/min). A CMP performed 24 hours post-MTX administration showed a SCr of 3.0 mg/dL (eGFR 20 mL/min) and a reduction in total urine output. MTX levels drawn at the same times showed an elevated MTX level of 13 μmol/L (Table 10.1.1). Urine alkalization with IV sodium bicarbonate was continued as needed and the leucovorin dose was increased (100 mg IV every 6 hours). After 72 hours of MTX administration, the SCr continued to rise and the patient developed nonoliguric kidney injury and progressive hypoxia due to the development of bilateral pleural effusions. The 48- and 72-hour MTX levels continued to remain elevated as well. Due to the delayed MTX clearance as a result of the AKI and resulting multiorgan failure, a single dose of glucarpidase (50 U/kg) was given on day 5. Leucovorin was held for 2 hours pre- and postadministration of glucarpidase to prevent inhibition of the glucarpidase enzymatic activity.

The patient's MTX levels rapidly cleared to undetectable levels within 24 hours of glucarpidase administration (Table 10.1.2) when analyzed by liquid chromatography mass spectrometry (LC/MS/MS) but remained elevated when analyzed by an immunoassay. His creatinine and urine output also began to improve 72 hours after glucarpidase administration. The patient was discharged 10 days after MTX administration and his SCr normalized (0.9 mg/dL) within 24 days. Following normalization, the patient completed 4 full cycles of R-MVP (Rituximab, MTX, Procarbazine, and Vincristine) treatment with reduced dose MTX (2 g/m^2) and accelerated leucovorin support (100 mg every 6 hours) without complications. A posttreatment MRI of the brain and CSF evaluation showed complete remission. The patient is scheduled to undergo consolidative high-dose chemotherapy and autologous hematopoietic cell transplantation to achieve a durable remission.

TABLE 10.1.1 Serum Creatinine (SCr) and Methotrexate (MTX) Levels After Administration of MTX

	Day 1	Day 2	Day 3	Day 4	Day 5	Day 6	Day 7	Day 8	Day 9
SCr (mg/dL)	0.7	2.2	3.3	3.5	3.7	3.5	3.1	2.8	2.5
MTX (μmol/L)			13	15	15	<0.005	<0.005	<0.005	

MTX was measured using LC/MS/MS. Glucarpidase was given on Day 5. *LC−MS/MS*, Liquid chromatography mass spectrometry.

TABLE 10.1.2 Plasma Methotrexate (MTX) Levels by Immunoassay and Liquid Chromatography Mass Spectrometry (LC−MS/MS) Before and After Glucarpidase Administration

Time (h)	MTX (μmol/L)	
	Immunoassay	LC−MS/MS
Preglucarpidase	15.0	15.0
0.25	5.0	0.2
1	3.0	0.1
6	1.6	0.05
24	0.9	<0.005
48	0.4	<0.005
72	0.3	<0.005

The immunoassay was the enzyme-multiplied immunoassays assay.
Source: Adapted from Wideman BC, Schwartz S, Jayaprakash N, Christensen R, Pui CH, Chauhan N, et al. Efficacy of glucarpidase (Carboxypeptidase G2) in patients with acute kidney injury after high-dose methotrexate therapy. Pharmacother: J Hum Pharmacol Drug Ther 2014;34(5):427−39; Al-Turkmani MR, Law T, Narla A, Kellogg MD. Difficulty measuring methotrexate in a patient with high-dose methotrexate−induced nephrotoxicity. Clin Chem 2010;56(12):1792−4; Mulder MB, Huisman R, Engels FK, van der Sluis IM, Koch BC. Therapeutic drug monitoring of methotrexate in plasma using ultra high-performance liquid chromatography−electrospray ionization−tandem mass spectrometry: necessary after administration of glucarpidase in methotrexate intoxications. Ther Drug Monit 2018;40(4):383−5 [1,2].

Discussion

MTX has a wide therapeutic index with doses ranging from 20 mg/m^2/week for the treatment of various malignant and nonmalignant conditions to high-dose regimens of 1000−33,000 mg/m^2 that incorporate leucovorin rescue [3]. High-dose MTX can be safely administered in patients with normal renal function when supported by vigorous hydration and urine alkalization. Bolus infusions of high-dose MTX infusions are generally associated with an increased risk of developing renal complications when compared with continuous infusion regimens. MTX undergoes active renal filtration and secretion by the kidneys [3]. MTX-induced renal injury is largely due to drug precipitation and accumulation of its metabolites, namely, 7-OH-MTX and 2,4-diamino-N(10)-methylpteroic acid (DAMPA) (Fig. 10.1.1). These are weak organic acid that are poorly soluble in acidic urine and are 10-fold less soluble than MTX. Therefore urine alkalization is essential with high-dose MTX regimens to prevent the development of renal toxicity [4]. An increase in urine pH from 6.0 to 7.0 increases the solubility of MTX by approximately eightfold. It is generally recommended that aggressive hydration should be initiated 12 hours prior to, during, and at least 24 hours after MTX infusion with intermittent urine pH measurements to maintain alkaline urine pH (≥ 7.5) using intravenous sodium bicarbonate supplementation [5].

MTX is an inhibitor of the enzyme dihydrofolate reductase (DHFR), and this inhibition is the basis of its antimetabolite activity. DHFR inhibition results in the depletion of serum folates that are required for purine synthesis, thereby resulting in its antitumor activity. The MTX rescue agent, leucovorin, restores the reduced folate pool after conversion to its active metabolites and helps to reduce the development of nonrenal toxicities associated with MTX administration. Elevated plasma concentrations of MTX have been directly correlated with development of renal toxicities [6,7].

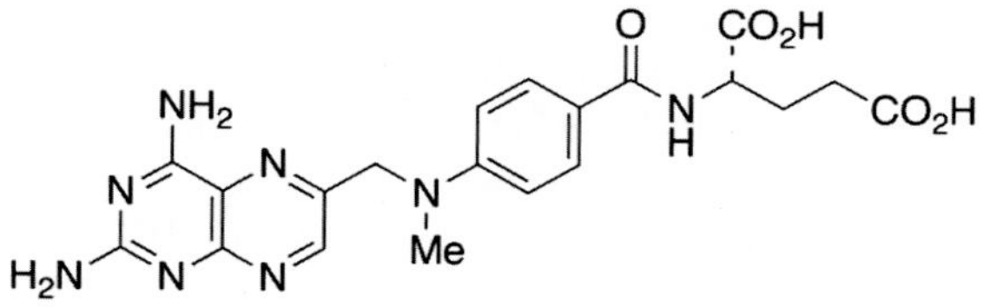

Methotrexate (MTX)

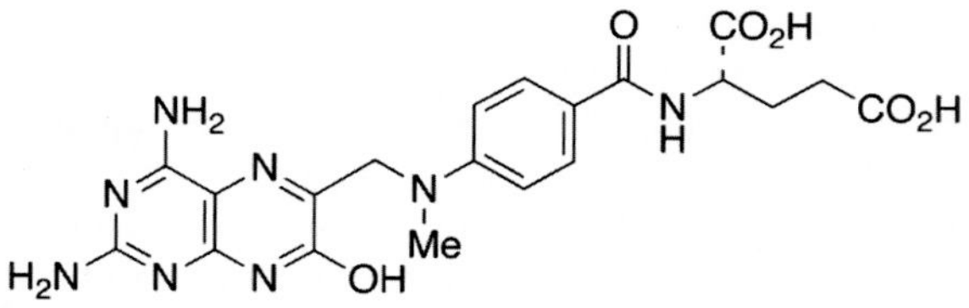

7- Hydroxy Methotrexate (7-OH MTX)

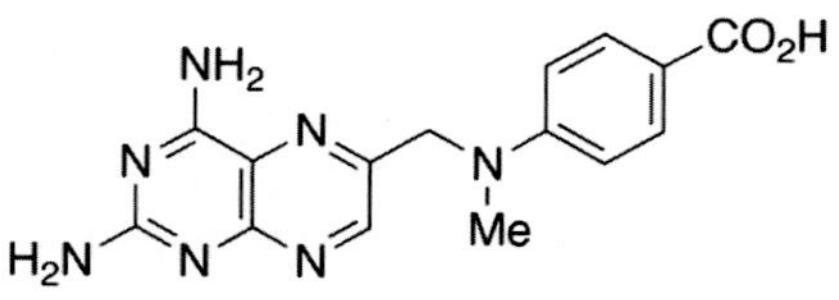

2,4,-diamino-N^{10}-methylpteroic acid (DAMPA)

FIGURE 10.1.1 The chemical structures of MTX, 7-OH MTX, and DAMPA. *DAMPA*, 2,4-Diamino-*N*(10)-methylpteroic acid; *MTX*, methotrexate.

In addition, sustained elevated plasma levels at 24 ($>5-10$ μM), 48 (>1 μM), and 72 hours (>0.1 μM) postdose have been shown to correlate with development of toxicities due to prolonged drug exposure. These toxicities can be minimized with pharmacokinetically dosed leucovorin based on the plasma MTX clearance rate [8–10].

MTX toxicity can be compounded when it is coadministered with drugs that compete with its renal clearance. Agents such as probenecid, proton pump inhibitors, salicylates, sulfamethoxazole, penicillins, and nonsteroidal antiinflammatory agents should generally be avoided in patients being treated with high dose MTX; however, data are conflicting with regards to the true clinical impact of these drug interactions [11–13].

Nausea, vomiting, and diarrhea are common side effects of MTX therapy, although these effects can be more pronounced in patients who develop MTX toxicity. Patients who develop MTX-induced renal injury are generally asymptomatic overall and most commonly present with nonoliguric renal injury marked by a rapid rise in serum creatinine during or after MTX administration [3,5]. Declines in renal function further exacerbate MTX toxicity by reducing the glomerular filtration rate, resulting in elevated serum levels and prolonged drug exposure. Moreover, prolonged drug exposure may also result in nonrenal side effects such as mucositis, prolonged myelosuppression, and hepatotoxicity [14–16].

For patients who develop MTX toxicity, conventional treatment options include aggressive hydration with urine alkalinization to maintain urine output and promote MTX clearance. Hemodialysis can be considered in patients at risk for severe toxicity due to prolonged elevations of MTX levels and can provide a 30%–80% decrease in plasma MTX levels [3]. Due to its high volume of distribution, a major limitation of dialysis-based removal of MTX is the rapid rebound in serum drug concentrations once dialysis is discontinued [17].

Glucarpidase, a recombinant bacterial carboxypeptidase enzyme, was approved by the US Food and Drug Administration (FDA) in 2012 for treatment of patients with delayed MTX clearance or AKI and elevated serum MTX levels (>10 μM/L) 48 hours after administration. Glucarpidase cleaves MTX into DAMPA and glutamate, both of which are inactive and nontoxic metabolites. A single dose of glucarpidase (50 U/kg IV) reduces plasma MTX concentrations by greater than 90% within 15 minutes of administration [1]. However, glucarpidase has no impact on intracellular and extravascular concentrations of MTX, so leucovorin rescue therapy must be continued in addition to aggressive hydration and urine alkalinization until renal recovery [4]. Leucovorin acts as a substrate for glucarpidase and may compete with MTX binding, so dosing should be held for 2 hours pre- and postglucarpidase administration. Reduction in plasma MTX levels with glucarpidase followed by timely administration of leucovorin (within 5–6 hours) allows for reduced competition between MTX and leucovorin while enhancing leucovorin transport to the intracellular space through a shared active transport mechanism [4]. Early use of glucarpidase may help to facilitate renal recovery and reduce the development of dose-limiting toxicity that may lead to treatment delay or discontinuation [1,18].

Monitoring serum drug levels is crucial in patients receiving high dose MTX treatment. Over the years, many different assay formats have been used to measure MTX in serum, plasma, CSF and other body fluids. One of the earliest methods to measure MTX was high-pressure liquid chromatography (HPLC) using either ultraviolet or fluorescent detection [19–22]. HPLC physically separates MTX from matrix components and structurally similar compounds, and these assays are sensitive, accurate, and selective. While still used today in many research laboratories, HPLC is generally not used clinically as the specimens require significant sample preparation and the analysis is much more time consuming than other types of automated assays. HPLC assays of MTX do not cross-react with 7-OH-MTX, the major MTX metabolite, or DAMPA, the cleavage product of MTX after glucarpidase treatment (Fig. 10.1.1). A number of studies have shown a relatively poor correlation between HPLC assays and immunoassays, which is most likely due to immunoassay interference and cross-reactivity with metabolites [23–25].

A number of enzymatic [26–28] and protein-binding assays [29] have been previously described and several have been adapted to high throughput automated clinical chemistry analyzers [30–32] and microtiter plate readers [28]. While many variations have been published, the most common approach is based on the inhibition of DHFR by MTX. DHFR catalyzes the reduction of dihydrofolate to tetrahydrofolate and in the process liberates $NADP^+$. The concentration of MTX is proportional to the increased absorbance at 340 nm that occurs with $NADP^+$ production. These assays have been demonstrated to be sensitive and specific but have some mild cross-reactivity with MTX metabolites [28]; however, they can still be used to measure MTX following glucarpidase administration [28]. While several kits are commercially available for research use, as of this writing none have FDA approval, which limits their use in the clinical laboratory.

The need for rapid measurement of routine MTX levels in the clinical setting led to the development of commercialized automated assays that utilize immunological techniques. Most immunoassays require limited sample preparation, can be fully automated, and produce results rapidly. Many different types of immunoassays have been described, including radioimmunoassays [33], heterogeneous or homogeneous immunoassays [34–36], fluorescence polarization immunoassays (FPIA) [33,37,38], and enzyme-multiplied immunoassays (EMIT) [39,40]. For many years the most widely used homogeneous immunoassay was the Abbott TDx (FPIA, Abbott Laboratories) which was sensitive, rapid, and easy to use. However, this assay has recently been withdrawn from the market and most users have switched to the homogeneous EMIT assay (ARK Diagnostics). While the immunoassays have good overall performance, they cross-react significantly with DAMPA (41% for FPIA and 100% with EMIT) [33,37,40] and cannot be used after glucarpidase administration. As shown in Table 10.1.2, the EMIT immunoassay falsely reported elevated MTX levels long after the MTX had been converted to DAMPA. The LC–MS/MS method showed the true MTX concentrations. In addition to the lack of specificity, immunoassays have often been shown to be susceptible to interferences from endogenous compounds present in the sample matrix [33].

More recently, assays using LC–MS/MS technology have been developed to measure MTX and its major metabolites DAMPA and 7-OH-MTX [41–45]. These assays are extremely sensitive and specific and can be used to monitor MTX levels after glucarpidase therapy [2,42,46–48]. While a properly developed mass spectrometry assay is not susceptible to interference from metabolites, all assays using mass spectrometry technology have a potential interference from ion suppression. This can occur when other ions in the specimen can possibly either suppress or enhance the ion of interest. It is imperative during method development and validation to rule out this type of interference. While this is the most sensitive and specific method for MTX analysis available, these instruments are very expensive and require a high degree of training to operate and troubleshoot. The cost and complexity renders it unsuitable for routine clinical use in most centers. However, the technology and software are constantly being simplified and mass spectrometry may replace many immunoassays in the near future.

References

[1] Widemann BC, Schwartz S, Jayaprakash N, Christensen R, Pui CH, Chauhan N, et al. Efficacy of glucarpidase (Carboxypeptidase G2) in patients with acute kidney injury after high-dose methotrexate therapy. Pharmacother: J Hum Pharmacol Drug Ther 2014;34(5):427–39.

[2] Al-Turkmani MR, Law T, Narla A, Kellogg MD. Difficulty measuring methotrexate in a patient with high-dose methotrexate–induced nephrotoxicity. Clin Chem 2010;56(12):1792–4.

[3] Widemann BC, Adamson PC. Understanding and managing methotrexate nephrotoxicity. Oncologist 2006;11(6):694–703.

[4] Howard SC, McCormick J, Pui CH, Buddington RK, Harvey RD. Preventing and managing toxicities of high-dose methotrexate. Oncologist 2016;21(12):1471–82.

[5] Glezerman IJE. Chemotherapy and kidney injury. Introduction to the American Society of Nephrology Onco-Nephrology Curriculum. American Society of Nephrology; 2016. p. 1–10.

[6] Chan H, Evans WE, Pratt CB. Recovery from toxicity associated with high-dose methotrexate: prognostic factors. Cancer Treat Rep 1977;61(5):797.

[7] Von DH, Penta JS, Helman LJ, Slavik M. Incidence of drug-related deaths secondary to high-dose methotrexate and citrovorum factor administration. Cancer Treat Rep 1977;61(4):745–8.

[8] Barrett JS, Mondick JT, Narayan M, Vijayakumar K, Vijayakumar S. Integration of modeling and simulation into hospital-based decision support systems guiding pediatric pharmacotherapy. BMC Med Inform Decis Mak 2008;8(1):6.

[9] Monjanel-Mouterde S, Lejeune C, Ciccolini J, Merite N, Hadjaj D, Bonnier P, et al. Bayesian population model of methotrexate to guide dosage adjustments for folate rescue in patients with breast cancer. J Clin Pharm Ther 2002;27(3):189–95.

[10] Dombrowsky E, Jayaraman B, Narayan M, Barrett JS. Evaluating performance of a decision support system to improve methotrexate pharmacotherapy in children and young adults with cancer. Ther Drug Monit 2011;33(1):99.

[11] Moore MJ, Erlichman C. Therapeutic drug monitoring in oncology. Clin Pharm 1987;13(4):205–27.

[12] Chan AJ, Rajakumar I. High-dose methotrexate in adult oncology patients: a case-control study assessing the risk association between drug interactions and methotrexate toxicity. J Oncol Pharm Pract 2014;20(2):93–9.

[13] Hall JJ, Bolina M, Chatterley T, Jamali F. Interaction between low-dose methotrexate and nonsteroidal anti-inflammatory drugs, penicillins, and proton pump inhibitors: a narrative review of the literature. Ann Pharmacother 2017;51(2):163–78.

[14] Frei E, Blum RH, Pitman SW, Kirkwood JM, Henderson IC, Skarin AT, et al. High dose methotrexate with leucovorin rescue: rationale and spectrum of antitumor activity. Am J Med 1980;68(3):370–6.

[15] Howell SB, Herbst K, Boss GR, Frei E. Thymidine requirements for the rescue of patients treated with high-dose methotrexate. Cancer Res 1980;40(6):1824–9.

[16] Stark AN, Jackson G, Carey PJ, Arfeen S, Proctor SJ. Severe renal toxicity due to intermediate-dose methotrexate. Cancer Chem Pharmacol 1989;24(4):243–5.

[17] Widemann BC, Balis FM, Murphy RF, Sorensen JM, Montello MJ, O'Brien M, et al. Carboxypeptidase-G2, thymidine, and leucovorin rescue in cancer patients with methotrexate-induced renal dysfunction. J Clin Oncol 1997;15(5):2125–34.

[18] Christensen AM, Pauley JL, Molinelli AR, Panetta JC, Ward DA, Stewart CF, et al. Resumption of high-dose methotrexate after acute kidney injury and glucarpidase use in pediatric oncology patients. Cancer 2012;118(17):4321–30.

[19] Nelson JA, Harris BA, Decker WJ, Farquhar D. Analysis of methotrexate in human plasma by high-pressure liquid chromatography with fluorescence detection. Cancer Res 1977;37(11):3970–3.

[20] Chen ML, Chiou WL. Sensitive and rapid high-performance liquid chromatographic method for the simultaneous determination of methotrexate and its metabolites in plasma, saliva and urine. J Chromatogr B: Biomed Sci Appl 1981;226(1):125–34.

[21] Collier CP, MacLeod SM, Soldin SJ. Analysis of methotrexate and 7-hydroxymethotrexate by high-performance liquid chromatography and preliminary clinical studies. Ther Drug Monit 1982;4(4):371–80.

[22] Cairnes DA, Evans WE. High-performance liquid chromatographic assay of methotrexate, 7-hydroxymethotrexate, 4-deoxy-4-amino-N10-methylpteroic acid and sulfamethoxazole in serum, urine and cerebrospinal fluid. J Chromatogr B: Biomed Sci Appl 1982;231(1):103–10.

[23] Howell SK, Wang YM, Hosoya R, Sutow WW. Plasma methotrexate as determined by liquid chromatography, enzyme-inhibition assay, and radioimmunoassay after high-dose infusion. Clin Chem 1980;26(6):734–7.

[24] So N, Chandra DP, Alexander IS, Webster VJ, Hughes DWG. Determination of serum methotrexate and 7-hydroxymethotrexate concentrations: method evaluation showing advantages of high-performance liquid chromatography. J Chromatogr B: Biomed Sci Appl 1985;337:81–90.

[25] Cosolo W, Drummer OH, Christophidis N. Comparison of high-performance liquid chromatography and the Abbott fluorescent polarization radioimmunoassay in the measurement of methotrexate. J Chromatogr B: Biomed Sci Appl 1989;494:201–8.

[26] Falk LC, Clark DR, Kalman SM, Long TF. Enzymatic assay for methotrexate in serum and cerebrospinal fluid. Clin Chem 1976;22(6):785–8.

[27] Werkheiser WC, Zakrzewski SF, Nichol CA. Assay for 4-amino folic acid analogues by inhibition of folic acid reductase. J Pharmacol Exp Ther 1962;137(2):162–6.

[28] Widemann BC, Balis FM, Adamson PC. Dihydrofolate reductase enzyme inhibition assay for plasma methotrexate determination using a 96-well microplate reader. Clin Chem 1999;45(2):223–8.

[29] Arons E, Rothenberg SP, da Costa M, Fischer C, Iqbal MP. A direct ligand-binding radioassay for the measurement of methotrexate in tissues and biological fluids. Cancer Res 1975;35(8):2033–8.

[30] Finley PR, Williams RJ. Methotrexate assay by enzymatic inhibition, with use of centrifugal analyzer. Clin Chem 1977;23(11):2139–41.

[31] Finley PR, Williams RJ, Griffith F, Lichti DA. Adaptation of the enzyme-multiplied immunoassay for methotrexate to the centrifugal analyzer. Clin Chem 1980;26(2):341–3.

[32] Pesce MA, Bodourian SH. Enzyme immunoassay and enzyme inhibition assay of methotrexate, with use of the centrifugal analyzer. Clin Chem 1981;27(3):380–4.

[33] Buice RG, Evans WE, Karas J, Nicholas CA, Sidhu P, Straughn AB, et al. Evaluation of enzyme immunoassay, radioassay, and radioimmunoassay of serum methotrexate, as compared with liquid chromatography. Clin Chem 1980;26(13):1902–4.

[34] Wannlund J, Azari J, Levine L, DeLuca M. A bioluminescent immunoassay for methotrexate at the subpicomole level. Biochem Biophys Res Commun 1980;96(1):440–6.

[35] Al-Bassam MN, O'Sullivan MJ, Bridges JW, Marks V. Improved double-antibody enzyme immunoassay for methotrexate. Clin Chem 1979;25(8):1448–52.

[36] Ferrua B, Milano G, Ly B, Guennec JY, Masseyeff R. An enzyme immunoassay design using labeled antibodies for the determination of haptens. Application to methotrexate assay. J Immunol Methods 1983;60(1-2):257–68.

[37] Pesce MA, Bodourian SH. Evaluation of a fluorescence polarization immunoassay procedure for quantitation of methotrexate. Ther Drug Monit 1986;8(1):115−21.

[38] Slørdal L, Prytz PS, Pettersen I, Aarbakke J. Methotrexate measurements in plasma: comparison of enzyme multiplied immunoassay technique, TDx fluorescence polarization immunoassay, and high pressure liquid chromatography. Ther Drug Monit 1986;8(3):368−72.

[39] Oellerich M, Engelhardt P, Schaadt M, Diehl V. Determination of methotrexate in serum by a rapid, fully mechanized enzyme immunoassay (EMIT). Clin Chem Lab Med 1980;18(3):169−74.

[40] Albertioni F, Rask C, Eksborg S, Poulsen JH, Pettersson B, Beck O, et al. Evaluation of clinical assays for measuring high-dose methotrexate in plasma. Clin Chem 1996;42(1):39−44.

[41] Schofield RC, Ramanathan LV, Murata K, Grace M, Fleisher M, Pessin MS, et al. Development and validation of a turbulent flow chromatography and tandem mass spectrometry method for the quantitation of methotrexate and its metabolites 7-hydroxy methotrexate and DAMPA in serum. J Chromatogr B 2015;1002:169−75.

[42] Kumar VS, Law T, Kellogg M. Liquid chromatography-tandem mass spectrometry (LC-MS-MS) method for monitoring methotrexate in the setting of carboxypeptidase-G2 therapy. Clin Appl Mass Spectrometry. Humana Press; 2010. p. 359−63.

[43] Roberts MS, Selvo NS, Roberts JK, Daryani VM, Owens TS, Harstead KE, et al. Determination of methotrexate, 7-hydroxymethotrexate, and 2,4-diamino-N10-methylpteroic acid by LC−MS/MS in plasma and cerebrospinal fluid and application in a pharmacokinetic analysis of high-dose methotrexate. J Liq Chromatogr Relat Technol 2016;39(16):745−51.

[44] Steinborner S, Henion J. Liquid−liquid extraction in the 96-well plate format with SRM LC/MS quantitative determination of methotrexate and its major metabolite in human plasma. Anal Chem 1999;71(13):2340−5.

[45] Nair H, Lawrence L, Hoofnagle AN. Liquid chromatography−tandem mass spectrometry work flow for parallel quantification of methotrexate and other immunosuppressants. Clin Chem 2012;58(5):943−5.

[46] Mitrovic D, Touw D, Tissing W. Treatment of high dose methotrexate toxicity with glucarpidase. J Clin Toxicol 2016;6(293) 2161-0495.

[47] Buchen S, Ngampolo D, Melton RG, Hasan C, Zoubek A, Henze G, et al. Carboxypeptidase G2 rescue in patients with methotrexate intoxication and renal failure. Br J Cancer 2005;92(3):480.

[48] Mulder MB, Huisman R, Engels FK, van der Sluis IM, Koch BC. Therapeutic drug monitoring of methotrexate in plasma using ultra high-performance liquid chromatography−electrospray ionization−tandem mass spectrometry: necessary after administration of glucarpidase in methotrexate intoxications. Ther Drug Monit 2018;40(4):383−5.

Further reading

Rule G, Chapple M, Henion J. A 384-well solid-phase extraction for LC/MS/MS determination of methotrexate and its 7-hydroxy metabolite in human urine and plasma. Anal Chem 2001;73(3):439−43.

Chapter 10.2

Methotrexate toxicity—case study

Alejandro R. Molinelli and Kristine R. Crews
Department of Pharmaceutical Sciences, St. Jude Children's Research Hospital, Memphis, TN, United States

Case description

A 9-year old girl initially presented to her pediatrician with a limp and pain in her left knee with no history of trauma. A plain X-ray showed a pathological fracture of the proximal tibia and she was referred to a pediatric oncologist. Her initial workup of blood tests and an abdominal ultrasound showed no abnormalities, but a chest X-ray revealed multiple lesions and an MRI of her left knee demonstrated an osteosarcoma in the proximal tibia which was later confirmed by biopsy. The patient began treatment with chemotherapy consisting of high-dose methotrexate (HDMTX), doxorubicin, cisplatin, ifosfamide, and etoposide. Her first course of HDMTX was administered at week three of therapy at a dose of 12 g/m^2 intravenously over 4 hours. Blood samples were collected in EDTA-containing tubes for MTX therapeutic drug monitoring. The plasma was separated and plasma MTX concentrations were measured locally within 1 hour of collection by using a commercially available immunoassay. Due to delayed MTX excretion indicated by an elevated plasma MTX concentration at 24 hours from the start of the MTX infusion (158.01 μmol/L), the patient received one dose of glucarpidase intravenously at a dose of 50 U/kg infused over 5 minutes. The patient also had a fourfold increase in serum creatinine from a baseline of 0.4 to 1.6 mg/dL (age specific reference range: 0.26–0.66 mg/dL). Supportive measures included continuous intravenous hydration, urine alkalinization with intravenous sodium bicarbonate, and leucovorin rescue. Following glucarpidase administration, plasma samples were collected for simultaneous analysis by immunoassay and by high performance liquid chromatography with UV detection. Her plasma MTX concentrations as measured by two different methods and her serum creatinine concentrations are shown in Table 10.2.1. Serum creatinine and urine output were monitored at least daily until complete MTX excretion was confirmed by two sequential plasma concentrations below the limit of quantification for the assay (≤0.06 μM at our hospital). The patient recovered and continued with treatment.

Discussion

Osteosarcoma is the most common type of bone cancer occurring in children and teens and the third most common cancer in teens after lymphomas and brain tumors. It occurs most often in the wide-end of long bones such as the femur, tibia, and humerus but can also occur in flat bones such as the pelvis and skull. Approximately 10%–20% of patients have metastases at diagnosis with the most common sites for metastasis being the lung and bones [1]. The presence or absence of metastasis at diagnosis is the only widely accepted prognostic factor available at diagnosis. The 5-year survival for patients with metastatic osteosarcoma ranges from 11% to 50% [2].

First-line chemotherapy for osteosarcoma (primary/neoadjuvant/adjuvant or metastatic) includes the use of cytotoxic agents only, in short intense regimens. Agents used include doxorubicin, cisplatin, MTX, ifosfamide, and etoposide, in different combinations over a period of approximately 40 weeks. Second-line therapy would include cytotoxic agents and sorafenib [3]. In the current case the patient received initial chemotherapy consisting of doxorubicin (37.5 mg/m^2/day IV over 1 hour daily for 2 days) and cisplatin (60 mg/m^2/day IV over 4 hours for 2 days). After the initial treatment lasting 2 days the next round of chemotherapy at week number three included HDMTX (12 g/m^2 IV over 4 hours with leucovorin rescue).

TABLE 10.2.1 Methotrexate Concentrations as Measured Utilizing an Immunoassay and a Chromatographic Method (UPLC-UV), and Serum Creatinine Concentrations

Time Post-MTX Administration	MTX Plasma Concentration by Immunoassay (μmol/L)	MTX Plasma Concentration by Chromatography (μmol/L)	Serum Creatinine (mg/dL)
4 h	948.10		0.4
24 h	158.01		1.6
32 h (postglucarpidase)	9.45		1.8
48 h	7.38	0.14	1.9
72 h	0.88	0.24	1.6
96 h	0.94	0.49	1.7
120 h	0.93	0.74	1.7
144 h	0.88	0.64	1.5
168 h	0.67	0.57	1.3
192 h	0.42	0.41	1.2
216 h	0.22	0.21	1.0
240 h	0.14		0.9
264 h	0.12		0.8
288 h	<0.06		0.9
312 h	<0.06		0.6

The reference range for creatinine at our institution for children ages 7–9 years is 0.26–0.66 mg/dL and the critical value is >3.00 mg/dL. *MTX*, Methotrexate.

HDMTX administration has been associated with severe nephrotoxicity in as many as 10% of patients [4]. The dose limiting toxicities of MTX are bone marrow suppression, ulcerative stomatitis, severe diarrhea, or acute nephrotoxicity. To prevent nephrotoxicity, patients must receive IV prehydration with sodium bicarbonate, preferably, prior to the scheduled administration of HDMTX and continued at a minimum of 100 mL/m^2/h until the plasma MTX concentration is <0.5 μmol/L and potentially longer if the patient has a history of toxicity, delayed MTX elimination, or the presence of an abnormal physiological fluid collection (e.g., intracranial fluid collection). Hydration rates are adjusted based on MTX concentrations and to maintain adequate urine output (at least 2 mL/kg/h).

Laboratory tests (that at a minimum should include a basic metabolic panel) should be drawn either the day before or the morning of HDMTX administration and twice daily thereafter. While following a patient receiving HDMTX, fluid input and output should be closely monitored to maintain adequate fluid balance and to ensure the patient is voiding at least every 2 hours. If a patient has not urinated in 4 hours, they may require diuresis. Since MTX precipitates in acidic urine, maintaining a urine pH of 6.5 or higher will increase MTX solubility, prevent drug precipitation in renal tubules, and decrease the chance of renal damage. Urine pH must be obtained prior to initiation of HDMTX and with each urine void. It is best to delay HDMTX until after the urine pH is ≥6.5, and the pH should be maintained between 6.5 and 9.0 until plasma MTX concentrations have declined to <0.5 μmol/L.

Plasma MTX concentrations higher than expected at designated times (Table 10.2.2) should be acted upon immediately. Initial interventions may include, but are not limited to, increasing the rate of IV fluids; obtaining an additional sample for MTX concentration; and decreasing the rate of, or stopping the MTX infusion. For patients receiving prolonged infusions of HDMTX, stopping the MTX infusion early in the setting of substantial decreases in renal function or other deterioration in clinical status should be considered.

Glucarpidase is a recombinant 390 amino acid glutamate carboxypeptidase enzyme produced in *Escherichia coli*. The enzyme hydrolytically cleaves the carboxyl-terminal glutamate residue from MTX and other folate analogues, thereby forming the noncytotoxic metabolites 4-amino-4-deoxy-*N*10-methylpteroic acid (DAMPA) and glutamic acid. In selected patient care settings, glucarpidase is a critical rescue medication. It is used as a rescue agent to reduce MTX

TABLE 10.2.2 Intervention Thresholds to Consider for Glucarpidase Administration for Osteosarcoma Patients Receiving Methotrexate (MTX) 10–12 g/m² IV Over 4 h

Time Post-MTX Infusion	MTX Concentration (μmol/L)	Interventions
24 h	>50.0 (start glucarpidase)	Leucovorin dosage individualized but commonly 500 mg/m² IV q6h. Hydration at 200 mL/m²/h. Urine alkalinization, check for nephrotoxic drugs, check creatinine, consider glucarpidase
48 h	>10.0 (consider glucarpidase)	
	>20.0 (start glucarpidase)	
72 h	>10.0 (start glucarpidase)	

toxicity in patients with severe delayed MTX elimination, MTX-induced nephrotoxicity, or accidental intrathecal MTX overdose. The use of glucarpidase is warranted in patients receiving short infusion of HDMTX (over 4 hours or less) and serum MTX concentration >50.0 μM at >24 hours after start of MTX infusion; or serum MTX concentration >10.0–20.0 μM at >48 hours after start of MTX infusion, especially when accompanied by an increase in serum creatinine.

Glucarpidase is most effective if given by 48 hours after HDMTX; the longer the delay between start of HDMTX and glucarpidase, the less effective it is because the MTX is polyglutamylated in tissues and thus less available to cleavage by glucarpidase in blood. Glucarpidase use should be accompanied with adequate hydration, urinary alkalinization, and concurrent leucovorin. As leucovorin is a substrate for glucarpidase that may compete with MTX for glucarpidase binding sites, it is recommended that the administration schedule for leucovorin be adjusted so that it is not administered within 2–4 hours prior to or within the 2 hours following glucarpidase dosing to minimize inactivation of leucovorin by glucarpidase.

After glucarpidase administration, all MTX immunoassays will overestimate the true MTX concentration, with the overestimation decreasing over time (as the DAMPA concentration decreases). Once the MTX concentration by immunoassay reaches 0.1 μmol/L the DAMPA concentration is likely to be negligible and the immunoassay results should be accurate [5].

In the current case the patient was observed to have delayed MTX clearance 24 hours following the start of the MTX infusion. The MTX concentration at 24 hour was 158.01 μmol/L, and it is recommended to use glucarpidase when MTX levels are higher than 50 μmol/L in osteosarcoma patients receiving the dosing of 10–12 g/m² IV over 4 hours. In addition, the patient had a four-fold change in creatinine from a baseline of 0.4–1.6 mg/dL (reference range for ages 7–9 years: 0.26–0.66 mg/dL). Upon the initiation of glucarpidase the presence of the MTX metabolite DAMPA cross-reacts with the MTX immunoassay (all commercial immunoassays) yielding inaccurate results. Concurrent MTX measurements utilizing the immunoassay and a chromatographic method are shown in Table 10.2.1. The MTX concentration between 24 and 48 hours dropped 99.9% from 158.01 to 0.14 μmol/L following the use of glucarpidase. Once the MTX concentration by both methods yields similar results, it is acceptable to continue using the immunoassay method alone. MTX levels in this patient were followed until two consecutive MTX results were below the limit of quantitation (<0.06 μmol/L). In some patients a modest rebound of MTX concentrations can occur after the administration of glucarpidase. This rebound is due to MTX previously bound to tissues being released back into the circulation. If a patient had fluid accumulation, it can also act as an MTX reservoir. In this case the MTX rebound can be observed in the chromatography results starting at 72 hours and continuing through 120 hours. MTX was completely cleared (two results below the limit of quantification) after 13 days. In this case the patient recovered and was able to continue therapy.

Note: The case presented in this text is specific for metastatic osteosarcoma. Because cancer treatment guidelines are updated continuously to reflect advances in scientific knowledge and the use of new drugs or treatment modalities, the treatment program presented in this case may differ from current treatment received by similar patients.

References

[1] Bielack SS, Kempf-Bielack B, Delling G, Exner GU, Flege S, Helmke K, et al. Prognostic factors in high-grade osteosarcoma of the extremities or trunk: an analysis of 1,702 patients treated on neoadjuvant cooperative osteosarcoma study group protocols. J Clin Oncol 2002;20(3):776—90.

[2] Kager L, Zoubek A, Pötschger U, Kastner U, Flege S, Kempf-Bielack B, et al. Primary metastatic osteosarcoma: presentation and outcome of patients treated on neoadjuvant Cooperative Osteosarcoma Study Group protocols. J Clin Oncol 2003;21(10):2011—18.

[3] Grignani G, Palmerini E, Dileo P, Asaftei SD, D'Ambrosio L, Pignochino Y, et al. A phase II trial of sorafenib in relapsed and unresectable high-grade osteosarcoma after failure of standard multimodal therapy: an Italian Sarcoma Group study. Ann Oncol 2012;23(2):508—16.

[4] Widemann BC, Adamson PC. Understanding and managing methotrexate nephrotoxicity. Oncologist 2006;11(6):694—703.

[5] Christensen AM, Pauley JL, Molinelli AR, Panetta JC, Ward DA, Stewart CF, et al. Resumption of high-dose methotrexate after acute kidney injury and glucarpidase use in pediatric oncology patients. Cancer 2012;118(17):4321—30.

Chapter 10.3

The importance of selecting an appropriate method for measuring methotrexate concentration after glucarpidase rescue: immunoassay or LC—MS/MS?

Fang Wu, Andrew W. Lyon and Martha E. Lyon
Division of Clinical Biochemistry, Department of Pathology & Laboratory Medicine, Saskatchewan Health Authority, Saskatoon, SK, Canada

Introduction

Methotrexate (MTX) is an antineoplastic agent used in the treatment of acute lymphoblastic leukemia, osteosarcoma, and cancers of lung, head, and neck. It exerts its anticancer effects through competitive inhibition of the enzyme dihydrofolate reductase and thus blocking the conversion of dihyrofolate to its active form (i.e., tetrahydrofolates) and ultimately inhibiting DNA synthesis. Intermediate to high doses of MTX are only tolerable when followed with leucovorin (5-formyltetrahydrofolate) rescue to salvage nontumor cells [1].

Severe renal and hepatic toxicity can occur after high dose MTX (HDMTX) therapy. Glucarpidase rescue is indicated for patients with HDMTX-induced acute kidney injury (AKI) and delayed MTX excretion during HDMTX chemotherapy [2,3]. Glucarpidase rapidly converts circulating MTX into the nontoxic metabolite e-deoxy-4-amino-N10-methylpeteroic acid (DAMPA) and glutamic acid [4]. Patients that undergo HDMTX therapies are required to be closely monitored so that toxic effects can be detected in a timely manner to avoid toxicity. Measurement of MTX post glucarpidase therapy by immunoassay typically is inappropriate because immunoassay methods cannot distinguish between MTX and DAMPA and thus significantly over-estimate the concentration of MTX [5].

Case report

An 11 year-old girl was diagnosed with osteosarcoma of the proximal right humerus. She underwent right forequarter amputation followed by neoadjuvant chemotherapy according to the Children's Oncology Group (COG AOST 0331) protocol. The chemotherapy used was a combination of HDMTX, cisplatin, and adriamycin. Before HDMTX administration, electrolytes, urea, creatinine, and liver enzymes were measured to assess renal and liver function. The results (Table 10.3.1) indicated normal renal and liver function of this patient prior to HDMTX therapy.

The patient received 17.4 g of IV MTX infused over 4 hours after her prechemotherapy hydration as well as IV sodium bicarbonate and leucovorin as per treatment protocol. Her urine output was monitored every 4 hours and initially seemed appropriate post chemotherapy. Blood MTX levels at 24, 36, 42, and 48 hours were monitored to determine when to administer leucovorin and/or glucarpidase. Routine blood work at 24 hours post HDMTX demonstrated that there were significant increases in her serum creatinine (from 33 to 224 mmol/L) and urea (from 3.1 to 12.2 mmol/L), indicating MTX-induced nephrotoxicity. Hepatotoxicity was also observed (significantly elevated liver

TABLE 10.3.1 Selected Clinical Chemistry Test Results Prior to High-Dose Methotrexate and Post High-Dose Methotrexate (MTX) Chemotherapy

	Pre-MTX	Post-MTX	Reference Range
Sodium (mmol/L)	140	126	135–146
Potassium (mmol/L)	3.9	6.1	3.5–5.1
Chloride (mmol/L)	104	94	100–110
Total CO_2 (mmol/L)	25	16	22–31
Urea (mmol/L)	3.1	12.2	3.7–7.0
Creatinine (μmol/L)	33	224	30–60
Anion gap (mmol/L)	11	15	8–16
Phosphate (mmol/L)	1.12	2.15	0.87–1.45
Aspartate Aminotransferase (U/L)	25	1344	10–50
Alanine Aminotransferase (U/L)	27	2905	5–45

TABLE 10.3.2 Comparison of Blood Methotrexate (MTX) Concentrations Between Immunoassays and Liquid Chromatography and Tandem Mass Spectrometry (LC–MS/MS)

		MTX Concentration (μmol/L)		
		ARK Immunoassay (The Author's Lab)	ARK Immunoassay (Reference Lab)	LC–MS/MS (Reference Lab)
24 h post high dose MTX		310	–	–
48 h post high dose MTX		470	–	–
Postglucarpidase	1	160	86	0.45
	2	80	47	0.16
	3	2.7	2.9	1.83
	4	2.9	2.5	2
	5	2.2	2.3	1.27
	6	1.3	1.5	0.78
	7	1.04	–	0.58
	8	0.24	–	0.14

enzymes). Decreased urine output and toxic levels of MTX confirmed HDMTX-induced AKI. Continuous renal replacement therapy (CRRT) was then started and her MTX levels dramatically decreased in 12 hours. Leucovorin was then continued at a higher dose, and glucarpidase was administered to reverse the effects of MTX given the MTX level was still in the toxic range.

Serial specimens post glucarpidase for MTX analysis were submitted to our laboratory. Our laboratory offers an immunoassay (ARK method) for the measurement of MTX, which is known to significantly cross react with DAMPA, a MTX metabolite produced during glucarpidase therapy. Recognizing this, specimens were also sent to a reference laboratory for the determination of MTX levels using liquid chromatography and tandem mass spectrometry (LC–MS/MS). A significant discrepancy between the results obtained from immunoassay and the LC–MS/MS assay were observed (Table 10.3.2). At the reference laboratory, ARK immunoassay testing was also requested for the same set of specimens for comparison purpose and lower MTX levels were observed. The difference between two immunoassay results (the authors' laboratory versus reference laboratory) is likely due to the delay of testing and in vitro MTX conversion by glucarpidase. Of note, ARK MTX assay is a competitive immunoassay in that the MTX in the specimen competes with MTX labeled with the enzyme glucose-6-phosphate dehydrogenase (G6PDH) for binding to the antibody reagent. G6PDH enzyme activity is directly proportional to the MTX concentration measured using the conversion of nicotinamide adenine dinucleotide (NAD) to NADH spectrophotometrically and the rate of change in absorbance monitored.

Post CRRT and serial medical management, this patient's urine output returned to normal (0.8–0.9 mL/kg/h), and her serum creatinine and urea levels also decreased. The patient was discharged with oral leucovorin with the plan to continue until MTX levels were less than 0.1 μmol/L.

Discussion

Most clinical laboratories only offer an immunoassay method for the quantification of MTX concentration, for example, ARK MTX assay used by the authors' laboratory. However, DAMPA (a MTX metabolite generated by glucarpidase) cross reacts considerably with this immunoassay, rendering it unsuitable for monitoring MTX levels post glucardipase rescue. In contrast, LC–MS/MS method should be used to avoid reporting falsely elevated MTX immunoassay values due to DAMPA interference that could confuse interpretation of the glucarpidase therapy. Clear communication between clinicians and laboratory is required in order to select an appropriate testing method. Collection of post glucardiase specimens on ice is recommended to avoid in vitro conversion of MTX to DAMPA, resulting in falsely low MTX concentrations. Typically, plasma MTX concentration can be measured by immunoassay 7–10 days after glucarpidase therapy (Table 10.3.2).

References

[1] Aquerreta I, Aldaz A, Giraldez J, Sierrasesumaga L. Methotrexate pharmacokinetics and survival in osteosarcoma. Pediatr Blood Cancer 2004;42:52–8.
[2] Ramsey LB, Balis FM, O'Brien MM, Schmiegelow K, Pauley JL, Bleyer A, et al. Consensus guideline for use of glucarpidase in patients with high-dose methotrexate induced acute kidney injury and delayed methotrexate clearance. Oncologist 2018;23:52–61.
[3] Green MR, Chamberlain MC. Renal dysfunction during and after high-dose methotrexate. Cancer Chemother Pharmacol 2009;63:599–604.
[4] Widemann BC, Sung E, Anderson L, Salzer WL, Balis FM, Monitjo KS, et al. Pharmacokinetics and metabolism of the methotrexate metabolite 2,4-diamino-N(10)-methylpteroic acid. J Pharmacol Exp Ther 2000;294:894–901.
[5] Gunther V, Mueller D, von Eckardstein A, Saleh L. Head to head evaluation of the analytical performance of two commercial methotrexate immunoassays and comparison with liquid chromatography-mass spectrometry and the former fluorescence polarization immunoassay. Clin Chem Lab Med 2016;54:823–31.

Chapter 10.4

Different cross-reactivity profiles of methotrexate immunoassays and the clinical management of methotrexate treatment

Yifei Yang* and Kiang-Teck J. Yeo

Pritzker School of Medicine, Section of Clinical Chemistry, Department of Pathology, The University of Chicago, Chicago, IL, United States

Case description

A 13-year-old male pediatric patient was recently diagnosed with nonmetastatic osteosarcoma and has just started his chemotherapy treatment with high dose methotrexate (>1 g/m^2), followed by leucovorin rescue therapy to minimize systemic toxicity. Following the high dose therapy protocol at the treated hospital, methotrexate is first administered via IV infusion and the initial serum methotrexate concentration was measured to be 75.6 μM. 24 hours post the methotrexate infusion, the leucovorin therapy was initiated at 7.5 mg dose every 6 hours, together with urine alkalization and hydration. To monitor the potential toxicity of methotrexate, its systemic clearance is monitored by serial drug measurement at different time points postinfusion. The leucovorin therapy was discontinued after the serum methotrexate concentration fell below 0.1 μM, 60 hours after the initial methotrexate dose, and the patient was discharged from the hospital. The clearance pattern of methotrexate in circulation measured by Siemens immunoassay EMIT Syva (Siemens Heather Diagnostics, Tarrytown, New York) is described in Table 10.4.1. Shortly after the first successful methotrexate therapy, the assay manufacturer (Siemens) announced a reagent formulation change to the immunoassay, in which the cross-reactivity profile of the assay would change based on a new batch of polyclonal antibody production. Based on the information provided by the product announcement bulletin, the assay will decrease its cross-reactivity to methotrexate metabolite 7-hydroxymethotrexate (7-OH-methotrexate), causing a 20% decrease in measured methotrexate concentrations. The clinical pharmacist who manages the patient methotrexate treatment and subsequent leucovorin rescue wondered how the reduction in cross-reactivity would change the measurement of methotrexate and the evaluation of its toxicity risk.

To investigate how reagents' different antibody formulations and their cross-reactivities can affect analytical results, we compared two FDA-approved assays from different manufacturers: Siemens and ARK Diagnostics (Freemont, California), with three different polyclonal antibody formulations (Syva old formulation, Syva new formulation, and ARK assay) on a Roche Cobas c501 analyzer. Although they share the same homogenous immunoassay design, their respective formulations are composed of different polyclonal antibody pools, with varying degrees of cross-reactivity to the methotrexate major metabolite, 7-hydroxymethotrexate.

Overall, the three formulations share acceptable analytical performances in their respective measuring range. Nevertheless, we observed considerable biases among them for measuring methotrexate in patient samples treated with high dose therapy. Pairwise analyses revealed that the mean proportional bias between Siemens new formulation relative to the old formulation was approximately −9%. For the ARK assay the proportional bias relative to the Siemens old assay is 12% across patient methotrexate serum concentration range.

---

*Current address: Department of Pathology, ARUP Laboratories, The University of Utah, Salt Lake City, UT, United States

TABLE 10.4.1 Patient Serum Methotrexate Concentrations at Different Times Post high dosets IV Infusion Measured by Three Different Assays formulations

t (h)	Siemens, Old Formulation (μM)	Siemens, New Formulation (μM)	ARK Assay (μM)
0	75.59	70.66	78.7
18	0.73	0.68	0.99
24	1.63	1.5	2.49
48	0.56	0.55	0.64
60	0.06	0.05	0.08

In order to determine how measurement bias may affect the toxicity risk assessment for individual patient and the appropriate cutoff limit to stop the leucovorin therapy, we investigated several patients' elimination pattern based on the serial methotrexate measurement across three methods. Specifically, for our patient, different results in Table 10.4.1 indicates that the methotrexate cutoff " < 0.1 μM" based on the Siemens old formulation would need to be modified to achieve the same leucovorin therapy duration.

Discussion

Methotrexate is a structural analog of folic acid and inhibits nucleotides synthesis needed for DNA and RNA synthesis, which is the basis for its clinical efficacy and toxicity as a chemo agent in both adult and pediatric neoplasms. Although the mechanism of action is not completely elucidated, methotrexate competitively inhibits dihydrofolate reductase (DHFR), an enzyme that participates in the tetrahydrofolate (THF) synthesis [1]. THF is needed for the de novo synthesis of the nucleoside thymidine, required for DNA synthesis. As folate is also essential for purine and pyrimidine base biosynthesis, RNA synthesis will be inhibited accordingly. The toxicity profile of methotrexate normally involves liver toxicity and hematopoiesis suppression [2]. For high dose methotrexate treatment, its therapeutic drug monitoring is crucial for its toxicity assessment and the management of the subsequent leucovorin rescue therapy in malignancy treatment. Leucovorin can be metabolized into purine/pyrimidine precursor without the activity of DHFR and, therefore, can restore certain DNA synthesis activity with the presence of methotrexate [3]. Due to the potential risk of myelosuppression, leucovorin is normally administered following high dose methotrexate to minimize the bone marrow suppression in patients. The duration of leucovorin therapy is managed by close assessment of the residual circulating methotrexate concentration following its initial infusion [4]. Based on treatment protocol at the local institution, the serum concentration of methotrexate is measured by enzyme immunoassays daily, starting 24 hours after the initial dose until sufficient clearance of methotrexate is guaranteed and reaches a preestablished cutoff level (<0.1 μM) [5].

At high dose, methotrexate has a half-life of 16–29 hours and is mainly eliminated through kidney and 16% methotrexate is eliminated as 7-OH-methotrexate [6]. Intracellularly, methotrexate primarily undergoes sequential conjugation to active polyglutamated forms. As the aqueous solubility of 7-OH-methotrexate is three- to fivefold lower than methotrexate, it is considered to be an inactive metabolite. The accumulation of 7-OH-methotrexate can become significant in high-dose methotrexate treatment [6]. When the polyclonal antibody formulation changes in immunoassay, their different cross-reactivities to 7-OH-methotrexate would lead to different quantification results for methotrexate in the same specimen.

When three assay formulations (Syva old formulation, Syva new formulation, and ARK assay) are compared together, their precision and analytical measuring ranges (AMR) were evaluated with spiked serum materials containing only methotrexate at four different methotrexate concentrations (Table 10.4.2). Accordingly, analytical performances of the three formulations are largely comparable, with similar precision within respective AMR ranges (Table 10.4.2). Nevertheless, when measuring patient samples containing different amounts of methotrexate and its major metabolite 7-OH-methotrexate, the three formulations exhibited discernable differences. The overall bias between Siemens new formulation relative to the old formulation was approximately −9%, and the proportional bias between ARK assay relative to the Siemens old assay is 12%. Considering the analytical precision data measured by the methotrexate quality

TABLE 10.4.2 The Analytical Precision of Three Assays at Four Different Methotrexate Concentrations (μM)

		QC Level I	QC Level II	QC Level III	QC Level IV
Siemens Syva, old formulation	Mean	0.07	0.40	1.30	9.61
	SD	0.005	0.01	0.05	0.42
	CV (%)	6.8	3.5	3.7	4.4
Siemens Syva, new formulation	Mean	0.08	0.38	1.12	8.43
	SD	0.005	0.01	0.07	0.41
	CV (%)	6.8	3.0	6.4	4.9
ARK methotrexate assay	Mean	0.11	0.43	1.05	9.63
	SD	0.009	0.02	0.09	0.96
	CV (%)	8.1	5.0	8.7	9.9

control materials, the overall bias observed among the three assays would be considered statistically significant. If we directly apply the proportional bias as conversion factor to modify the residual methotrexate cutoff (<0.1 μM), the proposed new threshold based on Siemens new formulation and ARK assay would be <0.09 and <0.11 μM, respectively.

As residual methotrexate concentration is used to determine the duration of leucovorin therapy, it is important to understand how the adjusted thresholds can affect the leucovorin treatment lengths. The methotrexate clearance pattern for our patient can be compared based on samples collected at different postdose time points (Table 10.4.1). Based on the adjusted methotrexate thresholds to stop leucovorin therapy, our patient would receive the same duration and doses for the leucovorin therapy using three different assay formulations.

The varying levels of cross-reactivity to methotrexate metabolites in the immunoassays can lead to different assessment of its toxicity risk and affect the clinical management of leucovorin rescue therapy based on our parallel evaluation of patients' postdose methotrexate level series. While biases do exist across different assay platforms, it is important to evaluate how the adjusted cutoff limit would impact the clinical management of patients on high dose methotrexate therapy.

References

[1] Dombrowsky E, et al. Evaluating performance of a decision support system to improve methotrexate pharmacotherapy in children and young adults with cancer. Ther Drug Monit 2011;33(1):99−107.

[2] Hospira, Inc. Methotrexate Injection, USP [Package Insert] U.S. Food and Drug Administration website. https://www.accessdata.fda.gov/drugsatfda_docs/label/2011/011719s117lbl.pdf. [accessed 22.02.20].

[3] Bedford Laboratories™. LEUCOVORIN CALCIUM INJECTION USP [Package Insert] U.S. Food and Drug Administration website. https://www.accessdata.fda.gov/drugsatfda_docs/label/2012/040347s010lbl.pdf. Revised 11/09/11 [accessed 22.02.20].

[4] Lambrecht L, et al. The role of the MTHFR C677T polymorphism in methotrexate-induced toxicity in pediatric osteosarcoma patients. Pharmacogenomics 2017;18(8):787−95.

[5] Goorin A, et al. Safety and efficacy of l-leucovorin rescue following high-dose methotrexate for osteosarcoma. Med Pediatr Oncol 1995;24(6):362−7.

[6] Baselt RC. Disposition of toxic drugs and chemicals in man. 2017.

Chapter 11

Cannabinoids

Hema Ketha[1] and Uttam Garg[2,3]

[1]Toxicology, Mass Spectrometry and Metals, Laboratory Corporation of America Holdings, Burlington, NC, United States, [2]Division of Laboratory Medicine, Pathology and Laboratory Medicine, Children's Mercy Hospital, Kansas City, MO, United States, [3]Pathology and Laboratory Medicine, University of Missouri School of Medicine, Kansas City, MO, United States

Introduction

Cannabis sativa, the plant source of most physiologically active component, delta-9-tetrahydrocannabinol (THC), has been used historically for 5 millennia, throughout the world. The term "cannabinoids" refers to the group of compounds that are derived from the cannabis sativa plant. Cannabis has found its way into medicinal, recreational, and spiritual aspects of medieval and modern human society. An estimated 200 million individuals aged 15–65 have used THC in 2010 alone [1]. The World Health Organization has classified cannabis as the world's most widely cultivated, trafficked, and abused illicit substance. THC has been used by physicians since the early 20th century for a variety of indications. Its common uses include use as an anxiolytic, antiemetic, and an appetite enhancer. The US federal government started to regulate cannabis use in the 1940s and by 1970 the US Congress added THC as a schedule I substance making it federally illegal and devoid of any purported medicinal value. Around the same era the illegal use and abuse of cannabis and related products exhibited a new increase which has not relented during the last many decades. Cannabis contains upward of 400 chemical components and more are being classified more recently. These compounds include amino acids, hydrocarbons, sugars, terpenes, simple and complex fatty acids, and more than 60 cannabinoids. When cannabis is smoked, pyrolysis can form hundreds of new cannabinoid compounds. A number of nonpsychoactive cannabinoid compounds, including cannabinol and cannabidiol (CBD), are being evaluated for alternate medicinal uses [2–6].

Other cannabis products such as hashish are also very popular in the illegal and naturopathic and alternate medical communities. Hashish is derived from the resin of pulp of young cannabis or marijuana buds or flowers. In various countries around the Indian Subcontinent, it is also referred to as charas or hash. Extracts of the resinous product is mixed into beverages termed "bhang." Hashish is commonly smoked in a pipe or in other smoking devices, including a bong [7]. Since hashish can be ineffective when used as a smoking agent by itself, it is widely smoked in combination with tobacco or cannabis leaves. The resin or pulp from the cannabis flowers and buds can also be consumed orally. The resin contains a mixture of THC and other cannabinoids that are derived from the stalked resin glands. Cannabis extracts are used in a variety of products, including teas, jelly, gummy bears, oils, lozenges, candy, and others. In confiscations across European Union (EU), the content of hashish in cannabis end product ranged from 4% to 18% [1,5]. The primary cannabis product popular in EU was hashish, whereas cannabis leaves are more common in North Americas [8,9]. Hashish is produced most widely in Afghanistan, Pakistan and the Kashmir region, Nepal, Lebanon, Turkey Greece, and Syria [10]. Until recently, due to increasing conflicts in the subcontinental region, Pakistan and Afghanistan were the largest suppliers for hashish to the EU [11–13]. Morocco has become the recent major player in hashish production [12,14,15].

Cannabinoids are the most widely abused group of illicit drugs around the world. In 1995 a massive shift in the state law occurred in the United States, wherein California became the first states to legalize THC for medical use. This set a landmark that divided state and federal legislature around cannabis as states legalized its use despite a federal ban on the product. Several states in the United States have legalized medical and/or recreational use of cannabis and its various products. There is considerable debate and heterogeneity in the state and federal laws that govern cannabis in the United States [16,17]. An exhaustive discussion of the scope of legal issues surrounding cannabis is outside the scope of this chapter. This chapter will discuss abuse, pharmacokinetics, toxicology, and laboratory testing aspects of cannabinoids that are relevant to the practice of clinical and forensic toxicology.

Cannabinoid compounds

THC is the most psychoactive component of cannabis sativa plant [7]. The common and popular notions that THC and cannabis use in general is harmless are being debated and a number of publications describing health risk of cannabis use have emerged [18–23]. THC is present in all parts of the cannabis plant [24]. The concentration can vary from ~1% to 12% depending on the portion of the cannabis plant in question. A user may be able to consume up to 22% of THC dose via smoking [1]. Dronabinol, or Marinol or Syndros, is the enantiomeric pure form of THC, (−)-trans-Δ^9-THC, found in cannabis. In contrast with Sativex, dronabinol does not include any other THC isomers or CBD. Dronabinol is used as an appetite stimulant in patients with HIV/AIDS and cancer [25,26]. The endogenous receptors for THC, central cannabinoid receptor (CB) 1, and peripheral CB2 have been characterized as transmembrane G-protein coupled receptors. CB1 receptor is activated by its endogenous agonists that act as retrograde neurotransmitters [27]. The physiological effect of THC, including its effect of appetite, cognition memory, and pain, is associated with the spatial arrangement of CB1 receptors in specific corresponding locations. Endogenous agonists include arachidonylethanolamide or anandamide and arachidonyl glycerol. Exogenous THC analogs such as SR141716 have also been developed. The most prominent physiological effects of cannabis use are tachycardia and euphoria [28]. Both the effects can be observed by THC consumption leading to binding to and activation of CB1 receptors. Synthetic cannabinoids such as K2 and spice act by binding to CB1 receptors but likely at a much higher potency. Many synthetic cannabinoids are relatively more toxic compared to THC as evidenced by fatalities and adverse effects of their use in the recent literature [29,30]. THC dose response curves have been studied. The curves for heart rate and subjective euphoria show a counterclockwise hysteresis. This indicates a delay between attainment of certain plasma concentration and physiological effects. The counterclockwise hysteresis indicates a long distribution phase, wherein THC distributes from the vascular site to the brain [31].

CBD is a cannabinoid with almost no psychoactive manifestations which has recently received immense public attention [32]. CBD formulation called Epidiolex was approved by the Food and Drug Administration in 2018 for the treatment epileptic disorders in the United States. It was approved on the basis of small randomized controlled trials to be used as an adjunct to other antiepileptic therapies for treatment of rare forms of epilepsy associated with Lennox–Gastaut syndrome or Dravet syndrome in patients 2 years of age and older [33]. Since its approval, CBD has been widely formulated into several daily used items such as food and beverages, oils, gums, candies, and other products such as deodorants. CBD is mostly unscheduled in the EU and CBD-containing hemp products in the United States are legally sold. There remains some confusion about the legal status of CBD. CBD extracted from marijuana is still illegal in the United States as of this writing (2019) but CBD derived from hemp was legalized by the 2018 Farm Bill [34]. One stipulation on hemp-derived CBD is that it must contain 0.3% of THC or lower to retain legal status. CBD has also been associated with several recent reports of toxic events. Nabiximols called Sativex in the United States is a drug approved in 2010 for use in supportive therapy in multiple sclerosis. Sativex was originally developed by the UK-based company GW Pharmaceuticals. It contains a combination of THC and CBD as the principle active cannabinoids [35]. Nabiximols is available as an oral mouth spray, wherein the effective dose delivered by each spray is 2.7 mg THC and 2.5 mg CBD. The strength of evidence supporting the effectiveness of Sativex for intended use has been debated. It is intended to alleviate neuropathic pain, spasticity, overactive bladder, and other symptoms of multiple sclerosis [36].

Pharmacokinetics and toxicokinetics

THC can be consumed by smoking or by oral consumption of cannabis products. Smoking cannabis leaves is the most common route of THC consumption. This route provides the most rapid route of THC absorption via the lungs directly into the brain. Smoking is the preferred route of THC consumption as the psychoactive effects of euphoria and pleasurable effects related to cannabis are immediately felt with the direct delivery of the active component to the brain. Users consume only a small fraction of THC present in the leaves when smoking the cannabis plant. Up to 30% of THC decomposes to other compounds due to pyrolysis [1,31]. In addition, half of the content can be lost in the smoke depending the person's smoking proficiency. Factors that affect the amount of THC absorbed via smoking include number, duration, and spacing of puffs, hold time, and inhalation volume. Depending on intersmoker variability causes a wide variability in dose availability and dose delivery. Quantifiable concentrations of THC can be seen in plasma seconds after the first puff of cannabis. Concentrations observed after inhalation of low THC dose cigarette at 1.75% THC or a high THC dose cigarette at 3.55% THC were 7.0 ± 8.1 (Mean ± SD) μg/L or 18.0 ± 12.0 μg/L. Peak concentrations have been documented prior to completion of smoking cycle about 9−10 minutes prior to the last puff sequence. Passive exposure to cannabis smoke can result in measurable plasma THC concentration. In one study, eight healthy study volunteers showed a serum concentration of 0.5 μg/L at 1.5 hours of passive exposure in a public shop [1,37]. Peak plasma concentration of 46−188 μg/L has been reported during a 10-minute smoking period in healthy adult volunteers who smoked a cigarette containing 8.8 mg of THC. Oral absorption is slower and can delay achievement of peak plasma THC levels by 1−5 hours [31]. In studies of oral delivery of Marinol, doses of 2.5−20 mg of THC resulted in 1.3−7.9 μg/L THC plasma levels at 1−2.5 hours. A buccal dose of 10.8 g THC (Sativex) resulted in average THC plasma levels of 6.1 μg/L in 2.5 hours and 3 μg/L of CBD at 2.8 hours. Oral administration of Marinol over a longer time frame of 1 week at 20 mg doses every 4−8 hours resulted in peak plasma levels at 2 hours after the 22nd dose of 43 μg/L. After intravenous administration of 4−5 mg of THC in health adults average peak plasma concentration of 62 μg/L was observed at 20 minutes [38].

Elimination half-lives of THC have been estimated to be as broad as 20−57 hours in infrequent users. However, the elimination half-life can be much longer up to 3−13 days in frequent users [37]. THC has a very large volume of distribution (3.4−10 L/kg) [31]. It is highly protein bound in circulation. It is estimated to be 97%−99% bound to alpha-1-acid glycoprotein. THC is highly lipid soluble. Therefore it is easily retained in the fat tissue. This results in a long circulating half-life in plasma of as high as 4.1 days. This physicochemical property of THC has an important bearing on laboratory testing. Once the user, with high body mass index and high overall fat content, has stopped smoking cannabis or using other orally consumed cannabis products, the slow, long release of the drug and appreciable enterohepatic recirculation can result in a positive THC test in plasma and for longer periods in urine [39].

THC is metabolized primarily to two monohydroxy compounds 11-OH-THC and 8-beta-OH-THC [31]. Both the compounds are active. It is unlikely that these compounds achieve high enough plasma concentration to exert any appreciable psychoactive effects. The overall concentration of 11-OH-THC is about 10% that of THC. The two cytochrome enzymes that are principal contributors to phase I metabolism of THC are P450-2C9 and 2C19. 11-COOH-THC [1,31,37] physiology of THC is complex. It is currently not classified as a stimulant, sedative, or a hallucinogen. Immediate effects of THC smoking are feelings of euphoria, relaxation, altered perception of reality and time, mood changes, and feelings of panic and paranoia. Toxic exposures are associated with tachycardia, hypertension, and excessive sedation [40]. In THC-associated deaths, ratio of heart to femoral blood concentration was found to be 0.3−3.1 (mean 1.5) for THC and 0.3−2.7 (mean 1.6) for 11-OH-THC [37]. Chronic THC use is associated with several detrimental effects, including psychosis, exacerbation of schizophrenia, lung and mouth cancer, reproductive and immunological effects. However, proving THC as the causative agent is very difficult [40].

Concentration of cannabinoid metabolites in urine is subject to variability among users. In controlled studies, while peak THC-COOH concentration in urine is associated with timing and dose, 10- to 12-fold intersubject variability was observed. In chronic users the THC-COOH concentration may vary significantly due to several physiological factors, including hydration status and renal function. It is common in routine toxicology practice to encounter a consult about prediction of timeline of cannabis use (recent vs chronic) based on urine THC metabolite ratios normalized to creatinine. Models have been developed for examining urine excretion profile of THC and its metabolites in nonchronic smokers. But such models in chronic smokers have not been successful. A ratio of 1.5 or greater between two positive urine THC-COOH tests normalized to creatinine has been suggested as an aid in differentiating new from prior drug use. Unfortunately, this ratio has not been validated thoroughly under controlled dosing conditions [1,31].

The physiological impact of THC exposure depends on several aspects, including time of onset of symptoms, duration of effect, and severity of clinical signs, depend upon the dose and the route of administration of the drug. It has been very difficult to define a toxic threshold dose for THC in humans due to variability in consumption of THC from smoked leaves, degree of purity for marijuana, and that the bioavailable dose depends upon the routes of exposure. Some information of impact of THC toxic exposures has been collected in other mammals such as dogs. A minimum lethal dose of 3 g/kg of THC has been reported for dogs. In dogs, clinical manifestations include incoordination, hypersalivation, depression, disorientation, hypothermia, mydriasis, bradycardia, vomiting, and tremors. Other presenting symptoms seen with marijuana ingestion in dogs are stupor, nystagmus, apprehension, vocalization, hyperexcitability, tachypnea, tachycardia, and hyperthermia [41].

Inhalation of 2−3 mg of THC or ingested doses of 5−20 mg of THC have been shown to impair attention, concentration, short-term memory, and executive functioning in adolescents and adults. At doses >7.5 mg/m^2 of THC, the adverse effects are more severe which include nausea, postural hypotension, delirium, panic attacks, anxiety, and myoclonic jerking [42,43]. The use of higher potency or highly concentrated products such as waxes or jellies can cause serious effects such as psychosis [44].

In small cohort-based studies, a dose-dependent response has been reported in children, wherein severity of symptoms and treatment required when they presented to the emergency department corresponded to the dose ingested. In children exposed to a dose of ~3.2 mg/kg of THC, minimal medical intervention and observation was needed for treatment of toxicity, whereas at 7.2 mg/kg of THC children had to be admitted to an inpatient setting with moderate medical intervention, and at 13 mg/kg of THC exposure exposed children had to be admitted to an intensive care unit with major medical interventions being a part of the treatment [45,46]. In this study, patients who have not been previously exposed to THC showed a greater degree of lethargy and somnolence and their duration of treatment and symptoms were longer [46,47].

Acute exposures in children occur in cases of accidental or exploratory consumption of cannabis products meant for adult consumption. In children exposed to THC, common symptoms include a combination of sleepiness and euphoria, irritability, and other behavioral changes. Sympathomimetic effects such as tachycardia and hypertension may be seen in the vital signs. This is particularly true for patients with depressed mental status, bradycardia. Slurred speech, nystagmus, ataxia, nausea, and vomiting have also been reported. Another symptom commonly reported in children with THC toxicity is dilated pupils. In exposures to concentrated products such as edible product, concentrated oils, or hashish presenting symptoms may include coma with apnea or depressed respirations [48,49].

In patient with THC toxicity a complete blood count (CBC) and other biochemical test are mostly normal. However, in managing a potential THC toxicity a CBC and a metabolic profile should be considered to rule out other underlying mechanisms of toxicity. Laboratory testing should include drug screen potentially with a confirmation test. However, results from confirmation testing may not be available for a few days. A qualitative gas chromatography (GC)—mass spectrometry(MS)-based urine drug screen should be able to confirm THC exposure in a matter of 1—1.5 hours where such testing is available as a STAT test. With the advent of liquid-chromatography (LC)-high-resolution mass spectrometry (HRMS)-based urine drug screen, an MS-based semiquantitative result may be available in 1—2 hours in some hospital settings. Patients presenting with chest pain in the setting of known THC toxicity, chest pain can be suggestive of myocardial ischemia or infarction that warrants a 12-lead electrocardiogram study and cardiac biomarker assessment. Of note, synthetic cannabinoid exposure may not present as a typical THC overdose as the potency, purity, and dose response of these unregulated drugs are highly variable. These "designer drugs" are not identified by most immunoassay urine drug screens. Urine GC—MS or LC—HRMS drug screen operated in a full-scan mode may be able to identify a novel synthetic cannabinoid provided the compound library that is used for identification has the structural information for the novel synthetic street drug. With newer advanced LC—HRMS instruments, structural analysis of retrospective mass fragmentation data is feasible but such analysis of unknown compounds is highly time consuming and tedious in nature.

Clinical management of THC toxicity is mostly supportive in nature. The type of supportive measure varies significantly by patient's age [45,46,50]. Life-threatening central nervous system depression from THC exposure is unique to the pediatric population. In this age group, acute THC exposure can present with profound depression, lethargy, and coma. Therefore maintaining airway, breathing, and circulation are priority. Since THC toxicity can present like other sedative hypnotic toxidromes, naloxone must be administered to support and treat a potential opioid toxicity. However, naloxone will not reverse THC toxicity. Seizures have been described in patients with coingestion of stimulates. Benzodiazepines are commonly used to treat seizures in such a case. In adults and adolescents with mild symptoms for THC exposure, treatment is with counseling and supportive care in a dimly lit room. In patients presenting with anxiety, benzodiazepine may be administered. Severe toxicity with THC exposure alone is rare in adults.

Analytical methods for cannabinoids

Like all other major classes of compounds relevant to a clinical and forensic toxicology laboratory, a combination of screening and confirmation tests is used to analyze cannabinoids and related compounds—particularly, THC and its major metabolite in urine, THC-COOH. Screening is commonly performed by immunoassay. Techniques such as thin layer chromatography are used in some laboratories but are becoming increasingly uncommon. With the advent of mass spectrometric methods in the toxicology laboratories, quantitative, targeted LC—MS/MS and semiquantitative, untargeted HRMS-based assays are being implemented widely in hospital-based and reference laboratory settings.

Urine is the matrix that is the most widely used for cannabinoid screening for clinical compliance testing. Other matrices used in postmortem, driving under the influence, pain management, and workplace drug-testing settings include blood, plasma, serum, and oral fluid. Until recently, urine was the specimen of choice for workplace drug-screening purposes. The federal workplace drug testing working group has recently recommended that oral fluid should be added as an additional matrix type in the federal workplace drug-testing program.

One important analytical variable to consider while selecting and screening immunoassay methods is the cross-reactivity of the glucuronide metabolites. Other considerations include cross-reactivity with commonly encountered over-the-counter medications and other interfering substances such as proteins and lipids. The most abundant THC metabolites present in urine are THC-COOH and glucuronides. Therefore for immunoassays in which the antibody was directed toward the free THC-COOH metabolite hydrolysis prior to sample processing was required. However, with antibodies which show enough cross-reactivity toward the free and glucuronidated metabolites, direct analysis of the urine sample without hydrolysis is possible with contemporary cannabinoid screening assays. Contemporary automated immunoassays offer screening cut-off of 10−50 ng/mL. These assays do not require any sample preprocessing and offer the laboratory fast turnaround time. However, immunoassay used for screening is prone to interferences that lead to false positives and in many cases false negatives. Most toxicology laboratories require a confirmation of all positive immunoassay screening results. Some immunoassays require a protein precipitation and an organic extraction step prior to quantitation but these assays are tedious and are less common in routine clinical and forensic toxicology laboratory settings.

Confirmation methods in toxicology laboratories are based on MS. In addition, the combination of screening and confirmation tests is expected to be of different chemical principals. Ideally, analytical sensitivity and specificity of the confirmation method should be higher compared to the screening methods and lower limit of quantitation. Older GC−MS-based methods are being replaced by LC−MS/MS or LC−HRMS-based methods. GC−MS methods require tedious derivatization steps and require long GC separation time. GC−MS methods can require 15−18 minutes of separation time, while LC−MS/MS analysis can be performed in under 5 minutes. As full-scan untargeted LC−HRMS assays become more common in toxicology laboratory settings, the utility of immunoassays for screening has been debated as LC−HRMS assays that can screen for hundreds of compounds in a single injection are being implemented in clinical laboratories [51].

The most abundant THC metabolites in urine are THC-COOH and its glucuronide and sulfate metabolites. For analysis of total THC-COOH concentration in urine samples, the glucuronide must be hydrolyzed under alkaline conditions. Until recently only a single supplier had a source for the THC-COOH glucuronide standard material. In addition, the efficiency of glucuronidase activity depends on the source of the glucuronidase (*Escherichia coli* or *Helix promatia*) [52]. Several methods for quantitation of THC and its metabolites in serum/plasma, whole blood, oral fluid, and hair have been published [52−57]. Oral fluid is a suitable alternative to serum or whole blood for drug of abuse testing in the clinical and forensic settings. Advantages of oral fluid use include ease of collection and simple matrix make up from an analytical standpoint. In addition, oral fluid is relatively harder to tamper with compared to urine as collection of the sample can be performed in completely supervised settings. On the other hand, oral fluid presents several disadvantages. Lag time between smoking or eating THC-containing products and collection of saliva sample can improve the detection in the oral fluid itself. But the saliva and the mucosal membranes can be saturated with THC and the concentration in saliva may not be correlated to circulating blood concentrations. In the case of THC measurements in oral fluid, saliva concentrations have been shown to not be a good predictor of circulating blood concentrations of THC. A possible reason for this lack of correlation is transmucosal THC absorption into blood [31,58].

Conclusion

THC is the most widely abused substance in the world. The dichotomy in laws governing THC use in countries and even across different states makes studies related to THC complicated to pursue. THC and/or cannabinoid extract are being used as pharmacotherapeutic agents most prominently for treatment of pain, as an appetite enhancer and as antiepileptic agent. Physiology of THC is complex. Endogenous CBs and its agonists are well studied. Illicitly produced and marketed synthetic cannabinoids act on the same receptors but with unpredictable physiological effects. THC toxicity is more common in children and adolescents. Treatment of toxicity can depend on the age of the patient. In adults, treatment is mostly supportive in nature, whereas the level of treatment needed is dependent of the dose to which the patient is exposed. Screening methods are commonly based on immunoassays and confirmation methods involve MS. HRMS-based screening methods have been developed but these are less commonly available.

References

[1] Kwong T, Magnani B, Rosano TG, Shaw L. In: Kwong T, Magnani B, Rosano TG, Shaw L, editors. Clinical toxicology laboratory: contemporary practice of poisoning evaluation. 2nd ed. American Association for Clinical Chemistry Inc. AACC Press; 2013.

[2] Kilmer B. Recreational cannabis—minimizing the health risks from legalization. N Engl J Med 2017;376(8):705–7. Available from: https://doi.org/10.1056/NEJMp1614783.

[3] Zuardi AW. History of cannabis as a medicine: a review. Rev Bras Psiquiatr 2006;28(2):153–7. Available from: https://doi.org/10.1590/s1516-44462006000200015.

[4] Turner CE, Elsohly MA, Boeren EG. Constituents of *Cannabis sativa* L. XVII. A review of the natural constituents. J Nat Prod 1980;43(2):169–234. Available from: https://doi.org/10.1021/np50008a001.

[5] Borroto Fernandez E, Peterseil V, Hackl G, Menges S, de Meijer E, Staginnus C. Distribution of chemical phenotypes (chemotypes) in European agricultural hemp (*Cannabis sativa* L.) cultivars. J Forensic Sci 2019. Available from: https://doi.org/10.1111/1556-?029.14242.

[6] Claussen U, Korte F. Concerning the behavior of hemp and of delta-9-6a, 10a-trans-tetrahydrocannabinol in smoking. Justus Liebigs Ann Chem 1968;713:162–5. Available from: https://doi.org/10.1002/jlac.19687130119.

[7] Mechoulam R. Marihuana chemistry. Science (80-) 1970;168(3936):1159–65. Available from: https://doi.org/10.1126/science.168.3936.1159.

[8] Botoeva G. Hashish as cash in a post-Soviet Kyrgyz village. Int J Drug Policy 2014;25(6):1227–34. Available from: https://doi.org/10.1016/j.drugpo.2014.01.016.

[9] Potter DJ. A review of the cultivation and processing of cannabis (*Cannabis sativa* L.) for production of prescription medicines in the UK. Drug Test Anal 2014;6(1-2):31–8. Available from: https://doi.org/10.1002/dta.1531.

[10] Balhara YPS, Mathur S. Bhang – beyond the purview of the narcotic drugs and psychotropic substances act. Lung India 2014;31(4):431–2. Available from: https://doi.org/10.4103/0970-2113.142109.

[11] Giroud C, Broillet A, Augsburger M, Bernhard W, Rivier L, Mangin P. Brief history of recent hemp cultivation in Switzerland and subsequent medico-legal problems resulting from hemp cultivation. Praxis (Bern 1994) 1999;88(4):113–21.

[12] Afsahi K, Darwich S. Hashish in Morocco and Lebanon: a comparative study. Int J Drug Policy 2016;31:190–8. Available from: https://doi.org/10.1016/j.drugpo.2016.02.024.

[13] Nguyen H, Malm A, Bouchard M. Production, perceptions, and punishment: restrictive deterrence in the context of cannabis cultivation. Int J Drug Policy 2015;26(3):267–76. Available from: https://doi.org/10.1016/j.drugpo.2014.08.012.

[14] Chouvy P-A, Afsahi K. Hashish revival in Morocco. Int J Drug Policy 2014;25(3):416–23. Available from: https://doi.org/10.1016/j.drugpo.2014.01.001.

[15] Chouvy P-A, Macfarlane J. Agricultural innovations in Morocco's cannabis industry. Int J Drug Policy 2018;58:85–91. Available from: https://doi.org/10.1016/j.drugpo.2018.04.013.

[16] Pacula RL, Smart R. Medical marijuana and marijuana legalization. Annu Rev Clin Psychol 2017;13:397–419. Available from: https://doi.org/10.1146/annurev-clinpsy-032816-045128.

[17] Stoecker WV, Rapp EE, Malters JM. Marijuana use in the era of changing cannabis laws: what are the risks? Who is most at risk? Mo Med 2018;115(5):398–404.

[18] Keyhani S, Steigerwald S, Ishida J, et al. Risks and benefits of marijuana use: a national survey of U.S. adults. Ann Intern Med 2018;169(5):282–90. Available from: https://doi.org/10.7326/M18-0810.

[19] Tashkin DP. Effects of marijuana smoking on the lung. Ann Am Thorac Soc 2013;10(3):239–47. Available from: https://doi.org/10.1513/AnnalsATS.201212-127FR.

[20] Martinasek MP, McGrogan JB, Maysonet A. A systematic review of the respiratory effects of inhalational marijuana. Respir Care 2016;61(11):1543–51. Available from: https://doi.org/10.4187/respcare.04846.

[21] Moore BA, Augustson EM, Moser RP, Budney AJ. Respiratory effects of marijuana and tobacco use in a U.S. sample. J Gen Intern Med 2005;20(1):33–7. Available from: https://doi.org/10.1111/j.1525-1497.2004.40081.x.

[22] Ghasemiesfe M, Barrow B, Leonard S, Keyhani S, Korenstein D. Association between marijuana use and risk of cancer: a systematic review and meta-analysis. JAMA Netw Open 2019;2(11):e1916318. Available from: https://doi.org/10.1001/jamanetworkopen.2019.16318.

[23] Reboussin BA, Wagoner KG, Sutfin EL, et al. Trends in marijuana edible consumption and perceptions of harm in a cohort of young adults. Drug Alcohol Depend 2019;205:107660. Available from: https://doi.org/10.1016/j.drugalcdep.2019.107660.

[24] Grof CPL. Cannabis, from plant to pill. Br J Clin Pharmacol 2018;84(11):2463–7. Available from: https://doi.org/10.1111/bcp.13618.

[25] Lucas CJ, Galettis P, Schneider J. The pharmacokinetics and the pharmacodynamics of cannabinoids. Br J Clin Pharmacol 2018;84(11):2477–82. Available from: https://doi.org/10.1111/bcp.13710.

[26] O'Connell BK, Gloss D, Devinsky O. Cannabinoids in treatment-resistant epilepsy: a review. Epilepsy Behav 2017;70(Pt B):341–8. Available from: https://doi.org/10.1016/j.yebeh.2016.11.012.

[27] Grotenhermen F. Pharmacokinetics and pharmacodynamics of cannabinoids. Clin Pharmacokinet 2003;42(4):327–60. Available from: https://doi.org/10.2165/00003088-200342040-00003.

[28] Pertwee RG. The diverse CB1 and CB2 receptor pharmacology of three plant cannabinoids: delta9-tetrahydrocannabinol, cannabidiol and delta9-tetrahydrocannabivarin. Br J Pharmacol 2008;153(2):199–215. Available from: https://doi.org/10.1038/sj.bjp.0707442.

[29] Alipour A, Patel PB, Shabbir Z, Gabrielson S. Review of the many faces of synthetic cannabinoid toxicities. Ment Health Clin 2019;9(2):93–9. Available from: https://doi.org/10.9740/mhc.2019.03.093.

[30] Diao X, Huestis MA. New synthetic cannabinoids metabolism and strategies to best identify optimal marker metabolites. Front Chem 2019;7:109. Available from: https://doi.org/10.3389/fchem.2019.00109.

[31] Huestis MA. Human cannabinoid pharmacokinetics. Chem Biodivers 2007;4(8):1770−804. Available from: https://doi.org/10.1002/cbdv.200790152.

[32] Crippa JA, Guimarães FS, Campos AC, Zuardi AW. Translational investigation of the therapeutic potential of cannabidiol (CBD): toward a new age. Front Immunol 2018;9:2009. Available from: https://doi.org/10.3389/fimmu.2018.02009.

[33] Sekar K, Pack A. Epidiolex as adjunct therapy for treatment of refractory epilepsy: a comprehensive review with a focus on adverse effects. F1000Research 2019;8. Available from: https://doi.org/10.12688/f1000research.16515.1.

[34] Mozaffarian D, Griffin T, Mande J. The 2018 farm bill-implications and opportunities for public health. JAMA 2019;321(9):835−6. Available from: https://doi.org/10.1001/jama.2019.0317.

[35] Keating GM. Delta-9-tetrahydrocannabinol/cannabidiol oromucosal spray (Sativex®): a review in multiple sclerosis-related spasticity. Drugs 2017;77(5):563−74. Available from: https://doi.org/10.1007/s40265-017-0720-6.

[36] Syed YY, McKeage K, Scott LJ. Delta-9-tetrahydrocannabinol/cannabidiol (Sativex®): a review of its use in patients with moderate to severe spasticity due to multiple sclerosis. Drugs 2014;74(5):563−78. Available from: https://doi.org/10.1007/s40265-014-0197-5.

[37] Baselt RC. In Disposition of toxic drugs and chemicals is man. 10th ed., Biomedical Publications.

[38] Huestis MA, Henningfield JE, Cone EJ. Blood cannabinoids. I. Absorption of THC and formation of 11-OH-THC and THCCOOH during and after smoking marijuana. J Anal Toxicol 1992;16(5):276−82. Available from: https://doi.org/10.1093/jat/16.5.276.

[39] Huestis MA, Mitchell JM, Cone EJ. Detection times of marijuana metabolites in urine by immunoassay and GC-MS. J Anal Toxicol 1995;19(6):443−9. Available from: https://doi.org/10.1093/jat/19.6.443.

[40] Volkow ND, Baler RD, Compton WM, Weiss SRB. Adverse health effects of marijuana use. N Engl J Med 2014;370(23):2219−27. Available from: https://doi.org/10.1056/NEJMra1402309.

[41] Fitzgerald KT, Bronstein AC, Newquist KL. Marijuana poisoning. Top Companion Anim Med 2013;28(1):8−12. Available from: https://doi.org/10.1053/j.tcam.2013.03.004.

[42] Devine ML, Dow GJ, Greenberg BR, et al. Adverse reactions to delta-9-tetrahydrocannabinol given as an antiemetic in a multicenter study. Clin Pharm 1987;6(4):319−22.

[43] Dow GJ, Meyers FH, Stanton W, Devine ML. Serious reactions to oral delta-9-tetrahydrocannabinol in cancer chemotherapy patients. Clin Pharm 1984;3(1):14.

[44] Andre C, Jaber-Filho JA, Bento RMA, Damasceno LMP, Aquino-Neto FR. Delirium following ingestion of marijuana present in chocolate cookies. CNS Spectr 2006;11(4):262−4. Available from: https://doi.org/10.1017/s1092852900020757.

[45] Wang GS, Roosevelt G, Heard K. Pediatric marijuana exposures in a medical marijuana state. JAMA Pediatr 2013;167(7):630−3. Available from: https://doi.org/10.1001/jamapediatrics.2013.140.

[46] Heizer JW, Borgelt LM, Bashqoy F, Wang GS, Reiter PD. Marijuana misadventures in children: exploration of a dose-response relationship and summary of clinical effects and outcomes. Pediatr Emerg Care 2018;34(7):457−62. Available from: https://doi.org/10.1097/PEC.0000000000000770.

[47] Claudet I, Mouvier S, Labadie M, et al. Unintentional cannabis intoxication in toddlers. Pediatrics 2017;140(3). Available from: https://doi.org/10.1542/peds.2017-0017.

[48] Caldicott DGE, Holmes J, Roberts-Thomson KC, Mahar L. Keep off the grass: marijuana use and acute cardiovascular events. Eur J Emerg Med 2005;12(5):236−44. Available from: https://doi.org/10.1097/00063110-200510000-00008.

[49] Bachs L, Morland H. Acute cardiovascular fatalities following cannabis use. Forensic Sci Int 2001;124(2−3):200−3. Available from: https://doi.org/10.1016/s0379-0738(01)00609-0.

[50] Wang GS, Roosevelt G, Le Lait M-C, et al. Association of unintentional pediatric exposures with decriminalization of marijuana in the United States. Ann Emerg Med 2014;63(6):684−9. Available from: https://doi.org/10.1016/j.annemergmed.2014.01.017.

[51] Colby JM, Lynch KL. Drug screening using liquid chromatography quadrupole time-of-flight (LC-QqTOF) mass spectrometry. Methods Mol Biol 2019;1872:181−90. Available from: https://doi.org/10.1007/978-1-4939-8823-5_17.

[52] Sempio C, Scheidweiler KB, Barnes AJ, Huestis MA. Optimization of recombinant beta-glucuronidase hydrolysis and quantification of eight urinary cannabinoids and metabolites by liquid chromatography tandem mass spectrometry. Drug Test Anal 2018;10(3):518−29. Available from: https://doi.org/10.1002/dta.2230.

[53] Hubbard JA, Smith BE, Sobolesky PM, et al. Validation of a liquid chromatography tandem mass spectrometry (LC-MS/MS) method to detect cannabinoids in whole blood and breath. Clin Chem Lab Med 2019. Available from: https://doi.org/10.1515/cclm-2019-0600.

[54] Scheidweiler KB, Newmeyer MN, Barnes AJ, Huestis MA. Quantification of cannabinoids and their free and glucuronide metabolites in whole blood by disposable pipette extraction and liquid chromatography-tandem mass spectrometry. J Chromatogr A 2016;1453:34−42. Available from: https://doi.org/10.1016/j.chroma.2016.05.024.

[55] Andersson M, Scheidweiler KB, Sempio C, Barnes AJ, Huestis MA. Simultaneous quantification of 11 cannabinoids and metabolites in human urine by liquid chromatography tandem mass spectrometry using WAX-S tips. Anal Bioanal Chem 2016;408(23):6461−71. Available from: https://doi.org/10.1007/s00216-016-9765-8.

[56] Aizpurua-Olaizola O, Zarandona I, Ortiz L, Navarro P, Etxebarria N, Usobiaga A. Simultaneous quantification of major cannabinoids and metabolites in human urine and plasma by HPLC-MS/MS and enzyme-alkaline hydrolysis. Drug Test Anal 2017;9(4):626−33. Available from: https://doi.org/10.1002/dta.1998.

[57] Klawitter J, Sempio C, Morlein S, et al. An atmospheric pressure chemical ionization MS/MS assay using online extraction for the analysis of 11 cannabinoids and metabolites in human plasma and urine. Ther Drug Monit 2017;39(5):556—64. Available from: https://doi.org/10.1097/FTD.0000000000000427.

[58] Menkes DB, Howard RC, Spears GF, Cairns ER. Salivary THC following cannabis smoking correlates with subjective intoxication and heart rate. Psychopharmacology (Berl) 1991;103(2):277—9. Available from: https://doi.org/10.1007/bf02244217.

Chapter 11.1

A review of impaired drivers under the influence of 5F-ADB

Amanda Chandler, Elizabeth Wehner, Brianna Peterson, Brian Capron and Fiona Couper
Toxicology Laboratory Division, Washington State Patrol, Seattle, WA, United States

Case description

Case 1

The driver was a 26-year-old female, with blood submitted from four driving under the influence (DUI) stops between October 29, 2015 and November 8, 2015. Driving behavior from the various stops included crashing into an embankment, slumped over at the wheel with a pipe in hand, a two-vehicle collision, and passed out while driving. Symptoms included poor coordination, droopy bloodshot eyes, lethargic with slurred speech, blank stares and eyelid tremors, and vomiting. "Spice" packages were noted on the floorboard of the car from one of the stops. Standard field sobriety tests (SFSTs) were not performed during three of the four stops, due to the need for medical treatment and refusal by the subject. The one stop when SFSTs were performed, there were four of six clues for horizontal gaze nystagmus (HGN) and three of four clues for the one leg stand (OLS) test. The toxicology reports for three of the four cases indicated "positive" for 5F-ADB and one indicated a concentration of 0.5 ng/mL.

Case 2

The driver was a 22-year-old male stopped for an inoperable headlight and veering across lanes of traffic until he drove over a curb. Field sobriety tests were not performed due to extreme impairment. Poison control was called and the subject was placed on a hold in the emergency room based on his behavior, which included a blank stare and bloodshot, watery, and droopy eyes while being uncommunicative or mumbling. He appeared to pass out while interacting with the officers and became unstable and frozen while standing. He was crying during the investigation and eventually calmed down after he got out of the car and some time had passed. The driver indicated that he had eaten "Spice" 2 hours prior to contact. The toxicology report indicated a 5F-ADB concentration of 0.3 ng/mL.

Case 3

The driver was a 26-year-old male with two blood draws submitted in February 2016 and July 2016. It had been noted on the submitted paperwork that the subject had more than 35 prior incidents with law enforcement, ranging from 911 calls by members of the public to multiple collisions. Impairment had often been noted at the point of contact but then was observed to diminish rapidly. An officer or civilian would reportedly make contact with the driver as he was passed out or slumped over, the driver would quickly awaken and attempt to drive off, often repeating this process a few times. Bloodshot, watery, and droopy eyes were frequently noted, along with poor coordination and balance. Presumed "Spice" packets and a pipe were found in his possession during several interactions with law enforcement. Field sobriety tests often yielded unimpaired results, presumably due to the quickness that symptoms would wear off. However, the stop during February 2016 indicated six of six clues for HGN, with the presence of Vertical Gaze Nystagmus, six of eight clues for the walk and turn (WAT), and three of four clues for the OLS. The toxicology reports for the February 2016 and July 2016 cases indicated results of 0.4 and 0.3 ng/mL of 5F-ADB, respectively.

Case 4

The driver was a 22-year-old male with two DUI incidents where blood was drawn. These incidents occurred within 1 week of each other. The first stop was made after a civilian called in to report an erratic driver. Upon contact the driver's vehicle was facing the wrong direction in the opposite lane of travel and the driver was observed leaning over the steering wheel, having difficulty keeping his bloodshot, watery eyes open. Suspected "Spice" flakes were noted inside the car and empty "Spice" bags were observed on the passenger seat. The second contact was made after the driver was called in for falling asleep in a fast food drive-thru. The officer noted that the driver mentioned that he "just can't stay awake" and later at the hospital, after being asked how the driver started using "Spice," he said "it is really addictive." Approximately 3 hours earlier, the driver had been contacted by police after he was observed passed out in his car in a deli parking lot. He was confused, lethargic, and sweating profusely. The officer deemed it necessary to initiate medical intervention and he was transported to the hospital. No field sobriety tests were performed, as impairment was extreme and obvious. The toxicology reports indicated 5F-ADB concentrations of 0.2 ng/mL from the first incident and 0.6 ng/mL from the second.

Case 5

The driver was a 60-year-old male who was stopped after nearly striking a patrol car. Limited information was provided regarding how the driver performed during the SFSTs, just that he "failed" them and had a preliminary breath test of 0.000. The driver admitted to using "Spice" and "Spice" packets were visible in the vehicle. The toxicology report indicated a 5F-ADB concentration of 1.31 ng/mL in the blood.

Discussion

Synthetic cannabinoids (SCs) were initially created as research compounds that would work therapeutically on the endogenous cannabinoid receptors CB1 and CB2. CB1 receptors are associated with reduced nausea, increased appetite, and euphoria, while CB2 receptors are associated with reduced inflammation and decreased pain perception [1]. However, it was discovered that the SC bound more strongly to the CB1 receptors than tetrahydrocannabinol (THC), the active component in marijuana, and research companies stopped investigating them. Subsequently, these compounds began to be synthesized illicitly and added to a plant material, claiming to be a "legal weed" or "legal high" [2]. Because of their effects, SC began to gain popularity as recreationally abused drugs in the early 2000s.

One such compound is methyl 2-(1-(5-fluoropentyl)-1H-indazole-3-carboxamido)-3,3-dimethylbutanoate or 5F-ADB [3], as seen in Fig. 11.1.1. It is an indazole-based SC with potent binding potential to the CB1 receptor [4]. The first published detection of 5F-ADB was in Japan in a postmortem case in late 2014 after an individual died subsequent to using a substance that contained the compound [5]. Further testing detected the compound in 10 more individuals who had died from an unexplained cause and had been determined to have inhaled the smoke of this substance.

While drug users began consuming SC for a variety of reasons (e.g., consume drugs "legally," avoid detection on workplace drug tests) [2], it was difficult to anticipate how these drugs would affect a person. The observed effects in users are not similar to those observed with THC use. The driving behaviors of individuals using SC, as noted by law enforcement officers, include a wide combination of effects seen from other common drug classes (Table 11.1.1).

The SC drug class is difficult to predict when it comes to the signs and symptoms that can occur after consumption. The drug recognition expert (DRE) case histories described in Table 11.1.1 indicate that upon evaluating these drivers,

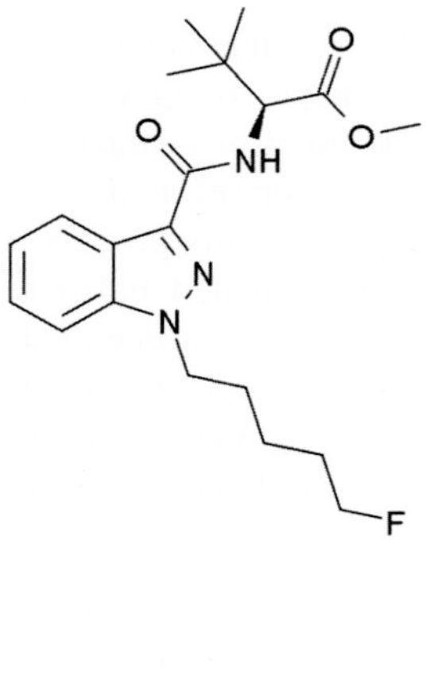

FIGURE 11.1.1

TABLE 11.1.1 Fourteen Featured Cases With Field Sobriety Test Results, Driving Behaviors, and Officer Determination of What was in the Driver's Blood

Case No.	HGN Six Clues	OLS Four Clues	WAT Eight Clues	LOC	Constricted/ Dilated	Driving Behaviors	Officer Determination	Toxicology Results for 5F-ADB (ng/mL)	Other Drugs Detected
1	NA	NA	NA	NA	NA	Passed out behind wheel, blocking traffic. Suspected SC packages in car	Synthetic cannabinoids	0.4	
2	NA	2	2	Y	Dilated	Call from gas station about "whacked out driver"	Stimulants	0.4	Carboxy-THC
3	6	1	3	Y	Normal	Admitted to smoking while driving. Performed better during DRE than during initial SFSTs	Depressants, cannabis	0.5	
4	6	3	5	Y	Normal	Hitting ramps, swerving, traveling at erratic speeds	NA	0.3	
5	0	3	3	Y	Dilated	Almost hit several road workers, almost collided head on with bus. Obvious impairment	Narcotic analgesics, cannabis	0.4	
6	0	2	4	Y	Dilated	Observed asleep at an intersection	Stimulants, cannabis	0.4	
7	4	2	4	N	Dilated	Observed to be incoherent. Pinpoint pupils on arrival but dilated during DRE exam. Cycled between calm and enraged	Depressants, cannabis	1.2	Carboxy-THC
8	0	1	2	Y	Constricted	Passed out behind wheel in parking lot with vehicle running. Disoriented	Cannabis	0.3	
9	0	1	NA	Y	Dilated	Found with vehicle stopped in the middle of road. Medical personnel said disoriented	Not impaired	0.3	
10	5	1	2	N	Dilated	Slow driving, drove into oncoming traffic. Smoking while stopped on roadway. Subdued	Cannabis	0.3	
11	2	1	4	Y	Dilated	Reported as unconscious driver in car. Drove away, poor driving. Slow movements, confused. Described as similar feeling to heroin	Stimulants	0.6	
12	0	2	3	N	Normal	Crossed over two lanes of traffic and drove into ditch	Narcotic analgesics, cannabis	0.4	
13	0	3	4	Y	Dilated	Stopped for lane travel, numerous calls made about erratic driver, running red lights, traveling slowly. Quick jerky movements	Cannabis	0.79	
14	0	3	3	Y	Dilated	Weaving in road way, below speed limit, crossing lane dividers	Stimulants	0.26	

DRE, Drug recognition expert; HGN, horizontal gaze nystagmus; LOC, lack of convergence; N, no; NA, not available; OLS, one leg stand; SC, synthetic cannabinoid; SFSTs, standard field sobriety tests; THC, tetrahydrocannabinol; WAT, walk and turn; Y, yes.
Source: Data obtained from DRE investigations submitted to the Washington State Patrol Toxicology Laboratory, Seattle, WA.

the officers commonly determined the cause of impairment to be either a stimulant or cannabis. Cannabis was sometimes expected based on scene observations, which often described drug paraphernalia consistent with marijuana use. However, many officers were at a loss as to what the contributing drug was. Their observations of the subjects did not point to a specific drug or a drug class based on the effects that they had been trained to evaluate. Routine testing, that did not include the detection of SC, yielded negative toxicology results. It was determined that further investigation would be necessary.

At times used as "legal marijuana" the desired effects may or may not be similar to those of marijuana. However, even though they are named cannabinoids, these compounds can cause a much different reaction, usually an adverse one. Once the adverse effects started to be noted and traced back to these compounds, state and federal legislation was created prohibiting their sale and use [6]. As a response, several SCs were synthesized where the compounds would be altered ever so slightly, to evade the law. A series of JWH compounds (JWH-018, JWH-073, etc.), named for John W. Huffman, the Clemson researcher who created them, turned into an alphanumeric soup of drugs, making up the enormous new realm of SC. Every new iteration of SC produced unexpected side effects, varying widely among compounds. The observable effects were especially unpredictable when multiple compounds were combined by manufacturers or the end users.

Starting in October 2015 law enforcement officers throughout Washington reported and witnessed an emergence of bizarre and unsafe behaviors, including dangerous driving [7]. Whether it was 911 call-ins regarding erratic driving or abnormal interactions officers had not previously experienced during typical traffic stops, SCs were creating concern and confusion. Law enforcement agencies use a set of tests to determine if the suspect driver is under the influence of alcohol or drugs. The three tests that are considered standard and are approved by National Highway Traffic Safety Association include the HGN test, the WAT test and the OLS test. Driving behaviors reported by civilians typically included high speed, erratic lane switching, or a driver asleep at the wheel. Upon arrival by an officer, it was often noted that drug paraphernalia was present. By then the driver was often asleep and difficult to arouse. After waking up the driver typically did not remember how they got there. Oftentimes, symptoms resolved within 30–60 minutes from point of contact, making it difficult for an officer to determine if the driver was impaired by something. Other effects observed included convulsions, blank stares, or incoherent speech. If it was concluded that a blood draw was needed, negative toxicology results were often returned. A new way of investigating these stops had to be developed in order to figure out how to request proper testing and obtain reports with results that explained the behavior.

Toxicology laboratory testing is capable of detecting a wide scope of illicit, prescription, and recreational drugs. However, it is impossible to utilize a single testing panel that includes all possible drugs, especially when the forms of synthetic drugs can be altered and developed quickly. Reference laboratories, many of which continually develop new test methods to detect these emerging drugs, can be a useful resource for toxicologists. In suspected DUI cases where a negative toxicology report has been issued, it is often a follow-up step to have blood samples sent out for supplemental testing using a panel of possible synthetic compounds. Information at the scene, such as package brands, can help one to guide what possible drug(s) might be in the individual's system, rather than testing the blood for all possible panels.

Upon receiving results from a reference laboratory, the toxicologist can update or produce a supplemental report and subsequently provide an explanation for the behavior observed by an officer. Limited dosing studies have been performed [1], but due to the fact that these compounds can vary structurally and each person can react to drugs differently, it is difficult to draw conclusions from them. Because of the novelty of these compounds and the lack of information available, it can be even more difficult for an officer or toxicologist to testify in court regarding these compounds and what their presence in an individual's blood might mean.

The Washington State Patrol Toxicology Laboratory started sending cases out to reference laboratories for SC testing in 2012. The cases sent out were usually cases where impairment had been indicated; yet, routine testing yielded negative results and/or there was evidence of SC use on the scene. In October 2015 the laboratory started receiving cases that described greater impairment than had been noted previously. Results from these specific cases showed the presence of the 5F-ADB compound, which had not yet been seen in Washington State driving cases prior to October 2015.

Sixty-five of the cases that were sent out to a reference laboratory for SC testing between October 2015 and March 2018 for 5F-ADB testing came back positive. The cases were comprised of death investigations (6 cases), DUIs (45), and DRE evaluations (14). Two of the death investigation cases were positive for only 5F-ADB. Other drugs detected in the remaining death investigation cases included benzodiazepines, opiates, and antidepressants. All of the death investigation cases were for male decedents, ages 35–59 (median 52). The concentrations of 5F-ADB detected ranged from 0.2 to 4.4 ng/mL, with a mean of 1.3 ng/mL and a median of 0.8 ng/mL. The reporting limit for 5F-ADB was

0.1 ng/mL. The majority of DUI and DRE cases (82%) were positive for only 5F-ADB. Other drugs detected in the remaining DUI and DRE cases included carboxy-THC, THC, and other SC. The majority of drivers were male (83%), ages 20–60 (median 29). The concentrations of 5F-ADB detected ranged from positive >0.1 to 1.31 ng/mL, with a mean of 0.47 ng/mL and a median of 0.4 ng/mL, including seven positive only cases.

The cases in this series describe the drivers' behavior at the point of contact or their poor driving performance, results from SFST, and the final result from laboratory testing. Of these cases the most common impairment types were poor driving to the point of collisions, passed out or asleep at the wheel, and lethargy. Other frequently cited observations included slurred speech, confusion, bloodshot watery eyes, and HGN.

References

[1] Adams WR, Logan BK. Pharmacodynamics and pharmacokinetics of synthetic cannabinoids. In: Presented at the 64th annual meeting of the American Academy of Forensic Sciences, Atlanta, GA; 2012.

[2] Castaneto MS, et al. Synthetic cannabinoids: epidemiology, pharmacodynamics, and clinical implications. Drug Alcohol Depend 2014;144:12–41. Available from: https://doi.org/10.1016/j.drugalcdep.2014.08.005.

[3] Rosenberg C. "Rules - 2016." 2016 – Notice of intent: temporary placement of six synthetic cannabinoids (5F-ADB, 5F-AMB, 5F-APINACA, ADB-FUBINACA, MDMB-CHMICA and MDMB-FUBINACA) Into Schedule I, 13 Dec. 2016, <www.deadiversion.usdoj.gov/fed_regs/rules/2016/fr1221.htm>.

[4] Banister SD, et al. Pharmacology of valinate and tert-leucinate synthetic cannabinoids 5F-AMBICA, 5F-AMB, 5F-ADB, AMB-FUBINACA, MDMB-FUBINACA, MDMB-CHMICA, and their analogues. ACS Chem Neurosci 2016;7(9):1241–54. Available from: https://doi.org/10.1021/acschemneuro.6b00137.

[5] Hasegawa K, et al. Identification and quantitation of 5-fluoro-ADB, one of the most dangerous synthetic cannabinoids, in the stomach contents and solid tissues of a human cadaver and in some herbal products. Forensic Toxicol 2014;33(1):112–21. Available from: https://doi.org/10.1007/s11419-014-0259-0.

[6] National Conference of State Legislatures. The federal government follows the lead of 17 states to Ban K2 [Press release]; 2019, November 24. Retrieved from <http://www.ncsl.org/press-room/federal-ban-on-k2.aspx>.

[7] Chandler A, Knoy L, Capron B, Peterson B, Couper F. Physiological indicators and driving behavior observed in suspected impaired driving cases positive for 5F-ADB. In: Poster presented at: Annual meeting of the Society of Forensic Toxicologists, Inc. Dallas, TX; 2016 October 19.

Chapter 11.2

Can cannabidiol use cause a false-positive tetrahydrocannabinol urine drug screen?

Mushal Noor and Y. Victoria Zhang

Department of Pathology and Lab Medicine, University of Rochester Medical Center, Rochester, NY, United States

Case description

The clinical lab received a call from a clinician: one of their patients had a positive urinary drug screen for delta-9-tetrahydrocannabinol (THC) metabolite. The patient denied the use of marijuana, cannabinoids, or any other illicit substances but did report using cannabidiol (CBD) oil once, several weeks ago. The clinician's inquiry was: could CBD use cause a positive or false-positive urine drug screen for THC? Also, could a one-time use of CBD or other cannabinoids result in a positive test result several weeks later?

Discussion

CBD is one of approximately 500 compounds found in *Cannabis sativa* and is the second most abundant compound the plant produces, following the psychoactive compound delta-9-THC. While CBD is a cannabinoid and shares some structural similarities with THC, it is nonpsychoactive and hence does not produce a "high." Unlike THC, CBD has very little effect on CB1 and CB2 receptors. The mechanism of action of CBD is not well understood, but it appears to involve several signaling systems: acting as an inverse agonist at several G protein–coupled receptors [1], an allosteric modulator of opioid receptors [2], and a partial agonist at serotonin 5-HT$_{1A}$ receptors [3].

CBD is perceived as less harmful than THC and comes in several different readily available forms, including beauty and health products, vapors, infused edibles, and the most common form: CBD oil. In recent years, there has been increased interest in CBD due to potential therapeutic use in the treatment of refractory epilepsy [4], for decreasing anxiety, improving sleep, and other neuroprotective effects [5]. CBD has also been used and investigated in pain, cancer, inflammation, and movement and mood disorders, including posttraumatic stress disorder. In 2018 the FDA approved a drug containing CBD (Epidiolex) for the treatment of seizures associated with two rare and severe forms of epilepsy, Lennox–Gastaut syndrome and Dravet syndrome. This is the first FDA-approved drug to contain a purified extract from the *Cannabis* plant.

In the United States the legal status of CBD depends on the source from which it is derived [6,7]. The Drug Enforcement Administration maintains that CBD extracted from marijuana is a Schedule I drug, and hence illegal. However, CBD extracted from hemp contains less than 0.3% THC and is federally lawful. Therefore as long as a hemp-derived CBD product contains less than 0.3% THC, it is legal to possess and distribute.

CBD has a good safety profile, with generally mild side effects. No lethal toxicity has been reported in the literature. Chronic use and high doses (up to 1500 mg per day) have been reported to be well tolerated by humans [8]. Unlike THC, CBD has no psychoactive effects. Its other effects are relatively minor and range from dry mouth and dizziness to diarrhea, drowsiness, headache, hypotension, decreased appetite, sleep disturbances, and psychomotor slowing. There is limited information, but some studies suggest that CBD may also inhibit hepatic drug metabolism [9]. In contrast, THC produces a wide range of psychoactive effects, including feeling "high," anxiety, paranoia, perceptual alterations, cognitive deficits (particularly in verbal recall), and exacerbates psychotic symptoms in patients with schizophrenia [10–12].

CBD itself would not result in a positive test for cannabinoids or their metabolites in drug screening assays. However, most CBD products are not well regulated and may have incorrect label information, contaminants and may even also contain THC above reported or legal levels [13,14]. If the CBD product contains THC at a sufficiently high concentration, it is a theoretical possibility that it may result in a positive urine drug test result for cannabinoids.

delta-9-THC is the main psychoactive compound in marijuana, the most widely used drug in the United States. It may be inhaled or ingested with subsequent metabolism to a variety of inactive chemicals, including delta-9-THC carboxylic acid (THC-COOH). THC itself has a clearance half-life of less than 30 minutes and is not detectable in urine. However, THC-COOH typically appears in the urine within 60 minutes to 4 hours. THC-COOH is inactive and is a very lipid soluble compound. Urine drug testing for marijuana or other cannabinoids is based on the detection of THC-COOH in urine.

The urine drug screen to detect cannabinoids is an enzyme immunoassay that not only detects THC-COOH but can also react with other cannabinoids present in the urine. Because of this cross-reactivity, immunoassay results are commonly reported in terms of "total cannabinoids" instead of THC-COOH only. The presence of THC-COOH at a level greater than the limit of quantitation (LOQ) indicates exposure to THC within 3 days after a single use, to approximately 30 days in heavy chronic users. The federal cutoff level for a positive urine cannabinoid screen is 50 ng/mL, although some laboratories may use a lower cutoff level of 20 ng/mL [15].

Immunoassay cross-reactivity to noncannabinoid compounds is very rare; however, confirmation methods such as gas chromatography/mass spectrometry (GC/MS) or liquid chromatography/tandem mass spectrometry (LC/MS/MS) can be used to confirm a positive immunoassay, especially in medicolegal cases, employee screening, or disputed test results. GC/MS or LC/MS/MS measure only THC-COOH and do not measure other cannabinoids; therefore they may yield a quantitative result that represents only 10%–50% of the measured value by immunoassays.

Marijuana use would result in a positive THC urine test, and the detection time since last use depends on a number of factors, including the route of administration, dose, frequency of use, individual metabolism, body mass index, and the LOQ. Cannabinoids are highly lipophilic and are extensively stored in body fat. Chronic marijuana use would result in THC accumulation in fatty tissues, resulting in slow elimination rates of marijuana metabolites, and positive urine drug screens days to weeks after cessation of use [15].

A single exposure to marijuana in nonusers may be detected in the urine only up to 72 hours. Regular smoking several times per week may result in urine specimens testing positive for 5–10 days. Chronic heavy marijuana use, in the form of daily smoking, can result in a positive test for 30 days or longer after cessation. Passive exposure to marijuana smoke will not result in a positive result for cannabinoids with a 50 ng/mL screening cutoff. Orally ingested marijuana may be detectable for 1–5 days.

In the case of our patient the positive THC test may have resulted from unreported cannabinoid use, as CBD use would be highly unlikely to result in a positive THC test. Even if the CBD oil hypothetically was contaminated with THC, the likelihood of detectable THC metabolite in patient's urine after several weeks had elapsed was extremely low. This case demonstrates a classic dilemma faced by a toxicologist in interpretation of cannabinoid urine drug screen in patients who report a concurrent CBD use. As CBD products are better understood, a clear analytical understanding of the impact of CBD products on cannabinoid urine dug tests will evolve.

References

[1] Laun AS, Shrader SH, Brown KJ, Song Z-H. GPR3, GPR6, and GPR12 as novel molecular targets: their biological functions and interaction with cannabidiol. Acta Pharmacol Sin 2019;40(3):300–8.

[2] Kathmann M, Flau K, Redmer A, Tränkle C, Schlicker E. Cannabidiol is an allosteric modulator at mu- and delta-opioid receptors. Naunyn Schmiedebergs Arch Pharmacol 2006;372(5):354–61.

[3] Russo EB, Burnett A, Hall B, Parker KK. Agonistic properties of cannabidiol at 5-HT1a receptors. Neurochem Res 2005;30(8):1037–43.

[4] Welty TE, Luebke A, Gidal BE. Cannabidiol: promise and pitfalls. Epilepsy Curr 2014;14(5):250–2.

[5] Shannon S, Opila-Lehman J. Cannabidiol oil for decreasing addictive use of marijuana: a case report. Integr Med (Encinitas, Calif) 2015;14(6):31–5.

[6] US Department of Justice, Drug Enforcement Administration. Title 21 United States Code of Federal Regulations, section 1308.35, 66 FR 51544, Oct. 9, 2001. https://www.ecfr.gov/cgi-bin/retrieveECFR?gp=&SID=758aab05ae8c73d4a74dadad886b36a7&mc=true&n=pt21.9.1308&r=PART&ty=HTML.

[7] US Department of Justice, Drug Enforcement Administration. Schedules of controlled substances: placement in schedule V of certain FDA-approved drugs containing cannabidiol; corresponding change to permit requirements, 2018. https://www.federalregister.gov/documents/2018/09/28/2018-21121/schedules-of-controlled-substances-placement-in-schedule-v-of-certain-fda-approved-drugs-containing.

[8] Bergamaschi MM, Queiroz RH, Zuardi AW, Crippa JA. Safety and side effects of cannabidiol, a *Cannabis sativa* constituent. Curr Drug Saf 2011;6(4):237–49.

[9] Iffland K, Grotenhermen F. An update on safety and side effects of cannabidiol: a review of clinical data and relevant animal studies. Cannabis Cannabinoid Res 2017;2(1):139–54.
[10] Boggs DL, Nguyen JD, Morgenson D, Taffe MA, Ranganathan M. Clinical and preclinical evidence for functional interactions of cannabidiol and $\Delta(9)$-tetrahydrocannabinol. Neuropsychopharmacology 2018;43(1):142–54.
[11] D'Souza DC, Perry E, MacDougall L, Ammerman Y, Cooper T, Wu YT, et al. The psychotomimetic effects of intravenous delta-9-tetrahydrocannabinol in healthy individuals: implications for psychosis. Neuropsychopharmacology 2004;29(8):1558–72.
[12] D'Souza DC, Abi-Saab WM, Madonick S, Forselius-Bielen K, Doersch A, Braley G, et al. Delta-9-tetrahydrocannabinol effects in schizophrenia: implications for cognition, psychosis, and addiction. Biol Psychiatry 2005;57(6):594–608.
[13] Bonn-Miller MO, Loflin MJE, Thomas BF, Marcu JP, Hyke T, Vandrey R. Labeling accuracy of cannabidiol extracts sold online. JAMA. 2017;318(17):1708–9.
[14] Hazekamp A. The trouble with CBD oil. Med Cannabis Cannabinoids 2018;1(1):65–72.
[15] Moeller KE, Kissack JC, Atayee RS, Lee KC. Clinical interpretation of urine drug tests: what clinicians need to know about urine drug screens. Mayo Clin Proc 2017;92(5):774–96.

Chapter 11.3

Clobazam intoxication due to cannabidiol consumption

Mikail Kraft[1], Ara Hall[2] and Uttam Garg[3,4]
[1]Pediatric Resident, Children's Mercy Hospital, Kansas City, MO, United States, [2]Epilepsy Section, Division of Neurology, Children's Mercy Hospital, Kansas City, MO, United States, [3]Department of Pathology and Laboratory Medicine, Children's Mercy Hospital, Kansas City, MO, United States, [4]University of Missouri, School of Medicine, Kansas City, MO, United States

Case description

An 11-year-old female with intractable generalized epilepsy presented to the emergency room (ER) with increased somnolence during the last 10 days. Her teachers and care assistants had been emailing her mother noting that the patient was extremely fatigued in the morning when she arrived at school, often took a midmorning nap, which she had not been doing previously. The patient's mother noticed that her child became drowsy after taking her morning medications. She was brought to the ER where her mother stated "this (drowsiness) is what they have been noticing at school. On the weekends I just think she needs a break so I let her nap. If she were older I would think she was drunk!"

The patient is being followed by neurology for idiopathic, intractable, generalized epilepsy with features of Jeavons syndrome. The syndrome is a form of generalized epilepsy characterized by the triad of eyelid myoclonus with and without absences, eye closure—induced seizures, EEG paroxysms, or both, and photosensitivity. She also has a history of mild developmental delay, congenital hypotonia, poor coordination, and gait differences. Her seizures started at 1 year of age, presented as generalized tonic colonic seizures, and presumed initially to be complex febrile seizures; she now has multiple events per week. Her seizures last 10–20 seconds. Her triggers include fatigue, sleep disruption and illness. Current seizures are more absence in nature, consisting of rapid eyelid flutter (eyelid myoclonus) with up rolling of the eyes or eyelid fluttering with staring. The postictal period consists of fatigue. She has had four prior hospitalizations (2014, 2012 × 2, and 2007). She has no surgical history. Genetic studies have been performed showing normal female (46,XX) with no known pathogenic variants. She was born at 40 weeks involving an uncomplicated pregnancy and vaginal delivery. Mild hypotonia was noted at birth.

Her current medications include valproic acid 325 mg BID (20 mg/kg/day), clobazam (Onfi) 7.5 mg BID (0.46 mg/kg/day), and topiramate 50 mg qHS (1.5 mg/kg/day). The parents report that she takes cannabidiol (CBD) oil, and approximately 10 days ago they increased the CBD dose from 8 to 40 drops BID. The brand and concentration of CBD product is unknown. She lives with her parents and her sibling at home. She attends regular school with Individualized Education Program, occupational therapy, and speech-language pathology. She is up-to-date on her vaccinations per ACIP.

Discussion

Cannabinoids are chemical compounds found in the plants of *Cannabis* genus. More than 100 cannabinoids have been isolated from these plants with particular clinical interest in delta-9-tetrahydrocannabinol (THC) and CBD. THC is a major component of marijuana and is known for its psychoactive/euphoric properties. Medically, it has been used as an appetite stimulant and antiemetic agent. It is available as a prescription drug under various trade names such as Dronabinol, Marinol and Syndros. THC has also been shown to help in other conditions, including pain, muscle spasticity, glaucoma, and insomnia. CBD is primarily derived from hemp due to its very low THC content (<0.3%). CBD has been used in the treatment of various conditions, including seizures, inflammation, pain, nausea, and anxiety. In recent years an interest in CBD as an adjunct therapy in the treatment of refractory epilepsy has grown [1–5]. In June 2018 the Food and Drug Administration approved Epidiolex, the first prescription medication to contain CBD for the treatment of certain types of seizures [6,7].

The patient described in this report had idiopathic and intractable generalized epilepsy with features of Jeavons syndrome. The patient was being treated with various antiseizure medications, including clobazam as an adjunct therapy. The major finding in this patient was altered mental status and somnolence. It was believed that the patient's clinical findings were secondary to her high levels of clobazam and its active metabolite N-desmethylclobazam (Table 11.3.1). Clobazam is a benzodiazepine class drug that is used as an anxiolytic agent and as an adjunctive treatment for seizures associated with Lennox–Gastaut syndrome. It has also been used as a monotherapy and adjunctive treatment for other forms of epilepsy. Clobazam, when given orally, is well absorbed with peak concentrations occurring 1–4 hours postdose. It is primarily metabolized in the liver by CYP3A4 and CYP2C19 (Fig. 11.3.1) with a long half-life of 36–42 hours [8]. Its active metabolite, N-desmethylclobazam, has an even longer half-life of 71–82 hours [8]. The drug's anxiolytic and anticonvulsant effects are due to its allosteric activation of the ligand-gated $GABA_A$ receptors on postsynaptic neurons [9,10]. Adverse and toxic effects of clobazam therapy include lethargy, sedation, slurred speech, somnolence, ataxia, and fatigue. Many of these features were noted in our patient.

Although the clobazam level was not tested before increasing CBD consumption, considering the patient's clobazam dose of 7.5 mg/BID and normal metabolism, the drug levels should have been within therapeutic range. High levels of clobazam and its active metabolite N-desmethylclobazam are likely due to an inhibition of clobazam metabolism by CBD. CYP3A4 and CYP2C19, involved in clobazam metabolism, are inhibited by CBD [8,11]; therefore coadministration of clobazam and CBD results in an increase of clobazam and its active metabolite N-desmethylclobazam. On the other hand, CYP3A4 inducers such as phenobarbital, phenytoin, and carbamazepine, when coadministered, decrease the concentrations of clobazam [12]. In an expanded access trial, patients with concomitant use of clobazam and CBD had an increase in clobazam and N-desmethylclobazam concentrations of 60% ± 80% and 500% ± 300%, respectively [8]. Despite a decrease in clobazam doses, N-desmethylclobazam concentrations remained high at week eight as compared to the baseline level. In another study a patient receiving 2000–2900 mg/day of CBD showed a threefold increase in tacrolimus concentration.

The case and cited literature highlight drug–drug interactions between CBD and clobazam, as well as other drugs. Dose adjustment and therapeutic drug monitoring may be necessary when CBD and other drugs metabolized by CYP3A4 and CYP2C19 are coadministered.

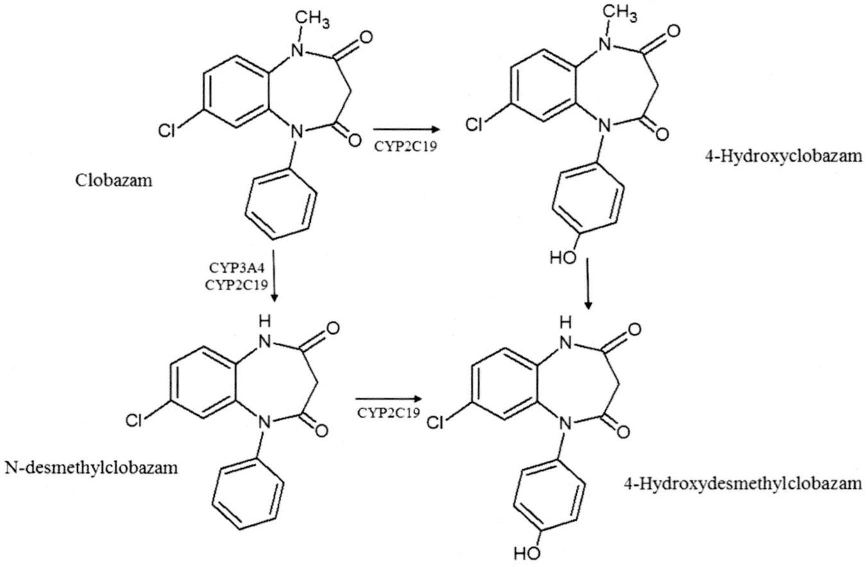

FIGURE 11.3.1 Metabolism of clobazam.

TABLE 11.3.1 Laboratory Values for Various Analytes in Our Patient.

Sodium	141 (135–145) mmol/L
Potassium	4.0 (3.5–5.2) mmol/L
Chloride	109 (99–122) mmol/L
Bicarbonate	21 (20–30) mmol/L
Anion Gap	11 (8–16) mmol/L
BUN	11 (5–20) mg/dL
Creatinine	0.48 (0.35–0.84) mg/dL
Glucose	78 (65–110) mg/dL
Calcium	9.6 (8.6–10.5) mg/dL
Protein total	6.1 (6.5–8.3) g/dL
Albumin	3.7 (2.5–5.1) g/dL
Bilirubin, total	0.7 (0.0–1.2) mg/dL
Bilirubin, direct	0.4 (1.0–0.4) mg/dL
Bilirubin, indirect	0.3 (0.0–1.2) mg/dL
AST	32 (12–50) U/L
ALT	31 (5–50) U/L
Alkaline phosphatase	244 (140–560) U/L
Valproic acid, total	71 (50–100) µg/dL
Topiramate	2.3 (5–20) µg/dL
Clobazam	**900 (30–300) ng/mL**
Desmethylclobazam	**12,256 (300–3000) ng/mL**

Reference ranges are given in parentheses.

References

[1] Ali S, Scheffer IE, Sadleir LG. Efficacy of cannabinoids in paediatric epilepsy. Dev Med Child Neurol 2019;61:13–18.
[2] De Caro C, Leo A, Citraro R, De Sarro C, Russo R, Calignano A, et al. The potential role of cannabinoids in epilepsy treatment. Expert Rev Neurother 2017;17:1069–79.
[3] Friedman D, Devinsky O. Cannabinoids in the treatment of epilepsy. N Engl J Med 2015;373:1048–58.
[4] Perucca E. Cannabinoids in the treatment of epilepsy: hard evidence at last? J Epilepsy Res 2017;7:61–76.
[5] Stockings E, Zagic D, Campbell G, Weier M, Hall WD, Nielsen S, et al. Evidence for cannabis and cannabinoids for epilepsy: a systematic review of controlled and observational evidence. J Neurol Neurosurg Psychiatry 2018;89:741–53.
[6] [No authors listed]. Cannabidiol (Epidiolex) for epilepsy. Med Lett Drugs Ther 2018;60:182–4.
[7] Sekar K, Pack A. Epidiolex as adjunct therapy for treatment of refractory epilepsy: a comprehensive review with a focus on adverse effects. F1000Res 2019;8.
[8] Geffrey AL, Pollack SF, Bruno PL, Thiele EA. Drug-drug interaction between clobazam and cannabidiol in children with refractory epilepsy. Epilepsia 2015;56:1246–51.
[9] Sankar R. GABA(A) receptor physiology and its relationship to the mechanism of action of the 1,5-benzodiazepine clobazam. CNS Drugs 2012;26:229–44.
[10] Vinkers CH, Olivier B. Mechanisms underlying tolerance after long-term benzodiazepine use: a future for subtype-selective GABA(A) receptor modulators? Adv Pharmacol Sci 2012;2012:416864.
[11] Jiang R, Yamaori S, Okamoto Y, Yamamoto I, Watanabe K. Cannabidiol is a potent inhibitor of the catalytic activity of cytochrome P450 2C19. Drug Metab Pharmacokinet 2013;28:332–8.
[12] Yamamoto Y, Takahashi Y, Imai K, Takahashi M, Nakai M, Inoue Y, et al. Impact of cytochrome P450 inducers with or without inhibitors on the serum clobazam level in patients with antiepileptic polypharmacy. Eur J Clin Pharmacol 2014;70:1203–10.

Chapter 11.4

Phencyclidine and marijuana exposure in utero

John O. Ogunbileje, Paul E. Young and Anthony O. Okorodudu
Department of Pathology, University of Texas Medical Branch at Galveston, Galveston, TX, United States

Case description

A 38-year-old pregnant woman with a recorded history of marijuana and phencyclidine (PCP) abuse presented to obstetrics clinic for follow-up of her pregnancy, which had been complicated by anemia and ABO incompatibility (the latter a cause of hemolytic disease of the fetus and newborn due to maternal IgG anti-A or anti-B antibodies crossing the placenta and attacking fetal red blood cells). The baby was delivered by cesarean section at 39 weeks gestation due to these complications and previous history of three cesarean sections. At birth the baby was in stable condition with weight (3685 g) and length (52.5 cm) within normal limits. However, within 24 hours after birth, she became jaundiced with unconjugated bilirubin peaking at 17.1 mg/dL. Treatment with phototherapy was subsequently initiated.

Due to the mother's long history of drug abuse, urine samples were obtained from her and the baby for the screening of illicit drugs using the immunoassay technique on Vitros 5600 Chemistry Analyzer (Ortho Clinical Diagnostics, Rochester, New York) within 24 hours of delivery. The baby's urine was positive for PCP 42 ng/mL (reference interval <25 ng/mL).

A meconium sample was also collected and screened by immunoassay. Meconium, as a drug test matrix, provides information about the potential length of drug exposure. The meconium sample was positive for both tetrahydrocannabinol (THC) and PCP using immunoassay but was insufficient for confirmatory reflex testing on liquid chromatography with tandem mass spectrometry (LC–MS/MS). The confirmation of the presence of PCP in the baby and mother's urine was, however, obtained using LC–MS/MS (Waters, Milford, Massachusetts).

Despite the presence of the PCP and THC in the baby's urine and meconium, the mother reported that the baby was feeding well every 2–3 hours. On examination the baby's heart rate was regular with a normal rhythm. Also, the baby did not present with any neurologic or respiratory problems after birth. There were no visible teratogenic effects or deformities. When questioned about the positive drug test results in her baby, the mother mentioned that she had a "drug problem" but only uses once a month. She reported that the last time she took PCP was approximately 8 days before she gave birth to her baby. This assertion does not support the reported half-life of PCP of 7–46 hours.

Discussion

PCP and THC are two harmful drugs that can lead to medical complications during pregnancy and after birth in the neonate. Substance abuse has profound impacts on the health of both mother and fetus [1]. Data shows that approximately 6% of pregnant women use illicit drugs at one point in their pregnancy, and this number increases to 18% in pregnant teens [2]. Some of the substances commonly abused by pregnant women include cocaine, opioids, THC, alcohol, tobacco, and PCP [3–5]. Exposure to these drugs during pregnancy has been associated with a number of complications, including intrauterine growth restriction, miscarriage, preterm delivery, placental abruption, intrauterine fetal demise, and other infant developmental defects [2,6]. Notably, most of these drugs are both water and lipid soluble. Also, as a result of their low molecular weights, they can cross the placenta, thus exposing the fetus to their toxic effects since the fetus lacks the appropriate enzymes to metabolize the drugs [4].

PCP is a hallucinogenic drug sought after because of its ability to produce euphoria and feelings of omnipotence [7]. This euphoria is linked to its affinity for the N-methyl-D-aspartate receptor complexes in the hippocampus, neocortex, basal ganglia, and limbic system of the central nervous system [7]. It was initially used as an anesthetic around the 1960s but was discontinued in the late 1970s due to its abuse and overdose with symptoms often mistaken for schizophrenia [7,8]. Recently, it has been observed that PCP is one of the main substances used to lace street marijuana to produce a greater "high" [7]. Smoking marijuana laced with PCP may be one possible source of PCP in the mother in our case. However, she is also THC and PCP dependent, thus creating the possibility that she consumed PCP via two sources. She admitted taking PCP 8 days before delivery but no record of when she last took THC. It has been reported that marijuana might be laced with 1–10 mg of PCP [7]. Also, a case of PCP detection in breast milk over 6 weeks after the use of an unknown quantity of PCP has been reported [9].

The main physical characteristic of PCP that increases its exposure to fetuses is that it is a weak base, making it lipid, water, and alcohol soluble [7]. Therefore this characteristic enables PCP to enter the placenta by simple diffusion [4]. In a study that examined 505 newborns exposed to illicit drugs in utero, they reported that PCP had a significant impact on the growth parameters. However, the effect was not as prominent as cocaine [4]. They reported that children exposed to PCP only present with a high-pitched cry, poor tracking, decreased attention, and 42% presented with low birthweight for gestational age [4]. The baby in our case study appeared to be stable. Still, she may benefit from close monitoring in the future for possible developmental or other dysfunctions associated with PCP or THC. The study of Walberg et al. might support the case of the baby we presented: They reported that they did not find a direct relationship of serum PCP concentration to either the clinical pattern of intoxication or to the history of the route of PCP use [7,10]. However, we must interpret the data from Walberg et al. with caution because sample analysis was carried out using only the enzyme immunoassay technique. Other factors that can determine the toxicity of PCP include nutritional state, coingestion of alcohol, and body habitus [7].

The presence of THC and PCP in the baby's meconium further confirmed the chronic consumption of these drugs by the mother. Meconium is believed to be a traditional gold standard for the determination of exposure of neonates to illicit substances; other biological matrices are urine, umbilical cord, and hair [5]. Meconium starts forming in the fetal intestine approximately 12 weeks after gestation making it an excellent matrix to determine drug exposure of the fetus in the last two trimesters of a mother's pregnancy [5]. This might partially explain why we were able to detect both PCP and THC in the baby's meconium but only a trace amount of THC in the baby's urine. The low concentration of hepatic cholinesterases in the neonate suggests that exposure to THC and PCP intrauterine may have subjected her to severe stress that might manifest in the future even if she appeared stable at delivery [4].

Further, illicit drug abuse sometimes mimics naturally occurring neurotransmitters [1]. For instance, marijuana has been reported to act as anandamides [1], a known endogenous agonists of the cannabinoid CB1 and CB2 receptors found in the central nervous system [11]. In the central nervous system, anandamides mediate the psychotropic effects of THC. This can further alter fetal growth, affect the mother's health, alter the baby's nutrition, and potentially lead to future behavioral problems.

Generally, children exposed to substance abuse tend to experience neonatal abstinence syndrome, and this might impact their ability to breastfeed effectively [12]. Therefore it is important that health-care providers understand the situation of these parents by communicating with them and following up with them on the best pathway to treat their drug dependences. The mother in our case has a history of depression and was only recently released from prison. All these events might complicate compliance with treatment of drug dependence. Therefore it will be important that the focus should not just be on the baby, but we should also encourage the mother to abstain from further illicit drug use. The mother needs to be abstinent because she has four other children that can be indirectly exposed to these drugs. Children born to drug-dependent mothers need a close follow-up at least until their teenage years to be able to detect any defect that might develop during their developmental age.

References

[1] Behnke M, Smith VC. Prenatal substance abuse: short- and long-term effects on the exposed fetus. Pediatrics 2013;131(3):e1009–1024.
[2] Gopman S. Prenatal and postpartum care of women with substance use disorders. Obstet Gynecol Clin North Am 2014;41(2):213–28.
[3] Zidkova M, Hlozek T, Balik M, Kopecky O, Tesinsky P, Svanda J, Balikova MA. Two cases of non-fatal intoxication with a novel street hallucinogen: 3-methoxy-phencyclidine. J Anal Toxicol 2017;41(4):350–4.
[4] Rahbar F, Fomufod A, White D, Westney LS. Impact of intrauterine exposure to phencyclidine (PCP) and cocaine on neonates. J Natl Med Assoc 1993;85(5):349–52.

[5] McMillin GA, Wood KE, Strathmann FG, Krasowski MD. Patterns of drugs and drug metabolites observed in meconium: what do they mean? Ther Drug Monit 2015;37(5):568–80.

[6] Wachsman L, Schuetz S, Chan LS, Wingert WA. What happens to babies exposed to phencyclidine (PCP) in utero? Am J Drug Alcohol Abuse 1989;15(1):31–9.

[7] Bey T, Patel A. Phencyclidine intoxication and adverse effects: a clinical and pharmacological review of an illicit drug. Cal J Emerg Med 2007;8(1):9–14.

[8] Lodge D, Mercier MS. Ketamine and phencyclidine: the good, the bad and the unexpected. Br J Pharmacol 2015;172(17):4254–76.

[9] NIDA. PCP (Phencyclidine). NIDA InfoFacts. U.S. Department of Health and Human Services. Washington (DC), USA; 2006:1–3. http://www.drugabuse.gov/PDF/Infofacts/PCP06.pdf.

[10] Walberg CB, McCarron MM, Schulze BN. Quantitation of phencyclidine in serum by enzyme immunoassay: results in 405 patients. J Anal Toxicol 1983;7(2):106–10.

[11] Luongo L, Maione S, Di Marzo V. Endocannabinoids and neuropathic pain: focus on neuron-glia and endocannabinoid-neurotrophin interactions. Eur J Neurosci 2014;39(3):401–8.

[12] Jansson LM, Velez ML, Butz AM. The effect of sexual abuse and prenatal substance use on successful breastfeeding. J Obstet Gynecol Neonatal Nurs 2017;46(3):480–4.

Chapter 12

Cardiac drugs

Hari Nair
Boston Heart Diagnostics, Framingham, MA, United States

Introduction

Cardiovascular system consists of the heart and an intricately long circulatory system consisting of veins, arteries, and capillaries. Heart disease and cardiovascular diseases (CVDs) constitute the most common cause of mortality in the United States; one in every four deaths occurs due to heart disease every year in the United States [1]. It is estimated that the prevalence of cardiovascular disease will continue to rise over the next few decades. While cardiotoxicity contributes to cardiovascular risk, the term cardiotoxicity is commonly used in the context of cancer care referring to the damage to heart muscle caused by chemotherapy drugs. But broadly cardiac-toxicity can be defined as the harm to cardiovascular health caused by chemical agents. Toxicity toward heart muscles and vascular system can be induced by the side effects of therapeutic doses or overdose of many therapeutic drugs (oncology or nononcology), illicit drugs, alcohol, or other agents.

When discussing cardiotoxicity, it may be useful to recognize that cardiotoxicity can result from direct impact to heart muscle, vasculature, or both, and the adverse effect of the agent can be primary or secondary to the action of the agent. Common forms of heart disease include, but are not limited to, conditions such as coronary artery disease, heart attack, arrhythmias, Heart valve disease, heart failure, congenital heart defects, and pericarditis. A number of known and unknown triggers can contribute to the many acute and chronic pathophysiological processes and events that occur in the cardiovascular system [2]. Adverse impact of agents can also depend on the genetic predisposition and the patient's history of heart disease and risk factors for heart disease. Therefore a brief review of the cardiovascular pathophysiology, acute and chronic risk factors, risk categories, and triggers for heart disease (lifestyle, genetic, and therapy-induced) might serve as a backdrop to the discussion of the central topic of cardiotoxicity.

Cardiotoxic agent can trigger heart disease or adverse event by causing dysfunction in cardiovascular functions such as cardiac output and tissue perfusion [3–5]. Every tissue in the body depends on the cardiac output (amount of work performed by the heart in response to body's need for oxygen) to receive oxygen and nutrients. Cardiovascular system accomplishes this, in part, by modulating heart rate and stroke volume. Any disfunction to this system can thus lead to morbidity and mortality. Similarly, other processes impacted can be those such as the microcirculation (which is partly controlled by endothelial products such as NO), sympathetic activation of the heart (autonomous nervous system), and regulation of circulatory volume (RAAS, ADH, and blood pressure). The adverse effect of the agents that can chronically or acutely impact cardiovascular system, include chronic heart failure, conduction abnormalities, conduction blocks, arterial hypertension, atherosclerosis, circulatory shocks, cardiomyopathy, and criterial and venous circulatory disorders [3–5].

Cardiovascular risk is a major factor determining the level of primary or secondary impact of the cardiotoxic agent. The primary risk factors for CVD include older age, male sex, race/ethnicity, abnormal lipid levels, high blood pressure, diabetes, and smoking [6]. Cardiovascular risk can be classified into nonmodifiable risk factors (age, gender, and heredity) and modifiable risk factors such as metabolic risk factors (hypertension, diabetes mellitus, metabolic syndrome, hyperlipidemias, and obesity/overweight) and lifestyle factors (smoking, physical activity, and diet). Academic and clinical understandings of novel risk factors such as lipoprotein (a), elevated homocysteine, inflammatory markers, prothrombotic factors and trace elements, and heavy metals that can impact cardiovascular system are emerging [7].

American College of Cardiology/American Heart Association (ACC/AHA) periodically publishes guideline on the primary prevention of CVD [8]. The 2019 guideline include information on risk factors such as dietary pattern, obesity and physical activity, T2DM, lipid levels, hypertension, and use of tobacco. The guideline also offers recommendations on the use and limitations of statin and low dose aspirin therapy, two of the widely administered regimes for ASCVD prevention. Although low dose aspirin is well established for secondary prevention of ASCVD by reducing the risk of atherothrombosis; recent studies have shown that aspirin therapy lacks net benefit for primary prevention. Known side effects of low dose aspirin include GI bleeding or peptic ulcer disease, bleeding from other sites, age >70 years, thrombocytopenia, coagulopathy, chronic kidney disease, and adverse drug–drug interactions (e.g., in the concurrent use with nonsteroidal antiinflammatory drugs, steroids, and anticoagulants). The guideline focuses on primary prevention of nine CVD conditions (acute coronary syndromes, myocardial infarction, stable or unstable angina, arterial revascularization, stroke/transient ischemic attack, and peripheral arterial disease) and two disease processes (heart failure and atrial fibrillation). The guideline includes the criteria for assessment of risk and recommendations for estimating the risk of ASCVD in asymptomatic adults.

In the perioperative population, it is estimated that cardiovascular complications are responsible for roughly one-half of all the mortality experienced by patients undergoing noncardiac surgery. The risk assessment for CVD and the risk assessment in the perioperative setting are described by Singh and Zeltser [9].

Lifestyle and genetic triggers

By predisposing the patient to adverse cardiovascular events, lifestyle, along with heredity, play an important role in cardiotoxicity. Lifestyle triggers include those mentioned in the ACC/AHA guidelines (dietary pattern, obesity, and physical activity) as well as strenuous activity and stressful events. To illustrate this later point in the first week of Iraqi missile attack on Israel during the onset of Iraq war, the rate of nonfatal myocardial infraction (MI) doubled compared to a year earlier [10]. Similarly, another study revealed that viewing a stressful soccer match more than doubles the risk of an acute cardiovascular event [11]. Lifestyle triggers for CVD—activities that result in short-term physiological changes that may lead directly to onset of acute CVD—is discussed in the review by Tofler and Muller [12]. Multicenter investigation of infarct size (MILIS) reported that common CVD triggers include emotional upset, moderate physical and heavy physical activity, lack of sleep, and overeating. Circadian variations such as morning peak in MI, sudden cardiac death, and that that up to 10% of ST-elevation MIs may be triggered by heavy physical exertion stroke point to the nonrandom nature of adverse events.

Large majority of cardiovascular diseases represent interaction between genetic and environmental risk factors (polygenic). Many polymorphisms have been identified as predisposing factors to cardiovascular risk and disease. Advances in our understanding of genetic and genomic underpinnings of cardiovascular disease is reviewed by O'Donnell and Nabel [13] and more recently by Kathiresan and Srivastava [14].

Understanding of the genetic variants that increase the risk of drug-induced drug toxicity can improve risk assessment in the patient management. For example, a systematic metaanalysis by Linschoten et al. [15] lists the polymorphisms identified that show potential to improve risk stratification (*CELF4* rs1786814, *RARG* rs2229774, *SLC28A3* rs7853758, *UGT1A6* rs17863783) and an intergenic variant (rs28714259). In yet another example, pregnant women who have been previously treated with Doxorubicin (DOX) may be particularly vulnerable to heart failure if they have a genetic predisposition to doxorubicin toxicity [16].

Drug-induced cardiotoxicity

As alluded to previously, mechanism of muscle damage caused by cardiotoxic agent can be direct or indirect. Agents causing direct muscle damage include alcohol, cocaine, glucocorticoids, lipid lowering drugs, antimalarial drugs, colchicine, and zidovudine. Antimalarials cause vascular myopathies. Colchicine is associated with vacuolar myopathies and zidovudine causes mitochondrial myopathies. Indirect muscle damage can trigger drug-induced coma with ischemic muscle compression, drug-induced hypokalemia triggered by diuretics, hyperkinetic states secondary to delirium tremens or seizures induced by alcohol, phenothiazine-induced dystonic states, and hyperthermia induced by cocaine. Alcohol binging can trigger hypokalemia, hypophosphatemia, and coma. Repeated intramuscular injections, particularly opiates, such as heroin or pentazocine can cause muscle damage. Intra-arterial injection is another cause of ischemic necrosis of muscle tissue.

Cancer treatment is a common source of cardiomyopathy, and a cause of long-term cardiovascular risk. Newer and better cancer therapies have prolonged cancer survival rates, but cancer treatment caused cardiotoxicity as well as predisposes the patient to long-term cardiovascular risk. Cardio-oncology [17–21] has emerged as a clinical discipline to help cancer patient complete cancer treatment without developing cardiotoxicity or cardiovascular risk. Patients are assessed for potential risk of developing heart conditions when undergoing certain types of cancer therapy or following radiation treatment to the chest. In addition to cancer therapy, cardiotoxicity can occur from classes of drugs such as antipsychotic, neuromyopathy, antimalarial, antiretroviral, lipid-lowering, inflammatory myopathy, and the use of Ipecac among others.

As mentioned earlier, agent of toxicity can adversely impact the heart, vascular system, or a combination of the two. Toxicity to heart can cause disturbances in the rhythm, myocardial ischemia, left ventricular (LV) dysfunction/heart failure, impairment of cardiac valves, induction of pericardial disease from immune triggered reactions, and hemorrhage. Toxicity to vascular system can occur from impact to arterial blood pressure and thromboembolic complications among others. Many drug classes such as β-blockers, sodium, potassium channel and pump inhibitors, anticancer drugs, NSAIDs, erythropoietin and analogs, glucocorticoids and female hormones among others can effect cardiovascular system directly or through drug-drug interactions. Drugs such as direct and indirect sympathomimetics, nicotine, tricyclic antidepressants, ethanol, cocaine, Ca^{2+} channel blockers, and maitotoxin affect both myocardium and vascular system.

Drugs that promote QT interval prolongation (*Torsade de pointes*) include antiarrhythmic drugs, calcium channel blockers, psychiatric drugs, antihistamines, antimicrobial and antimalarial drugs, serotonin agonists/antagonists, immunosuppressant (tacrolimus), antidiuretic hormone (vasopressin), and other agents such as adenosine, organophosphates, probucol, papaverine, and cocaine [22–23].

Drug–drug interaction

Another trigger for cardiotoxicity is drug–drug interactions which can cause acute or chronic cardiovascular toxicity. For example, angiotensin-converting enzyme (ACE) inhibitors are known to interact with antagonists at AT1 receptors for angiotensin II (causing increase in the risk of hypotension and hyperkalemia) and Neprilysin inhibitors causing angioedema. Similarly, alcohol can interact with antabuse drugs such as Disulfiram (used to treat alcohol addiction) and cause hypotension and tachycardia. When taken together, cardiotoxic effects of both cocaine and amphetamines are amplified leading to both short term (anxiety, tremors, seizure, or coma) and long-term (e.g., increased addiction and risk tolerance) cardiovascular risks. Common cardiac drugs—amiodarone, β-blockers, calcium channel blockers, and digoxin—can cause various adverse cardiovascular effects upon interaction with other drugs. Drug–drug interaction of common cardiac drugs with therapeutic and illicit drugs are listed in Table 12.1 along with the cardiovascular effects elicited by the interaction (adopted from Ref. [23]).

The list of known cardiotoxic agents is long and is likely to increase with the advent of newer medications for cancer and other illness. In the recent article "Surviving Cancer Without a Broken Heart," Caspi and Aronson [24] laid out the mechanisms, cardiotoxic impact, risk factors, and assessments as well as guidelines for monitoring and prevention of cardiotoxicity for the most commonly recognized chemotherapy agents that cause cardiotoxicity such as anthracyclines, proteasome inhibitors, human epidermal growth factor receptor 2 (HER2) targeted therapies, vascular endothelial growth factor inhibitors, and immune checkpoint inhibitors.

Chemotherapy associated myocardial toxicity

Cardiotoxicity induced by chemotherapy are generally differentiated as reversible, irreversible, acute, chronic, and late onset. Category type I describes irreversible damage while type II describe reversible damage. One difference between the two classes is that type I damage is usually caused by cumulative dosage while dosage is not directly related to the type II damage. Anthracycline toxicity is a key example of an agent causing type I damage.

Anthracyclines are used to treat many types of cancers such as leukemias, lymphomas, breast, stomach, uterine, ovarian, bladder cancer, and lung cancers. Anthracycline based regimen is used in about one million patients in the United States each year. Anthracycline therapy can result in clinical heart failure in up to 5% of high-risk patients [25]. Cumulative dosage, age, and preexisting conditions for cardiac disease are known risk factors for anthracycline toxicity. Because the clinical manifestations of reduction in LVEF from anthracycline-induced cardiotoxicity become evident only after a critical amount of myocardial damage has taken place, cardiac troponins can be used as sensitive indicator of early development of cardiotoxicity [26]. Endomyocardial biopsy is the gold standard for detection of acute doxorubin (an anthracycline)–induced cardiomyopathy. Heart failure typically occurs within a month to a year of last administration of anthracycline, it could occur as late as 6–10 years or later after treatment. Anthracycline related cardiomyopathy is typically treated with ACE inhibitors and β-blockers.

TABLE 12.1 Drug–Drug Interactions and Cardiovascular Effects

Drug 1	Drug 2	Relevant Cardiovascular Risks
ACEi	Antagonists at AT1 receptors for angiotensin II (sartans)	Hypotension and hyperkalemia
	Neprilysin inhibitors (sacubitril)	Angioedema
Disulfiram	Alcohol	Hypotension, tachycardia
Amphetamines	Cocaine	All risks described in relevant sections of this manuscript
Amiodarone	Nondihydropyridine Ca^{2+} channel blockers	Sinus arrest
	Some quinolones	Torsade de pointes
Drugs prolonging QT	Drugs lowering plasma potassium concentration (amphotericin B, β2-agonists, corticosteroids, loop and thiazide diuretics, theophylline, misuse, or overuse of laxatives)	Torsade de pointes
Amiodarone	Some? β-blockers (metoprolol and carvedilol)	Hypotension, bradycardia, asystole, and possibly ventricular fibrillation
	Sotalol	Torsade de pointes, hypotension
β-Blockers	Cholinomimetics	Bradycardia, AV blocks, and hypotension
	Nondihydropyridine Ca^{2+} channel blockers	Bradycardia, asystole, and sinus arrest
	Digoxin	Bradycardia, AV block
	Dronedarone	Bradycardia
	Antipsychotics-phenothiazines	Hypotension
	Propafenone	Profound hypotension and cardiac arrest
Some β-blockers	Some SSRi	Bradycardia, AV blocks, and hypotension
Calcium channel blockers	Azoles, clarithromycin, some HIV-protease inhibitors	Hypotension and/or bradycardia
Digoxin	Amiodarone	Dysrhythmias, also torsade de pointes
	Azoles, clarithromycin, some HIV-protease inhibitors	Dysrhythmias
	Nondihydropyridine Ca^{2+} channel blockers	Bradycardia, asystole, and sinus arrest
	Loop or thiazide diuretics, amphotericin B, corticosteroids	Dysrhythmias
	i.v. calcium	Dysrhythmias
	Propafenone	Dysrhythmias

Source: Adopted from Kumar V. et al., editors. Robbins and cotran pathologic basis of disease. 9th ed. Philadelphia, PA: Saunders, 2015.

Other notable cancer therapies associated with significant cardiac toxicity include Tyrosine kinase inhibitor therapy, trastuzumab, antimetabolites, alkylating agents, angiogenesis inhibitors, checkpoint inhibitors, Monoclonal antibodies—Mabs (e.g., HER2 targeted therapies), and radiation therapy. Mabs represent type II damage. Radiation therapy risks include coronary disease and cardiovascular abnormalities. Radiation therapy can cause fibrosis and thickening of pericardium leading to constrictive pericarditis. Modern radiotherapy techniques have substantially lowered risks compared to those used in earlier decades [26].

Cardiac monitoring is recommended in all patients prior to the initiation of chemotherapy. Patients receiving anthracyclines should have cardiac function monitoring at baseline—months 3, 6, and 9 and post initiation of treatment at 12–18 months. The patients should be counseled on the benefits of reducing cardiovascular risk through approaches

TABLE 12.2 Adverse Cardiovascular Effects of Different Classes of Chemotherapeutical Drugs

Chemotherapeutical Drug	Adverse Cardiovascular Effect
Anthracyclines	LVH, HF, myocarditis, arrhythmia
5-Fluorouracil	Ischemia, HF, pericarditis, cardiogenic shock
Taxanes (paclitaxel), vinca alkaloids	Sinus bradicardia, ventricular tachycardia, atrioventricular block, HF, ischemia
Cyclophosphamide	HF (neurohumoral activation), mitral regurgitation
Trastuzumab	HF, LVH, arrhythmia
Tamoxifen	Thromboembolism, cholesterol metabolism anomalies
Bevacizumab	Hypertension, thromboembolism
COX-2 specific inhibitors	Thromboembolism

HF, Heart failure; *LVH*, left ventricular hypertrophy.

such as lifestyle changes, statin therapy and blood pressure monitoring, and setting goals for blood pressure and cholesterol levels [27]. Table 12.2 lists most common classes of chemotherapy drugs that trigger cardiotoxicity and promote cardiovascular disease (adapted from Ref. [28]).

Cardiotoxicity associated with noncancer therapeutic drugs

Cardiotoxicity can accrue from a variety of noncancer therapies as mentioned earlier in the chapter. Thiazolidinediones (TZDs or glitazones), HIV therapy, drugs that cause QT prolongation, and drugs that cause vulvular disease [27,28]. TZDs are used to improve glycemic control in patients with type 2 diabetes. Glitazones can cause fluid retention, weight gain, and peripheral edema; risk factors include age, history of chronic diabetes, chronic renal failure, significant aortal and mitral valve disease, history of coronary disease, hypertension, and history of heart failure among others.

Antiretroviral therapies for HIV treatment have increased the survival of patients but has concurrently increased the risk of long-term cardiovascular diseases. Hyperlipidemia, hyperglycemia, and lipodystrophy are common side effects of protease inhibitors used in HIV therapy. Hyperlipidemia associated myocardial infarction is a key side effect of this class of drugs. HIV patients should be routinely screened for cardiovascular risk factors, including fasting lipid profile measurements before and after therapy. Statins (HMG-CoA reductase inhibitors) is used to treat hyperlipidemia. Protease inhibitors and many statins are metabolized by the cytochrome P450 system; therefore concurrent protease inhibitor therapy can increase plasma level of statins which can in turn increase the risk of drug toxicity. Pravastatin and atorvastatin that are not metabolized by the same P450 enzyme as protease inhibitor are therefore a more appropriate statin therapy in this setting. In addition, weight loss drugs of the past and some drugs for migraine head ache (e.g., ergotamine and methysergide) are linked to the development of vulvular structural abnormalities. Drugs that prolong QT have been discussed earlier in this chapter.

Cardiotoxicity associated with drugs of abuse

Alcohol and cocaine are the most common drugs of abuse that are associated with heart and cardiovascular toxicity. Heavy and prolonged use of alcohol is associated with dilated cardiomyopathy. Alcoholic cardiomyopathy is reversible with cessation of alcohol and standard heart failure therapy. Cocaine use poses severe risk for myocardial ischemia or infarction. A sympathomimetic drug, cocaine, blocks the reuptake of norepinephrine and dopamine at the postsynaptic receptor. The result is pronounced sympathetic activation leading to increased heart rate, blood pressure, myocardial contractility, coronary vasoconstriction, and propensity for acute thrombosis. Cocaine is detected by screening urine for benzoylecgonine which can be detected in urine for up to 24–48 hours after ingestion. Chest pain is a common presentation for patients suspected of cocaine-induced cardiotoxicity and is evaluated in the same way as the suspected acute coronary syndrome with the exception that ß-adrenergic antagonists may be avoided in the acute setting to avoid exasperation of vasoconstriction from unopposed α-agonism.

TABLE 12.3 Partial list of cardiac drugs indicated for TDM

Drug	Conditions Treated	Therapeutic Range (mg/L)	Toxic Concentration (mg/L)	Most Common Side Effects	Major Toxic Effects	Other Monitoring Required
Amiodarone	Life-threatening recurrent ventricular arrhythmias without response to adequate doses of other antiarrhythmics	0.5–2.0	>2.5	Photosensitivity, corneal microdeposits, thyroid abnormalities	Hypersensitivity, alveolar or interstitial pneumonitis, liver injury, worsened arrhythmias	Liver and thyroid functions, ECG, 2 PFTs, eye exam
Digitoxin	Heart failure, atrial flutter/fibrillation, and supraventricular tachycardia	0.01–0.03	>0.045	Anorexia, nausea, and vomiting	Mental status changes, vomiting, bradycardia, heart block	Serum K+; in acute overdose, serum [K+] decreases but increases in chronic overdose
Digoxin	Heart failure, atrial flutter/fibrillation, and paroxysmal atrial tachycardia	0.0005–0.002	>0.003	Nausea, vomiting, visual disturbances, weakness	AV block, premature cardiac contraction, arrhythmia, vomiting	Monitor [K+] and ECG
Disopyramide	Documented life-threatening ventricular arrhythmias	2.0–5.0	>7.0	Heart failure, hypotension	Apnea, arrhythmias, loss of consciousness	Serum glucose, renal function, ECG
Flecainide	Paroxysmal ventricular tachycardia, atrial fibrillation/flutter, supraventricular tachycardia	0.2–1.0	>1.0	Proarrhythmic effects	Nausea, vomiting, hypotension, syncope bradycardia, heart failure	Renal function, ECG
Lidocaine	Arrhythmias	1.5–5.0	>6.0	Excitatory or depressive CNS effects, allergic reaction, bradycardia	Cardiovascular depression, convulsions, and hypoxia	ECG
Mexiletine	Documented life-threatening ventricular arrhythmias (e.g., sustained ventricular tachycardias)	0.5–2.0	>2.0	Nausea, vomiting, heartburn, lightheadedness, tremors, coordination difficulties, changes in sleep habits	Nausea, hypotension, sinus bradycardia, paresthesia, seizures, left branch bundle block, asystole	ECG, liver function, CBC
Procainamide	Documented ventricular arrhythmias	4.0–8.0	>10	Hypotension, vomiting, lupus erythematosus-like syndrome, neutropenia	Ventricular extrasystoles and tachycardia	Monitor serum [NAPA], ECG, renal function
Propafenone	Life-threatening ventricular arrhythmias	<1.0	>4.86	Unusual taste, dizziness, constipation, dyspnea, nausea/vomiting, anxiety	Supraventricular tachycardia, atrial flutter, tinnitus, apnea	ECG, pacemaker function, renal, and hepatic functions
Quinidine	Conversion of (and decreased relapse into) atrial fibrillation and flutter; suppression of ventricular arrhythmias	2.0–5.0	>6.0	Diarrhea, headache, palpitations, rash, tremors	Ventricular arrhythmias, hypotension, vomiting, diarrhea, tinnitus	ECG

AV, Atrioventricular; TDM, therapeutic drug monitoring.
Source: Adapted from Jortani SA., Gheorghiade M., Valdes R. Standards of laboratory practice: cardiac drug monitoring. Clin Chem 1998;44(5):1096–109.

Endocrine therapy

Endocrine therapies (tamoxifen and aromatase inhibitors) are administered to reduce the risk of breast cancer in postsurgery patients for estrogen receptor and/or progesterone receptor positive breast cancer. The use of endocrine therapy increases the risk of composite adverse cardiovascular outcomes in part due to their role in reducing circulating estrogens that are cardio protective. Metaanalysis indicates that the risk adverse cardiovascular outcome is higher for aromatase inhibitors compared to tamoxifen [29].

In addition, there is widely varying myocardial infarction risks such as arterial thrombosis associated with oral contraceptives [30]. The risk is increased for patients older than 35 years, hypertensive and habituated to smoking. For patients with preexisting cardiovascular diseases the use of COX-2 inhibitors and NSAIDs pose increased risk of heart failure and renal failure.

Therapeutic drug monitoring of cardiac drugs

Cardiac drugs are administered to treat life-threatening conditions. However, when the dosage of these drugs exceed therapeutic limit and reaches the toxic limit, cardiac drug become cardiotoxic themselves. For example, antiarrhythmic drugs act of ion channels, but some of them have the potential to cause arrhythmias through other pathways. Toxicity from digoxin can result in premature ventricular contractions and hyperkalemias which increase the risk of arrhythmias [31]. Therefore for many cardiac drugs as with chemotherapy drugs, therapeutic monitoring—individualization of dosage by maintaining blood/plasma levels of drug concentration with a therapeutic window—is recommended and is a common practice. Therapeutic drug monitoring (TDM) is a three-step process that includes (1) the precise and reliable measure of the plasma concentration of a specific medicine, (2) the interpretation of the obtained concentration value according to the knowledge on the concentration-effect relationship, and (3) the calculation and proposal of an individual dose adjustment for that specific patient. Many cardiac drugs fit the indications for therapeutic drug monitoring [32–36]. Key cardiac drugs that are monitored for toxicity via TDM are listed in Table 12.3 (adopted from Ref. [32]). High dose methotrexate is an anticancer drug that is indicated for therapeutic drug monitoring [37,38].

Summary

Drug-induced cardiotoxicity can occur from the administration of a variety of therapeutic drugs (cancer and noncancer in therapeutic and/or overdose levels), illicit drugs, alcohol, or from drug–drug interactions. The susceptibility to cardiotoxicity can depend upon the history of cardiovascular diseases and risk factors such as lifestyle, biological and nonbiological triggers, and genetic predisposition. Prior to and after administration of therapeutic drugs, the dosage and the patient must be monitored closely via therapeutic drug monitoring to ensure that the drugs do not exceed toxic levels, while they remain within the therapeutic window. While on therapy with potential for cardiotoxicity, patient must be counseled on the need for cardiovascular risk reduction through lifestyle changes, adherence to medication, and monitoring of blood pressure.

Acknowledgment

The author thanks Professor Cynthia Taub, MD (Professor of Medicine—Cardiology at Albert Einstein College of Medicine) for reviewing and providing valuable insights during the preparation of this chapter.

References

[1] <https://www.cdc.gov/heartdisease/facts.htm>.
[2] Lobo SA, Fischer S. Cardiac risk assessment, [Updated 2019 May 7]. StatPearls [Internet]. Treasure Island, FL: StatPearls Publishing; 2019. Available from: <https://www.ncbi.nlm.nih.gov/books/NBK537146/>.
[3] Kumar V, et al., editors. Robbins and cotran pathologic basis of disease. 9th ed. Philadelphia, PA: Saunders; 2015.
[4] Boron WF, Boulpaep EL, editors. Medical physiology. 3rd ed. Philadelphia, PA: Elsevier; 2016.
[5] Lilly L, editor. Pathophysiology of heart disease. 6th ed. Philadelphia, PA: Lippincott; 2015.
[6] Guirguis-Blake JM, Evans CV, Senger CA, Rowland MG, O'Connor EA, Whitlock EP. Aspirin for the Primary Prevention of Cardiovascular Events: A Systematic Evidence Review for the U.S. Preventive Services Task Force. Evidence synthesis No. 131. AHRQ Publication No. 13-05195-EF-1. Rockville, MD: Agency for Healthcare Research and Quality; 2015.

[7] Alissa EM, Ferns GA. Heavy metal poisoning and cardiovascular disease J Toxicol 2011;21 Article ID 870125. Available from: https://doi.org/10.1155/2011/870125.
[8] 2019 ACC/AHA Guideline on the Primary Prevention of Cardiovascular Disease Mar 17, 2019 | Melvyn Rubenfire, MD, FACC.
[9] Singh S, Zeltser R. Cardiac risk stratification. [Updated 2019 Apr 28]. StatPearls [Internet]. Treasure Island, FL: StatPearls Publishing; 2019. Available from: <https://www.ncbi.nlm.nih.gov/books/NBK507785/>.
[10] Meisel SR, Kutz I, Dayan KI, Pauzner H, Chetboun I, Arbel Y, et al. Effect of Iraqi missile war on incidence of acute myocardial infarction and sudden death in Israeli civilians. Lancet 1991;338:660−1.
[11] Wilbert-Lampen Ute, David Leistner MD, Sonja Greven MD, Tilmann Pohl MS, Sebastian Sper MD, Völker C, et al. Cardiovascular events during world cup soccer. N Engl J Med 2008;358:475−83. Available from: https://doi.org/10.1056/NEJMoa0707427.
[12] Tofler Geoffrey H, Muller James E. Triggering of acute cardiovascular disease and potential preventive strategies. Circulation 2006;114:1863−72.
[13] O'Donnell CJ, Nabel EG. Genomics of cardiovascular disease. N Engl J Med 2011;365(22):2098−109.
[14] Kathiresan S, Srivastava D. Genetics of human cardiovascular disease. Cell. 2012;148(6):1242−57. Available from: https://doi.org/10.1016/j.cell.2012.03.001 PubMed PMID: 22424232; PubMed Central PMCID: PMC3319439.
[15] Linschoten M, Teske AJ, Cramer MJ, van der Wall E, Folkert W. Chemotherapy-related cardiac dysfunction: a systematic review of genetic variants modulating individual risk, Asselbergs Originally published 1 2018 <https://doi.org/10.1161/CIRCGEN.117.001753Circulation>: Genomic and Precision Medicine. 2018;11.
[16] Physiology (The American Physiological Society); Genetic predisposition to cardiovascular disease potentiates delayed doxorubicin cardiotoxicity during pregnancy; Pamela Fettig, Anthony Tobias, Lois Heller, Fred Apple and Leslie Sharkey Published Online:7 Mar 2006.
[17] Tajiri K, Aonuma K, Sekine I. Cardio-oncology: a multidisciplinary approach for detection, prevention and management of cardiac dysfunction in cancer patients. Jpn J Clin Oncol 2017;47(8):678−82. Available from: https://doi.org/10.1093/jjco/hyx068.
[18] Larsen CM, Mulvagh SL. Cardio-oncology: what you need to know now for clinical practice and echocardiography. Echo Res Pract 2017;4(1):R33−41. Available from: https://doi.org/10.1530/ERP-17-0013 PubMed PMID: 28254996; PubMed Central PMCID: PMC5435878.
[19] Lenneman CG, Sawyer DB. Cardio-oncology; an update on cardiotoxicity of cancer-related treatment. Circ Res 2016;118:1008−20.
[20] Chang HM, Moudgil R, Scarabelli T, Okwuosa TM, Yeh ETH. Cardiovascular Complications of Cancer Therapy. Best practices in diagnosis, prevention, and management: Part 1. J Am Coll Cardiol 2017;70:2536−51.
[21] Chang HM, Okwuosa TM, Scarabelli T, Moudgil R, Yeh ETH. Cardiovascular Complications of Cancer Therapy. Best practices in diagnosis, prevention, and management: Part 2. J Am Coll Cardiol 2017;70:2552−65.
[22] Yap YG, Camm AJ. Drug induced QT prolongation and torsades de pointes. Heart 2003;89(11):1363−72. Available from: https://doi.org/10.1136/heart.89.11.1363 PubMed PMID: 14594906; PubMed Central PMCID: PMC1767957.
[23] Mladěnka P, Applová L, Patočka J, Costa VM, Remiao F, Pourová J, et al. TOX-OER and CARDIOTOX Hradec Králové Researchers and Collaborators. Comprehensive review of cardiovascular toxicity of drugs and related agents. Med Res Rev 2018;38(4):1332−403. Available from: https://doi.org/10.1002/med.21476 PubMed PMID: 29315692; PubMed Central PMCID: PMC6033155.
[24] Caspi O, Aronson D. Surviving cancer without a broken heart. Rambam Maimonides Med J 2019;10(2):e0012. Available from: https://doi.org/10.5041/RMMJ.10366 PubMed PMID: 31002639; PubMed Central PMCID: PMC6474762.
[25] Henriksen PA. Anthracycline cardiotoxicity: an update on mechanisms, monitoring and prevention. Heart 2018;104:971−7.
[26] Wu AH. Cardiotoxic drugs: clinical monitoring and decision making. Heart 2008;94:1503−9.
[27] Thomas SA. Chemotherapy agents that cause cardiotoxicity. US Pharm 2017;42(9):HS24−33.
[28] Florescu M, Cinteza M, Vinereanu D. Chemotherapy-induced cardiotoxicity. Maedica (Buchar) 2013;8(1):59−67 PubMed PMID: 24023601; PubMed Central PMCID: PMC3749765.
[29] Matthews A, Stanway S, Farmer RE, Strongman H, Thomas S, Lyon AR, et al. Long term adjuvant endocrine therapy and risk of cardiovascular disease in female breast cancer survivors: systematic review. BMJ 2018;363. Available from: https://doi.org/10.1136/bmj.k3845 (Published 08 October 2018) Cite this as: BMJ 2018;363:k3845.
[30] Roach RE, Helmerhorst FM, Lijfering WM, Stijnen T, Algra A, Dekkers OM. Combined oral contraceptives: the risk of myocardial infarction and ischemic stroke. Cochrane Database Syst Rev 2015;(8). Available from: https://doi.org/10.1002/14651858.CD011054.pub2 Art. No.: CD011054.
[31] Meenakshisundaram R, Cannie DE, Shankar PR, Zadeh HZ, Bajracharya O, Thirumalaikolundusubramanian P. Chapter 8 — Cardiovascular toxicity of cardiovascular drugs. In: Ramachandran M, editor. Heart and toxins. Academic Press; 2015, ISBN: 9780124165953. p. 225−74.
[32] Jortani AS, Gheorghiade M, Valdes R. Standards of laboratory practice: cardiac drug monitoring. Clin Chem 1998;44(5):1096−109.
[33] Cooney L, Loke YK, Golder S, Kirkham J, Jorgensen A, Sinha I, et al. Overview of systematic reviews of therapeutic ranges: methodologies and recommendations for practice. BMC Med Res Methodol 2017;17:84.
[34] Schroeder LF, Guarner J, Amukele TK. Essential diagnostics for the use of world health organization essential medicines. Clin Chem 2018;64:1148−57.
[35] Rogers NM, Jones TE, Morris RG. Frequently discordant results from therapeutic drug monitoring for digoxin: clinical confusion for the prescriber. Intern Med J 2010;40:52−6.
[36] Rathore SS, Curtis JP, Wang Y. Association of serum digoxin concentration and outcomes in patients with heart failure. JAMA 2003;289:871−8.
[37] Al-Turkmani MR, Law T, Narla A, Kellogg MD. Difficulty measuring methotrexate in a patient with high-dose methotrexate−induced nephrotoxicity. Clin Chem 2010;56(12):1792−4. Available from: https://doi.org/10.1373/clinchem.2010.144824.
[38] Nair H, Lawrence L, Hoofnagle AN. Liquid chromatography−tandem mass spectrometry work flow for parallel quantification of methotrexate and other immunosuppressants. Clin Chem 2012;58(5):943−5. Available from: https://doi.org/10.1373/clinchem.2011.175067.

Chapter 12.1

Use of therapeutic plasma exchange to enhance propafenone elimination following intentional drug overdose

Yong Y. Han[1], Cindy George[2], Uttam Garg[3] and Gabor Oroszi[2,3]

[1]Department of Pediatrics, Division of Critical Care Medicine, Children's Mercy Hospital, Kansas City, MO, United States, [2]Apheresis Program, Children's Mercy Hospital, Kansas City, MO, United States, [3]Department of Pathology and Laboratory Medicine, Children's Mercy Hospital, Kansas City, MO, United States

Case description

After developing severe headache, nausea, and dizziness, a 16-year-old, 61.4 kg female confessed to her parents that ~7 hours earlier she had intentionally ingested between 20 and 25 of her father's 225 mg extended-release (ER) capsules of propafenone (Rythmol SR, GlaxoSmithKline, Philadelphia, PA) [4.5–5.6 g (73.3–91.6 mg/kg)] in a suicide attempt. Emergency medical services (EMS) was called, but while awaiting their arrival, she experienced either a syncopal episode or seizure for which bystander cardiopulmonary resuscitation (CPR) was briefly provided. When EMS arrived to the scene, she was found to have a pulse and appeared postictal. She was transported to a local emergency department where she was initially awake and verbally responsive with a sinus rhythm but with a PR interval at the upper limit of normal, mildly prolonged QRS duration, and mildly prolonged QTc interval on ECG (Fig. 12.1.1A; ~8 hours postingestion). During evaluation she suddenly developed a generalized tonic-clonic seizure and significantly prolonged QRS duration with a Brugada-like pattern (Fig. 12.1.1B). Lorazepam was administered for seizure and sodium bicarbonate and intravenous lipid emulsion therapy for propafenone toxicity. She did not require CPR or vasoactive agents. However, she was given a normal saline fluid bolus and was emergently intubated for airway protection. Her ECG morphology returned to sinus rhythm, but with slightly prolonged PR, QRS, and QTc intervals (Fig. 12.1.1C). Request to transfer to our pediatric intensive care unit (PICU) was made, and although she was maintaining adequate spontaneous circulation, our extracorporeal life support team was alerted and remained on "standby" mode in the event of precipitous cardiovascular collapse.

After arrival to our PICU, mechanical ventilatory support was continued and central venous and arterial accesses were obtained. Laboratory evaluation was not supportive of polypharmacy ingestion. There was concern for possible abrupt reemergence of life-threatening dysrhythmia, in part, because she had ingested ER capsules. Since propafenone is highly protein-bound and not efficiently removed by hemodialysis [1,2], the Apheresis Service was consulted to initiate urgent therapeutic plasma exchange (TPE). After obtaining parental consent, placement of a 12-French temporary hemodialysis catheter, and confirmation of a normal coagulation profile, a 1.6× plasma volume TPE using 5% albumin as replacement fluid was performed utilizing centrifugal apheresis technique (Spectra Optia, Terumo BCT, Lakewood, CO) with regional citrate anticoagulation (TPE duration ~1.5 hours). Post-TPE ECG revealed normalization of the PR interval and QRS duration but continued to show borderline prolonged QTc interval (Fig. 12.1.1D; ~19 hours postingestion). Although data were not available "real-time," serum propafenone levels were later found to have decreased over ~3.5 hours from 2.2 mcg/mL pre-TPE (~15 hours postingestion) to 0.7 mcg/mL post-TPE. She tolerated TPE without adverse effects. However, she later became hypotensive with development of metabolic acidosis requiring initiation of low-dose epinephrine infusion to maintain normotension and normal perfusion. She remained intubated overnight. By the following morning her ECG had normalized (Fig. 12.1.1E; ~32 hours postingestion), her metabolic acidosis had resolved, and she was successfully weaned off epinephrine infusion. She was

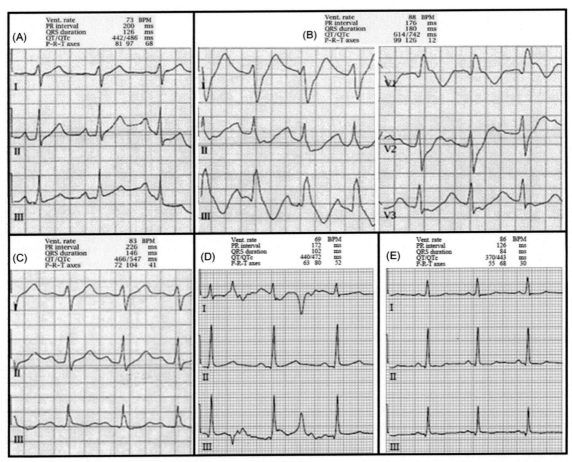

FIGURE 12.1.1 Serial, interval ECG's following intentional propafenone ingestion. (A, B, C): referring hospital ED (~8-hr post-ingestion); (D, E): our institution. A: initial presentation; B: Brugada-like pattern during sudden generalized tonic-clonic seizure; C: following administration of lorazepam, sodium bicarbonate, and lipid emulsion therapy. D: following therapeutic plasma exchange (~19-hr post-ingestion); E: normalization of ECG (~32-hr post-ingestion).

successfully extubated, was later transferred to the general care ward for continued medical observation, and was discharged 2 days later to an inpatient psychiatric facility.

Case discussion

Propafenone is Vaughan Williams class 1c antiarrhythmic that possesses local anesthetic properties, blocks the fast inward sodium channel in myocardial tissue and Purkinje fibers, and slows the rate of increase of phase 0 of the action potential [3–6]. It also exhibits weak beta-adrenergic receptor and calcium channel blocking activities [4–7] and exerts a negative inotropic effect on the myocardium [3–6,8]. On ECG, propafenone prolongs the PR interval and QRS duration and may prolong the QTc interval [3–6,9–10], but the latter may be difficult to interpret when QRS duration is prolonged [6]. Clinically, propafenone has been indicated for the treatment of paroxysmal atrial fibrillation, paroxysmal supraventricular tachycardia, and ventricular tachycardia [6].

The pharmacokinetics/pharmacodynamics of propafenone is complex. Propafenone is well-absorbed with serum concentrations peaking after oral ingestion 2–3 hours for immediate release (IR) tablets [5,10–12] and 3–8 hours for ER capsules [6]. However, propafenone undergoes extensive "first-pass" hepatic metabolism that reduces its bioavailability but simultaneously generates two intermediate, active metabolites (5-hydroxypropafenone and N-depropylpropafenone) before ultimate glucuronide and/or sulfate conjugate excretion [5,10–12]. This "first-pass" hepatic metabolism is saturable, leading to a nonlinear (disproportionate) increase in serum drug concentrations as dose is increased [5,6,10–12]. For example, a 150 mg IR tablet has an absolute bioavailability of 3.4%, while a 300 mg IR tablet has an absolute bioavailability of 10.6% [6]. Similarly, a threefold daily dose increase from 300 to 900 mg/day results in a 10-fold increase in mean steady-state drug concentrations [10]. The change in bioavailability stemming from this increased dosing regimen is not known but is expected to be higher. The relative bioavailability of propafenone from ER formulations is less than that of IR formulations such that the bioavailability of a 325 mg ER capsule BID regimen approximates that of a 150 mg IR tablet TID regimen [6]. However, mean exposure to the 5-hydroxypropafenone metabolite is about 20%—25% higher after ER capsule versus IR tablet ingestion [6]. The generation of the 5-hydroxypropafenone metabolite is particularly important, because it is pharmacologically more active and has a longer plasma half-life than the parent drug, contributing significantly to the overall clinical effects [12]. Genetic polymorphisms in hepatic cytochrome CYP2D6 delineate two distinct groups of metabolizers (extensive and poor), with ~7% of Caucasians belonging to the poor metabolizer group [13,14]. The elimination half-life for extensive metabolizers ranges 2—10 hours, while that for poor metabolizers ranges 10—32 hours [5,13,14]. Propafenone is extensively (85%—97%) protein-bound (mostly to alpha 1-acid glycoprotein) in plasma [5,11,15] and has a volume of distribution of 1.9—3 L/kg [5,6,11], making hemodialysis an inefficient modality for its removal [1,2].

Symptoms of mild to moderate propafenone toxicity include nausea, vomiting, dizziness, and blurred vision, while severe toxicity may lead to coma, generalized seizures, hypotension, and metabolic acidosis [5,6]. Adverse central nervous system effects have been reported to be more common as serum levels exceed 0.9 mcg/mL [5]. Inherent to its underlying electrophysiological properties, progressive cardiac conduction abnormalities with increasing PR interval and QRS complex durations, AV block, and ventricular dysrhythmias refractory to treatment become manifest [3—5,10—12]. Relevant to our case, a Brugada-like pattern on ECG has also been reported [16,17].

Although infrequent, fatalities have been reported [18—22], typically assumed, or attributed to be from cardiac causes. Postmortem blood levels from two men (46 and 43 years old) of 4.18 and 9.12 mcg/mL, respectively, have been reported after ingestion of unknown amounts of propafenone [18]. Another postmortem blood level from a 20-year-old female who had died ~6 hours after intentional ingestion of ~6 g of propafenone was found to be 12 mcg/mL [19]. A serum level of 2.5 mcg/mL ~10-hours postintentional ingestion of 1.5—3 g of propafenone was found in a 14-year-old female who had also coingested 1 g of metoprolol (serum level 2.6 mcg/mL) and later died [20]. A 68-year-old woman with history of paroxysmal atrial fibrillation on antiarrhythmic therapy died ~5 hours after presentation following failed resuscitation that included intracardiac pacing. Postmortem blood level of propafenone was found to be 5.27 mcg/mL [21]. Finally, although initially surviving her cardiac arrest, life-sustaining therapies were later discontinued for a 21-year-old female who had intentionally ingested ~4.5 g of propafenone; the serum drug level was not reported [22].

Cases of survival following cardiogenic shock or cardiac arrest have also been reported [17,23—26]. A 36-year-old female who had intentionally ingested ~4.9 g of propafenone was found to have a plasma level of 2.13 mcg/mL ~10-hours postingestion following prolonged cardiac arrest [23]. A 47-year-old male with a history of paroxysmal atrial fibrillation on antiarrhythmic therapy was found to have serum propafenone level 1.26 mcg/mL ~9—10 hours following cardiac arrest [24]. A 68-year-old male, also with a history of paroxysmal atrial fibrillation on antiarrhythmic therapy and presenting with cardiogenic shock, was found to have serum of propafenone level 2.8 mcg/mL [25]. Two women (44 and 56 year-old) developed cardiogenic shock following intentional ingestion of 4.5 and 3 g of propafenone, respectively; but serum drug levels were not reported [17]. Finally, a 15-year-old female survived cardiac arrest after intentional ingestion of 6 g of propafenone, but serum drug level was not reported [26]. In our patient, the amount of propafenone ingested (4.5—5.6 g) and the resultant ~15-hour postingestion serum drug level (2.2 mcg/mL) are comparable to the previously reported fatal and near-fatal cases.

Management of propafenone toxicity is primarily directed at providing supportive care to maintain adequate cerebro-cardiopulmonary function. Administration of sodium bicarbonate and serum alkalization may help stabilize early ventricular dysrhythmias [27], as was the case for our patient. Intravenous lipid emulsion therapy has also been reported to provide electrophysiological benefit [22,28], which our patient also received. Supported by experimental models of propafenone toxicity in rats [29,30], Bayram et al. recently reported the successful reversal of cardiac arrest that had been refractory to sodium bicarbonate with administration of insulin/glucose [26]. However, Chen and Yang did not observe benefit with insulin/glucose but rather attributed successful treatment of propafenone-induced cardiac arrest with calcium gluconate administration after failed improvement with sodium bicarbonate, insulin/glucose, temporary pacing, and even TPE [31]. Our patient did not receive insulin/glucose but coincidentally did receive calcium gluconate as part of her TPE procedure.

The American Society for Apheresis (ASFA) Guidelines designates the use of TPE for drug overdose/poisoning as a Category III indication (i.e., "Optimum role of apheresis therapy is not established. Decision-making should be individualized") [32]. The Guidelines state that "TPE may be used for the removal of drugs with a low volume of distribution

(<0.2 L/kg) and/or high plasma−protein binding (>80%)" [32], the latter of which is applicable to propafenone. In this regard, we have indeed found a 68% decrease in serum propafenone level following a 1.6 × plasma volume TPE based on the pre- and post-TPE drug levels collected 3.5 hours apart. Of importance, this reduction was likely contributed, in part, by ongoing hepatic drug metabolism. Using a mean elimination half-life of 6 hours [10], an elimination constant of 0.1155 can be determined, which in combination with a blood-sampling interval of 3.5 hours yields a calculated serum propafenone concentration of 1.5 mcg/mL had TPE not been performed. In this scenario, it can be estimated that TPE accounted for reduction of the remaining ∼0.8 mcg/mL, or ∼36% of drug removal. However, these figures might underestimate the contribution of TPE to drug elimination as ongoing absorption from the ER capsules may have continued during TPE.

We are aware of only two other case reports that have mentioned the use of TPE as adjunctive treatment for propafenone toxicity. Unal et al. reported successful management utilizing TPE to treat a 16-year-old female who had intentionally coingested 4.5 g of propafenone along with 5 mg of digoxin and 160 mg of warfarin [33]. However, serum propafenone levels were not determined and estimated extent of removal by TPE was not reported [33]. The previously mentioned report by Chen and Yang attributed successful treatment of cardiac arrest with calcium gluconate after failed improvement with TPE, the specific details of which were not disclosed [31]. Given that their patient was found to be significantly hypocalcemic during the pulseless electrical activity arrest following TPE, we speculate whether ionized calcium (iCa) levels had not been monitored and replaced accordingly during TPE. The potential to develop hypocalcemia during TPE is a well-known complication due to binding of calcium to albumin when 5% albumin is utilized as replacement fluid or to citrate when fresh frozen plasma is utilized as replacement fluid, or when citrate is used for regional anticoagulation [34]. At our institution, we standardly provide a calcium gluconate (or chloride) infusion with frequent monitoring of iCa levels during TPE to avoid iatrogenic hypocalcemia, as was the case for our patient.

In summary, we report for the first time that therapeutic plasma exchange can enhance the elimination of propafenone by removing an estimated 36%−68% of the drug (depending on the contribution from simultaneous hepatic metabolism) by performing a single, 1.6 × plasma volume exchange.

References

[1] Burgess ED, Duff HJ. Hemodialysis removal of propafenone. Pharmacotherapy 1989;9:331−3.

[2] Seto W, Trope AE, Gow RM. Propafenone disposition during continuous venovenous hemofiltration. Ann Pharmacother 1999;33:957−9.

[3] Ledda F, Mantell L, Manzinl S, Amerini S, Mugelli A. Electrophysiological and antiarrhythmic properties of propafenone in isolated cardiac preparations. J Cardiovasc Pharmacol 1981;3:1162−73.

[4] Kohlhardt M, Kohlhardt M. Basic electrophysiological actions of propafenone in heart muscle. In: Schlepper M, Olsson B, editors. *Cardiac arrhythmias: proceedings of the first international rythmonorm congress*. New York: Springer-Verlag; 1983. p. 91−101.

[5] Funck-Brentano C, Kroemer HK, Lee JT, Roder DM. Propafenone. N Engl J Med 1990;322:518−25.

[6] Product information: RYTHMOL SR (propafenone hydrochloride extended-release capsules). GlaxoSmithKline, Research Triangle Park, NC, 2018. [cited 2019 Aug 29]. Available from: <https://www.gsksource.com/pharma/content/dam/GlaxoSmithKline/US/en/Prescribing_Information/Rythmol_SR/pdf/RYTHMOL-SR-PI-PIL.PDF>.

[7] McLeod AA, Stiles GL, Shand DG. Demonstration of beta adrenoceptor blockade by propafenone hydrochloride: clinical pharmacologic, radioligand binding and adenylate cyclase activation studies. J Pharmacol Exp Ther 1984;228:461−6.

[8] Santinelli V, Arnese M, Oppo I, Matarazzi C, Maione S, Palma M, et al. Effects of flecainide and propafenone on systolic performance in subjects with normal cardiac function. Chest 1993;103:1068−73.

[9] Schamroth L, Myburgh DP, Schamroth CL, Scholtz ME, Pincus DR, Kawalsky DL. Oral propafenone in the suppression of chronic stable ventricular arrhythmias. Chest 1985;87:448−51.

[10] Connolly SJ, Kates RE, Lebsack CS, Harrison DC, Winkle RA. Clinical pharmacology of propafenone. Circulation 1983;68:589−96.

[11] Siddoway LA, Roden DM, Woosley RL. Clinical pharmacology of propafenone: pharmacokinetics, metabolism and concentration-response relations. Am J Cardiol 1984;54:9D−12D.

[12] Hollmann M, Brode E, Hotz D, Kaumeier S, Kehrhahn OH. Investigations on the pharmacokinetics of propafenone in man. Arzneimittelforschung 1983;33:763−70.

[13] Morike K, Magadum S, Mettang T, Griese EU, Machleidt C, Kuhlmann U. Propafenone in a usual dose produces severe side-effects: the impact of genetically determined metabolic status on drug therapy. J Intern Med 1995;238:469−72.

[14] Wagner F, Jahnchen E, Trenk M, Eichelbaum M, Harnasch P, Hauf G, et al. Severe complications of antianginal drug therapy in a patient identified as a poor metabolizer of metoprolol, propafenone, diltiazem, and sparteine. Klin Wochenschr 1987;65:1164−8.

[15] Gillis AM, Yee YG, Kates TE. Binding of antiarrhythmic drug to purified human alpha1-acid glycoprotein. Biochem Pharmacol 1985;34:4279−82.

[16] Arı ME, Ekici F. Brugada-phenocopy induced by propafenone overdose and successful treatment: a case report. Balk Med J 2017;34:473−5.

[17] Gil J, Marmelo B, Abreu L, Antunes H, Santos LFD, Cabral JC. Propafenone overdose: from cardiogenic shock to Brugada pattern. Arq Bras Cardiol 2018;110:292–4.
[18] Clarot F, Goullé JP, Horst M, Vaz E, Lacroix C, Proust B. Fatal propafenone overdoses: case reports and a review of the literature. J Anal Toxicol 2003;27:595–9.
[19] Maxeiner H, Klug E. Lethal suicidal intoxication with propafenone, after a history of self-inflicted injuries. Forensic Sci Int 1997;89:27–32.
[20] Kacirova I, Grundmann M, Kolek M, Vyskocilova-Hrudikova E, Urinovska R, Handlos P. Lethal suicide attempt with a mixed-drug intoxication of metoprolol and propafenone — a first pediatric case report. Forensic Sci Int 2017;278:e34–40.
[21] Zeljković I, Bulj N, Kolačević M, Čabrilo V, Brkljačić DD, Manola Š. Failure of intracardiac pacing after fatal propafenone overdose: a case report. J Emerg Med 2018;54:e65–8.
[22] ten? Tusscher BL, Beishuizen A, Girbes AR, Swart EL, van Leeuwen RW. Intravenous fat emulsion therapy for intentional propafenone intoxication. Clin Toxicol 2011;49:701.
[23] Ling B, Geng P, Tan D, Walline J. Full recovery after prolonged resuscitation from cardiac arrest due to propafenone intoxication: a case report. Medicine 2018;97:e0285.
[24] Ovaska H, Ludman A, Spencer EP, Wood DM, Jones AL, Dargan PI. Propafenone poisoning—a case report with plasma propafenone concentrations. J Med Toxicol 2010;6:37–40.
[25] Alsaad AA, Ortiz Gonzalez Y, Austin CO, Kusumoto F. Revisiting propafenone toxicity. BMJ Case Rep 2017. Available from: https://doi.org/10.1136/bcr-2017-219270.
[26] Bayram B, Dedeoglu E, Hocaoglu N, Gazi E. Propafenone-induced cardiac arrest: full recovery with insulin, is it possible? Am J Emerg Med 2013;31:457.e5–7.
[27] Brubacher J. Bicarbonate therapy for unstable propafenone-induced wide complex tachycardia. CJEM 2004;6:349–56.
[28] Jacob J, Heard K. Second case of the use of intravenous fat emulsion therapy for propafenone toxicity. Clin Toxicol 2011;49:946–7.
[29] Yi HY, Lee JY, Lee SY, Hong SY, Yang YM, Park GN. Cardioprotective effect of glucose-insulin on acute propafenone toxicity in rat. Am J Emerg Med 2012;30:680–9.
[30] Yi HY, Lee JY, Lee WS, Sung WY, Seo SW. Comparison of the therapeutic effect between sodium bicarbonate and insulin on acute propafenone toxicity. Am J Emerg Med 2014;32:1200–7.
[31] Chen X, Yang Z. Successful treatment of propafenone-induced cardiac arrest by calcium gluconate. Am J Emerg Med 2017;35:1209.e1–2.
[32] Padmanabhan A, Connelly-Smith L, Aqui N, Balogun RA, Klingel R, Meyer E, et al. Guidelines on the use of therapeutic apheresis in clinical practice — evidence-based approach from the Writing Committee of the American Society for Apheresis: The Eighth Special Issue. J Clin Apher 2019;34:171–354.
[33] Unal S, Bayrakci B, Yasar U, Karagoz T. Successful treatment of propafenone, digoxin and warfarin overdosage with plasma exchange therapy and rifampicin. Clin Drug Investig 2007;27:505–8.
[34] Weinstein R. Hypocalcemic toxicity and atypical reactions in therapeutic plasma exchange. J Clin Apher 2001;16:210–11.

Chapter 13

CNS depressants: benzodiazepines and barbiturates

Christine L.H. Snozek
Department of Laboratory Medicine and Pathology, Mayo Clinic in Arizona, Scottsdale, AZ, United States

Introduction

Two of the major drug classes that can inhibit central nervous system (CNS) activity are benzodiazepines and barbiturates. Although benzodiazepines have largely replaced the older barbiturates in clinical and recreational use, both drug classes share similarities and are of toxicological relevance. These drugs belong to a larger category of sedative/hypnotics, and induce a dose-dependent progression of effects from calming and drowsiness at lower doses, to unconsciousness, coma, and fatal respiratory depression at high doses [1]. Clinically, they are largely used to treat insomnia and seizures and to induce surgical anesthesia; benzodiazepines are also widely prescribed to relieve anxiety (anxiolytics).

Barbiturates came into medical use with the introduction of barbital and phenobarbital in the early 1900s, which then sparked development of several related drugs that were used commonly until the 1960–70s. Barbiturates are categorized according to the duration of their effect, which largely determines the clinical use of drugs within the class. Ultrashort-acting barbiturates such as thiopental produce immediate effects with very short duration and were used to induce or maintain surgical anesthesia. Short-acting barbiturates, such as pentobarbital, and intermediate-acting barbiturates such as butalbital act within 15–60 minutes and last several hours. They can be used for insomnia, preoperative sedation, and emergency management of seizures. Long-acting barbiturates include phenobarbital and have a duration of effect of a day or longer [1,2]. Few barbiturates still in somewhat common use in the United States include butalbital for treatment of migraine headaches and phenobarbital for management of seizure disorders. Anesthetic doses of barbiturates including pentobarbital are also used to relieve intracranial pressure from cerebral edema. Short- and intermediate-acting compounds have the highest abuse potential [3].

The first benzodiazepine marketed for clinical use was chlordiazepoxide, which received FDA clearance in 1960, and was quickly followed by diazepam and others. Benzodiazepines are also classified according to their duration of action, although some do not fit this scheme well due to active metabolites with very different half-lives than the parent drugs. Short-acting benzodiazepines include midazolam and triazolam, with half-lives of only a few hours. There are many intermediate-acting benzodiazepines, including alprazolam, lorazepam, and temazepam, with half-lives between 6 and 24 hours. Long-acting benzodiazepines such as diazepam have half-lives >24 hours [2]. Benzodiazepines are used extensively to treat anxiety and insomnia. Newer anticonvulsants of other drug classes have become the preferred agents for managing epilepsy, but some benzodiazepines are still used for emergency (lorazepam and diazepam) and long-term [clonazepam and clobazam] treatment of seizures [4]. Specific drugs within the class also have utility in managing withdrawal from alcohol [5] and other sedative/hypnotics [6], and as a preanesthetic (midazolam) [1]. Relative abuse risk for specific barbiturates is reflected in their classification as Schedule II (e.g., pentobarbital), Schedule III (e.g., butalbital), and Schedule IV (e.g., phenobarbital) drugs by the Drug Enforcement Agency [7].

Benzodiazepines and other drugs have largely supplanted barbiturates in current medical use. Much of the reason for this has to do with the lower risk of toxicity and dependence, as well as more selective regulation of CNS function. Most benzodiazepines are listed as Schedule IV drugs [7]. However, benzodiazepines do have abuse and addiction potential. The American Association of Poison Control Centers reported that fewer than 1700 cases logged in the National Poison Data System during 2016 were attributed to a single-drug exposure to a barbiturate.

In contrast, >74,000 single-drug exposures in the database were attributed to a benzodiazepine [8]. However, although two benzodiazepines (alprazolam and diazepam) were among the top 10 drugs associated with overdose deaths from 2010 to 2014, >95% of deaths related to those two benzodiazepines involved at least one other drug [9]. Benzodiazepines alone rarely cause fatal overdose but can potentiate the life-threatening effects of other drugs, for example, opioid-induced respiratory depression. For this reason, in 2016 the Centers for Disease Control guidelines were updated to recommend against prescribing benzodiazepines and opioids together [10].

Recreational drug users will often take benzodiazepines to potentiate the high and dampen withdrawal from opioids, or to reduce risk of seizures when using cocaine [3]. Several newer benzodiazepines have also emerged on the designer recreational drug scene. Some, such as etizolam, are available clinically in other countries, while others are novel drugs or active metabolites of older drugs [11,12]. Novel benzodiazepines are a common cause of emergency room visits and arrests for driving under the influence of drugs (DUID); like traditional benzodiazepines, the risks of morbidity and mortality are greatly increased when coingested with alcohol or other depressants [13].

In addition to recreational use, benzodiazepines and to a lesser degree barbiturates are of forensic interest for their potential use in facilitating crimes. Although the benzodiazepine flunitrazepam (Rohypnol, "roofies") is often mentioned in lay media as the stereotypical date-rape or drug-facilitated crime (DFC) drug, in reality the sedative/hypnotic and memory-impairing properties of most drugs in both classes provide the potential for their use in DFC [14]. A 2017 review noted that benzodiazepines as a class were the second most common drug (after alcohol) associated with DFC, with 3.5%–48.6% of victims in various studies testing positive for at least one benzodiazepine [15].

Mechanism of action

There is considerable structural diversity amongst both barbiturates and benzodiazepines, but each drug class has a common core structure. The general structure of barbiturates (Fig. 13.1) centers around a six-membered ring, with modifications commonly occurring at specific sites. Thiopental and other thiobarbiturates have a sulfur moiety replacing one or more of the ketone groups; methylated barbiturates such as methohexital and mephobarbital are alkylated at one of the ring nitrogens. The R groups on C_5 can include small alkyl groups, longer aliphatic chains, and aromatic rings [1,2].

The benzodiazepine core structure is more complex; the class name stems from the combination of a benzene ring with a seven-member diazepine ring. A phenyl ring attached at the C_5 position of the diazepine ring appears to be required for pharmacological activity [2]. Fig. 13.2 shows the general structure of benzodiazepines, with common substitutions included. Most benzodiazepines include a halogen, typically Cl or F, which can facilitate detection by some methodologies. The R_1 and R_2 groups can be simple hydrogen or ketone moieties or can be more complex including a variety of fused ring structures [1].

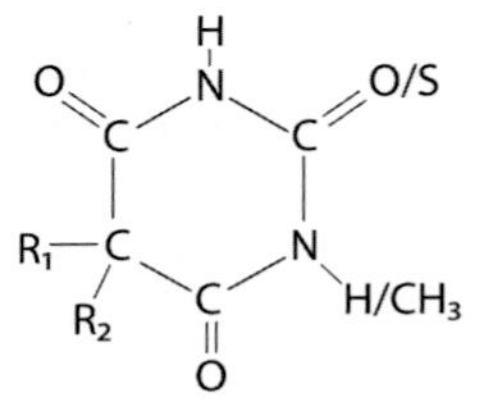

FIGURE 13.1 General structure of the barbiturate drug class.

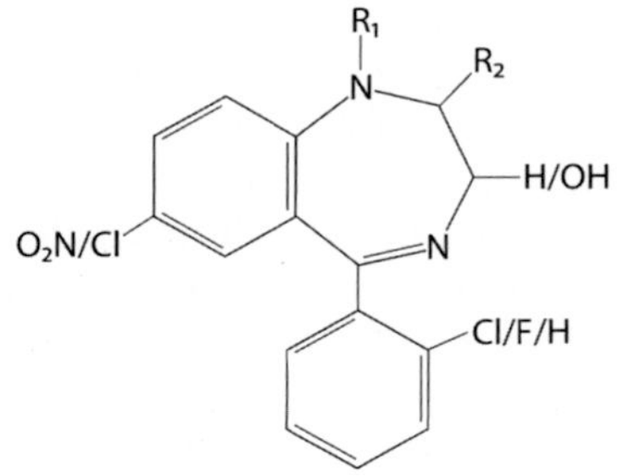

FIGURE 13.2 General structure of the benzodiazepine drug class.

Barbiturates and benzodiazepines both act by binding to gamma-aminobutyric acid (GABA) subtype A receptors, although each drug class has its own high-affinity GABA$_A$ binding site. The structurally unrelated "Z-drugs" including zolpidem and zopiclone bind at yet another site on GABA$_A$ but are often discussed together with benzodiazepines as "benzodiazepine-related drugs" due to similar hypnotic properties. GABA and its receptors comprise the primary inhibitory neurotransmission system in the CNS, and GABA$_A$ receptors in particular are thought to mediate the anticonvulsant and sedative properties of these drugs. GABA$_A$ receptors are heteropentameric chloride channels; opening of the receptor allows chloride to enter neurons resulting in a change in cell membrane potential that decreases neuronal activity. At therapeutic concentrations, barbiturates, and benzodiazepines are positive allosteric regulators of GABA$_A$ receptors, meaning they decrease the concentration of GABA required to open the channel [16]. At higher concentrations, barbiturates can act as agonists, leading to channel opening in the absence of GABA [1].

While classical benzodiazepines such as diazepam exhibit a broad range of pharmacological effects, many drugs within the class are more specific in the responses they elicit. Interaction with unique GABA$_A$ isoforms is thought to be responsible for this phenomenon: some studies suggest that the sedative effects are associated with drug binding to receptors containing an α1 subunit, whereas anxiolytic responses are attributed to α2 and α3 subunits [17,18]. Other studies attribute benzodiazepine selectivity to GABA$_A$ β subunit composition [16]. The anxiolytic properties of benzodiazepines are also thought to arise in part from effects on other systems, including response to norepinephrine neurotransmission [19]. In contrast to benzodiazepines, most barbiturates are fairly unselective and depress function throughout the CNS; one exception is phenobarbital which acts primarily as an anticonvulsant with less prominent sedative effects [1].

Chronic administration of benzodiazepines and barbiturates results in development of tolerance to some properties of the drugs. Tolerance to the mood-altering and sedative/hypnotic effects of barbiturates occurs to a greater degree than to the anticonvulsant or toxic effects, which contributes to the narrow therapeutic index of this drug class [1]. For benzodiazepines, tolerance develops rapidly to the sedative/hypnotic and anticonvulsant properties, although very little or no tolerance to the anxiolytic and memory-impairing properties occurs [17,18]. Prolonged use of barbiturates and benzodiazepines can also result in drug dependence, with precipitation of withdrawal symptoms such as anxiety, agitation, nausea, insomnia, tremors, and seizures upon cessation of use [2].

The pattern of undesirable side effects is similar between these drug classes and includes lingering drowsiness (e.g., residual daytime sleepiness or "hangover"), impaired performance on driving or similar tasks, vertigo, nausea, and vomiting. Interestingly, barbiturates and benzodiazepines both cause paradoxical effects in a subset of patients. Disinhibited behavior, including restlessness, excitement, delirium, and hostility, can occur, as can increased sensation of pain, tachycardia, and nightmares [1]. Benzodiazepines, in particular, have been associated with aggression and violence in a small percentage of patients; higher doses and concomitant use of alcohol appear to increase risk for this phenomenon [20]. Diazepam has the best studied link with aggressive behavior, although alprazolam and other benzodiazepines have been implicated in this response as well [21].

Pharmacokinetics and toxicokinetics

Barbiturates and benzodiazepines are well-absorbed when administered orally. They can also be administered intravenously, for example, to induce anesthesia or as emergency treatment for seizures. Other routes of administration, including intramuscular and rectal formulations, are less commonly used. Both drug classes are lipophilic and distribute throughout the body; they cross the blood—brain barrier readily and can also cross the placenta and be secreted in breast milk [1,2]. Barbiturates are teratogenic in animal studies, whereas benzodiazepines are not thought to be [2,22]. However, benzodiazepine withdrawal can be observed in neonates whose mothers take these drugs during pregnancy, and breastfeeding infants should be monitored for sedation and lethargy when benzodiazepines are used in lactating mothers [2].

Hepatic metabolism of barbiturates includes extensive varieties of oxidation, conjugation, hydroxylation, and dealkylation reactions. Except for phenobarbital and aprobarbital, only a small percentage of unchanged parent drug is found in urine for most barbiturates. Metabolism is slower in elderly patients and infants, and half-lives can be prolonged in pregnancy and chronic liver disease [1]. In addition to their own extensive metabolism, many barbiturates affect the metabolism of other compounds by inducing hepatic enzymes, including several cytochrome P450 (CYP) isoforms [23].

Unlike barbiturates, benzodiazepines do not induce hepatic enzyme activity; however, because CYP3A4 is involved in the metabolism of many of these drugs, inhibitors or inducers of that enzyme will affect benzodiazepine metabolism [3,23]. Benzodiazepines are extensively metabolized via both Phase I (oxidation) and Phase II (conjugation) reactions.

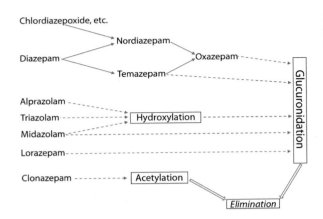

FIGURE 13.3 Metabolism of common benzodiazepines. Solid arrows denote that the metabolite has pharmacological activity; dashed arrows denote that the metabolite is inactive.

Many of the Phase I reactions form active metabolites; most Phase I metabolites also undergo Phase II conjugation (Fig. 13.3). Glucuronide conjugates are common metabolites for this drug class and are often the predominant form of benzodiazepines in urine. Among commonly prescribed benzodiazepines [24], lorazepam and clonazepam are somewhat unusual in terms of metabolism: lorazepam primarily undergoes glucuronide conjugation without prior oxidation, while clonazepam is largely eliminated as 7-aminoclonazepam without Phase II metabolism.

Several benzodiazepines related to diazepam have metabolites in common: N-demethylation of diazepam forms nordiazepam, whereas hydroxylation creates temazepam. Oxazepam can then be formed by either hydroxylation of nordiazepam or demethylation of temazepam. All of these compounds have pharmacological activity and are available commercially. Other benzodiazepines, including chlordiazepoxide and prazepam, can also be metabolized to nordiazepam and subsequently oxazepam. The long half-lives of these active metabolites can greatly increase the duration of action of diazepam and other benzodiazepines well beyond what would be predicted based on the pharmacokinetics of the parent drug alone [3].

The geriatric population is of specific concern with respect to benzodiazepines. Age-related changes in pharmacokinetics include altered drug distribution, longer half-lives, and decreased elimination with increasing age. In the elderly the effects of benzodiazepines such as weakness, ataxia, and dizziness exacerbate the already high risk of falls. There is also some evidence, though not definitive, that use of benzodiazepines accelerates cognitive decline and increases the likelihood of dementia in the elderly. Despite these risks, benzodiazepines remain among the most frequently prescribed medications in geriatric patients, with a high proportion of prescriptions for long-term use [25].

Treatment

Toxicity from barbiturate and benzodiazepine presents with features such as respiratory depression, altered mental status, slurred speech, involuntary eye motions (nystagmus), poor gait control (ataxia), lethargy and hypotension [26,27]. In overdose, barbiturates greatly depress the respiratory drive and cause hypotension through various effects on the peripheral nervous system and cardiovascular system. This latter effect can be more severe in patients with heart failure or hypovolemic shock and can result in marked oliguria or anuria. Benzodiazepines do not depress respiration to the same degree that barbiturates do, even in overdose; however, risk of hypoxia is increased if alcohol or other CNS depressants are also ingested, or if the patient has a preexisting condition such as chronic obstructive pulmonary disease [1]. Other symptoms reflective of coingestion of additional drugs may also be present: prescription formulations of barbiturates often include other medications [26], and multidrug use with benzodiazepines is common [3,9].

Management of overdose with either barbiturates or benzodiazepines is largely supportive and focused on preventing hypoxemia (due to respiratory depression) and hypotension. Although activated charcoal does reduce absorption of additional drug, there is debate as to whether this therapy improves clinical outcome, in particular, because of the increased risk of aspiration in patients with altered mental status. Renal elimination of barbiturates can be enhanced by alkalinizing urine; this is recommended for long-acting but not short-acting compounds. Hemodialysis and hemoperfusion are also options to enhance barbiturate elimination [26,27].

Flumenazil is a competitive inhibitor for binding at the benzodiazepine site on $GABA_A$. It can be used as a specific antidote to reverse benzodiazepine-induced sedation, but its use in overdose is controversial. Flumenazil has the potential to precipitate withdrawal and seizures [3,27] and exacerbates the condition for which the benzodiazepine is prescribed. There is no specific antidote for barbiturate toxicity, although administration of intravenous lipid emulsion shows some promise as a "sink" for sequestering these and other lipophilic drugs [26].

Serum concentrations of barbiturates and benzodiazepines do not correlate well to symptoms or outcome in overdose settings [3,26], thus quantitation is rarely necessary in emergency situations. With early intervention and in-hospital care, modern rates of barbiturate mortality from overdose are low [26]. Pulmonary complications or renal failure secondary to hypotension are the most likely causes for death after barbiturate intoxication [1]. Overdose with a benzodiazepine alone has a generally good prognosis; however, use with alcohol or other drugs that depress respiration greatly increases the risk for adverse outcomes and death [9].

Analytical methods and clinical-management implications

Therapeutic drug monitoring of phenobarbital and primidone (which metabolizes to phenobarbital) is common, although the use of these drugs in seizure management is declining [3,24]. Commercial immunoassays for measuring concentrations of these drugs in serum or plasma are readily available on a variety of automated analytical platforms. Few other barbiturates or benzodiazepines are measured in serum or plasma for the purpose of guiding therapeutic dosing, although some studies have suggested there is benefit to monitoring pentobarbital concentrations [28]. Pentobarbital is also sometimes measured in blood during evaluation of comatose patients, since the drug can induce a state that mimics brain death [29,30]. Therapeutic and toxic blood concentrations of select benzodiazepines and barbiturates are shown in Table 13.1 [31]. Relevant blood concentrations of designer benzodiazepines are less well-studied, although a few publications have investigated drug concentrations in relationship to parameters such as impairment in DUID cases [32].

Qualitative detection of benzodiazepines and barbiturates, for example, as part of drug of abuse screening or prescription medication compliance monitoring, is a common testing practice in most clinical and forensic laboratories. Commercial immunoassays, including those designed for automated platforms and point-of-care (e.g., cups or strips) testing, are widely available to screen for both drug classes in a variety of sample matrices. Most barbiturate immunoassays target secobarbital or a mixture of barbiturates [28], while most benzodiazepine immunoassays target various metabolites of diazepam, including nordiazepam, oxazepam, or temazepam [3]. Urine testing cutoffs of 200–300 ng/mL are

TABLE 13.1 Select Therapeutic and Toxic Concentrations for Central Nervous System Depressants [31]

Drug	Therapeutic	Toxic
Benzodiazepines		
Alprazolam	5–50 ng/mL	>100 ng/mL
Chlordiazepoxide	0.4–3 µg/mL	>3.5 µg/mL
Clobazam	30–300 ng/mL	>500 ng/mL
Clonazepam	20–80 ng/mL	>100 ng/mL
Diazepam	0.1–2.5 µg/mL	>3 µg/mL
Etizolam	8–20 ng/mL	>30 ng/mL
Flunitrazepam	5–15 ng/mL	>50 ng/mL
Lorazepam	80–250 ng/mL	>300 ng/mL
Temazepam	20–150 ng/mL	>1 µg/mL
Barbiturates		
Butalbital	1–5 µg/mL	>10 µg/mL
Pentobarbital	1–5 µg/mL	>10 µg/mL
Phenobarbital	10–30 µg/mL	>30 µg/mL
Primidone	4–12 µg/mL	>20 µg/mL
Thiopental	1–5 µg/mL	>7 µg/mL

common, although laboratories and providers should be aware that variable antibody cross-reactivity within each drug class can affect detection of specific drugs [33].

In particular, for benzodiazepines, the number of drugs and structural diversity within the drug class, as well as the wide range of potencies, create significant challenges for laboratories to detect all relevant compounds on screening or confirmation tests. Designer benzodiazepines also present a unique analytical challenge, both because the pattern of use changes rapidly and because the relevant compound for testing (e.g., predominant metabolite) is often unknown for novel agents [11]. Benzodiazepine immunoassays often cross-react to some degree with novel benzodiazepines [34], whereas many confirmation assays target specific compounds and will not detect other drugs.

In addition, many benzodiazepine immunoassays have poor or no cross-reactivity with glucuronide metabolites, which are the predominant forms of many of these drugs in urine. Pretreatment with a hydrolyzing agent such as glucuronidase, though time-consuming, can improve detection of some benzodiazepines [33]. However, sensitivity of some screening assays can be insufficient for monitoring high-potency benzodiazepines in pain management, even with this pretreatment [35]. Conversely, false-positive results due to specific medications have been described with screening assays for both barbiturates and benzodiazepines [33,36], although in general tests targeting these drug classes have fewer false-positives than immunoassays for other drugs of abuse.

Although immunoassays remain the workhorse of drug screening due to their availability and ease of use, many laboratories have shifted assays for applications such as drug of abuse and pain-management screening to mass spectrometry (MS) platforms. Current publications describing screening assays that include barbiturates and benzodiazepines tend to focus on multidrug panels, novel technologies, alternate matrices, or some combination of those factors (e.g., Ref. [37,38]). Improvements to traditional MS instruments and software have enabled validation of methods capable of simultaneously detecting dozens to hundreds of drugs in a single analytical run (e.g., Ref. [39]). Newer high-resolution (HRMS) techniques permit laboratories to collect both targeted and untargeted data, which can be followed by library matching to detect known compounds or prediction of novel compounds based on accurate mass [40,41]. HRMS has been particularly helpful in analysis of designer drugs including novel benzodiazepines. Untargeted data from clinical or forensic specimens can be queried as compound libraries expand, allowing retrospective identification of previously unrecognized drugs [42]. HRMS data can also be used to identify metabolites in biological specimens, to guide development of reference materials, and improve pharmacokinetic characterization of novel drugs [43].

Quantitative testing, whether for confirmation analysis or sensitive screening applications, is predominantly performed using MS platforms. Although the presence of specific atoms in most benzodiazepines facilitates the use of alternate detectors (electron capture and nitrogen–phosphorus) [2], these have largely been supplanted by MS in recent years. The confirmation of barbiturates by gas chromatography (GC–MS) generally requires derivatization for adequate sensitivity. Derivatization is generally time-consuming, but certain more rapid strategies have other limitations such as the inability to distinguish phenobarbital from mephobarbital when on-column methylation is performed [3]. Many benzodiazepines can be analyzed by GC–MS without derivatization; however, some of the more polar drugs within the class show better analytical performance on GC after derivatization. In addition, the thermal instability of some benzodiazepines (e.g., chlordiazepoxide) presents challenges for testing by GC techniques [2].

Liquid chromatography (LC) coupled to tandem mass spectrometry (MS/MS) or HRMS detectors offers significant advantages over GC for definitive detection of both barbiturates and benzodiazepines. Sample preparation on LC-based assays is often simpler than for GC; rapid extraction methods, including liquid–liquid, solid-phase, and online turbulent flow, have been described, as well as direct injection of urine samples with no preanalytical preparation [44–46]. The sensitivity on LC–MS platforms is often better than on GC–MS, without requiring derivatization [3]. HRMS, in particular, is useful in the setting of investigating novel designer drugs and/or metabolites, even in the absence of a reference standard prior to analysis [41,47]. Although matrix effects such as ion suppression can be an issue with LC methods, ready availability of deuterated internal standards for many drugs within each class helps alleviate this concern [48].

Even though only a small fraction of most barbiturates is eliminated unchanged, because of the high doses used, urine screens and confirmatory assays can successfully target parent drugs rather than metabolites. Confirmation testing for benzodiazepines in urine should incorporate metabolites; glucuronide hydrolysis can facilitate detection of many benzodiazepines (e.g., lorazepam, oxazepam), while others including alprazolam and clonazepam are found predominantly in urine as the Phase I metabolite (α-hydroxyalprazolam and 7-aminoclonazepam, respectively). Confirmation testing in blood generally targets parent drugs and active metabolites. In urine, most barbiturates can be detected 1–4 days after use, although long-acting compounds can be detected for weeks after last use. The detection window for benzodiazepines depends greatly on the specimen matrix, dose, frequency of use, and drug half-life. In urine the detection window ranges from a few days to several weeks. Detection windows in blood are shorter, generally <2–3 days depending on the half-life of each drug [3].

Oral fluid has emerged as a matrix of particular interest for benzodiazepine screening in settings such as pain management [49] and DUID [50]. Oral fluid has some advantages over blood or urine, notably that it can be collected noninvasively, the risk of adulteration is low, and collections can be witnessed without privacy concerns. Point-of-care screening devices are available for benzodiazepines and other potentially abused drugs and have shown utility in studies of traffic stops and similar applications [50–52]. Difficulties associated with oral fluid testing include small sample volumes, contamination with drugs taken orally or sublingually, and the need to detect relatively low drug concentrations urine [49]. Additional concerns for interpretation are that point-of-care devices cannot distinguish the use of licit versus illicit benzodiazepines, or whether the detected concentrations are therapeutic or likely to cause impairment. The detection window for most drugs in oral fluid is thought to be more similar to the shorter times associated with blood testing, although some studies have detected benzodiazepines in oral fluid for several days after ingestion of high doses [53]. Notably, some benzodiazepines including lorazepam and clonazepam (as 7-aminoclonazepam metabolite) are poorly detected in oral fluid compared to urine or blood, likely due to the chemical properties of these drugs [49].

Hair testing has been used to associate benzodiazepines and barbiturates with DFC after the detection window for matrices such as blood and urine has passed. An 8-year review of hair testing from suspected DFC victims detected benzodiazepines, Z-drugs, or phenobarbital in nearly 20% of cases, with sample collection occurring at a median of >1 month after the alleged crime [54]. Hair is also a useful matrix for other applications including postmortem testing [55] and evaluation of prenatal exposure [56]. However, studies have demonstrated that some benzodiazepines, for example, lorazepam, are not readily detected in hair, thus other matrices would be required for evaluating these drugs [57].

Meconium and, increasingly, cord tissue are also common sample types for evaluating drug exposure in utero. Both matrices offer long-detection windows, reflecting exposures throughout the third trimester and, in the case of meconium, extending into late second trimester. Benzodiazepines are frequently administered during labor, thus caution is required to ensure therapeutic use during delivery is not incorrectly interpreted as used during pregnancy [58]. Given the declining prevalence of barbiturate misuse and the limitations of these matrices (e.g., difficult sample preparation and small sample volumes), some laboratories are moving toward targeted confirmation of specific barbiturates such as butalbital and phenobarbital rather than including multiple drugs from the class [59].

Benzodiazepines are common contributors to both accidental and intentional drug-related deaths [60]. These drugs are readily detected in blood, but have also been measured in a wide variety of postmortem specimens including vitreous fluid [61], bone marrow [62], and intraosseous fluid [63] [64]. There are few recent publications addressing detection of barbiturates at autopsy, although occasional case reports highlight newer techniques for measuring these drugs in blood [65] or compare postmortem concentrations in other sample types [66]. Barbiturates are still in common use in veterinary medicine, for example, for euthanasia, and veterinary workers have used these agents to commit suicide [66].

References

[1] Goodman LS, et al. 11th ed. Goodman & Gilman's the pharmacological basis of therapeutics, xxiii. New York: McGraw-Hill; 2006. 2021 p.
[2] Levine B, American Association for Clinical Chemistry. 2nd ed. Principles of forensic toxicology, ix. Washington, DC: American Association for Clinical Chemistry; 2006. 418 p.
[3] Rifai NH, Rita A, Carl T W. Tietz textbook of clinical chemistry and molecular diagnostics. 6th ed. St Louis, MO: Elsevier; 2018.
[4] Antiepileptic Drugs. Available from: <https://emedicine.medscape.com/article/1187334-overview> [cited 18.10.18].
[5] Long D, Long B, Koyfman A. The emergency medicine management of severe alcohol withdrawal. Am J Emerg Med 2017;35(7):1005–11.
[6] Santos C, Olmedo RE. Sedative-hypnotic drug withdrawal syndrome: recognition and treatment. Emerg Med Pract 2017;19(3):1–20.
[7] Controlled Substance Schedules. Available from: <https://www.deadiversion.usdoj.gov/schedules> [cited 31.10.18].
[8] Gummin DD, et al. 2016 annual report of the American Association of Poison Control Centers' National Poison Data System (NPDS): 34th Annual Report. Clin Toxicol (Phila) 2017;55(10):1072–252.
[9] Warner M, et al. Drugs most frequently involved in drug overdose deaths: United States, 2010-2014. Natl Vital Stat Rep 2016;65(10):1–15.
[10] Dowell D, Haegerich TM, Chou R. CDC guideline for prescribing opioids for chronic pain—United States, 2016. MMWR Recomm Rep 2016;65(1):1–49.
[11] Kacinko SL, Papsun DM. The evolving landscape of designer drugs. Methods Mol Biol 2019;1872:129–35.
[12] Katselou M, et al. Metabolites replace the parent drug in the drug arena. The cases of fonazepam and nifoxipam. Forensic Toxicol 2017;35 (1):1–10.
[13] Logan BK, et al. Reports of adverse events associated with use of novel psychoactive substances, 2013–2016: a review. J Anal Toxicol 2017;41(7):573–610.
[14] *Recommended Minimum Performance Limits for Common DFC Drugs and Metabolites in Urine Samples.* Available from: <http://www.soft-tox.org/files/MinPerfLimits_DFC2017.pdf>. [cited 18.10.18].

[15] Anderson LJ, Flynn A, Pilgrim JL. A global epidemiological perspective on the toxicology of drug-facilitated sexual assault: A systematic review. J Forensic Leg Med 2017;47:46—54.

[16] Sigel E, Ernst M. The benzodiazepine binding sites of GABAA receptors. Trends Pharmacol Sci 2018;39(7):659—71.

[17] Licata SC, Rowlett JK. Abuse and dependence liability of benzodiazepine-type drugs: GABA(A) receptor modulation and beyond. Pharmacol Biochem Behav 2008;90(1):74—89.

[18] Vinkers CH, Olivier B. Mechanisms underlying tolerance after long-term benzodiazepine use: a future for subtype-selective GABA(A) receptor modulators? Adv Pharmacol Sci 2012;2012:416864.

[19] Mula M. Using anxiolytics in epilepsy: neurobiological, neuropharmacological and clinical aspects. Epileptic Disord 2016;18(3):217—27.

[20] Jones KA, et al. Benzodiazepines—their role in aggression and why GPs should prescribe with caution. Aust Fam Physician 2011;40 (11):862—5.

[21] Albrecht B, et al. Benzodiazepine use and aggressive behaviour: a systematic review. Aust N Z J Psychiatry 2014;48(12):1096—114.

[22] Crawford PM. Managing epilepsy in women of childbearing age. Drug Saf 2009;32(4):293—307.

[23] Drug Interactions. Available from: <https://drug-interactions.medicine.iu.edu/main-table.aspx> [cited 24.10.18].

[24] The Top 200 Drugs of 2018. Available from: <http://clincalc.com/DrugStats/Top200Drugs.aspx> [cited 24.10.18].

[25] Picton JD, Marino AB, Nealy KL. Benzodiazepine use and cognitive decline in the elderly. Am J Health Syst Pharm 2018;75(1):e6—e12.

[26] Barbiturate Toxicity. Available from: <https://emedicine.medscape.com/article/813155> [cited 19.10.18].

[27] Benzodiazepine Toxicity. Available from: <https://emedicine.medscape.com/article/813255> [cited 19.10.18].

[28] Humble RM, et al. Therapeutic drug monitoring of pentobarbital: experience at an Academic Medical Center. Ther Drug Monit 2015;37 (6):783—91.

[29] Meinitzer A, et al. Drugs and brain death diagnostics: determination of drugs capable of inducing EEG zero line. Clin Chem Lab Med 2008;46 (12):1732—8.

[30] Wijdicks EF. The diagnosis of brain death. N Engl J Med 2001;344(16):1215—21.

[31] Schulz M, et al. Therapeutic and toxic blood concentrations of nearly 1,000 drugs and other xenobiotics. Crit Care 2012;16(4):R136.

[32] Hoiseth G, Tuv SS, Karinen R. Blood concentrations of new designer benzodiazepines in forensic cases. Forensic Sci Int 2016;268:35—8.

[33] Magnani B, College of American Pathologists. Clinical toxicology testing: a guide for laboratory professionals, xviii. Northfield, IL: CAP Press; 2012. 286 p.

[34] Pettersson Bergstrand M, et al. Detectability of designer benzodiazepines in CEDIA, EMIT II Plus, HEIA, and KIMS II immunochemical screening assays. Drug Test Anal 2017;9(4):640—5.

[35] Darragh A, et al. KIMS, CEDIA, and HS-CEDIA immunoassays are inadequately sensitive for detection of benzodiazepines in urine from patients treated for chronic pain. Pain Physician 2014;17(4):359—66.

[36] Saitman A, Park HD, Fitzgerald RL. False-positive interferences of common urine drug screen immunoassays: a review. J Anal Toxicol 2014;38(7):387—96.

[37] Chepyala D, et al. Sensitive screening of abused drugs in dried blood samples using ultra-high-performance liquid chromatography-ion booster-quadrupole time-of-flight mass spectrometry. J Chromatogr A 2017;1491:57—66.

[38] McKenna J, et al. Toxicological drug screening using paper spray high-resolution tandem mass spectrometry (HR-MS/MS). J Anal Toxicol 2018;42(5):300—10.

[39] Dresen S, et al. Detection and identification of 700 drugs by multi-target screening with a 3200 Q TRAP LC—MS/MS system and library searching. Anal Bioanal Chem 2010;396(7):2425—34.

[40] Colby JM, Lynch KL. Drug screening using liquid chromatography quadrupole time-of-flight (LC-QqTOF) mass spectrometry. Methods Mol Biol 2019;1872:181—90.

[41] Colby JM, Thoren KL, Lynch KL. Suspect screening using LC-QqTOF Is a useful tool for detecting drugs in biological samples. J Anal Toxicol 2018;42(4):207—13.

[42] Partridge E, et al. A case study involving U-47700, Diclazepam and Flubromazepam—application of retrospective analysis of HRMS data. J Anal Toxicol 2018. p. bky039-bky039.

[43] Pettersson Bergstrand M, et al. Human urinary metabolic patterns of the designer benzodiazepines flubromazolam and pyrazolam studied by liquid chromatography-high resolution mass spectrometry. Drug Test Anal 2018;10(3):496—506.

[44] Alagandula R, Zhou X, Guo B. A dilute-and-shoot flow-injection tandem mass spectrometry method for quantification of phenobarbital in urine. Rapid Commun Mass Spectrom 2017;31(1):39—46.

[45] Drummer OH, Di Rago M, Gerostamoulos D. Analysis of benzodiazepines for drug-facilitated assaults and abuse settings (urine). Methods Mol Biol 2019;1872:23—39.

[46] Zhang X, et al. Rapid determination of nine barbiturates in human whole blood by liquid chromatography-tandem mass spectrometry. Drug Test Anal 2017;9(4):588—95.

[47] Mollerup CB, et al. Targeted and non-targeted drug screening in whole blood by UHPLC-TOF-MS with data-independent acquisition. Drug Test Anal 2017;9(7):1052—61.

[48] Perez ER, et al. Comparison of LC-MS-MS and GC-MS analysis of benzodiazepine compounds included in the drug demand reduction urinalysis program. J Anal Toxicol 2016;40(3):201—7.

[49] Petrides AK, et al. Monitoring opioid and benzodiazepine use and abuse: is oral fluid or urine the preferred specimen type? Clin Chim Acta 2018;481:75—82.

[50] Edwards LD, Smith KL, Savage T. Drugged driving in wisconsin: oral fluid versus blood. J Anal Toxicol 2017;41(6):523—9.
[51] Blencowe T, Vimpari K, Lillsunde P. Benzodiazepine whole blood concentrations in cases with positive oral fluid on-site screening test results using the DrugWipe® single for benzodiazepines. J Anal Toxicol 2011;35(6):349—56.
[52] Vindenes V, et al. Detection of drugs of abuse in simultaneously collected oral fluid, urine and blood from Norwegian drug drivers. Forensic Sci Int 2012;219(1—3):165—71.
[53] Nordal K, et al. Detection times of diazepam, clonazepam, and alprazolam in oral fluid collected from patients admitted to detoxification, after high and repeated drug intake. Ther Drug Monit 2015;37(4):451—60.
[54] Wang X, et al. Hair analysis in toxicological investigation of drug-facilitated crimes in Denmark over a 8-year period. Forensic Sci Int 2018;285:e1—e12.
[55] Hoiseth G, et al. Is hair analysis useful in postmortem cases? J Anal Toxicol 2018;42(1):49—54.
[56] Senczuk-Przybylowska M, et al. Diazepam and its metabolites in the mothers' and newborns' hair as a biomarker of prenatal exposure. J Physiol Pharmacol 2013;64(4):499—504.
[57] Kintz P, et al. Windows of detection of lorazepam in urine, oral fluid and hair, with a special focus on drug-facilitated crimes. Forensic Sci Int 2004;145(2—3):131—5.
[58] Colby JM. Comparison of umbilical cord tissue and meconium for the confirmation of in utero drug exposure. Clin Biochem 2017;50 (13—14):784—90.
[59] Scroggin TL, McMillin GA. Quantitation of cocaine and metabolites, phencyclidine, butalbital and phenobarbital in meconium by liquid chromatography-tandem mass spectrometry. J Anal Toxicol 2018;42(3):177—82.
[60] Jones AW, Holmgren A, Ahlner J. Post-mortem concentrations of drugs determined in femoral blood in single-drug fatalities compared with multi-drug poisoning deaths. Forensic Sci Int 2016;267:96—103.
[61] Metushi IG, Fitzgerald RL, McIntyre IM. Assessment and comparison of vitreous humor as an alternative matrix for forensic toxicology screening by GC-MS. J Anal Toxicol 2016;40(4):243—7.
[62] Wietecha-Posluszny R, et al. Human bone marrow as a tissue in post-mortem identification and determination of psychoactive Substances-Screening methodology. J Chromatogr B Anal Technol Biomed Life Sci 2017;1061-1062:459—67.
[63] Rodda LN, et al. Evaluation of intraosseous fluid as an alternative biological specimen in postmortem toxicology. J Anal Toxicol 2018;42 (3):163—9.
[64] Crichton ML, et al. Analysis of phenazepam and 3-hydroxyphenazepam in post-mortem fluids and tissues. Drug Test Anal 2015;7(10):926—36.
[65] de Almeida RM, et al. Hollow-fiber liquid-phase microextraction and gas chromatography-mass spectrometry of barbiturates in whole blood samples. J Sep Sci 2012;35(23):3361—8.
[66] Romain N, et al. Suicide by injection of a veterinarian barbiturate euthanasia agent: report of a case and toxicological analysis. Forensic Sci Int 2003;131(2—3):103—7.

Chapter 13.1

False-negative results in urine benzodiazepine immunoassay screening

Sheng Feng, Paul R. Hess, Ping Wang, Michael C. Malone and Leslie M. Shaw

Department of Pathology and Laboratory Medicine, Perelman School of Medicine at the University of Pennsylvania, Philadelphia, PA, United States

Case description

A 77-year-old woman with a past history of cardiomyopathy, congenital mitral stenosis, coronary artery disease, atrial fibrillation, and uncontrolled type 2 diabetes was prescribed lorazepam at 0.5 mg four times a day 20 years ago to reduce her anxiety when she was diagnosed with atrial fibrillation. She was weaned to 0.5 mg twice a day later on. She denied problems with sleeping. In addition, this patient has no record of substance abuse.

An initial urine drug screen was ordered and repeated monthly in this patient for compliance monitoring. The immunoassay for benzodiazepines has a general cutoff of 200 ng/mL. For four consecutive tests, urine samples from this patient were always negative for benzodiazepines (Table 13.1.1), although this patient insisted she was taking the medication according to her prescription and never missed a dose. To further investigate whether this patient actually adhered to her prescription, a direct benzodiazepine quantitative assay by a liquid chromatography—tandem mass spectrometry (LC—MS/MS) method was ordered. The assay yielded a lorazepam result of 1087 ng/mL in the absence of other benzodiazepines and metabolites, including diazepam, oxazepam, temazepam, nordiazepam, chlordiazepoxide, alprazolam, alpha-hydroxyalprazolam, clonazepam, 7-aminoclonazepam, midazolam, alpha-hydroxymidazolam, and zolpidem.

Due to the negative results for lorazepam on the immunoassay drug screen, the provider decided to monitor this patient routinely using the quantitative benzodiazepine test by LC—MS/MS every 2 months and later changed to every 3 months. Chart review of previous benzodiazepine screens and confirmations revealed that the patient always tested negative on the immunoassay screen but positive on the LC—MS/MS quantitative assay (Table 13.1.1).

Discussion and follow-up

Benzodiazepines are one of the most frequently prescribed medications in the United States. The number of adults filling a benzodiazepine prescription increased from 8.1 to 13.5 million between 1996 and 2013 [1]. According to the data from the National Institute on Drug Abuse, overdose deaths involving benzodiazepines increased from 1135 in 1999 to 8791 in 2015 [2]. Therefore more and more attention is being focused on the issue of benzodiazepines abuse. In addition, benzodiazepines have been implicated in drug-facilitated crimes [3,4]. Currently, many toxicology laboratories provide urine testing for benzodiazepines, although the detailed list of which ones may vary.

Lorazepam, brand name Ativan, is a high-potency benzodiazepine with short-acting characteristics. It is a potent sedative-hypnotic and antianxiety agent that binds the $GABA_A$ receptor with relatively high affinity [5]. Lorazepam dosing largely depends on the indication. For anxiolysis, dosing begins with 2—3 mg/day orally and the maximum daily dose should be less than 10 mg [6]. Lorazepam directly undergoes glucuronide conjugation without prior cytochrome p450 metabolism [7]. This unique feature makes it possible for the drug to be used by patients with hepatic dysfunction. In usual patients, rather than being immediately excreted, lorazepam glucuronide accumulates in plasma, exceeding the concentration of parent drug, with a longer half-life of approximately 16 hours. About 95% of a dose of lorazepam is accounted for in urine and feces over a period of 5 days. An estimated 75% is excreted in urine as lorazepam

TABLE 13.1.1 Immunoassay and Liquid Chromatography−Tandem Mass Spectrometry (LC−MS/MS) Results of the Patient

Months After First Test	Immunoassay Result for Benzodiazepine	Lorazepam Result by LC−MS/MS (ng/mL)
0	Negative	NT
1	Negative	NT
2	Negative	1484
3	Negative	NT
4	Negative	NT
5	Negative	1316
6	Negative	NT
7	Negative	NT
8	Negative	NT
9	Negative	1624
10	Negative	NT
11	Negative	NT
12	Negative	1330
13	Negative	NT
14	Negative	NT

NT: not tested.

glucuronide, and about 14% as minor metabolites, which include ring hydroxylation products and quinazolinone derivatives. Only negligible amounts of unchanged parent compound are detected in urine [8].

A typical drug test procedure includes an initial immunoassay screen and a quantitative confirmation by LC−MS/MS. However, the quantitative confirmation only analyzes those samples that tested positive on the initial immunoassay screen. Thus a potential false-negative result might be generated when the initial immunoassay screen fails to detect the targeted analyte in the urine.

In the current case, with consistently observed negative benzodiazepine immunoassay screens for the patient together with positive LC−MS/MS results, there are two plausible scenarios to explain this discrepancy. First, it is possible that the patient did not take lorazepam, and test results from the immunoassay were true negatives and the LC−MS/MS test captured an interference rather than the target analyte. However, this is highly unlikely given that the LC−MS/MS method has much greater specificity and sensitivity compared to immunoassay. In addition, this geriatric patient had no history of substance abuse or diversion. The second explanation is that the immunoassay failed to detect lorazepam while LC−MS/MS was able to detect it, which seems more reasonable according to the clinical context in this case.

The standard operating procedure accompanying the kit (Emit, Beckman Coulter) provides the following pertinent information: the test uses 200 ng/mL Lormetazepam as the general cutoff and does not incorporate an enzymatic hydrolysis step using β-glucuronidase. To produce a result approximately equivalent to this general cutoff, lorazepam has to be present at a concentration of 600 ng/mL or higher, and the immunoassay has no cross-reactivity towards lorazepam glucuronide. On the other hand, the confirmation quantitative test by LC−MS/MS includes sample preparation with a hydrolysis step using β-glucuronidase. In this case the patient had been prescribed 0.5 mg twice a day, totaling 1 mg per day, which is a relatively low dose. Unpublished data from Ameritox suggests that lorazepam urine testing results by LC−MS/MS for patients taking 1 mg per day lorazepam (939 tests obtained within 1 year) ranged from 40 to 2500 ng/mL. Quantitative analyses by the LC−MS/MS method revealed that lorazepam concentration in the urine of the Patients in this clinical vignette ranged from 553 to 2095 ng/mL, presumably existing as mostly lorazepam glucuronide before hydrolysis. This explains the discrepancy between immunoassay screening and quantitative LC−MS/MS assay.

Previous studies have suggested that hydrolysis using β-glucuronidase is a necessary step for immunoassay screening to adequately detect lorazepam, temazepam, and oxazepam, by reducing false-negatives [9]. However, currently, most toxicology laboratories perform immunoassay screening for benzodiazepines without hydrolysis. Furthermore, the enzymatic hydrolysis procedure might cause reductive transformation of oxazepam into nordiazepam, temazepam into diazepam, and lorazepam into delorazepam [10,11]. Therefore when those analytes are detected in hydrolyzed urines, results should be interpreted cautiously. Mari et al. reported a comparison between immunoassay screening tests and LC−MS/MS for urine detection of benzodiazepines and their metabolites, which pointed out the limitation of using immunoassay-only results for forensic purposes leading to false-positive and/or false-negative results [12].

In summary, for patients who are taking low-dose benzodiazepines, the initial immunoassay screening without enzymatic hydrolysis can yield false-negative results. The scenario presented in this case is of value to clinicians who are monitoring their patients' adherence to benzodiazepines. It is appropriate to directly monitor benzodiazepines using quantitative LC−MS/MS methodology without initial immunoassay screening in this situation, a practice that can reduce the chance of false-negatives.

References

[1] Bachhuber MA, Hennessy S, Cunningham CO, Starrels JL. Am J Public Health 2016;106:686.
[2] Lembke A, Papac J, Humphreys K. N Engl J Med 2018;378:693.
[3] Kintz P, Villain M, Cirimele V, Pepin G, Ludes B. Forensic Sci Int 2004;145:131.
[4] Negrusz A, Gaensslen RE. Anal Bioanal Chem 2003;376:1192.
[5] Matthew E, Andreason P, Pettigrew K, Carson RE, Herscovitch P, Cohen R, et al. Proc Natl Acad Sci U S A 1995;92:2775.
[6] Griffin III CE, Kaye AM, Bueno FR, Kaye AD. Ochsner J 2013;13:214.
[7] Olkkola KT, Ahonen J. Handb Exp Pharmacol 2008;335.
[8] Elliott HW. Br J Anaesth 1976;48:1017.
[9] Meatherall R. J Anal Toxicol 1994;18:385.
[10] Fu S, Lewis J, Wang H, Keegan J, Dawson M. J Anal Toxicol 2010;34:243.
[11] Fu S, Molnar A, Bowron P, Lewis J, Wang H. Anal Bioanal Chem 2011;400:153.
[12] Bertol E, Vaiano F, Borsotti M, Quercioli M, Mari F. J Anal Toxicol 2013;37:659.

Chapter 13.2

A suicide involving zolpidem

Bheemraj Ramoo[1,2], C. Clinton Frazee, III[1,2], Melissa Beals[1,2], Diane C. Peterson[3,4] and Uttam Garg[1,2]

[1]Department of Pathology and Laboratory Medicine, Children's Mercy Hospital, Kansas City, MO, United States, [2]University of Missouri-Kansas City School of Medicine, Kansas City, MO, United States, [3]Office of the Jackson County Medical Examiner, Kansas City, MO, United States, [4]Johnson County Department of Health and Environment/Medical Examiner's Office, Olathe, KS, United States

Case history

The subject was a 93-year-old male with a history of heart disease, hypertension, asthma, and gastroesophageal reflux disease. His prescription medications included digoxin, torsemide, potassium chloride, metoprolol, tamsulosin, warfarin, esomeprazole, atorvastatin, and albuterol. According to his grandson, the subject was unhappy about being unwillingly moved to an assisted living facility a few months prior to the incident. The subject was also told he was no longer allowed to drive. The subject was angry about these changes and threatened several times to take his entire bottle of zolpidem prescription at once. He reportedly had no history of suicidal attempts or ideation thereof. The subject was found unresponsive in his bed. A 90-count bottle of 10 mg zolpidem was located on a table next to subject's bed. The bottle was empty and two tablets found on the floor had markings consistent with the zolpidem prescription.

This case was an external examination only. No autopsy was performed. Femoral blood and vitreous fluid specimens were submitted for toxicology testing. The femoral blood was also used for comprehensive broad spectrum drug-screening that involved enzyme immunoassays (EIAs) for amphetamines, barbiturates, benzodiazepines, cannabinoids, cocaine metabolite, methadone, opiates, phencyclidine, and propoxyphene, and broad spectrum drug-screening for more than 200 drugs by gas chromatography–mass spectrometry (GC–MS). Drug screening by GC–MS involved liquid–liquid alkaline extraction using bicarbonate buffer (pH 11.0) and butyl acetate, and mass spectrometer operation in full scan mode. Presumptive identification of analytes was made by spectral library match and relative retention time comparison with reference standards. Confirmation and quantification were performed by a reference laboratory (NMS Labs, Willow Grove, PA).

The EIA drug screen was negative for drugs of abuse and no volatiles were detected. The GC–MS drug screen analysis indicated the presence of zolpidem, but no other drugs were detected. Postmortem femoral blood was then sent to NMS Labs for zolpidem confirmation and quantification by high-performance liquid chromatography/tandem mass spectrometry. Zolpidem in femoral blood was measured at a level of 4000 ng/mL.

Discussion

Zolpidem is a Schedule IV controlled substance under the Federal Controlled Substance Act. Zolpidem is a short-acting sedative hypnotic that has a rapid onset of action and a short elimination half-life of 1.4–4.5 hours, thus rendering it as an ideal drug for the treatment of insomnia. Zolpidem is sold under the brand names Ambien, Ambien CR, Edluar, Intermezzo and Zolpimist. It has been used as a hypnotic agent in the European countries since 1986 and in the United States since 1993 [1]. The drug is available as normal-release and extended-release tablets, and as a spray solution for intraoral administration. Zolpidem actions occur within 15 minutes following oral administration. Zolpidem is absorbed after oral administration and undergoes first-pass metabolism by the CYP 3A4 enzyme in the liver [2]. It is metabolized to pharmacologically inactive metabolites (Fig. 13.2.1) that are eliminated via renal (56%) and fecal (37%) excretion. Zolpidem is distributed into breast milk and crosses the placenta [3]. It is not readily removed by dialysis.

Toxicity related to the use of sleeping pills has increased in recent years due to their increased availability and abuse. Adverse effects of acute zolpidem toxicity include dizziness, amnesia, headache, nausea, tachycardia, ataxia,

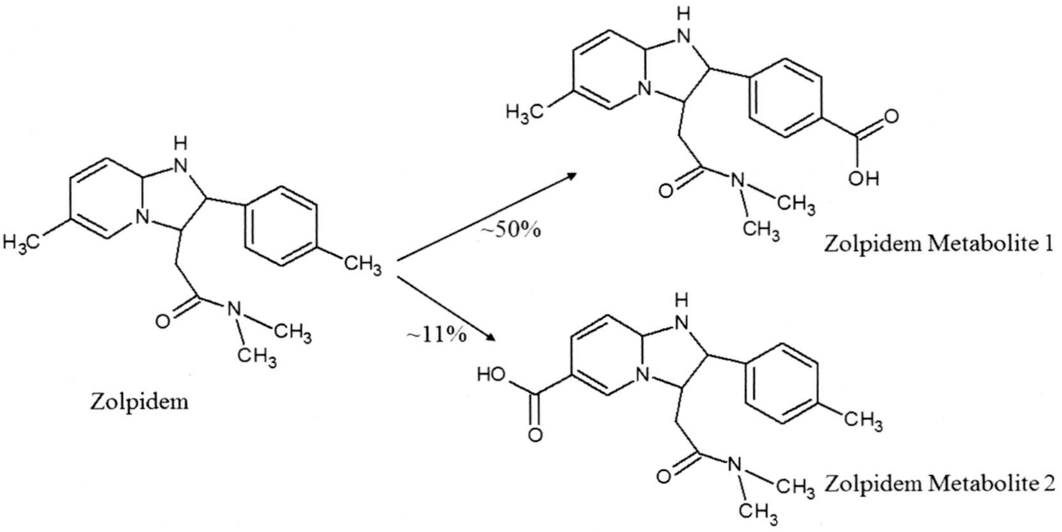

FIGURE 13.2.1 Metabolism of zolpidem.

slurred speech, emesis, hallucinations, and even death [2–4]. Long-term use of zolpidem can lead to a higher morbidity and mortality in insomniac patients. Patients who took zolpidem for more than a year commonly exhibited delirium, hallucinations, and abnormal behavior during sleep [4]. Adverse effects of zolpidem are potentiated by coingestion of ethanol, other sleep medications, and antidepressants. Incidents of driving under the influence and motor vehicle accidents due to zolpidem have become a relatively common occurrence. Ataxia, lethargy, slurred speech, confusion, and somnolence have been reported in subjects driving under the influence of zolpidem. In several large studies on intoxicated drivers, zolpidem average and median levels ranged from approximately 230 to 360 ng/mL [1].

Several fatalities involving zolpidem intoxication have been reported. In the majority of cases, zolpidem was administered orally and detected in combination with other drugs. In one such case, a 36-year-old female with a history of paranoid disorder, depression with panic episodes, and posttraumatic stress disorder was found dead in her secure home [3]. The decedent was prescribed risperidone, sertraline, and zolpidem (Ambien). Toxicological analyses of urine-detected caffeine, risperidone, and zolpidem. The concentrations of zolpidem in subclavian and iliac blood were 4500 and 7700 ng/mL, respectively [3]. The cause of death was determined to be acute zolpidem overdose, and the manner of death was suicide. Another fatality case reported involved the ingestion of zolpidem and acepromazine, a phenothiazine sedative used in veterinary practice. Zolpidem and acepromazine blood concentrations in this case were 3290 and 2400 ng/mL, respectively [5]. This report suggests a potentiation of the toxic effects of zolpidem and phenothiazines. An additional acute overdose is also reported in which a 68-year-old female was found dead at home after ingesting at least 30 tablets of 10 mg zolpidem [6]. Toxicological analyses detected zolpidem, meprobamate, and carisoprodol with blood concentrations of 4100, 19,300, and 2300 ng/mL, respectively [6]. A fatal case involving intravenous injection of zolpidem has been reported [7]. A male in his 20s self-administered zolpidem intravenously and was found dead on the floor of his apartment. A syringe containing dissolved crushed zolpidem tablets was found near the body. A tourniquet band and fresh track marks were on his right forearm. The concentration of zolpidem in femoral blood was 9550 ng/mL. This concentration far exceeds fatal levels of zolpidem reported, but it is consistent with the intravenous administration of zolpidem described in the case history.

Our case finding is unique because death is attributed to ingestion of zolpidem by itself with no mixed drug toxicity. The cause of death in this case was determined to be zolpidem overdose, and the manner of death was suicide.

References

[1] Baselt RC. Zolpidem. In: Baselt RC, editor. Disposition of toxic drugs and chemicals in man. Seal Beach, CA: Biomedical Publications; 2017. p. 2295–8.
[2] Hamad A, Sharma N. Acute zolpidem overdose leading to coma and respiratory failure. Intensive Care Med 2001;27:1239.
[3] Gock SB, Wong SH, Nuwayhid N, Venuti SE, Kelley PD, Teggatz JR, et al. Acute zolpidem overdose—report of two cases. J Anal Toxicol 1999;23:559–62.

[4] Jung M. Zolpidem overdose: a dilemma in mental health. Health Care Manag (Frederick) 2018;37:86—9.
[5] Tracqui A, Kintz P, Mangin P. A fatality involving two unusual compounds—zolpidem and acepromazine. Am J Forensic Med Pathol 1993;14:309—12.
[6] Winek CL, Wahba WW, Janssen JK, Rozin L, Rafizadeh V. Acute overdose of zolpidem. Forensic Sci Int 1996;78:165—8.
[7] Hasegawa K, Wurita A, Nozawa H, Yamagishi I, Minakata K, Watanabe K, et al. Fatal zolpidem poisoning due to its intravenous self-injection: postmortem distribution/redistribution of zolpidem and its predominant metabolite zolpidem phenyl-4-carboxylic acid in body fluids and solid tissues in an autopsy case. Forensic Sci Int 2018;290:111—20.

Chapter 14

Central nervous system stimulants

Erin Kaleta
Department of Laboratory Medicine and Pathology, Mayo Clinic, Scottsdale, AZ, United States

Introduction

Central nervous system (CNS) stimulants are drugs that increase activity of the CNS and compounds that exhibit sympathomimetic effects. Each sympathomimetic drug has a unique and characteristic mechanism of action on neurons, but these drugs either block monoamine reuptake or stimulate monoamine release. Generally, at low doses, these drugs increase focus, reduce appetite, enhance mood, and increase energy, but when taken at higher doses often cause agitation, delirium, can cause seizure, cardiotoxicity, and death. In addition to psychostimulation, CNS stimulants also cause activation of the sympathetic nervous system, and lead to hypertension, tachycardia, and hyperthermia. While there are some medical uses for CNS stimulants, they are limited, and therefore this drug class is most commonly encountered in clinical and forensic laboratories in cases of illicit use. The main compounds in this group include cocaine, amphetamine, methamphetamine, MDMA (3,4-methylenedioxymethamphetamine) (or ecstasy), and of most recent interest synthetic cathinones or "bath salts."

Cocaine

Cocaine (methyl benzoylecgonine) is a tropane alkaloid that is derived from the *Erythroxylon coca* plant, the leaves of which are dried and used to produce cocaine hydrochloride. Cocaine is rarely used in clinical medicine, with limited use for anesthesia and vasoconstriction during nasal surgery, but is more widely used for illicit recreational purposes [1]. Cocaine exists in two forms: hydrochloride salt and free base. The hydrochloride salt form of cocaine is a white powder which is ingested via insufflation (snorting), which is then absorbed by the mucus membranes of the nasal passage, or less commonly by intravenous injection. The salt form of cocaine is often "cut" with other similar powders, such as mannitol, lactose, or lidocaine [2]. The free base form is also known as "crack," which has a rock-like appearance, can be off-white to brown in color and derives its name from the crackling sound it makes when heated for smoking. Crack is prepared by neutralizing cocaine hydrochloride salt with a base, such as baking soda or ammonia [3]. Crack cocaine is typically smoked, as it has a lower vaporization point than the hydrochloride salt. The cocaine is then absorbed by the pulmonary vasculature, and into the blood stream.

Cocaine is ranked as one of the highest abused illicit substances, with 1.5 million cocaine users in the United States alone, according to the Substance Abuse Mental Health Service Administration (SAMHSA), and 18.3 million users globally according to the World Drug Report [4]. In 2014 there were 27.0 million Americans, aged 12 or older, that reported as current users of illicit substances. Of this number, 1.5 million reported using cocaine which includes about 350,000 users of crack cocaine and is equivalent to 0.6% and 0.1% of the population, respectively [5]. Cocaine has consistently ranked second or third among drugs involved in overdose deaths since 2010 and has increased from 1.4 per 100,000 to 1.8 per 100,000 in 2014. Cocaine is also frequently used concomitantly with other drugs, where one in three drug overdose deaths involving cocaine also involved heroin, and vice versa one in five drug overdose deaths involving heroin also involved cocaine [6]. According to the 2011 Drug Abuse Warning Network (DAWN) report, cocaine was the top illicit drug responsible for emergency department (ED) visits, resulting in 162 visits per 100,000 population, or approximately 40% of all illicit drug related ED visits [7].

Pharmacokinetics

Mechanism of action: The main pharmacological response to cocaine is due to inhibition of reuptake of neurotransmitters. Cocaine binds to the sodium binding site on transporter proteins that are responsible for reuptake of dopamine into presynaptic neurons prolonging its action in the synaptic cleft. This increase in dopamine is responsible for the euphoria associated with cocaine use. Similarly, cocaine also inhibits norepinephrine reuptake by a similar mechanism and is responsible for the sympathomimetic response of cocaine, inducing hyperthermia, tachycardia, and hypertension [3,8]. Over time, in response to cocaine, the postsynaptic neuron will decrease the number of dopamine receptors, causing insensitivity to cocaine, which may drive the user to compensate by increasing the cocaine dose [2]. Upregulation of dopamine transporters and cocaine receptors also occur, increasing the amount of cocaine needed for the same physiological response. This thereby produces the addiction associated with cocaine use.

Acute cocaine toxicity results in CNS stimulation and includes hypertension, hyperthermia, agitation, seizures, or coma, caused by catecholamine excess [9]. Cocaine toxicity is predominately due to cardiotoxicity cause by acute vasoconstriction and increased myocardial oxygen demand, predisposing patients to myocardial infarction [10]. Prolonged or repetitive vasoconstriction can result in endothelial damage, ultimately leading to increased atherosclerosis, exacerbating the cardiotoxicity of cocaine. In addition, cocaine binds to sodium channels, inhibiting sodium influx during depolarization. This may lead to life-threatening arrhythmias. These effects combined may cause sudden death following cocaine administration.

Routes of administration: Cocaine is typically ingested via insufflation, smoking or intravenously. Cocaine is generally not orally administered due to first pass metabolism. When smoked, cocaine absorbs rapidly over the alveoli of the lung into the blood, reaching peak plasma concentration in 2–5 minutes [11]. Similarly, when insufflated, cocaine absorbs readily in the mucus membranes, albeit more slowly than when smoked, with an average time-to-peak plasma concentration of 60 minutes. After ingestion the volume of distribution is 1.5–2 L/kg and is rapidly metabolized [12].

Metabolism: Metabolism of cocaine is quite complex. The major pathways are through ecgonine methyl ester and benzoylecgonine, both of which undergo enzymatic hydrolysis through benzoyl esterase and methyl esterase, respectively [2], or from spontaneous hydrolysis [13]. Benzoylecgonine is also produced via spontaneous hydrolysis that occurs at physiological and alkaline pH. Both ecgonine methyl ester and benzoylecgonine are metabolized to ecgonine. The half-life of cocaine is approximately 60 minutes, and the first appearance of benzoylecgonine can be seen in 15–30 minutes of cocaine administration. Benzoylecgonine has a longer elimination half-life of 4–6 hours. Due to its longer half-life, benzoylecgonine is often used in analytical methods for detection. Cocaine and its metabolites are excreted in urine as 1%–9% unchanged cocaine, 26%–54% benzoylecgonine, 18%–41% as ecgonine methyl ester, and 2%–3% as ecgonine, with variability depending on urine pH [2]. Within 3 days, 64%–69% of ingested cocaine has been eliminated and, depending on analytical methods, may be negative below the detection cut-off of the assay.

In the presence of ethyl alcohol, these esterases will also metabolize cocaine to cocaethylene, an active metabolite that has increased elimination half-life. Cocaethylene can increase euphoria associated with cocaine, leading to increased ingestion of the drug and effectively increasing associated toxicity [3]. In addition, cocaine undergoes metabolism by cytochrome P450 enzymes to produce norcocaine, whose downstream metabolites are hepatotoxic. Similarly, cocaethylene is metabolized to norethylcocaine. Cocaine, in the presence of ethyl alcohol, metabolizes to cocaethylene at a rate three times that of metabolism to benzoylecgonine, and metabolism to norethylcocaine could increase the hepatotoxicity when alcohol and cocaine use are combined, causing greater liver injury than either drug in isolation [2].

Treatment for toxicity: Treatment for cocaine toxicity is typically rapid cooling and fast-acting benzodiazepines. Next step is alpha-adrenergic blockers to relax muscles and decrease blood pressure, and calcium channel antagonists and sedatives when first line measures are inadequate [9]. The purpose of rapid cooling is to decrease temperature-induced vasodilation and decrease temperature to decrease protein denaturation. This will effectively decrease cardiac output, reduce myocardial oxygen demand, and mitigate the sympathomimetic effects of cocaine. Fast-acting benzodiazepines may also be administered and is effective in relieving hypertension, tachycardia, and vasoconstriction reducing the cardiotoxicity of cocaine. Additional research is underway to explore options for pharmacokinetic approaches as well as vaccines to generate cocaine-binding antibodies, and lipid emulsion sequestration of lipophilic drugs such as cocaine [9,14]. Chronic cocaine use has long-term cardiovascular effects caused by increased thrombus formation, myocardial ischemia, and increased atherosclerosis.

Amphetamines

Amphetamines are phenethylamine derivatives with sympathomimetic activity. Amphetamines are not naturally occurring substances and are synthesized for pharmaceutical and illicit use. This category of drugs includes amphetamine and methamphetamine, as well as newer designer amphetamines of which MDMA (ecstasy) and MDEA (3,4-methylenedioxyethylamphetamine; eve) are the most commonly abused [15]. Methamphetamine is a more lipophilic analog of amphetamine and is a more potent stimulant with a prolonged effect. Amphetamine and methamphetamine are chiral molecules and chirality is an important determinant of their pharmaceutical effects. The S(+)-enantiomers (*d*-amphetamine and *d*-methamphetamine) are more potent than the R(−)-enantiomers (*l*-amphetamine and *l*-methamphetamine). MDMA and MDEA are structurally similar to amphetamine, wherein the S(+) and R(−) forms have varying clinical effects, and the S(+) isomer is more amphetamine-like. All of these drugs have a similar sympathomimetic mechanism of action and are grouped together accordingly. Amphetamines are typically available as white crystalline powder and it is administered orally or through intravenous injection.

According to the 2014 National Survey on Drug Use and Health from SAMHSA, there were 1.6M people who identified as nonmedical users of amphetamine type stimulants, one-third of which reported using methamphetamine [5]. While the rate of users stayed the same across the preceding 10 years, the age-adjusted rate of methamphetamine overdose deaths more than doubled from 0.5 per 100,000 in 2010 to 1.2 per 100,000 in 2014. In comparison to other drugs, methamphetamine ranked seventh in 2014, behind heroin, cocaine, oxycodone, alprazolam, fentanyl, and morphine [6]. Methamphetamine use results in a significant burden to healthcare, as noted in the DAWN report for SAMHSA [16,7]. Similar to increase in deaths resulting from methamphetamine use, the DAWN report demonstrated a near doubling of ED visits related to methamphetamine, increasing from approximately 60,000 in 2009 to 100,000 in 2011. The majority of these patients were treated in the ED and released (64%), with only 16% being admitted to the hospital for prolonged treatment and monitoring, which is low when compared to all other drugs at a 25% admission rate.

Pharmacokinetics

Mechanism of action: In contrast to the mechanism of cocaine the primary mechanism of action for amphetamines is its impact on the presynaptic nerve and neurotransmitter vesicles within. Amphetamines are very similar in structure to monoamine neurotransmitters: serotonin, dopamine, and norepinephrine. Amphetamines are actively transported into the presynaptic nerve by reuptake transporters, acting as a competitive analog to their endogenous target. Once inside the cytosol, amphetamine again competes with monoamines, inhibiting transport to storage vesicle and ultimately increases transport of monoamines out of the neuron into the synapse, a process that is called "reverse transport." This same competition for reuptake receptors decreases the reuptake of these neurotransmitters, increasing the concentration in the synapse and driving hyperactivation of the postsynaptic nerve. [17,18] Amphetamine and methamphetamine stimulate the release of all monoamines, but primarily the effects experienced are due to dopamine release. MDMA/MDEA act preferentially on serotonin receptors, and while these drugs have similar stimulant effects, MDMA and MDEA are also classified as entactogens, or drugs that cause an increased emotional response [15,19].

Routes of administration: Amphetamine is typically available as illicit use of pharmaceutical preparations, and is in pill-form, which are then either taken orally, or crushed and dissolved for IV use. Due to the variety of manufacturers, the pharmacodynamic profile for amphetamines varies depending on the preparation but generally has a high bioavailability >90%, volume of distribution is 3−4 L/kg, time-to-peak concentration is approximately 4 hours with a elimination half-life of 12 hours [20]. Methamphetamine is the primarily abused drug of this group, due to its ease of manufacture leading to increased availability and decreased cost. Although small amounts of methamphetamine are available pharmaceutically, the majority is produced illegally. Methamphetamine is typically a water soluble white powder, ingested via oral route or IV. In this form, it is lipid soluble, with a 67% bioavailability, a volume of distribution of 3−7 L/kg, peak at 4 hours, and an elimination half-life of 10 hours. Methamphetamine is also available as a hydrochloride salt, which exists in larger crystals, and is often smoked. In this form the bioavailability increases to >90%, with immediate effects. The elimination half-life is approximately 11−12 hours [15].

MDMA and MDEA are ingested as either a pill-form or as a powder and can be taken orally or insufflated. The pharmacokinetics of MDMA and MDEA are nonlinear and are dose-dependent [21,22]. A typical dose is a 100 mg tablet with effects seen from 3 to 6 hours, with a time to peak of 2 hours, and an elimination half-life of approximately 9 hours. The volume of distribution is similar to other drugs in this group, approximately 6 L/kg.

Metabolism: Major methamphetamine metabolites include amphetamine, produced via demethylation, and hydroxymethamphetamine via hydroxylation [15]. Similarly, amphetamine is also hydroxylated to hydroxyamphetamine, norephedrine, and norhydroxyephedrine, and to a lesser extent, deaminated to phenylacetone, which is then oxidized to benzoic acid. A large portion of both parent drugs are eliminated in the urine. Amphetamine is a weak base, with a p*K*a of 9.9, and as a result, urinary elimination of amphetamines is largely pH dependent [23]. In acidic urine the plasma half-life greatly decreases from 12 to 7 hours, and in alkaline urine, it increases to 18−34 hours, as neutrals are more lipid soluble and are more readily reabsorbed in the renal tubules.

MDMA is metabolized to MDA by *N*-demethylation, and both of these metabolites are further processed by two pathways: (1) *O*-demethylation, methylation and glucuronidation and (2) *N*-dealkylation, deamination, and oxidation to be excreted as benzoic acid−glycine conjugates. MDEA follows a similar metabolic pathway [22]. The amount of drug excreted unchanged varies, depending on the dose, and is due to saturation of metabolic enzymes. Furthermore, the first step of metabolism, conversion to MDA, is quicker for the S(+) enantiomer than the R(−) form, leading to an imbalance of R/S isomers of each when excreted [3,24].

Treatment for toxicity: Toxicity with amphetamines is due to exacerbation of the main effects of these drugs on the CNS, and the adrenergic response. Increased agitation, body temperature, heart rate, blood pressure, and cardiac output are key features of toxicity from amphetamines [25]. More severe toxicity presents with rhabdomyolysis due to hyperthermia, coma, and/or seizures, and amphetamine-related fatality is caused by cerebral hemorrhage as a consequence of hypertension and acute cardiac failure [26]. The cause of death from amphetamine ingestion is reportedly caused by accidents and suicide resulting from extreme agitation and psychosis at high doses [27]. Toxicity from amphetamine use is not directly related to the dose taken, due to tolerance after repeated exposure. It can be difficult to distinguish acute intoxication with methamphetamine compared to cocaine, as the clinical effects are similar, but the duration of effects for methamphetamine are much longer than cocaine. Long-term effects from neurotoxicity of methamphetamine can be seen due to physical damage to dopaminergic and serotonergic terminals from oxidative stress [28].

Treatment of acute amphetamine toxicity involves reducing agitation and managing hyperthermia. Sedation is a first line treatment and involves the use of haloperidol or benzodiazepines to reduce the immediate effects of amphetamine intoxication. If orally ingested, activated charcoal may be given to decrease absorption [29] Elimination may be enhanced by acidifying urine, which decreases tubular reabsorption of amphetamines, but should be used sparingly, as this can exacerbate issues with metabolic acidosis [3].

Cathinones

Cathinones are designer drugs that are phenethylamine derivatives of a naturally occurring stimulant that is the principle psychoactive component of "khat," extracted from the leaves of the plant *Catha edulis* [30]. The leaves are brewed or chewed for stimulant effects, and the active ingredient has been identified as cathinone, which has effects similar to amphetamines and cocaine, with varying potency depending on the derivative. Designer drugs have been developed as derivatives of cathinone and are categorized as synthetic cathinones, and include methcathinone, mephedrone, methylenedioxypyrovalerone (MDPV), and methylone (MDMC). Each drug is structurally similar to amphetamines and similarly share the principle that the S(−) isomer is more potent than the R(−) form [30]. In 2013 at least 44 synthetic cathinones had been identified and is likely increasing due to the legal nature of designer drugs [30,31]. These drugs are marketed as "legal highs" and are frequently changing to avoid legal restrictions once each drug is identified [32]. In 2011 MDPV, mephedrone, and methylone were temporarily granted Schedule 1 status (no medical use, high abuse potential) in the United States, and permanent status was granted in 2012 [33]. This drives manufacturers to create alternatives with the same effects to avoid these restrictions.

These drugs are colloquially called "bath salts" due to their white crystalline appearance, and are available as powders, crystals or pressed into pill form [34,35]. Common street names include Bliss, Blizzard, Vanilla Sky, Meow Meow, White Rush, Explosion and Blue Magic, among others. Synthetic cathinones have been distributed as insect repellants, plant food/fertilizer, bath salts, or other products marketed as "not for human consumption" [35]. Prior to 2011 these drugs were considered legal and were distributed over the internet or in gas stations or head shops [36].

Synthetic cathinones produce an amphetamine-like sympathomimetic effect and are psychostimulatory, causing euphoria, mood enhancement, alertness, and hyperactivity [37−39]. Higher doses result in anxiety and agitation, tachycardia, and hypertension. Patients may also experience "excited delirium," which can cause increased delusions and aggressive

Synthetic cathinones were originally synthesized in the mid-1900s for a variety of clinical uses, including as antidepressants, appetite suppressants and stimulants, but were not marketed due to strong addiction potential [36]. However, reports of use appeared in Europe in the mid-2000s, and in the United States as early as 2009 [36,40−43]. Thereafter, the DAWN reported in 2011, of the 2.5 million ED visits involving drug use or abuse, 22,904 visits involved the use of cathinones, 52% of which were a combination of cathinones with other drugs of abuse [7,44] The use of cathinones peaked in 2011, with the number of calls to poison control centers totaling 6136, much higher than the 304 calls in the previous year. The frequency of cathinone-related calls has decreased since, with 2697 in 2012, 998 in 2013, and 587 in 2014 [45,46].

Recreational use of synthetic cathinones is less significant than other drugs of abuse. A study of data collected from the National Institute on Drug Abuse Monitoring the Future study shows that in 2017, fewer than 1% of high school seniors report use of cathinones, compared to 4.2% for cocaine use, 4.9% for MDMA use, and 1.1% for methamphetamine use [47]. These statistics all rely on self-reporting and may not be an accurate reflection of the current drug market. The global use of synthetic cathinones is likely increasing and has been reported by the European Monitoring Center for Drugs and Drug Addiction (EMCDDA) that users injecting cathinones are primarily those who also inject heroin or amphetamines, and are switching to cathinones or mixing them into the drug repertoire [4,48]. Global seizure of synthetic cathinones have been steadily increasing since hitting the market in 2010, reaching 1.3 t in 2014, which was triple the quantity seized in 2013, most of which was seized in western Europe [4]. To directly compare this quantity to other drugs, the total amount of MDMA seized in 2014 was 9 t, and the global cocaine seizure was 655 t globally.

Pharmacokinetics

Mechanism of action: Synthetic cathinones all have agonist activity against monoamine reuptake in the dopamine, serotonin, and norepinephrine pathways (paper 3). The mechanism of this varies slightly with each derivative, but as a group, these drugs block reuptake as well as stimulate the release of neurotransmitters (paper 1). Cathinone inhibits dopamine transport, but is less potent toward serotonin uptake. Mephedrone and methylone are nonselective, similar to cocaine, but also stimulate the release of serotonin mimicking MDMA activity [32,49,50]. MDPV is a potent reuptake inhibitor but is less effective on neurotransmitter release. The varying mechanisms of action of these drugs can have a synergistic effect, amplifying the reaction beyond the response of either drug alone. This may explain the strong physiological reactions reported to bath salts that may contain a combination of synthetic cathinones [30].

Routes of administration: Synthetic cathinones are commonly found in powder or crystalline form and are taken orally, via insufflation, rectally, or intravenously. When taken rectally or intravenously, effects are immediate, whereas insufflation, smoking, or vaping peaks within 1–30 minutes or within 1–2 hours if taken orally [51]. Symptoms last 2–3 hours after peak or can last longer depending on stomach contents and absorption [35,52,53]. A popular method of ingestion is "bombing" of "parachuting," where the powder is wrapped in paper and swallowed, and formed as a technique to avoid nasal irritation that comes with insufflation [54].

Metabolism: Synthetic cathinones are relatively new drugs with few pharmacokinetic studies available on controlled administration of these drugs to study participants. Several studies have been performed as a part of forensic toxicology, reviewing urine samples from patients who have reported cathinone use. MDPV, mephedrone, and methylone all undergo hepatic metabolism with cytochrome P450 2D6. MDPV undergoes O-demethylation to form 3,4-dihydroxypyrovalerone and O-methylation to form 4-hydroxy-3-methoxy-pyrovalerone, both of which undergo Phase II metabolism and are excreted as glucuronides and sulfates [55]. Mephedrone undergoes N-demethylation to form N-demethylmephedrone, hydroxylation to form 4′-hydroxymethylmephedrone and reduction of the ketone moiety to form 1-dihydromephedrone, which are all glucuronidated and excreted [56,57]. Methylone is similarly N-demethylated to form methylenedioxycathinone and demethylated followed by O-methylation of the hydroxyl groups to form 4-hydroxy-3-methoxymethcathinone, which is also glucuronidated and excreted [58]. Each of these drugs is almost entirely metabolized, with <5% of parent drug in patient urine.

Synthetic cathinones have a short half-life, where physiological effects are experienced for 2–3 hours after ingestion [59]. This rapid decline and short half-life may cause users to readminister, increasing the amount of drug ingested and may lead to large doses in a short amount of time (mephedrone big doses). This pattern of "stacking" (taking multiple doses at once) or "boosting" (taking multiple doses over time to prolong effects) can lead to increased toxicity.

Treatment for toxicity: Toxicity with synthetic cathinones reflects the sympathomimetic response; patients are typically agitated and often presenting with violent or aggressive delusions, which lead to self-inflicted injuries [36,46,60–62]. Hypertension, hyperthermia, tachycardia, and diaphoresis accompany these symptoms, as is similarly seen with other sympathomimetic drugs. Therapy and treatment for patients ingesting synthetic cathinones primarily include sedation with benzodiazepines or antipsychotics, to reduce the presenting agitation. Many symptoms are self-limited and resolve without intervention. Synthetic cathinones stimulate the release of dopamine, as well as reduce its reuptake, increasing the amount of dopamine in the synapse dramatically. Patients taking cathinones are often polydrug users [63], which may lead to increased neurotoxicity, particularly if taken with amphetamines. This increased dopamine can lead to delusions and agitation, or even sudden death [35,64].

Analytical considerations

The traditional means of screening patients for CNS stimulants is common to other types of drugs: initial screening with immunoassay, followed by a more specific confirmation using mass spectrometric methods. The specimen of choice is typically urine, as it is readily available and can provide a longer window of use compared to serum or saliva methods.

Routine urine immunoassay testing is mostly performed using laboratory based methods. These qualitative immunoassays are calibrated to designated cut-off concentrations of drug and/or metabolite that will flag positive or negative, relative to that specific cut-off. The window for detection of amphetamines is 1–3 days after use [65]. Immunoassays for amphetamines are directed against S(+)-amphetamine and S(+)-methamphetamine but are not exclusive for these isoforms, and cross-react with other substances [3,66,67]. The degree of cross-reactivity varies with assay manufacturer, but of particular interest is pseudoephedrine, MDA, MDMA, phentermine, and other similar sympathomimetic compounds, which are well known to cross-react with the amphetamines immunoassays [65,68]. Immunoassays for cocaine are developed against benzoylecgonine, the primary urinary metabolite of cocaine, because the parent drug is excreted only in small amounts. This metabolite has a half-life of 4–7 hours and can be detected in urine for 1–3 days after use [3,65]. Cocaine immunoassays are very specific since it has a unique chemical structure and are free from common cross-reactivity issues seen with other drugs [65,69,70].

Confirmatory testing follows immunoassay screening, to provide quantitative results, rule out false-positive and/or false negative results, and often has lower cut-offs than immunoassay testing. While immunoassay testing is routinely performed in most laboratory settings, confirmatory testing may be "sent out" to larger reference labs, which can create a delay in treatment and follow-up. Results from confirmatory testing typically do not impact immediate patient management, since the immunoassay result is able to provide detail on which drug type is causing the clinical scenario but is important for ongoing management, particularly in cases of suspected abuse. Confirmatory testing uses gas chromatography mass spectrometry (GCMS) or liquid chromatography mass spectrometry (LCMS) to provide specific detail on drugs in the urine.

GCMS is routinely used for amphetamines. It is important to note that amphetamines are small drugs, and as such, the resulting mass spectrum is uncomplicated, and not highly characteristic of the parent compound. Therefore derivatization is commonly used for amphetamines [3,71]. In addition, amphetamines are volatile substances that may be lost during dry-down evaporation of samples, extraction, or lost due to volatility at high temperatures in GCMS, adding to the advantages of derivatization. Derivatization is typically performed using heptafluorobutyric anhydride, pentafluoropropionic anhydride, trifluoroacetic anhydride, or 4-carbethoxyhexfluorobutyryl chloride [3]. While these methods identify the drug definitively, standard GCMS cannot differentiate D-amphetamine (illicit) from L-amphetamine (pharmaceutical). Additional chiral separation may be necessary if this differentiation is pertinent to clinical management [3,72,73]. LCMS offers the advantage by avoiding the need for derivatization, as it has softer ionization and provides more definitive fragmentation than GCMS methods [13] and allows for quantitation of amphetamine, methamphetamine, MDMA, and MDA in a single assay. Similar to GCMS, LCMS cannot distinguish D-isomer from L-isomer, requiring chiral separation in a separate assay if clinically warranted.

GCMS and LCMS are both used often for cocaine confirmation assays. When using GCMS, benzoylecgonine is derivatized via an acylation or silylation procedure (Forensic tox book) and is detected using selected ion monitoring [2]. Cocaine and metabolites are amenable to LCMS without derivatization. Solid phase extraction is typically performed and followed by reversed phase chromatography and multiple reaction monitoring tandem mass spectrometry [13,74]. While most assays focus on cocaine and benzoylecgonine, there may be additional interest in metabolites that are specific to crack cocaine by measuring anhydroecgonine methyl ester, a byproduct that indicates thermal degradation when cocaine is smoked [75,76]. In addition, cocaethylene is produced when cocaine and ethanol are used in combination, a metabolite with a much longer elimination half-life, and may indicate greater toxicity than cocaine alone [13,77].

Designer drugs

Detection of designer drugs, like synthetic cathinones, is much more complicated than the routine immunoassay-to-targeted mass spec methods. Some synthetic cathinones cross-react with amphetamine, methamphetamine or MDMA immunoassays, but it varies greatly by manufacturer and drug [54,78–80]. Individuals report that this inability to detect on routine screening methods is one of the reasons these drugs are used [37,42]. Mass spectrometric techniques require knowledge of the drug to be analyzed, and requires that the analyst keep pace with frequent changes within the drug class. This is particularly an issue for cathinones, which vary by small changes, and number in the hundreds [30,31].

GCMS is a commonly available technique for detection, but cathinones require derivatization to produce meaningful unique spectra, are difficult to derivatize [54]. Stability of derivatized cathinones in GCMS is also an issue, making quantitative analysis difficult. LCMS is an attractive technique for cathinone detection and does not require derivatization. Simple liquid—liquid extraction or solid phase extraction can be used as pretreatment, and there have been several studies on fragmentation pathways that can be used to identify the drugs [81—85]. These methods require knowledge about the drug to be analyzed for multiple reaction monitoring, but large menus can be built for the more commonly encountered cathinones. Detection by TOFMS is also an emerging option for screening, collecting fragmentation spectra from all eluting substances, and probing data for cathinones. While this is a broad screening method that can circumvent the issue of needing predetermined knowledge of each drug of interest, these instruments are expensive and not routinely found in clinical or toxicology laboratories [86,87].

Oral fluid testing

Oral fluid testing can have advantages over other methods of testing. Collecting saliva is a noninvasive procedure, and may be preferred for workplace testing, roadside testing, or in other situations where collecting other biological fluids is complicated. Oral fluid is related to plasma free drug concentration, so it is especially useful for identifying current intoxication. Limitations of oral fluid testing include small sample volume, low analyte concentration, oral contamination from smoking or insufflating, and a short detection window, similar to that of blood testing [3]. Studies have shown oral fluid testing for cocaine reflects plasma concentrations, with a cocaine half-life of approximately 3 hours and benzoylecgonine of 4—14 hours [88]. Similarly, studies involving disposition of methamphetamine have also shown that oral fluid reflects plasma levels, with the concentration being greater in oral fluid than plasma, likely due to the acidity of saliva [89]. Methamphetamine oral fluid testing provides a positive result up to 24 hours after dosing, and accumulates in oral fluid with repeat exposure. Less information is known about the deposition of synthetic cathinones in oral fluid, but due to chemical similarity, is likely to be similar to other amphetamine-like substances. Studies have demonstrated the ability to measure in this matrix [90—92]. This indicates that oral fluid can be used for identifying immediate impairment, similar to blood testing (Figs. 14.1—14.4).

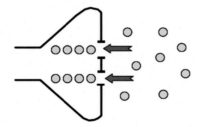
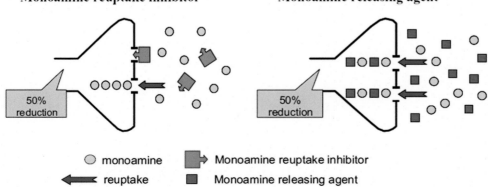

FIGURE 14.1 Mechanisms leading to an increase in monoamines in the synapse, produced by classical reuptake inhibition versus a releasing agent.
Source: Heal DJ, et al. Amphetamine, past and present—a pharmacological and clinical perspective. J Psychopharmacol 2013;27(6):52.

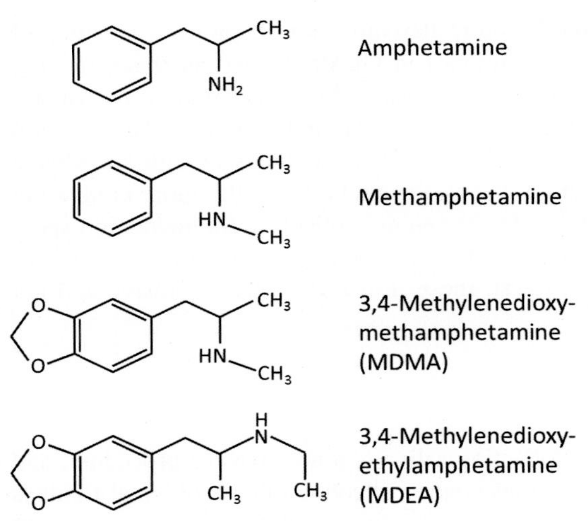

FIGURE 14.2 Structures of amphetamine-type stimulants.

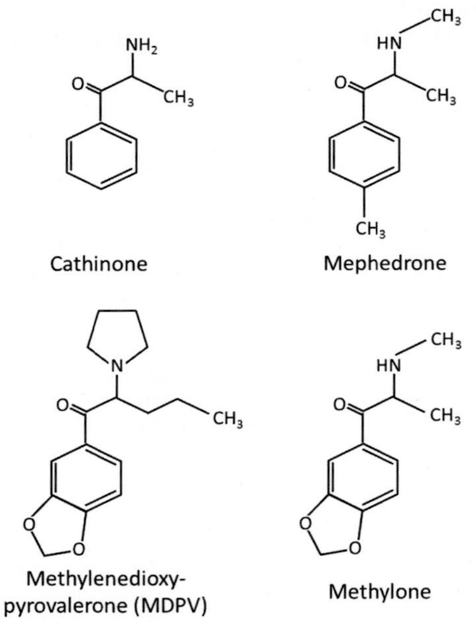

FIGURE 14.3 Structures of synthetic cathinones.

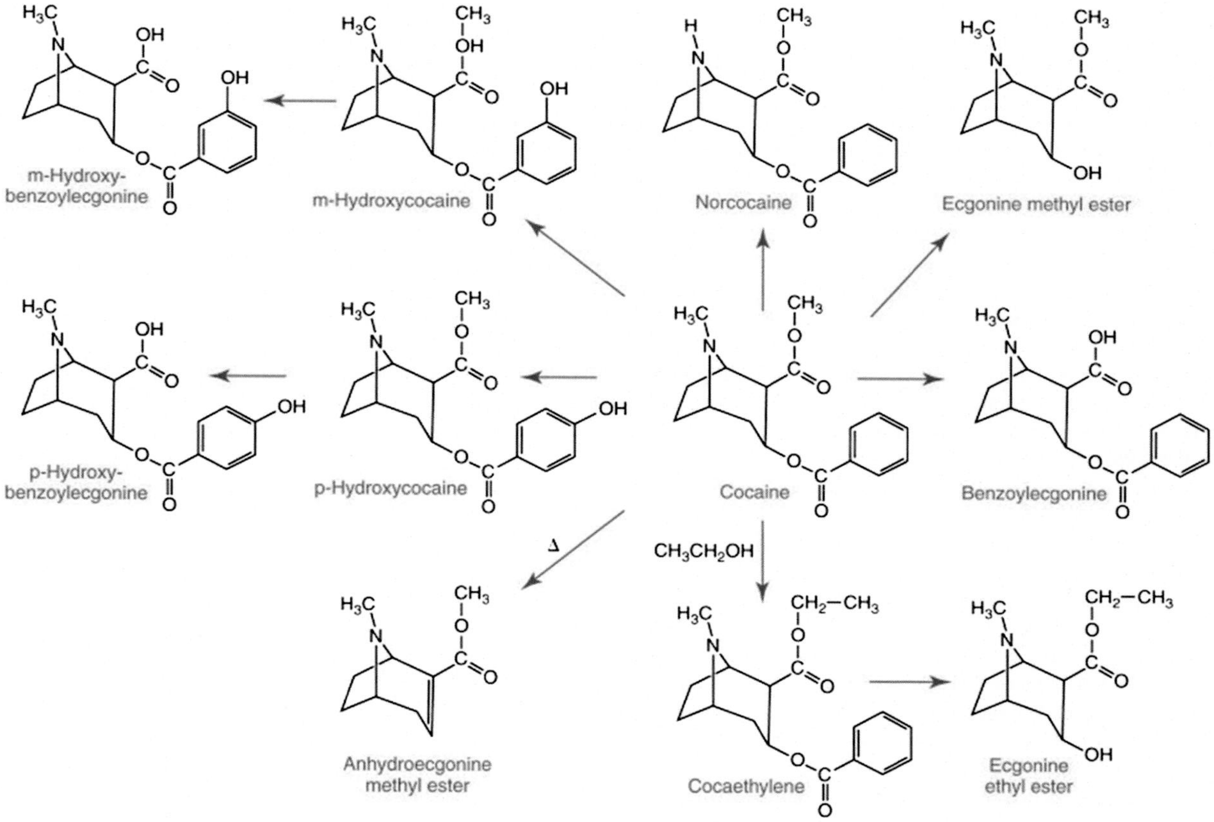

FIGURE 14.4 Cocaine metabolism.
Source: Langman L, Bechtel L, Holstege CP. Clinical toxicology. In: Burtis CA, Ashwood ER, Bruns DE, editors. Tietz textbook of clinical chemistry and molecular diagnostics. St. Louis, MO: Elsevier Saunders; 2012. p. 48.

References

[1] Long H, et al. Medicinal use of cocaine: a shifting paradigm over 25 years. Laryngoscope 2004;114(9):1625–9.
[2] Isenschmid DS. Cocaine. In: Levine B, editor. Principles of forensic toxicology. Washington, DC: AACC Press; 2006. p. 239–60.
[3] Langman L, Bechtel L, Holstege CP. Clinical toxicology. In: Burtis CA, Ashwood ER, Bruns DE, editors. Tietz textbook of clinical chemistry and molecular diagnostics. St. Louis, MO: Elsevier Saunders; 2012. p. 1109–88.
[4] United Nations Office on Drugs and Crime, *World Drug Report 2016* (United Nations publication, Sales No. E.16.XI.7)
[5] Center for Behavior Health Statistics and Quality. (2015). *Behavioral health trends in the United States: Results from the 2014 National Survey on Drug Use and Health* (HHS Publication No. SMA 15-4927, NSDUH Series H-50). Retrieved from http://www.samhsa.gov/data/
[6] Warner M, Trinidad JP, Bastian BA, Minino AM, Hedegaard H. Drugs most frequently involved in drug overdose deaths: United States, 2010-2014. Natl Vital Stat Rep 2016;65(10):15.
[7] Substance Abuse and Mental Health Services Administration, *Drug abuse warning network, 2011: National Estimates of Drug-Related Emergency Department Visits*. HHS Publication No. (SMA) 13-4760. DAWN Series D-39. Rockville, MD; 2013
[8] Sofuoglu M, Sewell RA. Norepinephrine and stimulant addiction. Addict Biol 2009;14(2):119–29.
[9] Connors NJ, Hoffman RS. Experimental treatments for cocaine toxicity: a difficult transition to the bedside. J Pharmacol Exp Ther 2013;347(2):251–7.
[10] Phillips K, et al. Cocaine cardiotoxicity: a review of the pathophysiology, pathology, and treatment options. Am J Cardiovasc Drugs 2009;9(3):177–96.
[11] Jenkins AJ, et al. Correlation between pharmacological effects and plasma cocaine concentrations after smoked administration. J Anal Toxicol 2002;26(7):382–92.
[12] Jones RT. Pharmacokinetics of cocaine: considerations when assessing cocaine use by urinalysis. NIDA Res Monogr 1997;175:221–34.
[13] Snozek CLH, Bjergum MW, Langman LJ. In: Langman LJ, Snozek CLH, editors. Cocaine and metabolites by LC-MS/MS, in LC-MS in drug analysis. New York: Humana Press; 2012. p. 91–104.

[14] Westfall TC, Westfall DP. Adrenergic agonists and antagonists. In: Brunton LL, Lazo JS, Parker KL, editors. Goodman & Gilman's the pharmacological basis of therapeutics. New York: McGraw-Hill Medical Publishing Division; 2006. p. 237–95.
[15] Moore KA. Amphetamines/sympathomimetic amines. In: Levine B, editor. Principles of forensic toxicology. Washington, DC: AACC Press; 2006. p. 277–96.
[16] Substance Abuse and Mental health Services Administration, Center for behavioral health statistics and quality. (June 19, 2014). *The DAWN Report: Emergency Department Visits Involving Methamphetamine: 2007 to 2011*. Rockville, MD
[17] Goodwin JS, et al. Amphetamine and methamphetamine differentially affect dopamine transporters in vitro and in vivo. J Biol Chem 2009;284(5):2978–89.
[18] Heal DJ, et al. Amphetamine, past and present—a pharmacological and clinical perspective. J Psychopharmacol 2013;27(6):479–96.
[19] Nichols DE. Differences between the mechanism of action of MDMA, MBDB, and the classic hallucinogens. Identification of a new therapeutic class: entactogens. J Psychoactive Drugs 1986;18(4):305–13.
[20] Markowitz JS, Patrick KS. The clinical pharmacokinetics of amphetamines utilized in the treatment of attention-deficit/hyperactivity disorder. J Child Adolesc Psychopharmacol 2017;27(8):678–89.
[21] de la Torre R, et al. Non-linear pharmacokinetics of MDMA ('ecstasy') in humans. Br J Clin Pharmacol 2000;49(2):104–9.
[22] de la Torre R, et al. Human pharmacology of MDMA: pharmacokinetics, metabolism, and disposition. Ther Drug Monit 2004;26(2):137–44.
[23] Poklis A, et al. Urinary excretion of d-amphetamine following oral doses in humans: implications for urine drug testing. J Anal Toxicol 1998;22(6):481–6.
[24] Fallon JK, et al. Stereospecific analysis and enantiomeric disposition of 3, 4-methylenedioxymethamphetamine (ecstasy) in humans. Clin Chem 1999;45(7):1058–69.
[25] Matsumoto RR, et al. Methamphetamine-induced toxicity: an updated review on issues related to hyperthermia. Pharmacol Ther 2014;144(1):28–40.
[26] Cruickshank CC, Dyer KR. A review of the clinical pharmacology of methamphetamine. Addiction 2009;104(7):1085–99.
[27] Logan BK, Fligner CL, Haddix T. Cause and manner of death in fatalities involving methamphetamine. J Forensic Sci 1998;43(1):28–34.
[28] Yang X, et al. The main molecular mechanisms underlying methamphetamine- induced neurotoxicity and implications for pharmacological treatment. Front Mol Neurosci 2018;11:186.
[29] Chyka PA, et al. Position paper: single-dose activated charcoal. Clin Toxicol (Phila) 2005;43(2):61–87.
[30] De Felice LJ, Glennon RA, Negus SS. Synthetic cathinones: chemical phylogeny, physiology, and neuropharmacology. Life Sci 2014;97(1):20–6.
[31] United Nations Office on Drugs and Crime, *The Challenge of New Psychoactive Substances, Global SMART Programme 2013* (United Nations publication; Retrived from https://www.unodc.org/documents/scientific/NPS_Report.pdf)
[32] Karila L, et al. Synthetic cathinones: a new public health problem. Curr Neuropharmacol 2015;13(1):12–20.
[33] Schedules of controlled substances: temporary placement of three synthetic cathinones into schedule I, 76 Fed. Reg. 204 (October 21, 2011). *Federal Register: The Daily Journal of the United States.*
[34] Graddy R, Buresh ME, Rastegar DA. New and emerging illicit psychoactive substances. Med Clin North Am 2018;102(4):697–714.
[35] Karch SB. Cathinone neurotoxicity ("the "3Ms"). Curr Neuropharmacol 2015;13(1):21–5.
[36] Spiller HA, et al. Clinical experience with and analytical confirmation of "bath salts" and "legal highs" (synthetic cathinones) in the United States. Clin Toxicol (Phila) 2011;49(6):499–505.
[37] Weaver MF, Hopper JA, Gunderson EW. Designer drugs 2015: assessment and management. Addict Sci Clin Pract 2015;10:8.
[38] Winstock A, et al. Mephedrone: use, subjective effects and health risks. Addiction 2011;106(11):1991–6.
[39] Vardakou I, Pistos C, Spiliopoulou C. Drugs for youth via Internet and the example of mephedrone. Toxicol Lett 2011;201(3):191–5.
[40] Camilleri A, et al. Chemical analysis of four capsules containing the controlled substance analogues 4-methylmethcathinone, 2-fluoromethamphetamine, alpha-phthalimidopropiophenone and N-ethylcathinone. Forensic Sci Int 2010;197(1-3):59–66.
[41] Bronstein AC, et al. 2008 Annual Report of the American Association of Poison Control Centers' National Poison Data System (NPDS): 26th annual report. Clin Toxicol (Phila) 2009;47(10):911–1084.
[42] Gunderson EW, et al. Substituted cathinone products: a new trend in "bath salts" and other designer stimulant drug use. J Addict Med 2013;7(3):153–62.
[43] Zawilska JB, Wojcieszak J. Designer cathinones—an emerging class of novel recreational drugs. Forensic Sci Int 2013;231(1-3):42–53.
[44] Substance Abuse and Mental Health Services Administration, *The DAWN report, Data spotlight: bath salts were involved in over 20,000 drug-related emergency department visits in 2011*. Sept 17 2013, samhsa.gov/data/sites/default/files/spot117-bath-salts-2013/spot117-bath-salts-2013.pdf, [accessed 28.02.20]
[45] Fact Sheet: synthetic cathinones. *Drug Policy Alliance*; Retrieved from https://www.drugpolicy.org/sites/default/files/DPA_Fact_Sheet_Synthetic_Cathinones_%28June%202016%29.pdf.
[46] Vazirian M, et al. Bath salts in the emergency department: a survey of emergency clinicians' experience with bath salts-intoxicated patients. J Addict Med 2015;9(2):94–8.
[47] NIDA. "Monitoring the future." *National Institute on Drug Abuse, National Institutes of Health, 31 Jan 2020*; Retrieved from https://www.drugabuse.gov/related-topics/trends-statistics/monitoring-future
[48] Perspective on drugs: Injection of synthetic cathinones. *European Monitoring Centre for Drugs and Drug Addiction, 28 May 2015*; Retrieved from http://www.emcdda.europa.eu/system/files/publications/2754/Synthetic%20cathinones_updated2015.pdf

[49] Baumann MH, et al. The designer methcathinone analogs, mephedrone and methylone, are substrates for monoamine transporters in brain tissue. Neuropsychopharmacology 2012;37(5):1192–203.

[50] Hadlock GC, et al. 4-Methylmethcathinone (mephedrone): neuropharmacological effects of a designer stimulant of abuse. J Pharmacol Exp Ther 2011;339(2):530–6.

[51] Johnson PS, Johnson MW. Investigation of "bath salts" use patterns within an online sample of users in the United States. J Psychoactive Drugs 2014;46(5):369–78.

[52] Brandt SD, et al. Analyses of second-generation 'legal highs' in the UK: initial findings. Drug Test Anal 2010;2(8):377–82.

[53] Schifano F, et al. Mephedrone (4-methylmethcathinone; 'meow meow'): chemical, pharmacological and clinical issues. Psychopharmacology (Berl) 2011;214(3):593–602.

[54] Kerrigan S. Improved detection of synthetic cathinones in forensic toxicology samples: thermal degradation and analytical considerations (Document number 249251). *U.S. Department of Justice*, 2015; Retrieved from https://www.ncjrs.gov/pdffiles1/nij/grants/249251.pdf.

[55] Anizan S, et al. 3,4-Methylenedioxypyrovalerone (MDPV) and metabolites quantification in human and rat plasma by liquid chromatography-high resolution mass spectrometry. Anal Chim Acta 2014;827:54–63.

[56] Pedersen AJ, et al. In vitro metabolism studies on mephedrone and analysis of forensic cases. Drug Test Anal 2013;5(6):430–8.

[57] Pozo OJ, et al. Mass spectrometric evaluation of mephedrone in vivo human metabolism: identification of phase I and phase II metabolites, including a novel succinyl conjugate. Drug Metab Dispos 2015;43(2):248–57.

[58] Kamata HT, et al. Metabolism of the recently encountered designer drug, methylone, in humans and rats. Xenobiotica 2006;36(8):709–23.

[59] Papaseit E, et al. Human pharmacology of mephedrone in comparison with MDMA. Neuropsychopharmacology 2016;41(11):2704–13.

[60] Lusthof KJ, et al. A case of extreme agitation and death after the use of mephedrone in The Netherlands. Forensic Sci Int 2011;206(1-3):e93–5.

[61] Maskell PD, et al. Mephedrone (4-methylmethcathinone)-related deaths. J Anal Toxicol 2011;35(3):188–91.

[62] Regan L, Mitchelson M, Macdonald C. Mephedrone toxicity in a Scottish emergency department. Emerg Med J 2011;28(12):1055–8.

[63] Angoa-Perez M, et al. Mephedrone does not damage dopamine nerve endings of the striatum, but enhances the neurotoxicity of methamphetamine, amphetamine, and MDMA. J Neurochem 2013;125(1):102–10.

[64] Vincent SG, et al. A murine model of hyperdopaminergic state displays altered respiratory control. FASEB J 2007;21(7):1463–71.

[65] Melanson SE. The utility of immunoassays for urine drug testing. Clin Lab Med 2012;32(3):429–47.

[66] Verstraete AG, Heyden FV. Comparison of the sensitivity and specificity of six immunoassays for the detection of amphetamines in urine. J Anal Toxicol 2005;29(5):359–64.

[67] Hsu J, et al. Performance characteristics of selected immunoassays for preliminary test of 3,4-methylenedioxymethamphetamine, methamphetamine, and related drugs in urine specimens. J Anal Toxicol 2003;27(7):471–8.

[68] Snyder ML, Melanson SE. Amphetamines. In: Magnani B, Bissell MG, Kwong TC, Wu AHB, editors. Clinical toxicology testing: a guide for laboratory professionals. Northfield, IL: CAP Press; 2012. p. 119–25.

[69] Moeller KE, Lee KC, Kissack JC. Urine drug screening: practical guide for clinicians. Mayo Clin Proc 2008;83(1):66–76.

[70] Carney S, et al. Evaluation of two enzyme immunoassays for the detection of the cocaine metabolite benzoylecgonine in 1,398 urine specimens. J Clin Lab Anal 2012;26(3):130–5.

[71] Kolbrich EA, et al. Major and minor metabolites of cocaine in human plasma following controlled subcutaneous cocaine administration. J Anal Toxicol 2006;30(8):501–10.

[72] Rasmussen LB, Olsen KH, Johansen SS. Chiral separation and quantification of R/S-amphetamine, R/S-methamphetamine, R/S-MDA, R/S-MDMA, and R/S-MDEA in whole blood by GC-EI-MS. J Chromatogr B: Analyt Technol Biomed Life Sci 2006;842(2):136–41.

[73] Cody JT. Determination of methamphetamine enantiomer ratios in urine by gas chromatography-mass spectrometry. J Chromatogr 1992;580(1-2):77–95.

[74] Robandt PP, Reda LJ, Klette KL. Complete automation of solid-phase extraction with subsequent liquid chromatography-tandem mass spectrometry for the quantification of benzoylecgonine, *m*-hydroxybenzoylecgonine, *p*-hydroxybenzoylecgonine, and norbenzoylecgonine in urine—application to a high-throughput urine analysis laboratory. J Anal Toxicol 2008;32(8):577–85.

[75] Smith RM. Ethyl esters of arylhydroxy- and arylhydroxymethoxycocaines in the urines of simultaneous cocaine and ethanol users. J Anal Toxicol 1984;8(1):38–42.

[76] Smith RM. Arylhydroxy metabolites of cocaine in the urine of cocaine users. J Anal Toxicol 1984;8(1):35–7.

[77] Isenschmid DS. Cocaine—effects on human performance and behavior. Forensic Sci Rev 2002;14(1-2):61–100.

[78] Toennes SW, Kauert GF. Excretion and detection of cathinone, cathine, and phenylpropanolamine in urine after kath chewing. Clin Chem 2002;48(10):1715–19.

[79] Truscott SM, et al. Violent behavior and hallucination in a 32-year-old patient. Clin Chem 2013;59(4):612–15.

[80] de Castro A, et al. Liquid chromatography tandem mass spectrometry determination of selected synthetic cathinones and two piperazines in oral fluid. Cross reactivity study with an on-site immunoassay device. J Chromatogr A 2014;1374:93–101.

[81] Namera A, et al. Comprehensive review of the detection methods for synthetic cannabinoids and cathinones. Forensic Toxicol 2015;33(2):175–94.

[82] Zuba D, Byrska B. Prevalence and co-existence of active components of 'legal highs'. Drug Test Anal 2013;5(6):420–9.

[83] Fornal E. Identification of substituted cathinones: 3,4-methylenedioxy derivatives by high performance liquid chromatography-quadrupole time of flight mass spectrometry. J Pharm Biomed Anal 2013;81-82:13–19.

[84] Jankovics P, et al. Identification and characterization of the new designer drug 4′-methylethcathinone (4-MEC) and elaboration of a novel liquid chromatography-tandem mass spectrometry (LC-MS/MS) screening method for seven different methcathinone analogs. Forensic Sci Int 2011;210(1-3):213—20.

[85] Bijlsma L, et al. Fragmentation pathways of drugs of abuse and their metabolites based on QTOF MS/MS and MS(E) accurate-mass spectra. J Mass Spectrom 2011;46(9):865—75.

[86] Smith JP, Sutcliffe OB, Banks CE. An overview of recent developments in the analytical detection of new psychoactive substances (NPSs). Analyst 2015;140(15):4932—48.

[87] Musah RA, et al. DART-MS in-source collision induced dissociation and high mass accuracy for new psychoactive substance determinations. Forensic Sci Int 2014;244:42—9.

[88] Scheidweiler KB, et al. Pharmacokinetics of cocaine and metabolites in human oral fluid and correlation with plasma concentrations after controlled administration. Ther Drug Monit 2010;32(5):628—37.

[89] Huestis MA, Cone EJ. Methamphetamine disposition in oral fluid, plasma, and urine. Ann NY Acad Sci 2007;1098:104—21.

[90] Williams M, Martin J, Galettis P. A validated method for the detection of 32 bath salts in oral fluid. J Anal Toxicol 2017;41(8):659—69.

[91] Miller B, Kim J, Concheiro M. Stability of synthetic cathinones in oral fluid samples. Forensic Sci Int 2017;274:13—21.

[92] Amaratunga P, Lorenz Lemberg B, Lemberg D. Quantitative measurement of synthetic cathinones in oral fluid. J Anal Toxicol 2013;37(9):622—8.

Chapter 14.1

Caffeine: massive accidental caffeine overdose treated with continuous veno-venous hemodiafiltration

Ruben Thanacoody[1,2]
[1]UK National Poisons Information Service, Newcastle Hospitals NHS Foundation Trust, Newcastle upon Tyne, United Kingdom, [2]Translational and Clinical Research Institute, Newcastle University, Newcastle upon Tyne, United Kingdom

Case description

Two 20-year-old male university students developed severe acute toxicity following accidental ingestion of a potentially lethal 30 g (instead of the intended 30 mg) dose of caffeine arising from a 1000-fold dose calculation error during a university chemistry experiment.

At presentation to the emergency department, the first student (case 1) developed nausea, hematemesis, agitation, tremor, and diarrhea. His heart rate was 84/minute, blood pressure 171/71 mmHg. Electrocardiography showed sinus rhythm with a normal QRS interval (118 ms) and prolonged QTc interval (586 ms). The second student (case 2) developed nausea, hematemesis, tremor, blurred vision, epigastric pain, and diarrhea. His heart rate was 64/minute, blood pressure 94/69 mmHg. Electrocardiography showed sinus rhythm with a normal QRS interval (106 ms) and prolonged QTc interval (536 ms). Initial laboratory results are shown in Table 14.1.1. Both students were transferred to critical care for monitoring and management.

Case 1 was treated with potassium replacement and intravenous ondansetron for protracted vomiting. Continuous veno-venous hemodiafiltration (CVVHDF) was initiated at 7 hours postingestion to enhance caffeine elimination and continued until 43 hours postingestion. The serum caffeine concentration declined from a peak concentration of 367 to 1.8 mg/L. The blood pressure and QTc interval normalized over the first 24 hours. The patient developed rhabdomyolysis with a peak creatine kinase activity of 6891 U/L at 29 hours postingestion.

Case 2 also received potassium replacement, intravenous ondansetron for vomiting and diazepam for agitation. Cardiac monitoring revealed multifocal premature ventricular contractions and nonsustained runs of ventricular tachycardia which was treated with intravenous magnesium sulfate. CVVHDF was initiated at 9 hours postingestion and continued until 45 hours postingestion. The serum caffeine concentration declined from a peak concentration of 352 to 1.5 mg/L. Severe hypertension (BP 190/90 mmHg) was initially treated with glyceryl nitrate infusion on day 2 followed by labetalol infusion on days 3 and subsequently oral metoprolol. Abdominal pain and muscle pain was treated with fentanyl. The creatine kinase activity peaked 27 hours post ingestion at 14717 U/L.

Both patients made a full recovery without any permanent sequelae.

Plasma caffeine concentrations were measured in our university toxicology department by liquid chromatography—mass spectrometry (LC—MS) (Fig. 14.1.1).

Discussion and follow-up

Caffeine (1,3,7-trimethylxanthine) is a natural product commonly found in foodstuffs, beverages and medicinal products. Recently, energy drinks containing larger amounts of caffeine or dietary supplements containing anhydrous

TABLE 14.1.1 Blood Results on Admission		
	Case 1	Case 2
White cell count (10^9/L)	14.23	11.43
Serum potassium (mmol/L)	2.7	2.4
Serum bicarbonate (mmol/L)	20	17.7
Arterial blood pH	7.39	7.39
Blood lactate (mmol/L)	6.9	6.9
ALT (U/L)	163	136
Serum creatine kinase (U/L)	899	1369
Serum creatinine (mmol/L)	67	78

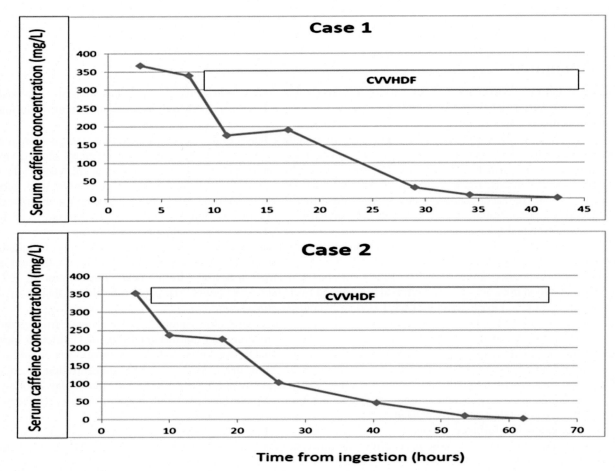

FIGURE 14.1.1 Caffeine concentrations.

caffeine have become increasingly used as performance enhancers. Severe side effects have been reported after ingestion of 6–10 g (100–150 mg/kg body weight) of caffeine, although patients have survived after ingestion of up to 100 g caffeine with supportive care.

Caffeine is structurally similar to adenosine. Nonselective antagonism of the adenosine receptor subtypes A1 and A2 by caffeine at high doses can result in seizures. Caffeine also causes beta-receptor agonism and inhibition of the enzyme phosphodiesterase leading to increased adrenergic effects. Sympathomimetic effects include anxiety, tremor, irritability, tachyarrhythmias (β_1-stimulation); vasodilation and hypotension, hypokalemia, leukocytosis, and glycogenolysis (all β_2-stimulation); and lipolysis (β_3-stimulation). In toxic doses, caffeine directly releases calcium from intracellular stores, which may also increase the susceptibility to arrhythmias. Rhabdomyolysis can result from a combination of ischemia, seizures and tetany mediated by intracellular calcium release from the endoplasmic reticulum.

In healthy adults, caffeine is completely and rapidly absorbed following ingestion, with the mean peak plasma concentration occurring within 15−120 minutes. Peak caffeine serum concentrations after drinking three cups of coffee (250 mg caffeine) are around 5−10 mg/L. Caffeine undergoes demethylation by cytochrome P450 isoform CYP1A2 to paraxanthine (80%), theobromine (11%), and theophylline (4%) by the hepatic cytochrome P_{450} CYP1A2 system. The mean elimination half-life of caffeine in healthy adults is 4−5 hours but varies widely between patients and may be prolonged in overdose, in those with hepatic impairment, and in pregnancy.

Although plasma caffeine concentration might be useful in confirming the diagnosis, it is not routinely available in the clinical setting. Analytic methods used for measuring caffeine and its metabolites include immunoassay with monoclonal antibodies, liquid chromatography with ultraviolet detection, liquid chromatography coupled to a mass spectrometry detector (LC−MS) and high performance liquid chromatography. We used LC−MS in our university-based toxicology laboratory to determine plasma caffeine concentrations using a published method [1]. In the absence of LC−MS, methods using high-performance liquid chromatography also have high sensitivity [2].

The clinical severity of caffeine toxicity is generally determined from the sympathomimetic effects. Hypokalemia can be very severe, resulting from β_2-stimulation leading to movement of potassium into cells. Hypokalemia can cause electrophysiological effects, including delayed cardiac repolarization manifest as QT prolongation on the ECG, which may increase the risk of arrhythmias. Lactic acidosis is also correlated with the severity of poisoning. It results from a combination of tissue hypoxia and sympathetic stimulation causing increased glycogenolysis and lipolysis, leading to the formation of pyruvate which is then converted to lactate [3,4].

Management of caffeine toxicity is largely supportive and symptomatic. Activated charcoal can be administered to prevent absorption but is often precluded by protracted nausea and vomiting. Continuous cardiac monitoring is necessary. Intravenous fluids should be used together with aggressive correction of serum potassium. Hemodynamically stable arrhythmias can usually be treated by correction of hypoxia, metabolic acidosis with sodium bicarbonate and treatment of hypokalemia and hypomagnesaemia. Short-acting beta-blockers (e.g., esmolol) may be useful if arrhythmias persist. Advanced cardiac life support protocols are used for hemodynamically unstable arrhythmias. Lipid emulsion has been used successfully in a few cases of caffeine overdose for severe cardiovascular instability [5,6] but the mechanism for its potential beneficial effect is unclear.

Caffeine has a low protein binding of around 35% and low volume of distribution of 0.6−0.8 L/kg, suggesting that its clearance might be enhanced by extracorporeal elimination techniques such as hemodialysis. There are few reported cases of use of hemodialysis in caffeine toxicity. In one case of caffeine overdose in a 20-year-old woman, cardiac arrest occurred but she survived after advanced cardiac life support, esmolol infusion, and hemodialysis. Serum caffeine concentrations measured around 6 and 18 hours after overdose (pre/postdialysis) were 240.8 and 150.7 mg/L [7]. Similarly, a patient survived after a 100 g caffeine overdose after institution of hemodialysis followed by CVVHDF. CVVHDF was less effective than hemodialysis in enhancing caffeine clearance [8]. In our 2 cases, CVVHDF was used successfully with rapid falls in caffeine concentrations. However, the elimination caffeine half-lives of around 4−6 hours are comparable to those in normal healthy volunteers but shorter than the caffeine half-life of up to 16 hours in overdose.

References

[1] Ptolemy AS, Tzioumis E, Thomke A, Rifai S, Kellogg M. Quantification of theobromine and caffeine in saliva, plasma and urine via liquid chromatography-tandem mass spectrometry: a single analytical protocol applicable to cocoa intervention studies. J Chromatogr B 2010;878(3-4):409−16.

[2] Lopez-Sanchez RDC, Lara-Diaz VJ, Aranda-Gutierrez A, Martinez-Cardona JA, Hernandez JA. HPLC method for quantification of caffeine and its three major metabolites in human plasma using fetal bovine serum matrix to evaluate prenatal drug exposure. J Anal Methods Chem 2018;2018:2085059.

[3] Schmidt A, Karlson-Stiber C. Caffeine poisoning and lactate rise: an overlooked toxic effect? Acta Anaesthesiol Scand 2008;52:1012−14.

[4] Morita S, Yamagiwa T, Aoki H, et al. Plasma lactate concentration as an indicator of plasma caffeine concentration in acute caffeine poisoning. Acute Med Surg 2014;1:159−62.

[5] Schmidt M, Farna H, Kurcova I, Zakharov S, Fric M, Waldauf P, et al. Successful treatment of supralethal caffeine overdose with a combination of lipid infusion and dialysis. Am J Emerg Med 2015;33:738.
[6] Muraro L, Longo L, Geraldini F, Bortot A, Paoli A, Boscolo A. Intralipid in acute caffeine intoxication: a case report. J Anesth 2016;30:895–9.
[7] Laskowski LK, Henesch JA, Nelson LS, Hoffman RS, Smith SW. Start me up! Recurrent ventricular tachydysrhythmias following intentional concentrated caffeine ingestion. Clin Toxicol (Phila) 2015;53:830–3.
[8] Fausch F, Uehlinger DE, Jakob S, Pasch A. Haemodialysis in massive caffeine intoxication. Clin Kidney J 2012;5:150–2.

Chapter 14.2

Delayed presentation and conservative management of an intraarterial injection of crushed amphetamine/dextroamphetamine salts

Cornelia McDonald[1] and Justin Arnold[2]

[1]*Department of Emergency Medicine, University of Alabama at Birmingham, Birmingham, AL, United States,* [2]*Department of Internal Medicine, Division of Emergency Medicine, University of South Florida, Tampa, FL, United States*

Case description

A 37-year-old right hand dominant Caucasian female with a medical history of bipolar disorder, depression, hepatitis C, and intravenous (IV) drug abuse presented to a tertiary medical center emergency department with pain and discoloration to her right upper extremity (RUE). The onset of the patient's symptoms was 15 hours prior to arrival to the hospital. She disclosed that she had attempted inject crushed amphetamine/dextroamphetamine tablets dissolved in water into a vein she had located in her right antecubital fossa. Immediately after injecting the patient felt a burning sensation throughout her forearm, and about 20 minutes later, she started having more severe diffuse arm pain and skin changes distal to the injection site. The patient's vital signs on arrival to the emergency department were unremarkable with a temperature of 97.5°F, heart rate of 86 beats per minute, respiratory rate of 18 breaths per minute, blood pressure of 139/86 mmHg, and oxygen saturation of 98% on room air. In triage, she was noted to have a radial pulse to palpation but the right hand was swollen with fingertips having a purple discoloration.

Her physical exam was notable for skin of the right forearm, hand, and fingers that had a molted appearance with numerous violaceous patches that was also cool to the touch (Fig. 14.2.1). Her capillary refill was less than 3 seconds and but she luckily had palpable radial and ulnar pulses in the RUE. Motor exam of RUE was limited secondary to pain. Given her story and physical exam on arrival, it became evident that the patient had inadvertently injected herself arterially with her drug of choice rather than intravenously. The intraarterial (IA) injection of her prepared amphetamine/dextroamphetamine solution had caused acute ischemic effects to her upper extremity, and this mechanism was likely secondary to arteriospasm versus arterial thrombosis distal to the injection site. An emergent RUE arterial and venous duplex ultrasound was obtained in the emergency department and did not show any evidence of flow-limiting stenosis or other abnormalities that would require anticoagulation or immediate surgical or intravascular intervention. Initial laboratory evaluation was notable for a normal complete blood count, basic metabolic panel including renal function, and a creatine phosphokinase (CPK) level of 2233 U/L (normal < 190 U/L) suggesting muscle injury, possibly from ischemia. Emit II Plus Urine Drug Screen Assay (Siemens Healthineers USA) was qualitatively positive for amphetamines.

Previously reported IA injections of other sympathomimetics frequently describe patients presenting earlier in their onset of symptoms and undergoing treatment with arteriodilators, most commonly phentolamine, to improve distal blood flow [1–4]. However, at the time of this patient's presentation the alpha-adrenergic antagonist phentolamine was unavailable due to national shortages. The patient was subsequently admitted to the hospital for compartment and pulse checks, pain control, IV fluids, and elevation of her affected extremity.

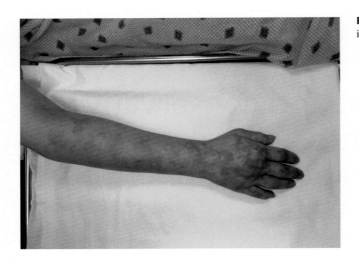

FIGURE 14.2.1 Right forearm and hand 15 hours post-intraarterial injection.

On day 2 of admission, the patient's CPK reached a peak serum concentration of 3362 U/L. A commuted tomography (CT) scan of her right forearm and hand revealed ill-defined evidence of myonecrosis of the flexor and extensor muscles of the forearm and thenar eminence. While she had evidence of rhabdomyolysis and CT evidence of diffuse distal extremity muscle injury on imaging, her physical exam on this day had actually demonstrated clinical improvement. The plastic surgery consult service was consulted and they noted normal capillary refill and poor demarcation of the myonecrosis on CT and agreed with continued conservative management and close monitoring.

The patient continued to improve and was discharged home on day 6 with decreased overall swelling and pain and resolving rhabdomyolysis. She was discharged home with an oral pain regimen consisting of gabapentin 100 mg three times a day (TID), ibuprofen 800mg TID, and Tramadol 50 mg TID as needed. She was seen in clinic at 1 and 2 weeks following discharge from the hospital with continued improvement and near-resolution of all symptoms nearly 3 weeks after her injury.

Discussion

Amphetamines are a broad class of lipophilic and weakly basic xenobiotics that exhibit their clinical effect by increasing release of dopamine and norepinephrine from presynaptic storage vesicles and by blocking reuptake of catecholamine within the central nervous system (CNS). As a class, they serve as a CNS stimulant that increases wakefulness, alertness, energy, reduces hunger and provides an overall sense of well-being. Amphetamines can be ingested, injected, or smoked and inhaled and are readily absorbed through all routes.

Amphetamine and in combination with its enantiomer dextroamphetamine specifically have a rapid onset of action within 30–60 minutes and a duration of effects lasting 4–6 hours when taken orally at therapeutic doses. It has a V_d of 3.5–4.6 L/kg and is metabolized through hepatic glucuronidation and CYP450 monooxygenase with a therapeutic half-life in adults of 10 and 13 hours for D-amphetamine and L-amphetamine, respectively (Fig. 14.2.2). IV or IA kinetics are not well delineated, but given avoidance of first pass metabolism via injection, a much more rapid onset of symptoms is anticipated. As a class they are metabolized to deaminated and hydroxylated metabolites and excreted in the urine where they are most often easily detected. Elimination is predominantly renal in nature and is highly pH dependent with acidic urine significantly increasing drug elimination.

Our patient's urine was analyzed with the Emit II Plus Urine Drug Screen Assay (Siemens Healthineers USA) which screened for amphetamines, barbiturates, benzodiazepines, buprenorphine, cannabis, cocaine, heroin, methadone, opiates, and oxycodone. The amphetamine assay, specifically, was a monoclonal antibody immunoassay that detects D-amphetamine, D-methamphetamine, L-amphetamine, L-methamphetamine, D,L-methamphetamine, methylenedioxyamphetamine, methylenedioxymethamphetamine, and methylenedioxyethylamine in human urine. The use of monoclonal antibodies in this assay is subject to less false-positive reactions involving amphetamine-like compounds when compared to similar polyclonal assays. The Emit II Plus immunoassay can detect amphetamines in the urine within 3 hours after and up to 24–48 hours following amphetamine use [5,6]. The limit of detection for our laboratory's assay was 1000 ng/mL for D-amphetamine and 8651 ng/mL for L-amphetamine, the two primary metabolites of the patient's amphetamine/dextroamphetamine.

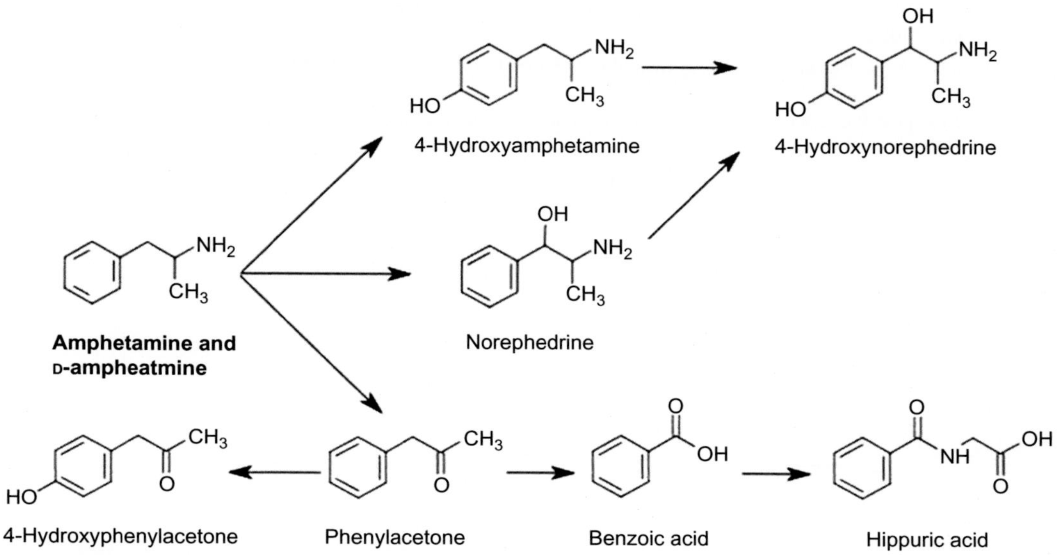

FIGURE 14.2.2 Metabolism of amphetamine and its enantiomer dextroamphetamine.

Given the patient provided a reliable history and clinical examination consistent with inadvertent IA injection of her crushed amphetamine/dextroamphetamine salts, qualitative immunoassay-based urine drug screening was felt to be sufficient and quantitative GC/MS testing which was not available on-site was not felt to be clinically useful and was deferred. Had there been diagnostic uncertainty regarding the injected substance, GC/MS testing to determine the specific compound could have been pursued by the treating team.

This case was unique when compared to previously reported IA injection of sympathomimetic xenobiotics given the delay in presentation and the lack of availability of the most appropriate treatment medication due to hospital shortage. Given palpable pulses and the lack of any flow limiting lesion on vascular ultrasound it was felt persistent drug-induced vasospasm was unlikely and aggressive arteriodilator therapy would not likely provide additional benefit. In fact, phentolamine could theoretically contribute to further damage by mechanism of ischemic reperfusion injury associated with a rapid improvement in tissue perfusion following the ischemic event.

Although this patient experienced pain as well as significant skin changes and localized myonecrosis, she recovered fully without any pharmacologic or surgical intervention. Clinicians should now consider conservative management as a possible treatment option in patients that present with myonecrotic injuries associated inadvertent IA injection of sympathomimetic agents. The decision to utilize arteriodilators in this clinical scenario can be guided by rapid evaluation the degree of arteriorspasm and amount of myonecrosis that has and is anticipated to occur. Clinicians are encouraged to supplement their clinical examination with arterial ultrasound versus CT angiography in addition to evaluation of serum CPK, lactate, basic metabolic panel as well as urine immunoassay screening and possible GC/MS confirmation of the offending agent(s).

In conclusion, we report a successful case of conservative treatment of an inadvertent IA injection of crushed amphetamine/dextroamphetamine salts. Although arteriodilators such as phentolamine have been traditionally been considered first-line therapies, rapid clinical assessment, laboratory evaluation, and imaging may aid clinicians in determining if pharmacologic or conservative therapies are the most appropriate treatment choices for their specific patient.

References

[1] Birkhahn H, Heifetz M. Accidental intra-arterial injection of amphetamine: an unusual hazard of drug addiction. Br J Anaesth 1973;45 (7):761–3. Available from: https://doi.org/10.1093/bja/45.7.761.
[2] Markovchick V, Burkhart KK. The reversal of the ischemic effects of epinephrine on a finger with local injections of phentolamine. J Emerg Med 1991;9(5):323–4. Available from: https://doi.org/10.1016/0736-4679(91)90374-o.
[3] Sen S, Chini EN, Brown MJ. Complications after unintentional intra-arterial injection of drugs: risks, outcomes, and management strategies. Mayo Clin Proc 2005;80(6):783–95. Available from: https://doi.org/10.4065/80.6.783.

[4] Hinterberger JW, Kintzi HE. Phentolamine reversal of epinephrine-induced digital vasospasm. How to save an ischemic finger. Arch Fam Med 1994;3(2):193—5 PMID: 7994442.
[5] NIDA research monograph 73 Hawks RL, Chiang CN, editors. Urine testing for drugs of abuse. Rockville, MD: National Institute on Drug Abuse (NIDA); 1986.
[6] American Society of Hospital Pharmacists. AHFS drug information '87, American Society of Hospital Pharmacists, Inc; 1987. p. 1105—7.

Chapter 14.3

Now you see it, now you don't: ecstasy or not?

Jayson V. Pagaduan[1,2,*], Marianne Benyon[1,2,*] and Sridevi Devaraj[1,2]
[1]Department of Pathology and Immunology, Baylor College of Medicine, Houston, TX, United States, [2]Section of Clinical Chemistry, Department of Pathology, Texas Children's Hospital, Houston, TX, United States

Case description

A 37-year-old female, in her third trimester of pregnancy was admitted to the Pavilion for Women at Texas Children's Hospital (PFWTCH), Houston Texas. The patient reported no complications with her pregnancy and denied any significant exposure to alcohol, tobacco, drugs, or other known teratogens during pregnancy. It is TCH policy to perform urine drug screen when an expectant mother has no prior record of prenatal care.

The patient was placed under general anesthesia. Before delivery, urine was collected and was sent for urine drug-screen test to the clinical laboratory. Our laboratory uses TOX/SEE (Bio-Rad Laboratories) for urine-drug screening. Results showed positive for 3,4-methylenedioxymethamphetamine (MDMA), also known as Ecstasy, and methamphetamine (MET). After an uneventful delivery and 4.5 hours after the previous urine drug screen, urine was collected again and drug screen was repeated. The test showed negative for MET and indeterminate for MDMA. Only the second urine sample was sent to reference laboratory, ARUP, for urine screen with reflex to confirmation.

TOX/SEE is a CLIA waved test which utilizes competitive immunoassay to detect presence of drug. Given that the control line is positive, the absence of colored line next to the drug name is interpreted as negative while the presence of the colored line is presumed positive. The TOX/SEE concentration cut-off for MET and MDMD are 1000 ng/mL and 500 ng/mL. At TCH, two technologists interpret the result and a discrepant interpretation will be considered indeterminate.

ARUP Laboratories confirmatory drug-screen test is a two-step process. The first drug-screen is done using immunoassay. In their assay, MDMA is reported together with the amphetamines. The indicated immunoassay amphetamine concentration cut-off is 300 ng/mL. If sample tested positive, LC/MS or GC/MS would be performed. But since the sample tested negative for any drugs in their panel, the sample was not further analyzed.

Discussion

Review of the patient's medical history revealed that 25 mg of ephedrine was given to the patient subcutaneously before delivery. The first urine was collected about 3.5 hours after ephedrine was administered. The urine drug screen showed positive for MDMA and MET. About 8 hours after ephedrine was given, another urine sample was collected and urine drug screen result was interpreted as negative for MET and indeterminate for MDMA.

Ephedrine is a sympathomimetic amine that is commonly included in over the counter medicine for cold cough, asthma, and even dietary supplements [1–3]. In clinical obstetrics, ephedrine is used during labor to mitigate arterial hypotension and ensure uterine perfusion [4]. Ephedrine and MDMA are both MET derivatives. We hypothesize that ephedrine may cause the positive urine drug screen for MET and MDMA. Ephedrine is mostly excreted unchanged in urine [3,5]. Pharmacokinetics studies showed that ephedrine has a plasma half-life of 3–6 hours depending on the urinary pH [6]. Furthermore, ephedrine is presumed to cross the placental barrier [7].

---

*Authors contributed equally.

TABLE 14.3.1 Urine Drug Screen Results After Spiking With Known Ephedrine

Ephedrine in Urine (μg/mL)	TOX/SEE Results
5	Negative for all drugs
250	Positive only for MET
500	Positive for MET, indeterminate for MDMA
1000	Positive for MET and MDMA

MDMA, 3,4-Methylenedioxymethamphetamine; *MET*, methamphetamine.

Urine drug screen is easy to perform however, the multitude of possible cross-reacting substances result in false positive necessitating the need to perform confirmatory testing for positive urine drug screen results [8]. To confirm the hypothesis that ephedrine resulted in positive urine drug screen, we spiked urine that is negative from any drugs with different ephedrine concentrations. The ephedrine-spiked urine samples were tested on TOX/SEE. At our institution we use the TOX/SEE urine drug screen cartridge that can detect 12 classes of drugs such as amphetamines, METs, barbiturates, benzodiazepine, cocaine/benzoylecgonine, methadone, opiates, phencyclidine, tetrahydrocannabinoids, tricyclic antidepressants, MDMA, and oxycodone. The results were interpreted 5 minutes after sample application by visual inspection of the bands.

The results (see Table 14.3.1) indicated positive for MET only when the ephedrine concentration is at least 250 μg/mL. While at 500 μg/mL, the result was interpreted as positive for MET and indeterminate for MDMA due to the very faint MDMA band. Unequivocal positive urine drug screen for MET and MDMA was observed when ephedrine concentration was 1000 μg/mL. The concentrations that led to positive MET and MDMA in our spiked-urine samples were way above the cut-off for the TOX/SEE. The package insert for TOX/SEE indicated cross-reactivity with other drugs but no cross-reactivity cut-off was reported for ephedrine.

The negative result for MET and MDMA from ARUP may be attributed to the lower concentration of ephedrine in the second urine sample. In this case the first urine sample should have been sent for confirmatory testing since the first urine sample was unequivocally positive compared to the second urine sample. However, it is also important to note that urine drug screen kits from different vendors may have different cut-off and cross-reactivity, resulting in discrepant results. Therefore it is important to know the limitations of the assay used.

Urine drug screen for expectant mothers are usually performed when the patient admitted use of illicit drugs or suspicion of drug abuse by clinician. If urine drug screen is part of the protocol in the hospital, we recommend that urine drug test must be done, if possible, before any medication is given to the mother to decrease the incidence of false positive urine drug screen results. By doing so, the need for confirmatory testing may be reduced by decreasing false positive results. If not possible, review of medications taken prior to urine collection may be done to give clinicians possible causes of positive results while waiting for confirmatory results. The latter suggestion requires a collaboration with manufacturers and laboratorians to report interfering drugs. However, with the advent of increasing variety of synthetic drugs, there is a possibility that synthetic drugs may still be missed even when confirmatory testing is done. As laboratorians, we can also make the clinicians aware of iatrogenic agents that could interfere with urine drug screening.

Key points

1. Immunoassay urine drug screening method is susceptible to false positive results, thus requiring confirmatory testing before releasing the results to the patient.
2. Ephedrine could result in false positive for MDMA and MET in TOX/SEE urine drug screen.
3. Review of medications and patient history are vital to glean clues that could explain the urine drug-screen results.
4. We strongly suggest, if possible, to collect urine for drug screening before any medications are given to avoid false-positive results due to drugs administered after the patient is admitted.

References

[1] Powers ME. Ephedra and its application to sport performance: another concern for the athletic trainer? J Athl Train 2001;36:420–4.
[2] Cooper RJ. Over-the-counter medicine abuse—a review of the literature. J Subst Use 2013;18:82–107.

[3] Pickup ME, May CS, Ssendagire R, Paterson JW. The pharmacokinetics of ephedrine after oral dosage in asthmatics receiving acute and chronic treatment. Br J Clin Pharmacol 1976;3:123–34.
[4] Ducros L, Bonnin P, Cholley BP, Vicaut E, Benayed M, Jacob D, et al. Increasing maternal blood pressure with ephedrine increases uterine artery blood flow velocity during uterine contraction. Anesthesiology 2002;96:612–16.
[5] Tseng YL, Shieh MH, Kuo FH. Metabolites of ephedrines in human urine after administration of a single therapeutic dose. Forensic Sci Int 2006;157:149–55.
[6] Csajka C, Haller CA, Benowitz NL, Verotta D. Mechanistic pharmacokinetic modelling of ephedrine, norephedrine and caffeine in healthy subjects. Br J Clin Pharmacol 2005;59:335–45.
[7] Ngan Kee WD, Khaw KS, Tan PE, Ng FF, Karmakar MK. Placental transfer and fetal metabolic effects of phenylephrine and ephedrine during spinal anesthesia for cesarean delivery. Anesthesiology 2009;111:506–12.
[8] Brahm NC, Yeager LL, Fox MD, Farmer KC, Palmer TA. Commonly prescribed medications and potential false-positive urine drug screens. Am J Health Syst Pharm 2010;67:1344–50.

Chapter 14.4

What is in the cocaine? The vessels never lie—cocaine-induced vasculitis

Alexandra Rapp and Anthony O. Okorodudu
Department of Pathology, The University of Texas Medical Branch, Galveston, TX, United States

Case description

A 45-year-old female with a past medical history significant for type II diabetes mellitus, hypertension, vasculitis of unknown type with no family history of autoimmune vasculitis, untreated hepatitis C, bipolar disorder, and cocaine abuse presented with purpuric skin lesions on the nose, bilateral ears, and bilateral upper and lower extremities. Two days prior to presentation, the patient had been in pain secondary to arthritis, so she used cocaine from a friend for self-management. Overnight following the exposure to cocaine, she developed skin lesions. She reported similar skin lesions following cocaine self-administration in the past. Regarding her history of vasculitis, she has had recurrent episodes of vasculitis of unknown etiology for the past 3 years treated with corticosteroids (dose and duration unknown), which she stopped 1 month ago due to running out of medication and not renewing her indigent care. The episodes have been associated temporally with cocaine use.

On presentation the patient's vital signs were stable. Physical exam demonstrated multiple large, purpuric lesions with bullae, mild hemorrhage, and crusting located on the bilateral upper and lower extremities, bilateral ears, and nose.

Given the patient's history of cocaine abuse, the patient's urine was sent for a comprehensive urine drug-screen involving automated enzyme immunoassays for amphetamines, opiates, phencyclidine, cocaine, benzodiazepines, propoxyphene, cannabinoids, methadone, and barbiturates. Immunoassay resulted as presumptively positive for cocaine and opiates, the latter consistent with her currently identified prescriptions of morphine and tramadol. The urine was then used for a more comprehensive confirmatory test via liquid chromatography-mass spectrometry that tests for approximately 40 drugs and/or their metabolites. The results were positive for morphine and tramadol, again consistent with the patient's current prescriptions for pain. The results were also positive for benzoylecgonine, the metabolite of cocaine, at a concentration of >300,000 ng/mL prior to creatinine normalization, which demonstrated that the urine was not diluted. This result was consistent with the patient's stated history of recent cocaine use.

Because of the patient's temporal relationship of developing vasculitic skin lesions after cocaine use in the context of a history of vasculitis of unknown type with no family history, we highly suspected the patient to have cocaine-induced vasculitis.

Discussion

Cocaine is an alkaloid extract from the *Erythroxylum* coca plant that is abused for its euphoric and stimulatory effects and has significant adverse impact on the cardiovascular system, including direct vasoconstrictor effects [1]. In addition to stimulating vasoconstriction and vasospasm, it has been discovered to cause a wide variety of vasculitis and vasculitis-like syndromes. Cerebral vasculitis was one of the first types of vasculitis associated with cocaine use documented in the medical literature as a JAMA case report. The patient was a 22-year-old, previously healthy man with no significant past medical history other than cocaine use who was found to have cerebral angiogram findings consistent with cerebral vasculitis. After a thorough work-up including serological testing, which revealed no other vasculitic etiology, his treating team diagnosed him with cocaine-induced cerebral vasculitis [2]. Multiple other cases of cerebral vasculitis associated with cocaine use have been reported. However, these vasculitis-like presentations were discovered to not just be limited to the central nervous system, with the first case documented consisting of a presentation of skin vasculitis, hypokalemia, and acute renal failure in cocaine-induced rhabdomyolysis [3].

Returning to the presentation of cerebral vasculitis associated with cocaine use, the main limitation of this diagnosis had previously been relying on angiographic findings without histologic confirmation of vasculitis. The importance of this histologic verification lies in differentiating vasculitis versus cocaine-induced vasospasm, which present similarly with necrotic, destructive, and/or ischemic lesions but requires markedly different treatments. In 1991 a case of acute encephalopathy following intravenous and intranasal cocaine was the first documented case of histologically confirmed cerebral vasculitis. In this case, brain biopsy demonstrated vascular changes of primarily small arteries, including lymphocytic infiltration, endothelial thickening, and deposition of proteinaceous amorphous material within and around vessel walls—all confirming the diagnosis of cocaine-induced vasculitis. In light of this diagnosis, high-dose intravenous steroids were administered, and the patient subsequently improved [4].

With histopathology, multiple different vasculitic syndromes have been described. First was a case of cocaine-induced eosinophilic granulomatosis with polyangiitis confirmed with skin and muscle biopsy of a patient who after freebase cocaine-smoking developed bronchoconstriction, arthritis, microhematuria, pruriginous skin rash, and mononeuritis multiplex [5]. Gertner and Hamlar reported the first documented case of a necrotizing granulomatosis vasculitis associated with nasal destruction and an oronasal fistula in a chronic cocaine user [6].

In addition, there is an ANCA-associated vasculitis attributed to levamisole adulteration. Levamisole, an anthelmintic drug, present in about 71% of cocaine on the US streets, can cause an ANCA-associated vasculitis characterized by cutaneous, renal, and pulmonary manifestations. Biopsy typically reveals a small vessel vasculitis that can be distinguished from granulomatosis with polyangiitis, eosinophilic granulomatosis with polyangiitis, and microscopic polyangiitis by its temporal relationship to cocaine use. Ninety percent of patients present with cutaneous lesions include purpura, necrosis, abscesses, and bullae, similar to our above case presentation. Arthralgias occur in 31%—83%, and constitutional symptoms are present 72% of the time [7]. This form of vasculitis as well as levamisole-related pseudovasculitis—characterized by the absence of granulomas or leukocytoclasia on biopsy—is important to differentiate from the above diagnoses because drug cessation is highly curative in the former and thus can avoid exposing affected patients to potentially dangerous side effects of corticosteroids and immunosuppressants [7—9]. Pathogenesis of this form of vasculitis has recently been attributed to neutrophil extracellular traps, with implications for further treatment [7].

In conclusion, based on the presentation of purpuric lesions with bullae and arthralgia associated with vasculitis temporally related to cocaine use from a potentially questionable source, the patient we present likely has levamisole-associated vasculitis. While the patient has received steroids in the past, it is likely that drug cessation alone can be used to effectively treat her vasculitis, as relapse in such cases is attributed directly to continued cocaine use. Thus patient and physician education is required to effectively manage these patients while minimizing cost and side effects of immunosuppressive medications, thus improving patients' quality of life.

References

[1] Boghdadi MS, Henning RJ. Cocaine: pathophysiology and clinical toxicology. Heart Lung 1997;26:466—83.
[2] Kaye BR, Fainstat M. Cerebral vasculitis associated with cocaine abuse. JAMA 1987;258(15):2104—6.
[3] Enriquez R, Palacios FO, Gonzalez CM, Amoros FA, Cabezuelo JB, Hernandez F. Skin vasculitis, hypokalemia and acute renal failure in rhabdomyolysis associated with cocaine. Nephron 1991;59(2):336—7.
[4] Fredericks RK, Lefkowitz BS, Challa VR, Troost BT. Cerebral vasculitis associated with cocaine abuse. Stroke 1991;22(11):1437—9.
[5] Orriols R, Munoz X, Ferrer J, Huget P, Morell F. Cocaine-induced Churg-Strauss vasculitis. Eur Respir J 1996;9(1):175—7.
[6] Gertner E, Hamlar D. Necrotizing granulomatous vasculitis associated with cocaine use. J Rheumatol 2002;29(8):1795—7.
[7] Jin Q, Kant S, Alhariri J, Geetha D. Levamisole adulterated cocaine associated ANCA vasculitis: review of literature and update on pathogenesis. J Community Hosp Intern Med Perspect 2018;8(6):339—44.
[8] Fan T, Macaraeg J, Haddad TM, Bacon H, Le D, Maza M, et al. A case report on suspected levamisole-induced pseudovasculitis. WMJ 2017;116(1):37—9.
[9] Subesinghe S, van Leuven S, Yalakki L, Sangle S, D'Cruz D. Cocaine and ANCA associated vasculitis-like syndromes — a case series. Autoimmun Rev 2018;17(1):73—7.

Chapter 14.5

Polysubstance abuse and rhabdomyolysis

Paul E. Young, John O. Ogunbileje and Anthony O. Okorodudu
Department of Pathology, University of Texas Medical Branch, Galveston, TX, United States

Case description

A 23-year-old man with 1-day history of agitation, passage of tea colored urine and progressively decreased urine volume presented to the emergency room in the company of law enforcement officers. He had been an inmate in the county jail for the past 2 days and was brought in restraints owing to his agitated state. He reported usage of marijuana, methamphetamine, and Xanax (alprazolam) "bars" 2 days prior to presentation and stated that methamphetamine is his preferred drug. According to his drug use history, he uses methamphetamine "all day, every day." While in jail, his urine became dark and the volume gradually decreased, prompting presentation. On physical examination, he was severely dehydrated, anxious, agitated and hallucinating. His vital signs on presentation include: BP 137/88, pulse 104 beats per minute, temperature 37.9°C (100.3°F), respiratory rate 18 breaths per minute and SpO_2 97% on room air. Initial laboratory results include potassium 4.8 mmol/L [reference interval (RI) 3.5–5.0 mmol/L], sodium 153 mmol/L (RI 135–145 mmol/L), chloride 120 mmol/L (RI 98–108 mmol/L), bicarbonate 15 mmol/L (RI 23–31 mmol/L), total protein 8.6 g/dL (RI 6.3–8.2 g/dL), albumin 5.5 g/dL (RI 3.5–5.0 g/dL), creatinine 2.60 mg/dL (RI 0.60–1.25), BUN 50 mg/dL (RI 7–23 mg/dL), total bilirubin 2.1 (RI 0.1–1.1 mg/dL), AST 119 U/L (RI 13–40 U/L), ALT 80 U/L (RI 9–51 U/L), anion gap 18 (RI 2–16), and estimated glomerular filtration rate (eGFR) 30.7 mL/min/1.73 m^2. He had no prior history of kidney disease. Past medical/psychiatric history was significant for depression, bipolar disorder, post-traumatic stress disorder (PTSD) since age 12 years and polysubstance abuse (methamphetamine, cocaine, heroin, alcohol, marijuana, and benzodiazepines) since age 15 years, with documented alcohol withdrawal seizures in the past. A diagnosis of acute renal failure (ARF) and severe dehydration was made. He was started on intravenous fluids and worked up for the etiology of renal failure.

Patient follow-up

The physical examination and laboratory results are consistent with dehydration, anion gap metabolic acidosis and ARF of unclear etiology. The presence of tea colored urine and decreased urine volume prompted suspicion for rhabdomyolysis [1]. Creatine kinase (CK) activity was ordered and reported at 5305 U/L (RI 33–194 U/L). Because the patient had no other risk factors for rhabdomyolysis, including trauma, the etiology was suspected to be secondary to illicit drug abuse. One study reported that over 80% of cases of rhabdomyolysis were caused by alcohol and drugs of abuse, with one third of these patients developing ARF [1].

Urine toxicology drug screen by immunoassay on the Vitros 5600 analyzer (Ortho-clinical diagnostics, Rochester, NY) revealed positive results for benzodiazepines, amphetamine, methamphetamine, tetrahydrocannabinol (THC), and cocaine. Liquid chromatography tandem mass spectrometry (Waters, Milford, MA), confirmed the presence of alprazolam (Xanax), amphetamine, methamphetamine, and cocaine. The concentration of methamphetamine and amphetamine, its metabolite, were the highest, with values of >15,000 ng/mL and >40,000 ng/mL, respectively. Given the high sensitivity and specificity of THC immunoassay, coupled with the patient's admission of marijuana use, it was deemed unnecessary to confirm THC using mass spectrometry, as this would not affect therapy in any way. In addition, ethanol was not tested since the patient had a documented history of alcoholism and he confirmed drinking a lot of alcohol prior to his arrest. He stated that he drinks a fifth of liquor and a six-pack daily.

A further diagnosis of rhabdomyolysis complicated by ARF secondary to illicit drug abuse was made. Following treatment with intravenous fluids, the kidney function improved remarkably. Twenty-four hours after presentation, BUN and creatinine decreased to 15 mg/dL (RI 15−23 mg/dL) and 0.81 mg/dL (RI 0.60−1.25 mg/dL), respectively, while eGFR increased to 118.1 mL/min/1.73 m^2. CK activity steadily trended downward to 4506 and 2257 U/L at 8 and 32 hours, post admission. His urine output also improved and ARF resolved completely at the time of discharge from the hospital.

Prior to discharge, the patient complained of generalized body pain, depressed mood and suicidal ideations which are consistent with methamphetamine withdrawal [2,3]. He refused further diagnostic tests and indicated that he wished to be left alone so he could take his life. He had made three attempts at suicide in the past by overdosing on pills and hanging himself. He was referred to psychiatry on account of high risk of suicide.

Discussion

Polysubstance abuse

Alcohol, marijuana, and tobacco are some of the most commonly abused drugs in the United States, especially among teenagers [4]. Studies have shown that teenagers who start off using more socially acceptable drugs, albeit illegal-for-age, such as alcohol and tobacco eventually progress to the use of more illicit drugs such as marijuana, methamphetamine, and cocaine [4]. Our patient started using illicit drugs in his teenage years and progressed steadily to the use of several illicit drugs. His longest period of abstinence from drug use was for 18 months between ages 17 and 19 years.

Substance abuse and dependence may be triggered or worsened by chronic medical and/or psychiatric illnesses. Our patient had a traumatic experience at the age of 12 years, which led to a diagnosis of PTSD. When probed about the events leading to the diagnosis of PTSD, the patient refused to discuss the subject matter.

Coingestion of multiple drugs of abuse exposes patients to greater toxic effects since many of these drugs work synergistically. Cocaine and alcohol for instance combine in the body to form a compound called cocaethylene, which is more toxic than either drug consumed alone due to the long half-life of cocaethylene (about five times the half-life of cocaine). The risk of death from cocaethylene is 18−25 times greater than that of cocaine [5].

Our patient consumed alcohol, cocaine, methamphetamine, Xanax, and marijuana. The latter four were confirmed by immunoassay and/or liquid chromatography tandem mass spectrometry. Alcohol was not tested because of the known history of alcoholism, coupled with the patient's admission of usage. He has had documented alcohol withdrawal seizures in the past. Agitation, anxiety and hallucination which the patient presented with could be explained by alcohol withdrawal [6], owing to an enforced abstinence from alcohol for the past 2 days in the county jail. He was given oxazepam taper for management of alcohol withdrawal.

Rhabdomyolysis

Rhabdomyolysis is a syndrome characterized by injury to skeletal muscle with release of the muscle contents into plasma [1,7−10]. The muscle contents spilled into plasma include myoglobin, CK, lactate dehydrogenase, phosphate, and potassium [1]. Myoglobin is an early marker for rhabdomyolysis and the dark urine as seen in our patient is an indication of myoglobinuria [1]. The half-life of myoglobin in plasma is short because of its rapid clearance within 6 hours by metabolism in the liver and elimination by the kidneys [1,7]. For this reason, normal serum myoglobin level does not rule out rhabdomyolysis [1,7]. The liver metabolism of myoglobin to bilirubin may explain the increased bilirubin levels in this patient. CK which has a plasma half-life of 1.5 days, is the most abundant enzyme in muscles and is a sensitive marker for rhabdomyolysis [1,7]. After muscle damage, CK activity peaks between 24 and 48 hours and activity level greater than 1000 U/L is generally considered diagnostic of rhabdomyolysis [1,7,9,10].

Trauma is one of the most recognized causes of rhabdomyolysis, especially given that early reports of rhabdomyolysis in the English literature were made during World War II in British people who suffered trauma during bomb blasts [1,9]. Drugs of abuse are however the leading cause of rhabdomyolysis in patients who present to the emergency room, especially in patients who are agitated [1]. Cocaine, amphetamines, heroin and ethanol are among the more common drugs involved [1,7,8]. Our patient has a history of use of all of these drugs. The mechanism by which these drugs of abuse cause rhabdomyolysis vary widely.

Cocaine is directly toxic to muscle [1]. Being a potent vasoconstrictor, chronic cocaine use may lead to muscle ischemia [1]. Additionally, substances used to adulterate cocaine may be myotoxic. About 24% of cocaine abusers develop rhabdomyolysis [1]. Rhabdomyolysis may or may not present with muscle pain, swelling or tenderness [1,9]. In one study, only about 4% of cocaine users with rhabdomyolysis presented with muscle pain/tenderness [9].

Amphetamine abusers typically present in an agitated state, as in our patient. Applying physical restraints to these patients increases risk of exertional rhabdomyolysis [8]. Also, the characteristic choreiform movement seen in amphetamine abusers may predispose muscle to easy breakdown [1]. Like cocaine, amphetamines are vasoconstrictors, thus chronic use may lead to ischemia of muscle cells. Amphetamine users typically have reduced sleep hours, this may also increase their risk of rhabdomyolysis [1].

Ethanol is directly toxic to the muscle cells. Ethanol intoxication leads to immobilization, which further increases risk of rhabdomyolysis [1]. Severe alcohol withdrawal leading to seizures also increases risk of muscle breakdown [1].

Generally, drug abusers tend to have decreased food and fluid intake. This translates to diminished ATP production, dehydration and increased risk of rhabdomyolysis [1].

Kidney failure in rhabdomyolysis is secondary to the toxicity of myoglobin and the obstruction caused by myoglobin and urate crystals in the renal tubules [1]. As these toxic compounds filter through the kidney, they damage the tubules resulting in the characteristic dark colored urine and ARF [1]. This process is worsened by dehydration. It is estimated that about 30% of patients with rhabdomyolysis will develop renal failure and about 10% of all ARF cases are caused by rhabdomyolysis [1].

Disseminated intravascular coagulation may occur due to release of thromboplastin from damaged muscle cells [1]. Very high plasma potassium levels may predispose patients to arrhythmias [1]. Other complications of rhabdomyolysis include metabolic acidosis, liver damage, hyperuricemia, hypocalcemia, and compartment syndrome [1].

The management of rhabdomyolysis involves identification of the etiology of muscle damage, correcting it and aggressively hydrating with intravenous crystalloids to rescue the kidneys by flushing out the toxic compounds [1,8]. The prognosis of well-treated rhabdomyolysis is very good as the repair mechanism of skeletal muscles is very efficient [7].

References

[1] Richards JR. Rhabdomyolysis and drugs of abuse. J Emerg Med 2000;19:51–6.
[2] Marshall BDL, Werb D. Health outcomes associated with methamphetamine use among young people: a systematic review. Addiction 2010;105:991–1002.
[3] McGregor C, Srisurapanont M, Jittiwutikarn J, Laobhripatr S, Wongtan T, White JM. The nature, time course and severity of methamphetamine withdrawal. Addiction 2005;100:1320–9.
[4] Moss HB, Chen CM, Yi H. Early adolescent patterns of alcohol, cigarettes, and marijuana polysubstance use and young adult substance use outcomes in a nationally representative sample. Drug Alcohol Dependence 2014;136:51–62.
[5] Andrews P. Cocaethylene toxicity. J Addict Dis 1997;16(3):75–84.
[6] Maldonado JR, Sher Y, Ashouri JF, Hills-Evans K, Swendsen H, Lolak S, et al. The "Prediction of Alcohol Withdrawal Severity Scale" (PAWSS): systematic literature review and pilot study of a new scale for the prediction of complicated alcohol withdrawal syndrome. Alcohol 2014;48:375–90.
[7] Poels PJE, Gabreels FJM. Rhabdomyolysis: a review of the literature. Clin Neurol Neurosurg 1993;95:175–92.
[8] Curry SC, Chang D, Connor D. Drug and toxin-induced rhabdomyolysis. Ann Emerg Med 1989;18:1068–84.
[9] Welch RD, Todd K, Krause GS. Incidence of cocaine-associated rhabdomyolysis. Ann Emerg Med 1991;20:154–7.
[10] Counselman FL, McLaughlin EW, Kardon EM, Bhambhani-Bhavnani AS. Creatine phosphokinase elevation in patients presenting to the emergency department with cocaine-related complaints. Am J Emerg Med 1997;15:221–3.

Chapter 14.6

Unexpected finding of amphetamine and methamphetamine in a patient on monoamine oxidase inhibitor

Geza S. Bodor
Department of Pathology, University of Colorado School of Medicine, UC Health Laboratories, Aurora, CO, United States

Case description

A 36-year-old female patient, diagnosed with posttraumatic stress disorder had been treated with tricyclic antidepressants for major depressive disorder (depression). She had been self-medicating with medical marijuana in the past but she would like to be completely drug free; therefore she requested to be included in a drug rehabilitation program and be treated as an inpatient. Prerequisite for such treatment is signing a treatment contract that stipulates participation in random urine drug testing and requires the patient's agreement to remain free of nonprescribed drugs. Her urine drug screens have been consistently negative until recently, when she was found positive for amphetamines by the Syva EMIT II. Plus (assay cut off: 1000 ng/mL). Repeat amphetamine screen was performed on a fresh urine specimen and it also showed amphetamines concentration >1000 ng/mL by immunoassay. Amphetamine confirmation by liquid chromatography tandem mass spectrometry (LC–MS/MS) reported amphetamine >2000 ng/mL and methamphetamine >2000 ng/mL. The treating physician requested consultation for interpretation of the unexpected findings because the patients had shown excellent compliance in the past, still denied using any amphetamine-like substance and was motivated to be completely drug free. The review of the patient's medical record and further consultation with her clinician disclosed recent change in her antidepressant medication. Her prior tricyclic antidepressant medication was changed to selegiline approximately one week before the first positive urine drug screen.

Because the use of illicit amphetamine or methamphetamine by the patient would constitute violation of her treatment contract resulting in expulsion from the program, the clinician wanted to know if the presence of amphetamine and methamphetamine is a true or false-positive finding and if it can prove noncompliance.

Discussion

Selegiline (L-deprenyl) is a substituted phenethylamine, the propargyl derivative of L-methamphetamine. Selegiline has been used for the treatment of Parkinson's disease since the late 1970s [1]. Its monoamine oxidase (MAO) enzyme inhibiting properties have been known since the same time [2] prompting its therapeutic utilization for the treatment of depression. Transdermal patch, providing 6–12 mg selegiline in 24 hours, has replaced oral administration because of similar efficacy and diminished side effects [3]. Because of selegiline's specificity for MAO-B enzyme, only minimal numbers of tyramine induced hypertensive attacks have been reported with transdermal selegiline application leading to better tolerance by patients and improved adherence to treatment.

Selegiline is available under many brand names for human and veterinary use. Its chemical structure and its metabolites are depicted on Fig. 14.6.1. The three major metabolites of selegiline are L-desmethylselegiline, L-amphetamine, and L-methamphetamine, which have been reported from treated patients [4] and in an overdose case [5]. More recently Kalasz et al. reviewed the metabolism of selegiline [6] and reported deprenyl-*N*-oxide and over 40 minor metabolites in addition to the well documented three major metabolites. All the major metabolites are the levorotatory isomers, an important fact when interpreting results for patient compliance with illicit drug use policy.

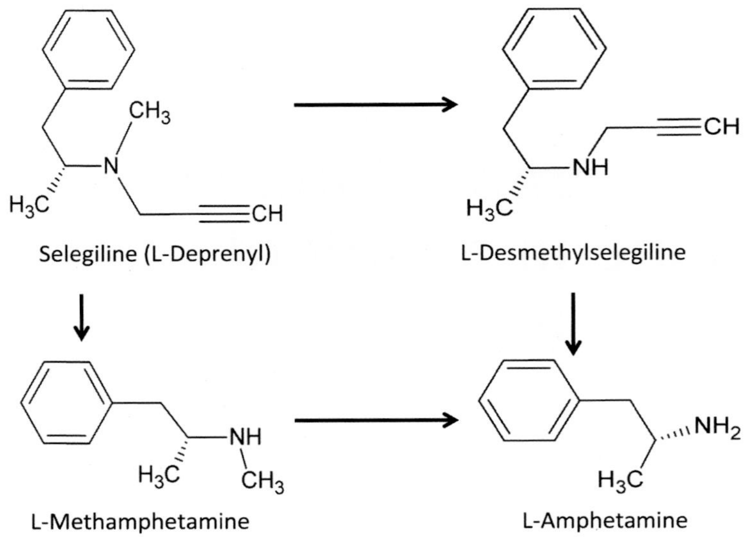

FIGURE 14.6.1 Chemical structure of selegiline and its metabolites. Arrows indicate the irreversible metabolic reactions leading to methamphetamine and amphetamine production in the human body.

Because the presence of amphetamine and methamphetamine in patient samples could result from amphetamines abuse or from taking prescription selegiline, testing for the major metabolites is important to rule out erroneous conclusion of amphetamines abuse.

Testing for the parent drug selegiline is not routinely done in clinical laboratories. It is most likely that prescribed selegiline will be considered when it causes unexpected positive urine amphetamines screen, although this will not happen with all subjects who take selegiline. The reason for this is the exclusive production of levorotatory metabolites and the variable cross reactivity of amphetamine immunoassays with L-amphetamine or L-methamphetamine. Therapeutic amphetamine preparations either use the D-isomers (such as Dexedrine) or a racemic mixture (such as Adderall) of both stereoisomers. Street amphetamines are the racemic mixture. The commercial urine amphetamine immunoassays are designed to preferentially detect D-amphetamine or D-methamphetamine and three of the reagents with high market penetration in the United States, the Syva EMIT II. Amphetamine/methamphetamine and the Cedia As or A/E assays, exhibit less than 10% cross reactivity with the L-forms. On the other hand, the immunoassay screen in our laboratory, the Syva EMIT II. Plus Amphetamines assay, exhibits approximately 41% cross reactivity with L-methamphetamine. In our case this significant cross reactivity is the explanation for the positive screening result.

Immunoassays don't have the selectivity to quantitate the various amphetamine-like substances they are present simultaneously in the sample. For accurate detection and identification chromatography-mass spectrometry techniques are required. Gas chromatography-mass spectrometry (GC-MS) of the past is being replaced by LC–MS/MS as the method of choice for identifying phenethylamine derivatives and their metabolites.

Several LC–MS/MS methods have been described in the literature capable of measuring selegiline metabolites [5,7]. While these methods possess the necessary specificity and sensitivity to detect even low concentrations of the parent drug or its metabolites, the routine confirmation tests by LC–MS/MS usually lack the capability of separately quantitating chiral isomers. For this, additional chiral chromatography is required. The proper work up of a patient with unexpected positive amphetamine screen, therefore, must include identification of the compounds present along with the distribution of the chiral isomers.

Other phenethylamine derivative drugs than selegiline can also be metabolized into amphetamine and/or methamphetamine. Cody reviewed the licit and illicit drugs that metabolized into amphetamine and methamphetamine [8] and noted that not all of them will produce exclusively levorotatory compounds but a racemic mixture. These metabolites can be detected by urine immunoassay screens more readily requiring confirmation testing for accurate interpretation. In addition to the many stimulants obtained by prescription or from illicit sources, one must keep this list of phenethylamine drugs in mind when interpreting amphetamine test results.

Conclusion of our case

After reviewing the literature with the clinician he concluded that the presence of amphetamine and methamphetamine could be explained by the administration of selegiline. Chiral testing for amphetamines was offered but the clinician declined it on the basis of the patient's motivation to become drug free and her history of strict compliance in the past. The patient remained positive for amphetamine and methamphetamine and negative for all other drugs tested during her treatment.

A common question from clinicians in similar situation is if the positive amphetamines result is a "false positive" or not. This question is an unfortunate outcome of simplistic interpretation of drug testing. In our case both the immunoassay and the LC—MS/MS confirmation detected true amphetamine and methamphetamine, therefore the result is true positive in the analytical sense. Amphetamine and methamphetamine were produced in the body; therefore this is a physiologically true positive result. However, the positive result is not due to illegal use of amphetamines. We must make this distinction clear to the clinicians when consulting on drug test results.

References

[1] Csanda E, et al. Experiences with L-deprenyl in Parkinsonism. J Neural Transm 1978;43(3—4):263—9.
[2] Magyar K, Knoll J. Selective inhibition of the "B form" of monoamine oxidase. Pol J Pharmacol Pharm 1977;29(3):233—46.
[3] Bied AM, Kim J, Schwartz TL. A critical appraisal of the selegiline transdermal system for major depressive disorder. Expert Rev Clin Pharmacol 2015;8(6):673—81.
[4] Kronstrand R, et al. Quantitative analysis of desmethylselegiline, methamphetamine, and amphetamine in hair and plasma from Parkinson patients on long-term selegiline medication. J Anal Toxicol 2003;27(3):135—41.
[5] Fujita Y, et al. Detection of levorotatory methamphetamine and levorotatory amphetamine in urine after ingestion of an overdose of selegiline. Yakugaku Zasshi 2008;128(10):1507—12.
[6] Kalasz H, et al. Metabolism of selegiline [(−)-deprenyl)]. Curr Med Chem 2014;21(13):1522—30.
[7] Katagi M, et al. Simultaneous determination of selegiline-N-oxide, a new indicator for selegiline administration, and other metabolites in urine by high-performance liquid chromatography-electrospray ionization mass spectrometry. J Chromatogr B Biomed Sci Appl 2001;759(1):125—33.
[8] Cody JT. Precursor medications as a source of methamphetamine and/or amphetamine positive drug testing results. J Occup Environ Med 2002;44(5):435—50.

# Chapter 14.7

A death due to acute nicotine intoxication

Tiffany Hollenbeck[1], Marius Tarau[1], Lindsey Haldiman[1] and Uttam Garg[2,3]
[1]Office of the Jackson County Medical Examiner, Kansas City, MO, United States, [2]Department of Pathology and Laboratory Medicine, Children's Mercy Hospital, Kansas City, MO, United States, [3]University of Missouri School of Medicine, Kansas City, MO, United States

Case description

A 23-year-old-male was found by his roommate unresponsive, covered in vomit, in the bathroom of their residence. Emergency medical services were dispatched to the scene and resuscitative attempts were made for approximately 30 minutes before the subject was pronounced deceased. His roommate reported that he had seen the subject about 1 hour and 45 minutes prior and he did not appear to be intoxicated at the time. Of note, the subject had a bag which contained commercially available "stress relief-eucalyptus and spearmint," bath salts, sleeping pills, aspirin, some other drugs and three empty and apparently unused syringes which were therefore not collected for analysis (Fig. 14.7.1). One almost full package of cigarettes was found near the decedent but no other nicotine products were identified at the scene. Autopsy revealed marked pulmonary edema, and congestion of the kidneys and liver. There was no evidence of trauma identified.

The heart blood was used for volatile and drug screening. Volatile screening for ethanol, methanol, isopropanol and acetone was performed by headspace gas chromatography with flame ionization detector. Comprehensive broad-spectrum drug-screening included enzyme immunoassays (EIAs) for amphetamines, barbiturates, benzodiazepines, cannabinoids, cocaine metabolite, methadone, opiates, phencyclidine, and propoxyphene, and analysis for >200 drugs by gas chromatography mass spectrometry (GC–MS). Drug screening by GC–MS involved liquid–liquid alkaline extraction with bicarbonate buffer (pH 11.0) and butyl acetate followed by mass spectral analysis in full scan mode. Presumptive identification of analytes was made by a spectral library match and relative retention time comparison with reference standards.

Volatile screen was positive for ethanol at a concentration of 13 mg/dL. The EIA drug screen was positive for amphetamine drug class. However, mass spectrometry testing was negative for a number of sympathomimetic amines including amphetamine, methamphetamine, ephedrine, methylenedioxyamphetamine, methylenedioxyethylamphetamine, methylenedioxymethylamphetamine, methylephedrine, pseudoephedrine, norpseudoephedrine, phentermine, and phenylpropanolamine. GC–MS was positive for nicotine and cotinine (a metabolite of nicotine). Femoral blood was sent to a reference laboratory for a quantification of nicotine and cotinine. Nicotine was quantified at a concentration of 2400 ng/mL and cotinine at a level of 210 ng/mL (limit of detection for nicotine and cotinine is 5.0 and 10.0 ng/mL, respectively). Gastric contents were sent to a reference laboratory for quantification of nicotine and cotinine. Nicotine was quantified at a concentration of 2,300,000 ng/mL and cotinine was quantified at a concentration of 3500 ng/mL. A novel psychoactive substances screen for bath salts and other psychoactive drugs performed by National Medical Services (NMS) laboratory (Horsham, PA) was negative. Toxicology results are summarized in Table 14.7.1.

Discussion

Though a number of drugs and chemicals were found on the scene and detected on drug screening, the most remarkable finding in this case was extremely high levels of nicotine. Nicotine is a lipid soluble alkaloid substance which is extracted from the leaves of the *Nicotiana* plant. It is one of the most toxic and popular drugs of abuse.

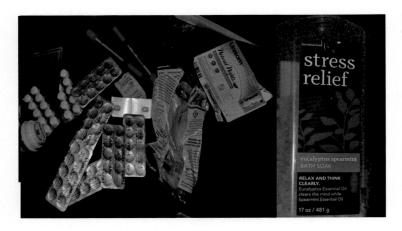

FIGURE 14.7.1 Picture of drugs, syringes, and essential oil found on the scene.

TABLE 14.7.1 Postmortem Toxicology Results on Heart Blood.

Enzyme Immunoassay	Results
Amphetamine	**Positive**
Barbiturates	Negative
Benzodiazepines	Negative
Cannabinoids	Negative
Cocaine metabolite	Negative
Methadone	Negative
Opiates	Negative
Phencyclidine	Negative
Propoxyphene	Negative
GC-FID volatile screening	
Ethanol	13 mg/dL
Acetone	<5 mg/dL
Methanol	<5 mg/dL
Isopropanol	<5 mg/dL
GC–MS Screening for >200 Drugs and Toxins	**Confirmation and Quantification**
Nicotine	Nicotine—2400 ng/mL
Cotinine	Cotinine—210 ng/mL
	Negative for sympathomimetic amines and novel psychoactive substances

EIA, Enzyme immunoassay; *GC-FID*, gas-chromatography with flame ionization detector; *GC–MS*, gas chromatography mass spectrometry.

Most commonly, nicotine is used in tobacco products such as cigarettes, transdermal patches and chewing tobacco. It is also found in some pesticides at very high concentrations; for example, "Black Leaf 40" contains 40% nicotine sulfate [1,2]. Cigarette tobacco contains approximately 1.5% nicotine or 10–15 mg of nicotine per cigarette. Chewing tobacco contains 2.5%–8% nicotine. In recent years, nicotine use via electronic cigarettes (e-cigarettes) has increased considerably. E-cigarettes were first introduced in 2004 and have gained significant popularity in recent years. These are the devices that heat and vaporize a solution which contains nicotine dissolved in propylene glycol and/or glycerin to increase vapor and flavor. E-cigarettes have a battery and a heating device. In automatic e-cigarettes, puffing activates a heating coil and vaporizes the nicotine containing liquid. A typical e-cigarette is shown in Fig. 14.7.2. E-cigarettes are

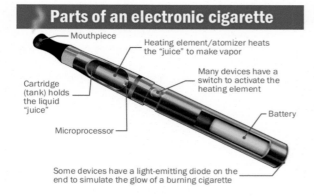

FIGURE 14.7.2 Parts of electronic cigarette. https://www.usfa.fema.gov/downloads/pdf/publications/electronic_cigarettes.pdf.

advertised to satisfy nicotine addiction without all of the health hazards of inhaling tar and other carcinogens known to be present in traditional cigarettes. However, the harm of nicotine e-cigarettes is well established [3,4]. Liquid nicotine, used to fill e-cigarettes, contains very high concentrations of nicotine, up to 36.6 mg/mL in a refill solution [5]; typical concentrations are 10–20 mg/mL available in 5–10 mL bottles. The amount of nicotine present in these solutions is sufficient to cause a serious injury or even death. The minimal estimated lethal dose of nicotine is about 30–60 mg in an adult [1].

Nicotine is absorbed rapidly from all the routes including inhalation, oral, intramuscular, intravenous and dermal, and reaches the brain quickly [1,2]. It has a half-life of approximately 40–60 minutes with volume of distribution of 1–3 L/kg [1]. The mechanism of toxicity is through stimulation of nicotinic cholinergic receptors. Effects of nicotine on the body are complex, with small doses causing a stimulatory effects leading to nausea, vomiting, dizziness, miosis, tachycardia, hypertension, sweating, and salivation. High doses results in depressive effects such as prostration, convulsions, respiratory paralysis, and cardiac arrhythmias [1]. Although rare, fatalities involving nicotine toxicity are not uncommon. Fatalities are primarily from accidental exposure but deaths from intentional overdose have been reported [1]. Accidental exposures include dermal exposure to concentrated nicotine solutions, tobacco harvesting and oral ingestion of tobacco containing products by young children [1,2]. A number of cases of intentional overdose from ingestion of liquid nicotine have been reported [6–12]. In the past intentional overdoses primarily occurred from ingesting concentrated nicotine-containing insecticide solutions. In recent years, attempted suicide by ingesting e-cigarette solutions have been reported [6,7,12–14]. Three men, who intentionally ingested e-cigarette solutions, but survived, devolved nausea, emesis, tachycardia, and hypertension [1]. In three deaths involving oral ingestion or intravenous injections of e-cigarette solutions, nicotine doses ranged from 400 to 11,000 mg with nicotine levels ranging from >1000 to 5500 ng/mL [6,12,13]. Nicotine level of 2400 ng/mL in our case parallels with the levels reported in these cases. In our case, although liquid nicotine was not located on the scene, based on a literature review and pre- and postmortem findings, the most likely method of sustaining such a high nicotine level in such a short period of time is ingestion of a concentrated form of nicotine, as found in e-cigarette refills.

References

[1] Baselt RC. Nicotine. In: Baselt RC, editor. Disposition of toxic drugs and chemicals in man. Seal Beach, CA: Biomedical Publications; 2017. p. 1521–5.
[2] Benzowitz NL. Nicotine. In: Olson KR, editor. Poisoning and drug overdose. New York: McGraw Hill; 2018. p. 337–9.
[3] Hughes A, Hendrickson RG. An epidemiologic and clinical description of e-cigarette toxicity. Clin Toxicol (Phila) 2019;57:287–93.
[4] Tegin G, Mekala HM, Sarai SK, Lippmann S. E-cigarette toxicity? South Med J 2018;111:35–8.
[5] Goniewicz ML, Gupta R, Lee YH, Reinhardt S, Kim S, Kim B, et al. Nicotine levels in electronic cigarette refill solutions: a comparative analysis of products from the U.S., Korea, and Poland. Int J Drug Policy 2015;26:583–8.
[6] Chen BC, Bright SB, Trivedi AR, Valento M. Death following intentional ingestion of e-liquid. Clin Toxicol (Phila) 2015;53:914–16.
[7] Eberlein CK, Frieling H, Kohnlein T, Hillemacher T, Bleich S. Suicide attempt by poisoning using nicotine liquid for use in electronic cigarettes. Am J Psychiatry 2014;171:891.
[8] Jalkanen V, Varela V, Kalliomaki J. Case report: two severe cases of suicide attempts using nicotine containing e-cigarette liquid. Duodecim 2016;132:1480–3.
[9] Kim JW, Baum CR. Liquid nicotine toxicity. Pediatr Emerg Care 2015;31:517–21 quiz 522–4.
[10] Miller A. Nicotine poisoning increase due to e-cigarettes. CMAJ 2014;186:E367.

[11] Morley S, Slaughter J, Smith PR. Death from ingestion of e-liquid. J Emerg Med 2017;53:862—4.
[12] Thornton SL, Oller L, Sawyer T. Fatal intravenous injection of electronic nicotine delivery system refilling solution. J Med Toxicol 2014;10:202—4.
[13] Bartschat S, Mercer-Chalmers-Bender K, Beike J, Rothschild MA, Jubner M. Not only smoking is deadly: fatal ingestion of e-juice-a case report. Int J Legal Med 2015;129:481—6.
[14] Solarino B, Rosenbaum F, Riesselmann B, Buschmann CT, Tsokos M. Death due to ingestion of nicotine-containing solution: case report and review of the literature. Forensic Sci Int 2010;195:e19—22.

Chapter 14.8

Cocaethylene—ethanol adding fuel to cocaine's fire

Alexandra Rapp and Anthony Okorodudu
Department of Pathology, University of Texas Medical Branch, Galveston, TX, United States

Case description

A 56-year-old female with a past medical history significant for combined systolic-diastolic heart failure (ejection fraction 15%—20%, NYHA class III, stage C) status-post implantable cardioverter—defibrillator, hypertension, hyperlipidemia, tobacco and cocaine abuse, and frequent hospitalizations for congestive heart failure (CHF) presented with acute worsening of shortness of breath. For the past four days, the patient noted progressive worsening of baseline shortness of breath accompanied by cough productive of white sputum, along with recent exposure to sick contacts. She reported compliance with her home prescription Lasix 40 mg BID and denied illicit drug use.

On presentation the patient's vital signs were stable, and her oxygen saturation was 100% on 2 L nasal cannula of oxygen. Physical exam was notable for jugular venous distension up to mid-neck and crackles right mid-lung downward as well as decreased breath sounds at the right lung base. Chest X-ray revealed bilateral pleural effusions and mild pulmonary edema, consistent with her history of CHF, as well as a right lower lung opacity concerning for pneumonia. Brain natriuretic peptide was elevated at 4530 pg/mL, consistent with the patient's history of CHF, though it was not substantially higher than the patient's baseline.

Given the patient's history of cocaine abuse, the patient's urine was sent for a comprehensive urine drug-screen involving automated enzyme immunoassays for amphetamines, opiates, phencyclidine, cocaine, benzodiazepines, propoxyphene, cannabinoids, methadone, and barbiturates. Immunoassay resulted as presumptively positive for cocaine. The urine was then used for a more comprehensive confirmatory test via liquid chromatography—mass spectrometry that tests for approximately forty illicit substances and/or their metabolites. The results were positive for benzoylecgonine, the metabolite of cocaine, at a concentration of 2113 ng/mL prior to creatinine normalization and 1635 ng/mL with creatinine normalization. In addition, ethyl sulfate and ethyl glucuronide were also detected at 430 and 781 ng/mL with creatinine normalization, respectively, consistent with recent alcohol consumption. Although our laboratory does not test for cocaethylene, a metabolite formed by the metabolism of cocaine and ethanol, we strongly suspected this substance to be in the patient given the detection of cocaine and ethanol metabolites.

Discussion

Cocaine is an alkaloid extract from the *Erythroxylum coca* plant that is abused for its euphoric and stimulatory effects [1]. In addition to these desired effects, cocaine has significant adverse impacts on the cardiovascular, pulmonary, and immunologic systems. Cocaine, combined with other stimulants, was responsible for over 1 million illicit drug-related emergency room visits in 2010 [2,3].

Cocaethylene is an unnatural alkaloid of the coca leaf formed by the transesterification of cocaine by liver carboxylesterase in the presence of ethanol [4,5]. Like cocaine, cocaethylene blocks dopamine transporters on dopaminergic presynaptic nerve terminals in the brain, leading to increased postsynaptic stimulation, which in turn produces a euphoric sensation. However, cocaethylene lacks the serotonergic-related inhibitory mechanism, thereby causing a more intense euphoria than cocaine [4], which may explain the increasing trend of co-abusing cocaine and ethanol to achieve this even more desirable "high."

Unfortunately, with cocaethylene's potentiated positive effects also comes a much greater risk of toxicity and lethal outcomes. In 1991 this enhanced lethality of cocaethylene was noted in experiments with Swiss-Webster mice, with the LD_{50} of cocaethylene being 60.7 mg/kg (males) and 63.8 mg/kg (females) versus cocaine's LD_{50} of 93.0 mg/kg [6]. In canine models, cocaethylene was shown to increase hypertension and systemic vascular resistance and to decrease myocardial function, slow cardiac conduction, and cause arrhythmias [7]. Since then, cocaethylene's increased cardiovascular toxicity—relative to cocaine—in humans has been noted [4]. In addition, cocaethylene has a plasma half-life of three to five times that of cocaine, which may be implicated in its lethality, and is also associated with seizures, liver damage, and a compromised immune system [8,9]. The immunologic impairment appears to be due to decreased splenic weight and a suppression of splenocyte mitogen-stimulatory production of gamma-interferon, TNF, and IL-2, which is more severe for cocaethylene than cocaine [9]. In addition, cocaethylene is responsible for many acute and chronic lung diseases via inhibition of cellular proliferation of alveolar epithelial type II cells [10]. Lastly, cocaethylene carries an 18-to-25-fold increase in sudden death compared to cocaine alone [5,8].

Referring back to our case presentation, the patient was treated for both pneumonia and CHF exacerbation, with significant improvement in symptoms following administration of antibiotics and IV diuresis. It is important to note during this admission that there was a decrease in baseline ejection fraction from 15%–20% to 10%–15%. Given that cocaethylene is known to cause increased blood pressure and systemic vascular resistance and decreased myocardial function, perhaps the concomitant use of cocaine and ethanol contributed to this specific CHF exacerbation. In addition, cocaethylene may have lowered the patient's immunity and pulmonary function, contributing to her propensity to develop pneumonias. It is important to note that the patient was admitted less than a month later with an almost identical presentation, further supporting this hypothesis.

In conclusion, this case highlights the dangerous effects of combining cocaine and alcohol, the latter potentiating both the positive and negative effects of the former—adding fuel to cocaine's "fire." With the growing prevalence of cardiac conditions, this case also demonstrates how multi-substance abuse can exacerbate chronic medical conditions, which not only places a significant financial strain on our health care system but also greatly impairs an individual's quality of life.

References

[1] Boghdadi MS, Henning RJ. Cocaine: pathophysiology and clinical toxicology. Heart Lung 1997;26:466–83.
[2] Tseng W, Sutter ME, Albertson TE. Stimulants and the lung: review of the literature. Clin Rev Allergy Immunol 2014;46(1):82–100.
[3] Tacker DH, Okorodudu AO. Evidence for injurious effect of cocaethylene in human microvascular endothelial cell. Clin Chim Acta 2004;345 (1-2):69–77.
[4] Landry MJ. An overview of cocaethylene, an alcohol-derived, psychoactive, cocaine metabolite. J Psychoact Drugs 1992;24(3):273–6.
[5] Pennings EJ, Leccese AP, Wolff FA. Effects of concurrent use of alcohol and cocaine. Addiction 2002;97:773–83.
[6] Hearn WL, Rose S, Wagner J, Ciarleglio A, Mash DC. Cocaethylene is more potent than cocaine in mediating lethality. Pharmcol Biochem Behav 1991;39(2):531–3.
[7] Wilson LD, French S. Cocaethylene's effects on coronary artery blood flow and cardiac function in a canine model. J Toxicol Clin Toxicol 2002;40(5):535–46.
[8] Andrews P. Cocaethylene toxicity. J Addictive Med 1997;16(3):75–84.
[9] Pirozhkov SV, Watson RR, Chen GJ. Ethanol enhances immunosuppression induced by cocaine. Alcohol 1992;9(6):489–94.
[10] Bazuaye-Ekwuyasi EA, Ogunbileje JO, Kaphalia BS, Eltorky MA, Okorodudu AO. Comparative effects of cocaine and cocaethylene on alveolar epithelial type II cells. Toxicol Mech Methods 2015;25(8):604–13.

Chapter 14.9

A fatality caused by acute methamphetamine intoxication due to intravaginal absorption

Melissa Beals[1,2], Megan Weitzel[1,2], Diane C. Peterson[3,4], C. Clinton Frazee, III[1,2] and Uttam Garg[1,2]

[1]Department of Pathology and Laboratory Medicine, Children's Mercy Hospital, Kansas City, MO, United States, [2]University of Missouri School of Medicine, Kansas City, MO, United States, [3]Office of the Jackson County Medical Examiner, Kansas City, MO, United States, [4]Johnson County Department of Health and Environment/Medical Examiner's Office, Olathe, KS, United States

Case history

A 28-year-old female was found deceased lying prone on the floor of her apartment bedroom. The subject had a history of chronic kidney disease, celiac disease, and drug abuse. She was last known alive approximately 3 days prior. The body displayed mild decomposition, including localized skin slippage, green discoloration of the abdomen, purple discoloration and marbling of the face, and mummification of the hands and feet. The residence was secure and no foul play was suspected. Two prescription medications, losartan potassium and metronidazole, were found on the scene.

The subject's primary care physician refused to sign the death certificate, stating that although the decedent had 40% kidney function, her renal disease should not have caused death. The case was deferred to the Jackson County Medical Examiner. The external examination of the body was nondescript, with only a few superficial abrasions and small healing wounds present. Upon internal examination, the lungs were mildly edematous with the right and left pleural cavities each containing 25 mL of purge-type fluid. Consistent with the decedent's history of stage IV chronic kidney disease, the kidneys appeared atrophic with histologically demonstrable glomerular sclerosis. More notable, a clear plastic bag was found inside the decedent's vaginal cavity. The bag contained approximately 12 mL of a yellow-tinged liquid and a transparent crystalline material. Heart blood, urine, vitreous fluid, gastric contents, liver tissue, brain tissue, and the intact plastic bag were submitted for toxicological analysis.

Preliminary drug screen testing was performed using heart blood. Following methanol dissolution, enzyme multiplied immunoassay technique (EMIT) was used to screen blood for cocaine metabolite, cannabinoids, opiates, benzodiazepines, phencyclidine, amphetamines class drugs, barbiturates, methadone, and propoxyphene. EMIT results indicated the presence of amphetamine class drugs, but no other drugs of abuse were detected. Comprehensive drug screening for >200 drugs was performed using alkaline liquid-liquid extraction with n-butyl acetate, followed by gas chromatography–mass spectrometry (GC/MS). The GC/MS drug screen analysis indicated the presence of methamphetamine and amphetamine. No other drugs were detected. Volatile testing by gas chromatography/flame ionization detection was negative for ethanol, methanol, isopropanol, and acetone. Sympathomimetic amine screening performed at NMS Labs (Horsham, PA) by LC/MS/MS quantified D,L-ephedrine, D,L-phenylpropanolamine, D,L-amphetamine, and D,L-methamphetamine at concentrations of 43, 100, 2841, and 50,517 ng/mL, respectively.

Discussion

Methamphetamine, the N-methyl derivative of amphetamine, is a central nervous system stimulant with a high potential for abuse. Its clandestine production and ease of synthesis have made illicit methamphetamine readily available. Though seldom prescribed due to concerns relating to human toxicity and the propensity for addiction, D-methamphetamine has been used clinically since 1930 to treat narcolepsy, obesity, and attention deficit disorder [1].

Methamphetamine dominates the global market for synthetic drugs, and international methamphetamine use exceeds that of all other drugs of abuse except cannabis [2]. The D-methamphetamine isomer has greater central stimulant action and weaker peripheral sympathomimetic activity than the L-isomer, which is used as a nasal decongestant and in certain non-prescription inhalers [1,3]. Hence, differentiating the D,L stereoisomers can be a useful tool for determining illicit methamphetamine abuse, especially in probationary drug screen testing.

Through interactions with 5-HT transporters, monoamine transporters, and N-methyl-D-aspartate receptors, methamphetamine affects serotonergic, noradrenergic, and glutamatergic systems. Methamphetamine increases extracellular dopamine levels by reversing the transport of dopamine and displacing dopamine from vesicular stores [4]. Following methamphetamine administration, users experience euphoria, disinhibition, increased energy and productivity, and reduced anxiety [5]. Methamphetamine toxicity affects the nervous and cardiovascular systems, and severe sympathomimetic effects include hypertension, tachycardia, agitation, hyperthermia, seizures, altered mental status, and violent behavior [2,6]. An overdose of methamphetamine can cause cerebrovascular hemorrhage, seizures, hyperpyrexia, renal failure, hepatic failure, cardiac arrest, and death [6].

Methamphetamine can be inhaled or smoked, swallowed, nasally insufflated, intravenously injected, and rectally or vaginally absorbed [1,2,6–8]. The drug is readily absorbed through various listed routes and undergoes extensive distribution across high lipid content tissues [2,6]; thus acute toxicity can easily occur intentionally or accidentally by absorption. Intentional absorption of methamphetamine might include "parachuting," or the ingestion of methamphetamine which have been packaged in a partially-opened plastic bag or a wrapper that then slowly dissolves. In this manner, parachuting mimics the sustained-release mechanism used with prescription and over-the-counter drug preparations [9]. Accidental absorption can occur from body *packing*, which is the ingestion of packaged drugs with the intent of smuggling and with the expectation of defecating the packages intact [10], or from body *stuffing*, which is the ingestion of packaged or unpackaged drugs in an attempt to hide them from law enforcement and avoid arrest [11].

In one study, a female body *stuffer* who was in police custody admitted to concealing plastic bags of methamphetamine in her vagina. After going into seizures and subsequent hospitalization, her serum methamphetamine concentration was measured at 3100 ng/mL with amphetamine at 110 ng/mL [8]. She recovered from the event. In another body stuffing case, a young woman who was in county jail custody for suspicion of a felony drug offense was pronounced dead after being found unconscious and unresponsive in the holding cell. During the autopsy, plastic packages containing what was later identified as methamphetamine were found in the decedent's vaginal cavity. The methamphetamine and amphetamine levels in the decedent's subclavian blood were measured at 42,600 and 1300 ng/mL, respectively [7]. Our case denotes some of the higher postmortem methamphetamine (50,517 ng/mL) and amphetamine (2,841 ng/mL) levels found in literature and is significant by virtue of the mode of administration through vaginal absorption. Though the motive for intravaginal stuffing with this decedent is unknown, the acute level of methamphetamine in her system resulted from an apparent direct mucosal absorption of the drug. The cause of death was determined to be methamphetamine overdose. The manner of death was ruled as an accident.

References

[1] Baselt RC. D-Methamphetamine. In: Baselt RC, editor. Disposition of toxic drugs and chemicals in man. Seal Beach, CA: Biomedical Publications; 2017. p. 1319–22.
[2] Schep LJ, Slaughter RJ, Beasley DM. The clinical toxicology of metamfetamine. Clin Toxicol (Phila) 2010;48:675–94.
[3] Baselt RC. L-Methamphetamine. In: Baselt RC, editor. Disposition of toxic drugs and chemicals in man. Seal Beach, CA: Biomedical Publications; 2017. p. 1322–3.
[4] Nordahl TE, Salo R, Leamon M. Neuropsychological effects of chronic methamphetamine use on neurotransmitters and cognition: a review. J Neuropsychiatry Clin Neurosci 2003;15:317–25.
[5] Krasnova IN, Cadet JL. Methamphetamine toxicity and messengers of death. Brain Res Rev 2009;60:379–407.
[6] Cruickshank CC, Dyer KR. A review of the clinical pharmacology of methamphetamine. Addiction 2009;104:1085–99.
[7] Jones P, Mutsvunguma R, Prahlow JA. Accidental death via intravaginal absorption of methamphetamine. Forensic Sci Med Pathol 2014;10:234–8.
[8] Kashani J, Ruha AM. Methamphetamine toxicity secondary to intravaginal body stuffing. J Toxicol Clin Toxicol 2004;42:987–9.
[9] Hendrickson RG, Horowitz BZ, Norton RL, Notenboom H. "Parachuting" meth: a novel delivery method for methamphetamine and delayed-onset toxicity from "body stuffing". Clin Toxicol (Phila) 2006;44:379–82.
[10] Gupta M, Bailey S, Lovato LM. Bottoms up: methamphetamine toxicity from an unusual route. West J Emerg Med 2009;10:58–60.
[11] West PL, McKeown NJ, Hendrickson RG. Methamphetamine body stuffers: an observational case series. Ann Emerg Med 2010;55:190–7.

Chapter 15

Designer drugs

Gregory Janis
MedTox Laboratories, Laboratory Corporation of America Holdings, St. Paul, MN, United States

Introduction

Designer drugs can be best defined as chemical agents clandestinely synthetized for use as psychoactive agents mimicking traditional drug of abuse. Designer drugs have arisen from serendipity, others originated as research tools, while others were specifically rationalized and subsequently synthesized for their potential efficacy or as a mechanism to avoid legal controls or detection. While designer drugs mimic historically abused drugs, their array of activity often deviates from their intended pharmacology creating entirely new spectrums of both physical and psychological effects. Likewise, the toxicity of designer drugs can significantly depart from the toxicity of their better characterized corresponding traditional drugs of abuse. With the evolution of each class of designer drugs and each individual designer drug, the users, the legal community, and the medical community must learn to recognize the active agent and understand its myriad of effects.

Designer drugs have been a component of drug culture since the 1960s. In that decade the man who would become known as the Godfather of Ecstasy, Alexander "Shasha" Shulgin, began his quest to find and develop synthetic versions of the hallucinogens, mescaline, and later tryptamine. As chronicled in two novel length books, Shulgin details the chemical synthesis, dosing, observed effects, and design rational of over 200 different compounds [1,2]. While most of these compounds lack significant psychoactive activity, a handful of these compounds are potent hallucinogens. The more potent compounds found an occasional foothold as clandestinely produced designer drugs. 2,5-Dimethoxy-4-methylamphetamine and 2,5-dimethoxy-4-bromophenethylamine (2C-B) are examples of Shulgin's synthetic handiwork which were later identified as street drugs on the designer drug scene (see Fig. 15.1 for structures). Independent of Sasha Shulgin, the phenylpiperidine opioids meperidine and fentanyl spawned separate lineages of designer opiates in the 1970s. 1-Methyl-4-phenyl-4-propionoxypiperidine (MPPP) was clandestinely synthesized as a nonregulated analog of meperidine. Unfortunately due to poorly controlled synthetic routes, the synthesis of MPPP was contaminated with 1-methyl-4-phenyl-1,2,3,6-tetrahydropyridine (MPTP). The contaminant MPTP was metabolized to 1-methyl-4-phenyl-pyridinium; a potent toxin to dopaminergic neurons (Fig. 15.2). MPPP users inadvertently exposed to MPTP developed rapid onset, irreversible Parkinson's disease; these stark tragedies brought the dangers of designer drugs to the general population [3,4]. Also in that same decade, clandestinely produced fentanyl analogs began appearing in the heroin using population. Alpha methylfentanyl was the first to appear in southern California in the late 1970s sold as a china white heroin [5,6]. In 1981 when the DEA moved to classify alpha methylfentanyl as a schedule I compound, alternative

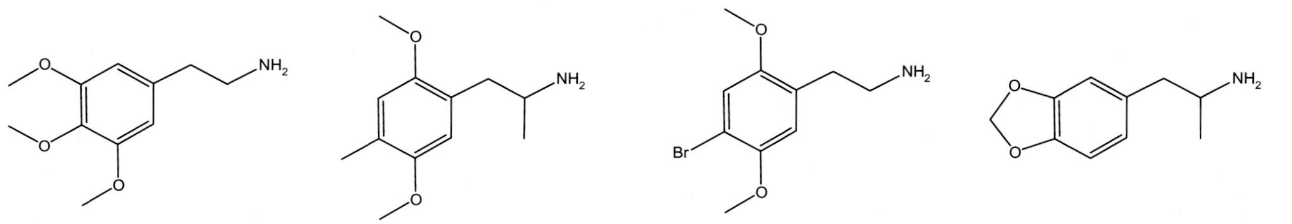

Mescaline 2,5-dimethoxy-4-methylamphetamine (DOM) 2,5-dimethoxy-b-bromophenethylamine (2C-B) MDA

FIGURE 15.1 Structure of mescaline and synthetic analogs.

Toxicology Cases for the Clinical and Forensic Laboratory. DOI: https://doi.org/10.1016/B978-0-12-815846-3.00015-6

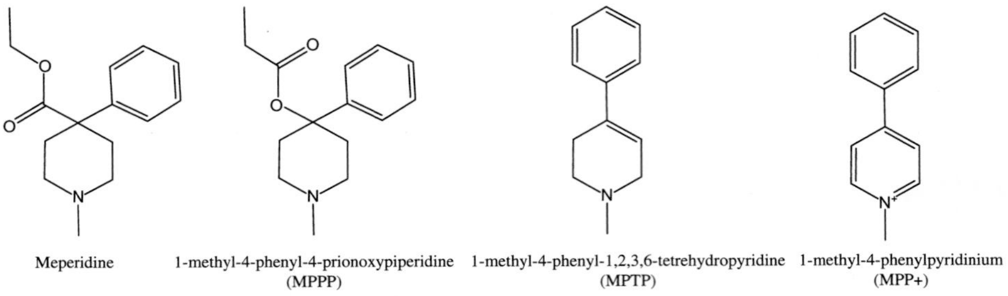

FIGURE 15.2 Structure of meperidine, its synthetic analog MPPP, the synthetic contaminant MPTP, and the toxic metabolite MPP+. *MPPP*, 1-Methyl-4-phenyl-4-propionoxypiperidine; *MPP+*, 1-methyl-4-phenylpyridinium; *MPTP*, 1-methyl-4-phenyl-1, 2,3,6-tetrahydropyridine.

Meperidine 1-methyl-4-phenyl-4-prionoxypiperidine (MPPP) 1-methyl-4-phenyl-1,2,3,6-tetrehydropyridine (MPTP) 1-methyl-4-phenylpyridinium (MPP+)

fentanyl analogs were already on the street. This pattern of cat-and-mouse scheduling followed by the emergence of new drug analogs foreshadowed the rapid cycle of emerging compounds which defines designer drugs.

As was seen with MPPP and alpha methylfentanyl, evading legal restriction continues to be one of the main driving forces that promote waves of designer drugs. The federal government combats emerging designer drugs with two main legal strategies designed to enable a nimble and rapid legal response. The Attorney General is empowered with emergency scheduling powers to classify a substance as a schedule I drug with no legitimate use and a high potential for abuse if the substance is previously unscheduled and possesses an imminent hazard to public safety [21 U.S.C. 811(h)]. While the Office of the Attorney General is faced with the burden of proving that a chemical poses an "imminent hazard" based upon patterns of abuse and toxicity, emergency scheduling has been the most effective tool for rapidly responding to emerging designer drugs. Between January of 2000 and September of 2018, emergency scheduling powers have been exercised to temporarily schedule more than 60 individual compounds in response to the ongoing evolution of designer drugs (Title 21 CFR Part 1308.11).

The Federal Analogue Act is another legal strategy combating designer drugs. It enables a substance to be considered as being a controlled substance if the compound (1) is similar in structure to a scheduled compound, (2) is shown to possess similar activity on the central nervous system, and (3) is represented or intended to be used to obtain the CNS effects. However, prosecutions under the Federal Analogue Act are complex. The basic premise of what constitutes structural similarity can be argued from multiple perspectives and cases can frequently pit one expert against another debating structural similarity. In addition, the concept of intent when prosecuting under the analog act is often deliberated in the courts. Cases hinging on the Federal Analogue Act have been appealed as high as the Supreme Court on multiple occasions. As a result, many prosecutors are reluctant to use the Federal Analogue Act as their lone basis of prosecution.

In addition to legal motivations the desire of a user to conceal their drug use also fuels their use of designer drugs. In general, designer drugs are not detected by mandatory drug tests performed for parole, rehabilitation, or workplace drug testing. Point of care devices used for rapid drug screening based on immunoassay technologies are rarely capable of detecting newly emerging designer drugs, unless by luck they possess a degree of cross reactivity with specific designer drugs. Laboratory-based screening methods utilizing either more sophisticated immunoassay screening technologies or targeted mass spectrometry−based screening techniques are better able to keep up with emerging drugs due to the increased flexibility of these assays. However, targeted lab-developed tests can only detect what the test developers thought to include as assay targets. Newly emerged drugs may escape being targeted simply because the assay developers are unaware of either the drug species or the proper metabolic targets indicative of drug exposure. Untargeted assays, such as time-of-flight (TOF) mass spectrometry, increase the ability to detect designer drugs as these methods collect full spectrum data. However, full-scan data are still reliant on either software algorithms or a human analyst knowing what drug substance or drug metabolite to mine the data for. In addition, biological samples contain a myriad of unidentifiable peaks of unknown origin making the process of sorting through untargeted data to find drug-related information complex and time consuming.

Finally, when discussing the motivating factors driving waves of new designer drugs, the simple facts of profit and logistics cannot be overlooked. Clandestine laboratories target designer drugs that are cheaply and easily synthesized. When combined with the high potency of most designer drugs, the potential profitability for synthetic designer drugs far exceeds most traditional drugs such as heroin or cocaine. For instance, a kilogram of fentanyl is currently 20 times more profitable than a kilogram of heroin due to the higher potency allowing for dramatically more individual doses contained within each kilogram. In addition, interdiction methods such as drug sniffing dogs are generally constructed to intercept traditional contraband and are blind to most designer drugs transported in smaller, more easily concealed physical packages.

Synthetic cannabinoids

The most recent wave of designer drugs began in earnest with an influx of synthetic cannabinoids (Fig. 15.3). In the early 2000s an obscure research tool appeared as a component of smoking blends in London. The drug, JWH-018, is one of dozens of similar compounds developed at Clemson University by Prof. John Huffman as a research tool for studying cannabinoid receptors. JWH-018 and many of the other developed research tools have potency at the cannabinoid receptor exceeding that of delta9-tetrahydrocannabinol (delta9-THC), the primary active chemical constituent in marijuana [7,8]. In addition, JWH-018 and some of the similarly potent, related synthetic cannabinoids are relatively easily synthesized enabling moderately trained synthetic chemists to produce large quantities of the drugs. These synthesized compounds were then sprayed onto innocuous plant materials along with various perfuming agents and most frequently sold as potpourri. Originally branded as K2 or Spice, these products were labeled as "not for human consumption" but were smoked to obtain cannabis-like effects. Prior to legal restrictions, the availability of synthetic cannabinoids was staggering with nearly every tobacco shop, head shop, and truck stop peddling concoctions containing synthetic cannabinoids.

While synthetic cannabinoids were frequently referred to as being "synthetic marijuana," users' experiences with synthetic cannabinoids varied, ranging from being comparable to marijuana to being completely negative and described as "bad trips" full of undesired effects. In addition to the marijuana-like high, users report negative effects of synthetic cannabinoids to include anxiety, paranoia, memory loss, nausea, agitation, hypertension, tachycardia, and chest pain [9]. Some of these negative effects from synthetic cannabinoids are shared with marijuana, but the reported negative effects are generally more intense. In extreme cases, users of synthetic cannabinoids have required emergency medical attention to treat cardiac events, nephrotoxicity, seizures, severe agitation, and psychotic events [10,11]. Without an available antagonist, the treatment of acute synthetic cannabinoid intoxication is generally supportive. Benzodiazepines are administered to combat debilitating anxiety, irritability, and seizures [12,13]. Similarly, antipsychotics are administered to combat acute psychotic events [14]. While withdrawal symptoms are not usually associated with marijuana use, synthetic cannabinoid withdrawal has been reported following chronic use. Withdrawal symptoms are similar to the negative effects seen with acute administrations and include anxiety, irritability, and agitation. The same strategy of supportive care is utilized to combat acute toxicity and withdrawal symptoms.

Multiple factors may play into the reported higher intensity of the negative effects of synthetic cannabinoids. For one, delta9-THC is a partial agonist at the neuronal CB1 receptor, while many synthetic cannabinoids are full agonists at this receptor. In addition, marijuana contains both CB1 and CB2 receptor agonists, as well as CB1 receptor antagonists such as cannabidiol. These antagonists present in marijuana are hypothesized to attenuate the CB1 receptor activation from THC. In contrast, synthetic cannabinoids exclusively possess agonist activity at cannabinoid receptors. Furthermore, the potent CB1 receptor activity of synthetic cannabinoids is shared by some of the metabolites of synthetic cannabinoids in addition to the parent drugs themselves; this phenomenon is not observed with marijuana where metabolites of THC lack receptor activity.

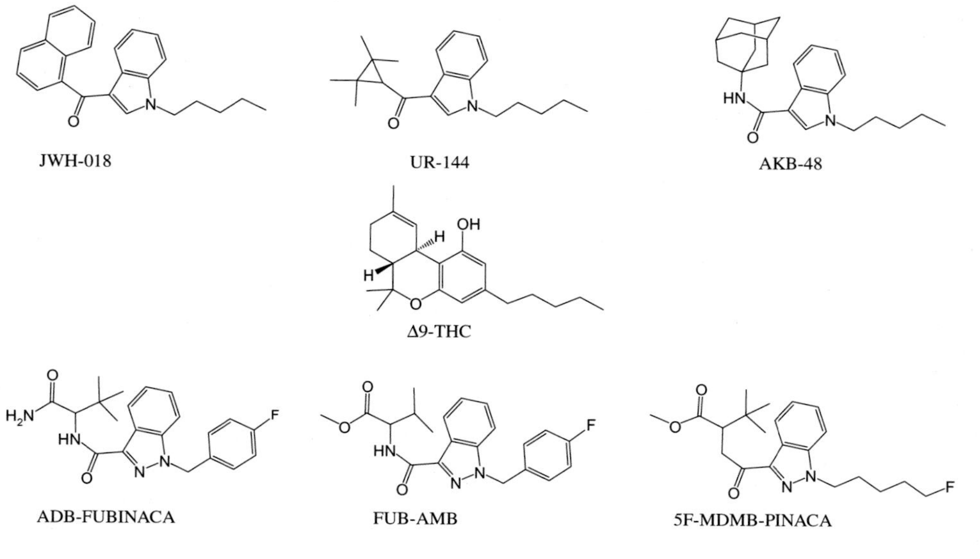

FIGURE 15.3 Structure of THC and a select group of synthetic cannabinoids.

Synthetic cannabinoids produce notoriously different experiences dependent on the user. While some of the variability in user experience is likely the result of personal variability as well as dose, no doubt a substantial component of the reported variation is the result of the constantly changing active ingredients found in synthetic cannabinoid products. The exact chemical species of synthetic cannabinoids available on the street have continued to evolve since their first appearance in the mid and late 2000s. It is likely that emergency scheduling of synthetic cannabinoids accelerated the rapid proliferation of new chemicals present in smoking blends. Until 2011 when the DEA scheduled the first round of clandestinely produced synthetic cannabinoids, smoking blends containing these chemicals were widely and easy available at tobacco shops, head shops, record stores, truck stops, and convenience stores. Following a period of rapid evolution of both active ingredients and legal responses, broader laws restricting synthetic cannabinoids and a greater awareness of the dangers of synthetic cannabinoids have dramatically reduced their prevalence and availability as a street drug. At the same time, the appearance of new synthetic cannabinoids as street drugs has also slowed. In the past year FUB-AMB, ADB-FUBINACA, and 5F-MDMB-PINACA have been the dominant synthetic cannabinoids (DEA Emerging Threat Report, Third Quarter 2018). Earlier dominant species of synthetic cannabinoids are essentially absent from the current clandestine drug marketplace.

Detection of synthetic cannabinoids in biological samples is complicated by multiple factors. First, the appearance of new synthetic cannabinoids replacing existing cannabinoids has generally occurred at a faster rate than other designer drugs. In the period coinciding with the federal scheduling of the abused initial synthetic cannabinoids, JWH-018 and JWH-073, new compounds would appear, disappear, and be subsequently replaced in a period of a few months. As a result of this rapid turnover, analytical laboratories and manufacturers of reference materials struggled to keep up with this rapidly changing landscape. The situation was made more complicated by both the route of administration and the metabolism of synthetic cannabinoids. The act of smoking synthetic cannabinoid products generates pyrolysis products of certain active drugs [15]. Metabolic considerations additionally complicate attempts to identify synthetic cannabinoids in biological specimens. In urine, synthetic cannabinoids are exclusively found as phase II glucuronide conjugates of phase I metabolites. Unchanged parent molecules are generally not detectable. Thus the laboratory must either have knowledge of the phase I metabolism of the drug species of interest or the lab must be able to predict the phase I metabolism. The laboratory must also either target detecting the phase II conjugates or subject the urine sample to enzymatic or chemical hydrolysis procedures and then target the unconjugated phase I metabolites. Analysis is additionally complicated by the availability of reference standards. Reference materials for metabolites are slower to be developed than standards than the parent drugs, again due to the need for understanding the metabolic pathways that transform the dosed drugs.

Like THC, synthetic cannabinoids are highly lipophilic drugs. In broad screening methodologies, synthetic cannabinoids will elute off of a reverse phase High Performance Liquid Chromatography (HPLC) chromatographic system after other drugs of abuse. Like most xenobiotics, the phase I metabolites of synthetic cannabinoids are generally more polar than the corresponding parent molecules and will elute earlier. Metabolic transitions generally consist of hydroxylation reactions on aliphatic chains or demethylation or deamination reactions on compounds where those functional groups are available for metabolic attack. Multiple, sequential phase I metabolic transitions do not appear to be common. Thus each synthetic cannabinoid generally possesses one or two dominant phase I metabolites that can be targeted in urine analyses. Unfortunately, general rules of analysis across compounds are not easily applied to blood analyses for synthetic cannabinoids. Most of the early compounds appear to circulate primarily as parent drug. This is not the case for the more recent compounds, some of which only appear to circulate as phase I metabolites. Thus knowledge of the metabolism of synthetic cannabinoids is a prerequisite to monitoring assays in both blood and urine.

Designer opiates

Designer opiates that, as discussed earlier, first appeared as abused drugs in the 1970s resurfaced in this decade. A resurgence of heroin associated with opiate prescription abuse was a fertile field for designer opiates to reenter the drug of abuse marketplace. Abusers and addicts of opiate prescriptions occasionally would turn to heroin as the supply of opiate pills became unreliable or cost prohibitive. Since the potency of fentanyl and fentanyl derivatives far exceeds that of heroin, small quantities of fentanyl compounds could be sold as counterfeit heroin or used to increase the potency of low-grade heroin. The high potency also enabled small but profitable quantities of drugs to be easily smuggled into the United States; bags of fentanyl were and still are shipped hidden inside of packages transported by normal mail routes from manufacturing sites in China. Heroin is not the only drug to have been replaced by clandestinely produced fentanyl and its analogs. Through legally purchased manufacturing equipment, fentanyl compounds and binder agents have been found pressed into pills of identical shape and markings to legitimate pharmaceuticals. Street pills of hydrocodone, oxycodone, and alprazolam have all been identified as containing fentanyl or fentanyl derivatives. This presents a very dangerous situation for users of pharmaceuticals illicitly sold on the street. A pharmaceutical with legitimate and appropriate markings carries the perception of being a known entity of both composition and dose. However, the potency of the expected entity and the actual potency of the counterfeit pharmaceutical is unpredictable and likely has little in common with the potency of the pharmaceutical it is mimicking in appearance.

No matter what the form, the high potency of designer opiates has resulted in frequent overdose situations. The Centers for Disease Control and Prevention (CDC) estimates that nearly 30,000 deaths occurred in 2017 involving fentanyl and fentanyl derivatives. This number is nearly twice the number of deaths resulting from heroin and roughly equivalent to the number of deaths resulting from pharmaceutical opiates and heroin combined (NIH Overdose Death Rates August 2018 report [16]). Once an overdose situation is underway, the most effective course of action is to administer the opioid antagonist naloxone that effectively reverses the respiratory depression associated with opiate overdoses. In response to the increasing number of opiate-related deaths, first responders as well as community activist in many locations now carry naloxone as a rescue medication. However, due to the high potency of many designer opiates, multiple administrations of naloxone may be required to achieve a rescue [17,18]. While naloxone has proven to be a very effective tool for reducing opiate overdoses and saving the life of opiate users, its use is not entirely benign. Naloxone reverses all of the effects of opiates, not just respiratory depression. Thus in a habituated opiate user, the administration of naloxone will precipitate opiate withdrawal. Physical withdrawal symptoms such as nausea, muscle cramping, and diarrhea as well as agitation, anxiety, and drug craving often follow a naloxone rescue in an addict. Continued supporting care of the rescued user is imperative to ensure that they do not succumb to the intense drive to redose to alleviate the symptoms of opiate withdrawal. In addition, continued supportive care of the rescued user is necessary to ensure that naloxone is onboard until the abused opiates are metabolized and excreted. Naloxone has a half-life of between 30 and 80 minutes, after that a user may resuccumb to the respiratory depression effect of longer lasting opiate agonists as circulating naloxone levels drop. Thus additional naloxone doses may be necessary.

Laboratory detection of designer opiates is most challenged by the potency of the compounds and like all designer drugs, by the changing nature of the drug du jour. Most fentanyl derivatives appear to have some reactivity to antibodies targeting fentanyl [19]. Thus immunoassay screening procedures are capable of providing nondefinitive evidence that fentanyl derivatives are present. Confirmational testing is best performed by Liquid chromatography - tandem mass spectrometry (LC-MS/MS). Fentanyl and most of the derivatives generate a strong signal by electrospray ionization tandem mass spectrometry. In addition, most fentanyl derivatives circulate as parent drugs are primarily excreted without phase II metabolism and are at least partially excreted as unchanged drug. Thus knowledge of the metabolism of new compounds and reference standards for the metabolites are beneficial but are not absolutely required for detecting fentanyl derivatives in urine or blood.

Simple reverse phase chromatographic separations are usually adequate for most fentanyl derivatives unless a definitive discrimination of positional isomers is required. Multiple tandem mass spectrometry (MS/MS) transitions are recommended in an attempt to discriminate positional isomers, but collision-induced fragmentation patterns may overlap sufficiently to hinder an MS/MS-based identification. Thus carefully thought out and powerful chromatographic separations may be required to discriminate isomers. This is the case when attempting to identify the exact structure of methylated or halogenated derivatives where the functional groups can be relocated with little impact on the pharmacology of the drug and little impact on the behavior of the compound within an analytical system. Examples of a few of these challenging isomers and their relationship to the structure of fentanyl are shown in Fig. 15.4.

Cathinones

Cathinone derivatives are the third most prevalent class of designer drugs of this decade. The cathinone derivative methcathinone first gained prominence in the Soviet states with soldiers returning from the Soviet Afghanistan War. Since that point cathinone, derivatives have developed into a broad and diverse class of stimulants. Some of these drugs maintain the basic structure and pharmacology of traditional stimulants, while others appear to more resemble 3,4-Methylenedioxymethamphetamine (MDMA). Like synthetic cannabinoids, most cathinone derivatives could at one point be found sold legally at head shops, truck stops, and the occasional convenience store. Like methamphetamine, cathinones potentiate catecholamines from noradrenergic, serotonergic, and dopaminergic nerves [20]. Cathinones are generally abused for their stimulant and euphoric properties with some compounds producing MDMA-like empathogenic effects. Chronic use or acute toxicity is associated with paranoid excitatory delirium, extreme agitation, paranoia, and unpredictable violent behaviors [21,22]. Treatment for cathinone intoxication is primarily supportive.

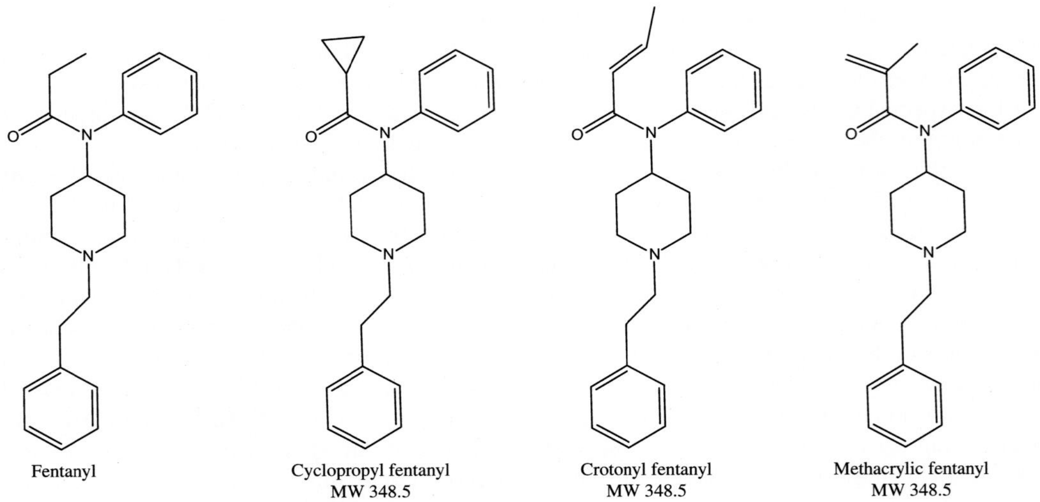

FIGURE 15.4 Structure of fentanyl and three challenging to distinguish analogs.

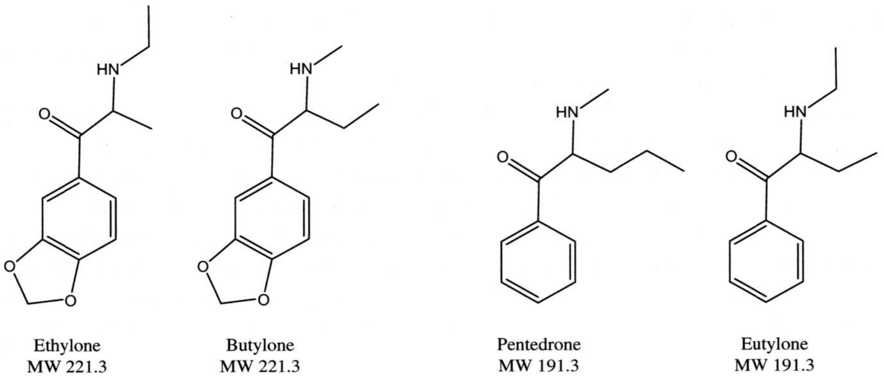

FIGURE 15.5 Two isometric pairs of challenging cathinone derivatives.

Ethylone
MW 221.3

Butylone
MW 221.3

Pentedrone
MW 191.3

Eutylone
MW 191.3

Fentanyl

Cyclopropyl fentanyl
MW 348.5

Crotonyl fentanyl
MW 348.5

Methacrylic fentanyl
MW 348.5

Laboratory detection of cathinones in biological samples is reasonably straightforward. Parent drugs typically circulate and are excreted at least in part without metabolic transformation. With cathinones as with most designer drugs, the challenge is knowing what drugs to target and obtaining reference materials for the targeted drugs. The smaller size of most cathinones enables their analysis in biological samples to be performed by either LC−MS/MS or gas chromatography mas spectrometry (GC-MS) without great difficulty. However, the existence of multiple cathinone isomers can complicate the analysis when using either technique. Many of the isomers may have parallel fragmentation patterns and only subtle differences in chromatographic characteristics. Thus analytical methods must be powerful enough to definitively identify any given compound at the exclusion of both known and unknown isomers. A single example of isomers that can be challenging to discriminate is shown in Fig. 15.5.

Other designer drugs

Hallucinogenic designer drugs have maintained a low-level presence since Shulgin's early experiments. Drugs based upon the structure of mescaline, including those designed by Shulgin, can still be found sold through online retailers. Methyl and dimethoxy derivatives of mescaline have been utilized to avoid legal restriction and detection. Further development of the base structure has led to highly potent hallucinogens such as the NBOMe series of compounds as well as Bromo-dragonfly, Fig. 15.6. The potencies of these compounds are on par with Lysergic acid diethylamide (LSD) and thus they are often sold as counterfeit LSD on blotter papers. However, their activity at serotonin receptors subtypes appears to be more diverse resulting in both potent hallucinogenic activity at serotonin 5-HT-2A receptors as

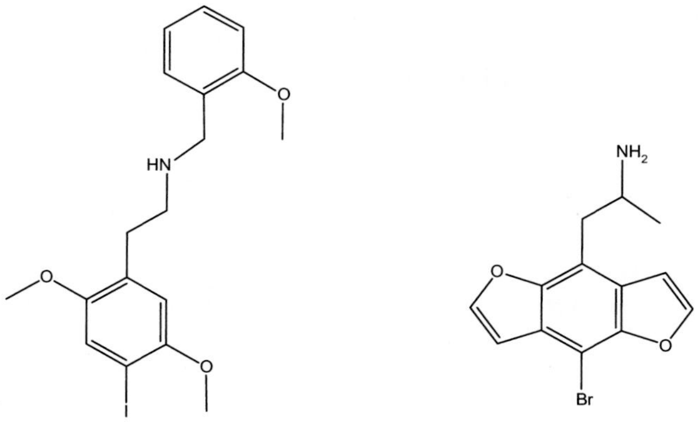

FIGURE 15.6 Potent hallucinogens 25I-NBOMe and Bromo-dragonfly.

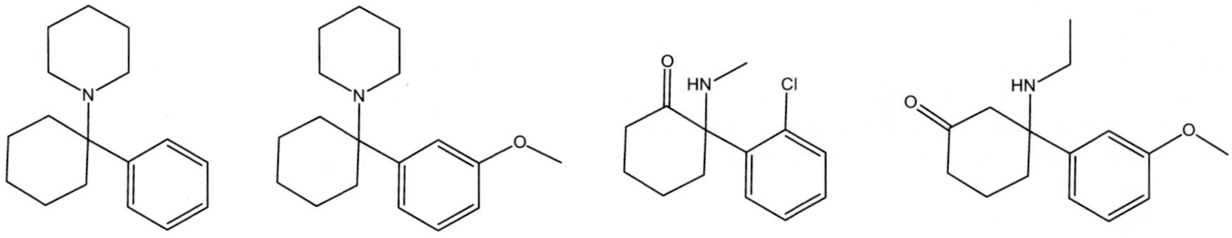

FIGURE 15.7 PCP and dissociative analogs.

well as other serotonin receptors resulting in toxic serotonergic syndrome. In addition to the intended hallucinations, agitation, aggression, and seizures have been reported following the use of these compounds. In extreme cases, severe vasoconstriction has occurred resulting in the development of gangrene and subsequently necessary amputations following even single exposures. In addition, collapsing capillary beds has resulted in examples of bleeding from the eyes, ears, and mouths in overdosing individuals. There are no specific treatments to counteract the effects of these designer hallucinogens. Liquid replacement and perhaps the administration of benzodiazepines as supportive care are the best available options until the effects subside.

Designer drugs mimicking ketamine and phencyclidine (PCP) have also found their way into the club drug scene (Fig. 15.7). Methoxetamine and methoxy PCP are two of the more frequently encountered analogs in this class. Little scientific information exists on these compounds, but the class appears to possess all the physical and psychological effects of dissociative anesthetics acting through postulated noncompetitive N-methyl-D-aspartate (NMDA) receptor blockage [23,24]. At sufficient quantities the desired dissociative effects can transform into a state of dissociative psychosis.

Due to the large number of designer hallucinogens and PCP analogs but a general lack of a dominant species, assays intended to detect hallucinogens and dissociative agents can be cumbersome to design, perform, and maintain in a relevant state. For this purpose, Liquid chromatography - time of flight mass spectrometry (LC-TOF) procedures are of high utility. The procedure should be designed to run in full scan collecting information on the molecular ions of all species. The methodology should also collect information on the molecular fragmentation of all species. With this information a library of known compounds and metabolites can then be built. In addition, new compounds can be added to the library as they emerge. Using a full-scan data collection technique, previously collected data on historical samples can be mined for the presence of previously unknown compounds once these drugs or their metabolites are described in the literature.

Conclusion

In conclusion, designer drugs continue to be a complex and ever-evolving challenge. Their presence and prevalence stress the medical community, unsuspecting drug users, vulnerable families, and a society ill equipped to respond to an ever-changing and poorly understood drug threat. In order to best serve the needs of these groups, laboratories must constantly stay educated on drugs trends and the metabolism of these novel agents. Laboratory methods must remain nimble to target relevant compounds and minimize the risk of overlooking emerging threats.

References

[1] Shulgin A, Shulgin A. PiHKAL: a chemical love story. Berkeley, CA: Transform Press; 1991.
[2] Shulgin A, Shulgin A. TiHKAL: the continuation. Berkeley, CA: Transform Press; 1997.
[3] Lagnston JW, Ballard P. Parkinson's disease in a chemist working with 1-methyl-4-phenyl-1,2,5,6-tetrahydropyridine. N Engl J Med 1983;309(5):310.
[4] Lagnston JW, Ballard P, Tetrud JW, Irwin I. Chronic parkinsonism in humans due to a product of meperidine-analog synthesis. Science 1983;219(4587):979—80.
[5] Ayres WA, Starsiak MJ, Sokolay P. The bogus drug: three methyl & alpha methyl fentanyl sold as "China White". J Psychoact Drugs 1981;13(1):91—3.
[6] Gillespie TJ, Gandolfi AJ, Davis TP, Morano RA. Identification and quantitation of alpha-methylfentanyl in post mortem specimens. J Anal Toxicol 1982;6(3):139—42.
[7] Artwood BK, Huffman J, Straiker A, Mackie K. JWH018 a common constituent of "spice" herbal blends is a potent and efficacious cannabinoid CB receptor agonist. Br J Pharmacol 2010;160(3):585—93.
[8] Compton DR, Gold LH, Ward SJ, Balster RL, Martin BR. Aminoalkylindole analogs: cannabimimetic activity of a class of compounds structurally distinct from delta-9-tetrahydrocannabinol. J Pharmacol Exp Ther 1992;263(3):1118—26.
[9] Castaneto MS, Gorelick DA, Desrosiers NA, Hartman RL, Pirard S, Huestis MA. Synthetic cannabinoids: epidemiology, pharmacodynamics, and clinical implications. Drug Alcohol Depend 2014;144:12—41.
[10] Heath TS, Burroughs Z, Thompson AJ, Tecklenburg FW. Acute intoxication caused by a synthetic cannabinoid in two adolescents. J Pediatr Pharmacol Ther 2012;17(2):177—81.
[11] Roberto AJ, Lorenzo A, Li KJ, Young J, Mohan A, Pinnaka S, et al. First-episode of synthetic cannabinoid-induced psychosis in a young adult, successfully managed with hospitalization and risperidone. Case Rep Psychiatry 2016;2016:7257489.
[12] Hermanns-Clausen M, Kneisel S, Szabo B, Auwärter V. Acute toxicity due to the confirmed consumption of synthetic cannabinoids: clinical and laboratory findings. Addiction 2013;108(3):534—44.
[13] Cooper ZD. Adverse effects of synthetic cannabinoids: management of acute toxicity and withdrawal. Curr Psychiatry Rep 2016;18(5):52.
[14] Ozer U, Ceri V, Evren C. Capgras syndrome after use of synthetic cannabinoids: an adolescent case. J Psychiatry Neurol Sci 2016;29:374—8.
[15] Kaizaki-Mitsumoto A, Hataoka K, Funada M, Odanaka Y, Kumamoto H, Numazawa S. Pyrolysis of UR-144, a synthetic cannabinoid, augments an affinity to human CB1 receptor and cannabimimetic effects in mice. J Toxicol Sci 2017;42(3):335—41.
[16] National Institute on Drug Abuse. Overdose death rates, <https://www.drugabuse.gov/related-topics/trends-statistics/overdose-death-rates>; 2018.
[17] Faul M, Lurie P, Kinsman J, Dailey M, Crabaugh C, Sasser S. Multiple naloxone administrations among emergency medical service providers is increasing. Prehosp Emerg Care 2017;4(21):411—19.
[18] Skolnick P. On the front lines of the opioid epidemic: rescue by naloxone. Eur J Pharmacol 2018;835:147—53.
[19] Helander A, Stojanovic K, Villén T, Beck O. Detectability of fentanyl and designer fentanyls in urine by 3 commercial fentanyl immunoassays. Drug Test Anal 2018;1297—1304.
[20] Kalix P. The releasing effect of the isomers of the alkaloid cathinone at central and peripheral catecholamine storage sites. Neuropharmacology 1986;25(5):499—501.
[21] Penders TM, Gestring RE, Vilensky DA. Excited delirium following use of synthetic cathinones (bath salts). Gen Hosp Psychiatry 2012;34(6):647—50.
[22] Kasick DP, McKnight CA, Klisovic E. "Bath salt" ingestion leading to severe intoxication delirium: two cases and a brief review of the emergence of mephedrone use. Am J Drug Alcohol Abuse 2012;38(2):176—80.
[23] Horsley RR, Lhotkova E, Hajkova K, Jurasek B, Kuchar M, Palenicek T. Detailed pharmacological evaluation of methoxetamine (MXE), a novel psychoactive ketamine analogue—behavioural, pharmacokinetic and metabolic studies in the Wistar rat. Brain Res Bull 2016;126(Pt 1):102—10.
[24] Halberstadt AL, Slepak N, Hyun J, Buell MR, Powell SB. The novel ketamine analog methoxetamine produces dissociative-like behavioral effects in rodents. Psychopharmacology (Berl) 2016;233(7):1215—25.

Chapter 15.1

A death involving extremely high levels of 3,4-methylenedioxymethamphetamine

Leo Johnson[1,2], C. Clinton Frazee, III[1,2], Diane C. Peterson[3] and Uttam Garg[1,2]
[1]Department of Pathology and Laboratory Medicine, Children's Mercy Hospital, Kansas City, MO, United States, [2]University of Missouri School of Medicine, Kansas City, MO, United States, [3]Johnson County Department of Health and Environment/Medical Examiner's Office, Olathe, KS, United States

Case description

A 20-year-old white female was found unresponsive in bed by her boyfriend the morning after they had used party drugs (marijuana, alprazolam, cocaine, and ecstasy) the night before. The subject was observed to have a white foam froth around her mouth and nose upon discovery. General examination of the body noted superficial scars and several healing abrasions but revealed no obvious signs of foul play or significant injury. The subject's medical history was unknown. The body was transported to the Jackson County Medical Examiner's Office, Kansas City, MO. Per protocol, both internal and external examinations were performed and no evidence of injury was detected. Heart blood, subclavian blood, and liver tissue were submitted for toxicological analyses. The heart blood was screened for volatiles (ethanol, methanol, isopropanol, and acetone) by headspace gas chromatography with flame ionization detector (GC—FID). The heart blood was also used for comprehensive broad-spectrum drug-screen testing that included enzyme immunoassays (EIAs) for amphetamines, barbiturates, benzodiazepines, cannabinoids, cocaine metabolite, methadone, opiates, phencyclidine and propoxyphene, and analysis for >200 drugs by gas chromatography—mass spectrometry (GC—MS). Drug-screen testing by GC—MS was performed using liquid—liquid alkaline extraction with bicarbonate buffer (pH 11.0) and butyl acetate followed by mass spectral analysis in full-scan mode. Presumptive identification of analytes was made by a spectral library match and relative retention time comparison with reference standards.

The EIA drug screen was positive for benzoylecgonine (cocaine metabolite), cannabinoids, benzodiazepines, opiates, and amphetamines class drugs. However, urine was negative for opiates and, thus, quantification of opiates in the blood was not pursued. Because cannabinoids toxicity is extremely rare, quantification of the cannabinoid active metabolites was not performed. Heart blood volatiles screen testing by GC—FID was negative for ethanol, methanol, isopropanol, and acetone. The heart blood GC—MS drug screen revealed the presence of alprazolam, levamisole, cocaine, and 3,4-methylenedioxymethamphetamine (MDMA). Heart blood alprazolam level quantitated at 606 ng/mL by ultrahigh liquid chromatography with photodiode array detector. GC—MS was used to confirm cocaine and metabolites. Concentrations of cocaine, benzoylecgonine, and ecgonine methyl ester were 3394, 3650, and 1512 ng/mL, respectively. Heart blood was submitted to a reference laboratory for amphetamine confirmation and quantification by high-performance liquid chromatography—tandem mass spectrometry (LC—MS/MS). Methamphetamine, MDMA, and 3,4-methylenedioxyamphetamine (MDA) were present at the concentrations of 6.8, 97,000, and 640 ng/mL, respectively. Levamisole is frequently used as a cutting agent with cocaine and was not quantitated. Results of toxicology finding are summarized in Table 15.1.1.

Discussion

Although levels of alprazolam and cocaine were in the toxic range, the most striking finding in this case was the extremely high level of MDMA at a concentration of 97,000 ng/mL. MDMA, commonly known as ecstasy and Molly, is a ring-substituted derivative of methamphetamine (Fig. 15.1.1). It was first synthesized in the early 1900s as an appetite suppressant and later in the 1970s as an adjunct psychotherapeutic agent. In the 1970s and 1980s, MDMA became a

TABLE 15.1.1 Postmortem Toxicology Results on Heart Blood	
EIA	Results
Amphetamine	**Positive**
Barbiturates	Negative
Benzodiazepines	**Positive**
Cannabinoids	**Positive**
Cocaine metabolite	**Positive**
Methadone	Negative
Opiates	**Positive**
Phencyclidine	Negative
Propoxyphene	Negative
GC–FID Volatile Screening	
Ethanol	<10 mg/dL
Acetone	<5 mg/dL
Methanol	<5 mg/dL
Isopropanol	<5 mg/dL
GC–MS Screening	**Confirmation and Quantification**
Alprazolam	MDMA—97,000 ng/mL
Levamisole	Cocaine—3394 ng/mL
Cocaine	Alprazolam—606 ng/mL
MDMA	

EIA, Enzyme immunoassay; GC–FID, gas chromatography with flame ionization detector; GC–MS, gas chromatography–mass spectrometry; MDMA, 3,4-methylenedioxymethamphetamine.
Note: Bold indicates the positive results.

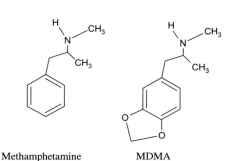

FIGURE 15.1.1 Structures of methamphetamine and MDMA.

Methamphetamine   MDMA

popular street drug being used at dance parties and raves. In 1985 it was listed as a schedule 1 drug by the US Drug Enforcement Agency.

Typical recreational doses of MDMA range from 50 to 150 mg per tablet, with some variation depending upon the origin of the tablets [1]. These dosages usually produce blood levels of approximately 100–250 ng/mL. Like methamphetamine and cocaine, MDMA can exist as free base or as salt. However, MDMA has a higher boiling point than cocaine and methamphetamine and cannot to be inhaled in its free base form. MDMA is readily absorbed from the gastrointestinal tract with peak levels at 2–3 hours, and a half-life of 4–12 hours [2,3]. MDMA is metabolized in the liver by CYP2D6, to 3,4-methylenedioxymethylamphetamine (MDA), an active metabolite which has a longer half-life ($t_{1/2} \simeq$ 16–38 hours). As higher MDMA concentrations are reached and CYP2D6 activity becomes saturated, small changes in dose can increase the risk for drug toxicity and death. This could have been the reason for the very high MDMA concentration in our case. MDMA increases the release and inhibits the reuptake of serotonin [1,4]. MDMA can also enhance the release of dopamine and noradrenaline. The physiological effects of MDMA usage include marked increase in wakefulness, sexual arousal, endurance, and energy. The psychological effects of the drug have been characterized as giving one a sense of euphoria, greater sociability, and sensory perception, as well as a "closeness" with other people. The adverse effects at toxic levels include paranoia, restlessness, insomnia, disorientation, teeth grinding, sweating, nausea, tachycardia, and hyperthermia at toxic levels.

Numerous cases of MDMA toxicity have been reported in the literature, most of them involving multidrug toxicity. In one study from the United Kingdom in which 81 deaths involving MDMA were reported, only 6 were from MDMA by itself (Office for National Statistics, the United Kingdom). Five adults who ingested one to three tablets of MDMA were treated in a hospital for severe toxicity. Despite treatment, these subjects died with blood MDMA levels ranging from 100 to 1300 ng/mL [5]. An 18-year-old female who ingested 150 mg MDMA died of ventricular fibrillation, with an MDMA level of 1000 ng/mL and an ethanol level of 0.04 g/dL [6]. In other reports of fatalities, MDMA levels ranged from 600 to 9300 ng/mL [3,5,7–11]. In our case the level of 97,000 ng/mL is much higher than the reported cases. However, peripheral blood levels of MDMA may have been lower as the drug undergoes postmortem redistribution. In addition to obvious overdose situations, there have been deaths reported that were due to "ecstasy-induced" depression which became severe enough to cause risk-taking behavior and/or suicide [4,5,9,11]. An unusual case of a death due to an apparent allergic reaction following MDMA ingestion has also been reported [12].

The cocaine concentration of 3394 ng/mL found in our case is also in the toxic range and central nervous system (CNS) stimulant effects of cocaine and MDMA may have been synergistic to cause of death. An alprazolam concentration of 606 ng/mL was also in the toxic range (therapeutic range 10–60 ng/mL). Alprazolam is a CNS depressant. It is difficult to postulate its effect on MDMA and cocaine-induced toxicity, but the alprazolam may have contributed to higher dosing and tolerance of these CNS stimulants. In the presented case the manner of death was ruled as accidental due to multidrug intoxications of MDMA, cocaine, and alprazolam.

In conclusion, our case presents a fatality involving MDMA at a level of 97,000 ng/mL, which is one of the highest levels reported to date.

References

[1] Kalant H. The pharmacology and toxicology of "ecstasy" (MDMA) and related drugs. CMAJ 2001;165:917–28.
[2] Mas M, Farre M, de la Torre R, Roset PN, Ortuno J, Segura J, et al. Cardiovascular and neuroendocrine effects and pharmacokinetics of 3,4-methylenedioxymethamphetamine in humans. J Pharmacol Exp Ther 1999;290:136–45.
[3] Baselt RC. Methylenedioxymethamphetamine. In: Baselt RC, editor. Disposition of toxic drugs and chemicals in man. Seal Beach, CA: Biomedical Publications; 2017. p. 1380–3.
[4] Meyer JS. 3,4-Methylenedioxymethamphetamine (MDMA): current perspectives. Subst Abuse Rehabil 2013;4:83–99.
[5] Henry JA, Jeffreys KJ, Dawling S. Toxicity and deaths from 3,4-methylenedioxymethamphetamine ("ecstasy"). Lancet 1992;340:384–7.
[6] Dowling GP, McDonough III. ET, Bost RO. 'Eve' and 'ecstasy'. A report of five deaths associated with the use of MDEA and MDMA. JAMA 1987;257:1615–17.
[7] Chandra YG, Shetty AR, Jayanth SH, Hugar BS, Praveen S, Harish S. A death due to ecstasy—a case report. Med Leg J 2016;84:46–8.
[8] Crifasi J, Long C. Traffic fatality related to the use of methylenedioxymethamphetamine. J Forensic Sci 1996;41:1082–4.
[9] Dams R, De Letter EA, Mortier KA, Cordonnier JA, Lambert WE, Piette MH, et al. Fatality due to combined use of the designer drugs MDMA and PMA: a distribution study. J Anal Toxicol 2003;27:318–22.
[10] Elliott SP. MDMA and MDA concentrations in antemortem and postmortem specimens in fatalities following hospital admission. J Anal Toxicol 2005;29:296–300.
[11] Klys M, Rojek S, Wozniak K, Rzepecka-Wozniak E. Fatality due to the use of a designer drug MDMA (ecstasy). Leg Med (Tokyo) 2007;9:185–91.
[12] Sauvageau A. Death from a possible anaphylactic reaction to ecstasy. Clin Toxicol (Phila) 2008;46:156.

Chapter 15.2

Rhabdomyolysis associated with laboratory-confirmed FUB-AMB use

Stephen L. Thornton[1] and Roy Gerona[2]
[1]University of Kansas Hospital Poison Control Center, Kansas City, KS, United States, [2]Department of Laboratory Medicine, University of California, San Francisco, San Francisco, CA, United States

Case description

A 25-year-old female with history of scoliosis and depression presented to the emergency department complaining of back pain and bilateral lower extremity weakness. These symptoms began on awakening 8 hours prior to arrival and were not consistent with prior episodes of back pain. She denied ingestion, seizures, injury, or immobilization. She also denied any bowel or bladder problems. Her vital signs were heart rate of 73 bpm, blood pressure of 120/71 mmHg, respiratory rate of 17, and a temperature of 36.8°C. She was anxious but oriented and not agitated. She was given intravenous lorazepam and ketorolac without relief of her symptoms. Physical examination noted effort weakness in her lower extremities with intact sensation, no swelling, mild tenderness, and normal reflexes. Significant pain was present with movement of her legs. Laboratory evaluation revealed a normal complete blood count and basic chemistry, including a normal serum creatinine of 0.84 mg/dL (normal 0.4−1 mg/dL). A seven-panel urine drug immunoassay screen was positive only for benzodiazepines. A serum ethanol level was negative. She continued to have severe pain and would not able to ambulate. It was difficult to determine if her weakness and inability to ambulate were due to pain or weakness. MRIs of the thoracic and lumbar spine were ordered and were interpreted as normal. Neurology was consulted and after evaluation they felt an acute neurologic problem was unlikely and to consider musculoskeletal causes. At this point, rhabdomyolysis was considered and a creatinine kinase (CK) was ordered. It resulted at 5265 U/L (normal 21−215 U/L). She was started on normal saline intravenous fluids and admitted to the hospital. Her CK peaked at 17,991 U/L on hospital day 2. Her serum creatinine remained normal throughout her hospitalization.

While in the hospital, she admitted to smoking what she stated was marijuana the night before her symptoms began. She continued to deny any seizures or trauma. Serum samples from her admission were obtained and screened by liquid chromatography—quadrupole time-of-flight mass spectrometry (QTOF 6550, LC 1260, Agilent)—and detected methyl 2-(1-(4-fluorobenzyl)-1H-indazole-3-carboxamido)-3-methylbutanoate (AMB-FUBINACA or FUB-AMB) at a concentration of 4.2 ng/mL. The screen was negative for 549 other substances. Also detected were formula matches to predicted metabolites of FUB-AMB—$C_{21}H_{22}FN_3O_4$ and $C_{20}H_{20}FN_3O_3$—which could not be confirmed due to lack of reference standards. The patient could not provide a name or sample of what she had smoked. She was discharged after a 3-day hospitalization with her pain greatly improved, an improving CK of 4972 U/L, and a normal serum creatinine of 0.74 mg/dL.

Case discussion

FUB-AMB is an indazole synthetic cannabinoid structurally related to AB-FUBINACA. Synthetic cannabinoids are a collection of chemicals that, as a rule, have potent effects on the various cannabinoid receptors [1]. Since approximately 2010 they have been widely marketed as "K2," "Spice," and a multitude of other "herbal potpourris" and are now frequently distributed over the internet [2]. Many are controlled substances and FUB-AMB was made a schedule I controlled substance by the Drug Enforcement Administration in 2017 [3].

The use of synthetic cannabinoids has been associated with multiple toxic effects. Tachycardia, agitation/anxiety, and gastrointestinal distress are commonly reported [4]. More serious effects such as psychosis, myocardial infarctions, acute kidney injury, and strokes have all been associated with synthetic cannabinoid use [5].

Like other synthetic cannabinoids, FUB-AMB is a potent chemical, with estimates of it being 85 times more potent than Δ9-tetrahydrocannabinol [6]. Consistent with this potency are the reports of severe toxicity associated with its use. FUB-AMB was the causative agent in the "Zombie outbreak" in New York City in 2016 [7]. This outbreak resulted in 18 patients (mean age 36.8 years) being transported to health-care facilities for treatment of a severe delirium. Interestingly, investigators were not able to identify FUB-AMB itself in the biological sample but rather only its metabolites were identified. This contrasts with our case which is the only case in the medical literature where both FUB-AMB and its purported metabolites were identified in a biological sample. This may imply a potentially large exposure to FUB-AMB in this case. FUB-AMB is believed to have a short half-life due to the rapid serum hydrolysis of similar synthetic cannabinoids [8]. Detection of the parent compound at least 8 hours from exposure could be explained by a large initial level. This potentially large exposure may have led to the rhabdomyolysis. Unlike in this case, rhabdomyolysis was not reported as a complication of the "zombie" outbreak. Rhabdomyolysis has previously been attributed to synthetic cannabinoid use; however, those reports were limited by lack of analytical confirmation and presence of other known causes of rhabdomyolysis (agitation, hyperthermia, and seizures) which were not present in this case [9–11]. In this case the exact mechanism by which FUB-AMB caused muscle injury is not known and further investigation is warranted.

Health-care practitioners should be aware of the risk of rhabdomyolysis from FUB-AMB even in the absence of classic risk factors.

References

[1] Tai S, Fantegrossi WE. Pharmacological and toxicological effects of synthetic cannabinoids and their metabolites. Curr Top Behav Neurosci 2017;32:249–62.

[2] Brents LK, Prather PL. The K2/Spice phenomenon: emergence, identification, legislation and metabolic characterization of synthetic cannabinoids in herbal incense products. Drug Metab Rev 2014;46(1):72–85.

[3] Drug Enforcement Administration, Department of Justice. Schedules of controlled substances: temporary placement of FUB–AMB into schedule I. Temporary amendment; temporary scheduling order. Fed Regist 2017;82(212):51154–8.

[4] Courts J, Maskill V, Gray A, Glue P. Signs and symptoms associated with synthetic cannabinoid toxicity: systematic review. Australas Psychiatry 2016;24(6):598–601 Epub 2016 Aug 24.

[5] Tait RJ, Caldicott D, Mountain D, Hill SL, Lenton S. A systematic review of adverse events arising from the use of synthetic cannabinoids and their associated treatment. Clin Toxicol (Phila) 2016;54(1):1–13.

[6] Banister SD, Moir M, Stuart J, et al. Pharmacology of indole and indazole synthetic cannabinoid designer drugs AB-FUBINACA, ADB-FUBINACA, AB-PINACA, ADB-PINACA, 5F-AB-PINACA, 5F-ADB-PINACA, ADBICA, and 5F-ADBICA. ACS Chem Neurosci 2015;6:1546–59.

[7] Adams AJ, Banister SD, Irizarry L, Trecki J, Schwartz M, Gerona R. "Zombie" outbreak caused by the synthetic cannabinoid AMB-FUBINACA in New York. N Engl J Med 2017;376(3):235–42.

[8] Andersson M, Diao X, Wohlfarth A, Scheidweiler KB, Huestis MA. Metabolic profiling of new synthetic cannabinoids AMB and 5F-AMB by human hepatocyte and liver microsome incubations and high-resolution mass spectrometry. Rapid Commun Mass Spectrom 2016;30:1067–78.

[9] Adedinsewo DA, Odewole O, Todd T. Acute rhabdomyolysis following synthetic cannabinoid ingestion. N Am J Med Sci 2016;8(6):256–8.

[10] Durand D, Delgado LL, de la Parra-Pellot DM, Nichols-Vinueza D. Psychosis and severe rhabdomyolysis associated with synthetic cannabinoid use: a case report. Clin Schizophr Relat Psychoses 2015;8(4):205–8.

[11] Sweeney B, Talebi S, Toro D, Gonzalez K, Menoscal JP, Shaw R, et al. Hyperthermia and severe rhabdomyolysis from synthetic cannabinoids. Am J Emerg Med 2016;34(1):121.e1–2.

Chapter 15.3

Clinical and pathological findings in fatal cases involving the ingestion of methylone

Diane M. Boland

Miami-Dade Medical Examiner Department, Toxicology Laboratory, Miami, FL, United States

Case descriptions

Case 1

The decedent was a 32-year-old male who, according to family, was in good health and had no documented medical concerns. He was known to smoke three packets of cigarettes per day, drink vodka daily, and had a history of using "Mollies," cocaine, and marijuana. On the date of his death, the decedent consumed "Mollies" and vodka while at a holiday party with his wife. Upon returning to his residence, the decedent became ill. Fire rescue responded and transported the decedent to the emergency department. He was reportedly diaphoretic, agitated, confused, violent, and unable to follow commands. He was also speaking to himself and experiencing visual hallucinations. Hospital records indicate that the decedent was acidotic, exhibited tachycardia with hypotension, and had a rectal temperature of 107.9°F. Initial urine toxicology screen at the hospital was positive for amphetamines and cannabinoids. The decedent expired while being treated. Autopsy revealed cerebral edema and diffuse petechial hemorrhages of the right cerebral hemisphere of the brain, pulmonary edema and congestion, biventricular dilatation of the heart, and an ecchymosis of the tongue. Histologic sections of the brain demonstrated extensive intraparenchymal and perivascular hemorrhages limited primarily to the cortex, consistent with an acute infarct. Postmortem toxicology testing utilizing peripheral blood revealed the presence of methylone at a concentration of 1.5 mg/L. Other drugs identified were amphetamine, diphenhydramine, doxylamine, ibuprofen, and nordiazepam. The cause of death was determined to be right hemispheric cerebral Infarct due to acute methylone toxicity, with the manner of death as accident.

Case 2

The decedent was a 30-year-old white male who was never married, did not have any children, and lived with a roommate. According to a friend, the decedent was known to take a drug called "G" mixed with alcohol, although it was not clear if he consumed any on the evening of his death. After a night of partying the decedent began to vomit and, subsequently, became unresponsive. Fire rescue responded and transported him to the emergency department where he arrived in cardiac arrest and, after medical efforts, expired. Significant autopsy findings include cardiomegaly (heart: 530 g), epicardial plaque, redness and softening of the left ventricular free wall, left ventricular hypertrophy, hepatosplenomegaly (liver: 2190 g, spleen: 310 g), capsular scar, and multiple hemorrhagic lesions of the liver and right renal cyst. Toxicology revealed low levels of amphetamine in antemortem urine and saphenous vein blood. Atropine, lidocaine, and etomidate were detected in saphenous vein blood, and methylone was detected in saphenous vein blood at a concentration of 0.29 mg/L. The cause of death was determined to be methylone toxicity, with the manner of death as accident.

Case 3

The decedent was a 21-year-old white male who was not under the care of a medical doctor but, according to family, smoked cigarettes, drank alcohol socially, and used marijuana occasionally. Per family and friends, he was reportedly visiting South Florida to attend a 3-day music festival. On the day of his death, his friends stated that he was "popping Mollies" and using lysergic acid diethylamide (LSD), a mixture referred to as "Sassy." During the day of the incident, the decedent was behaving normally, but at some point he left his group of friends to smoke with an unknown male. Shortly after his return, he started acting erratically, speaking to someone who was not there, and rolling on the ground and waving his hands around. Fire rescue responded and transported him to the emergency department. According to the attending physician, the decedent was extremely aggressive and required multiple security guards and emergency room staff to restrain him. For his protection the decedent was sedated and intubated. Initial diagnosis included drug overdose, rhabdomyolysis, hyperthermia (107.6°F body temperature), lactic acidosis, hyperkalemia, possible aspiration, and supraventricular tachycardia. Cardiopulmonary resuscitation was met with negative results and death was pronounced less than 24 hours after his admission. Autopsy disclosed minor injuries and a hemorrhagic diathesis. Toxicology analysis identified methylone at a concentration of 1.2 mg/L and the presence of 9-carboxy-tetrahydrocannabinol. The cause of death was determined to be acute methylone toxicity with the manner of death as accident.

Discussion

Methylone, a cathinone derivative and beta-keto analog of methylenedioxymethamphetamine (MDMA), is an entactogen and stimulant marketed as bath salts, plant food, research chemicals, or as a replacement for MDMA in "Molly" capsules [1]. It is appealing to users because it produces euphoria and increased sociability; however, it can be extremely dangerous as it can also cause paranoia, agitations, hallucinations, and psychotic behavior [2]. The Miami-Dade Medical Examiner Department (MDME) has investigated 72 cases since 2011, in which methylone was identified by the toxicology laboratory during a routine drug screen. Of the death investigation cases received over the last 6 years, 46 were classified as homicides, 8 were classified as suicides (gunshot wounds and hangings), 16 were classified as accidents, and 2 were classified as undetermined. Of the accidental deaths, seven were overdoses; however, only three of the accidental deaths were directly related to the ingestion of methylone and its toxic effects.

Routine toxicological analyses were performed on postmortem specimens, or antemortem specimens when available. All cases received a volatile analysis by headspace gas chromatography with flame ionization detection. Urine specimens were analyzed by immunoassay for amphetamines, barbiturates, opiates, benzoylecognine, oxycodone, benzodiazepines, and phencyclidine. Urine was also subjected to a comprehensive drug screen using liquid–liquid extraction followed by analysis using gas chromatography mass spectrometry. Quantitation of methylone in blood was performed using solid-phase extraction followed by analysis using liquid chromatography tandem mass spectrometry.

Due to its relatively recent introduction as an illicit drug, blood concentrations of methylone have not been widely reported in fatalities. Pearson et al. describe three fatalities involving methylone toxicity with peripheral blood concentrations of 0.84, 3.3, and 0.56 mg/L [3]. The decedents exhibited symptoms consistent with sympathomimetic toxicity, including elevated body temperature, metabolic acidosis, rhabdomyolysis, acute renal failure, and disseminated intravascular coagulation (DIC) [3]. Another case report by Warrick et al. involves a decedent exhibiting serotonin syndrome, DIC, and renal failure; however, a blood concentration was not reported [4]. Cawrse et al. described a fatality of a young man who collapsed suddenly while jogging. Peripheral and central blood concentrations of methylone were 0.67 and 0.74 mg/L, respectively, and the cause of death was sudden cardiac death associated with methylone use [5]. Most recently, Barrios et al. reported a fatality in which the decedent ingested methylone orally and, soon after, developed difficulty breathing accompanied by polypnea, as well as cardiopulmonary arrest. Peripheral and central blood concentrations of methylone were 3.13 and 6.64 mg/L, respectively, and methylone was also qualitatively identified in the stomach contents [6]. The fatalities in Miami-Dade County that include methylone in the cause of death exhibited blood methylone concentrations (0.29, 1.2, and 1.5 mg/L) comparable to the previously published case reports. In those cases where terminal event, fire rescue, and/or emergency department information were available, similarities were also noted.

References

[1] Drug Enforcement Administration. 3,4-Methylenedioxymethcathinone (methylone). Office Diversion Control, Drug & Chemical Evaluation Section; 2013.
[2] Baselt RC, editor. Disposition of toxic drugs and chemicals in man. 9th ed. Seal Beach, CA: Biomedical Publications; 2011.
[3] Pearson JM, Hargraves TL, Hair LS, Massucci CJ, Frazee CC, Garg U, et al. Three fatal intoxications due to methylone. J Anal Toxicol 2012;36(6):444−51.
[4] Warrick BJ, Wilson J, Hedge M, Freeman S, Leonard K, Aaron C. Lethal serotonin syndrome after methylone and butylone ingestion. J Med Toxicol 2012;8(1):65−8.
[5] Cawrse BM, Levine B, Jufer RA, Fowler DR, Vorce SP, Dickson AJ, et al. Distribution of methylone in four postmortem cases. J Anal Toxicol 2012;36:434−9.
[6] Barrios L, Grison-Hernando H, Boels D, Bouquie R, Monteil-Ganiere C, Clement R. Death following ingestion of methylone. Int J Leg Med 2016;130:381−5.

Chapter 15.4

A death involving a "bath salt" methylenedioxypyrovalerone and tramadol

Uttam Garg[1,2], Clinton Frazee[1,2] and Diane Peterson[3]
[1]Department of Pathology and Laboratory Medicine, Children's Mercy Hospital, Kansas City, MO, United States, [2]University of Missouri School of Medicine, Kansas City, MO, United States, [3]Office of the Jackson County Medical Examiner, Kansas City, MO, United States

Case description

A 41-year-old white male was found unresponsive in his bed. He had a history of hypertension, anxiety, bipolar disorder, chronic ethanol abuse, and methamphetamine use. He had recently stopped using methamphetamine and had begun using "bath salts" approximately 2 weeks prior to his death. He had been "high" and awake for the last 3 days. He had head and neck cyanosis, superficial ulcers of the mucosa of the upper and lower lips, and linear discontinuous healing of superficial excoriations, as well as an apparent needle puncture site with adjacent ecchymosis on the skin of the arms. Coronary arteries were mildly to markedly narrowed by atheroma. There was moderate pulmonary edema, and mucosal erosions at the mid and lower aspects of the esophagus. Histologically, there was no definitive contraction band necrosis.

Femoral blood, heart blood, vitreous fluid, urine, liver tissue, brain tissue, gastric contents, and a packet labeled "Blue Magic 350 mg" were submitted for toxicological analysis. The femoral blood was used volatile screen (ethanol, methanol, isopropanol, and acetone) by headspace gas chromatography with flame ionization detector. The femoral blood was also used for a comprehensive broad spectrum drug-screening that utilized enzyme immunoassays (EIAs) for amphetamines, barbiturates, benzodiazepines, cannabinoids, cocaine metabolite, methadone, opiates, phencyclidine and propoxyphene, and drug-screen testing for >200 drugs by gas chromatography–mass spectrometry (GC–MS). Drug-screen testing by GC–MS employed a liquid–liquid alkaline extraction using bicarbonate buffer (pH 11.0) and butyl acetate followed by mass spectrometer detection analysis in full-scan mode. Presumptive identification of analytes was made by spectral library match and relative retention time comparison with reference standards. The packet submitted, "Blue Magic 350 mg," had a zip lock closure with the inscription "Bath Salts, Novelty Bath Salts, Not for human consumption" on the back aspect of the packet. The packet contained an unknown white powder. Approximately 10 mg of white powder was dissolved in deionized water and analyzed by EIA and GC/MS.

The volatiles screen was negative for ethanol (<10 mg/dL), methanol (<5 mg/dL), isopropanol (<5 mg/dL), and acetone (<5 mg/dL). The nine panel EIA drug screen was positive for benzodiazepine. The GC–MS drug screen for >200 drugs indicated the presence of diphenhydramine, methylenedioxypyrovalerone (MDPV), and tramadol. Diphenhydramine peak was very small and was not quantified. Benzodiazepine confirmation by HPLC/UV showed the presence of alprazolam at a concentration of 26 ng/mL. Femoral blood was submitted to a reference laboratory (NMS Labs, Willow Grove, PA) for the quantification of tramadol and MDPV. Tramadol and O-desmethyltramadol concentrations were 9000 and 320 ng/mL, respectively. MDPV concentration was 130 ng/mL. Results are summarized in Table 15.4.1. The packet labeled "Blue Magic 350 mg" tested positive for MDPV.

Discussion

Designer drugs are synthetic compounds that mimic the effects of other legal or illicit drugs. As drug analogs, their primary purpose is to circumvent legal ramifications and allude detection. MDPV is a synthetic cathinone with structural

TABLE 15.4.1 Postmortem Toxicology Results on Femoral Blood.

EIA	Results
Amphetamine	Negative
Barbiturates	Negative
Benzodiazepines	**Positive**
Cannabinoids	Negative
Cocaine metabolite	Negative
Methadone	Negative
Opiates	Negative
Phencyclidine	Negative
Propoxyphene	Negative
GC–FID volatile screening	
Ethanol	<10 mg/dL
Acetone	<5 mg/dL
Methanol	<5 mg/dL
Isopropanol	<5 mg/dL
GC–MS Screening	**Quantification**
Diphenhydramine	Alprazolam—26 ng/mL
MDPV	Tramadol—8000 ng/mL
Tramadol	*O*-Desmethyltramadol—320 ng/mL

EIA, Enzyme immunoassay; *GC–FID*, gas chromatography with flame ionization detector; *GC–MS*, gas chromatography–mass spectrometry; *MDPV*, methylenedioxypyrovalerone.
Note: Bold indicates the positive results.

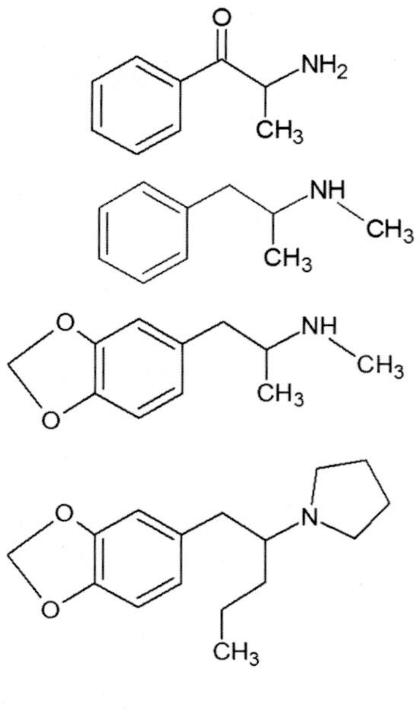

Cathinone

Methamphetamine

MDMA

MDPV

FIGURE 15.4.1 Chemical structures of cathinone, methamphetamine, MDMA, and MDPV. *MDMA*, 3,4-Methylenedioxymethamphetamine; *MDPV*, methylenedioxypyrovalerone.

similarities to methamphetamine and 3,4-methylenedioxymethamphetamine (Fig. 15.4.1). Cathinone itself is a naturally occurring amphetamine analog found in the leaves of *Catha edulis* (Khat), a plant indigenous to Northeast Africa and the Arabian Peninsula [1]. In recent years, synthetic cathinones have grown in popularity and are typically sold as "bath salts" or "plant food" [2]. Other commonly used trade names for synthetic cathinones include "Ivory Wave," "White Lightening," "Bliss," "Cloud Nine," "Lunar Wave," and "Vanilla Sky." They are labeled "not for human consumption" in an attempt to avoid penalty under the Analogue Enforcement Act and are distributed to the general public through street-level drug distributors, head shops, smoke shops, gas stations, and internet retailers [1]. These compounds are typically sold as granulated crystal salts and modes of administration vary. Though they are most commonly snorted, they can be consumed orally or rectally or injected intravenously or intramuscularly. What is more, synthetic cathinones are water soluble and can be easily dissolved in beverages. Commonly reported synthetic cathinones include α-pyrrolidinovalerophenone, mephedrone, methoxetamine, methylone, pentedrone, and MDPV.

MDPV was first synthesized in 1967, and its abuse was reported around 2005 [1,3]. Doses of 5–20 mg produce sympathomimetic (stimulant) effects that can last for 1–5 hours. In addition to its stimulant effects, MDPV can produce effects that mirror those of hallucinogenics with users experiencing anxiety, confusion, paranoia, psychosis, myoclonus, insomnia, severe agitation, diaphoresis, tachycardia, and hypertension. As a result, MDPV overdose can lead to severe hypertension, tachycardia, and serotonin syndrome resulting in death [1–5]. Unsurprisingly, concomitant use of MDPV with other stimulants leads to greater monoamine toxicity [6] and a number of toxicological cases with an array of symptoms have been described [1,5,7–9]. In most of these cases a mixed drug intoxication had been reported. For example, the New York City Medical Examiner conducted a 3-year retrospective analysis of deaths in which cathinones were detected. In 15 cases, cathinones were a contributory cause of death and 2 deaths were attributed solely to cathinone intoxication [10].

Although both MDPV and tramadol were found in the present case and tramadol was detected in a toxic to lethal concentration range, the decedent's history of being awake for 3 days is primarily consistent with MDPV (stimulant) use. The average tramadol level reported in 11 acutely overdosed patients was 7869 ng/mL (range 3910–14,950 ng/mL) [11]. Two adults who survived after consuming 3–6 g tramadol had levels of 8663 and 9500 ng/mL [12]. Seventeen adults who believed to have died from tramadol overdose had levels of 1100–12,000 ng/mL [13]. These findings show a clear overlap between toxic and lethal concentration of tramadol. Based upon these reported observations, the tramadol level seen in our case is likely to cause respiratory depression, hypertension, and tachycardia. In combination with a sympathomimetic drug such as MDPV, tramadol toxic effects could have been potentiated. Of note, oral mucosal (aphthous) ulcers seen in this case have not been previously reported as being a potential adverse effect of either tramadol or "bath salts." The cause of death in this case was ruled as tramadol overdose and MDPV intoxication, and manner of death was accident.

References

[1] Banks ML, Worst TJ, Rusyniak DE, Sprague JE. Synthetic cathinones ("bath salts"). J Emerg Med 2014;46:632–42.
[2] Thornton MD, Baum CR. Bath salts and other emerging toxins. Pediatr Emerg Care 2014;30:47–52 quiz 53–55.
[3] Baselt RC. Methylenedioxypyrovalerone. In: Baselt RC, editor. Disposition of toxic drugs and chemicals in man. Seal Beach, CA: Biomedical Publications; 2017. p. 1384–5.
[4] Baumann MH, Partilla JS, Lehner KR, Thorndike EB, Hoffman AF, Holy M, et al. Powerful cocaine-like actions of 3,4-methylenedioxypyrovalerone (MDPV), a principal constituent of psychoactive 'bath salts' products. Neuropsychopharmacology 2013;38:552–62.
[5] Stanciu CN, Penders TM, Gnanasegaram SA, Pirapakaran E, Padda JS, Padda JS. Withdrawn: the behavioral profile of methylenedioxypyrovalerone (MDPV) and alpha-pyrrolidinopentiophenone (PVP)—a systematic review. Curr Drug Abuse Rev 2016;(9):1–5.
[6] Angoa-Perez M, Kane MJ, Francescutti DM, Sykes KE, Shah MM, Mohammed AM, et al. Mephedrone, an abused psychoactive component of 'bath salts' and methamphetamine congener, does not cause neurotoxicity to dopamine nerve endings of the striatum. J Neurochem 2012;120:1097–107.
[7] Antonowicz JL, Metzger AK, Ramanujam SL. Paranoid psychosis induced by consumption of methylenedioxypyrovalerone: two cases. Gen Hosp Psychiatry 2011;33:640.e5–6.
[8] Penders TM, Gestring R. Hallucinatory delirium following use of MDPV: "bath salts". Gen Hosp Psychiatry 2011;33:525–6.
[9] Thornton SL, Gerona RR, Tomaszewski CA. Psychosis from a bath salt product containing flephedrone and MDPV with serum, urine, and product quantification. J Med Toxicol 2012;8:310–13.
[10] deRoux SJ, Dunn WA. "Bath salts" the New York City medical examiner experience: a 3-year retrospective review. J Forensic Sci 2017;62:695–9.
[11] Khosrojerdi H, Alipour Talesh G, Danaei GH, Shokooh Saremi S, Adab A, Afshari R. Tramadol half life is dose dependent in overdose. Daru 2015;23:22.
[12] Baselt RC. Tramadol. In: Baselt RC, editor. Disposition of toxic drugs and chemicals in man. Seal Beach, CA: Biomedical Publications; 2017. p. 2147–50.
[13] Tjaderborn M, Jonsson AK, Hagg S, Ahlner J. Fatal unintentional intoxications with tramadol during 1995–2005. Forensic Sci Int 2007;173:107–11.

Chapter 15.5

Clonazolam abuse: a report of two cases

Heath A. Jolliff
Department of Emergency Medicine, Doctors Hospital, OhioHealth, Columbus, OH, United States

Case descriptions

Case #1

A 20-year-old male with a past medical history of anxiety and depression presented to the emergency department (ED) via emergency medical services (EMS) after being found slumped over in his car. He and a friend admitted to ingesting several tablets of a drug called "Pinzor" shortly before arrival (Figs. 15.5.1 and 15.5.2). They purchased this drug online for recreational use. The patient stated that he ingested three 0.7 mg pills prior to being found unconscious in his car. Upon arrival in the ED, he had stable vital signs but became progressively hypotensive (BP 79/46 mmHg) and bradycardic (51 bpm). His blood pressure was responsive to IV fluids and he did not require atropine. He was drowsy, although arousable, and had 3 mm pupils. His medical work-up was unremarkable except for a urine drug of abuse screen that was positive for benzodiazepines and negative for opiates, amphetamines, oxycodone, phencyclidine, and cannabinoids. The causative drug was subsequently identified as clonazolam, a novel synthetic benzodiazepine (Fig. 15.5.3). The patient was admitted to the ICU and was subsequently discharged the following morning after a return of normal mental status, stable vitals, and a mental health evaluation.

Case #2

An 18-year-old male with no past medical history arrived via EMS after falling asleep in the same car as the patient in case #1. He admitted to taking two 0.7 mg "Pinzor" tablets prior to being found by EMS. He stated that he became very drowsy and was awakened by EMS and transported to the ED with patient #1. In the ED, he was alert and oriented. His vitals in contrast to the other patient were significant for tachycardia (135 bpm) and hypertension (174/107 mmHg) but otherwise stable. His mentation was normal throughout his ED course except for some restlessness and anxiety. His work-up was remarkable for a creatinine (1.10 mg/dL), a thyroid stimulating hormone (TSH) (517 mU/mL), and a positive urine drug of abuse screen for benzodiazepines, cannabinoids, and opiates. The patient was admitted to the ICU and had an uneventful hospital course. His creatinine and vital signs normalized after IV hydration and he was discharged the following day.

Discussion

The hospital urine immunoassay drug screen for both patients was positive for benzodiazepines.

The purchased pills were sent to a reference lab and analyzed by GC/MS revealing a single major peak with an accurate mass of 354.0779, consistent with $C_{17}H_{12}ClN_5O_2$, the molecular formula for clonazolam. This identification was confirmed by verification of retention time and GC−MS analysis relative to a clonazolam reference standard (Agilent 5975; Santa Clara, CA).

Samples of both patients' urine were sent to the same reference lab as the pills. A screening analysis was performed using liquid chromatography quadrupole time-of-flight mass spectrometry (LC−QTOF) (Xevo G2-S with an Acquity I-class UPLC, Waters; Milford, MA) to establish the presence of drugs and their accurate mass. The method has been previously described and validated against an in-house database for approximately 1200 drugs and their common metabolites, including medicinal, therapeutic, and novel psychoactive substances.

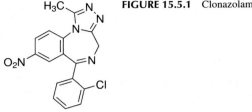

FIGURE 15.5.1 Clonazolam tablets.

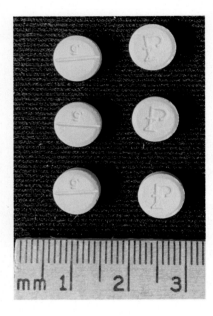

FIGURE 15.5.2 "Pinzor" lable.

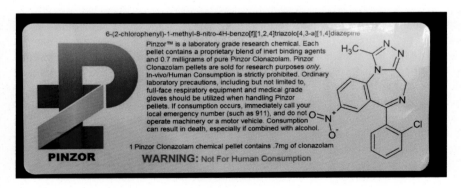

FIGURE 15.5.3 Clonazolam structure.

Confirmatory analysis for clonazolam was performed by liquid chromatography tandem mass spectrometry (LC−MS/MS) (Agilent 6430 with a 1200 Series HPLC, Santa Clara, CA). Confirmatory methods for urine were validated according to SWGTOX guidelines, including determination of linear range, limits of detection and quantitation, precision, accuracy, stability, and interferences [1]. Patient #1 had a urinary concentration of clonazolam of 15.3 ng/mL. Patient #2 had a urinary concentration of clonazolam of 30.7 ng/mL.

Clonazolam was synthesized in the 1970s as a potent antidepressant but was never evaluated in humans. It has emerged in the illicit market as a designer benzodiazepine. Information on its public health impact is scarce. Over the past few years, multiple synthetic benzodiazepines have been identified, including etizolam, flubromazepam, flubromazolam, bromazepam, and now clonazolam [2,3]. Clonazolam is a potent triazolo-analog of clonazepam and also known as clonitrazolam. As little as 0.5 mg of oral clonazolam has been shown to cause sedation and amnesia [2].

At the time of the patients' presentation in the ED, we believed that these were the first reported cases of clonazolam ingestion and subsequent toxicity reported in the United States. The local Poison Center and inpatient Toxicology service were consulted from the ED and had no knowledge of this particular drug. Online searches revealed that clonazolam was a designer benzodiazepine intended for "research use" produced in Europe and only available for online purchase. A PubMed search for "clonazolam" resulted in only one publication in which a case report of flubromazolam, another designer benzodiazepine, ingestion was described [4]. Of note there were multiple online social forums in which the use of clonazolam was described. Since the time of the patients presentation, three more review articles on clonazolam have been published [5–7].

Our case series is notable due to the markedly different presentations of the two patients in regards to vital signs and mental status. Whether this was a dose-dependent relationship of clonazolam or due to a combination of the clonazolam, cannabinoids, and opiates in the second patient is uncertain. We hope this case series will raise awareness for the use of a new designer benzodiazepine in recreational abuse, as well as to encourage further studies of this drug.

References

[1] Clonazolam. Washington, DC; 2016. pp. 1–5. <SWGDRUG.org/monographs.htm>.

[2] Moosmann B, King LA, Auwärter V. Designer benzodiazepines: a new challenge. World Psychiatry 2015;14(2):248. Available from: https://doi.org/10.1002/wps.20236.

[3] Huppertz LM, Bisel P, Westphal F, Franz F, Auwärter V, Moosmann B. Characterization of the four designer benzodiazepines clonazolam, deschloroetizolam, flubromazolam, and meclonazepam, and identification of their in vitro metabolites. Forensic Toxicol 2015;1–8. Available from: https://doi.org/10.1007/s11419-015-0277-6.

[4] Łukasik-Głębocka M, Sommerfeld K, Teżyk A, Zielińska-Psuja B, Panieński P, Żaba C. Flubromazolam – a new life-threatening designer benzodiazepine. Clin Toxicol (Phila) 2015;54(1):66–8. Available from: https://doi.org/10.3109/15563650.2015.1112907.

[5] Meyer MR, Bergstrand MP, Helander A, Beck O. Identification of main human urinary metabolites of the designer nitrobenzodiazepines clonazolam, meclonazepam, and nifoxipam by nano-liquid chromatography-high-resolution mass spectrometry for drug testing purposes. Anal Bioanal Chem 2016;408(13):3571–91. Available from: https://doi.org/10.1007/s00216-016-9439-6.

[6] Pope JD, Choy KW, Drummer OH, Schneider HG. Novel benzodiazepines (clonazolam and flubromazolam) identified in candy-like pills. J Appl Lab Med 2018;3(1). Available from: https://doi.org/10.1373/jalm.2017.025387.

[7] Dowling G, Kavanagh PV, Eckhardt HG, et al. An approach to shortening the timeframe between the emergence of new compounds on the drugs market and the availability of reference standards: the microscale syntheses of nitrazolam and clonazolam for use as reference materials, utilizing polymer-supported reagents. Drug Test Anal 2018;10(7):1198–208. Available from: https://doi.org/10.1002/dta.2383.

Chapter 16

Hallucinogens—psychedelics and dissociative drugs

Mark Petersen[1], Uttam Garg[2,3] and Hemamalini Ketha[1]

[1]Department of Toxicology, Clinical Mass Spectrometry and Metals Testing, Center for Esoteric Testing, Laboratory Corporation of America Holdings, Burlington, NC, United States, [2]Department of Pathology and Laboratory Medicine, Children's Mercy Hospital, Kansas City, MO, United States, [3]University of Missouri School of Medicine, Kansas City, MO, United States

Introduction

Hallucinogens are compounds that produce psychoactive manifestations including hallucinations, dissociative behavior, and aberrations in perception of the user's reality and their state of connectivity to tangible reality [1]. These compounds are recreationally used and abused for the unusual perceptual and cognitive distortions and "out-of-body" experiences. Hallucinogens are a broad class of drugs that can be classified into several classes based on chemical structure and mechanisms of action. Sometimes the term psychedelic is used interchangeably with the term hallucinogen. National Institute on Drug Abuse classifies hallucinogens into two primary types—classical hallucinogens and dissociatives. Classical hallucinogens may also be referred to as psychedelics. The term hallucinogen is coined based on the Latin word *alucinari* that means "a wandering mind," whereas the term psychedelic finds its roots in ancient Greek that implies the manifestations of mind and the soul. In this chapter, psychedelics will be considered as classic hallucinogens. The two categories of hallucinogens that will be primarily focused on in this chapter are classic hallucinogens or psychedelics and dissociatives [2].

Psychedelics are classified as drugs that cause changes in thought, visual and auditory hallucinations, and an altered state of consciousness. The most commonly used and abused psychedelics are lysergic acid diethylamide (LSD), tryptamines, and phenethylamines [3]. Dissociatives produce a feeling of being detached from one's self, emotions, or tangible reality. The most common dissociative drugs include phencyclidine (PCP, also called angel dust), ketamine, and nitrous oxide. These two categories vary greatly in their structures and mechanisms of action but do exhibit important similarities in the cognitive effects of the user's experience. Other categories of drugs that are sometimes included with hallucinogens include entactogens, such as methylenedioxymethamphetamine, and atypical hallucinogens, such as cannabis. Although hallucinations are side effects of toxic exposures to many other drugs, including amphetamines, these are not classified as hallucinogens.

In this chapter, we review the pharmacology, toxicokinetics, and guidance for management of hallucinogen toxicity. We also describe analytical methods used in a toxicology laboratory used for the identification and quantitation of psychedelic and dissociative drugs.

Psychedelic drugs or classical hallucinogens

Psychedelics can further be classified into two structural subclasses: tryptamines and phenethylamines. Tryptamines and phenethylamines are referred to as serotonergic psychedelics because their mechanism of action is mediated by serotonin receptors. Serotonin (5-hydroxytryptamine, abbreviated as 5-HT) receptors are found many places in the body. Dense concentrations of 5-HT receptors are located in the gastrointestinal tract, blood platelets, and the central nervous system. 5-HT is formed from tryptophan, which is converted to 5-hydroxytryptophan and then to 5-HT [4,5]. 5-HT is metabolized by monoamine oxidase, forming 5-hydroxyindoleacetic acid that is excreted in urine. 5-HT is thought to have a role in peristalsis, vomiting, platelet aggregation, and hemostasis, as an inflammatory mediator, sensitization of nociceptors, and microvascular control. 5-HT is also thought to have many functions in the central nervous system, including appetite regulation, sleep, mood disorders, and anxiety. Clinical conditions associated with disturbed 5-HT function include migraine, carcinoid syndrome, mood disorders, and anxiety. Studies have suggested that the mechanism of action is through activation of 5-HT$_{2A}$ receptor. The potency of tryptamine has been linked with affinity for 5-HT$_{2A}$ receptor. For example, LSD has been shown to bind with high affinity to at least eight different serotonergic receptor subtypes [6]. By comparing receptor binding of tryptamines with phenethylamines, common sites of action are observed to be at the 5-HT$_{2A}$ and 5-HT$_{2C}$ receptors. In studies where subjects were administered psilocybin, simultaneous administration of risperidone blocked the hallucinogenic effects of the psilocybin. Risperidone is an antagonist as both 5-HT$_{2A}$ receptors as well as D2 dopamine receptors [7]. Haloperidol, which antagonizes D2 receptors but not 5-HT$_{2A}$, failed to prevent hallucinations when administered concurrently to the psilocybin.

Hallucinogens produce rapid tolerance, which is also believed to be linked to the 5-HT$_{2A}$ receptor [8]. 5-HT$_{2A}$ downregulation has been demonstrated in rats given LSD, psilocybin and phenethylamines [9]. However, mescaline administration did not produce downregulation, so there are likely multiple mechanisms of tolerance formation. The location of the hallucinogen-producing receptors is still being studied, but there appears to be a role played by the locus coeruleus (LC). The LC is a dense cluster of norepinephrine-containing neurons located in the pons and is responsible for the majority of the noradrenergic projections to the forebrain. The LC receives and integrates input from all major sensory systems and sends information to all areas of the cortex, including the sensory cortex. LSD and mescaline have been observed to decrease spontaneously in rat LC neurons but to paradoxically enhance the excitation of these cells by sensory stimulation. Hallucinogens are theorized to work on receptors modulating the LC rather than the LC itself. Simultaneous reduction of spontaneous neuronal firing with enhancement of sensory responsiveness means that in the presence of hallucinogens the LC is more responsive to sensory input. Another theory, proposed by Vollenweider and Geyer, hypothesizes that the hallucinogenic effects are produced by a disruption of normal information processing in a circuit that includes the prefrontal cortex, the striatum, and the thalamus. In this hypothesis, hallucinogens interfere with gating of sensory information passing through this circuit, leading to information overload at the cortex [10].

Hallucinogens are not physiologically addictive. Use does not result in a physical dependence or withdrawal symptoms when usage is ceased [11–13]. Psychological dependence or cravings are rare compared with other drugs. Hallucinogens have not been observed to be effective as positive reinforcements in self-administration studies. The most common adverse experience is a "bad trip"—an acute anxiety or panic attack in response to the drug's effects. Bad trips are believed to be a product of several factors, including effects of the drug, the user's emotional state, and the environment in which it is used. Talking the user through the trip is usually sufficient, although in extreme cases, hospitalization may be necessary. Although believed to be rare, an effect associated with hallucinogens is the flashback. A flashback is defined by the Diagnostic and Statistical Manual of Mental Disorders-IV as the reexperiencing, following cessation of use of a hallucinogen, of one or more of the perceptual symptoms that were experienced while intoxicated with the hallucinogen. HPPD—hallucinogen persisting perception disorder—is characterized by flashbacks severe enough to cause major disturbance or impairment. The use of other psychoactive agents, such as marijuana, may trigger these flashbacks. The most severe adverse reaction is a psychotic breakdown. While uncommon, users that have been previously diagnosed with a psychotic disorder, such as schizophrenia, are believed to be at an elevated risk of experiencing a psychotic breakdown.

Phenethylamines are monoaminergic psychedelics [14]. Structurally, they are similar to catecholamines and have been observed to produce effects similar to those of both psychedelics and stimulants. The stimulant effect, which is observed at low doses, is theorized to be achieved via catecholamine systems. At higher doses, psychedelic effects may be produced through action on serotonergic systems. Mescaline is a naturally occurring phenethylamine psychedelic, while many chemically similar synthetic forms have been discovered and are much more potent. Phenethylamines produce sympathomimetic effects, such as elevated heart rate and dilated pupils. Toxic effects include hyperactivity, tremors, seizures, and death [15]. Other naturally occurring monoaminergic psychedelics include myristin, elemicin, and safrole.

Mescaline is produced by the peyote cactus, which is native to the Southwestern United States and Mexico. The crown of the cactus is cut and dried to create what are known as a mescal button or peyote buttons. Ancient peyote buttons have been and determined to be 5700 years old by radiocarbon dating. These buttons can be chewed raw or cooked and eaten. Mescaline can also be extracted and consumed as a relatively pure powder. Today mescaline is still consumed as part of religious rituals in Mexico and by the Native American Church of the United States and Canada. Pure mescaline was first isolated in 1896 and first synthesized in 1919. Usage of synthetic mescaline is not prevalent due to the high cost of synthesis and general lack of demand, possibly due in part to its tendency to induce nausea. Far more potent phenethylamines are available in illicit markets, further reducing demand for synthetic mescaline. Studies have suggested that long-term peyote use does not result in cognitive deficits and may prove effective in the treatment of alcoholism [17–19].

There are many synthetic psychedelic phenethylamines and discovery of new variants continues today [20]. Many were pioneered by Dr. Alexander Shulgin and detailed in his book PiHKAL: A Chemical Love Story. Examples of synthetic phenethylamines include 2,5-dimethoxy-4-methylamphetamine (known as DOM), 25I-NBOMe (N-bomb), and Bromo-DragonFLY. DOM, which is also referred to as Serenity, Tranquility, and Peace (STP) produces altered visual perception, slowed passage of time and increased sexual drive and pleasure. It may also produce altered states of consciousness and transcendental experiences. DOM has a slower onset than similar drugs such as LSD that may cause an inexperienced user to redose before it takes effect. N-bomb is a highly potent hallucinogen, with effective dosing in the submilligram level. Duration of effects varies with method of administration. A 2014 WHO report attributed 14 deaths to N-bomb usage. Some of these deaths were attributed to drug-induced injurious behavior, while others have been due to seizures, cardiac arrest, and renal failure. Because it is so potent, simply handling the drug can be extremely dangerous. Bromo-DragonFLY is also active at submilligram levels. The most common method of administration is oral via blotters, similar to LSD [21].

Tryptamine is a monoamine alkaloid, structurally similar to the amino acid tryptophan. The core chemical structure of tryptamine is an indole ring connected to an amine group by an ethyl side chain. Substituted tryptamines replace one or more of tryptamine hydrogen atoms with another group. Serotonin is an example of a substituted tryptamine. The class of drugs known as tryptamines also falls in this category. Psilocybin is a naturally occurring tryptamine that is found in numerous species of mushrooms around the world. These species include the genera *conocybe*, *copelandia*, *panaeolus*, *psilocybe*, and *stropharia*. The mushrooms are dried and eaten in amounts ranging from 1 to 5 g depending on the species. Psilocybin is the major hallucinogenic component of these mushrooms along with its related compound psilocin. Psilocybin is enzymatically converted via dephosphorylation by first-pass hepatic metabolism to the 5-HT$_{2A}$ agonist psilocin, which is the active form of the compound. Like peyote, the use of hallucinogenic mushrooms in religious rituals dates back thousands of years. Algerian rock paintings dating back to 3500 BCE show a shaman with an animal head and mushrooms sprouting from its skin. The Aztecs referred to these mushrooms as *teonanacatl*, meaning "flesh of the gods." In 1938 Richard Schultes of Harvard University traveled to Oaxaca and collected mushrooms used in rituals by the Mazatec peoples. This inspired Gordon Wasson to visit Oaxaca in 1953 and again in 1955. On his visits, Wasson participated in ritual mushroom consumption led by Mazateca curandera Maria Sabina. Wasson chronicled his experience in an article that ran in Life magazine entitled "Seeking the Magic Mushroom." Wasson's experience inspired Harvard professor Timothy Leary to experiment with mushrooms during his own visit to Mexico. Leary would go on to start the Harvard Psilocybin Research Program with Richard Alpert conducting experiments with graduate students and faculty. Leary conducted what is known as the Good Friday experiment in which he administered a 30 mg dose of psilocybin to psychedelic-naive divinity students on a religious chapel on Good Friday [22,23]. The time and setting were chosen purposefully to amplify religious sentiments among a group of religiously inclined people. The psilocybin was significantly more effective at producing experiences that were deemed to be meaningful in a religious capacity than an active placebo. Leary also conducted a prison experiment where prisoners were given psilocybin as part of a group-therapy program. Upon release the recidivism rates among test subjects were significantly lower than that of the general prison population, although follow-up studies have cast these results in doubt. Psilocybin was first isolated in 1958 by Albert Hoffman. Although various studies have suggested that psilocybin use is largely benign and exhibits the least risk of harm to self or others when compared to commonly abused drugs, users are still at risk of strong fear and/or anxiety reaction and potentially HPPD. Psilocybin has shown potential as a possible treatment of anxiety and alcoholism as well as smoking cessation [24].

Like psilocybin, DMT (*N*-dimethyltryptamine) is a naturally occurring tryptamine. DMT is found in a number of plants native to South America. It is used in the making of the hallucinogenic drink ayahuasca, which means "vine of the soul" in Quechua. Ayahuasca is typically made with stalks from the DMT-containing leaves of the *Psychotria viridis* and/or *Diplopterys cabrerana*, as well as the stalks of the *Banisteriopsis caapi* vine. DMT is generally inert when taken orally but produces psychoactivity in ayahuasca. This may be due to the beta-carboline harmine that is found in the stalks and acts as a monoamine oxidase inhibitor, blocking the breakdown of the DMT and potentially allowing it to reach the brain and exert hallucinogenic effects. Ayahuasca has long been used in religious rituals in Brazil, and although DMT is illegal in the United States, churches have been granted the legal right to use ayahuasca in religious ceremonies. DMT has been found in the urine of nondrug users as well in the brain of rats, although the function of endogenous DMT is not certain. Unlike other psychedelics, DMT has not been shown to be tolerance inducing. Ayahuasca has been investigated and shows potential as a treatment for alcoholism and addiction as well as depression.

Prior to isolating psilocybin in 1958, Albert Hofmann had been studying the substance ergot. Ergot is produced by the fungus *Claviceps purpurea* that can infect rye and wheat. Ergot is extremely toxic and the consumption of it leads to a condition known as ergotism. Ergotism was fairly common in the Middle Ages and typically results in headache, vomiting, diarrhea due to its convulsive effects as well as gangrene of the fingers and toes. Due to these convulsive effects, ergot is pharmaceutically useful for inducing childbirth and reducing postbirth uterine hemorrhaging by producing uterine contractions. The core structure of all ergot alkaloids is lysergic acid. Albert Hofmann was synthesizing new lysergic acid derivatives when he absorbed a small amount of LSD through his fingertips. Hoffman described his experience as such:

"Last Friday, April 16, 1943, I was forced to interrupt my work in the laboratory in the middle of the afternoon and proceed home, being affected by a remarkable restlessness, combined with a slight dizziness. At home I lay down and sank into a not unpleasant intoxicated-like condition, characterized by an extremely stimulated imagination. In a dreamlike state, with eyes closed (I found the daylight to be unpleasantly glaring), I perceived an uninterrupted stream of fantastic pictures, extraordinary shapes with intense, kaleidoscopic play of colors. After some two hours this condition faded away."

Hofmann believed these effects to be due to his exposure to LSD. Three days later he dissolved 250 μg of LSD in water and drank it. Less than an hour later, he began to experience intense changes in perception. With his assistant escorting him, Hofmann rode his bicycle home. On his way, he began experiencing anxiety and paranoia. However, upon arriving home, this gave way to pleasurable sensations accompanied by colorful hallucinations. These events are now remembered as "Bicycle Day." Timothy Leary and Richard Alpert experimented heavily with LSD, while the Central Intelligence Agency investigated its utility as a potential tool for espionage and mind control. Therapeutic studies have shown LSD to potentially be effective in treating alcoholism.

LSD is an extremely potent hallucinogen, effective at the microgram level. It is ingested orally with only about 1% of the substance reaching the brain. It has very low toxicity and thus a wide therapeutic index. Mild sympathomimetic effects occur including dilated pupils and slightly elevated heart rate, blood pressure, and body temperature. Side effects of LSD may include headache, nausea, vomiting, tremors, dizziness, and muscle weakness. Desirable effects attributed to LSD use include lucid thoughts, intensified sensations, distorted visual imagery, hallucinations, and transcendental states. Undesirable effects include panic, fear, anxiety, paranoia, isolation, and mood changes. The experience of the LSD "trip" can be divided into four phases. Phase one is the onset of effects, lasting from 30 to 60 minutes. During this, phase visual effects begin to occur. Strange visions may be seen with one's eyes closed and the perception of colors may be intensified. Stage two is the plateau phase that lasts for the next 2 hours or so. At this point the user's subjective sense of time slows down. Stage three is the peak that occurs after about 3 hours and lasts for about 3 hours. During this phase the user feels as if they are in another world. A continuous stream of bizarre and distorted images may be seen, which can either be a pleasant or terrifying experience. Synesthesia may occur, where stimulation of one sensory pathway leads to an experience in another, for example, hearing colors or feeling sounds. Stage four is the comedown. All together the subjective effects of LSD can last up to 12 hours. The high from LSD can either be pleasant (a good trip) or disturbing (a bad trip). This seems to be the product of several factors, including the dose, the user's personality, expectations, emotional state, previous experiences, and settings. LSD does not result in any physical dependence and very low psychological dependence. LSD is rapidly tolerance forming (within days), although these effects are reversible. The long-term effects of LSD usage were unclear. HPPD is associated with LSD more so than other drugs although studies have suggested widely ranging prevalence estimates [1,3]. In addition to its action as a serotonergic agonist, LSD has also been observed to be active at both dopaminergic and adrenergic receptor sites. LSD was investigated as a potential psychological aid in helping neurotic patients recover repressed thoughts and feelings. This was done by one of two methods, either giving the patient small but gradually increasing doses or by administering one large dose. These studies ceased when LSD was made illegal by the Controlled Substances Act in 1970, but interest has picked back up in the past few years.

Dissociative drugs

Dissociatives are noncompetitive N-methyl-D-aspartate (NMDA) receptor antagonists. NMDA receptors are ionotropic glutamate receptor channels. Glutamate is an excitatory neurotransmitter formed mainly by the Krebs cycle intermediate α-oxoglutarate by the action of gamma aminobutyric acid-aminotransferase. NMDA receptors are connected directly to cation channels that are highly permeable to Ca^{2+} and blocked by Mg^{2+}. When opened, Na^+ and Ca^{2+} ions are conducted into the cell, which can activate second messenger systems. These ionotropic glutamate NMDA receptors depolarize the membrane of the postsynaptic cell causing an excitatory response. In addition to glutamate, NMDA receptors require glycine as a coagonist. Dissociatives are noncompetitive NMDA receptor antagonists, meaning that they prevent activation by glutamate by binding a site other than the glutamate site [25].

PCP is a dissociative psychedelic that was first tested in the 1950s as a potential anesthetic [26]. The anesthetic response produced by PCP was unusual in that although subjects showed no response to nociceptive stimuli, their mental state was not the relaxed unconsciousness typical of barbiturates and other traditional anesthetics known at the time. Instead the subjects were observed to exhibit trance-like or catatonic-like states characterized by vacant facial expressions, fixed and staring eyes, and maintenance of muscle tone. Initial studies showed promise due to the absence of respiratory depression associated with barbiturates thus giving it a high therapeutic index. However, many patients experienced problematic reactions. Common postoperative symptoms included blurred vision, dizziness, and mild disorientation, while more severe reactions included hallucinations, severe agitation, and violent behavior. This led to the termination of clinical use of PCP in 1965. However, use as a recreational drug persisted [27]. Street PCP is generally obtained as powder or pill and can be taken orally, intranasally, or injected intravenously or intramuscularly. The most common method of administration is smoking. PCP is applied to leafy material such as marijuana or tobacco and smoked. Effects of PCP at a subanesthetic dosage include euphoria, feeling detached from one's body, sensations of floating, numbness, and dream-like states. Other effects include drowsiness, paranoia, apathy, negativism, inebriation, and hostility. Patients on PCP exhibit marked cognitive dysfunction, characterized by difficulty maintaining concentration or focus and halting speech. Although PCP was abandoned as an anesthetic, studies of related compounds continued.

Ketamine was first synthesized in 1962 and proved to be useful as an anesthetic, particular in children and those at risk of hypotension or with respiratory conditions. Low doses of ketamine can also be used for postoperative pain management, and ketamine has shown to be a rapidly acting antidepressant. Nonmedical use of ketamine began in the 1970s [28]. While the prevalence of PCP usage is low compared to other drugs of abuse, ketamine usage is believed to be on the rise due in part to popularity at raves. Ketamine is commonly used in medical settings, so medical and veterinary practitioners often have access to it. Commercially, ketamine is distributed as an injectable liquid; however, the street version is often evaporated and snorted as a powder or compressed into a pill and taken orally. Low doses of ketamine result in intoxication, euphoria, and numbness, while high doses can cause agitation, catalepsy, amnesia, analgesia, and seizures. Experiences with high doses of ketamine are termed the "K-hole", which, similar to LSD trips, can either be pleasurable or terrifying experiences. This is a state of extreme dissociation with visual and auditory hallucinations. Reported experiences include the complete loss of one's sense of time, out-of-body experiences, and the distortion of body shape or size.

In addition to being noncompetitive NMDA receptor antagonists, PCP and ketamine have both been shown to activate midbrain dopamine cell firing and stimulate dopamine release. Both have been shown to be addictive in animal models by self-administration and conditioned place-preference studies. Due to the psychotomimetic effects, PCP and ketamine have been used in the development of animal models for schizophrenia [29]. Repeated PCP treatment in animal studies led to a reduction of dopamine activity in the prefrontal cortex. Monkeys fed regular doses of PCP exhibit deficits in prefrontal cortex associated tasks even after administration is halted. This is useful as an animal model as the prefrontal cortex is thought to be dysfunctional in schizophrenic patients [30].

Dextromethorphan (DXM) is a cough suppressant found in many over-the-counter cold and flu medications. In addition to its antitussive properties, DXM also acts as an NMDA receptor agonist at high doses. Recreational use of DXM can induce similar effects to ketamine and PCP. Because of its presence in many cough syrups, DXM is often much more accessible to teens than other drugs [31]. Other medications present in these cough syrups, such as acetaminophen, can cause adverse effects such as renal failure at dosages necessary to produce dissociative effects. DXM also acts similar to a selective serotonin reuptake inhibitor. Sudden cessation of recreational usage can produce physical and psychological withdrawal effects.

Toxicokinetics and clinical management of overdose

LSD undergoes N-demethylation, N-deethylation, and hydroxylation to form N-desmethyl-LSD, 2-OXO-LSD, and 2-OXO-3-OH-LSD. Hydroxylated metabolites further undergo glucuronidation. 2-OXO-3-OH-LSD and the glucuronidated products are the major LSD metabolites in urine. Whereas ketamine undergoes N-demethylation to form norketamine which in turn undergoes dehydrogenation to form dehydronorketamine. These metabolites further undergo hydroxylation and conjugation before excreted in urine. Ketamine is excreted in urine primarily unchanged along with small quantities of other demethylated and conjugated metabolites.

Classic hallucinogens can cause users to see images, hear sounds, and feel sensations that seem real but do not exist. The effects generally begin within 20–90 minutes and can last as long as 12 hours in some cases (LSD) or as short as

15 minutes in others (synthetic DMT). Hallucinogen users refer to the experiences brought on by these drugs as "trips." If the experience is unpleasant, users sometimes call it a "bad trip." Along with hallucinations, other short-term general effects include increased heart rate, nausea, intensified feelings and sensory experiences (such as seeing brighter colors), and changes in sense of time (e.g., the feeling that time is passing by slowly). In high doses, dissociative drugs can cause memory loss, panic and anxiety, seizures, psychotic symptoms, amnesia, inability to move, mood swings, and trouble breathing. More research is needed on the long-term effects of dissociative drugs. It is known that repeated use of PCP can result in addiction. Other long-term effects may continue for a year or more after use stops, including speech problems, memory loss, weight loss, anxiety, depression, and suicidal thoughts.

Most classic hallucinogens may produce extremely unpleasant experiences at high doses, although the effects are not necessarily life threatening. However, serious medical emergencies and several fatalities have been reported from a phenethylamine derivative 25I-NBOMe. Of note, several of 2,5-dimethoxy derivatives of phenethylamine, also termed 2C-B compounds, have been shown to have hallucinogenic properties.

Overdose is more likely with some dissociative drugs. High doses of PCP can cause seizures, coma, and death. In addition, taking PCP with depressants such as alcohol or benzodiazepines can also lead to coma. Benzodiazepines, such as alprazolam (Xanax), are prescribed to relieve anxiety or promote sleep. However, users of both classic hallucinogens and dissociative drugs also risk serious harm because of the profound alteration of perception and mood these drugs can cause. Users might do things they would never do in real life, such as jump out of a window or off a roof, for instance, or they may experience profound suicidal feelings and act on them. With all drugs, there is also a risk of accidental poisoning from contaminants or other substances mixed with the drug. Users of psilocybin also run the risk of accidentally consuming poisonous mushrooms that look like psilocybin. Taking poisonous mushrooms can result in severe illness or possible death.

Specific short-term effects of some hallucinogens include increased blood pressure, breathing rate, or body temperature, loss of appetite, dry mouth, sleep problems, spiritual experiences, feelings of relaxation, uncoordinated movements, excessive sweating, panic, paranoia—extreme and unreasonable distrust of others, psychosis—disordered thinking detached from reality, and bizarre behaviors.

Physical symptoms of PCP intoxication are summarized by the mnemonic RED DANES: rage, erythema, dilated pupils, delusions, amnesia, nystagmus, excitation, and skin drying. Treatment is typically supportive care, although benzodiazepines may be administered to control agitation or seizures or antipsychotics to treat psychotic symptoms. If the drug was taken orally, activated charcoal may absorb PCP and lessen renal effects.

Signs of a ketamine overdose may include difficulty breathing, heightened blood pressure nausea/vomiting, severe confusion, chest pain, irregular heart rate, paralysis, violence, terrors possibly related to hallucinations, seizures, extreme sedation, and loss of consciousness or coma. When injected, ketamine can be lethal at high concentration. Treatment is supportive care while monitoring vitals and potentially benzodiazepines for seizures and anxiety [32,33].

Clinical management of hallucinogen overdose is mainly supportive in nature. Patients who are intoxicated with a hallucinogen or a dissociative drug have a low threshold for erratic behavior. As discussed earlier, patients acting out a behavior such as jumping out of a window can indirectly lead to fatality caused by hallucinogen consumption. Therefore keeping the patient calm is an important aspect of prehospital care in hallucinogen overdose. In the emergency department, hyperthermia, dehydration, and any cardiac changes must be addressed rapidly. Benzodiazepines are commonly used for managing agitated delirium that many patients who have consumed hallucinogens present with. Rhabdomyolysis in the setting of LSD consumption must also be managed with a rapid response.

Analysis of psychedelics, dissociatives, and other hallucinogens

Common analytical methods for identification of hallucinogens include immunoassay screens and chromatographic methods with spectrophotometric or mass spectrometric detection methods [34–36]. Due to low prevalence of PCP use in the general population, many clinical laboratories may not even include LSD and PCP as a part of their general drug screen [37]. The physician encountering the patient may have to request the screen to be performed when a hallucinogen use is suspected. Immunoassays tend to show false-positive results for some hallucinogens [38,39]. It should be noted that enzyme-multiplied immunoassay technique–based PCP immunoassay has been shown to be prone to false positives from some commonly used OTC and prescription drugs. The drugs that cause false positive on a PCP immunoassay drug screen are DXM, tramadol, alprazolam, clonazepam, and carvedilol [40].

Most qualitative gas chromatography–mass spectrometry (GC–MS) methods for urine drug screening will be able to identify LSD or PCP use in an exposed patient [41,42]. Quantitative GC–MS (with flame ionization or with nitrogen-selective detection) and LC−MS/MS methods have been published for commonly used hallucinogens, including LSD and its metabolite 2-OXO-3-OH-LSD [43−45] PCP [46,47], ketamine and norketamine [48,49], phenethylamines, and tryptamines [50,51].

Conclusion

Serotonergic compounds that can be classified as psychedelics or hallucinogens have strong psychoactive physiological manifestations. These compounds can alter perception and mood and have an impact on several cognitive functions. LSD is generally considered safe to use but because of its effects on perception and cognitive function, users can engage in unsafe behaviors. Adverse effects of LSD are mostly believed to be due to the physiological impact on perception. The "new-age" designer phenethylamines, on the other hand, have a narrow safety index, and several deaths have been reported due to toxic exposures to the new synthetic psychedelic agents [15]. Psychedelics are reemerging and are being used in very low doses in a practice referred to as psychedelic microdosing [52,53]. Very sparse literature available claims a positive effect of psychedelic microdosing on mood state and cognitive processes such as concentration [53]. A systematic review of clinical trials published in the last 25 years showed that LSD along with some other psychedelics may have beneficial effects for treatment-resistant depression, anxiety and depression associated with life-threatening diseases, and tobacco and alcohol dependence. This chapter describes pharmacokinetic, clinical and analytical aspects of commonly encountered hallucinogens and psychedelics in a toxicology laboratory.

References

[1] Nichols DE. Psychedelics. Pharmacol Rev 2016;68(2):264−355. Available from: https://doi.org/10.1124/pr.115.011478.
[2] Abraham HD, Aldridge AM, Gogia P. The psychopharmacology of hallucinogens. Neuropsychopharmacology 1996;14(4):285−98. Available from: https://doi.org/10.1016/0893-133X(95)00136-2.
[3] Nichols DE. Hallucinogens. Pharmacol Ther 2004;101(2):131−81. Available from: https://doi.org/10.1016/j.pharmthera.2003.11.002.
[4] Nichols DE. Chemistry and structure-activity relationships of psychedelics. Curr Top Behav Neurosci 2018;36:1−43. Available from: https://doi.org/10.1007/7854_2017_475.
[5] Winter JC. Hallucinogens as discriminative stimuli in animals: LSD, phenethylamines, and tryptamines. Psychopharmacology (Berl) 2009;203(2):251−63. Available from: https://doi.org/10.1007/s00213-008-1356-8.
[6] Passie T, Halpern JH, Stichtenoth DO, Emrich HM, Hintzen A. The pharmacology of lysergic acid diethylamide: a review. CNS Neurosci Ther 2008;14(4):295−314. Available from: https://doi.org/10.1111/j.1755-5949.2008.00059.x.
[7] Vollenweider FX, Vollenweider-Scherpenhuyzen MF, Babler A, Vogel H, Hell D. Psilocybin induces schizophrenia-like psychosis in humans via a serotonin-2 agonist action. Neuroreport 1998;9(17):3897−902. Available from: https://doi.org/10.1097/00001756-199812010-00024.
[8] Baumeister D, Barnes G, Giaroli G, Tracy D. Classical hallucinogens as antidepressants? A review of pharmacodynamics and putative clinical roles. Ther Adv Psychopharmacol 2014;4(4):156−69. Available from: https://doi.org/10.1177/2045125314527985.
[9] Buchborn T, Lyons T, Knöpfel T. Tolerance and tachyphylaxis to head twitches induced by the 5-HT 2A agonist 25CN-NBOH in mice. Front Pharmacol 2018;9:17. Available from: https://doi.org/10.3389/fphar.2018.00017.
[10] Vollenweider FX. Brain mechanisms of hallucinogens and entactogens. Dialogues Clin Neurosci 2001;3(4):265−79.
[11] Halberstadt AL. Recent advances in the neuropsychopharmacology of serotonergic hallucinogens. Behav Brain Res 2015;277:99−120. Available from: https://doi.org/10.1016/j.bbr.2014.07.016.
[12] Halberstadt AL, Geyer MA. Effect of hallucinogens on unconditioned behavior. Curr Top Behav Neurosci 2018;36:159−99. Available from: https://doi.org/10.1007/7854_2016_466.
[13] Halberstadt AL, Geyer MA. Multiple receptors contribute to the behavioral effects of indoleamine hallucinogens. Neuropharmacology 2011;61(3):364−81. Available from: https://doi.org/10.1016/j.neuropharm.2011.01.017.
[14] Fantegrossi WE, Murnane KS, Reissig CJ. The behavioral pharmacology of hallucinogens. Biochem Pharmacol 2008;75(1):17−33. Available from: https://doi.org/10.1016/j.bcp.2007.07.018.
[15] Dean BV, Stellpflug SJ, Burnett AM, Engebretsen KM. 2C or not 2C: phenethylamine designer drug review. J Med Toxicol 2013;9(2):172−8. Available from: https://doi.org/10.1007/s13181-013-0295-x.
[16] Koelle GB. The pharmacology of mescaline and D-lysergic acid diethylamide (LSD). N Engl J Med 1958;258(1):25−32. Available from: https://doi.org/10.1056/NEJM195801022580106.
[17] Cassels BK, Saez-Briones P. Dark classics in chemical neuroscience: mescaline. ACS Chem Neurosci 2018;9(10):2448−58. Available from: https://doi.org/10.1021/acschemneuro.8b00215.
[18] Albaugh BJ, Anderson PO. Peyote in the treatment of alcoholism among American Indians. Am J Psychiatry 1974;131(11):1247−50. Available from: https://doi.org/10.1176/ajp.131.11.1247.
[19] Johnson MW, Hendricks PS, Barrett FS, Griffiths RR. Classic psychedelics: An integrative review of epidemiology, therapeutics, mystical experience, and brain network function. Pharmacol Ther 2019;197:83−102. Available from: https://doi.org/10.1016/j.pharmthera.2018.11.010.

[20] Hill SL, Thomas SHL. Clinical toxicology of newer recreational drugs. Clin Toxicol (Phila) 2011;49(8):705−19. Available from: https://doi.org/10.3109/15563650.2011.615318.

[21] Kyriakou C, Marinelli E, Frati P, et al. NBOMe: new potent hallucinogens—pharmacology, analytical methods, toxicities, fatalities: a review. Eur Rev Med Pharmacol Sci 2015;19(17):3270−81.

[22] Geiger HA, Wurst MG, Daniels RN. DARK classics in chemical neuroscience: psilocybin. ACS Chem Neurosci 2018;9(10):2438−47. Available from: https://doi.org/10.1021/acschemneuro.8b00186.

[23] Reingardiene D, Vilcinskaite J, Lazauskas R. Hallucinogenic mushrooms. Medicina (Kaunas) 2005;41(12):1067−70.

[24] Johnson MW, Griffiths RR. Potential therapeutic effects of psilocybin. Neurother J Am Soc Exp Neurother 2017;14(3):734−40. Available from: https://doi.org/10.1007/s13311-017-0542-y.

[25] Mion G, Villevieille T. Ketamine pharmacology: an update (pharmacodynamics and molecular aspects, recent findings). CNS Neurosci Ther 2013;19(6):370−80. Available from: https://doi.org/10.1111/cns.12099.

[26] Johnson KM, Jones SM. Neuropharmacology of phencyclidine: basic mechanisms and therapeutic potential. Annu Rev Pharmacol Toxicol 1990;30(1):707−50. Available from: https://doi.org/10.1146/annurev.pa.30.040190.003423.

[27] Morris H, Wallach J. From P.C.P. to MXE: a comprehensive review of the non-medical use of dissociative drugs. Drug Test Anal 2014;6(7−8):614−32. Available from: https://doi.org/10.1002/dta.1620.

[28] Li L, Vlisides PE. Ketamine: 50 years of modulating the mind. Front Hum Neurosci 2016;10:612. Available from: https://doi.org/10.3389/fnhum.2016.00612.

[29] Jones CA, Watson DJG, Fone KCF. Animal models of schizophrenia. Br J Pharmacol 2011;164(4):1162−94. Available from: https://doi.org/10.1111/j.1476-5381.2011.01386.x.

[30] Linn GS, O'Keeffe RT, Schroeder CE, Lifshitz K, Javitt DC. Behavioral effects of chronic phencyclidine in monkeys. Neuroreport 1999;10(13):2789−93. Available from: https://doi.org/10.1097/00001756-199909090-00017.

[31] Reissig CJ, Carter LP, Johnson MW, Mintzer MZ, Klinedinst MA, Griffiths RR. High doses of dextromethorphan, an NMDA antagonist, produce effects similar to classic hallucinogens. Psychopharmacology (Berl) 2012;223(1):1−15. Available from: https://doi.org/10.1007/s00213-012-2680-6.

[32] Warner LL, Smischney N. Accidental ketamine overdose on induction of general anesthesia. Am J Case Rep 2018;19:10−12. Available from: https://doi.org/10.12659/ajcr.906205.

[33] Stewart CE. Ketamine as a street drug. Emerg Med Serv 2001;30(11) 30, 32, 34 passim.

[34] Kerrigan S, Mott A, Jatzlau B, et al. Designer psychostimulants in urine by liquid chromatography−tandem mass spectrometry. J Forensic Sci 2014;59(1):175−83. Available from: https://doi.org/10.1111/1556-4029.12306.

[35] Kerrigan S, Banuelos S, Perrella L, Hardy B. Simultaneous detection of ten psychedelic phenethylamines in urine by gas chromatography-mass spectrometry. J Anal Toxicol 2011;35(7):459−69.

[36] Kerrigan S, Mellon MB, Banuelos S, Arndt C. Evaluation of commercial enzyme-linked immunosorbent assays to identify psychedelic phenethylamines. J Anal Toxicol 2011;35(7):444−51. Available from: https://doi.org/10.1093/anatox/35.7.444.

[37] Crane CA, Easton CJ, Devine S. The association between phencyclidine use and partner violence: an initial examination. J Addict Dis 2013;32(2):150−7. Available from: https://doi.org/10.1080/10550887.2013.797279.

[38] Landy GL, Kripalani M. False positive phencyclidine result on urine drug testing: a little known cause. BJPsych Bull 2015;39(1):50. Available from: https://doi.org/10.1192/pb.39.1.50.

[39] Sena SF, Kazimi S, Wu AHB. False-positive phencyclidine immunoassay results caused by venlafaxine and O-desmethylvenlafaxine. Clin Chem 2002;48(4):676−7. Available from: http://clinchem.aaccjnls.org/content/48/4/676.abstract.

[40] Rengarajan A, Mullins ME. How often do false-positive phencyclidine urine screens occur with use of common medications? Clin Toxicol 2013;51(6):493−6. Available from: https://doi.org/10.3109/15563650.2013.801982.

[41] Clarkson ED, Lesser D, Paul BD. Effective GC-MS procedure for detecting iso-LSD in urine after base-catalyzed conversion to LSD. Clin Chem 1998;44(2):287−92. Available from: http://clinchem.aaccjnls.org/content/44/2/287.abstract.

[42] Sklerov JH, Kalasinsky KS, Ehorn CA. Detection of lysergic acid diethylamide (LSD) in urine by gas chromatography-ion trap tandem mass spectrometry. J Anal Toxicol 1999;23(6):474−8. Available from: https://doi.org/10.1093/jat/23.6.474.

[43] Libong D, Bouchonnet S, Ricordel I. A selective and sensitive method for quantitation of lysergic acid diethylamide (LSD) in whole blood by gas chromatography-ion trap tandem mass spectrometry. J Anal Toxicol 2003;27(1):24−9. Available from: https://doi.org/10.1093/jat/27.1.24.

[44] Sklerov JH, Magluilo Jr. J, Shannon KK, Smith ML. Liquid chromatography-electrospray ionization mass spectrometry for the detection of lysergide and a major metabolite, 2-oxo-3-hydroxy-LSD, in urine and blood. J Anal Toxicol 2000;24(7):543−9. Available from: https://doi.org/10.1093/jat/24.7.543.

[45] Berg T, Jorgenrud B, Strand DH. Determination of buprenorphine, fentanyl and LSD in whole blood by UPLC-MS-MS. J Anal Toxicol 2013;37(3):159−65. Available from: https://doi.org/10.1093/jat/bkt005.

[46] Poklis JL, Guckert B, Wolf CE, Poklis A. Evaluation of a new phencyclidine enzyme immunoassay for the detection of phencyclidine in urine with confirmation by high-performance liquid chromatography-tandem mass spectrometry. J Anal Toxicol 2011;35(7):481−6. Available from: https://doi.org/10.1093/anatox/35.7.481.

[47] Feng J, Wang L, Dai I, Harmon T, Bernert JT. Simultaneous determination of multiple drugs of abuse and relevant metabolites in urine by LC-MS-MS. J Anal Toxicol 2007;31(7):359−68. Available from: https://doi.org/10.1093/jat/31.7.359.

[48] Moore KA, Sklerov J, Levine B, Jacobs AJ. Urine concentrations of ketamine and norketamine following illegal consumption. J Anal Toxicol 2001;25(7):583–8. Available from: https://doi.org/10.1093/jat/25.7.583.

[49] Legrand T, Roy S, Monchaud C, Grondin C, Duval M, Jacqz-Aigrain E. Determination of ketamine and norketamine in plasma by micro-liquid chromatography-mass spectrometry. J Pharm Biomed Anal 2008;48(1):171–6. Available from: https://doi.org/10.1016/j.jpba.2008.05.008.

[50] Bjornstad K, Beck O, Helander A. A multi-component LC-MS/MS method for detection of ten plant-derived psychoactive substances in urine. J Chromatogr B Anal Technol Biomed Life Sci 2009;877(11-12):1162–8. Available from: https://doi.org/10.1016/j.jchromb.2009.03.004.

[51] Yritia M, Riba J, Ortuno J, et al. Determination of N,N-dimethyltryptamine and beta-carboline alkaloids in human plasma following oral administration of Ayahuasca. J Chromatogr B Anal Technol Biomed Life Sci 2002;779(2):271–81.

[52] Yanakieva S, Polychroni N, Family N, Williams LTJ, Luke DP, Terhune DB. The effects of microdose LSD on time perception: a randomised, double-blind, placebo-controlled trial. Psychopharmacology (Berl) 2019;236(4):1159–70. Available from: https://doi.org/10.1007/s00213-018-5119-x.

[53] Kuypers KP, Ng L, Erritzoe D, et al. Microdosing psychedelics: more questions than answers? An overview and suggestions for future research. J Psychopharmacol 2019;33(9):1039–57. Available from: https://doi.org/10.1177/0269881119857204.

Chapter 16.1

Ketamine—a review of published cases

Manoj Tyagi[1], Amit Bansal[2] and Alina G. Sofronescu[3]
[1]Align Laboratories, Clinical Chemistry Consultant, Charlotte, NC, United States, [2]Department of Medicine, Rochester Regional Health, Rochester, NY, United States, [3]Department of Pathology and Microbiology, University of Nebraska Medical Center, Omaha, NE, United States

Clinical cases description

Acute ketamine toxicity

Acute ketamine toxicity [1], reported by Licata et al. [2], involves an 18-year-old white male who was found half-laying on the driver's seat of a locked car, parked in the remote mountainous and uninhabited area. Death was thought to have occurred 12–16 hours before the discovery of the body. Three fresh needle marks were found on the anterior/external quadrant of the left buttock, and ketamine vials were found on the crime scene. An equivalent of 1 g of the total dose was estimated to be administered. The postmortem examination excluded any trauma or spontaneous pathology as a primary cause of death. Death was attributed to massive pulmonary edema. Norketamine, a metabolite of ketamine, was detected in all fluids and tissues examined (notably in blood, urine, bile, brain, liver, and kidneys). The cause of death was determined to be due to ketamine overdose based on pathology and toxicology findings. It was presumed that two or more doses of ketamine were administered, the last dose shortly before death. The finding of norketamine in fluids and tissues and the high level of ketamine in the liver and kidneys led investigators to think that a long period of time elapsed between the first administration of ketamine and the victim's death.

Chronic ketamine toxicity

Chronic ketamine toxicity, reported by Tao et al. [3], involves a 34-year-old female who resided in an apartment with her husband and was discovered unconscious with no respirations or heartbeat shortly after taking a bath and then drinking a cup of coffee. She was brought to the hospital where she was pronounced dead in the emergency room. During the investigation a coffee cup was collected, and ketamine injection vials were found in the refrigerator of her house. Reconnaissance investigation was carried out, and an autopsy and toxicological analyses were performed. The victim was reported to be in good general health with no previous medical history. Her symptoms initially started about a year ago with a headache and insomnia. One day, she became unconscious suddenly and was found to have arrested respiration and heartbeat. Similar symptoms recurred on three or four subsequent occasions for which she was taken to hospital but medical examination, including electrocardiogram (ECG), failed to find any evidence of illness in the vital organs. It was found that the husband had married the victim because he wanted to become a postgraduate student of her father. A few years before the reported incident of the victim's death, the husband had fallen in love with another young girl. The husband was a pediatric surgeon who had the knowledge about ketamine's toxicity as well as had easy access to the drug. About a year before her death, the husband began preparing a cup of tea or coffee for his wife after her bath. This was something new he started doing. The husband's fingerprints were found on the surface of empty vials of ketamine collected from a garbage can, as well as on the cup of coffee used by his victim shortly before she became unconscious and had a respiratory arrest. When police detained the husband and investigated him as a suspect, he confessed that he had been using ketamine to poison his wife. Initially, he used to add one vial (100 mg) of ketamine to his wife's coffee cup. When he found no noticeable symptoms of toxicity, he started adding two vials (200 mg) to her drink. This led to arrested respiration and pulse for which she got hospitalized each time, but hospital doctors were not able to reach any conclusion besides that they suspected unusual cardiac malfunction. Finally, the husband added three vials (300 mg) to the coffee and the lady died.

Autopsy findings did not show any injuries on the surface or inner tissue suggesting no assault with strangulation or mechanical asphyxia. Lungs were found to be edematous and congested (right 775 g; left 700 g) and had hemorrhagic spots on the surface. The heart (220 g) had proportionate chambers and was otherwise unremarkable but had leukoplakia and hemorrhagic spots on the surface. The thickness of the left heart wall was 1.3 cm and the right was 0.4 cm. All valves were of normal size and in good working condition. The surface of the kidneys (right 220 g; left 225 g) was smooth. The brain (1550 g) was congested with no trauma. Microscopic examination of the lungs confirmed pulmonary congestion and edema. There were some heart-failure cells in the interstitial and alveolar cavity filled with edema liquid, cellulose, and erythrocytes. Histology of the heart showed widespread scattered cardiac muscle fibrosis around the small arteries. The newly occurred necrosis foci had some infiltrated inflammatory cells. The endothelium of the coronary artery was slightly thickened. The brain showed signs of chronic hypoxic changes, including glial and formation of starch bodies and some edema characteristics. The small arteries in heart and brain showed shedding of endothelium and hyaline degeneration of smooth muscle. To summarize, the autopsy revealed cerebral edema, overinflation of lungs, pulmonary edema accompanied by hyperemia, interalveolar, and interstitial hemorrhages, all of these signs pointed to an asphyxial death, but there was no evidence of mechanical asphyxia or autoerotic asphyxia in this case. High concentration of ketamine was found as follows: 3.8 μg/mL in cardiac blood, 21 μg/mL in stomach contents, and 1.2 μg/mL in urine. Ketamine was also detected in the remnants and sediments of the coffee cup. This case describes acute as well as chronic toxicity secondary to ketamine.

Pediatric ketamine toxicity

Pediatric ketamine overdose, reported by Villelli et al, involves a 10-month-old previously healthy girl who was brought to the hospital for unresponsiveness [4]. The patient's brother and grandmother were also found unresponsive. The 10-month-old infant was in critical condition requiring intubation. On examination, she had a Glasgow Coma Scale score of 6 T. She had extensor posturing in her right upper extremity, withdrawing in her left upper extremity and withdrawing in her lower extremities bilaterally. Her lower extremities revealed increased muscular tone bilaterally. Her pupils were reacting sluggishly and were 3 mm in diameter. Blood work revealed a serum glucose level of 400 mg/dL, sodium bicarbonate level of 8 mmol/L, PCO_2 greater than 52 mmHg and a base excess of −20 mmol/L. The carbon monoxide levels were normal, and serum acetaminophen, salicylate, and ethanol concentrations were undetectable. A urine drug screen testing was positive for benzodiazepines, ketamine, and levetiracetam. A urine ketamine and metabolite confirmation test detected high levels of ketamine and norketamine. The child received benzodiazepine and levetiracetam in the hospital but never received any ketamine or ketamine derivative. Imaging studies were performed, and they revealed cytotoxic edema involving the bilateral cerebellar hemispheres. Cerebellar swelling obstructed the foramen magnum, effaced the fourth ventricle, leading to obstructive hydrocephalus, thereby requiring an external ventricular drain and surgical decompression of the posterior fossa. She was very sick initially, but her neurological exam slowly improved and she was discharged to an inpatient rehabilitation unit after about 15 days of hospital admission. The child made a remarkable recovery. The cause and circumstances of the incident were initially unknown, but a police investigation revealed that family members were involved in the illegal production of a recreational drug composed of tobacco leaves soaked in ketamine (called "Kommon"). The child and other two family members were served eggs cooked in the same dishes that were used to prepare the drug.

Discussion

Ketamine was introduced as an anesthetic in the 1960s and continues to be used therapeutically both as an anesthetic and also increasingly in pain management [5]. In addition to use in humans, ketamine is widely used as an anesthetic in veterinary medicine. It is also used in research as a pharmacological model of schizophrenia and bipolar depression approaches.

In recent years, ketamine has increasingly been used as a recreational drug. Though it is now a controlled substance in many countries in the world due to its psychotropic properties, its use as a recreational drug has become a worldwide phenomenon. In addition to the risk of acute toxicity associated with ketamine use, long-term ketamine toxicity can produce significant adverse effects and consequently lead to a burden on health-care resources [6].

In the clinical cases above, we highlighted the changes in different organs and toxicity on the central nervous system and other organs. We hereby described three cases of ketamine overdose, one each from the following categories: acute ketamine toxicity, chronic ketamine toxicity, and pediatric ketamine overdose.

Mechanism of action

Ketamine acts by noncompetitive antagonism of N-methyl-D-aspartic acid (NMDA) receptor. It also interacts with opioid receptors, monoamine, cholinergic, purinergic, and adrenoreceptor systems as well as having local anesthetic effects. Prolonged antidepressant effects of ketamine are thought to be due to downstream postdrug effects such as activity-induced increase in structural synaptic connectivity [7].

Pharmacokinetics and pharmacodynamics

Ketamine is a highly lipid-soluble substance that undergoes rapid breakdown and redistribution to peripheral tissues. It is extensively metabolized in the liver by N-demethylation and ring hydroxylation pathways [8]. The main metabolite is norketamine that is one-third to one-fifth as potent as ketamine as an anesthetic. Ketamine is excreted in feces and urine as norketamine and as hydroxylated derivatives. It has a cumulative effect and a gradual resistance builds upon repeated administration of ketamine [9].

Ketamine stimulates the cardiovascular system leading to tachycardia, increase in blood pressure and increase in cardiac output, principally mediated through sympathetic nervous system. It produces airway relaxation by acting on various receptors and inflammatory cascades and bronchial smooth muscles. It increases salivation and muscle tone. Ketamine has cataleptic, amnestic, analgesic, and dose-dependent anesthetic actions. Ketamine leads to a dissociative state in which the patient appears awake but is detached from the surroundings [10].

Pharmacodynamic drug interactions

The combination of ketamine with midazolam (or propofol) for induction and maintenance of anesthesia is a useful technique, particularly in frail patients with hemodynamic compromise. However, it is known that ketamine and propofol have distinctly different modes of action. Midazolam is used to produce sleepiness or drowsiness and relieve anxiety before surgery or certain procedures. It is also used as an anesthesia to produce a loss of consciousness before and during surgery. Midazolam is often used to suppress emergence phenomena due to ketamine and can also be used in combination for sedation or induction of anesthesia. Midazolam has an infraadditive effect on ketamine hypnosis, and no effect on the ketamine dose required to suppress movement to a noxious stimulus. However, these drugs have opposing effects on hemodynamic variables. At an optimal propofol:ketamine dose ratio of 3:1, there is minimal change in both heart rate and blood pressure when used for anesthesia induction.

Ketamine and barbiturates or diazepam should not be mixed together in the same syringe, as they are chemically incompatible. Other cerebral depressant medications prolong ketamine's effects and delay recovery.

Ketamine as anesthetic and use in pain management

Ketamine is widely used to manage pain, as its use has diminished the need of opiates and their side effects in postoperative period [11]. Ketamine provides sedation by acting as both a functional and electrophysiological dissociative agent in the brain. Ketamine is currently one of the most common intravenous induction agents used to achieve unconsciousness and apnea, together with propofol and etomidate, and it is used in various types of surgery (e.g. neurosurgery, amputations, obstetric and gynecologic interventions). The continued presence of a gag reflex under ketamine anesthesia makes operations inside the mouth problematic, although teeth can be wired for stabilization using ketamine. Bronchoscopy cannot be done under ketamine unless neuromuscular blockers are used. Furthermore, due to its side effects, ketamine is rarely used in adult patients in critical care units [12] A major benefit of using ketamine is that, unless high doses are used or smaller doses are given rapidly, the patient's airway remains open and he or she breathes spontaneously. Ketamine is highly lipophilic and results in a rapid onset of sedation and analgesia through stimulation of NMDA and opioid receptors, with a duration of action of 0.5–2 hour [13].

Importantly, ketamine decreases catecholamine reuptake and preserves respiratory drive and reflexes. The sympathomimetic effects of ketamine also make it a viable alternative for patients with cardiovascular or pulmonary instability. However, ketamine is associated with many side effects. The ketamine-induced cataleptic state is accompanied by nystagmus with pupillary dilation, salivation, lacrimation, and spontaneous limb movements with increased overall muscle tone, possible severe confusion, and hallucinations [14].

In therapeutic doses, ketamine can exert a mild sympathomimetic effect on the cardiovascular system with slight increases in blood pressure and heart rate. Hypotension and bradycardia are rare. Ketamine increases myocardial oxygen demand and may increase the risk of acute coronary syndrome, particularly in patients with additional risk factors for cardiovascular disease. Therefore ketamine use in patients with cardiovascular risk should be done with caution.

Ketamine was reported to increase intracranial pressure (ICP), limiting its use in patients with intracranial pathology. Ketamine may be a reasonable supplemental agent in patients receiving exceptionally high doses of traditional agents or patients suffering from hemodynamic compromise.

As a general anesthetic in children, ketamine anesthesia is considered the standard of care for children in many developing countries and medical facilities lacking amenities for the safe administration of inhalational anesthesia. Intramuscular ketamine is ideal for use in children who require repeated painful procedures. This accessible ketamine administration route is particularly useful in small infants. However, ketamine as a sole anesthetic agent is less successful in small babies, and repeated doses may lead to apnea. It is not recommended for sedation in children less than three months old [15].

Pharmacologic management of complex regional pain syndrome

Ketamine activates the sympathetic nervous system causing a systemic release of catecholamines, inhibition of parasympathetic output, and inhibition of norepinephrine reuptake at peripheral nerves. However, in high doses, it has direct myocardial depressant properties. Patients undergoing low-dose ketamine infusions may experience tachycardia, increases in pulmonary arterial pressure, increase in systemic blood pressure, and increases in myocardial oxygen consumption. Therefore all patients undergoing ketamine infusion therapy require continuous monitoring of their cardiovascular parameters (heart rate, blood pressure, and oxygenation) [16].

The pharmacology of ketamine and its use in outpatient anesthesia

In the acute, outpatient setting, the most common reasons to sedate patients are for fracture reductions/treatments, joint relocations, laceration repairs, outpatient/office procedures and lumbar punctures. There are very low number of complications occurring after ketamine administration (0.7% of cases). Therefore ketamine is considered one of the safest anesthetics.

Ketamine antagonizes the excitatory amino acids at the NMDA receptor and may reduce the neuronal damage that occurs in patients with an intracranial injury. However, ketamine is infrequently used for the management of head-injury patients because of its detrimental effects on cerebral blood flow and ICP. Some newer studies outlining the use of ketamine in head-injury patients have been published, they suggest minimal effects on ICP and patient outcome when using specific multiagent protocols. Although these preliminary studies show promising results regarding ketamine involvement in patients with intracranial injury, further studies are warranted.

References

[1] Bokor G, Andderson PD. Ketamine: an update on its abuse. J Pharm Pract 2014;27:582–6.
[2] Licata M, Pierini G, Popoli G. A fatal ketamine poisoning. J Forensic Sci 1994;39:1314–20.
[3] Tao Y, Chen XP, Qin ZH. A fatal chronic ketamine poisoning. J Forensic Sci 2005;50:173–6.
[4] Villelli N, Hauser N, Gianaris T, et al. Severe bilateral cerebellar edema from ingestion of ketamine: case report. J Neurosurg Pediatr 2017;20:393–6.
[5] Matuszko G, et al. Extracellular matrix alterations in the ketamine model of schizophrenia. Neuroscience 2017;350:13–22.
[6] Liu Q, et al. Effects of co-administration of ketamine and ethanol on the dopamine system via the cortex-striatum circuitry. Life Sci 2017;179:1–8.
[7] Kurdi MS, Theerth KA, Deva RS. Ketamine: current applications in anesthesia, pain, and critical care. Anesth Essays Res 2014;8(3):283–90. Available from: https://doi.org/10.4103/0259-1162.143110.
[8] Reves JG, Glass PS, Lubarsky DA, McEvoy MD, Ruiz RM. Intravenous anaesthetics. In: Miller RD, editor. Miller's Anaesthesia. 7th ed. USA: Churchill Livingstone; 2010. p. 719–71.
[9] Dundee JW, Wyant GM. Intravenous anaesthesia. New York: Churchill Livingstone; 1988. p. 135–59. Ketamine.
[10] Kolawole IK. Ketamine hydrochloride: a useful but frequently misused drug. Niger J Surg Res 2001;3:118–25.
[11] Bell RF, et al. Perioperative ketamine for acute postoperative pain. Cochrane Database Syst Rev 2006;(1):CD004603.
[12] Devlin JW, et al. Clinical practice guidelines for the prevention and management of pain, agitation/sedation, delirium, immobility, and sleep disruption in adult patients in the ICU. Crit Care Med 2018;46(9):e825–73.

[13] Stevenson C. Ketamine: a review. Update Anaesth 2005;20:25–9.
[14] Weiner AL, et al. Ketamine abusers presenting to the emergency department: a case series. J Emerg Med 2000;18(4):447–51.
[15] Hodges SC, Walker IA, Bosenberg AT. Paediatric anaesthesia in developing countries. Anaesthesia 2007;62(Suppl 1):26–31.
[16] Sacchetti A, et al. Procedural sedation in the community emergency department: initial results of the ProSCED registry. Acad Emerg Med 2007;14(1):41–6.

Chapter 16.2

Tryptamine trauma: N,N-dipropyltryptamine-associated trauma and rhabdomyolysis

Stephen L. Thornton[1] and Roy G. Gerona[2]

[1]Department of Emergency Medicine, University of Kansas Health System, Kansas City, KS, United States, [2]Department of Laboratory Medicine, University of California-San Francisco, San Francisco General Hospital, San Francisco, CA, United States

Case description

A 17-year-old male with history of attention deficit hyperactivity disorder and opioid abuse was noted by family to be hallucinating. He then became agitated and violent. This was reported to have occurred after he had smoked marijuana. He had a physical altercation and fell and struck his head. He began having seizure-like activity. Emergency medical services were called and administered midazolam 5 mg IV. Upon arrival at the emergency department, he no longer had seizuring but was unresponsive, and subsequently, he was endotracheally intubated using etomidate and rocuronium. His initial vital signs were a pulse of 155 bpm, blood pressure of 128/59 mmHg, and temperature of 98.6°F. His heart rate increased to 169 bpm after intubation. He was started on dexmedetomidine that improved his blood pressure and heart rate. His physical exam demonstrated multiple abrasions to his upper extremities and a scalp laceration without active hemorrhage. The only prescription medication he actively took was lisdexamfetamine. Laboratory evaluation showed an initial pH of 6.96, leukocytosis of 25.6 K/UL, serum creatinine of 1.75 mg/dL (normal 0.3–1.0 mg/dL), and creatine kinase (CK) of 404 U/L (normal <232 U/L). His urine drug screen was positive for amphetamine, opiates, and benzodiazepines but negative for marijuana. He was admitted to the trauma intensive care unit. His pH improved to 7.33, and he was extubated after approximately 4 hours of mechanical ventilation. However, he remained agitated and confused. This altered mental status persisted for the next 4 days, and he required lorazepam and haloperidol along with dexmedetomidine infusion for sedation. He was also given levetiracetam and no further witnessed seizures.

During his hospitalization, he was found to be in rhabdomyolysis. Over the first 24 hours his CK rose to 77,400 U/L. It peaked at 83000 U/L, 48 hours after admission. He was treated with aggressive normal saline hydration and sodium bicarbonate infusion, and his CK declined to 12,295 U/L upon discharge. His creatinine never went higher than 1.75 mg/dL and was 0.69 mg/dL on discharge. He was hospitalized for a total of 6 days. Upon follow-up 2 weeks after discharge, he had a normal CK and creatinine. Upon further questioning, his family suggested that he may have taken a substance called "N,N-dimethyltryptamine (DMT)" that he obtained off the internet. He eventually admitted then to using a substance that he had purchased off the internet though he was not certain its exact identity.

Admission serum and urine samples were obtained and analyzed by liquid chromatography time-of-flight mass spectrometry (TOF 6230, LC 1260, Agilent) using a library of 550 drugs including 285 novel psychoactive substances. N,N-Dipropyltryptamine (DPT) was detected in both serum and urine at concentrations of 1740 and 971 ng/mL, respectively. All other substances detected could be explained by prescription history.

Discussion

DPT is a rarely encountered tryptamine with no published case of human toxicity or overdose. Like all tryptamines, DPT is characterized by the indole ring structure, a fused double ring composing of a pyrrole ring and a benzene ring, joined to an amino group by two carbon side chain [1]. Tryptamines can be synthetic or naturally occurring. DPT is an example of a synthetic tryptamine and has the nickname of "The Light." It was first described in the medical literature 1973 and initially was proposed to have a possible role in psychotherapy [2]. Examples of naturally occurring tryptamines include psilocin, the psychoactive component in magic mushrooms, and DMT that is found in the South American plant *Psychotria viridis* and was reported by the family as the ingested substance in this case [3,4]. In all, there are dozens of tryptamines available and many are misused as novel psychoactive substances for their hallucinogenic effect [5]. It is due to their potential for misuse that multiple tryptamines are controlled substances in the United States and Europe [6,7]. DPT is not a controlled substance in the United States but is in Europe.

Tryptamines as a class are considered potent serotonergic drugs. While human data is lacking, animal studies suggest DPT has potent effects at both 5-HT2A and 5-HT1A receptors, likely explaining its hallucinogenic properties. [8]. Clinically, tryptamines cause hallucinations along with some sympathomimetic effects such as agitation, tachycardia, and hypertension [9]. Cases of rhabdomyolysis and seizures have been reported with other tryptamines thought not DPT [10,11] In this case the cause of the seizures could have been the DPT or the head trauma. The duration of the clinical effects is dependent on the tryptamine ingested and the dose. Fatalities have occasionally been associated with tryptamine use [12]. Treatment for tryptamine toxicity is based on symptomatic care with chemical sedation and hydration key in preventing complications such as rhabdomyolysis. Benzodiazepines seem effective in treating the hallucinations and agitation [13].

While there are no previous case reports of *N,N*-diisopropyltryptamine toxicity in the medical literature, in this case, symptoms appear similar to those reported with more widely used tryptamines such as alpha-methyltryptamine and DMT. Clinicians should be aware of the use of DPT and its potential to cause toxicity.

References

[1] Nichols DE. Chemistry and structure-activity relationships of psychedelics. Curr Top Behav Neurosci 2018;36:1–43.
[2] Soares DBS, Duarte LP, Cavalcanti AD, Silva FC, Braga AD, Lopes MTP, et al. *Psychotria viridis*: chemical constituents from leaves and biological properties. An Acad Bras Cienc 2017;89(2):927–38.
[3] Soskin RA, Grof S, Richards WA. Low doses of dipropyltryptamine in psychotherapy. Arch Gen Psychiatry 1973;28(6):817–21.
[4] Passie T, Seifert J, Schneider U, Emrich HM. The pharmacology of psilocybin. Addict Biol 2002;7(4):357–64.
[5] Araújo AM, Carvalho F, Bastos Mde L, Guedes de Pinho P, Carvalho M. The hallucinogenic world of tryptamines: an updated review. Arch Toxicol 2015;89(8):1151–73.
[6] U.S. Department of Justice. Drug Enforcement Administration Diversion Control Division. scheduling actions controlled substances regulated chemicals; 2018. <https://www.deadiversion.usdoj.gov/schedules/orangebook/orangebook.pdf>. [Accessed 28.11.18].
[7] EMCDDA. Substances and classifications table; 2018. <http://www.emcdda.europa.eu/attachements.cfm/att_65482_EN_Substances%20and%20classifications%20Nov08.xls>. [Accessed 28.11.18].
[8] Fantegrossi WE, Reissig CJ, Katz EB, Yarosh HL, Rice KC, Winter JC. Hallucinogen-like effects of *N,N*-dipropyltryptamine (DPT): possible mediation by serotonin 5-HT1A and 5-HT2A receptors in rodents. Pharmacol Biochem Behav 2008;88(3):358–65.
[9] Tittarelli R, Mannocchi G, Pantano F, Romolo FS. Recreational use, analysis and toxicity of tryptamines. Curr Neuropharmacol 2015;13(1):26–46.
[10] Alatrash G, Majhail NS, Pile JC. Rhabdomyolysis after ingestion of "foxy," a hallucinogenic tryptamine derivative. Mayo Clin Proc 2006;81(4):550–1.
[11] Kamour A, James D, Spears R, Cooper G, Lupton DJ, Eddleston M, et al. Patterns of presentation and clinical toxicity after reported use of alpha methyltryptamine in the United Kingdom. A report from the UK National Poisons Information Service. Clin Toxicol (Phila) 2014;52(3):192–7.
[12] Boland DM, Andollo W, Hime GW, Hearn WL. Fatality due to acute alpha-methyltryptamine intoxication. J Anal Toxicol 2005;29(5):394–7.
[13] Smolinske SC, Rastogi R, Schenkel S. Foxy methoxy: a new drug of abuse. J Med Toxicol 2005;1(1):22–5.

Chapter 16.3

Case reports involving the use of lysergic acid diethylamide

Diane M. Boland

Miami-Dade Medical Examiner Department, Toxicology Laboratory, Miami, FL, United States

Case descriptions

Case 1

The decedent was a 24-year old male with no significant past medical or psychiatric history who, shortly after using lysergic acid diethylamide (LSD), leapt from a fifth-floor residential balcony. The decedent's uncle, who was also using illicit substances, witnessed the event and attempted, without success, to stop the decedent during his sprint toward the balcony. Investigation of the residence revealed drug paraphernalia, including multiple capsules of unknown substance, suspected blotter paper, and probable homegrown cannabis. Entrance to the balcony is through a bedroom and via a sliding glass door. The balcony is enclosed by partial-length railing. Autopsy examination revealed multiple, fatal traumatic injuries. Toxicological examination revealed the presence of LSD and metabolites of cocaine in the urine and chest blood. The cause of death was blunt force injuries of the head and torso with the manner of death initially listed as a suicide. Further investigation revealed that the decedent had no prior history of suicide attempts or ideations and no evidence to suggest that he wanted to end his life. Because LSD is one of the most potent hallucinogenic agents known and toxicology testing confirmed that the decedent had an acute LSD intoxication, it was concluded that LSD precipitated the decedent's erratic behavior and subsequent jump from the balcony. Furthermore, the decedent could not have been aware of his actions nor the consequences they would produce. Based on the totality of information, the manner of death was amended to an accident.

Case 2

The decedent was a 22-year-old male who was in apparent good health with no known clinically diagnosed mental health conditions. On the date of his death, he was at a lake in an apartment complex when he was heard yelling "Carlos" and nonsensical words repeatedly. The decedent was witnessed in his underwear acting strangely and irrationally, and upon approach by a security guard, jumped into the lake and began swimming without difficulty back and forth. Police divers entered the water to retrieve him, at which point he became uncooperative and combative. He suddenly stopped moving and was subsequently removed from the water without ever being submerged. Fire rescue arrived and found the decedent pulseless, apneic, and asystolic. He was transported to the emergency department where he arrived unresponsively. The decedent was asystolic and hypothermic (94.8°F), with laboratory results revealing lactic acidosis and multiple electrolyte abnormalities. He could not be resuscitated and was pronounced approximately 30 minutes after arrival. The autopsy revealed cardiomegaly (470 g) and cardiac dilatation, severe pulmonary congestion and edema, cholesterolosis of the gallbladder, and 3 mL of fluid in the sphenoid sinus. Toxicology testing revealed LSD and its metabolite in aortic blood. Also identified in iliac vein, blood was diphenhydramine, tetrahydrocannabinol and metabolites, flumazenil, ibuprofen, and naloxone. A white strip paper found in the decedent's phone case was tested and identified as LSD. The decedent's actions were determined to be consisted of excited delirium syndrome. Excited delirium syndrome is characterized by delirium with agitation, including panic, fear, shouting, violence, and hyperactivity, sudden cessation of struggle, signs of hyperthermia, unexpected strength, respiratory arrest, and death [1]. Although the decedent's body temperature was low (likely due to being in the water), the decedent's overall actions and behaviors are consistent with an acute intoxication of LSD, including symptoms of excited delirium syndrome. Therefore the cause of death was determined to be excited delirium syndrome associated with LSD use, with the manner of death as an accident.

Discussion

LSD is a semisynthetic potent hallucinogen derived from diethylamide and lysergic acid found in the ergot fungus on rye and other grains [2]. It is manufactured as a tartrate salt that is odorless, colorless, and tasteless [3]. Dosage forms are most commonly "blotters" or "paper squares" that are adorned with artwork or designs. These sheets of decorated absorbent paper are perforated so they can be torn into small squares, typically 7 mm, each containing a single dose that is applied to the tongue where the drug is rapidly absorbed [4]. Other less common forms of LSD include liquid, tablets (microdots), thin gelatin squares (window panes), capsules, impregnated sugar cubes, gel wraps, and candy [3].

LSD was discovered in 1938 by Albert Hoffman, a Swiss natural products chemist who accidentally ingested a small amount of the synthesized drug while researching pharmacologically active derivatives of lysergic acid [2]. The effects of LSD that Hoffman reported included "restlessness, dizziness, a dream-like state and an extremely stimulated imagination" [5]. Due to the profound pharmacological effects reported by Hoffman, LSD became widely researched for decades and a better understanding of its interaction with the brain's serotonin neurotransmitter system evolved. Although the mode of action of LSD is not fully understood, it is thought to bind and activate 5-hydroxytryptamine subtype 2 receptor that interferes with inhibitory systems resulting in sensory−perceptual disturbances [4]. Small doses as low as 20 μg may illicit visual disturbances in the form of geometric shapes or figures in patterns, flashes of intense color, and/or stable objects dissolving or moving [3]. Perception of time may be altered or slowed, and synesthesia, or cross-sensory perception, may also occur in which voices or music evokes perception of particular colors or shapes. Additional effects may include intensified emotions, sudden and dramatic mood swings, depersonalization, and impairment of attention, concentration, and motivation [3]. Adverse effects are dependent on the dose and may include dilated pupils, elevated body temperature, increased heart rate and blood pressure, loss of appetite, profuse sweating, sleeplessness, dry mouth and tremors [3]. LSD and its effect on users are highly variable. Individual reactions are dependent on the dose, the mind-set of the user, and the external circumstances surrounding the user while under the influence of the drug. Each LSD experience can therefore produce very different outcomes, from frightening "bad trips" to deeply meaningful and positive experiences [5]. Because of LSD's hallucinogenic properties, it has been, and continues to be, a drug of abuse. According to the National Forensic Laboratory Information System, reporting of LSD from federal, state, and local forensic laboratories has increased from 1064 reports in 2011 to 4287 reports in 2017 [6,7]. In 2018 alone, from January to June, forensic laboratories identified 373 reports [3]. To date, there have been no reports of fatal overdoses involving LSD as it is considered one of the least toxic drugs used recreationally. The Miami-Dade Medical Examiner, however, has identified LSD in two fatalities, both of which include the drug as the cause of death or contributing to the cause of death.

Routine toxicological analyses were performed on postmortem specimens. All cases received a volatile analysis by headspace gas chromatography coupled to a flame ionization detector. Urine specimens were analyzed by enzyme multiplied immunoassay technique for amphetamines, opiates, benzoylecognine, oxycodone, benzodiazepines, and cannabinoids. When urine was not available, blood was analyzed by enzyme-linked immunosorbent assay for benzodiazepines, benzoylecgonine, opiates, oxycodone, and cannabinoids. The blood was also subjected to a comprehensive drug screen using solid-phase extraction followed by gas chromatography−nitrogen phosphorous detection and gas chromatography−mass spectral analysis.

The two medical examiner cases reported here support that LSD is an extremely potent hallucinogen that produces unpredictable outcomes for individual users. Case 1 was a healthy young man who, out of character, jumped to his death off of a fifth-floor balcony. The manner of death was originally considered suicide; however, based on the lack of previous suicidal ideations, the presence of LSD in the blood, and the documented hallucinogenic effects of LSD, the manner was amended to an accident. Case 2, also a healthy individual, began to experience the sympathomimetic effects of LSD that manifested in excited delirium syndrome, which ultimately caused his death. These cases illustrate two very different scenarios involving fatal outcomes and the ingestion of LSD. Although LSD is considered relatively safe, it can be implicated in the cause and manner of death especially if the circumstances suggest that the drug precipitated the terminal event.

References

[1] Takeuchi A, Ahern TL, Henderson SO. Excited delirium. West J Emerg Med 2011;12(1):77–83.
[2] Passie T, Halpern JH, Sticktenoth DO, Emrich HM, Hintzen A. The pharmacology of lysergic acid diethylamide: a review. CNS Neurosci Ther 2008;14:295–314.
[3] Office of Diversion Control, Drug & Chemical Evaluation Section. Drug enforcement administration: D-lysergic acid diethylamide. Office of Diversion Control, Drug & Chemical Evaluation Section; 2018.
[4] European Monitoring Centre for Drugs and Drug Addiciton. Lysergide (LSD) drug profile. <http://www.emcdda.europa.eu/publications/drug-profiles/lsd>; 2017.
[5] Drug Policy Alliance. LSD fact sheet; 2017.
[6] Drug Enforcement Administration, Diversion Control Division. National forensic laboratory information system. In: 2011 Annual drug report.
[7] Drug Enforcement Administration, Diversion Control Division. National forensic laboratory information system. In: 2017 Annual drug report.

Chapter 17

Therapeutic drug monitoring of immunosuppressants

Sami Albeiroti[1], Vincent Buggs[2], Bjoern Schniedewind[3], Kimia Sobhani[4], Uwe Christians[3] and Kathleen A. Kelly[2,5]

[1]*Sutter Health Shared Laboratory, Livermore, CA, United States,* [2]*UCLA Medical Center, Geffen School of Medicine, Los Angeles, CA, United States,* [3]*iC42 Clinical Research and Development, Department of Anesthesiology, University of Colorado Anschutz Medical Campus, Aurora, CO, United States,* [4]*Cedars-Sinai Medical Center, Los Angeles, CA, United States,* [5]*Pathology and Laboratory Medicine, Clinical Chemistry and Toxicology Geffen School of Medicine, University of California at Los Angeles, Los Angeles, CA, United States*

Case

A 59-year-old patient with cryptogenic cirrhosis was given an orthotopic liver transplantation. The patient was started on a standard immunosuppressant regimen of 1.0 mg BID tacrolimus and 250 mg BID mycophenolate mofetil. Acute kidney injury developed shortly after transplantation, and this was treated with hemodialysis, as can be seen by the rise and fall of serum creatinine level (Fig. 17.1). After a period of stabilization the serum creatinine level remained at approximately 1.0 mg/dL creatinine. To prevent further kidney injury by calcineurin inhibitor (CNI) therapy with tacrolimus, the immunosuppressant regimen was changed to discontinue the mycophenolate mofetil, reduce tacrolimus to 0.5 mg BID, and add 0.5 mg BID everolimus. The most recent serum creatinine level (0.93 mg/dL) showed a decrease.

Introduction

It has been more than a half century since the first successful kidney transplant was performed between monozygotic twins [1]. Since then, vast improvements in organ transplantation have occurred despite the need for continuous immunosuppression to prevent graft rejection. In order to optimize and individualize immunosuppressive therapy, reliable and precise methods are required for drug monitoring. Not all immunosuppressive drugs are routinely monitored such as corticosteroids, which are dosed based on empirical guidelines. Several methods have been developed to measure azathioprine; however, this antiproliferative agent is seldom monitored in transplant patients [2–4]. The immunosuppressive drugs, cyclosporin A (CsA), tacrolimus, sirolimus, everolimus, and mycophenolic acid (MPA) are routinely monitored for the following reasons: (1) there is a clear relationship between drug concentration and clinical response; (2) the drug has a narrow therapeutic index; (3) there is a high degree of inter- and intra-patient variability due to interactions with food, disease and nonadherence; (4) the pharmacological response is difficult to distinguish from unwanted side effects; (5) there is a risk of poor or noncompliance because the drug is administered for the lifetime of the graft or patient; and (6) there are significant drug–drug interactions.

Combination immunosuppressive therapy is widely used today to achieve more efficient immunosuppression and avoid the nephrotoxicity associated with higher doses of CNIs (CsA and tacrolimus). The most common combinations of immunosuppressive drugs used in solid organ transplant patients are tacrolimus and MPA (see the case presented earlier), followed by CsA and MPA [5]. CsA and sirolimus or tacrolimus and sirolimus are also used together. Corticosteroids are typically included in all immunosuppressive drug regimens [6]. CsA inhibits transport of a MPA metabolite from the liver into bile, which results in lower MPA concentrations when the drugs are used in combination [6,7]. The combination of CsA and sirolimus or tacrolimus and sirolimus results in higher than expected blood concentrations of sirolimus [6,8]. This can result in unpredictable drug concentrations and emphasizes the need for therapeutic drug monitoring.

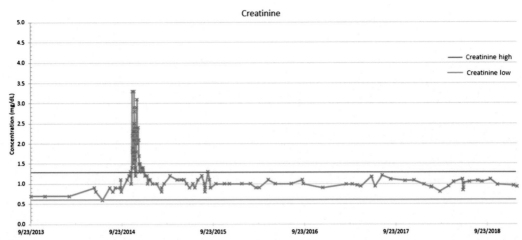

FIGURE 17.1 Creatinine levels were measured in serum from 2013 to 2018 (>150 samples) using a creatininase enzymatic method approved for the cobas c 701/702 system (Roche Diagnostics GmbH, Mannheim, Germany). The reference interval is denoted by the blue line (high, 1.3 mg/dL) and the green line (low, 0.6 mg/dL).

TABLE 17.1 Drugs Used in Solid Organ Transplantation for Immunosuppression.

Class of Drug	Drug Name	Brand Names
Antimetabolites	Mycophenolate mofetil	CellCept, generic formulations
	Mycophenolate sodium	Myfortic, generic formulations
Calcineurin Inhibitors	Cyclosporin A	Sandimmune, Neoral, generic formulations
	Tacrolimus (FK-506)	Prograf, Advagraf, Envarsus, XR, generic formulations
mTOR Inhibitors	Sirolimus (Rapamycin)	Rapamune, generic formulations
	Everolimus	Zortress

mTOR, Mammalian target of rapamycin.

Until recently, most laboratories measured immunosuppressive drugs by semiautomated and automated immunoassays. The development of automated sample handling systems and improvements in mass spectrometers have made liquid chromatography−mass spectrometry (LC−MS) systems more attractive to laboratories that monitor immunosuppressive drugs. This chapter focuses on methods to measure FDA-approved immunosuppressive drugs, including methods for measuring everolimus, a promising new immunosuppressive agent (Table 17.1).

Calcineurin inhibitors

CsA and tacrolimus are two CNI that function to suppress the immune response by blocking activation and proliferation of CD4 + and CD8 + T lymphocytes. Specifically, these drugs inhibit interleukin-2 (IL-2) production [9,10]. The complex of major histocompatibility and peptide antigens results in T lymphocyte activation of calcium/calmodulin-dependent serine/threonine phosphatase calcineurin. Calcineurin activation causes dephosphorylation of the nuclear factor of activated T lymphocytes (NF-AT) and nuclear translocation of NF-AT. NF-AT binds proinflammatory cytokine genes such as IL-2 in the nucleus, resulting in upregulated gene transcription and proliferation of T lymphocytes [11]. CsA and tacrolimus diffuse across cell membranes and form complexes with their respective cytoplasmic binding proteins—immunophilin, cyclophilin, and FK506-binding protein-12 [12,13]. These complexes of drug-immunophilin inhibit calcineurin activity to prevent nuclear translocation of NF-AT, which in turn downregulates cytokine gene transcription. The result is a block in CD4 + and CD8 + lymphocyte activation [14−16].

TABLE 17.2 Types of Immunoassays Used for Measuring Whole Blood Immunosuppressants.

Immunoassay[a]	Drug Applications	Manufacturer
ACMIA	CsA, tacrolimus, sirolimus	Siemens Healthineers
CMIA	CsA, tacrolimus, sirolimus	Abbott Laboratories
CEDIA Plus	CsA, MPA	Thermo Scientific
EMIT 2000	CsA, tacrolimus	Siemens Healthineers
CLIA	CsA	Siemens Healthineers
Enzyme Assay	MPA	Roche Diagnostics
QMS	Everolimus, tacrolimus	Thermo Scientific

[a]ACMIA, antibody conjugated magnetic immunoassay; CMIA, chemiluminescent microparticle immunoassay; CEDIA, cloned enzyme donor immunoassay; EMIT, enzyme-multiplied immunoassay technique; CLIA, chemiluminescent immunoassay; QMS, quantitative microsphere system.

CNI are the mainstay of immunosuppression in solid organ transplantation [17]. However, CNI are associated with adverse effects, namely, renal and cardiovascular [18,19]. Furthermore, CNI are considered the leading cause of nephrotoxicity, which can contribute to renal deterioration [20]. The addition of MPA has been shown to reduce the adverse effects of CNI while providing optimal immunosuppression and long-term graft function [21]. Recently, the addition of mTOR inhibitors, sirolimus and especially everolimus, to the immunosuppressive regimen have shown a potential for reducing adverse renal effects [22,23]. Reduction of CNI following liver transplantation has shown to be a viable means to prevent renal deterioration [24,25].

Cyclosporin A

CsA is a cyclic polypeptide that is derived from fungal cultures of *Tolypocladium inflatum Gams* in 1970 [26]. CsA is approved in the United States for solid organ transplantation of kidney, liver, heart, and bone marrow. CsA is administered orally and intravenously (Sandimmune). Neoral is a microemulsion formulation of CsA with more reproducible absorption characteristics following oral administration [27]. CsA is concentrated in red blood cells and bound to lipoproteins in plasma.

Cyclosporin A immunoassays

The importance of immunosuppressive therapies following solid organ transplantation and narrow therapeutic window of CNI to avoid adverse effects has made it critical for laboratories to provide rapid and precise drug concentrations. There are currently five types of immunoassays used for measuring whole blood CsA in the United States (Table 17.2). Immunoassays are used by approximately 84% of all laboratories (2018 College of American Pathologists Immunosuppressive Drug Monitoring Survey), although many have significant cross-reactivity with CsA metabolites. Consensus panels specify that the analytical method for CsA should be specific for the parent compound [28–30]. Antibody conjugated magnetic immunoassay (ACMIA) and chemiluminescent microparticle immunoassay (CMIA) have reported the least cross-reactivity with CsA metabolites.

LC–MS/MS methods for cyclosporin A

Liquid Chromatography-Mass Spectrometry/Mass Spectrometry (LC–MS/MS) methods are specific for CsA parent compound and are now considered the reference method for validating immunoassays. Sample preparation usually involves a protein precipitation step followed by solid-phase extraction using columns containing C18 or another solid-phase material. Sample extraction using disposable solid-phase columns can be performed manually or automated using liquid handling systems [31,32]. Alternatively, liquid–liquid extraction can be performed manually or can be semiautomated [32].

Several methods using online solid-phase extraction have also been described [33,34]. CsA is a large neutral molecule that does not readily protonate under electrospray ionization and forms Na^+, K^+, and NH_4^+ adducts in the mobile phase. LC−MS methods mainly measure a sodium adduct (m/z 1224.6) since this produces the strongest signal [33,35]. The majority of LC−MS/MS methods monitor an ammonium adduct transition (m/z 1220→1203) since ammonium adducts are relatively unstable and easy to fragment [36,37].

Analytical issues

Immunoassay instruments are easy to operate and can be obtained through reagent agreements instead of purchasing capital equipment. Another advantage is that medical technologists do not need specialized training to operate immunoassay instruments in contrast to mass spectrometry instruments. Immunoassay methods also have the advantage of rapid turnaround times. Turnaround times can be a critical issue, especially in an outpatient setting where it is desirable for the drug concentration to be available when the patient is seen in clinic. The later can require a 1−2 hours turnaround time for immunosuppressive drug testing.

The biggest drawback using CsA immunoassays is metabolite cross-reactivity. None of the immunoassays are completely immune from cross-reactivity with CsA metabolites. After transplantation, CsA levels are closely monitored for individualizing CsA dose. Shortly after transplantation, metabolites are typically present in the highest concentrations complicating dosing regimens [38]. As expected, the magnitude of metabolite cross-reactivity contributes to the positive bias when comparing immunoassays with LC or LC−MS methods. Mean concentrations of CsA were shown to be approximately 12%, 13%, 17%, 22%, and 40% higher than results obtained by LC−UV when measured by ACMIA, enzyme-multiplied immunoassay technique (EMIT), cloned enzyme donor immunoassay (CEDIA Plus), fluorescent polarization immunoassay (FPIA) on the TDx, and FPIA on the AxSYM, respectively [39−43]. Thus it is important to consider metabolite cross-reactivity and the degree of positive bias when selecting an immunoassay for monitoring CsA.

A potential problem with all CsA immunoassays that require a manual extraction step is poor laboratory technique, which can significantly contribute to overall assay imprecision. Careful attention to detail and good technique can minimize variations at this critical preanalytical step. This can be a problem with any of the whole blood immunosuppressive drug assays requiring a manual extraction step.

The advent of C2 monitoring for CsA (CsA levels 2 hours postdosing) results in concentrations that range from 660 to 1700 μg/L, depending on the type of graft and time after transplantation [44]. This can be a problem for some immunoassays with calibration curves that do not have sufficient analytical linearity to measure C2 concentrations without sample dilution. Sample dilution can lead to inaccuracies in quantitation since CsA metabolites may not dilute in a linear fashion. Differences in the time needed for CsA metabolites to reequilibrate after dilution can vary depending on the immunoassay and dilution protocol [45,46]. Some of the immunoassay manufacturers have eliminated the potential for dilution error by providing a separate calibration curve for C2 monitoring (ACIMA, CEDIA Plus, and ADVIA Centaur).

Clearly, the major advantage of LC−MS/MS methods over immunoassays is their excellent specificity and lack of interference from CsA metabolites. MS detection methods have excellent correlation with UV detection methods ($r = 0.998$) [42]. It is important that the linearity of the method is high enough to cover C2 monitoring of CsA; otherwise, the method can produce dilution errors, as described above for immunoassays. Most methods are typically linear from 10 to 2000 μg/L [47]. Another advantage is that other immunosuppressive drugs can be measured in the same sample, which reduces labor costs, analysis time, and sample volume requirements [48]. Turnaround times can be a drawback for LC−MS/MS methods, especially when expedited test results are needed. Run time per sample can be long, ranging from 5 to 15 minutes, depending on the method [47]. Liquid handling systems and solid-phase extraction in a 96-well plate format can help improve sample throughput and reduce turnaround times, especially at institutions that perform large testing volumes. Other potential solutions to improve sample throughput is the use of LC−MS/MS systems equipped with online sample extraction and measurement of multiple immunosuppressive drugs in the same sample [36,37].

The selection of internal standard is critical and can potentially contribute to assay imprecision. Cyclosporin D, deuterium-labeled (d12) CsA, and ascomycin are commonly used internal standards. Ascomycin is often used when monitoring CsA and other immunosuppressive drugs are in the same sample. Some metabolites in blood can interfere with nondeuterated internal standards such as cyclosporin D by increasing peak areas for certain transitions [49,50]. The use of cyclosporin D as an internal standard results in lower CsA results compared with deuterium-labeled (d12) CsA as the internal standard [51]. Ascomycin produces the highest imprecision when compared to cyclosporine D and deuterium-labeled CsA as internal standard [49]. Thus internal standards must be carefully selected and evaluated when developing LC−MS/MS methods to monitor CsA either separately or simultaneously with other immunosuppressive drugs.

Despite the specificity of LC−MS/MS methods for measuring CsA, the majority of laboratories in the United States are still using immunoassays. The lack of widespread acceptance of LC−MS/MS methods for measuring CsA is mostly based on the high initial equipment costs and the need for specialized training in chromatographic techniques and mass spectrometry detection methods. Unfortunately, this type of specialized training is only available in a limited number of clinical laboratories and is not part of many medical technology training programs. As newer LC−MS/MS systems emerge that are easier to operate and less expensive to purchase, more clinical laboratories may consider switching to these systems for measuring immunosuppressive drugs.

Tacrolimus

Tacrolimus (FK-506) is a macrolide antibiotic isolated from the fungus *Streptomyces tsukubaensis* [52]. Tacrolimus (brand name Prograf) was approved for use in liver transplantation in 1994 and kidney transplantation in 1997 in the United States. Tacrolimus is approximately 100 times more potent than CsA. It has been shown to be associated with less acute and chronic rejection, better long-term graft survival, and improved renal function [53]. Tacrolimus monitoring is an integral part of any organ transplant program due to the variability of blood concentrations and narrow therapeutic index. For these reasons, tacrolimus is rapidly replacing CsA. Similar to CsA, the majority of tacrolimus is found within red blood cells and is bound in plasma to albumin, α1-acid glycoprotein, and lipoproteins [54].

Tacrolimus immunoassays

Tacrolimus can be measured by manual ELISA and automated immunoassay. The ELISA assay is manual and is used by only a few laboratories in the United States. Approximately 84% of the laboratories measure tacrolimus by immunoassay (2018 College of American Pathologists Immunosuppressive Drug Monitoring Survey). The ACMIA is fully automated, whereas the CMIA, CEDIA Plus, and Syva EMIT 2000 are semiautomated methods requiring a separate whole blood pretreatment step.

The CMIA is the most popular immunoassay (used by 59% of all laboratories in the United States) and is available on Architect *i* systems. The CMIA uses the same mouse monoclonal antitacrolimus antibody that was used in the microparticle enzyme immunoassay, which is no longer commercially available. The Syva EMIT 2000 has received FDA approval and is used by only 4% of laboratories. The Microgenics CEDIA for tacrolimus is also not widely used despite being available for measuring tacrolimus on several different instruments. The fully automated ACMIA has been available since 2006 and is the second most popular tacrolimus immunoassay. Recently, the ACMIA assay reagents have been improved. The correlation between ACMIA and CLIA are close to identical with the ACMIA comparing well with LC−MS/MS [55].

LC−MS/MS methods for tacrolimus

LC−MS/MS methods are used by only 15% laboratories in the United States. LC−MS/MS methods satisfy the recommendations in consensus documents regarding specificity for parent compound [30]. However, the cost of mass spectroscopy instrumentation and the need for special operating skills appear to affect the use of LC−MS/MS methods. LC−MS/MS methods for tacrolimus have been shown to correlate better with biochemical data from transplant patients followed longitudinally than results by immunoassay [56]. Although the original LC−MS method only measured tacrolimus, other immunosuppressive drugs such as CsA, sirolimus, and everolimus can be measured in the same sample because of similar solubility in alcohols, acetonitrile, ethers, and halogenated hydrocarbons [57,58]. There are numerous protocols to measure tacrolimus by LC−MS/MS with a lower limit of quantitation around 0.2 μg/L [36,59−62]. LC−MS methods mainly monitor a sodium adduct (*m/z* 826) in positive ion mode, while LC−MS/MS methods monitor an ammonium adduct transition (*m/z* 821 → 768) [35−37]. Kushnir et al. compared the specificity of LC−MS and LC−MS/MS for measuring tacrolimus and other immunosuppressive drugs and concluded as expected, that LC−MS/MS is the superior method for verifying drug identity and avoiding interferences [63].

Analytical issues

All of the immunoassays have significant cross-reactivity with tacrolimus metabolites. The ELISA, CMIA, and EMIT cross-react with M-II (31-*O*-demethyl), M-III (15-*O*-demethyl), and M-V (15,13-di-*O*-demethyl) metabolites of tacrolimus [64,65]. The CEDIA has significant cross-reactivity with M-I (13-*O*-demethyl) but does not cross-react with M-II or M-III.

Cross-reactivity of the CEDIA with M−V has not been examined [66]. The ACMIA is expected to have metabolite cross-reactivity similar to the EMIT since both assays use the same monoclonal antibody. Metabolite cross-reactivity in patients with good liver function is typically not a problem because metabolite concentrations are relatively low compared to parent drug [67]. However, metabolites tend to accumulate with reduced liver function and immediately after liver transplant, resulting in significant metabolite cross-reactivity and falsely elevated tacrolimus concentrations [68].

The ACMIA, CMIA, and EMIT produce tacrolimus results that are 0.12 μg/L lower, 0.92 μg/L higher and 1.50 μg/L higher than results obtained by LC−MS/MS, respectively [69]. The CMIA produces results that are 1.21 μg/L higher and 0.47 μg/L lower than results obtained by the ACMIA and EMIT, respectively. The ACMIA was shown to produce results that are 1.66 μg/L lower than results by the EMIT and produced a falsely high tacrolimus result for a patient not receiving tacrolimus [70]. A multisite study found that the CMIA produces results similar to results by the ACMIA and 0.51−1.63 μg/L higher than results obtained by LC−MS/MS [71]. Recently the ACMIA reagents were improved showing a superior correlation with LC−MS/MS than that of CMIA or EMIT [55]. The CEDIA produces results that are 19%−22% higher than results obtained by LC−MS/MS [66,72]. Calibration error may contribute to some of the overall positive bias.

The recommended therapeutic range for whole blood tacrolimus concentrations after kidney and liver allograft transplants is 5.0−20 μg/L by LC−MS [73]. When tacrolimus is used with other immunosuppressive agents such as sirolimus, the desired target tacrolimus concentration can be reduced to 2.0−4.0 μg/L. Because of this, the European Consensus Conference on Tacrolimus Optimization recommends a limit of quantitation of 1 μg/L for tacrolimus in order to provide reliable concentrations during low-dose tacrolimus therapy [74]. Thus it is important for laboratories to use a tacrolimus testing method that has acceptable performance at low concentrations and make transplant services aware of the imprecision at the limit of quantitation. The limit of quantitation (between-day CV $<20\%$) of the ACMIA was shown to range from 2.5 to 4.0 μg/L, whereas the CMIA ranged from 0.5 to 0.8 μg/L [69,71,73]. The CEDIA and EMIT have a limit of quantitation of 3.6 and 3.0 μg/L, respectively [72,75]. The reagents for ACMIA have recently been changed and studies show that the limit of quantitation (between-day CV $<20\%$) has improved 0.8 μg/L [55].

Assay standardization and reagent lot variations have been shown to be a major source of within laboratory imprecision for tacrolimus immunoassays [76]. It is recommended that laboratories sequester a large supply of a single lot of reagents and carefully monitor new lots for bias and imprecision before beginning patient testing.

LC−MS/MS methods for measuring tacrolimus do not have metabolite cross-reactivity and have limits of quantitation suitable for monitoring tacrolimus. LC−MS methods produce variability in tacrolimus testing due to poor chromatography, variability in calibration curves, and the absence of isotope-labeled internal standards [65]. LC−MS/MS methods typically require the in-house preparation of calibrators, controls, and internal standard, which can contribute to variability. A commercially available kit containing freeze-dried calibrators and controls, and a solution of internal standard is currently available (MassTrak Immunosuppressants Kit), but less than 2% of the labs currently use this kit due to cost (2018 College of American Pathologists Immunosuppressive Drug Monitoring Survey). The kit performed well in a multicenter evaluation and has the potential to help harmonize tacrolimus testing across laboratories [69]. Another potential source of variability is the use of ascomycin as internal standard. Ascomycin is not structurally identical to tacrolimus (ethyl analog) and may behave differently and not completely compensate for variable recovery during sample extraction and matrix effects that produce ion suppression.

Mycophenolic acid

MPA is a fermentation product of *Penicillium* species [77,78]. The 2-morpholinoethyl ester of MPA, mycophenolate mofetil (brand name CellCept), was developed for oral and intravenous use to improve bioavailability [79]. Mycophenolate mofetil received FDA approval in 1995. A sodium salt of MPA, mycophenolate sodium (brand name Myfortic), is also available as delayed-release (enteric-coated) tablets for oral use. MPA has largely replaced azathioprine in organ transplantation.

MPA noncompetitively inhibits the enzymatic activity of inosine monophosphate dehydrogenase (IMPDH), the rate-limiting step in the production of guanosine nucleotides [80]. Guanosine nucleotides are needed for DNA synthesis and cellular proliferation and are synthesized by the IMPDH pathway and an alternate salvage pathway. The salvage pathway is not present in lymphocytes, so MPA selectively inhibits lymphocyte proliferation [81,82]. Two isoforms of IMPDH exist and MPA selectively inhibits the one expressed predominantly by activated lymphocytes [83]. Almost all MPA ($>99\%$) is located in plasma, so MPA concentrations are measured in serum and plasma samples [84]. The use of plasma from EDTA-anticoagulated whole blood allows the same sample to be used to measure other immunosuppressive drugs using LC−MS/MS methods.

When MPA was originally approved for use as mycophenolate mofetil, therapeutic drug monitoring was not considered necessary. However, studies have found wide variations in total drug exposure (as high as 10-fold) following a fixed dose, suggesting that individualized dosing may be beneficial [85,86]. Interpatient variability and significant drug interactions during combination immunosuppressive therapy recently prompted the recommendation to monitor blood levels [87]. Trough levels of MPA are a relatively good indicator of total drug exposure [88]. The generally accepted therapeutic range for MPA is 1.0–3.5 mg/L, which can be easily measured by current analytical methods with good precision [89–91].

Circulating MPA is mostly bound to albumin and concentrations of free (unbound) MPA range from 1.2% to 2.5% of total MPA concentrations [84]. Free MPA can be increased in certain clinical conditions, such as hypoalbuminemia, hyperbilirubinemia, and uremia [92]. Several studies have shown that the immunosuppressive effects of MPA and clinical toxicity correlate better with free MPA rather than total MPA concentrations [84]. This is especially true for children because they have higher concentrations of MPA that are more variable due to differences in glucuronosyltransferase activity [93]. In contrast, a few studies have found that free MPA is not superior to total MPA in predicting clinical outcomes in transplant patients [94]. Monitoring of total MPA is recommended but not the separate monitoring of free MPA or MPA metabolites [95].

Immunoassays for mycophenolic acid

Currently only 61 laboratories in the United States measure MPA, which is one-tenth the number of laboratories that measure tacrolimus. Approximately half of the laboratories measure MPA by automated immunoassay or enzymatic assay. Six laboratories use the EMIT and only eight use CEDIA to measure MPA. Fifteen laboratories use the Roche enzymatic immunoassay for measuring total MPA. The Roche enzymatic immunoassay uses MPA in the sample to inhibit IMPDH II, the enzyme that normally catalyzes the conversion of inosine monophosphate to xanthosine monophosphate. During this reaction, NAD^+ is converted to NADH, and the rate of NADH generation is monitored. The concentration of MPA in the sample is inversely related to NADH.

LC–UV and LC–MS/MS methods for MPA

The laboratories using HPLC or LC–MS/MS for measurement of MPA, only 18% use LC–UV while the majority use LC–MS/MS. LC–MS/MS methods have excellent specificity and sensitivity for parent compound [96–100]. Free MPA can be measured after the removal of protein-bound MPA by ultrafiltration or precipitation. However, free MPA is typically in low concentration (approximately 2% of total) and is more appropriately quantitated using LC–MS/MS methods [85]. Although LC–MS/MS methods have been described to measure multiple immunosuppressants in the same sample, MPA is structurally different and requires separate calibrators and sample preparation techniques [96,101–104]. LC–MS/MS methods for MPA typically monitor an ammonium adduct transition (m/z 338→207) in selected reaction monitoring mode utilizing positive electrospray ionization and MPA carboxybutoxy ether as internal standard.

Analytical issues

As previously described for other immunosuppressive drugs, the major benefits of MPA immunoassays are simple testing format, rapid turnaround times, and minimal startup costs. The major limitation is metabolite cross reactivity, which can translate into significant assay bias. The two major metabolites are the 7-O-glucuronide (MPAG) and acyl-glucuronide (AcMPAG) conjugates of MPA. MPAG does not have immunosuppressive activity, whereas AcMPAG has anti-IMPDH activity [85,105]. The EMIT, Roche enzymatic assay, and CEDIA have 30%, <5%, and 158% cross-reactivity with AcMPAG, respectively [106–108]. Bias due to metabolite cross-reactivity for the EMIT and CEDIA can be further exaggerated in transplant patients with impaired renal function due to the accumulation of AcMPAG [109,110].

The EMIT has been shown to overestimate MPA concentrations by 10%–30% when compared with LC–UV methods [109,111–113]. Other studies have shown that the positive bias of the EMIT ranges from 19% to 61% compared with LC–MS/MS methods and is dependent on whether the patient is also receiving sirolimus or CsA [110,114]. The Roche enzymatic immunoassay produces results that are only 0.26–0.45 mg/L lower than results obtained by LC–MS/MS [115]. Further studies across multiple testing centers have confirmed that the Roche enzymatic immunoassay produces MPA results that are in close agreement with those produced by LC–MS/MS [115,116].

LC–UV and LC–MS/MS methods for MPA are specific for parent compound and are used for validating immunoassays [98–100,117]. LC–UV methods typically do not have a limit of quantitation suitable for monitoring free MPA unless the sample is extensively concentrated before analysis [118]. When using an LC−MS/MS method to measure MPA in conjunction with other immunosuppressive drugs, it is important to carefully validate the method for drug recovery and eliminate any assay bias. Another concern when monitoring MPA by LC−MS/MS is the selection of internal standard. Due to structural similarity, the carboxybutoxy ether of MPA is typically used as internal standard. However, there are reports that MPA carboxybutoxy ether is subject to in-source ion fragmentation to MPA, which may result in significant bias when measuring free MPA concentrations [119,120].

Mammalian target of rapamycin (mTOR) inhibitors

The mTOR inhibitors or so-called proliferation signal inhibitors sirolimus and everolimus are macrocyclic lactones. Sirolimus (rapamycin) is a lipophilic molecule isolated from *Streptomyces hygroscopicus*. It was identified in the early 1970s and was approved by the FDA in 1999 for use with CsA to reduce acute kidney rejection [121]. Sirolimus is available for both oral and intravenous administration. Everolimus is a chemically modified version of sirolimus to improve pharmacokinetic and bioavailability. It contains a 2-hydroxyethyl group at position 46 of sirolimus, which increases the hydrophilic properties [122]. Everolimus was approved by the FDA in April 2010 for oral use in adult kidney transplant patients and for adult liver transplant patients in 2013.

mTOR inhibitors readily cross the lymphocyte plasma membrane and bind the intracellular immunophilin, FK506 binding protein-12 [123]. Complexes of siroliumus or everolimus with immunophilin are highly specific inhibitors of the mammalian target of rapamycin (mTOR), a cell cycle serine/threonine kinase involved in the protein kinase B signaling pathway. T lymphocyte proliferation is caused by blocking cell cycle progression into S phase [124].

mTOR inhibitors can be used to minimize or replace the use of CNI inhibitors in patients that experience declining renal allograph function. It was shown that kidney allograph patients at risk of renal dysfunction benefited from conversion from CNI inhibitors to mTOR inhibitors based on an improvement in renal glomerular filtration rate (mean increase of 38%) and serum creatinine (mean reduction of 17%) [125]. Recently, the combination of everolimus and CNI have been investigated for adult liver transplant patients. These studies showed that the early addition of everolimus produced suitable immunosuppression and minimized adverse effects [22,23].

Sirolimus

Sirolimus synergizes with CNI to produce profound immunosuppression of T lymphocytes. Sirolimus is primarily found in red blood cells (96%), with a small amount in plasma and lymphocytes/granulocytes [126]. Almost all of the sirolimus in plasma is bound to lipoproteins.

Sirolimus immunoassays

Therapeutic monitoring of sirolimus is critical because the administered dose is a poor predictor of total drug exposure due to patient variables. Because of the long drug half-life, daily monitoring of sirolimus is typically not necessary. Weekly monitoring may be needed shortly after transplantation, followed by monthly monitoring. Target concentrations for sirolimus range from 4.0 to 12.0 μg/L when used with a CNI [127]. Similar to tacrolimus, these relatively low whole blood concentrations can be a challenge analytically for some immunoassays. As combination immunosuppressant therapies continue to evolve, target concentrations for sirolimus may become even lower, further challenging the analytical performance of immunoassays.

According to the most recent College of American Pathologist Immunosuppressive Drug Monitoring Survey, there are 249 laboratories performing sirolimus testing. Approximately 55% of the laboratories in the United States measure whole blood sirolimus by CMIA (ARCHITECT *i* instruments) [128]. Another 14% of the laboratories use the ACMIA (Dimension instruments) and only a handful use the CEDIA.

LC−MS/MS methods for sirolimus

LC−UV methods for sirolimus suffer from the same drawbacks as described previously for CsA. Approximately 28% of laboratories measure sirolimus using LC−MS/MS methods. LC−MS/MS methods typically have limits of quantitation below 0.3 μg/L [33,35,129−132]. LC−MS methods mainly monitor a sodium adduct (*m/z* 936.6), while LC−MS/MS methods monitor an ammonium adduct transition (*m/z* 931→864) in selected reaction monitoring mode [35,36]. These assays have acceptable total imprecision that ranges from 0.96% to 10.8%, with a reported accuracy of

92%–108% [36]. As indicated previously for CNI, some of these methods use online extraction methods to improve turnaround times. A rapid LC–MS/MS method with turbulent flow technology has recently been introduced with markedly reduced turnaround times compared to traditional LC–MS methods [133].

Analytical issues

As with other immunoassays, the major problem with sirolimus immunoassays is metabolite cross-reactivity. The CMIA has 37% cross-reactivity with 11-hydroxy-sirolimus and 20% cross-reactivity with 41-O-desmethyl-sirolimus [134]. The CEDIA has 44% cross-reactivity with 11-hydyroxy-sirolimus and 73% cross-reactivity with 41-O- and 32-O-desmethyl-sirolimus [135]. However, immunoassay metabolite cross-reactivity may be less of an issue from a clinical standpoint since the distribution of metabolites in whole blood is similar among patients and relatively stable over long periods of time [136].

Sirolimus results using the ACMIA were shown to be 0.95 μg/L higher than results obtained by LC/MS–MS [137]. At the authors' institution, sirolimus results obtained by CMIA are 0.74 μg/L higher than results obtained by LC–MS. A multisite evaluation found that the CMIA produced results that were 14%–39% higher across three sites compared with results obtained by LC–MS/MS [134]. The CEDIA produces whole blood sirolimus results with a mean positive bias of 20.4%, compared with results obtained by LC–UV and LC–MS/MS methods [138].

The ACMIA, CMIA, and CEDIA have a limit of quantitation of 2.0, ≤1.5, and 3.0 μg/L, respectively [134,137,139,140]. Although these limits of detection are adequate for monitoring sirolimus, it is important that laboratories experimentally determine their own limit of quantitation (using whole blood samples) and do not rely on package insert information or published data.

As noted earlier for the CNI, a major advantage of LC–MS/MS sirolimus methods is better sensitivity and specificity compared to immunoassays. A drawback is the longer result turnaround times compared to immunoassay, but this problem can be partially addressed with online sample extraction. The use of in-house calibrators and controls, and different internal standards can result in variable results across laboratories [141]. Hopefully, the variability across laboratories using LC–MS/MS methods can be reduced when kits containing all the critical reagents become commercially available.

Another potential problem with LC–MS/MS methods for sirolimus is the choice of internal standard. Many of the published methods currently use desmethoxy-rapamycin as internal standard. A recent study compared sirolimus results by LC–MS/MS using deuterium-labeled-sirolimus and desmethoxy-rapamycin as internal standards and found matrix effects that produced variable suppression/enhancement of the desmethoxy-rapamycin response [141]. This can result in a positive bias for sirolimus when compared with results obtained using a deuterium-labeled internal standard. The interbatch assay imprecision is also higher using desmethoxy-rapamycin as internal standard. Thus internal standards need to be carefully validated in LC–MS/MS methods when quantitating immunosuppressive drugs and, whenever possible, deuterium labeled internal standards should be used after verifying that they do not contain a significant amount of unlabeled drug.

Everolimus

Everolimus was approved for use in adult kidney transplant patients in the United States 2010. In a large international, multicenter study, everolimus was found to preserve kidney function using lower doses of cyclosporine [142]. In liver transplant recipients, everolimus was shown to improve renal function when combined with low-dose tacrolimus and to decrease the rates of hepatocellular carcinoma recurrence [22,143]. Everolimus partitions between red blood cells and plasma with approximately 75% of everolimus in RBC [23]. Almost all of the everolimus in plasma is bound to proteins [23]. Everolimus metabolites are in relatively low concentrations when monitoring trough blood levels [144]. Evaluation of everolimus metabolites (39-O-des-methyl-, 16-O-des-methyl-, 24-hydroxy-, 25-hydroxy-, 46-hydroxy-, and 11-hydroxy-everolimus) measured by LC–MS/MS found that their distribution was similar between liver and kidney transplant samples. However, the metabolite-to-parent drug ratio was higher in kidney transplant patients, particularly 24-hydroxy-everolimus [145]. The recommendation is to monitor trough levels of everolimus since they correlate well with the level of immunosuppression and rate of adverse effects [146]. In transplantation settings the therapeutic range of everolimus is 3–8 ng/mL when used in combination CNIs and glucocorticoids; however, in CNI-free regimens, the everolimus target trough range should be 6–10 ng/mL [147].

Everolimus immunoassays

The quantitative microsphere system (QMS) everolimus immunoassay (Thermo Fisher Scientific) is the only FDA-approved immunoassay for monitoring whole blood everolimus concentrations. The QMS everolimus assay is a particle-enhanced turbidometric assay that uses calibrators adjusted based on the LC−MS/MS results of a training set of trough samples from kidney transplant patients, which minimizes the effect of the assay's cross-reactivity with everolimus metabolites. The assay can be used with different automated clinical chemistry analyzers; as a result, varying, and sometimes contradictory, data are available about the performance of the assay depending on the evaluated analyzer. The assay has an adequate limit of quantitation and was shown to be 0.75 ng/mL in one report. Negative and positive bias was shown when QMS assay was correlated with LC−MS/MS, depending on the study and the analyzer [148]. However, in a long-term study, analyzing the performance of the QMS assay, the immunoassay was shown to correlate very well with LC−MS/MS with low interlaboratory variability [149]. Although the immunoassay compared favorably with LC−MS/MS, it demonstrated underestimation in kidney transplant samples and overestimation in liver transplant samples at low therapeutic values of 3.0 ng/mL. Overall, everolimus in liver transplant samples showed better correlation with LC−-MS/MS than kidney transplant samples [145]. It is important to note that the immunoassay has 46% cross-reactivity with sirolimus, which is not surprising given the structural similarity between molecules [150].

LC−MS/MS methods for everolimus

LC−UV methods have been developed to measure everolimus, but they only have a limit of quantitation of 2.0 µg/L [146]. Fully validated LC−MS/MS analysis is currently the preferred standard in everolimus monitoring. At least nine LC−MS/MS methods have been developed to measure everolimus with limits of quantitation around 0.25 µg/L [146,151,152]. LC−MS methods monitor a sodium adduct (m/z 980.6) in positive ion mode. The LC−MS/MS methods monitor an ammonium adduct transition (m/z 975 → 908) in selected reaction monitoring mode. A deuterium labeled everolimus is available and highly recommended since ascomycin underestimates everolimus concentrations when used as internal standard [153−155]. For LC−MS/MS assays, recommended whole blood pretreatment includes enhancement of cell lysis by adding distilled water, freezing the samples, or incubating in the presence of ammonium bicarbonate, followed by $ZnSO_4$ treatment and organic extraction [147]. Compared to the QMS immunoassay, there is higher interlaboratory variability among laboratories using LC−MS/MS [149].

Concluding remarks

Therapeutic monitoring of immunosuppressive drugs is essential for optimizing therapeutic effectiveness and minimizing unwanted adverse effects following organ transplantation. Although it has been firmly established that immunoassays suffer from significant metabolite cross-reactivity, the majority of laboratories in the United States still use immunoassays to measure most immunosuppressive drugs. The exception is MPA, which is commonly monitored by LC−UV, LC−MS/MS, and enzymatic assay. Immunoassays have gained widespread use and will continue to dominate the market because they are simple to operate, have fast turnaround times, have low start-up costs, and do not require specialized training. These considerations overshadow the excellent sensitivity, specificity, and versatility of LC−MS/MS methods for immunosuppressive drug monitoring. The high cost of LC−MS/MS systems is a major obstacle, especially for hospitals with limited capital equipment budgets. Furthermore, it is difficult to recruit testing personal with experience in chromatographic separation techniques and mass spectrometry. However, LC−MS/MS methods are rapidly improving to become an economically viable and practical approach for monitoring immunosuppressive drugs.

References

[1] Merrill JP, Murray JE, Harrison JH, Guild WR. Successful homotransplantation of the human kidney between identical twins. JAMA 1956;160:277−82.

[2] Bruunshuus I, Schmiegelow K. Analysis of 6-mercaptopurine, 6-thioguanine nucleotides and 6-thiuric acid in biological fluids by high-performance liquid chromatography. Scand J Clin Invest 1989;49:779−84.

[3] Kreuzenkamp-Jansen CW, De Abreu RA, Bokkerink JPM, Trijbels JMF. Determination of extracellular and intracellular thiopurines and methylthiopurines with HPLC. J Chromatogr 1995;672:53−61.

[4] Rabel SR, Stobaugh JF, Trueworthy R. Determination of intracellular levels of 6-mercaptopurine metabolites in erythrocytes utilizing capillary electrophoresis with laser-induced fluorescence detection. Anal Biochem 1995;224:315−22.

[5] Wolfe RA, Roys EC, Merion RM. Trends in organ donation and transplantation in the United States, 1999-2008. Am J Transpl 2010;10(4 part 2):961−72.

[6] Filler G, Lepage N, Delisle B, Mai I. Effect of cyclosporine on mycophenolic acid area under the concentration-time curve in pediatric kidney transplant recipients. Ther Drug Monit 2001;23:514—19.

[7] van Gelder T, Klupp J, Barten MJ, Christians U, Morris RE. Comparison of the effects of tacrolimus and cyclosporine on the pharmacokinetics of mycophenolic acid. Ther Drug Monit 2001;23:119—28.

[8] Undre NA. Pharmacokinetics of tacrolimus based combination therapies. Nephrol Dial Transpl 2003;18:i12—15.

[9] Shibasaki F, Hallin U, Uchino H. Calcineurin as a multifunctional regulator. J Biochem 2002;131:1—15.

[10] Siekierka JJ, Hung SHY, Poe M, Lin CS, Sigal NH. A cytosolic binding protein for the immunosuppressant FK506 has peptidyl-prolyl isomerase activity but is distinct from cyclophilin. Nature 1989;341:755—7.

[11] Schreiber SL, Crabtree GR. The mechanism of action of cyclosporin A and FK-506. Immunol Today 1992;13:136—42.

[12] Flanagan WM, Corthesy B, Bram RJ, Crabtree GR. Nuclear association of a T-cell transcription factor blocked by FK-506 and cyclosporin A. Nature 1991;352:803—7.

[13] Clipstone NA, Crabtree GR. Identification of calcineurin as a key signalling enzyme in T-lymphocyte activation. Nature 1992;357:695—7.

[14] Schreiber SL. Chemistry and biology of immunophilins and their immunosuppressive ligands. Science 1991;251:283—7.

[15] Gummert JF, Ikonen T, Morris R. Newer immunosuppressive drugs: a review. J Am Soc Nephrol 1999;10:1366—80.

[16] Jørgensen KA, Koefoed-Nielsen PB, Karamperis N. Calcineurin phosphatase activity and immunosuppression. A review on the role of calcineurin phosphatase activity and the immunosuppressive effect of cyclosporin A and tacrolimus. Scand J Immunol 2003;57:93—8.

[17] Calne RY. Immunosuppression in liver transplantation. N Engl J Med 1994;331(17):1154—5.

[18] Ojo AO, Held PJ, Port FK, Wolfe RA, Leichtman AB, Young EW, et al. Chronic renal failure after transplantation of a nonrenal organ. N Engl J Med 2003;349:931—40.

[19] Rubin A, Sánchez-Montes C, Aguilera V, Juan FS, Ferrer I, Moya A, et al. N Engl J Med 2013;349:931—40.

[20] Bechstein WO. Neurotoxicity of calcineurin inhibitors: impact and clinical management. Transpl Int 2000;13:313—26.

[21] US Department of Health and Human Services. OPTN/SRTR 2011 annual data report: liver, <http://srtr.transplant.hrsa.gov/annual_reports/2011/pdf/03_%20liver_12.pdf>; 2011 [accessed 01.12.18].

[22] De Simone P, Nevens F, De Carlis L, Metselaar HJ, Beckebaum S, Saliba F, et al. Everolimus with reduced tacrolimus improves renal function in de novo liver transplant recipients: a randomized controlled trial. Am J Transpl 2012;12(11):3008—20.

[23] Kovarik JM, Kahan BD, Kaplan B, Lorber M, Winkler M, Rouilly M, et al. Longitudinal assessment of everolimus in *de novo* renal transplant recipients over the first post-transplant year: pharmacokinetics, exposure-response relationships, and influence on cyclosporine. Clin Pharmacol Ther 2001;69:48—56.

[24] Cillo U, Saracino L, Vitale A, Bertacco A, Salizzoni M, Lupo F, et al. Very early introduction of everolimus in de novo liver transplantation: results of a multicenter, prospective, randomized trial. Liver Transpl 2019;25:242—51.

[25] De Simone P, Carrai P, Coletti L, Ghinolfi D, Petruccelli S, Precisi A, et al. Everolimus vs mycophenolate mofetil in combination with tacrolimus: a propensity score-matched analysis in liver transplantation. Transpl Proceed 2018;50:3615—20.

[26] Borel JF, Feurer C, Gubler HU, Stahelin H. Biological effects on cyclosporin A: a new antilymphocytic agent. Agents Actions 1976;6:468—75.

[27] Vonderscher J, Meinzer A. Rationale for the development of Sandimmune Neoral. Transpl Proc 1994;26:2925—7.

[28] Kahan B, Shaw L, Holt D, Grevel J, Johnston A. Consensus document: Hawk's Cay meeting on therapeutic drug monitoring of cyclosporine. Clin Chem 1990;36:1510—16.

[29] Shaw L, Yatscoff R, Bowers L, Freeman D, Jeffery J, Keown P, et al. Canadian Consensus Meeting on cyclosporine monitoring: report of the consensus panel. Clin Chem 1990;36:1841—6 464.

[30] Ollerich M, Armstrong VW, Kahan B, Shaw L, Holt DW, Yatscoff R, et al. Lake Louise consensus conference on cyclosporin monitoring in organ transplantation: report of the consensus panel. Ther Drug Monit 1995;17:642—54.

[31] Annesley TM, Clayton L. Simple extraction protocol for analysis of immunosuppressant drugs in whole blood. Clin Chem 2004;50:1845—8.

[32] Brignol N, McMahon LM, Luo S, Tse FLS. High-throughput semi-automated 96-well liquid/liquid extraction and liquid chromatography/mass spectrometric analysis of everolimus (RAD 001) and cyclosporin a (CsA) in whole blood. Rapid Commun Mass Spectrom 2001;15:898—907.

[33] Vidal C, Kirchner GI, Wunsch G, Sewing K-F. Automated simultaneous quantification of the immunosuppressants 40-*O*-(2-hydroxyethyl)rapamycin and cyclosporine in blood with electrospray-mass spectrometric detection. Clin Chem 1998;44:1275—82.

[34] Jemal M. High-throughput quantitative bioanalysis by LC/MS/MS. Biomed Chromatogr 2000;14:422—9.

[35] Deters M, Kirchner G, Resch K, Kaever V. Simultaneous quantification of sirolimus, everolimus, tacrolimus and cyclosporine by liquid chromatography-mass spectrometry (LC-MS). Clin Chem Lab Med 2002;40:285—92.

[36] Korecka M, Shaw L. Review of the newest HPLC methods with mass spectrometry detection for determination of immunosuppressive drugs in clinical practice. Ann Transpl 2009;14:61—72.

[37] Sallustio BC. LC—MS/MS for immunosuppressant therapeutic drug monitoring. Bioanalysis 2010;2:1141—53.

[38] Ryffel B, Foxwell BM, Mihatsch MJ, Donatsch P, Maurer G. Biologic significance of cyclosporine metabolites. Transpl Proc 1988;20:575—84.

[39] Steimer W. Performance and specificity of monoclonal immunoassays for cyclosporine monitoring: how specific is specific? Clin Chem 1999;45:371—81.

[40] Schutz E, Svinarov D, Shipkova M, Niedmann P-D, Armstrong VW, Wieland E, et al. Cyclosporin whole blood immunoassays (AxSYM, CEDIA, and Emit): a critical overview of performance characteristics and comparison with HPLC. Clin Chem 1998;44:2158—64.

[41] Hamwi A, Veitl M, Manner G, Ruzicka K, Schweiger C, Szekeres T. Evaluation of four automated methods for determination of whole blood cyclosporine concentrations. Am J Clin Pathol 1999;112:358—65.

[42] Terrell AR, Daly TM, Hock KG, Kilgore DC, Wei TQ, Hernandez S, et al. Evaluation of a no-pretreatment cyclosporin A assay on the Dade Behring Dimension RxL clinical chemistry analyzer. Clin Chem 2002;48:1059—65.
[43] Butch AW, Fukuchi AM. Analytical performance of the CEDIA cyclosporine PLUS whole blood immunoassay. J Anal Toxicol 2004;28:204—10.
[44] Oellerick M, Armstrong VW. Two-hour cyclosporine concentration determinations: an appropriate tool to monitor neoral therapy? Ther Drug Monit 2002;24:40—6.
[45] Morris RG, Holt DW, Armstrong VW, Griesmacher A, Napoli KL, Shaw LM. Analytic aspects of cyclosporine monitoring, on behalf of the IFCC/IATDMCT Joint Working Group. Ther Drug Monit 2004;26:227—30.
[46] Holt DW, Johnston A, Kahan BD, Morris RG, Oellerich M, Shaw LM. New approaches to cyclosporine monitoring raise further concerns about analytical techniques. Clin Chem 2000;46:872—4.
[47] Salm P, Taylor PJ, Lynch SV, Warnholtz CR, Pillans PI. A rapid HPLC-mass spectrometry cyclosporin method suitable for current monitoring practices. Clin Biochem 2005;38:667—73.
[48] Taylor PJ. Therapeutic drug monitoring of immunosuppressant drugs by high-performance liquid chromatography-mass spectrometry. Drug Monit 2004;26:215—19.
[49] Taylor PJ, Brown SR, Cooper DP, Salm P, Morris MR, Pillans PI, et al. Evaluation of 3 internal standards for the measurement of cyclosporin by HPLC-mass spectrometry. Clin Chem 2005;51:1890—3.
[50] Vogeser M, Spöhrer U. Pitfall in the high-throughput quantification of whole blood cyclosporin A using liquid chromatography-tandem mass spectrometry. Clin Chem Lab Med 2005;43:400—2.
[51] Taylor PJ. Internal standard selection for immunosuppressant drugs measured by high-performance liquid chromatography tandem mass spectrometry. Ther Drug Monit 2007;29:131—2.
[52] Goto T, Kino T, Hatanaka H, Nishiyama M, Okuhara M, Kohsaka M, et al. Discovery of FK-506, a novel immunosuppressant isolated from *Streptomyces tsukubaensis*. Transpl Proc 1987;19:4—8.
[53] First MR. Tacrolimus based immunosuppression. J Nephrol 2004;17:25—31.
[54] Zahir H, Nand RA, Brown KF, Tattam BN, McLachlan AJ. Validation of methods to study the distribution and protein binding of tacrolimus in human blood. J Pharmacol Toxicol Methods 2001;46:27—35.
[55] Kaneko T, Fujioka T, Suzuki Y, Nagano T, Sato Y, Asakura S, et al. Comparison of whole-blood tacrolimus concentrations measured by different immunoassay systems. J Clin Lab Anal 2018;32:e22587. Available from: https://doi.org/10.1002/jcla.22587.
[56] Napoli KL. Is microparticle enzyme-linked immunoassay (MEIA) reliable for use in tacrolimus TDM? Comparison of MEIA to liquid chromatography with mass spectrometric detection using longitudinal trough samples from transplant recipients. Ther Drug Monit 2006;28:491—504.
[57] Alak AM. Measurement of tacrolimus (FK506) and its metabolites: a review of assay development and application in therapeutic drug monitoring and pharmacokinetic studies. Ther Drug Monit 1997;19:338—51.
[58] Christians U, Jacobsen W, Serkova N, Benet LZ, Vidal C, Sewing KF, et al. Automated, fast and sensitive quantification of drugs in blood by liquid chromatography-mass spectrometry with on-line extraction: immunosuppressants. J Chromatogr B Biomed Sci Appl 2000;748:41—53.
[59] Koster RA, Dijkers ECF. Robust, high-throughput LC-MS/MS method for therapeutic drug monitoring of cyclosporine, tacrolimus, everolimus, and sirolimus in whole blood. Ther Drug Monit 2009;31:116—25.
[60] Wang S, Magill JE, Vicente FB. A fast and simple high-performance liquid chromatography/mass spectrometry method for simultaneous measurement of whole blood tacrolimus and sirolimus. Arch Pathol Lab Med 2005;129:661—5.
[61] Lensmeyer GL, Poquette MA. Therapeutic monitoring of tacrolimus concentrations in blood: semi-automated extraction and liquid chromatography-electrospray ionization mass spectrometry. Drug Monit 2001;23:239—49.
[62] Keevil B, McCann S, Cooper D, Morris M. Evaluation of a rapid micro-scale assay for tacrolimus by liquid chromatography-tandem mass spectrometry. Ann Clin Biochem 2002;39:487—92.
[63] Kushnir MM, Rockwood AL, Nelson GJ, Yue B, Urry FM. Assessing analytical specificity in quantitative analysis using tandem mass spectrometry. Clin Biochem 2005;38:319—27.
[64] Iwasaki K, Shiraga T, Matsuda H, Nagase K, Tokuma Y, Hata T, et al. Further metabolism of FK506 (tacrolimus). Identification and biological activities of the metabolites oxidized at multiple sites of FK506. Drug Metab Dispos 1995;23:28—34.
[65] Wallemacq P, Armstrong VW, Brunet M, Haufroid V, Holt DW, Johnston A, et al. Opportunities to optimize tacrolimus therapy in solid organ transplantation: report of the European Consensus Conference. Ther Drug Monit 2009;31:139—52.
[66] Microgenics Corporation. CEDIA® tacrolimus assay (package insert). Fremont, CA: Microgenics Corporation; 2005.
[67] Staatz CE, Taylor PJ, Tett SE. Comparison of an ELISA and an LC/MS/MS method for measuring tacrolimus concentrations and making dosage decisions in transplant recipients. Ther Drug Monit 2002;24:607—15.
[68] Gonschior A, Christians U, Winkler M, Linck A, Baumann J, Sewing K. Tacrolimus (FK506) metabolite patterns in blood from liver and kidney transplant patients. Clin Chem 1996;42:1426—32.
[69] Napoli KL, Hammett-Stabler C, Taylor PJ, Lowe W, Franklin ME, Morris MR, et al. Multi-center evaluation of a commercial kit for tacrolimus determination by LC/MS/MS. Clin Biochem 2010;43:910—20.
[70] Bazin C, Guinedor A, Barau C, Gozalo C, Grimbert P, Duvoux C, et al. Evaluation of the ARCHITECT tacrolimus assay in kidney, liver, and heart transplant recipients. J Pharm Biomed Anal 2010;53:997—1002.
[71] Wallemacq P, Goffinet J-S, O'Morchoe S, Rosiere T, Maine GT, Labalette M, et al. Multi-site analytical evaluation of the Abbott ARCHITECT tacrolimus assay. Ther Drug Monit 2009;31:198—204.

[72] Westley IS, Taylor PJ, Salm P, Morris RG. Cloned enzyme donor immunoassay tacrolimus assay compared with high-performance liquid chromatography-tandem mass spectrometry and microparticle enzyme immunoassay in liver and renal transplant recipients. Ther Drug Monit 2007;29:584–91.

[73] Busuttil RW, Klintmalm GBG, Lake JR, Miller CM, Porayko M. General guidelines for the use of tacrolimus in adult liver transplant patients. Transplantation 1996;61:845–7.

[74] Amann S, Parker TS, Levine DM. Evaluation of 2 immunoassays for monitoring low blood levels of tacrolimus. Ther Drug Monit 2009;31:273–6 10.

[75] LeGatt DF, Shalapay CE, Cheng SB. The EMIT 2000 tacrolimus assay: an application protocol for the Beckman Synchron LX20 PRO analyzer. Clin Biochem 2004;37:1022–30.

[76] Steele BW, Wang E, Soldin SJ, Klee G, Elin RJ, Witte DL. A longitudinal replicate study of immunosuppressive drugs. Arch Pathol Lab Med 2003;127:283–8.

[77] Microgenics Corporation. CEDIA® sirolimus assay (package insert). Fremont, CA: Microgenics Corporation; 2004.

[78] Quinn CM, Bugeja VC, Gallagher JA, Whittaker PA. The effect of mycophenolic acid on the cell cycle of *Candida abicans*. Mycopathologia 1990;111:165–8.

[79] Lee WA, Gu L, Miksztal AR, Chu N, Leung K, Nelson PH. Bioavailability improvement of mycophenolic acid through amino ester derivatization. Pharm Res 1990;7:161–6.

[80] Franklin TJ, Cook JM. The inhibition of nucleic acid synthesis by mycophenolic acid. Biochem J 1969;113:185–204.

[81] Wu JC. Mycophenolate mofetil: molecular mechanisms of action. Perspect Drug Discov Des 1994;2:185–204.

[82] Eugui EM, Allison AC. Immunosuppressive activity of mycophenolate mofetil. Ann N Y Acad Sci 1993;685:309–29.

[83] Allison AC, Eugui EM. Purine metabolism and immunosuppressive effects of mycophenolate mofetil (MMF). Clin Transpl 1996;10:77–84.

[84] Nowak I, Shaw L. Mycophenolic acid binding to human serum albumin: characterization and relation to pharmacodynamics. Clin Chem 1995;41:1011–17.

[85] van Gleder T, Shaw LM. The rationale for and limitations of therapeutic drug monitoring for mycophenolate mofetil in transplantation. Transplantation 2005;80:S244–53.

[86] Brunet M, Cirera I, Martorell J, Vidal E, Millán O, Jiménez O, et al. Sequential determination of pharmacokinetics and pharmacodynamics of mycophenolic acid in liver transplant patients treated with mycophenolate mofetil. Transplantation 2006;81:541–6.

[87] van Gelder T, Meur YL, Shaw LM, Oellerich M, DeNofrio D, Holt C, et al. Therapeutic drug monitoring of mycophenolate mofetil in transplantation. Ther Drug Monit 2006;28:145–54.

[88] Mahalati K, Kahan B. Pharmacological surrogates of allograft outcome. Ann Transpl 2000;5:14–23.

[89] Shaw LM, Holt DW, Oellerich M, Meiser B, van Gelder T. Current issues in therapeutic drug monitoring of mycophenolic acid: report of a roundtable discussion. Ther Drug Monit 2001;23:305–15.

[90] Oellerich M, Shipkova M, Schütz E, Wieland E, Weber L, Tönshoff B, et al. Pharmacokinetic and metabolic investigations of mycophenolic acid in pediatric patients after renal transplantation: implications for therapeutic drug monitoring. Ther Drug Monit 2000;22:20–6.

[91] Shaw L, Korecka M, Aradhye S, Grossman R, Bayer L, Innes C, et al. Mycophenolic acid area under the curve values in African American and Caucasian renal transplant patients are comparable. J Clin Pharmacol 2000;40:624–33.

[92] Kaplan B, Meier-Kriesche H, Friedman G, Mulgaonkar S, Gruber S, Korecka M, et al. The effect of renal insufficiency on mycophenolic acid protein binding. J Clin Pharmacol 1999;39:715–20.

[93] Filler G, Bendrick-Peart J, Christians U. Pharmacokinetics of mycophenolate mofetil and sirolimus in children. Ther Drug Monit 2008;30:138–42 10.109.

[94] Weber LT, Shipkova M, Armstrong VW, Wagner N, Schütz E, Mehls O, et al. The pharmacokinetic-pharmacodynamic relationship for total and free mycophenolic acid in pediatric renal transplant recipients: a report of the German Study Group on mycophenolate mofetil therapy. J Am Soc Nephrol 2002;13:759–68.

[95] Kuypers DRJ, Meur YL, Cantarovich M, Tredger MJ, Tett SE, Cattaneo D, et al. Consensus report on therapeutic drug monitoring of mycophenolic acid in solid organ transplantation. Clin J Am Soc Nephrol 2010;5:341–58.

[96] Streit F, Shipkova M, Armstrong VW, Oellerich M. Validation of a rapid and sensitive liquid chromatography-tandem mass spectrometry method for free and total mycophenolic acid. Clin Chem 2004;50:152–9.

[97] Saunders DA. Simple method for the quantitation of mycophenolic acid in human plasma. J Chromatogr B Biomed Sci Appl 1997;704:379–82.

[98] Teshima D, Kitagawa N, Otsubo K, Makino K, Itoh Y, Oishi R. Simple determination of mycophenolic acid in human serum by column-switching high-performance liquid chromatography. J Chromatogr B Anal Technol Biomed Life Sci 2002;780:21–6.

[99] Sparidans RW, Hoetelmans RM, Beijnen JH. Liquid chromatographic assay for simultaneous determination of abacavir and mycophenolic acid in human plasma using dual spectrophotometric detection. J Chromatogr B Biomed Sci Appl 2001;750:155–61.

[100] Renner UD, Thiede C, Bornhauser M, Ehninger G, Thiede HM. Determination of mycophenolic acid and mycophenolate mofetil by high-performance liquid chromatography using postcolumn derivatization. Anal Chem 2001;73.

[101] Ceglarek U, Casetta B, Lembcke J, Baumann S, Fiedler GM, Thiery J. Inclusion of MPA and in a rapid multi-drug LC-tandem mass spectrometric method for simultaneous determination of immunosuppressants. Clin Chim Acta 2006;373:168–71.

[102] Kuhn J, Prante C, Kleesiek K, Götting C. Measurement of mycophenolic acid and its glucuronide using a novel rapid liquid chromatography-electrospray ionization tandem mass spectrometry assay. Clin Biochem 2009;42:83–90.

[103] Brandhorst G, Streit F, Goetze S, Oellerich M, Armstrong VW. Quantification by liquid chromatography tandem mass spectrometry of mycophenolic acid and its phenol and acyl glucuronide metabolites. Clin Chem 2006;52:1962–4.
[104] Annesley TM, Clayton LT. Quantification of mycophenolic acid and glucuronide metabolite in human serum by HPLC-tandem mass spectrometry. Clin Chem 2005;51:872–7.
[105] Schutz E, Shipkova M, Armstrong VW, Wieland E, Oellerich M. Identification of a pharmacologically active metabolite of mycophenolic acid in plasma of transplant recipients treated with mycophenolate mofetil. Clin Chem 1999;45:419–22.
[106] Shipkova M, Schutz E, Armstrong VW, Niedmann PD, Weiland E, Oellerich M. Overestimation of mycophenolic acid by EMIT correlates with MPA metabolite. Transpl Proc 1999;31:1135–7.
[107] Brandhorst G, Marquet P, Shaw LM, Liebisch G, Schmitz G, Coffing MJ, et al. Multicenter evaluation of a new inosine monophosphate dehydrogenase inhibition assay for quantification of total mycophenolic acid in plasma. Ther Drug Monit 2008;30:428–33. Available from: https://doi.org/10.1097/FTD.
[108] Thermo Scientific. CEDIA mycophenolic acid immunoassay (MDA) package insert. Fremont, CA: Thermo Scientific, 2017.
[109] Weber LT, Shipkova M, Armstrong VW, Wagner N, Schutz E, Mehls O, et al. Comparison of the EMIT immunoassay with HPLC for therapeutic drug monitoring of mycophenolic acid in pediatric renal-transplant recipients on mycophenolate mofetil therapy. Clin Chem 2002;48:517–25.
[110] Prémaud A, Rousseau A, Le Meur Y, Lachâtre G, Marquet P. Comparison of liquid chromatography-tandem mass spectrometry with a commercial enzyme-multiplied immunoassay for the determination of plasma MPA in renal transplant recipients and consequences for therapeutic drug monitoring. Ther Drug Monit 2004;26:609–19.
[111] Beal JL, Jones CE, Taylor PJ, Tett SE. Evaluation of an immunoassay (EMIT) for mycophenolic acid in plasma from renal transplant recipients compared with a high-performance liquid chromatography assay. Ther Drug Monit 1998;20:685–90.
[112] Schütz E, Shipkova M, Armstrong VW, Niedmann PD, Weber L, Tönshoff B, et al. Therapeutic drug monitoring of mycophenolic acid: comparison of HPLC and immunoassay reveals new MPA metabolites. Transpl Proc 1998;30:1185–7.
[113] Westley IS, Sallustio BC, Morris RG. Validation of a high-performance liquid chromatography method for the measurement of mycophenolic acid and its glucuronide metabolites in plasma. Clin Biochem 2005;38:824–9.
[114] Prémaud A, Rousseau A, Picard N, Marquet P. Determination of mycophenolic acid plasma levels in renal transplant recipients co-administered sirolimus: comparison of an enzyme multiplied immunoassay technique (EMIT) and liquid chromatography-tandem mass spectrometry. Ther Drug Monit 2006;28(2):274–7.
[115] Decavele A-SC, Favoreel N, Heyden FV, Verstraete AG. Performance of the Roche Total Mycophenolic Acid® assay on the Cobas Integra 400®, Cobas 6000® and comparison to LC–MS/MS in liver transplant patients. Clin Chem Lab Med 2011;49:1159–65 [Epub ahead of print].
[116] van Gelder T, Domke I, Engelmayer J, Fijter HD, Kuypers D, Budde K, et al. Clinical utility of a new enzymatic assay for determination of mycophenolic acid in comparison with an optimized LC-MS/MS method. Ther Drug Monit 2009;31:218–23.
[117] Tsina I, Kaloostian M, Lee R, Tarnowski T, Wong B. High-performance liquid chromatographic method for the determination of mycophenolate mofetil in human plasma. J Chromatogr B Biomed Appl 1996;681:347–53.
[118] Mandla R, Line P-D, Midtvedt K, Bergan S. Automated determination of free mycophenolic acid and its glucuronide in plasma from renal allograft recipients. Ther Drug Monit 2003;25:407–14.
[119] Patel CG, Mendonza AE, Akhlaghi F, Majid O, Trull AK, Lee T, et al. Determination of total mycophenolic acid and its glucuronide metabolite using liquid chromatography with ultraviolet detection and unbound mycophenolic acid using tandem mass spectrometry. J Chromatogr B 2004;813:287–94.
[120] Atchison CR, West AB, Balakumaran A, Hargus SJ, Pohl LR, Daiker DH, et al. Drug enterocyte adducts: possible causal factor for diclofenac enteropathy in rats. Gastroenterology 2000;119:1537–47.
[121] Miller JL. Sirolimus approved with renal transplant indication. Am J Health Syst Pharm 1999;56:2177–8.
[122] Sedrani R, Cottens S, Kallen J, Schuler W. Chemical modification of rapamycin: the discovery of SDZ RAD. Transpl Proc 1998;30:2192–4.
[123] Abraham RT, Wiederrecht GJ. Immunopharmacology of rapamycin. Annu Rev Immunol 1996;14:483–510.
[124] Kimball PM, Derman RK, Van Buren CT, Lewis RM, Katz S, Kahan BD. Cyclosporine and rapamycin affect protein kinase C induction of intracellular activation signal, activator of DNA replication. Transplantation 1993;55:1128–32.
[125] Sahin S, Gürkan A, Uyar M, Dheir H, Turunç V, Varilsuha C, et al. Conversion to proliferation signal inhibitors-based immunosuppressive regimen in kidney transplantation: to whom and when? Transplan Proc 2011;43:837–40.
[126] Yatscoff R, LeGatt D, Keenan R, Chackowsky P. Blood distribution of rapamycin. Transplantation 1993;56:1202–6.
[127] Holt DW, Denny K, Lee TD, Johnston A. Therapeutic monitoring of sirolimus: its contribution to optimal prescription. Transpl Proc 2003;35:S157–61.
[128] Kahn SE, Vazquez D, Meyer P, Dickson D, Kenney D, Edwards M. Analytical evaluation of the Abbott ARCHITECT sirolimus assay. Clin Chem 2008;54:A14 (abstr).
[129] Koal T, Deters M, Casetta B, Kaever V. Simultaneous determination of four immunosuppressants by means of high speed and robust on-line solid phase extraction-high performance liquid chromatography-tandem mass spectrometry. J Chromatogr B 2004;805:215–22.
[130] Holt DW, Lee T, Jones K, Johnston A. Validation of an assay for routine monitoring of sirolimus using HPLC with mass spectrometric detection. Clin Chem 2000;46:1179–83.
[131] Poquette MA, Lensmeyer GL, Doran TC. Effective use of liquid chromatography-mass spectrometry (LC/MS) in the routine clinical laboratory for monitoring sirolimus, tacrolimus, and cyclosporine. Ther Drug Monit 2005;27:144–50.

[132] Streit F, Armstrong VW, Oellerich M. Rapid liquid chromatography-tandem mass spectrometry routine method for simultaneous determination of sirolimus, everolimus, tacrolimus, and cyclosporin A in whole blood. Clin Chem 2002;48:955—8.
[133] Wang S, Miller A. A rapid liquid chromatography-tandem mass spectrometry analysis of whole blood sirolimus using turbulent flow technology for online extraction. Clin Chem Lab Med 2008;46:1631—4.
[134] Schmid RW, Lotz J, Schweigert R, Lackner K, Aimo G, Friese J, et al. Multi-site analytical evaluation of a chemiluminescent magnetic microparticle immunoassay (CMIA) for sirolimus on the Abbott ARCHITECT analyzer. Clin Biochem 2009;42:1543—8.
[135] Wilson D, Johnston F, Holt D, Moreton M, Engelmayer J, Gaulier JM, et al. Multi-center evaluation of analytical performance of the microparticle enzyme immunoassay for sirolimus. Clin Biochem 2006;39:378—86.
[136] Holt DW, McKeown DA, Lee TD, Hicks D, Cal P, Johnston A. The relative proportions of sirolimus metabolites in blood using HPLC with mass-spectrometric detection. Transpl Proc 2004;36:3223—5.
[137] Westley IS, Morris RG, Taylor PJ, Salm P, James MJ. CEDIA(R) sirolimus assay compared with HPLC-MS/MS and HPLC-UV in transplant recipient specimens. Ther Drug Monit 2005;27:309—14.
[138] Mullett WM. Determination of drugs in biological fluids by direct injection of samples for liquid-chromatographic analysis. J Biochem Biophys Methods 2007;70:263—73.
[139] Mahalati K, Kahan BD. Clinical pharmacokinetics of sirolimus. Clin Pharmacokinet 2001;40:573—85.
[140] Cervinski MA, Duh S-H, Hock KG, Gray J, Wei TQ, Kilgore DC, et al. Performance characteristics of a no-pretreatment, random access sirolimus assay for the Dimension® RxL clinical chemistry system. Clin Biochem 2009;42:1123—7.
[141] O'Halloran S, Ilett KF. Evaluation of a deuterium-labeled internal standard for the measurement of sirolimus by high-throughput HPLC electrospray ionization tandem mass spectrometry. Clin Chem 2008;54:1386—9.
[142] Motzer RJ, Escudier B, Oudard S, Hutson TE, Porta C, Bracarda S, et al. Phase 3 trial of everolimus for metastatic renal cell carcinoma: final results and analysis of prognostic factors. Cancer 2010;116:4256—65.
[143] Cholongitas E, Mamou C, Rodríguez-Castro KI, Burra P. Mammalian target of rapamycin inhibitors are associated with lower rates of hepatocellular carcinoma recurrence after liver transplantation: a systematic review. Transpl Int 2014;27(10):1039—49.
[144] Kirchner GI, Winkler M, Mueller L, Vidal C, Jacobsen W, Franzke A, et al. Pharmacokinetics of SDZ RAD and cyclosporin including their metabolites in seven kidney graft patients after the first dose of SDZ RAD. Br J Clin Pharmacol 2000;50:449—54.
[145] Buggs V, Schniedewind B, Albeiroti S, Huss J, Sobhani K, Christians U, et al. Performance of the Thermo Scientific QMS® everolimus (EVER) assay based on transplant type, metabolite differences and assay platform following everolimus immunosuppression. [poster presentation]. In: Paper presented at: 69th American Association of Clinical Chemistry Annual Scientific Meeting, 2017; San Diego, CA; 2017.
[146] Kirchner GI, Meier-Wiedenbach I, Manns MP. Clinical pharmacokinetics of everolimus. Clin Pharmacokinet 2004;43:83—95.
[147] Shipkova M, Hesselink DA, Holt DW, Billaud EM, van Gelder T, Kunicki PK, et al. Therapeutic drug monitoring of everolimus: a consensus report. Ther Drug Monit 2016;38(2):43—69.
[148] Shu I, Wright AM, Chandler WL, Bernard DW, Wang P. Analytical performance of QMS everolimus assay on ortho Vitros 5,1 FS fusion analyzer: measuring everolimus trough levels for solid organ transplant recipients. Ther Drug Monit 2014;36(2):264—8.
[149] Schniedewind B, Niederlechner S, Galinkin JL, Johnson-Davis KL, Christians U, Meyer EJ. Long-term cross-validation of everolimus therapeutic drug monitoring assays: the Zortracker study. Ther Drug Monit 2015;37(3):296—303.
[150] Dasgupta A, Davis B, Chow L. Evaluation of QMS everolimus assay using Hitachi 917 analyzer: comparison with liquid chromatography/mass spectrometry. Ther Drug Monit 2011;33:149—54.
[151] Baldelli S, Murgia S, Merlini S, Zenoni S, Perico N, Remuzzi G, et al. High-performance liquid chromatography with ultraviolet detection for therapeutic drug monitoring of everolimus. J Chromatogr B 2005;816:99—105.
[152] Salm P, Taylor PJ, Lynch SV, Pillans PI. Quantification and stability of everolimus (SDZ RAD) in human blood by high-performance liquid chromatography-electrospray tandem mass spectrometry. J Chromatogr B Anal Technol Biomed Life Sci 2002;772:283—90.
[153] Taylor PJ, Franklin ME, Graham KS, Pillans PI. A HPLC-mass spectrometric method suitable for the therapeutic drug monitoring of everolimus. J Chromatogr B 2007;848:208—14.
[154] Boernsen KO, Egge-Jacobsen W, Inverardi B, Strom T, Streit F, Schiebel H-M, et al. Assessment and validation of the MS/MS fragmentation patterns of the macrolide immunosuppressant everolimus. J Mass Spectrom 2007;42:793—802.
[155] Hoogtanders K, van der Heijden J, Stolk LM, Neef C, Christiaans M, van Hooff J. Internal standard selection for the high-performance liquid chromatography tandem mass spectroscopy assay of everolimus in blood. Ther Drug Moni 2007;29:673—4.

Further reading

Alnouti Y, Li M, Kavetskaia O, Bi H, Hop CECA, Gusev AI. Method for internal standard introduction for quantitative analysis using on-line solid-phase extraction LC—MS/MS. Anal Chem 2006;78:1331—6.
Kirchner GI, Vidal C, Jacobsen W, Franzke A, Hallensleben K, Christians U, et al. Simultaneous on-line extraction and analysis of sirolimus (rapamycin) and ciclosporin in blood by liquid chromatography-electrospray mass spectrometry. J Chromatogr B: Biomed Sci Appl 1999;721:285—94.
Napoli KL, Kahan BD. Sample clean-up and high-performance liquid chromatographic techniques for measurement of whole blood rapamycin concentrations. J Chromatogr B Biomed Appl 1994;654:111—20.
Taylor PJ. Matrix effects: the Achilles heel of quantitative high-performance liquid chromatography-electrospray-tandem mass spectrometry. Clin Biochem 2005;38:328—34.

Taylor PJ, Salm P, Lynch SV, Pillans PI. Simultaneous quantification of tacrolimus and sirolimus, in human blood, by high-performance liquid chromatography-tandem mass spectrometry. Ther Drug Monit 2000;22:608−12.

Whitman DA, Abbott V, Fregien K, Bowers LD. Recent advances in high-performance liquid chromatography/mass spectrometry and high-performance liquid chromatography/tandem mass spectrometry: detection of cyclosporine and metabolites in kidney and liver tissue. Ther Drug Monit 1993;15:552−6.

Wild DG, editor. The immunoassay handbook. 3rd ed. Elsevier B.V; 2006.

Zhou L, Tan D, Theng J, Lim L, Liu YP, Lam KW. Optimized analytical method for cyclosporine A by high-performance liquid chromatography-electrospray ionization mass spectrometry. J Chromatogr B Biomed Sci Appl 2001;754:201−7.

Chapter 18

Opioids

Jessica A. Hvozdovich, Meagan L. Wisniewski and Bruce A. Goldberger
Department of Pathology, Immunology and Laboratory Medicine, University of Florida College of Medicine, Gainesville, FL, United States

Introduction

The opium poppy (*Papaver somniferum* L.) has been used as a medicinal plant throughout Europe and Asia for millennia and cultivated for the opium produced by the plant's seedpods. Its ability to induce analgesia has been a mainstay of patient treatment and continues to be the basis of modern pain management and anesthesia. Opium contains many pharmacologically active alkaloids such as morphine, codeine, and thebaine. These natural compounds are referred to as opiates, while opioid is an umbrella term that also includes synthetic and semisynthetic analytes with similar pharmacological effects involving CNS depression. Their structures sometimes, but not always, mimic the endogenous opioid peptides known as enkephalins, endorphins, and dynorphins. This class is often referred to as narcotics, a term used in legal contexts that is broadly applied to any sleep-inducing drug with addictive potential. Diacetylmorphine, commonly known as heroin, and fentanyl are two opioids that are prominent in the current illicit drug market; however, both are deemed necessary for palliative care, their legal status varying between countries. For example, fentanyl is a prescribed controlled substance both in the United States and the United Kingdom (DURAGESIC, ACTIQ), while diacetylmorphine is only indicated for pain treatment in the United Kingdom (Diamorphine). This family of drugs has a dichotomous reputation, both a critical component of modern medicine and the source of substance use disorders, overdoses, and many criminal activity.

Mechanism of action

Mechanisms that impede neurotransmission are the basis of opioid analgesia, preventing nociceptive signals from being relayed to the brain. Compounds that bind to and activate opioid receptors are agonists, while antagonists bind to receptors without activation. Antagonists that have higher affinities for the receptors are able to prevent and/or displace agonists, thereby halting agonist action. Mixed agonist—antagonists can activate or block opioid receptors depending on the receptor type, selectivity, and conditions at the binding site.

Classical receptor types include μ, δ, and κ receptors, also referred to as MOP (Mu OPioid receptor), DOP (Delta OPioid receptor), and KOP (Kappa OPioid receptor), respectively. Most exogenous opioids exert their analgesic effects by activating MOP, yet crossover activation of DOP and/or KOP is common. The N/OFQ (Nociceptin/Orphanin FQ, formerly ORL-1) receptor (NOP or Nociceptin OPioid receptor) is also implicated in pain modulation. The opioid receptors can form homo- or heterodimers, such as MOP—DOP or DOP—KOP, altering their affinity and activation. Putative receptor subtypes for MOP, DOP, and KOP have been pharmacologically defined based on responses to agonists; however, splice variants, heterodimer formation, and differences in intracellular signaling may be largely responsible for the observed variation. The opioid receptors are G protein—coupled receptors (GPCR) which form seven alpha-helices that span the plasma membrane. Once an agonist is bound to the GPCR, a conformational change occurs in the second and third intracellular loops that cause the exchange of guanosine diphosphate (GDP) for guanosine triphosphate (GTP) by the associated G protein. Alternatively, the conformational change of the GPCR can cause phosphorylation of the intercellular domain of the GPCR and subsequent recruitment of beta-arrestin. Both of these pathways lead to an intracellular signaling cascade. The downstream effect of signaling in supraspinal regions has been shown in preclinical models to rely on blocking GABA release, resulting in decreased neuron excitability. The downstream effect in the spinal regions is blocking or enhancing ion channels, thereby altering the excitation of neurons. Peripherally, immune cells are known to release endogenous opioids in response to inflammation, and in this case, exogenous opioids modulate analgesia; however, the mechanism for this is still largely unresolved.

The analgesic effect associated with opioids is predominantly mediated by MOP, as are the common side effects of euphoria, miosis, reduced GI motility, respiratory depression, and physical dependence. A mutation (Arg181Cys) in the intracellular second loop between the third and fourth transmembrane domain of the MOP has been shown to cause poor (heterozygous) or no (homozygous) response to administered opioids, implicating its critical role. DOP activation is associated with analgesia, antidepressant effects, physical dependence, and cough suppression. KOP activation is associated with spinal and supraspinal analgesia, miosis, and diuresis; however, it does not seem to contribute to respiratory depression. Studies in rodent and primate models have shown activation of NOP induces antinociception in several pain modalities while mitigating side effects such as respiratory depression.

The common side effects of exogenous opioids are caused by several mechanisms. Mood alteration (e.g., euphoria, tranquility) has been linked to supraspinal dopamine increases resulting from GABA inhibition. Hyperalgesia, or increased pain sensitivity, is thought to involve alterations in spinal or supraspinal signaling with NMDA receptor activation likely playing a role. Conversely, tolerance is an expected result of repetitive drug use where drug response decreases with continued administration. Tolerance and hyperalgesia may share similar pathways; however, nociception should respond to an increased dose of opioids in tolerant individuals. Opioid dependence is common once a state of tolerance is reached; discontinuing exogenous opioids or exposure to an antagonist at this point leads to cellular excitability at all levels, resulting in agitation, hyperalgesia, hypertension, diarrhea, and the effects of released pituitary and adrenomedullary hormones. This is known as opioid withdrawal syndrome and is considered the only evidence of dependence. The terms addiction and dependence are sometimes used interchangeably, but addiction is more specifically a chronic substance use disorder associated with compulsive drug seeking and continued administration despite harmful effects. Tolerance and dependence are two of many potential indicators of an opioid use disorder (OUD). Additional criteria for diagnosis include using more than intended, continued desire to quit or failed attempts to regulate use, preoccupation with drug acquisition and use, cravings, neglecting personal and professional obligations, and risky use. It should be stressed that especially in clinical settings, tolerance and dependence do not imply addiction, and these symptoms are expected results of continued opioid treatment. Respiratory depression is a potentially fatal side effect of opioid use, and anoxic brain damage is often listed as a primary cause of death secondary to opioid administration. Signaling through MOP and DOP decreases the body's response to high CO_2 partial pressure and hypoxia and states which would normally increase respiratory rate and tidal volume. Acute respiratory distress syndrome is linked to opioid use either through breathing after a period of apnea (spontaneous or due to antagonist treatment) or attempting inspiration with a closed glottis. The result is pulmonary edema and frothy sputum evident in the airways or endotracheal tube, the characteristic "foam cone." Hypotension is associated with opioid use; however, cardiovascular effects are largely dependent on the type of opioid and dose.

Common to all of the exogenous opioids are a structure consisting of a phenolic ring, a nitrogen protonated at physiologic pH, and a hydrophobic domain. A variable linker domain is associated with further moieties that allow the compound to bind with specific receptors. Antagonists are theorized to bind in such a way to prevent the conformational change of a receptor required to initiate signaling cascades. Full agonists provide the best analgesia results in patients. Full antagonists are used to reverse the side effects of the agonists at specific sites (peripherally for constipation relief or centrally for respiratory depression reversal) and to treat opioid and alcohol dependency. Full and partial agonists are currently used to treat OUDs to mitigate withdrawal symptoms that hinder abstinence. Mixed agonist−antagonists often consist of a KOP agonist and a MOP antagonist. While this subclass has a better safety profile for dependence and respiratory depression, it is also associated with undesirable behavioral and cardiovascular effects with limited analgesic properties compared to full agonists. Development of an ideal opioid that limits these potentially fatal side effects while providing effective pain relief is still ongoing, with mixed agonist−antagonists showing the most promise.

Pharmacokinetics and pharmacodynamics

Opioid absorption occurs in most tissues; thus many routes of administration are employed in both clinical and recreational settings. Subcutaneous, intramuscular, and intravenous injection allow for direct access to systemic circulation and provide more rapid onset of pharmacological activity with the potential for effects to have increased intensity. Dermal absorption is possible for more lipophilic opioids such as fentanyl, which is available in patches. Oral routes are highly subject to first-pass metabolism in the liver, and while such formulations are convenient to prescribe, pills are also an easy target for diversion and illicit manufacturing. Subdermal implants provide a continuous opioid release and are available to treat opioid dependency for improved adherence and safety compared to daily oral and sublingual formulations. Insufflation and inhalation are routes of administration associated with recreational use, bypassing the dangers associated with injection, including bloodborne pathogens transmitted through needle sharing. Clinical formulations are subject to misuse. In addition to using more than directed, products may be crushed, dissolved, or tampered with to insufflate or inject the opioid to enhance its effects.

Distribution is dependent on an individual's physiology and the chemical properties of the opioid. Bioavailability is influenced by a compound's lipophilicity and the extent that first-pass metabolism is circumvented. Volume of distribution and plasma protein binding are variable among members of the opioid class that concentrate in highly perfused organs. Lipophilic compounds also accumulate in fatty tissues, skeletal muscle, and more easily traverse the blood--brain barrier. For example, heroin is much more lipophilic than its deacetylated metabolites, 6-acetylmorphine (6-AM), and morphine, but heroin itself has minimal pharmacological activity and a very short half-life. Thus optimal transport is achieved by the parent drug, while its amphoteric metabolites produce the desired effects.

Metabolism occurs primarily in the liver where polar, and sometimes more active, metabolites are produced to promote conjugation and excretion. The main route of excretion is via the kidneys, making urine an ideal matrix for drug monitoring and toxicological testing. Liver and renal impairment may result in reduced metabolism and excretion. The subsequent buildup of parent compounds that cannot be metabolized, active metabolites that cannot be excreted, or a combination thereof, may result in overdose at concentrations thought to be within the drug's therapeutic range. Conversely, tolerant patients and longtime recreational users may regularly administer doses that would be fatal to a naïve individual. As with all xenobiotics, a universal lethal dose cannot be estimated and must be assessed on a case-by-case basis.

In addition to analgesia and anesthesia, opioids are indicated for antidiarrheal and antitussive therapy. Opioids also form the basis of treating opioid overdose and opioid dependency. Adverse effects are numerous and include the nervous, pulmonary, gastrointestinal, cardiac, genitourinary, and neuroendocrine systems. Therefore opioid use by patients with preexisting conditions warrants close monitoring. As CNS depressants, opioids may exacerbate the effects of other CNS depressants, resulting in an increased risk of respiratory depression when coadministered. Drug interactions should be considered when determining therapeutic doses. Treatment of significant respiratory depression and apnea include airway, breathing, and circulation control (ABC), and in most cases, administration of an opioid antagonist. Continued opioid use can lead to tolerance of both the positive and negative effects, necessitating increased doses to provide analgesia while preventing fatal respiratory depression. However, the unknown potency and content of illicit drug supplies threaten even tolerant, experienced users.

A summary of the pharmacokinetic and pharmacodynamic properties of opioids is presented in Table 18.1.

Opioid use disorder and treatment

The psychoactive effects of opioids, such as sedation and euphoria, make this class of drugs desirable for recreational and self-medication purposes. Opioid use, especially without professional supervision and trustworthy drug manufacturers, comes with a high risk of developing an addiction. Drug use disorders, including OUD, are considered a chronic disease, and as such treatment is a long, challenging process with anticipated periods of relapse. Detoxification is necessary to address physical dependence, and while opioid withdrawal syndromes are not typically lethal, they are overwhelmingly unpleasant and part of the drive to continue use despite negative ramifications.

Withdrawal begins within a few hours of the last dose and can persist for days with drug clearance determining the intensity and duration. If methadone, buprenorphine, or other long-acting maintenance agonists or partial agonists are used to substitute short-acting opioids such as heroin the withdrawal effects may persist but are milder. With opioid agonist therapy, drug cravings diminish over time as the intense highs and lows stabilize during recovery. The initial maintenance dose is the amount needed to reduce withdrawal to a manageable level and is decreased during the detoxification period. Opioid cross-tolerance can therefore be used to treat the patient, and it is noteworthy that lower doses of maintenance opioids are not shown to be inherently beneficial. In fact, higher methadone doses are associated with less illicit use because tolerance to euphoria is maintained. Once tolerance to methadone sedation is established, patients are typically relieved by its mild stimulating effects as they readjust to normal life. A more dangerous method that has been attempted is rapid antagonist-precipitated withdrawal performed under general anesthesia for a quicker detoxification process without discomfort. This is not widely performed due to its risks and lack of long-term benefits; however, it is possible that removing the need for daily drug administration during detoxification may prevent relapse after a patient leaves a clinic. Buprenorphine implants are a more modern technique to minimize withdrawal while circumventing the logistical issues of classic medication-assisted therapies and helping patients adhere to treatment plans. Group support offering emotional assistance improves success rates in monitored clinics. A protracted withdrawal syndrome can persist for months as the body reacclimatizes to the absence of a compound that has altered the neuroendocrine system and lifestyle of the recovering individual. Hospital discharge following an overdose without continued maintenance

TABLE 18.1 Summary of the Pharmacokinetic and Pharmacodynamic Properties of Opioids.

Drug	Potency Relative to Morphine	Mode of Action	Major Metabolites[a]	Half-Life (h)	Duration of Analgesia (h)	V_d (L/kg)	Protein Binding (%)	Addictive Potential
Buprenorphine	25–50	partial μ agonist κ antagonist	norbuprenorphine, buprenorphine	2–4 PAR 18–49 SL	4–8	1.4–6.2	96	Low
Codeine	0.1	weak μ and δ agonist	codeine, morphine, norcodeine	1.2–3.9	3–4	2.5–3.5	7–25	Moderate
Fentanyl	100–200	strong μ agonist	despropionylfentanyl, norfentanyl, hydroxyfentanyl hydroxynorfentanyl	3–12	1–1.5 IV >12 ER	3–8	79	High
Heroin	1–5	strong μ agonist	6-acetylmorphine, morphine, normorphine	0.03–0.10	3–5	25	<5	High
Hydrocodone	1–2	μ agonist	hydromorphone, norhydrocodone hydrocodol, hydromorphol	3.4–8.8	3–5 IV 12 ER	3.3–4.7	25	Moderate
Hydromorphone	7–10	strong μ agonist	hydromorphol, hydromorphone	3–9 IV 10–22 ER	4–5 IV 12–24 ER	2–4	19	High
Methadone	1	strong μ agonist	EDDP, EDMP, methadone, methadol normethadol	15–55	4–6	4–7	87	High
Morphine	1	strong μ agonist weak κ and δ agonist	morphine, normorphine	1.3–6.7	4–5 IV 8–24 ER	2–5	35	High
Naloxone	NA	strong μ, κ, δ antagonist	naloxone, nornaloxone, naloxol	0.5–1.3	NA	2.6–2.8	46	NA
Oxycodone	1–2	μ agonist	noroxycodone, oxymorphone, oxycodone	3–6	4–6 IV 12 ER	1.8–3.7	45	Moderate
Oxymorphone	8–15	strong μ agonist	6-oxymorphol, oxymorphone	4–12	4–6 IV 12 ER	2–4	10–12	High
Tramadol	0.1–0.2	μ agonist alternative mechanisms	O-monodesmethyltramadol N,O-didesmethyltramadol N-desmethyltramadol	4.3–6.7	4–6 IV 12–24 ER	2.6–29	15–20	Low

ER, Extended release; *IV*, intravenous/immediate release; *PAR*, parenteral; *SL*, sublingual.
[a] *Most metabolites and remaining parent drug are conjugated prior to elimination.*

Adapted from Kerrigan S, Goldberger BA. Chapter 15: Opioids. In: Levine B, editor. Principles of forensic toxicology. 3rd ed. Washington, DC: AACC Press; 2010. p 225–44 [1]; Ropero-Miller JD, Goldberger BA, Reisfield GM. Chapter 11: Opioids. In: Kwong TC, Magnani B, Rosano TG, Shaw, LM, editors. The clinical toxicology laboratory: contemporary practice of poisoning evaluation. 2nd ed. Washington, DC: AACC Press; 2013. p. 155–77 [2]; Baselt RC. Disposition of toxic drugs and chemicals in man. 11th ed. Seal Beach, CA: Biomedical Publications; 2017 [3].

therapy can be dangerous if compulsive use is continued with lowered tolerance after a period of abstinence. Comprehensive long-term treatment is necessary, but often cost prohibitive and unavailable to individuals with the highest rates of OUD. Recently released inmates are a population with a particularly high risk of overdose.

Overprescribing of pharmaceutical opioids coupled with decreased cost and increased potency of street drugs have contributed to an international opioid crisis. Harm reduction strategies have been a controversial component of crisis management since the start of the epidemic. Prioritizing the lives of people with substance use disorders and mitigating the risks associated with drug use can be misconstrued as facilitating and promoting illegal activity. Needle exchange services, distribution of drug checking strips, and supervised injection facilities are programs that are met with skepticism, though research has indicated they reduce risk of infectious disease transmission, inform users about the contents of their drugs (particularly the presence of fentanyl) so they are more likely to perform safer administration practices and prevent overdose when medical professionals are available to intervene in the event of respiratory depression. Naloxone distribution programs and their increased availability have been better received by policymakers. However, all of these methods have the potential to connect individuals with treatment programs and initiate long-term recovery.

Analytical methods

Sample preparation

Opioid metabolism commonly involves conjugation, necessitating sample preparation with a hydrolysis component for "free and total" measurements. This analysis is routinely performed for morphine and quantitates both conjugated and unconjugated morphine, "free" referring to the unconjugated molecules analyzed without hydrolysis and "total" representing both the conjugated metabolites released via hydrolysis and the unconjugated molecules together. This is done in order to approximate how recently the drug was administered. Higher quantities of free drugs imply that there was no significant time for metabolism to occur or the conjugation process was hindered in an individual. Enzymatic hydrolysis with an incubation period is often preferred, as some acids can degrade the 6-AM metabolite of heroin. In addition, liquid chromatography (LC) methods are able to analyze conjugated metabolites without hydrolysis if deuterated standards are available. Unlike traditional gas chromatography (GC) protocols, LC methods do not typically require derivatization to improve volatility and stability for analytes of interest, decreasing sample preparation time and cost. Limitations such as incomplete hydrolysis and introduced interferences that influence the accuracy of results are also circumvented. Different assays may be optimized for the most efficient extraction and analysis of different opioid types (i.e., metabolites, parent compounds, low and high concentration ranges).

Both liquid–liquid extraction (LLE) and solid-phase extraction (SPE) have been employed for sample cleanup. With LLE the basic opioid drug is extracted into an organic solvent from its polar matrix by increasing the sample pH to minimize ionization of unconjugated analytes and promote their exchange to the more lipophilic layer. SPE operates on similar principles of pH adjustment to promote absorption onto a conditioned solid phase. Hydrophobic sorbents are commonly used to remove nonpolar, unionized drugs from polar matrices. Rinsing with additional solvents allows for enhanced cleanup prior to elution with less analyte loss and a concentrated quantity of the analyte of interest. SPE is often faster than LLE and better at removing interferences for cleaner results, but the cartridge expense can be prohibitory. There are many examples of optimized extraction methods in the literature. For rapid screening, protein precipitation or "dilute and shoot" methods may be sufficient, though they produce more wear on the instrument and will require additional downtime for system maintenance long-term.

Screening

Immunoassay techniques are commonly used for opioid screening in forensic toxicology laboratories. They are rapid and nonspecific, therefore cross-reactions may occur with metabolites and structurally similar compounds depending on the particular antibody. Conversely, analogs and novel psychoactive substances might not react with assays developed for classic drug panels and require the development of more specific assays. Fentanyl analogs are of particular concern.

GC techniques have been used for more comprehensive targeted drug screens coupled with detectors including flame ionization detectors, nitrogen phosphorous detectors, or mass spectrometers (MS). High-resolution mass spectrometry (HRMS) methods including LC-quadrupole time of flight (LC-QTOF) and triple TOF methods are newer screening techniques that are becoming more accessible to forensic toxicology laboratories. Depending on the acquisition methods that are applied, retrospective data mining is possible with untargeted data collection through HRMS, allowing toxicologists to reanalyze data and identify emerging substances in samples tested before the novel compounds were known to the forensic community.

The widespread prescription and misuse of opioids have created an increase in demand for opioid tests that can provide rapid presumptive results onsite. Applications include prescription compliance monitoring, workplace drug testing, and roadside testing in suspected drugged driving cases. Lateral-flow immunoassays are available to detect opioids in urine for clinical use and portable devices that test oral fluid are undergoing validation for roadside use. Both of these methods provide qualitative testing options, as do at-home test strips that are widely available. Fentanyl test strips have been used to test drug supplies for the presence of fentanyl or fentanyl analogs before consumption as they are often unknown, undesired adulterants. Increasing test strip availability among individuals that use drugs has been advocated as a harm reduction strategy to prevent overdose deaths attributed fentanyl contamination, especially in populations such as opioid-naïve cocaine users.

Confirmation

Confirmatory methods are designed for the unequivocal identification of target analytes. If a sample's results were presumptively positive for opioids using the screening methods previously described, they would move on to this more specific testing. GC—mass spectrometry (GC—MS) is the classic "gold standard" of identification in forensic toxicology. Retention time and fragmentation data are compared to known standards and compound libraries to identify and quantitate the amount of analyte in a sample with excellent specificity. Derivatization is commonly used to stabilize heat-labile compounds, improve separation, prevent interferences, and increase the volatility of analytes of interest for improved response. LC separation is much gentler, making analyte stability during separation less critical. As was previously mentioned, bulky conjugated metabolites can also be separated via LC for direct analysis without time-consuming hydrolysis steps. LC allows this to be done with even more efficiency, resulting in shorter runtimes combined with simpler sample preparation. LC coupled with a tandem mass spectrometer (MS/MS) is a means of achieving subpicogram detection limits for targeted methods. Interferences from coeluting compounds may be minimized by optimizing separation. Funding permitted, toxicology laboratories are moving toward integrating LC—MS/MS systems into their testing schema as they become more accessible.

Interpretation

Each case must be evaluated individually as there are many factors to be considered along with the drug identities and quantities. The analytical data is interpreted in context with patient information such as the history of drug use, the presence of tolerance, and any observed effects of impairment. There are no true set points for lethal concentrations. Enzyme mutations and impaired elimination pathways can cause typical doses to be improperly metabolized or excreted, resulting in the accumulation of active compounds, lowering the expected therapeutic window with potentially fatal consequences. Cooccurring substances also complicate toxicological interpretation. In cases of fatal overdose the medical examiner makes the cause of death determinations and should work closely with toxicologists for guidance while forming their opinion. Blood samples are used to determine the amount of drug producing an effect on the body at the time of collection or death. In postmortem work, peripheral and central sources may differ due to redistribution, and peripheral samples such as the femoral vein are generally preferred over sites such as the heart or body cavities in cases of trauma. Liver samples tend to correlate with blood concentrations depending on the compound. Urine is commonly used for screening because of the accumulation of xenobiotics in this matrix prior to excretion. While quantitation in urine is not able to be interpreted reliably, there is a greater chance of qualitatively detecting substances that have been used over time. Similarly, alternative matrices such as hair and bone are used to determine drug history but are challenging to work with, and most toxicology laboratories do not provide this testing regularly.

There are several possible sources of opioids that contribute to adverse clinical outcomes. Accidental or intentional misuse of prescription opioids, acquisition of illicitly manufactured or diverted products, and inadvertent exposures are some of the most common occurrences. Opioids are available as over the counter products or prescriptions, from both clinicians and veterinarians. Pharmaceutical evidence such as packaging, unused products, and receipts acquired from a scene may be used to determine how the drugs were acquired. Contaminants and synthesis byproducts identified with toxicology testing may also contribute to source identification for illicit products. For example, illicitly manufactured fentanyl tends to contain larger quantities of acetyl fentanyl than pharmaceutical fentanyl and medical examiner findings indicate that this analog is more likely a synthesis byproduct than an intentional additive. Due to structural similarities, some compounds are more challenging to identify due to shared metabolites and short metabolic half-lives.

Heroin, morphine, and codeine are a classic example. Heroin's rapid metabolism to 6-AM and morphine contributes to the likely underreporting of heroin overdoses. The 6-AM metabolite is diagnostic for heroin, but it is only detectable if a sample is collected within a few hours of exposure. Further complications arise when heroin products contain acetylcodeine, an analyte that metabolizes to codeine and is partially converted to morphine. Morphine detected in the urine can also be produced from poppy seed ingestion; however, workplace testing and compliance cutoffs have been increased to prevent such exposures from being misidentified as illicit drug use. A challenge associated with novel psychoactive substances is the separation and identification of isomers. These compounds may produce the same mass spectral signatures, making alternative testing such as nuclear magnetic resonance, UV–Vis, and IR necessary to identify subtle conformational and structural differences if they cannot be chromatographically separated.

Commonly encountered opioids

Morphine

Morphine is a model full MOP receptor agonist and is the primary analgesic alkaloid produced by the opium poppy. It is used clinically for moderate-to-severe pain treatment. Morphine's CNS depressant effects are separate from its psychostimulant abilities associated with dopamine release, the root of this compound's addictive potential. Concentrations of morphine can be challenging to interpret. Blood results best correlate with drug-induced effects, while urine, bile, liver, kidney, and other samples indicate prior exposure.

Oral bioavailability is approximately 25% due to effective first-pass metabolism in the liver. Parenteral administration provides more rapid absorption and distribution. Morphine not only is predominantly metabolized by glucuronidation but also undergoes demethylation, sulfonation, and N-oxide formation. Morphine-6-glucuronide is an active metabolite that is more potent than the parent compound, and while it is more polar, intramolecular folding improves its ability to cross the blood–brain barrier. Meanwhile, morphine-3-glucuronide may produce excitatory effects, though it has a limited affinity to opioid receptors. Even with this activity, the polar metabolites are readily excreted in the urine and approximately 90% of a dose is removed within 24 hours by normally functioning kidneys.

Free and total tests are performed to determine the amount of unbound versus conjugated morphine, with the "total" sample undergoing hydrolysis in addition to the standard extraction procedure for the removal of glucuronides added during metabolism to promote excretion. In overdose deaths that have occurred rapidly, there is a higher ratio of free-to-total morphine, with more metabolism correlating with a longer survival period. Morphine is also an active metabolite of both codeine and heroin. If these parent compounds or intermediate metabolites, such as 6-AM, are not present due to complete metabolism, results may incorrectly implicate morphine as the administered compound. Tolerance, cooccurring substances, and other circumstances complicate interpretation.

Heroin

Diacetylmorphine, also known as heroin, is a semisynthetic opioid produced by acetylating the two hydroxyl groups of morphine. This structural addition allows the molecule to more efficiently traverse the blood–brain barrier. While parent heroin has poor opioid receptor affinity, it is rapidly deacetylated by blood esterases to the more active 6-AM and subsequently to morphine. Thus its pharmacological profile and effects are similar to morphine with high risks of physical dependence and severe withdrawal symptoms. Its rapid onset after intravenous administration produces a brief period of euphoria lasting a few minutes with sedation lasting a few hours, followed by the sickly feelings of early withdrawal. Inhalation of heroin by heating it on a spoon or foil, called "chasing the dragon," is another common route of administration. The development of tolerance to euphoria and the associated reward mechanisms occur rapidly. Heroin excretion occurs primarily in the urine as free and conjugated morphine. The parent compound is rarely detectable in biological samples, and while the presence of 6-AM is conclusive for heroin, this intermediate metabolite also has a short half-life. As previously mentioned, morphine alone cannot implicate that an individual has administered heroin. This likely results in underreporting of heroin-involved overdoses. Hair of regular users accumulates lipophilic parent drug over time and is more likely to contain 6-AM. However, hair testing is not a common component of standard drug screening as it requires extensive sample preparation and its interpretation is challenging.

While heroin is classified as a Schedule I drug in the United States with no approved medical uses and a high potential for misuse, it is prescribed as diamorphine in Canada and the United Kingdom. It is considered to be twice the potency of morphine with similar pharmacological effects. Diamorphine achieves rapid peak plasma concentrations and has extended duration due to its active metabolites, but its therapeutic advantages have been disputed. Clinical trials have concluded that hydromorphone is just as effective for severe pain treatment and cannot be distinguished from heroin in double-blind studies. Illicit heroin may contain a variety of adulterants, diluents, contaminants, and synthetic artifacts with variable purity. It is also frequently taken in combination with other drugs, intentionally and unintentionally. Heroin and cocaine is a common combination known as "speedball," but individuals who are physically or psychologically dependent may take whatever is cheap and available in an attempt to maintain homeostasis and prevent withdrawal symptoms. The unpredictability of street heroin and desperation of those struggling with addiction increases the risk of overdose.

Fentanyl and fentanyl analogs

Fentanyl is a synthetic MOP receptor agonist originally developed for anesthesia. In clinical settings, it is commonly administered as an epidural dose for postoperative pain or analgesia during labor and is indicated for the treatment of chronic pain. Its fast action and potency are useful for medical purposes, but also dangerous when misused. In the United States, fentanyl is classified as a Schedule II substance due to its physical dependence liability and ability to rapidly produce tolerance. The lipophilicity of fentanyl at physiological pH allows it to rapidly traverse the blood–brain barrier, enhancing the euphoria experienced and increasing the risk of respiratory depression. This characteristic allows for transdermal administration through patch formulations for continuous pain relief. However, patches also present unique challenges because heating that occurs with fevers, heated blankets, saunas, and physical exertion increases the rate of release. They are not always applied as directed and are misused by intentionally heating them, removing the reservoir and injecting the contents, or chewing the patch to allow the full dose to be transmucosally absorbed. Transmucosal absorption is also employed with oralet formulations described as lollipops indicated for the treatment of breakthrough cancer pain. Fentanyl and its more potent analog sufentanil are preferred over other opioids for anesthesia because unlike morphine, histamine release, and the resulting bronchoconstriction and vasodilation do not occur and cardiovascular effects are much less pronounced for an improved safety profile. Metabolism via n-dealkylation occurs in the liver and is catalyzed by CYP3A4 resulting in the renal excretion of parent fentanyl and norfentanyl.

Illicit fentanyl is both diverted from legitimate sources and illicitly manufactured by clandestine laboratories. Its increased potency compared to heroin makes it ideal for smuggling, allowing for smaller quantities of product to be transported and distributed. Analysis of heroin and cocaine samples has detected fentanyl as an adulterant. Counterfeit prescription opioid pills containing fentanyl have also been seized. Fentanyl's rapid onset and potent effects present risks to drug users, especially opioid naïve individuals that may not know their drug of choice, such as cocaine, is contaminated. Overdose is more likely when a typical quantity is administered, but additional or more potent products are present for which the user has less or no tolerance. Tolerant individuals have been known to seek out batches associated with overdoses in hopes that it may produce euphoria in their altered neuroendocrine systems. However, several user reports have stated that fentanyl is undesirable and the potency does not correlate with positive effects.

Fentanyl derivatives have also emerged in the illicit market, contributing to the issues of novel psychoactive substance testing in forensic toxicology and drug chemistry laboratories. This includes the use of time and resources to develop methods for analytes that only briefly appear in casework, waiting for standards to be developed, and the need for continuous monitoring to search for new analogs. Many emerging substances are not detected by traditional testing methods. The underreporting that results masks the true extent of the problem. Accurate assessments of the drug market are needed to inform public health efforts for proper resource allocation. Carfentanil, a Schedule II substance indicated for the sedation of large animals, made a sustained appearance in several parts of the world. Its incredible potency, reportedly 10,000 times that of morphine, caused widespread panic. Only a few milligrams can produce a rapid fatal overdose. Other prominent novel opioids have included 4-ANPP (a fentanyl starting material and metabolite), acetyl fentanyl (a suspected synthesis material or byproduct from illicit fentanyl), cyclopropylfentanyl, methoxyacetylfentanyl, butyrylfentanyl, among many others with new analogs appearing on an almost monthly basis. The mechanisms and effects of the analogs are not well studied. In February of 2018 the US Drug Enforcement Administration placed all illicit fentanyl analogs under temporary emergency scheduling to prosecute traffickers and discourage the production of other analogs created to bypass scheduling laws.

Naloxone

Naloxone is an antagonist with high MOP receptor affinity, allowing it to competitively block the effects of opioid agonists. However, its lack of agonist effect does not provide any relief from withdrawal, limiting its use to detoxification instead of maintenance. It is especially useful when opioid overdose is suspected for fast-acting reversal of respiratory depressant effects. The NARCAN formulation is marketed as an opioid overdose reversal medication and is available as a nasal spray or an injectable syringe for intravenous, intramuscular, and subcutaneous administration. Multiple doses and continued administration with surveillance may be required to maintain respiratory function if the potency and duration of action of the opioid agonist exceed that of naloxone. It is commonly carried by first responders and prescribed to individuals who have recently survived an overdose, but NARCAN is now also available for purchase without a prescription to promote bystander administration and safer drug-use practices.

Buprenorphine

Buprenorphine is a semisynthetic mixed agonist—antagonist derived from thebaine. It has a high MOP receptor affinity with slow dissociation, providing long-lasting analgesia. This affinity also allows it to competitively bind and block the effects of heroin, making it useful in maintenance treatment. Due to minimal withdrawal and a lower overdose potential, a trained physician can treat patients with buprenorphine outside of a methadone clinic environment, improving treatment access. Misuse and diversion are still possible, but buprenorphine's addictive potential is lowered with the sublingual Suboxone formulation, a combination of buprenorphine and naloxone. While naloxone has no pharmacological activity orally, if the pill is dissolved for attempted injection the mild high and potential respiratory depression of buprenorphine is prevented by naloxone. A ceiling effect is reached and if a patient's physical dependence is not relieved the full agonist methadone is recommended. Because buprenorphine is only a partial MOP agonist, it may induce withdrawal symptoms after switching from methadone. Besides maintenance treatment, it is also used clinically to antagonize the respiratory depression of fentanyl during anesthesia. *N*-Dealkylation to norbuprenorphine followed by glucuronidation is the primary metabolic pathway, though most of a dose is eliminated unchanged in the feces.

Conclusion

Public perception of illicit drug use is continuously shifting. Toxicologists must strive to provide objective data and research that informs policy change as well as individual decision making for the benefit of public health. Today, addiction is described as a disease rather than a moral shortcoming, challenging years of stigma that has contributed to the current crisis. Treating those who struggle with OUDs as patients rather than criminals is the first step in combatting this tragedy, prioritizing compassionate problem-solving over blind enforcement. The idea that recreational drug use is inherently immoral continues to be challenged. Consider how intoxication by some substances is culturally acceptable while others are not. How can we work to create safer, consistent, more feasible regulations that are not misconstrued as promotion of potentially life-threatening practices? There is no easy answer or clear solution. Criminalization and prohibition movements have historically resulted in detrimental effects, leading to the trafficking of unregulated products, merging drug use with more serious violent crime, and viewing individuals with illness as undesirables. As professionals we must promote innovative, research-driven solutions as we attempt to curb an epidemic that has claimed millions of lives internationally. Providing data to vigilantly track emerging opioids and contributing to epidemiological studies that assess the extent of the opioid crisis are efforts that encourage professional involvement. While the widespread misuse of opioids has many underlying cultural and infrastructural causes that fall outside the scope of our work as toxicologists, our contributions will hopefully lead to solutions that better society.

References

[1] Kerrigan S, Goldberger BA. Opioids. In: Levine B, editor. Principles of forensic toxicology. 3rd ed. Washington, DC: AACC Press; 2010. p. 225–44. Chapter 15.
[2] Ropero-Miller JD, Goldberger BA, Reisfield GM. Opioids. In: Kwong TC, Magnani B, Rosano TG, Shaw LM, editors. The clinical toxicology laboratory: contemporary practice of poisoning evaluation. 2nd ed. Washington, DC: AACC Press; 2013. p. 155–77. Chapter 11.
[3] Baselt RC. Disposition of toxic drugs and chemicals in man. 11th ed. Seal Beach, CA: Biomedical Publications; 2017.

Chapter 18.1

Heroin or not: a case for timing of specimen collection

Chelsea Milito and Y. Victoria Zhang
Department of Pathology and Lab Medicine, University of Rochester Medical Center, Rochester, NY, United States

Case description

A 54-year-old male went to his outpatient clinic for urine drug screening. The patient's medical history includes heroin abuse, hypertension, paraplegia, and anemia of chronic disease. Two samples of urine on the same day were sent to the toxicology lab at the hospital for urine drug screening testing and confirmation for any positive screen results. One sample was labeled "from bag" with a collection time of 13:00 and the other labeled "catheter" with a collection time of 13:15. The sample taken at 13:00 tested positive for opiates and acetaminophen on the screening assay and the opiate confirmatory assay showed morphine and codeine. The sample taken at 13:15 tested positive for opiates, acetaminophen, fentanyl/norfentanyl, quinine/quinidine, lidocaine, and cyclobenzaprine. The opiate confirmatory assay for this second sample showed 6-monoacetylmorphine (6-MAM), morphine, and codeine. Other screening results were otherwise negative. Test results for these two samples are shown in Table 18.1.1.

Discussion

Although this case did not include a symptomatic overdose, it is important to know how drug overdoses present clinically to help create a better interface between the laboratory and the clinical staff. Heroin belongs to the opioid class of drugs, which includes a number of legal formulations used for the management of chronic and/or severe pain as well as many illicit ones. Heroin, when taken in excess, can cause an overdose presenting with pinpoint pupils, decreased respiratory rate and effort, decreased level of consciousness, seizures, and death. This is not limited to the illicit forms, as the legal formulations can easily be abused as well and patients who are taking opioid medications should receive close follow-up with their provider to make sure they are not overusing them [1,2]. The patient described here had a history of heroin abuse that his physician was aware of. The physician was monitoring him for heroin use which is why he ordered the urine drug screen for him.

Urine drug testing at our institution is performed by a screening immunoassay followed by a confirmatory test utilizing liquid chromatography—tandem mass spectrometry (LC—MS/MS) for any positive results upon request. Urine drug screens using immunoassays are a very effective method of detection for many opioid medications, both legal and illicit [3]. Antibodies in the immunoassay drug screens are commonly developed to cross-react with several drugs that share the same epitope in the patient's urine. The antibody for opioids is designed to detect morphine and other opioids that are structurally similar, with a varying degree of cross-reactivity as the compounds get less similar to morphine [4]. There are forms of opioids that are too dissimilar to the structure of morphine to cross-react with the antibody utilized in the screening assay, such as oxycodone. In this case the screening test will give a negative result if only oxycodone or another nonreactive opiate formulation is present. Many laboratories have additional assays available to detect those opiates when providers need to order them. The screening test only gives the provider a presumptive positive or a presumptive negative result, as the immunoassay can cause false-positive and false-negative results. However, a subsequent confirmatory assay, most often performed with LC—MS/MS, can aid in clarifying what specific types of opiates the patient has been taking and can give quantitative information.

TABLE 18.1.1 Results of Urine Drug Screen and Confirmation Testing Ordered on the Patient.

	Screening Results			Confirmation Results	
Tested Compounds	Cutoff Values (ng/mL)	Sample From 13:00	Sample From 13:15	Sample From 13:00	Sample From 13:15
Acetaminophen	1000	Positive	Positive		
Barbiturate	200	Negative	Negative		
Cocaine/metabolites	300	Negative	Negative		
Benzodiazepine	200	Negative	Negative		
Opiates	300	Positive	Positive	Morphine, codeine	6-Monoacetylmorphine, morphine, codeine
Propoxyphene	300	Negative	Negative		
THC metabolites	50	Negative	Negative		
Fentanyl		Negative	Positive		
Urine drug screen (unknowns)			Quinine/quinidine, lidocaine, and cyclobenzaprine		

THC, 9-Tetrahydrocannabinol.

LC–MS/MS can provide very specific and definitive answers for compounds in patient samples. However, there are other preanalytical factors that affect the test results. One of the factors is the timing of sample collection from the patient. Depending on the sample collection timing relative to when the patient has last taken the drug, the parent compound may no longer be detectable in the sample, and the assays will either show the metabolites or give a negative test result. Understanding the metabolism of the compound of interest is essential for correct interpretation of the results.

For the patient that was investigated in this instance, it is important to understand heroin metabolism. Heroin (6,12-diacetyl morphine) is hydrolyzed to 6-MAM and subsequently to morphine. Both of these compounds are pharmacologically active metabolites of heroin. A small fraction of morphine can also be converted into hydromorphone prior to being excreted by the kidneys as glucuronides. When testing patients for heroin use, the laboratories detect 6-MAM rather than the parent compound of heroin because the half-life of heroin ranges from 2 to 8 minutes [5–7]. 6-MAM does not have a long half-life either, ranging from 16 to 24 minutes, while the half-life of morphine ranges from 67 to 110 minutes. Therefore it's mostly morphine that is providing the sustained physiologic effects of taking a dose of heroin. When testing for heroin use though, if a patient only has morphine positivity, it is impossible to say if they actually took heroin or took another morphine-containing compound. 6-MAM positivity in a sample definitively indicates the use of heroin as it is a unique metabolite of heroin.

Given this information, it is clear why it did not make sense to the laboratory that this patient would test positive for 6-MAM only in the second sample sent by the provider. If the samples were only drawn 15 minutes apart, how would the patient have developed 6-MAM positivity during that time, when he would presumably not have been able to take heroin while he was at his physician's office?

Further investigation included a conversation with the provider which indicated that the patient, due to his paraplegia, had a condom catheter with a bag that he carried around with him. The first sample, that was labeled "from bag," had been drawn from the urine that was in this bag. However, due to the fact that the provider had no way of knowing if the patient had tampered with what was in the bag prior to coming to the office, nor how old the urine in the bag was, he additionally had his nurse catheterize the patient to get a second, fresh sample that he then labeled "catheter." This additional information on the timing of sample collection made the two results from the two samples more coherent. The first sample, although marked as having been taken 15 minutes prior to the second, was actually taken at an unknown time because the provider had no way to ascertain what was in the catheter bag the patient had with him, whereas the test results from the second sample revealed that the patient had used heroin recently and the provider knew exactly when that sample was collected.

When such seemingly discrepant results are brought to the attention of a toxicologist a clear understanding of drug disposition and metabolism, communication with the provider can alleviate dilemmas about test results that were critical for confirming illicit drug use, such as those in our case. This is another example to illustrate the importance of the lab interacting with the providers to provide the most effective lab support for the patient care.

References

[1] Peglow SL, Binswanger IA. Preventing opioid overdose in the clinic and hospital: analgesia and opioid antagonists. Med Clin North Am 2018;102(4):621–34.
[2] Dasgupta N, Funk MJ, Proescholdbell S, Hirsch A, Ribisl KM, Marshall S. Cohort study of the impact of high-dose opioid analgesics on overdose mortality. Pain Med 2016;17(1):85–98.
[3] Mahajan G. Role of urine drug testing in the current opioid epidemic. Anesth Analg 2017;125(6):2094–104.
[4] Milone MC. Laboratory testing for prescription opioids. J Med Toxicol 2012;8(4):408–16.
[5] Rook EJ, Huitema AD, van den Brink W, van Ree JM, Beijnen JH. Population pharmacokinetics of heroin and its major metabolites. Clin Pharmacokinet 2006;45(4):401–17.
[6] Gyr E, Brenneisen R, Bourquin D, Lehmann T, Vonlanthen D, Hug I. Pharmacodynamics and pharmacokinetics of intravenously, orally and rectally administered diacetylmorphine in opioid dependents, a two-patient pilot study within a heroin-assisted treatment program. Int J Clin Pharmacol Ther 2000;38(10):486–91.
[7] Rentsch KM, Kullak-Ublick GA, Reichel C, Meier PJ, Fattinger K. Arterial and venous pharmacokinetics of intravenous heroin in subjects who are addicted to narcotics. Clin Pharmacol Ther 2001;70(3):237–46.

Chapter 18.2

A death involving fentanyl-laced pills

Bheemraj Ramoo[1,2], Robert B. Pietak[3], C. Clinton Frazee, III[1,2] and Uttam Garg[1,2]

[1]Department of Pathology and Laboratory Medicine, Children's Mercy Hospital, Kansas City, MO, United States, [2]University of Missouri School of Medicine, Kansas City, MO, United States, [3]Office of the Jackson County Medical Examiner, Kansas City, MO, United States

Case history

A 35-year-old white male was found lying supine on the left side of the bed with his right foot hanging off the edge. He was wearing a pair of gym shorts and a t-shirt. The subject was in a state of full rigor mortis and livor mortis was present on his posterior side. The lividity was purple in color and blanching. There was foam present around the decedent's mouth, but no signs of injury or trauma.

The decedent was last known to be alive around midnight the day he was found. According to his girlfriend, she and the decedent went to bed at the same time, and the following morning she found him unresponsive. She called 911, EMS responded and found him deceased. His girlfriend mentioned that he was known to buy prescription pills from the street rather than through a doctor and pharmacy. She described the pills as "little green ones." Per his girlfriend, the decedent was also known to snort cocaine. However, no medications or illicit drugs were found at the scene.

Femoral blood, heart blood, vitreous, urine, liver, brain tissue, and gastric contents were submitted to the laboratory for toxicological analysis. No pill specimen was available for analysis. Heart blood was screened for volatiles (ethanol, methanol, isopropanol, and acetone) by headspace gas chromatography with flame ionization detector (GC-FID). The heart blood was also used for comprehensive broad-spectrum drug-screening that involved enzyme immunoassays (EIAs) for amphetamines, barbiturates, benzodiazepines, cannabinoids, cocaine metabolite, methadone, opiates, phencyclidine and propoxyphene, and broad-spectrum drug-screening for >200 drugs by GC—mass spectrometry (GC—MS). Drug screening by GC—MS involved liquid—liquid alkaline extraction using bicarbonate buffer (pH 11.0) and butyl acetate, and mass spectrometer operation in full scan mode. Presumptive identification of analytes was made by spectral library match and relative retention time comparison with reference standards. Confirmation was performed by a reference laboratory (NMS Labs, Willow Grove, Pennsylvania).

The initial EIA drug screen was negative for the nine drugs of abuse mentioned previously. The volatiles screen was negative for ethanol, methanol, isopropanol, and acetone. However, the drug screen by GC—MS was positive for fentanyl only. No cocaine or cocaine metabolites as indicated by case history were detected in any of the drug screens. Postmortem femoral blood was sent to NMS Labs for fentanyl confirmation and quantification by high-performance liquid chromatography/tandem MS. Fentanyl and its metabolite norfentanyl were reported at the concentrations of 22 and 7.1 ng/mL, respectively. The results are summarized in Table 18.2.1.

Discussion

Fentanyl is a synthetic narcotic analgesic that is 50—100 times more potent than morphine [1]. It is marketed under various trade names such as Abstral, Actiq, Duragesic, Fentora, Ionsys, Lazanda, Sublimaze, Matrifen, PecFent, and Subsys [2]. It is majorly used as an adjunct therapy to surgical anesthesia and in pain management. Fentanyl can be administered intravenously or intramuscularly or through transdermal patches, buccal tablets, nasal spray, and sublingual spray. Sublingual sprays such as Subsys are used for the management of breakthrough pain in cancer patients. The most popular use of fentanyl is for the management of chronic pain through transdermal patches which provide a slow and controlled release of the drug. Patches contain 2.5—10 mg of fentanyl and provide a dose of 20—100 μg/h for up to 72 hours [2]. Plasma concentration of 0.6—2 ng/mL provides effective analgesia without respiratory depression.

TABLE 18.2.1 Postmortem Toxicology Results on Heart Blood.

EIA	Results
Amphetamine	Negative
Barbiturates	Negative
Benzodiazepines	Negative
Cannabinoids	Negative
Cocaine metabolite	Negative
Methadone	Negative
Opiates	Negative
Phencyclidine	Negative
Propoxyphene	Negative
GC-FID volatile screening	
Ethanol	<10 mg/dL
Acetone	<10 mg/dL
Methanol	<10 mg/dL
Isopropanol	<10 mg/dL
GC–MS screening	Fentanyl
LC–MS/MS (femoral blood)	Fentanyl—22 ng/mL
	Norfentanyl—7.1 ng/mL

EIA, Enzyme immunoassay; *GC-FID*, gas chromatography with flame ionization detector; *GC–MS*, gas chromatography–mass spectrometry; *LC–MS/MS*, liquid chromatography/ tandem mass spectrometry.

Approximately 0.4%–6% of fentanyl is eliminated unchanged in urine [2]. Fentanyl has a half-life of 3–12 hours and a volume of distribution of 3.2–5.6 L/kg. It is metabolized to norfentanyl, hydroxyfentanyl, hydroxynorfentanyl, and despropionylfentanyl. With the exception of despropionylfentanyl, these metabolites are eliminated through the urine.

Fentanyl and its analogs, for example, acetyl fentanyl, alfentanil, butyryl fentanyl, carfentanil, 4-fluorobutyrylfentanyl, furanyl fentanyl, 4-methoxybutyrylfentanyl, ocfentanil and sufentanil, quickly cross the blood–brain barrier and potentiate CNS effects. Overdose can result in respiratory depression, hypotension, sedation, confusion, unconsciousness, seizures, coma, and even death [1,2]. In the past, most fentanyl overdoses and deaths resulted from abuse and/or misuse of transdermal patches or intravenous injections. Transdermal patch abuse includes the use of multiple patches to get doses higher than prescribed and may entail chewing or smoking the patches [3,4]. At current time, overdoses from fentanyl and its analogs are happening from the nonpharmaceutical products produced in clandestine laboratories. Street names for fentanyl or fentanyl-laced heroin include Apache, China Girl, China White, Dance Fever, Friend, Goodfella, Jackpot, Murder 8, TNT, and Tango and Cash [5]. These nonpharmaceutical forms of fentanyl are sold in various forms including powder, spiked on blotter paper, mixed with or substituted for heroin and other drugs, and as tablets that mimic prescription pills [5]. For example, fentanyl has been shown to be a major ingredient in "counterfeit" hydrocodone and oxycodone tablets. In addition to being mixed with or substituted as other opioids, fentanyl is even being sold falsely labeled as methylenedioxymethamphetamine (MDMA or ecstasy). One possible explanation for the increased use of fentanyl within the clandestine drug market is its low production cost. Heroin costs about $65,000 per kilogram wholesale, for example, whereas illicit fentanyl cost roughly $3,500 per kilogram [6].

In recent years, fentanyl encounters have increased exponentially. The number of fentanyl encounters reported by the DEA National Forensic Laboratory Information System (NFLIS) from 2014 to 2017 was 4768, 14,440, 34,199, and 56,530, respectively [7]. This appears to correlate with the surge in the number of fentanyl-related deaths [1,8,9]. It is estimated that more than 100 Americans die every day from opioid overdose [9]. Therefore deaths from fentanyl overdose are not uncommon. But the abuse and manner of deaths in many cases are unusual. We described a case of misuse of a fentanyl transdermal patch (ingestion by chewing on the patch) followed by complications of aspiration of the patch [3].

FIGURE 18.2.1 Heroin and fentanyl-laced pills (https://www.dea.gov/galleries/drug-images/fentanyl).

Increased illegal distribution of fentanyl-laced drugs is another troubling trend which is contributory to high morbidity and mortality [10–12]. As an example, heroin and fentanyl-laced pills are shown in Fig. 18.2.1. Many times, victims who die from fentanyl overdose may have been unaware of the presence of the drug if it was used to lace other drugs such as heroin, oxycodone, and hydrocodone. In postmortem cases, blood fentanyl concentrations ranged from 3 to 28 ng/mL with an average of 8.3 ng/mL [2]. The fentanyl blood level in this case was 22 ng/mL and well above the average observed concentration in reported fatalities. The findings in this case indicate that the subject died as a result of severe respiratory depression due to fentanyl intoxication. Death investigation findings indicated that the victim may not have been aware that he was using a fentanyl-laced product. The manner of death was ruled accidental.

References

[1] Centers for Disease Control and Prevention (CDC). Drug and opioid involved overdose deaths—United States, 2013-2017; 2019. [last modified 03.01.19].
[2] Baselt RC. Fentanyl. In: Baselt RC, editor. Disposition of toxic drugs and chemicals in man. Seal Beach, CA: Biomedical Publications; 2017. p. 883–6.
[3] Carson HJ, Knight LD, Dudley MH, Garg U. A fatality involving an unusual route of fentanyl delivery: chewing and aspirating the transdermal patch. Leg Med (Tokyo) 2010;12:157–9.
[4] Palmer RB. Fentanyl in postmortem forensic toxicology. Clin Toxicol (Phila) 2010;48:771–84.
[5] National Institute of Drug Abuse, NIDA. Fentanyl. <https://www.drugabuse.gov/publications/drugfacts/fentanyl>; 2016 [accessed 28.01.19].
[6] Frank RG, Pollack HA. Addressing the fentanyl threat to public health. N Engl J Med 2017;376:605–7.
[7] U.S. Drug Enforcement Administration, Diversion Control Division. Retrieved from the NFLIS Public Resource Library at <https://www.nflis.deadiversion.usdoj.gov/Resources/NFLISPublicResourceLibrary.aspx> [accessed 03.06.20].
[8] Jannetto PJ, Helander A, Garg U, Janis GC, Goldberger B, Ketha H. The fentanyl epidemic and evolution of fentanyl analogs in the United States and the European Union. Clin Chem 2019;62:242–53.
[9] Scholl L, Seth P, Kariisa M, Wilson N, Baldwin G. Drug and opioid-involved overdose deaths—United States, 2013-2017. MMWR Morb Mortal Wkly Rep 2018;67:1419–27.
[10] Boddiger D. Fentanyl-laced street drugs "kill hundreds". Lancet 2006;368:569–70.
[11] Bode AD, Singh M, Andrews J, Kapur GB, Baez AA. Fentanyl laced heroin and its contribution to a spike in heroin overdose in Miami-Dade County. Am J Emerg Med 2017;35:1364–5.
[12] Turock MK, Watts DJ, Mude H, Prestosh J, Stoltzfus J. Fentanyl-laced heroin: a report from an unexpected place. Am J Emerg Med 2009;27:237–9.

Chapter 18.3

Opioid metabolism: impact of concomitant medications and genetic variations on the interpretation of drug tests

Hila Shaim, Paul E. Young and Anthony O. Okorodudu
Department of Pathology, University of Texas Medical Branch, Galveston, TX, United States

Case 1 description

A 65-year-old man presented to pain clinic to establish care on account of his chronic back pain. He had recently moved to the area. He reported a history of multiple back injuries with diagnoses of lumbosacral spondylosis, central lumbar intervertebral disc herniation (L5−S1) and spinal stenosis of the lumbar region, status post−surgical fusion of L4 and L5 vertebrae done seventeen years prior. He had suffered from chronic low back pain since the age of 16. He described the pain as constant and throbbing, with radiation down the left lower limb to the foot, and down the right lower limb to the knee, with associated numbness and weakness. Medications from his previous physician(s) include fentanyl patch 50 μg/h every 72 hours, duloxetine 60 mg daily, pregabalin (Lyrica) 100 mg daily, modafinil 4 mg daily, oxycodone 10 mg as needed and clonazepam 1 mg twice daily. Since moving to the area, he had not refilled his prescriptions and had run out of some of his medications. Per clinic guidelines, the patient was asked to submit to a routine urine drug test prior to signing a pain contract and possible prescription of opioids. Particular attention is paid to drugs with addiction potential, especially opioids and illicit drugs such as cocaine and methamphetamine in this test. The urine drug test by liquid chromatography−tandem mass spectrometry (LC−MS/MS) was positive for oxycodone (208 ng/mL), noroxycodone (3454 ng/mL), norfentanyl (131 ng/mL), and 7-aminoclonazepam (282 ng/mL). Oxymorphone, a minor metabolite of oxycodone, was not detected. Why was oxymorphone not detected? Is this pattern consistent with the use of oxycodone as prescribed? In the discussion section, we will cover some of the possibilities.

Case 2 description

A 73-year-old woman presented to pain clinic for follow-up of her chronic pain. She had a history of lumbago secondary to a low back injury sustained 30 years ago, post−laminectomy syndrome (L3−S1), cervical spondylosis, myofascial pain syndrome, and migraine. Her back pain is dull and achy, located in the midline and associated with occasional radiation down both legs. Her regular medications include hydrocodone−acetaminophen 10−325 mg as needed, 5% lidocaine patch 700 mg daily, topiramate 50 mg twice daily, escitalopram 10 mg daily, and celecoxib 100 mg twice daily. She was asked to submit to a routine urine drug test to ensure compliance with her prescription and to rule out the use of other drugs. Urine drug test by LC−MS/MS was positive for hydrocodone (1648 ng/mL) and norhydrocodone (1310 ng/mL). Hydromorphone, a minor metabolite of hydrocodone, was not detected. Review of previous urine drug tests in the patient's record revealed similar result patterns. Why was hydromorphone not detected in this patient's urine? Is this pattern consistent with her use of the hydrocodone as prescribed?

Discussion

Opioid metabolism

Hydrocodone and oxycodone are semisynthetic opioids administered orally as analgesic medications for the treatment of moderate to severe pain. Both drugs produce their analgesic effect by activating the μ-opioid receptors. Hydrocodone and oxycodone themselves are low efficacy agonists at these receptors. They undergo hepatic metabolism via cytochrome P450 3A4 (CYP3A4) (major pathway) and cytochrome P450 2D6 (CYP2D6) (minor pathway) into their major metabolites norhydrocodone and noroxycodone and their minor metabolites hydromorphone and oxymorphone, respectively [1,2].

Oxymorphone has a 40- to 45-fold higher μ-opioid receptor binding affinity than oxycodone and is 14 times more potent than oxycodone [3,4]. Similarly, hydromorphone is reportedly responsible for the analgesic effects of hydrocodone and is 30 times more potent than hydrocodone [5–7].

Polypharmacy poses a major challenge in the management of patients with chronic pain especially those requiring opioids [8]. Many commonly prescribed and over the counter medications may interfere with the activity of the CYP isoenzymes that are involved in the metabolism of opioids. Some of these medications may induce the CYP isoenzymes, while others may inhibit the activity of these enzymes [5]. Therefore, before prescribing opioids to a patient, care providers must carefully review all the medications the patient is taking at the time. Periodical review of patient's medications should be done at intervals to assess for any changes.

Selective serotonin reuptake inhibitors such as duloxetine and escitalopram are inhibitors of CYP2D6. In our cases as discussed previously, patients 1 and 2 were taking duloxetine and escitalopram, respectively. In addition to escitalopram, patient 2 was also taking the COX-2 inhibitor, celecoxib, also a known inhibitor of CYP2D6 [5]. Although a weak inhibitor of CYP2D6, escitalopram has been reported to increase the peak plasma concentration of CYP2D6 substrates by 40% (FDA data sheet). As a result, in patients taking a combination of these drugs, the in vivo formation of hydromorphone and oxymorphone may be decreased, leading to concentrations below the limit of detection for these drugs in urine. On the other hand, reduced formation of the metabolites (hydromorphone and oxymorphone) may lead to the accumulation of the parent drugs hydrocodone and oxycodone, thus increasing their toxicity in these patients. The same phenomenon applies to the inactive prodrug codeine. CYP2D6 inhibitors also potentially inhibit the conversion of codeine to its active metabolite morphine resulting in a buildup of codeine and increased potential for toxicity [5].

Furthermore, patients 1 and 2 were taking modafinil and topiramate, respectively. Both medications are inducers of the CYP3A4 enzyme that is responsible for the conversion of hydrocodone and oxycodone to their major metabolites norhydrocodone and noroxycodone respectively [5,9]. Induction of this enzyme can boost its enzymatic activity, resulting in a rapid elimination of the parent drugs with increased formation of norhydrocodone and noroxycodone. Again, this will also occur with codeine, inducing its conversion to the inactive metabolite norcodeine [10].

When a CYP2D6 inhibitor is taken in addition to a CYP3A4 inducer, as in the index clinical cases, the urine toxicology screen is expected to show decreased levels of hydromorphone and oxymorphone with increased ratios of norhydrocodone to hydrocodone and noroxycodone to oxycodone [5,7]. Besides altering the expected urine drug testing profile ratios, this impaired biotransformation is also thought to result in phenotypic differences in drug response, potentially minimizing their analgesic benefits [3,11]. Some studies however report no effect of the alterations to the oxycodone drug profile caused by CYP2D6 inhibitors on achievement of clinical analgesia [2,11]. This is probably due to the analgesic effect of oxycodone itself.

The activity of the CYP isoenzymes is genotype dependent, thus plasma concentrations of these medications and their metabolites are affected to varying degrees. Genetic polymorphism of the CYP isoenzymes also plays a role in metabolism of opioids. For example, 7% of Caucasians have little to no CYP2D6 enzymatic activity. Consumption of hydrocodone and oxycodone in these patients will result in nonformation of hydromorphone and oxymorphone, respectively [1,3,6].

Time after drug consumption should also be taken into account when interpreting drug test results. Cone et al. [1] depicted the detection of hydrocodone and its metabolites in the urine following a single dose administration (20 mg). Urine concentration of norhydrocodone is higher than hydrocodone at almost any given time point. Similar levels of hydrocodone and norhydrocodone, as seen in case two, are only seen very early after drug consumption. Higher ratios of hydrocodone to norhydrocodone and oxycodone to noroxycodone can also be seen at other time points by taking CYP3A4 inhibitors and in patients who are poor metabolizers of these drugs. This will also potentially increase the hydromorphone/oxymorphone concentration [2,12–14]. In patients taking codeine, it may cause toxicity due to the increase in the CYP2D6 pathway, resulting in higher levels of morphine which may be confused with morphine consumption based on the codeine to morphine ratio in the urine [15].

Lastly, the metabolites hydromorphone and oxymorphone undergo phase II metabolism via glucuronidation soon after formation, converting to hydromorphone-3-glucuronide and oxymorphone-3-glucuronide respectively. Because these metabolites are glucuronidated, lab methods used to detect them must include a digestion step to release the "free" drug.

In summary, drug–drug interactions and genetic polymorphisms may potentially result in nonresponse which may be overlooked or neglected in clinical practice. Toxicology drug testing and expert interpretation by the clinical toxicologist plays a very key role in bringing to light such unique situations. When necessary, the clinical toxicologist may recommend adjustment or stopping of medications.

References

[1] Cone EJ, et al. Prescription opioids. II. Metabolism and excretion patterns of hydrocodone in urine following controlled single-dose administration. J Anal Toxicol 2013;37(8):486–94.
[2] Kummer O, et al. Effect of the inhibition of CYP3A4 or CYP2D6 on the pharmacokinetics and pharmacodynamics of oxycodone. Eur J Clin Pharmacol 2011;67(1):63–71.
[3] Stamer UM, et al. CYP2D6 genotype dependent oxycodone metabolism in postoperative patients. PLoS One 2013;8(3):e60239.
[4] Samer CF, et al. The effects of CYP2D6 and CYP3A activities on the pharmacokinetics of immediate release oxycodone. Br J Pharmacol 2010;160(4):907–18.
[5] Gudin J. Opioid therapies and cytochrome p450 interactions. J Pain Symptom Manage 2012;44(6 Suppl.):S4–14.
[6] Hutchinson MR, et al. CYP2D6 and CYP3A4 involvement in the primary oxidative metabolism of hydrocodone by human liver microsomes. Br J Clin Pharmacol 2004;57(3):287–97.
[7] Smith HS. Opioid metabolism. Mayo Clin Proc 2009;84(7):613–24.
[8] Taylor Jr. R, et al. Economic implications of potential drug-drug interactions in chronic pain patients. Expert Rev Pharmacoecon Outcomes Res 2013;13(6):725–34.
[9] Nallani SC, et al. Dose-dependent induction of cytochrome P450 (CYP) 3A4 and activation of pregnane X receptor by topiramate. Epilepsia 2003;44(12):1521–8.
[10] Caraco Y, Sheller J, Wood AJ. Pharmacogenetic determinants of codeine induction by rifampin: the impact on codeine's respiratory, psychomotor and miotic effects. J Pharmacol Exp Ther 1997;281(1):330–6.
[11] Monte AA, et al. The effect of CYP2D6 drug-drug interactions on hydrocodone effectiveness. Acad Emerg Med 2014;21(8):879–85.
[12] Hagelberg NM, et al. Voriconazole drastically increases exposure to oral oxycodone. Eur J Clin Pharmacol 2009;65(3):263–71.
[13] Gronlund J, et al. Effect of telithromycin on the pharmacokinetics and pharmacodynamics of oral oxycodone. J Clin Pharmacol 2010;50(1):101–8.
[14] Barakat NH, et al. Relationship between the concentration of hydrocodone and its conversion to hydromorphone in chronic pain patients using urinary excretion data. J Anal Toxicol 2012;36(4):257–64.
[15] Gasche Y, et al. Codeine intoxication associated with ultrarapid CYP2D6 metabolism. N Engl J Med 2004;351(27):2827–31.

Chapter 18.4

Emergence and sudden disappearance of pink: U-47700 in southeast Michigan

Milad Webb[1,2] and Jeffrey Jentzen[1,2]

[1]Wayne County Medical Examiner's Office, Detroit, MI, United States, [2]Department of Pathology, University of Michigan Health System, Ann Arbor, MI, United States

Case description

The decedent, a 23-year-old white female with a history of substance abuse, including heroin, cocaine, marijuana, alcohol, and benzodiazepines, was found unresponsive in a room in her mother's basement. Drug paraphernalia was found near the body and there is no evidence of foul play.

Scene examination: The scene took place in a residential home. The house was clean and well kept. The decedent was located in a furnished bedroom in the basement.

The body is identified as being consistent with a 23-year-old white female. She was laying in the prone position with her pelvis and legs on the bed and her torso and extremities laying over the edge of the bed (Fig. 18.4.1A). The decedent appeared to be reaching for a small box that was under the bed. The body was cool to the touch (80°F) and in full rigor. Lividity was consistent with the position of the body and blanched with pressure. Tardieu spot was present on the upper extremities. A possible injection site was identified on the left antecubital fossa. There were no injuries or trauma identified to account for death. On the bed is a folder with Narcotic Anonymous information, worksheets, and a notebook with personal letters. Once the decedent was removed, it was discovered that the small box under the bed contained a syringe and a spoon (Fig. 18.4.1B).

Death investigation: Per investigation, the decedent had recently been at a drug rehabilitation facility for 90 days of treatment. Her mother had suspicion that she was again started using drugs three weeks prior to the death when she discovered a syringe in her room. The decedent had left the house to attend a Narcotics Anonymous meeting and returned home at approximately 2300 hours. A camera setup in the home captured the decedent going through some luggage at 0330 hours then heading back to her room in the basement. When she did not leave her room that day, her mother went down to look for her at approximately 2100 hours and found her unresponsive.

She called 911 and emergency medical services arrived on the scene to pronounce her dead by protocol.

Postmortem examination: External examination of the decedent found no injury or trauma to account for death. An injection site was confirmed on the left antecubital fossa. Internal examination revealed pulmonary edema and no other significant abnormalities.

Postmortem toxicology: Postmortem toxicology was performed on the decedent's peripheral blood (iliac blood) and urine and was positive for U-47700, fentanyl (2.8 ng/mL), and norfentanyl (0.73 ng/mL).

Discussion

Sold under the street name "Pink," U-47700 (3,4-dichloro-N-[2-(dimethylamino)cyclohexyl]-N-methylbenzamide) is a selective agonist of the μ-opioid receptor. The compound can be used orally, insufflated, or injected intravenously. It was originally developed by the Upjohn Company in the 1970s (Fig. 18.4.2). The patent applications submitted on July 4, 1978 indicated that the drug was intended for use "…for alleviating pain in animals…." The drug is known to be 7–8 fold more potent than morphine [1,2].

FIGURE 18.4.1 The decedent was found in a basement bedroom of a secure residence. She was discovered unresponsive on her bed, reaching for a box on the ground. It was revealed that the box contained drug paraphernalia, a spoon, and a syringe.

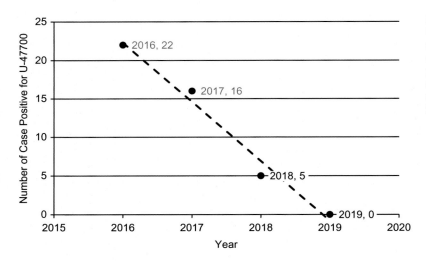

FIGURE 18.4.2 Case confirmed positive for U-47700 from 2016 through February 2019 in southeast Michigan. There was an obvious precipitous decline in the appearance of U-47700 after an initial peak. Concurrently, there was decline in the appearance of carfentanil, and fentanyl derivatives, namely, as furanyl fentanyl. During the same period, fentanyl use has increased exponentially.

No formal studies to evaluate the effects of the drug on humans have been performed. Based on anecdotal reports from users, the drug appears to have a typical opioid profile with analgesia, sedation, euphoria, as well as constipation, itching, and respiratory depression [2–4]. In initial reports, drug users were unaware when the compound was added to or substituted for other illicit preparations. Concentrations and purity of the compound are found to vary widely, and the same quantity of powder may illicit significantly different effects [1].

There are several different subtypes of opioids including naturally occurring compounds (e.g., morphine, codeine), semisynthetic compounds (e.g., hydrocodone, oxycodone), and synthetic compounds (fentanyl, U-47700). All opioids interact with the opioid receptor on neurons. This interaction produces analgesia as well as euphoria. These compounds have multiple side effects including somnolence and profound respiratory depression. Although generally safe when used for short periods and under the supervision of licensed medical providers, because of the potential for abuse and dependence, opioid use has become the preeminent health crisis of this generation [5]. Since 2007, overdose deaths associated with opioid use have skyrocketed without much optimism for future change.

As a result of increasing fatalities noted in 2015 and early 2016, U-47700 received emergency scheduling and was placed into the Schedule 1 of the Controlled Substances Act [6] as of November 2016.

In Wayne County, Michigan (the location of Detroit, Michigan), there were 43 confirmed cases (demographics available in Table 18.4.1) of U-47700 from 2016 to 2018 (Fig. 18.4.2 and Tables 18.4.2 and 18.4.3). The appearance of the compound was highly correlated with mixed drug toxicities, and not a single case was due to single drug toxicity. In 2016 appearance of U-47700 was highly correlated to the appearance of carfentanil (82% of cases), whereas in 2017 fentanyl was the most dominant cotoxicity (94%). In 2018 all cases of U-47700 were accompanied by fentanyl toxicity. Trends show a significant decrease of U-47700 and carfentanil, with an ever-increasing presence of fentanyl as the most dominant opioid of choice.

TABLE 18.4.1 Demographics Data for Cases Positive for U-47700 From 2016 to 2018 in Southeast Michigan.

Demographics: 43 Total Cases From 2016 to 2018		
Age	Mean	43.0
	Median	46
Race	White	28 (65%)
	Black	14
	Other	1
Sex	Male	30 (70%)
	Female	12

The typical user was identified as a white male in their early 40s. Of the 15 nonwhite individuals, only 2 were female. Among black users, 86% were male. Among white users, 63% were male. Average age of black decedents was 55.1 years versus 37.3 years for white decedents. Similar age disparity was seen among black males (55.8 vs 36.7) and black females (51.5 vs 38.3) versus their white counterparts.

TABLE 18.4.2 Distribution of Mix Drug Toxicity with Cases Positive for U-47700 From 2016 to 2018 in Southeast Michigan.

Compound	Number of Cases (%)
Fentanyl	34 (79%)
Cocaine	27 (63%)
Fentanyl derivative	25 (58%)
Carfentanil	19 (44%)
THC	18 (42%)
Ethanol	14 (33%)
Benzodiazepine	14 (33%)
Heroin	13 (30%)
Methamphetamine	8 (19%)
Barbiturate	2 (5%)

THC, 9-Tetrahydrocannabinol.

Blood concentrations of U-47700 in 16 postmortem blood samples were 17–490 ng/mL (average 253 ng/mL). Distribution of U-47700 in postmortem samples revealed a 1.8 ratio (central/peripheral blood) and a liver to peripheral blood distribution of 8.9 L/kg [4]. By standard, a liver/peripheral blood distribution ratio of <5 L/kg indicates a negligible propensity toward postmortem redistribution and 20–30 L/kg indicates significant postmortem distribution. Carfentanil (4-[{1-oxopropyl}-phenyl-amino]-1-[2-phenylethyl]-4-piperidinecarboxylic acid methyl ester) is a synthetic opioid and analogue of fentanyl. It was first synthesized by Janssen Pharmaceuticals (Belgium) in 1974 under the brand name Wildnil and approved in 1986 for veterinary use as an intramuscular tranquilizer for large animals [7]. It is not approved for human use except in the research setting as radiolabeled [11C]-carfentanil, where it is used to map μ-opioid receptors in humans by positron emission tomography. The relative analgesic potency of carfentanil is approximately 10,000 times that of morphine, 4000 times that of heroin, and 20–30 times that of fentanyl. Lethal dose in humans has not been studied, and currently, there is no known safe concentration in humans.

TABLE 18.4.3 Distribution of Mix Drug Toxicity With Cases Positive for U-47700 in 2016 in Comparison to 2017 in Southeast Michigan.

2016		2017	
Compound	Number of Cases (%)	Compound	Number of Cases (%)
Carfentanil	18 (82%)	Fentanyl	15 (94%)
Fentanyl	14 (64%)	Cocaine	11 (69%)
Cocaine	13 (59%)	Fentanyl derivative	11 (69%)
Fentanyl derivative	13 (59%)	THC	5 (31%)
THC	10 (45%)	Ethanol	5 (31%)
Heroin	10 (45%)	Benzodiazepine	5 (31%)
Benzodiazepine	9 (41%)	Methamphetamine	5 (31%)
Ethanol	6 (27%)	Heroin	3 (19%)
Methamphetamine	3 (14%)	Carfentanil	1 (6%)
Barbiturate	2 (9%)	Barbiturate	0

THC, 9-Tetrahydrocannabinol.

In consideration of the investigation, postmortem examination, and postmortem toxicology, the death was certified as fentanyl and U-47700 toxicity and the manner of death was accident. U-47700 is a synthetic opioid, approximately eightfold more potent than morphine. In southeast Michigan the presence of the compound was highly correlated with cotoxicities of carfentanil, fentanyl derivative, such as furanyl fentanyl, and fentanyl. Following a peak in 2015–16, there has been a rapid decline in detected cases.

Opioid abuse and dependence is a crisis of epidemic proportions. Synthetic drugs, such as U-47700 and others like it, are bound to appear and disappear with the whims of their users. It is the job of the medical, public health, and law enforcement communities to do their very best to keep people safe. It is a challenging proposition; however, with improved detection, identification, and epidemiology the task is possible.

References

[1] Rambaran KA, Fleming SW, An J, Burkhart S, Furmaga J, et al. U-47700: a clinical review of the literature. J Emerg Med 2017;53(4):509–19.
[2] Rohrig TP, Miller SA, Baird TR. U-47700: a not so new opioid. J Anal Toxicol 2018;42(1):e12–14.
[3] Ruan X, Chiravuri S, Kaye AD. Comparing fatal cases involving U-47700. Forensic Sci Med Pathol 2016;12(3):369–71.
[4] McIntyre IM, Gary RD, Joseph S, Stabley R. A fatality related to the synthetic opioid U-47700: postmortem concentration distribution. J Anal Toxicol 2016;158–60.
[5] Brown RE, Sloan PA. The opioid crisis in the United States: chronic pain physicians are the answer, not the cause. Anesth Analg 2017;125(5):1432–4.
[6] Controlled Substances Act. 2018.
[7] Leen JLS, Juurlink DN. Carfentanil: a narrative review of its pharmacology and public health concerns. Can J Anesth 2019;66(4):414–21.

Chapter 19

Toxic herbals and plants in the United States

Amitava Dasgupta
Department of Pathology and Laboratory Medicine, University of Texas McGovern Medical School at Houston, Houston, TX, United States

Introduction

According to the World Health Organization (WHO), approximately 80% of the world population relies on herbal medicines. In addition, many patients take herbal medicines concurrently with conventional drugs. For example, in the United States, the concurrent use of herbals and conventional drugs occurs in 20–30% patients [1]. As a result, a relatively safe herbal product such as St. John's wort may cause treatment failure due to drug–herb interactions [2]. In the United States, herbal remedies are sold according to the 1994 Dietary Supplement Health and Education Act where herbals are classified as food supplements. In addition to herbal remedies, food supplements such as vitamins, minerals, amino acids, extracts, and metabolites are also sold under this law. Although manufacturers of herbal remedies are not allowed by law to claim any medical benefits from using such products, at the same time they are not under surveillance by the Food and Drug Administration (FDA). Therefore dietary supplements can be manufactured and sold without demonstrating safety and efficacy to FDA as required for drugs. Unlike drugs that are highly regulated by the FDA, for herbal supplements, FDA cannot take steps against a toxic supplement unless patients have already been harmed. Furthermore, no government agency has the authority to force the manufacturer to provide proof of efficacy of any herbal supplement [3]. There is a common perception that herbal supplements are safe and effective. Although many herbal supplements are safe, scientific studies have shown significant toxicity and even fatality associated with certain herbal medicines.

There are also many toxic plant species. One common example is oleander which grows freely in the Southern part of the United States and also in many other countries. All parts of this plant are toxic and accidental ingestion may cause severe toxicity and even death. Similarly, lily of the valley plant that grows in a relatively cooler climate especially in Northern parts of the United States is also toxic. Despite toxicity, extracts of both oleander and lily of the valley are used in some herbal preparations. Other poisonous plants include castor bean, monkshood, foxglove, and Jimson weed. This chapter focuses on toxic herbal supplements and plants.

Popularity and safety of herbal supplements

The popularity of using herbal supplements is steadily increasing among the general population in the United States where it has been estimated that roughly 20,000 herbal products are available. The 10 most commonly used herbal supplements are echinacea, ginseng, ginkgo biloba, garlic, St. John's wort, peppermint, ginger, soy, chamomile, and kava [4]. Population survey also indicates that one-third to one-half Americans take dietary supplements. Users are more likely to be women, non-Hispanic whites, and more financially secure than nonusers. The annual sale of herbal supplements has increased from $9.6 billion in 1994 to $36.7 billion in 2014 [5].

In general, vitamin and calcium supplements are relatively safe if consumed as per the prescribed dosage. Moreover, some herbal supplements are safe and effective. GRAS stands for "generally recognized as safe," a designation used by the FDA to label a food additive as safe for human consumption. GRAS exemptions are granted for substances that are generally recognized among experts as safe food additives or if a substance was used as a food prior to

TABLE 19.1 Examples of Herbal Supplements With Known Toxicity

Organ Toxicity	Specific Herbal Supplement	Reported Death
Liver	Kava, comfrey, chaparral, germander, coltsfoot, mistletoe, pennyroyal, skullcap, Lipo Kinetix, kombucha tea, also known as Manchurian fungus tea), green tea extract, some weight loss products	Reported fatality with kava, coltsfoot, germander, and pennyroyal oil
Cardiac toxicity	Ma huang (ephedra), oleander containing herbs, lily of the valley containing herbs, Chinese medicine Chan Su, thunder god vine, aconite containing Chinese herbs/monkshood	Reported fatality with ma huang, oleander, Chan Su, Thunder God Vine and aconite containing Chinese herbs
Nephrotoxicity	Aristolochic acid containing herbs indicated for weight loss such as wild ginger, snake weed, and birthroot, sassafras, calamus, juniper berry, horse chestnut, wormwood oil	No known fatality
Neurological	Wormwood oil	No known fatality
Hypertension	Yohimbe bark	No known fatality
Metabolic acidosis	Oil of wintergreen and methyl salicylate containing Chinese medicines	May be fatal

January 1958. GRAS status may be granted to herbal supplements after evaluating the safety record provided by the manufacturer. FDA may grant "GRAS" status by saying no comments to the claim or deny the requested GRAS status if FDA officials determine that the herbal supplement does not have an adequate safety record. In this case, they notify the manufacturer their decision and cause of action. Many herbal products sold today in the United States fall in the gray zone and do not have GRAS status, for example, flaxseed oil, which is generally considered as a safe laxative, does not have a GRAS status.

Certain herbal supplements are also effective in treating certain conditions. For example, silymarin, active ingredient of milk thistle is effective in treating liver diseases including fatty liver disease, chronic liver disease, cirrhosis of liver, and hepatocellular carcinoma [6]. Insomnia is one of the most common sleep disturbances defined as poor sleep quality, difficulty to fall asleep, difficulty to maintain sleeping, and early awakening. However, herbal supplements such as valerian, hop, and jujube are effective to alleviate symptoms of insomnia [7].

Toxic herbal supplements: an overview

However, many herbal supplements also have known toxicity. The most common toxicity associated with the use of herbal supplements is liver toxicity. Drug-Induced Liver Injury Network has determined that between 2003 and 2013, herbal related liver injury increased from 8% to 20% in the United States [8]. Severe hepatotoxicity may result from use of herbs such as kava, chaparral, comfrey, germander, pennyroyal, and mistletoe. Several deaths have been reported due to use of herbal antidepressant kava. Another major toxicity associated with use of herbal supplement is cardiac toxicity. Herbal supplements containing oleander extract or lily of the valley extract may cause severe cardiac toxicity. Although ephedra alkaloid is banned by the FDA, some Chinese herbal supplements still available in the United States may contain ephedra. Taking such products may cause cardiac toxicity. Toxic herbs are listed in Table 19.1 including herbs associated with fatality.

Herbs toxic to the liver

Most commonly reported herbal supplement causing liver toxicity is kava (Table 19.2). Toxicity after use of kava is mostly attributable to kavalactones where six kavalactones, methysticin, 7,8-dihydromethysticin, kavain, 7,8-dihydrokavain, yangonin, and desmethoxyyangonin, account for approximately 96% of the total kavalactones. Commercial products are usually ethanol or acetone extracts, while traditional extracts used in the South Pacific have been water extracts. A major survey of 93 clinical case reports was conducted by the WHO in 2007 where 7 patients died and 14 patients had liver transplants. Eight cases were found to have a close association between the use of kava

TABLE 19.2 Active Ingredients of Common Toxic Herbs/Plants

Toxic Herb/Plant	Active Ingredient
Kava	Six kavalactones—methysticin, 7,8-dihydromethysticin, kavain, 7,8-dihydrokavain, yangonin, and desmethoxyyangonin—account for approximately 96% of the total kavalactones
Comfrey	Comfrey contains as many as 14 pyrrolizidine alkaloids, including 7-acetylintermedine, 7-acetyllycopsamine, echimidine, intermedine, lasiocarpine, lycopsamine, myoscorpine, symlandine, symphytine, and symviridine
Chaparral	Nordihydroguaiaretic acid
Pennyroyal oil	Pulegone
Coltsfoot	Sesquiterpenoids, such as tussilagone
Kombucha/Manchurian tea	Usnic acid
Concentrated green tea extract	Epigallocatechin 3-gallate in high concentration may cause liver damage but drinking green tea is safe.
Ma huang	Ephedra alkaloids
Oleander plant/herbals containing extract	Oleandrin
Lily of the valley plant/herbals containing extract	Convallatoxin
Chan Su	Bufalin and related compounds
Monkshood plant/herbals containing extract	Aconite
Thunder God Vine	Celastrol
Herbal weight loss products	Some Chinese weight loss products contain nephrotoxic aristo cholic acid
Wormwood oil	3,5-dicaffeoyl-epi-quinic acid
Yohimbe	Yohimbine, an indole alkaloid
Oil of wintergreen	Methyl salicylate
Castor bean	Ricin
Jimson weed (Datura stramonium)	Various alkaloids including atropine, hyoscyamine, and scopolamine
Poison hemlock	2-Pentyl-3,4,5,6-tetrahydropyridine, 5-hydroxy-2-pentylpiperidine, 2-pentylpiperidine
Foxglove	Digitoxin, purpurea A, lanatoside, purpurea B and maybe a trace amount of digoxin

and liver dysfunction because the patients recovered on withdrawal of kava. Fifty-three cases were classified as having a possible relationship, but they could not be fully assessed due to insufficient data or other potential causes of liver damage. Ketola et al. reported a case of suicide due to overdose of kavalactones using i.v. injection together with administration of ethanol. The authors confirmed the presence of kavalactones in femoral blood using both gas chromatography/mass spectrometry (MS) and liquid chromatography (LC) combined with MS [9]. Regular consumption of kava has been associated with increased concentrations of γ−glutamyltransferase suggesting potential hepatotoxicity. Escher et al. described a case in which severe hepatitis was associated with kava use. A 50-year-old man took three to four kava capsules daily for 2 months (maximum recommended dose—three capsules). Liver function tests showed a 60−70-fold increase in AST and ALT. Tests for viral hepatitis (HAV, HBV, and HCV) were all negative as were the tests for CMV and HIV. The patient eventually received a liver transplant [10].

Escher et al. presented a case of a 56-year-old woman who had a two-week history of fatigue, nausea and increasing jaundice after taking herbal supplement for anxiety, prescribed and provided by a naturopath (Kava 1800 Plus, Eagle Pharmaceuticals, Castle Hill, NSW; one tablet thrice daily, labeled as containing kavalactones 60 mg, *Passiflora incarnata* 50 mg, and *Scutellaria lateriflora* 100 mg). She was diagnosed with benign monoclonal gammopathy 12 months ago. However, a repeated bone marrow biopsy did not suggest the presence of multiple myeloma. A trans-jugular liver biopsy performed on the fifth day of admission showed nonspecific severe acute hepatitis with panacinar necrosis and collapse of hepatic lobules. Over the subsequent week, the patient's condition deteriorated and she received a liver transplant on day 17. Unfortunately, the procedure was complicated by massive bleeding that did not correct following implantation of the donor liver, and the patient died of progressive blood loss, hypotension and circulatory failure. Subsequent analysis of the supplement confirmed the presence of kava and her cause of death was attributed to fulminant hepatic failure due to use of kava [11].

Pyrrolizidine alkaloids are associated with severe liver toxicity. There are over 6000 plant species that contain pyrrolizidine alkaloids are more than 300 pyrrolizidine alkaloids have been characterized [12]. Poisoning caused by these toxins is associated with acute and chronic liver damage. *Tussilago farfara* (coltsfoot), *Petasites hybridus* (common butterbur), *Senecio vernalis* (eastern groundsel) and *Symphytum officinale* (comfrey) are traditional herbal supplements that contain toxic pyrrolizidine alkaloids [13]. Coltsfoot is often found in combination with other products and is used as a cough remedy. There are several reports of death due to the use of coltsfoot supplement [14]. Sesquiterpenoids, such as tussilagone, are the major active compound of coltsfoot [15]. Comfrey is a toxic herbal supplement due to the presence of toxic pyrrolidine alkaloids. Comfrey contains as many as 14 pyrrolizidine alkaloids, including 7-acetylintermedine, 7-acetyllycopsamine, echimidine, intermedine, lasiocarpine, lycopsamine, myoscorpine, symlandine, symphytine, and symviridine. In addition to liver toxicity, these toxins also exhibit genotoxicity and carcinogenicity but mechanism of such toxicity is not fully understood [16]. Russian comfrey is even more toxic than comfrey that is found in Europe and Asia because it contains more potent hepatotoxins. The liver disease caused by comfrey is called veno-occlusive disease which may lead to liver cirrhosis and eventually liver failure [17]. A 55-year-old woman taking 1600 mg per day of germander became jaundiced after 6 months and was admitted to the hospital. On admission, she showed highly elevated bilirubin of 13.9 mg/dL. Her liver enzymes were also elevated (ALT:1500, AST: 1180 U/L). However, prothrombin time was normal. Results of serological test for hepatitis A, B, and C were negative. A liver biopsy showed bridging necrosis, and collapse, lobular and portal-tract inflammatory infiltration with polymorphonuclear and mononuclear cells, and mild portal tract fibrosis. Germander therapy was discontinued and hepatitis resolved in 2 months [18].

Fatal hepatitis in a 68-year-old woman was related to use of the herbal weight loss product Tealine which contained hepatotoxic germander extract [19]. Chaparral is a plant found in Southwestern United States and Northern Mexico and leaves of this plant are used as an herbal supplement indicated for treating a wide variety of symptoms from cold sore to muscle pain and is also used as an antioxidant, anti-HIV and anticancer herbal product. It can be taken as dried extract in capsule or tablet form. Leaves, stems, and bark in bulk are also available for brewing tea. The active ingredient of chaparral is nordihydroguaiaretic acid and has antiinflammatory and antioxidant properties However, this product can cause severe hepatotoxicity. Several reports of Chaparral-associated hepatitis have been reported. In one case report, a 45-year-old woman took Chaparral 160 mg/day for 10 weeks presented to the hospital with jaundice, anorexia, fatigue, nausea, and vomiting. Liver enzymes and other liver function tests showed abnormally high values indicating severe liver injury. [20]. Sheikh et al. reviewed 18 cases of illness from using chaparral and found that 13 cases of hepatotoxicity related to chaparral. Clinical presentation of these cases is significant elevation of liver enzymes and other liver markers 3–52 weeks after ingestion of chaparral and after stopping use of chaparral liver enzymes return to normal values. However, the authors also commented that chaparral induced liver injury may also be irreversible and may cause liver failure requiring liver transplant [21].

Pennyroyal is a widely available herb that has long been used as an abortifacient despite its potentially lethal hepatotoxic effects. Anderson et al. reported four cases of pennyroyal ingestion where one patient died, one received N-acetylcysteine, and two ingested minimally toxic amounts of pennyroyal and were not treated with N-acetylcysteine. In the fatal case, postmortem examination of a serum sample, which had been obtained 72 hours after the acute ingestion, identified 18 ng of pulegone per mL and 1 ng of menthofuran per mL. Review of 18 previous case reports of pennyroyal ingestion documented moderate to severe toxicity in patients who had been exposed to at least 10 mL of pennyroyal oil [22].

Lipo Kinetix has been promoted for weight loss by increasing metabolism and as an alternative to exercise. This product contains phenylpropanolamine (which is banned by FDA in all over the counter cold medications), caffeine, yohimbine, diiodothyronine, and sodium usniate. In 2002, seven patients who were using Lipo Kinetix developed acute hepatitis which was due to usnic acid [23]. Moreover, usnic acid is present in kombucha tea (also known as Manchurian Mushroom or Manchurian Fungus tea), which is prepared by brewing kombucha mushroom in sweet black tea. Acute liver damage due to drinking of this tea has been reported [24].

Epigallocatechin 3-gallate (EGCG), the major catechin present in green tea, is an effective antioxidant. Consumption of green tea is not associated with liver damage in humans, and green tea infusion and green tea−based beverages are considered safe. The tolerable upper intake level of 300 mg EGCG/person/day is proposed for food supplements. However, taking green tea extract where EGCG intake level is much higher than 300 mg/day (upper tolerable level is usually considered as 600 mg/day, liver toxicity may be encountered) [25]. Since 2006, there have been more than 50 reports in the medical literature of clinically apparent acute liver injury with jaundice attributed to green tea extracts [26].

In 2001 FDA issued warning regarding liver toxicity from use of comfrey, in 2001 liver toxicity associated with use of Lipo Kinetix (weight loss product), in 2002 liver toxicity due to use of kava, in 2009 liver failure due to use of weight loss product hydroxycut, in 2011 waning against use of uprizing because it contained anabolic steroid methyl drostanolone and in 2013 acute liver failure due to use of weight loss product OxyElite Pro [27].

Cardiotoxic herbs

Some plants as well as herbal products derived from such plants are cardiotoxic. The oleanders are evergreen ornamental shrubs with various colors of flowers that belong to the Dogbane family and grow in the Southern parts of the United States from Florida to California, Australia, India, Sri Lanka, China, and also other parts of the world. All parts of oleander plants are toxic. Drinking oleander tea may be fatal. There is a case report of death of a 30-year-old woman after consuming herbal tea she believed she prepared from eucalyptus. Unfortunately, she prepared herbal tea using oleander leaves that are very toxic. However, despite aggressive treatment and cardiac massage in the hospital, her cardiac rhythm deteriorated to an agonal rhythm and then to asystole. Despite inserting a transthoracic pacemaker, the patient died. Postmortem examination revealed serum digoxin level of 6.4 ng/mL. However, the patient never received digoxin. The cause of death was accidental oleander poisoning [28].

Unfortunately, despite toxicity, oleander is still used in some herbal medicines. Boiling or drying the plant does not inactivate the toxins. Oleander contains oleandrin which is similar in structure to the cardioactive drug digoxin and is responsible for cardiac toxicity after accidental ingestion of oleander leaves or taking herbal supplement containing oleander plant extract. False serum digoxin levels after ingesting oleander plant have been reported [29,30]. Because of the structural similarity with digoxin, severe oleander poisoning can be treated with the same antidote (Digibind) as digoxin poisoning. However, Digibind is expensive and there is a severe shortage of such antidotes in countries such as Sri Lanka causing death from the oleander poisoning [31].

The lily family is composed of 280–300 genera made up of 4000–4600 different species but only 90 genera representing approximately 525 species are found in North America. Lilies are popular decorative plants and are also found in floral arrangements. Lily of the valley usually grows in the cooler climate of North America and also in Europe and certain parts of Asia. The entire plant is toxic, containing cardiac glycosides causing digitalis-like toxicity. The principle toxic cardiac glycoside found in lily of the valley is convallatoxin. There are several reports of lily of the valley poisoning [32]. Lethal poisoning of lily of the valley ingestion in a dog has also been reported [33]. Despite toxicity, lily of the valley extract is used in herbal medicine despite as a tonic for heart [34]. Convallatoxin, the major cardiac glycoside present in lily of the valley, has digitalis-like properties by inhibition of the Na, K-ATPase, thereby causing a positive inotropic effect [35]. Convallatoxin has structural similarity with digoxin, and as a result rapid detection of convallatoxin in human serum using luminescent oxygen channeling technology based digoxin assay (LOCI Digoxin assay) has been reported [36]. Fink et al. also compared five digoxin assays for rapid detection of convallatoxin [37].

Poisoning from foxglove (*Digitalis purpurea*) may be accidental, intentional or due to drinking of herbal tea. A 55-year-old woman presented to the emergency department with nausea, vomiting, generalized weakness and lightheadedness eight hours after drinking "comfrey" herbal tea. Physical examination did not reveal any abnormal findings. Laboratory studies showed elevated potassium 7.6 mEq/L and initial electrocardiogram revealed a first-degree atrioventricular conduction block with interatrial conduction delay and premature atrial contractions. Chest radiograph showed mild bilateral pulmonary congestion. Initial therapy for hyperkalemia included intravenous calcium gluconate, insulin, dextrose, and sodium bicarbonate. A serum digoxin level was 151.2 ng/mL was also observed although the patient was not receiving digoxin. The patient developed ventricular fibrillation and despite aggressive therapy and treatment with eight vials of Digibind she expired on hospital in day 7. A botanist confirmed that the plant was foxglove [38].

Another case of foxglove poisoning due to ingestion of herbal tea in an 85-year-old man produced a serum digoxin level of 1.8 ng/mL and potassium of 5.4 mEq/L. The serum digitoxin concentration on the second day was 59.0 ng/mL [39]. Glycoside composition of foxglove extract contains digitoxin, purpurea A, lanatoside, purpurea B and maybe a trace amount of digoxin. Rich et al. presented a case where a 22-year-old man presented to the emergency room after intentional overdose of a homemade foxglove extract. The patients showed a serum digoxin level of 7.82 ng/mL (FPIA) and serum digitoxin level of 172.5 ng/mL. The patient was treated initially with 10 vials of Digibind (800 mg FAB) and then 24 hours later with an additional 400 mg of Digibind. The patient was eventually discharged indicating that Digibind can be used as an antidote in treating foxglove poisoning [40].

Chinese medicine Chan Su is prepared from the dried white secretion of the auricular glands and the skin glands of Chinese toads (Bufo melanostictus Schneider or Bufo bufo gargarzinas Gantor). Chan Su is also a major component of traditional Chinese medicines Liu-Shen-Wan and Kyushin [41]. These medicines are used in China for the treatment of tonsillitis, sore throat, furuncle, palpitation, etc. because of their anesthetic and antibiotic action. Traditional use of Chan Su given in small doses also includes stimulation of myocardial contraction, antiinflammatory effect and pain relief. The pharmacological effect of Chan Su is due to bufalin and cinobufagin but bufalin is also very toxic. A 25-year-old woman with a previous history of miscarriage was pregnant again and drank a bowl of herbal tea (approximately 100 mL) as prescribed by a Chinese herbalist. She complained about nausea and then vomited with increasing abdominal pain and was admitted to the hospital. Approximately 30 minutes after admission, the patient became pale, bradycardiac, and progressively hypotensive. After administration of atropine, sinus tachycardia developed followed by cardiopulmonary arrest. Despite all efforts, she died 2.5 hours after admission. Extensive postmortem toxicology analysis revealed an apparent digoxin level of 4.9 ng/mL and mouse bioassay showed that herbal tea was very toxic. Mass spectrometric analysis identified bufadienolides as toxic component of herbal tea which originated from Chinese remedy Chan Su. Cause of death was due to ingestion of toxic herbal tea containing Chan Su [42].

Although not related to Chan Su, a 23-year-old man died from heart failure after ingestion of a West Indian aphrodisiac "Love Stone." Chemical analysis showed the presence of a controlled substance bufotenine as well as bufalin and related compounds found in Chan Su [43]. Bufalin cross reacts with digoxin immunoassays. Therefore Chan Su poisoning can indirectly be determined by using digoxin immunoassays in a patient not taking digoxin [44].

Aconite is derived from a perennial plant genera Aconitum (monkshood) and is traditionally used for treating many illnesses (pain reducer, antirheumatic, neurological problems, etc.) and also for detoxification. Aconite is also used as an active ingredient in Homeopathic remedies. Because in Homeopathic remedies aconite is used in extremely small amount, toxicity from use of aconite containing Homeopathic remedies is unlikely. However, severe toxicity and even death may occur from using herbal supplement containing much higher amount of aconite compared to homeopathic remedies. Aconite is a very toxic plant alkaloid which is well known for causing arrhythmias and even death. Aconite is also very toxic to the nervous system. The lethal dose of aconite in human is 2 mg to 6 mg [45,46]. One healthy 25-year-old man died after eating wild berries during a nature walk. At autopsy, the nail beds were noted to be cyanosed. Internal examination showed severe congestion of all organs. Histologic examination revealed bilateral massive intrapulmonary hemorrhage and edema. During postmortem analysis, urine toxicology screen was negative. However, his friends were able to identify the berries as *Aconitum napellus* (monkhood) indicating possibility of aconitine poisoning. A LC tandem MS (LC−MS/MS) analyses confirmed the presence of aconitine which was in the postmortem femoral blood sample at a concentration of 3.6 ng/mL and in the urine sample at a concentration of 149 ng/mL. The cause of death was aconite poisoning [47].

Ephedra is a small perennial shrub with thin stems which rarely grows over one foot. Some of the better-known species include *Ephedra sinica* and *Ephedra equisetina* (collectively called ma huang) from China. Ephedrine is the predominant active compound found in the ephedra plants although other compounds such as pseudoephedrine, norephedrine, and phenylpropanolamine are also found. Ephedra is a major component of ma huang, a Chinese weight loss product. Many reports of toxicity from consuming ephedra-containing products have been published. Haller and Benowitz evaluated 140 reports of ephedra related toxicity that were submitted to the FDA between June 1997 and March 31, 1999. The authors conclude that 31% of the cases were definitely related to ephedra toxicity and another 31% were possibly related. Hypertension was the single most frequent adverse reaction followed by palpitation, tachycardia, stroke, and seizure. Ten events resulted in death and 13 events caused permanent disability. The authors conclude that use of a dietary supplement that contains ephedra may pose a serious health risk [48]. Other than death ephedra can increase blood pressure and can cause cardiac arrhythmia and stroke. On April 12, 2004 the FDA prohibited the sale of ephedra-containing dietary weight loss supplements in the United States [49].

Thunder God Vine (lei gong teng prepared from the plant *Triptergium wilfordii*) has been used in traditional Chinese medicine for over 2000 years for local treatment of arthritis and inflammatory tissue swelling and treating cancer. This supplement has also been indicated for treating rheumatoid arthritis. Celastrol, an active compound extracted from the root bark of Thunder God Vine, was found to inhibit tumor cell growth and promote apoptosis in several tumor cell lines [50]. Unfortunately, Thunder God Vine can cause severe adverse reactions and also is poisonous if it is not carefully extracted from the skinned roots. Other parts of the plant such as leaves, flowers, and skin of the root are very toxic to human and may cause death if ingested. A 36-year-old man was admitted to the hospital with severe diarrhea and vomiting for 3 days. Three days before his admission to the hospital he consumed an herbal supplement. The person died 15 hours after admission to the hospital due to shock, hypotension and cardiac damage. The herbal supplement the patient was taking was identified as thunder god vine [51].

Herbs with nephrotoxicity

In 1993 rapidly progressing kidney damage was reported in a group of young women who were taking pills containing Chinese herbs while attending a weight loss clinic in Belgium. It was discovered that one prescription Chinese herb has been replaced by another Chinese herb containing aristolochic acid, a known toxin to kidney [52]. Later, there were many reports of kidney damage due to use of herbal supplements contaminated with aristolochic acid in the medical literature. National Kidney foundation stated that wormwood plant, sassafras, Chinese herbal medicine Tung Shueh and horse chestnut may be toxic to kidney. There are several herbal supplements that are known to cause hematuria and urine proteinuria. Examples of these herbs are aloe juice from leaf, kava, and saffron many other herbs such as calamus, chaparral, horse chestnut seed, and wormwood oil may cause nephrotoxicity [53].

Other toxic herbs

The medical use of the wormwood plant *Artemisia absinthium* L. dates back to at least Roman times, while during the last century, this tradition was seemingly on the decline due to fears of absinthism, a syndrome allegedly caused by the wormwood-flavored spirit absinthe and more specifically as a result of thujone, a monoterpene ketone often present in the essential oil of wormwood which is a neurotoxin [54]. Recently, active ingredient of wormwood oil has been determined to be 3, 5-dicaffeoyl-epi-quinic acid [55].

Extracts of the bark of the central African tree Pausinystalia yohimbe contain yohimbine, an indole alkaloid, which is used to treat erectile dysfunction. The reported side effects of over-the-counter preparations of yohimbine include gastrointestinal upset, anxiety, increased blood pressure, headache, agitation, rash, tachycardia, and frequent urination [56].

Methyl salicylate is a major component of oil of wintergreen which is prepared by distillation from wintergreen leaves. Methyl salicylate has an analgesic effect and is used in many over-the-counter analgesic creams or gels designed only for topical use. Methyl salicylate if ingested is very poisonous. Although aspirin is acetylsalicylate which is structurally close to methyl salicylate, acetylsalicylate after ingestion rapidly breaks down into salicylic acid in our blood and also by the liver. Salicylic acid is responsible for the analgesic effect of acetylsalicylate. In contrast, methyl salicylate is very irritating to the gastric mucosa and very little methyl salicylate is broken down to salicylic acid by the enzymes in our blood. The majority of methyl salicylate is converted into salicylic acid by the liver.

Oil of wintergreen contains 98% methyl salicylate and ingestion of 5 mL of oil of wintergreen (1 teaspoon) is equivalent to 7000 mg of aspirin ingestion and for a child less than 6 years of age one teaspoon or less of oil of wintergreen has been implicated in several well-documented deaths. Methyl salicylate is a relatively common cause of poisoning of a child [57]. Many Chinese medicines and medicated oils contain high amounts of methyl salicylate Chinese medicated oils such as white flower oil and red flower oil contain high amounts of methyl salicylate and severe salicylate toxicity due to ingestion of white or red flower oil has been reported in adults including one death due to ingestion of red flower oil [58]. Parker et al. reported another case where a 58-year-old Vietnamese woman who lived alone in an apartment in San Diego died. Her brother discovered the body and called the emergency service. The paramedics confirmed death at the site and the police officer discovered a coffee cup containing a clear liquid with menthol-like odor, and several empty medicine bottles including one Chinese herbal medicine bottle on a bedside table. All of them were submitted to the Medical Examiner's Office for investigation. High amounts of salicylic acid were found in the blood and gastric content of the decedent. The Chinese herbal medicine bottles were identified as Koong Hung Yick Far oil containing 67% oil of wintergreen and Po Sum oil containing 15% menthol. Methyl salicylate was also present in the gastric content of the decedent confirming that the cause of death was ingestion of methyl salicylate-containing Chinese-medicated oil [59].

Poisonous plants

There are many poisonous plants in the United States and detail discussion on this topic is beyond the scope of this chapter. In this section, toxicity due to ingestion of common toxic plants is addressed.

Ingestion of jimson weed (*Datura stramonium*) may cause severe toxicity because it contains a variety of alkaloids including atropine, hyoscyamine, and scopolamine. Other species of Datura are also toxic and patients usually present with the symptoms of atropine intoxication. Elevated levels of liver enzyme may be observed in these patients through an atropine induced hepatotoxicity. In one patient, AST, ALT, CK, and LDH were 1829, 2052, 6970, and 1087 U/L on the second day of admission where initial concentrations were 370, 304, 5739, 1014 U/L. The mechanism of elevation in CK concentration is not clear [60]. The patient recovered with supportive treatment. The use of antidote in the management of poisoning with jimson weed is controversial. Physostigmine can be used as an antidote but its use has been associated with significant problems such as seizures, bradyarrhythmias, and cholinergic crisis [61]. A previously healthy 2-year-old boy was brought to the emergency department by paramedics after unexplained loss of consciousness and difficulty breathing. He ate low-lying green leafy plant about 6–8 in. tall on the walk home from preschool with his sister. An erythematous rash was present on his right arm, neck, and face. High-flow oxygen via bag valve mask improved his saturations to 100%. He was endotracheally intubated for respiratory failure using atropine 0.3 mg (0.02 mg/kg), etomidate 4.5 mg (0.3 mg/kg), and rocuronium 9 mg (0.6 mg/kg). Some remnants of the plant were brought from the child's house by his father for examination. An Internet search by the health care providers preliminarily identified the plant as poison hemlock, and this was confirmed by a master gardener at the Oregon Poison Center. The patient received supportive therapy and was discharged home that evening after a complete neurologic recovery [62].

Major alkaloids found in poison hemlock include 2-pentyl-3,4,5,6-tetrahydropyridine and 5-hydroxy-2-pentylpiperidine and 2-pentylpiperidine [63].

Ricin is a protein toxin derived from the castor bean plant *Ricinus communis*. The ricin content of castor beans has been reported to be 1%–5%. The castor oil produced from the beans is popularly used as a purgative. Because ricin is inactivated during oil extraction under heated conditions, it is not expected to be found in the conventional castor bean oil or its related products. However, more than 1000 ricin poisoning cases secondary to intentional castor bean consumption have been reported in the literature [63]. A male teenager with a history of depression and suicide attempts was presented to the emergency department after intending to die by ingesting 200 castor beans mixed with juice in a blender. Eight hours after ingestion, he developed weakness, light-headedness, nausea, and several vomiting episodes but was no longer suicidal and sought medical treatment. The patient was admitted and his initial urine drug screen using immunoassays was negative. However, a comprehensive untargeted drug screen test showed the presence of ricinine, a biomarker for ricin poisoning. Three days after ingestion, the patient was medically cleared for psychiatric disposition. The authors commented that oral ricin ingestion is most likely the route of exposure that emergency physicians may encounter in their daily practice. The orally ingested ricin initially damages the GI system, causing GI tract symptoms (e.g., nausea, vomit, and bloody diarrhea), leading to severe dehydration, visceral organ damage (e.g., liver, kidney, and spleen), and possible death in severe cases [64].

Conclusion

Despite toxicity people use herbal remedies. However, popular herbal remedies such as kava can cause fatality. Major toxicity reported after use of herbal medicines is hepatotoxicity. However, cardiac toxicity and nephrotoxicity after use of herbal medicines have also been reported. Plant poisoning is also another concern especially accidental exposure of toxic plants to toddlers.

References

[1] Choi JG, Eom SM, Kim J, Kim SH, Huh E, Kim H, et al. A comprehensive review of recent studies on herb-drug interaction: a focus on pharmacodynamic interaction. J Altern Complement Med 2016;22:262–79.

[2] Madabushi R, Frank B, Drewelow B, Derendorf H, Butterweck V. Hyperforin in St. John's wort drug interactions. Eur J Clin Pharmacol 2006;62:225–33.

[3] Pray WS. Orin Hatch and the dietary supplement health and education act: Pandora's box revisited. J Child Neurol 2012;27:561–3.

[4] Bent S. Herbal medicine in the United States: review of efficacy, safety and regulation. J Gen Intern Med 2008;23:854–9.

[5] Navarro VJ, Khan I, Bjornsson E, Seeff LB, Serrano J, Hoofnagle JH. Liver injury from herbal and dietary supplements. Hepatology 2017;65:363–73.

[6] Federico A, Dallio M, Loguercio C. Silymarin/silybin and chronic liver disease: a marriage of many years. Molecules 2017;22:E191.

[7] Palmieri G, Contaldi P, Fogliame G. Evaluation of effectiveness and safety of a herbal compound in primary insomnia symptoms and sleep disturbances not related to medical or psychiatric causes. Nat Sci Sleep 2017;9:163–9.

[8] Avigan M, Mozersky RP, Seeff LB. Scientific and regulatory perspectives in herbal and dietary supplements associated hepatotoxicity in the United States. Int J Mol Sci 2016;17:331.
[9] Ketola RA, Viinamäki J, Rasanen I, Pelander A, Goebeler S. Fatal kavalactone intoxication by suicidal intravenous injection. Forensic Sci Int 2015;249:e7–e11.
[10] Escher M, Desmeules J. Hepatitis associated with kava, a herbal remedy. Br Med J 2001;322:139.
[11] Gow PJ, Connelly NJ, Hill RL, Crowley P, Angus PW. Fatal fulminant hepatic failure induced by a natural therapy containing kava. Med J Aust 2003;178:442–3.
[12] Neuman MG, Cohen L, Opris M, Nanau RM, Hyunjin J. Hepatotoxicity of pyrrolizidine alkaloids. J Pharm Pharm Sci 2015;18:825–43.
[13] Şeremet OC, Bărbuceanu F, Ionică FE, Margină DM, GuȚu CM, Olaru OT, et al. Oral toxicity study of certain plant extracts containing pyrrolizidine alkaloids. Rom J Morphol Embryol 2016;57:1017–23.
[14] Dailey A, Johns Cupp M. In: Johns Cupp M, editor. Coltsfoot in "Toxicology and clinical pharmacology of herbal products. Human Press; 2000.
[15] Cao K, Xu Y, Zhao TM, Zhang Q. Preparation of sesquiterpenoids from *Tussilago farfara* L. by high-speed counter-current chromatography. Pharmacogn Mag 2016;12(48):282–7.
[16] Mei N, Guo L, Fu PP, Fuscoe JC, Luan Y, Chen T. Metabolism, genotoxity, and carcinogenicity of comfrey. J Toxicol Env Health B Crit Rev 2010;7-8:509–26.
[17] Stickel L, Seitz HK. The efficacy and safety of comfrey. Public Health Nutr 2000;3(4A):501–8.
[18] Laliberte L, Villeneuve JP. Hepatitis after use of germander, a herbal remedy. Can Med Assoc J 1996;154:1689–92.
[19] Mostefa-Kara N, Pauwels A, Pines E, Biour M, Levy VJ. Fatal hepatitis after herbal tea. Lancet 1992;340:674.
[20] Alderman S, Kailas S, Goldfarb S, Singaram C, Malone DG. Cholestatic hepatitis after ingestion of Chaparral leaves: confirmation by endoscopic retrograde cholangiopancreatography and liver biopsy. J Clin Gastroenterol 1994;19:242–7.
[21] Sheikh NM, Philen RM, Love LA. Chaparral associated hepatotoxicity. Arch Intern Med 1997;157:913–19.
[22] Anderson IB, Mullen WH, Meeker JE, Khojasteh-Bakht SC, Oishi S, Nelson SD, et al. Pennyroyal toxicity: measurement of toxic metabolite levels in two cases and review of the literature. Ann Intern Med 1996;124:726–34.
[23] Favreau JT, Ryu ML, Braunstein G, Orshansky G, Park SS, Coody GL, et al. Severe hepatotoxicity associated with use of dietary supplement. Ann Intern Med 2002;136:590–5.
[24] Gedela M, Potu KC, Gali VL, Alyamany K, Jha LK. A case of hepatotoxicity related to kombucha tea consumption. S D Med 2016;69:26–8.
[25] Dekant W, Fujii K, Shibata E, Morita O, Shimotoyodome A. Safety assessment of green tea based beverages and dried green tea extracts as nutritional supplements. Toxicol Lett 2017;277:104–8.
[26] Mazzanti G, Di Sotto A, Vitalone A. Hepatotoxicity of green tea: an update. Arch Toxicol 2015;89:1175–91.
[27] Navarro VJ, Lucena MI. Hepatotoxicity induced by herbal and dietary supplements. Semin Liver Dis 2014;34:172–93.
[28] Haynes BE, Bessen HA, Wightman WD. Oleander tea: herbal draught of death. Ann Emerg Med 1985;14:350–3.
[29] Dasgupta A, Datta P. Rapid detection of oleander poisoning using digoxin immunoassays: comparison of five assays. Ther Drug Monit 2004;26:658–63.
[30] Dasgupta A, Welsh KJ, Hwang SA, Johnson M, Actor JK. Bidirectional (negative/positive) interference of oleandrin and oleander extract on a relatively new Loci digoxin assay using Vista 1500 analyzer. J Clin Lab Anal 2014;28:16–20.
[31] Eddleston M, Senarathna L, Mohammed F, Buckley N, Juszczak E, Sheriff MH, et al. Deaths due to the absence of an affordable antitoxin for plant poisoning. Lancet 2000;362:1041–4.
[32] Alexandre J, Foucault A, Coutance G, Scanu P, Milliez P. Digitalis intoxication induced by an acute accidental poisoning of lily of the valley. Circulation 2012;125:1053–5.
[33] Moxley RA, Schneider NR, Steingger DH, Carlson MP. Apparent toxicosis associated with lily of the valley (*Convallaria majalis*) ingestion in a dog. J Am Vet Med Assoc 1989;195:485–7.
[34] Haass LF. *Convallaria majalis* (lily of the valley) (also known as our lady's tears, ladder to heaven). J Neurol Neurosurg Psychiatry 1995;59:367.
[35] Choi DH, Kang DG, Cui X, Cho KW, et al. The positive inotropic effect of the aqueous extract of *Convallaria keiskei* in beating rabbit atria. Life Sci 2006;79:1178–85.
[36] Welsh KJ, Huang RS, Actor JK, Dasgupta A. Rapid detection of the active cardiac glycoside convallatoxin of the lily of the valley using LOCI digoxin assay. Am J Clin Pathol 2014;142:307–12.
[37] Fink SL, Robey TE, Tarabar AF, Hodsdon ME. Rapid detection of convallatoxin using five digoxin immunoassays. Clin Toxicol (Phila) 2014;52:659–63.
[38] Wu IL, Yu JH, Lin CC, Seak CJ, Olson KR, Chen HY. Fatal cardiac glycoside poisoning due to mistaking foxglove for comfrey. Clin Toxicol (Phila) 2017;55:670–3.
[39] Dickstein E, Kunkel FW. Foxglove tea poisoning. Am J Med 1980;69:167–9.
[40] Rich SA, Libera JM, Locke RJ. Treatment of foxglove extract poisoning with digoxin-specific fab fragment. Ann Emerg Med 1993;22:1904–7.
[41] Morishita S, Shoji M, Oguni Y, Ito C, Higuchi M, Sakanashi M. Pharmacological actions of "Kyushin" a drug containing toad venom: cardiotonic and arrhythmogenic effects and excitatory effect on respiration. Am J Chin Med 1992;20:245–6.
[42] Ko R, Greenwald M, Loscutoff S, Au A, Appel BR, Kreutzer RA, et al. Lethal ingestion of Chinese tea containing Chan Su. West J Med 1996;164:71–5.

[43] Barry TL, Petzinger G, Zito SW. GC/MS comparison of the West Indian aphrodisiac "Love Stone" in the Chinese medicine Chan Su: bufotenine and related bufadienolides. J Forensic Sci 1996;41:1068–73.
[44] Dasgupta A. Therapeutic drug monitoring of digoxin: impact of endogenous and exogenous digoxin-like immunoreactive substances. Toxicol Rev 2006;25:273–81.
[45] Wang Y, Wang S, Liu Y, Yan L, Dou G, Gao Y. Characterization of metabolites and cytochrome P 450 isoforms involved in the microsomal metabolism of aconite. J Chromatogr B 2006;844:292–300.
[46] Fu M, Wu M, Qiao Y, Wang Z. Toxicological mechanisms of aconite alkaloids. Pharmazie 2006;61:735–41.
[47] Pullela R, Young L, Gallagher B, Avis SP, Randell EW. A case of fetal aconite poisoning by monkshood. J Forensic Sci 2008;53:491–4.
[48] Haller CA, Benowitz NL. Adverse and central nervous system events associated with dietary supplements containing ephedra alkaloids. N Eng J Med 2000;343:1833–8.
[49] Seamon MJ, Clauson KA. Ephedra: yesterday, DSHEA, and tomorrow-a ten year perspective on the dietary supplement health and education act of 1994. J Herb Pharmacother 2005;5:67–86.
[50] Yu X, Zhou X, Fu C, Wang Q, Nie T, Zou F, et al. Celastrol induces apoptosis of human osteosarcoma cells via the mitochondrial apoptotic pathway. Oncol Rep 2015;34:1129–36.
[51] Chou WC, Wu CC, yang PC, Lee YT. Hypovolemic shock and mortality after ingestion of *Tripterygium wilfordii* hook F: a case report. Int J Cardiol 1995;49:173–7.
[52] Vanhaelen JL, Vanhaelen-Fastre R, Nut P, vanhaelen-Fastre R, Vanhaelen M, et al. Rapidly progressive interstitial renal fibrosis in young women: association with slimming regimen including Chinese herb. Lancer 1993;341:387–91.
[53] Blowey DL. Nephrotoxicity of over the counter analgesics, natural medicines, and illicit drugs. Adolesc Med 2005;16:31–43.
[54] Lachenmeier DW. Wormwood (*Artemisia absinthium* L.)—a curious plant with both neurotoxic and neuroprotective properties? J Ethnopharmacol 2010;131:224–7.
[55] Nam SY, Han NR, Rah SY, Seo Y, Kim HM, Jeong HJ. Anti-inflammatory effects of *Artemisia scoparia* and its active constituent, 3,5-dicaffeoyl-epi-quinic acid against activated mast cells. Immunopharmacol Immunotoxicol 2018;40:52–8.
[56] Myers A, Barrueto Jr. F. Refractory priapism associated with ingestion of yohimbe extract. J Med Toxicol 2009;5:223–5.
[57] Davis JE. Are one or two dangerous? Methyl salicylate exposure in toddlers. J Emerg Med 2007;32:63–9.
[58] Chan TY. Ingestion of medicated oils by adults: the risk of severe salicylate poisoning is related to the packaging of these products. Hum Exp Toxicol 2002;21:171–4.
[59] Parker D, Martinez C, Stanley C, Simmons J, McIntyre IM. The analysis of methyl salicylate and salicylic acid from Chinese herbal medicine ingestion. J Anal Toxicol 2004;28:214–16.
[60] Ertekin V, Selimogly MA, Altinkaynak S. A combination of unusual presentation of *Datura stramonium* intoxication in a child: rhabdomyolysis and fulminant hepatitis. [letter]. J Emerg Med 2005;28:227–8.
[61] Francis PD, Clarke CF. Angel trumpet lily poisoning in five adolescents: clinical findings and management. J Paediatr Child Health 1999;35:93–5.
[62] West PL, Horowitz BZ, Montanaro MT, Lindsay JN. Poison hemlock-induced respiratory failure in a toddler. Pediatr Emerg Care 2009;25:761–3.
[63] Radulović N, Dorđević N, Denić M, Pinheiro MM, Fernandes PD, Boylan F. A novel toxic alkaloid from poison hemlock (*Conium maculatum* L., Apiaceae): identification, synthesis and antinociceptive activity. Food Chem Toxicol 2012;50:274–9.
[64] Bozza WP, Tolleson WH, Rivera Rosado LA, Zhang B. Ricin detection: tracking active toxin. Biotechnol Adv 2015;33:117–23.

Further reading

Lopez Nunez OF, Pizon AF, Tamama K. Ricin poisoning after oral ingestion of castor beans: a case report and review of the literature and laboratory testing. J Emerg Med 2017;53:e67–71.

Chapter 19.1

Toxicity due to kava tea consumption in conjunction with alcohol and multiple antidepressants

Theresa Swift[1], Brian Wright[1] and Hema Ketha[2]
[1]Department of Pathology, University of Michigan Health System, Ann Arbor, MI, United States, [2]Center for Esoteric Testing, Laboratory Corporation of America Holdings, Burlington, NC, United States

Case presentation

A 30-year-old male was brought to the emergency room due with an altered mental status by his parents. The patient had a history of chronic anxiety, obsessive compulsive disorder and alcohol abuse. His anxiety and obsessive compulsive disorder were being treated with multiple medications prescribed by his psychiatrist. His prescribed medications included quetiapine, propranolol, naltrexone, mirtazapine, escitalopram, clomipramine, and clonazepam. The patient's parents stated that the patient had been acting normally throughout the day, until approximately 2 hours before his presentation to the hospital when he appeared sweaty, had slurred speech, was confused and lethargic. He had displayed ataxia and his eyes were rolling back into his head. At that time, his parents quickly brought him to the emergency room.

In the emergency room the patient had normal vital signs (temperature: 36.4°C, blood pressure: 121/78 mmHg, heart rate: 87 beats/minute, respiratory rate: 20 respirations/minute SpO_2: 99%). He had a normal heart rate with a regular rhythm. He appeared diaphoretic and lethargic, but he was easily awakened by voice. The patient was extremely anxious and appeared intoxicated upon presentation. His pupils were dilated bilaterally and his speech was delayed. The patient had normal breath sounds, but was breathing slowly. He had no abdominal mass, distension, nor tenderness.

The patient admitted that he did not remember which medication or how much he took. He also admitted to drinking about four drinks of vodka earlier in the day. The patient had no history of suicide attempts nor psychiatric admissions, but he has had multiple admissions for alcohol detoxification. As the patient was observed in the emergency room he became more awake, alert and anxious. The patient was given intravenous fluids and he requested something for his anxiety for which he was given lorazepam and olanzapine. The patient was not given charcoal since the ingestion had occurred over an hour prior to his arrival in the emergency department.

In order to assess the cause of patient's presenting symptoms, his medications were counted and it did not appear that any of the medication was missing or that excessive medication was taken. However, empty bottles of kava root supplement were found within his medication box. The patient admitted to taking the kava root supplement 2–6 times a day to help with his anxiety in addition to his other medications. The patient denied any other illicit drug use (Table 19.1.1).

Urine drug testing results were consistent with the drugs prescribed to the patient, and with what he admitted.

The medical team concluded that that the patient's presentation was due to a combination of kava tea intoxication consumed along with antidepressant medications. Once he was medically cleared, he was voluntarily admitted to the psychiatric unit for medication adjustment, but he signed himself out the next day.

TABLE 19.1.1 Laboratory Results for the Patient Presenting With Kavain Toxicity Consumed in Conjunction With Alcohol and Antidepressants

Ethanol	167 mg/dL
Acetaminophen	<10 µg/mL
Salicylate	<3 mg/dL
Sodium	137 mmol/L
Potassium	4.3 mmol/L
CO_2	31 mmol/L
Urea nitrogen	12 mg/dL
Creatinine	1.17 mg/dL
AST	27 IU/L
ALT	35 IU/L
Bilirubin, total	0.5 mg/dL
Glucose	120 mg/dL
Calcium	9.7 mg/dL
Venous pH	7.35
pCO_2, venous	58 mmHg
pO_2, venous	24 mmHg
HCO_3, venous	32 mmol/L
Total hemoglobin	15.4 g/dL
Urine drug screen (immunoassay)	Amphetamine: positive, barbiturate: negative, benzodiazepine: negative, cannabinoid: negative, cocaine: negative, opiate: negative, oxycodone: negative
Qualitative GC–MS urine drug confirmation	Kavain, mirtazapine citalopram, quetiapine, clomipramine, and caffeine

Discussion

Kava is derived from the roots of the *Piper methysticum* plant which is native to islands in the western and south pacific. In Polynesian cultures, kava has been consumed as a drink in ceremonies and recreationally for centuries. In the traditional preparation the root of the plant is either chewed or grinded into a water based suspension. The preparation sold in the United States, Canada, Australia, and Europe is synthetic or prepared from an organic extraction using ethanol or acetone and is in pill form. The pharmacologically active ingredients known as kavalactones are found concentrated in the ribosomes, roots, and root stems of the kava plants. The further from the root of the kava plant the concentration of kavalactones decreases. The dried root of the kava plant is 4%–8% kavalactones [1].

There are approximately 15 kavalactones but 96% of the pharmacologic activity comes from 6 kavalactones; kavain, dihydrokavain, methysticin, dihydromethysticin, yangonin, and demethylxyyangonin. The kava is well absorbed orally, and only trace amounts of the unchanged kavain and 4′-OH kavain are found in urine [1].

Kava can be used in the treatment of patients with anxiety disorders, stress, and insomnia. These active kavalactones can cause relaxation, talkativeness, and euphoria while maintaining mental clarity. Kavalactones pass through the blood brain barrier and can cause behavioral changes at low concentrations. Kavalactones enhance ligand binding to the $GABA_a$ receptor by inhibiting the activation of neurons in the reticular formation and the limbic system. (Stickel) They also block voltage gated Na^+ and Ca^{2+} channels. Blockage of the voltage gated cation channels can decrease neuronal function and can cause sedation. They also interact with monoamine systems by blocking the reuptake of norepinephrine and inhibiting MAO_b. Blocking norepinephrine can help increase alertness [2,3].

Kava has the potential for overdose but it is rare. A 37-year-old male drank several cups of an herbal tea. After eating the man was very dizzy, unable to stand and began vomiting. He was taken to an emergency room where his heart rate and blood pressure were elevated and his pupils were dilated. His only notable lab was an elevated glucose level. The patient's condition improved over the next few hours while he was being observed. The patient admitted to preparing the herbal tea containing kava which was too strong. The kava was the only medication he was taking [4].

Kava is well tolerated but it can have possible side effects include headache, dizziness, drowsiness, and diarrhea. Kava dermatopathy has been reported in 45% of regular kava users and 78% of heavy kava users. Kava dermatopathy is an eruption on the skin that begins as a powdery dryness that progresses to a nonerythematous desquamating keratosis. This dermopathy is not permanent. If a kava user stops consuming kava the dermopathy will resolve.

The most serious of the potential side effects that can occur with kava consumption is hepatotoxicity. From 1990 through 2002 there were 93 case reports of presumed hepatotoxicity associated with kava that were reviewed by the World Health Organization [5]. These cases occurred in the United States, Canada, and Europe. From these cases, 14 patients received a liver transplant and 7 patients died. Eight of the cases had a probable relationship between a liver disorder and the use of kava. Fifty-three of the cases had a possible relationship between a liver disorder and the use of kava. All of the seven deaths were classified as having a possible relationship between the kava and liver disorder. Out of the 93 reviewed cases, only 5 of the cases were prepared using water extracts, the remainder were prepared with an organic extract [5].

In a case of a 50-year-old male, who presented to his physician with complaints of fatigue and jaundice. The patient had been taking kava supplement for 2 months. His liver function tests were markedly elevated and his hepatitis testing was negative. The patient was admitted to the hospital and within 2 days he developed stage 4 encephalopathy and was intubated. He received a liver transplant 2 days later and made a full recovery. His liver showed severe hepatocellular necrosis [6].

Another case involves a 22-year-old female who went to her physician complaining of a three-week history of nausea, fatigue and she had recently become jaundiced. Her only daily medication was an oral contraceptive and kava. She took rizatriptan and acetaminophen occasionally. She had been taking kava for four months before the onset of symptoms. The patient was eventually admitted to the hospital for worsening jaundice and encephalopathy. She tested negative for hepatitis A, B, C, herpes simplex virus, and cytomegalovirus. During her stay at the hospital, her liver failure and encephalopathy worsened. She became comatose and received a liver transplant. The patient was released from the hospital 12 weeks after the transplant; however she died 6 months after the transplant from aspergillin sepsis [7].

After the reported cases of hepatotoxicity lead to changes in the legal status of kava. In the United States an advisory was issued by the FDA regarding the potential for hepatotoxicity but has remained available as an over the counter supplement. In Europe, countries took various actions to regulate the sale of kava. For example, in Germany, kava products were removed from the market in 2002, but in 2014 it became available again, but only as a prescription. As of 2018 Poland was the only country with a ban on the product [5].

There is a large potential for drug interaction with kava. In the United States, kava is sold as an over the counter supplement. People of all educational backgrounds use alternative therapies. They may choose to use them because they did not respond well to traditional medications or they experienced severe adverse side effects, or the withdraw from these medications can be difficult. One national survey suggests that 42% of Americans do not inform their physician of the use of alternative therapies, like kava [8].

Without informing physicians of what medications are being taken prescription or supplement it there is a large potential for a interaction. In one case report a 53-year-old male who was prescribed alprazolam, cimetidine and terazosin took kava for three days prior to being hospitalized for being in a semicomatose state. He denied overdosing on any of the medications. It was believed that the additive effect between the kava and alprazolam that put induced his semicomatose state [9].

Alcohol also appears to have an effect on kava users. One study compared the effects of a placebo, kava, ethanol, and ethanol and kava. The study subject subjectively rated their own level of sedation, cognition, coordination and intoxication. Kava had no measurable effect on sedation, cognition nor coordination. The effect of the combination of the kava and the ethanol was greater on the measured abilities when compared with alcohol alone [10].

To be able to diagnose a possible overdose or drug interaction with kava, communication of all medications taken with a physician is essential. Physicians should also ask their patients if they are taking any supplements in addition to prescribed medication or illegal substances. In most circumstances, there are not any lab abnormalities that a kava user would have. Immunoassay drug screens do not detect kavalactones. Depending on the GC–MS method GC–MS analysis may not detect any of the kavalactones and few hospitals have the instrumentation.

References

[1] Baselt RK. Disposition of toxic drugs and chemicals in man. 6th ed. Foster City, CA: Biomedical Publications. p. 772−3.
[2] Spinella M. The importance of pharmacological synergy in psychoactive herbal medicines. Altern Med Rev 2002;7(2):130−7.
[3] Singh YN, Singh NN. Therapeutic potential of kava in the treatment of anxiety disorders. CNS Drugs 2002;16(11):731−43.
[4] Perez J, Holmes JF. Altered mental status and ataxia secondary to acute kava ingestion. J Emerg Med 2005;28(1):49−51.
[5] World Health Organization. Assessments of the risk of hepatotoxicity with kava products. Geneva: WHO Document Production Services; 2007.
[6] Escher M, Giostra E, et al. Hepatitis associated with kava, a herbal remedy for anxiety. BMJ 2001;322:139.
[7] Brauer RB, Stangl, et al. Acute liver failure after administration of herbal tranquilizer kava-kava (*Piper methysticum*). J Clin Psychiatry 2003;64(2):216−18.
[8] Spinella M. Herbal medicines and epilepsy: the potential for benefit and adverse effects. Epilepsy Behav 2001;2(6):524−32.
[9] Almeida JC, Grimsley EW. Coma from the health food store: interaction between kava and alprazolam. Ann Intern Med 1996;125(11):940−1.
[10] Foo H, Lemon J. Acute effects of kava, alone or in combination with alcohol, on subjective measures of impairment and intoxication and on cognitive performance. Drug Alcohol Rev 1997;16(2):147−55.

Chapter 19.2

Kratom, a novel herbal opioid in a patient with benzodiazepine use disorder

Heather M. Stieglitz[1] and Steven W. Cotten[2]
[1]Department of Pathology, The Ohio State University, Columbus, OH, United States, [2]Department of Pathology and Laboratory Medicine, University of North Carolina at Chapel Hill, Chapel Hill, NC, United States

Case presentation

A 22-year old male with a medical history of depression, attention-deficit/hyperactivity disorder, seizure, and traumatic brain injury presented to the psychiatry department for treatment of benzodiazepine use disorder. The patient appeared well-nourished and clean at presentation. His psychomotor function appeared retarded, but language was well-formed and intact and his affect was blunted, calm, and cooperative. He denied self-harm, delusions, paranoia, and auditory or visual hallucinations. The patient's prior urine drug screen analyses by immunoassay for methadone, opiates, barbiturates, benzodiazepine, and cocaine were all negative. He had a single prior positive urine drug screen result for amphetamine and cannabinoids. After disclosing that he had begun experimenting with drugs purchased on the Internet, his provider ordered a urine novel psychoactive drug screen that was positive for mitragynine (kratom) at a cutoff of 50 ng/mL measured by high performance liquid chromatography time of flight mass spectrometry, (LC-TOF/MS) which was subsequently confirmed by liquid chromatography tandem mass spectrometry with a cutoff of 10 ng/mL. The clinician discouraged the patient from purchasing drugs from online retailers and encouraged continued work toward recovery. The patient returned for five additional follow-up visits over the subsequent 6 months with similar presentation. The patient reported no adverse events related to his continued kratom use and kratom was consistently detected in his urine each at visit.

Discussion

Kratom (*Mitragyna speciosa*) is a psychoactive plant indigenous to certain regions of Southeast Asia where it has been traditionally used for its stimulatory and analgesic effects. In recent years, Kratom has gained popularity in the Western countries as a novel psychoactive substance that can be purchased from smoke shops or online retailers where it has been marketed as an "herbal supplement," "alternative medicine," "opium substitute," and a "legal high" [1]. Traditional methods of consuming kratom by populations where kratom trees grow naturally are by chewing fresh leaves or preparing tea. Typical routes of exposure by people who purchase kratom online are orally by ingesting dried leaves, which may be loose or in capsules, powders which can be added to food or beverages (so-called toss-and-wash method), or by smoking the leaves [2]. Brewed tea from kratom leaves may also be purchased in bars or cafes in regions where kratom remains uncontrolled. The desired effects of kratom reported by users include mental and physical stimulation, mood elevation, and euphoria at lower doses (1–5 g) and sedation and analgesia at higher doses (5–15 g) [3]. Kratom is also commonly used for self-treatment of opioid addiction and withdrawal [4]. Undesired effects are associated with both short- and long-term use and include tachycardia, nausea, constipation, insomnia, temporary erectile dysfunction, xerostomia, pruritus, anorexia, skin darkening, and hair loss.

Kratom contains over 40 different alkaloids, of which the most abundant is mitragynine, followed by paynantheine, speciogynine, and speciocilatine, that differ in composition among the different plant varieties. Mitragynine is an indole alkaloid exhibiting partial agonist activity at the μ-opioid receptor while paynantheine, speciogynine, and speciociliatine are weak competitive antagonists of this receptor [5,6]. None of these alkaloids share structural similarity to opiates [7].

According to in vitro functional assays using transfected HEK cells, mitragynine showed displacement of the full opioid agonist [d-Ala^2, N-Me-Phe^4, Gly-ol^5]-enkephalin (DAMGO) from the human μ-opioid receptor with an EC_{50} = 339 nm suggesting 100-fold less potency than morphine which showed an EC_{50} of 3 nm in this assay [5,6]. The minor alkaloid, 7-hydroxymitragynine, also shows partial agonist activity with the μ-opioid receptor at a potency of 10-fold less than that of morphine as demonstrated by an EC_{50} of 35 nm in the same functional assay [5]. Both 7-hydroxymitragynine and mitragynine are competitive antagonists for the δ- and, to a lesser extent, κ-opioid receptors. Signaling at the μ-opioid receptor is mediated through G protein activation rather than via β-arrestin recruitment, which may contribute to the lack of respiratory depression that is typically observed with other μ-opioid receptor agonists such as morphine or heroin [6]. Mitragynine has also been reported to interact with non-opioid central nervous system receptors, however, the importance of these interactions in mediating kratom's physiological effects remains unknown [6].

Pharmacology

The onset of kratom effects begins at about 10 minutes and full effects are experienced 30–60 minutes following ingestion, and last 5–7 hours [3]. This is consistent with the observed mitragynine T_{max} (time to reach maximum plasma concentration) of 49.8 minutes determined in a small pharmacokinetic study of 10 adult male, chronic kratom users who were given kratom tea for 7 days [8]. This study also demonstrated that mitragynine elimination is consistent with first-order kinetics and the terminal half-life in humans was estimated to be 23.24 hours (standard deviation = 16.07 hours) [8]. The apparent volume of distribution of mitragynine is 38.04 L/kg and an exponential decline was observed in the pharmacokinetic time profiles in most patients evaluated suggesting mitragynine follows an oral two-compartment model [8]. Mitragynine undergoes phase I and II metabolism in humans producing several hydroxylated and carboxylic acid metabolites that are then conjugated to glucuronides and sulfates [9,10]. The concentration of mitragynine metabolites in urine depends on the metabolite detected and analysis method but has generally been measured in the range of 1.2 to >50,000 ng/mL [7,9].

Studies involving kratom extracts have demonstrated potent inhibition of CYP3A4, CYP2D6, and P-glycoprotein which may make it susceptible to interactions with drugs that are metabolized by these enzymes [11–13].

Toxicity

The toxicity of kratom in humans is incompletely understood and most of the data available is from animal studies or case series of self-reported kratom use. These data, however, are subject to limitations including inconsistent results in animal studies, the presence of other health conditions or other drugs in human patient case reports, and the lack of demonstrating the presence of kratom in patient samples at the time of presentation. Nevertheless, several undesired effects following kratom use have been documented and required medical attention. Between 2011 and 2017, 1599 exposures of kratom were reported to the US Poison Control Center, and over 90% were then treated at a health care facility with 31.8% requiring hospital admission [14].

The most common adverse effects following self-reported kratom exposure include agitation, drowsiness, confusion, seizures, tremor, tachycardia, hypertension, nausea, and vomiting [4,14,15]. Drug-induced liver injury has also been attributed to kratom use in cases where other etiologies have been ruled out and signs and symptoms coordinate with kratom use and discontinuation. In some reports the biochemical injury pattern appears consistent with hepatocellular insult, for example, elevated bilirubin, aspartate aminotransferase, and alanine aminotransferase [16], while in other cases, the pattern suggests kratom use can involve an intrahepatic cholestatic process with hyperbilirubinemia [17–19]. In one report, elevated liver enzymes decreased with N-acetylcysteine administration suggesting one mechanism of kratom toxicity in the liver may involve glutathione depletion similar to acetaminophen toxicity [16]. Endocrine abnormalities have also been reported with kratom use and include low testosterone, mildly elevated prolactin, and hypothyroidism, although the mechanism by which kratom is associated with these clinical effects has not been elucidated [20–22].

Withdrawal symptoms have been described with chronic, heavy kratom use suggesting some users may experience physical dependence [2,4,23]. These symptoms include irritability, diarrhea, increased urination, loss of appetite, muscle symptoms, diaphoresis, and mood disturbances. Neonatal abstinence syndrome, a clinical diagnosis most frequently associated with maternal opioid use, has also been reported with kratom use during pregnancy [24–28]. In these cases, newborns were successfully treated with supportive care, morphine, or morphine with benzodiazepines.

While fatalities have been reported with kratom exposure, the lack of a defined lethal mitragynine concentration in humans and simultaneous detection of other drugs in the deceased have made it difficult to determine the contribution of kratom to death.

Kratom analysis in a toxicology laboratory

The lack of structural similarity between kratom alkaloids, mitragynine and 7-hydroxymitragynine and their metabolites with opiates, morphine and codeine, limit the cross-reactivity of kratom compounds in traditional opiate immunoassay drug screens used in the clinical setting. Therefore drug panels that specifically test for mitragynine are required to identify kratom use in patients. Currently, kratom testing in the clinical setting is limited mostly to reference toxicology laboratories that use LC−TOF/MS or tandem mass spectrometry (LC−MS/MS) and can achieve a limit of detection in the 1−10 ng/mL range in urine [9,29]. Qualitative and/or quantitative testing for mitragynine is available in urine, serum/plasma, or whole blood either alone and as part of multidrug panels.

Some of the analytical challenges faced with mitragynine detection using mass spectrometry methods reported in the literature include identifying appropriate and stable metabolites to measure and determining the necessity of enzymatic hydrolysis in sample extraction [30]. One report suggests 17-hydroxymitragynine is too unstable for quantitative analysis; however, mitragynine along with the less psychoactive kratom alkaloids, speciociliatine, speciogynine, and paynantheine are suitable for quantitation [30]. Another report discusses whether an enzymatic hydrolysis step is needed for sample preparation for accurate mitragynine measurement. This study determined in postvalidation studies that in samples with mitragynine >1000 ng/mL, most of the compound was not conjugated to glucuronide, and therefore a hydrolysis step may not be needed depending on the clinical needs of the method [7]. Other chromatographic and/or mass based techniques that have been developed for the detection of mitragynine and its metabolites include liquid chromatography with ultraviolet detection [31], thin layer chromatography [32], gas chromatography mass spectrometry [33], ion mobility spectrometry [34], and direct analysis in real time-mass spectrometry (DART-MS) [35]; however, these methods are not routinely employed in the clinical setting.

Immunochromatographic techniques using monoclonal antibodies raised against mitragynine haptens in lateral flow and ELISA formats have also been described to detect mitragynine, although these methods are not as analytically sensitive as the chromatographic methods and have been limited to research settings [36−38]. One commercially available ELISA assay has been developed to test for mitragynine in whole blood, post mortem blood, or urine with a limit of detection of 0.54 ng/mL. This assay shows the highest reactivity with mitragynine (100%) and has some reactivity with the *o*-desmethyl mitragyinine (18.1%), and 7-hydroxymitragynine (0.4%) compounds as well. A CEDIA method has also been developed to detect kratom in human urine and is available for use with automated clinical analyzers [39]. This assay also shows the highest cross reactivity with mitragynine (100%), and limited cross reactivity with the other kratom alkaloids, 7-hydroxymitragynine (0.14%), paynantheine (0.2%), and speciociliatine (0.2%).

Legality

Kratom remains an uncontrolled substance in the United States at the federal level. In November of 2018, the Department of Health and Human Services recommended to the Drug Enforcement Administration (DEA) that mitragynine and 7-hydroxymitragynine should be banned due to high potential for abuse; however, the DEA has withdrawn scheduling of kratom alkaloids but still lists kratom as a Drug and Chemical of Concern [40]. While kratom remains legal at the federal level, some states and cities have made their own legislation listing kratom as a schedule I substance including Alabama, Arkansas, Indiana, Tennessee, Vermont, Wisconsin, the District of Columbia, Denver, and San Diego [41]. In the European Union, *M. speciosa* and/or mitragynine and/or 7-hydroxymitragynine is controlled in several Member States including Denmark, Latvia, Lithuania, New Zealand, Poland, Romania, and Sweden. Australia, Malaysia, Myanmar, and Thailand also control kratom use and possession [42].

Conclusion

Kratom has become a popular opioid alternative due to its legal status in many regions, availability online, and desired stimulatory, euphoric, and analgesic effects. However, clinicians should be aware that although kratom remains licit in many areas, there is risk for addiction and toxicity in patients who use kratom either alone or in combination with other prescription or illicit substances. Routine urine drug screening panels used in most clinical settings are not equipped to detect kratom alkaloids; therefore clinicians should question patients about all substances they may be using including those marketed as herbal and natural supplements in order to select appropriate testing.

References

[1] Hillebrand J, Olszewski D, Sedefov R. Legal highs on the internet. Subst Use Misuse 2010;45:330–40. Available from: https://doi.org/10.3109/10826080903443628.

[2] Cinosi E, Martinotti G, Simonato P, Singh D, Demetrovics Z, Roman-Urrestarazu A, et al. Following "the roots" of kratom (*Mitragyna speciosa*): the evolution of an enhancer from a traditional use to increase work and productivity in Southeast Asia to a recreational psychoactive drug in Western countries. Biomed Res Int 2015;2015. Available from: https://doi.org/10.1155/2015/968786.

[3] Warner ML, Kaufman NC, Grundmann O. The pharmacology and toxicology of kratom: from traditional herb to drug of abuse. Int J Leg Med 2016;130:127–38. Available from: https://doi.org/10.1007/s00414-015-1279-y.

[4] Grundmann O. Patterns of Kratom use and health impact in the US—results from an online survey. Drug Alcohol Depend 2017;176:63–70. Available from: https://doi.org/10.1016/j.drugalcdep.2017.03.007.

[5] Kruegel AC, Grundmann O. The medicinal chemistry and neuropharmacology of kratom: a preliminary discussion of a promising medicinal plant and analysis of its potential for abuse. Neuropharmacology. 2018;134:108–20. Available from: https://doi.org/10.1016/j.neuropharm.2017.08.026.

[6] Kruegel AC, Gassaway MM, Kapoor A, Váradi A, Majumdar S, Filizola M, et al. Synthetic and receptor signaling explorations of the mitragyna alkaloids: mitragynine as an atypical molecular framework for opioid receptor modulators. J Am Chem Soc 2016;138:6754–64. Available from: https://doi.org/10.1021/jacs.6b00360.

[7] Le D, Goggin MM, Janis GC. Analysis of mitragynine and metabolites in human urine for detecting the use of the psychoactive plant kratom. J Anal Toxicol 2012;36:616–25. Available from: https://doi.org/10.1093/jat/bks073.

[8] Trakulsrichai S, Sathirakul K, Auparakkitanon S, Krongvorakul J, Sueajai J, Noumjad N, et al. Pharmacokinetics of mitragynine in man. Drug Des Dev Ther 2015;9:2421–9. Available from: https://doi.org/10.2147/DDDT.S79658.

[9] Lee MJ, Ramanathan S, Mansor SM, Yeong KY, Tan SC. Method validation in quantitative analysis of phase I and phase II metabolites of mitragynine in human urine using liquid chromatography-tandem mass spectrometry. Anal Biochem 2018;543:146–61. Available from: https://doi.org/10.1016/j.ab.2017.12.021.

[10] Philipp AA, Wissenbach DK, Weber AA, Zapp J, Maurer HH. Phase I and II metabolites of speciogynine, a diastereomer of the main Kratom alkaloid mitragynine, identified in rat and human urine by liquid chromatography coupled to low- and high-resolution linear ion trap mass spectrometry. J Mass Spectrom 2010; <https://onlinelibrary.wiley.com/doi/abs/10.1002/jms.1848> (accessed 15.04.19).

[11] Kong WM, Chik Z, Ramachandra M, Subramaniam U, Aziddin RER, Mohamed Z. Evaluation of the effects of *Mitragyna speciosa* alkaloid extract on cytochrome P450 enzymes using a high throughput assay. Molecules 2011;16:7344–56. Available from: https://doi.org/10.3390/molecules16097344.

[12] Hanapi NA, Ismail S, Mansor SM. Inhibitory effect of mitragynine on human cytochrome P450 enzyme activities. Pharmacogn Res 2013;5:241–6. Available from: https://doi.org/10.4103/0974-8490.118806.

[13] Rusli N, Amanah A, Kaur G, Adenan MI, Sulaiman SF, Wahab HA, et al. The inhibitory effects of mitragynine on *P*-glycoprotein in vitro. Naunyn-Schmiedeberg's Arch Pharmacol 2019;392:481–96. Available from: https://doi.org/10.1007/s00210-018-01605-y.

[14] Post S, Spiller HA, Chounthirath T, Smith GA. Kratom exposures reported to United States poison control centers: 2011–2017. Clin Toxicol (Phila) 2019;1–8. Available from: https://doi.org/10.1080/15563650.2019.1569236.

[15] Nelsen JL, Lapoint J, Hodgman MJ, Aldous KM. Seizure and coma following kratom (*Mitragynina speciosa* Korth) exposure. J Med Toxicol 2010;6:424–6. Available from: https://doi.org/10.1007/s13181-010-0079-5.

[16] Mousa MS, Sephien A, Gutierrez J, O'Leary C. *N*-Acetylcysteine for acute hepatitis induced by kratom herbal tea. Am J Therapeutics 2018;25:e550. Available from: https://doi.org/10.1097/MJT.0000000000000631.

[17] Antony A, Lee T-P. Herb-induced liver injury with cholestasis and renal injury secondary to short-term use of kratom (*Mitragyna speciosa*). Am J Ther 2018. Available from: https://doi.org/10.1097/MJT.0000000000000802.

[18] Kapp FG, Maurer HH, Auwärter V, Winkelmann M, Hermanns-Clausen M. Intrahepatic cholestasis following abuse of powdered kratom (*Mitragyna speciosa*). J Med Toxicol 2011;7:227–31. Available from: https://doi.org/10.1007/s13181-011-0155-5.

[19] Dorman C, Wong M, Khan A. Cholestatic hepatitis from prolonged kratom use: a case report. Hepatology. 2015;61:1086–7. Available from: https://doi.org/10.1002/hep.27612.

[20] Sheleg SV, Collins GB. A coincidence of addiction to "Kratom" and severe primary hypothyroidism. J Addict Med 2011;5:300–1. Available from: https://doi.org/10.1097/ADM.0b013e318221fbfa.

[21] LaBryer L, Sharma R, Chaudhari KS, Talsania M, Scofield RH. Kratom, an emerging drug of abuse, raises prolactin and causes secondary hypogonadism: case report. J Investig Med High Impact Case Rep 2018;6. Available from: https://doi.org/10.1177/2324709618765022.

[22] Singh D, Murugaiyah V, Hamid SBS, Kasinather V, Chan MSA, Ho ETW, et al. Assessment of gonadotropins and testosterone hormone levels in regular *Mitragyna speciosa* (Korth.) users. J Ethnopharmacol 2018;221:30–6. Available from: https://doi.org/10.1016/j.jep.2018.04.005.

[23] UNODC. Bulletin on narcotics—1975 Issue 3—002; n.d. <https://www.unodc.org/unodc/en/data-and-analysis/bulletin/bulletin_1975-01-01_3_page003.html> (accessed 15.03.19).

[24] Eldridge WB, Foster C, Wyble L. Neonatal abstinence syndrome due to maternal kratom use. Pediatrics 2018;142. Available from: https://doi.org/10.1542/peds.2018-1839.

[25] Murthy P, Clark D. An unusual cause for neonatal abstinence syndrome. Paediatr Child Health 2019;24:12–14. Available from: https://doi.org/10.1093/pch/pxy084.

[26] Davidson L, Rawat M, Stojanovski S, Chandrasekharan P. Natural drugs, not so natural effects: neonatal abstinence syndrome secondary to "kratom". Neonatal Perinat Med 2018. Available from: https://doi.org/10.3233/NPM-1863.
[27] Mackay L, Abrahams R. Novel case of maternal and neonatal kratom dependence and withdrawal. Can Fam Physician 2018;64:121−2.
[28] Cumpston KL, Carter M, Wills BK. Clinical outcomes after Kratom exposures: a poison center case series. Am J Emerg Med 2018;36:166−8. Available from: https://doi.org/10.1016/j.ajem.2017.07.051.
[29] Strickland EC, Cummings OT, Mellinger AL, McIntire GL. Development and validation of a novel all-inclusive LC-MS-MS designer drug method. J Anal Toxicol 2019;43:161−9. Available from: https://doi.org/10.1093/jat/bky087.
[30] Basiliere S, Bryand K, Kerrigan S. Identification of five *Mitragyna* alkaloids in urine using liquid chromatography-quadrupole/time of flight mass spectrometry. J Chromatogr B Anal Technol Biomed Life Sci 2018;1080:11−19. Available from: https://doi.org/10.1016/j.jchromb.2018.02.010.
[31] Mudge EM, Brown PN. Determination of alkaloids in Mitragyna speciosa (kratom) raw materials and dietary supplements by HPLC-UV: single-laboratory validation, first action 2017.14. J AOAC Int 2018;102:322−4.
[32] Kowalczuk AP, Łozak A, Zjawiony JK. Comprehensive methodology for identification of Kratom in police laboratories. Forensic Sci Int 2013;233:238−43. Available from: https://doi.org/10.1016/j.forsciint.2013.09.016.
[33] Philipp AA, Meyer MR, Wissenbach DK, et al. Monitoring of kratom or krypton intake in urine using GC-MS in clinical and forensic toxicology. Anal Bioanal Chem 2011;400:127−35 <https://link.springer.com/article/10.1007%2Fs00216-010-4464-3> (accessed 15.04.19).
[34] Fuenffinger N, Ritchie M, Ruth A, Gryniewicz-Ruzicka C. Evaluation of ion mobility spectrometry for the detection of mitragynine in kratom products. J Pharm Biomed Anal 2017;134:282−6. Available from: https://doi.org/10.1016/j.jpba.2016.11.055.
[35] Lesiak AD, Cody RB, Dane AJ, Musah RA. Rapid detection by direct analysis in real time-mass spectrometry (DART-MS) of psychoactive plant drugs of abuse: the case of *Mitragyna speciosa* aka "kratom". Forensic Sci Int 2014;242:210−18. Available from: https://doi.org/10.1016/j.forsciint.2014.07.005.
[36] Limsuwanchote S, Wungsintaweekul J, Keawpradub N, Putalun W, Morimoto S, Tanaka H. Development of indirect competitive ELISA for quantification of mitragynine in kratom (*Mitragyna speciosa* (Roxb.) Korth.). Forensic Sci Int 2014;244:70−7. Available from: https://doi.org/10.1016/j.forsciint.2014.08.011.
[37] Limsuwanchote S, Putalun W, Tanaka H, Morimoto S, Keawpradub N, Wungsintaweekul J. Development of an immunochromatographic strip incorporating anti-mitragynine monoclonal antibody conjugated to colloidal gold for kratom alkaloids detection. Drug Test Anal 2017. Available from: https://doi.org/10.1002/dta.2354.
[38] Smith LC, Lin L, Hwang CS, Zhou B, Kubitz DM, Wang H, et al. Lateral flow assessment and unanticipated toxicity of kratom. Chem Res Toxicol 2019;32:113−21. Available from: https://doi.org/10.1021/acs.chemrestox.8b00218.
[39] CEDIA TM mitragynine (kratom) assay. Package insert; 2019. <https://www.thermofisher.com/document-connect/document-connect.html?url = https%3A%2F%2Fassets.thermofisher.com%2FTFS-Assets%2FCDD%2FPackage-Inserts%2F10026620-CEDIA-Mitragynine-Kratom-Assay-EN.pdf&title = Q0VESUEgTWl0cmFneW5pbmUgKEtyYXRvbSkgQXNzYXkgUGFja2FnZSBJbnNlcnQgW0VOXQ == > (accessed 12.04.19).
[40] Drugs of Abuse, a DEA resource guide 2017 edition; 2017. <https://www.dea.gov/sites/default/files/sites/getsmartaboutdrugs.com/files/publications/DoA_2017Ed_Updated_6.16.17.pdf#page = 84> (accessed 15.03.19).
[41] Kratom Legality Map.; n.d. <http://speciosa.org/home/kratom-legality-map/> (accessed 15.04.19).
[42] Kratom (*Mitragyna speciosa*) drug profile; n.d. <http://www.emcdda.europa.eu/publications/drug-profiles/kratom#headersection> (accessed 15.03.19).

Chapter 19.3

Accidental death involving psilocin from ingesting "magic mushroom"

Uttam Garg[1,2], Jeff Knoblauch[1,2], C. Clinton Frazee, III[1,2], Adrian Baron[3] and Mary Dudley[3]

[1]Department of Pathology and Laboratory Medicine, Children's Mercy Hospital, Kansas City, MO, United States, [2]University of Missouri School of Medicine, Kansas City, MO, United States, [3]Office of the Jackson County Medical Examiner, Kansas City, MO, United States

Case history

The decedent was a 34-year-old white male visiting from out of town and staying at a friend's house. A witness noticed that the subject had few alcoholic drinks and was acting strange last night at the dinner. The group returned home around 0100 and was seen alive around 0230. At that time, he was in the bed and eating an old pizza with mushrooms. When the subject was not seen by 1400 the next day, the witness went to check on him. He discovered the decedent unresponsive and cold to the touch. The witness observed a prescription bottle belonging to a girl with whom the subject spent a night one week prior. He also found a bag of mushrooms on the nightstand. The witness said that he got scared and flushed the mushrooms down the toilet before calling 911. EMS arrived at the scene and noted vomit around the subject's mouth and on the pillow next to his head. There were no signs of injury or foul play. The investigator also noted that the aforementioned prescription bottle had two different kinds of pills present. In addition, the investigator discovered what appeared to be a mushroom stem floating in the toilet (Fig. 19.3.1). The mushroom stem and the pizza found on the scene were collected by Crime Scene Investigator (Fig. 19.3.2).

The decedent was brought to the Jackson County Medical Examiner's Office, Kansas City. The autopsy showed mild cerebral edema and pulmonary congestion. The gastric contents of the subject contained intact slices of undigested mushrooms. The scene investigation and history insinuated the ingestion of hallucinogenic mushrooms. Postmortem femoral blood, vitreous fluid, and urine samples were submitted for toxicological analysis. The femoral blood was screened for volatiles (ethanol, methanol, isopropanol, and acetone) by headspace gas chromatography with flame ionization detector. The blood was also used for comprehensive broad spectrum drug-screening that involved enzyme immunoassays (EIAs) for amphetamines, barbiturates, benzodiazepines, cannabinoids, cocaine metabolite, methadone, opiates, phencyclidine, and propoxyphene, and drug-screening for >200 drugs by gas chromatography mass spectrometry (GC−MS). Drug screening by GC−MS involved liquid-liquid alkaline extraction using bicarbonate buffer (pH 11.0) and butyl acetate, and mass spectrometer operation in full scan mode. Presumptive identification of analytes was made by spectral library match and relative retention time comparison with reference standards.

Results of toxicological analyses are summarized in Table 19.3.1. EIA drug screen testing indicated the presence of benzodiazepine, cannabinoids, and opiates. An 18 analyte benzodiazepine panel was performed by LC−MS/MS and yielded alprazolam and diazepam at therapeutic or below therapeutic concentrations. An eight analyte panel of free opiates was performed by GC−MS. Hydromorphone was suspected from the decedent's clinical history, but no free opiates were detected. A LC−MS/MS urine screen for psilocybin as total psilocin confirmed its presence. However, blood quantitation of psilocin was not available at the time of testing. Volatile screen testing for isopropanol, methanol, ethanol and acetone was positive for ethanol at 169 mg/dL.

Discussion

Psilocybin (*O*-phosphoryl-4-hydroxy-*N,N*-dimethyltryptamine) is a tryptophan indole-based alkaloid found in genus *Psilocybe*, *Panaeolus*, *Conocybe*, *Gymnopilus*, *Stropharia*, *Pluteus*, and *Panaeolina* mushrooms. They are commonly

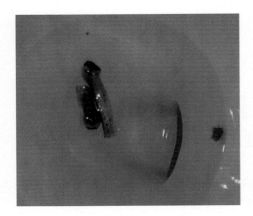

FIGURE 19.3.1 Mushroom stems found floating in the toilet bowl.

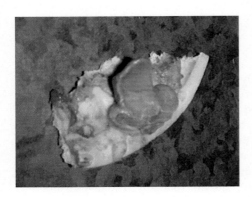

FIGURE 19.3.2 Pizza, with mushroom, found on the scene.

referred as magic, hallucinogenic, psychedelic, psychoactive, and sacred or saint mushrooms. Psilocybin concentration in these mushrooms varies from 0.1% to 2%. Typical hallucinogenic dose is ~10 mg of psilocybin, which can be found in 1 g of fresh *Psilocybe cubensis* mushroom [1]. This variety of mushroom is popular among recreational users of the drug.

Psilocybin is fairly nontoxic. Its toxicity is conferred from its active metabolite psilocin which acts as a $5HT_{2A}$, $5HT_{2C}$, and $5HT_{1A}$ agonist. These agonist properties of psilocin are being explored for its potential medicinal uses, particularly in the area of substance abuse and psychiatry [2–5]. Metabolism of psilocybin is shown in Fig. 19.3.3 [1,6]. The major clinical manifestations of psilocin toxicity include visual and auditory hallucinations, anxiety, paranoid delusions, disorientation, ataxia, and tachycardia. Most of these effects are acute and may last from 2 to 12 hours. Psilocin differs from other psychoactive drugs because it does not induce tolerance, dependency or addiction [6]. Psilocin has a half-life of 1.8–4.5 hours, volume of distribution of 2.5–5.0 L/kg, and oral bioavailability of ~53% [1]. Psilocin can be measured by HPLC with UV, electrochemical or mass spectrometry detection [1]. Immunoassay has also been described for the assay of psilocin [7]. Enzymatic hydrolysis is used to cleave glucuronide group from psilocin for total psilocin determination.

Several cases of psilocybin toxicity have been reported. One case involved intentional intoxication with psilocybin mushrooms in Poland. An 18-year-old man ingested *Psilocybe semilanceata* and experienced Wolff–Parkinson–White syndrome, arrhythmia, and myocardial infarction [8]. In this case, psilocybin was the only agent involved. In another case, a young man ingested 4–5 g of *Psilocybe* mushrooms and was arrested for erratic driving. The serum levels of free and total psilocin were 0.23 and 1.76 μg/mL respectively [9]. A 10-year-old boy developed ataxia, aphasia, and hallucinations after ingesting magic mushrooms [10]. As the symptoms of psilocybin toxicity are reversible and mushroom consumption often happens in remote geographic areas, its toxicity is possibly underreported.

Although toxicity involving ingestion of "magic mushroom" is not uncommon, fatalities due to psilocybin toxicity are rare. Flogstad reported three cases of mushroom poisoning with one fatality [11]. A 6-year-old child who accidentally ingested *Psilocybe* mushrooms developed hyperthermia, seizures and eventually died [12]. Another case involved a "mushroom party" in France. A 22-year-old man harvested and prepared psilocybin mushrooms for his friends.

TABLE 19.3.1 Postmortem Toxicology Results

EIA	Results
Amphetamine	Negative
Barbiturates	Negative
Benzodiazepines	**Positive**
Cannabinoids	**Positive**
Cocaine metabolite	Negative
Methadone	Negative
Opiates	**Positive**
Phencyclidine	Negative
Propoxyphene	Negative
GC–FID volatile screening	
Ethanol (Blood)	169 mg/dL
Ethanol (Vitreous)	207 mg/dL
Acetone	<5 mg/dL
Methanol	<5 mg/dL
Isopropanol	<5 mg/dL
LC–MS/MS	
Alprazolam	44 ng/mL
Diazepam	51 ng/mL
Psilocin	**Positive**

EIA, Enzyme immunoassay; *GC–FID*, gas chromatography with flame ionization detector.

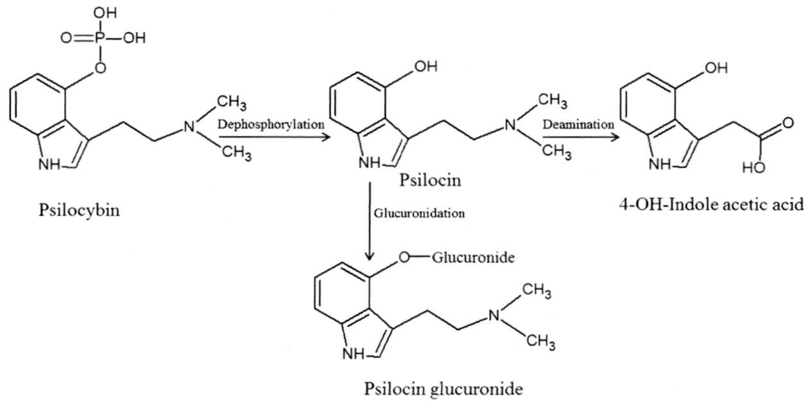

FIGURE 19.3.3 Metabolism of psilocybin.

He had already ingested some mushrooms while picking them. When he met his friends in the bar, he was already intoxicated from the mushrooms. He brought some of the mushrooms with him, ate some in front of his friends, and then invited them to join in. Later, the group consumed tea which was prepared from 20 to 40 of the mushrooms. All subjects had various euphoric and symptomatic experiences. After about an hour and a half, the lead subject went into convulsions and fell into a coma. The patient was taken to an emergency room where he could not be resuscitated and was pronounced dead. Postmortem toxicological results drawn 36 hours after death revealed a blood concentration of

4 μg/mL of psilocin. All other toxicology tests were negative [13]. Psilocin level in this case seems to be the highest reported. Lim et al. reported a case of transplant recipient who ingested *Psilocybe* mushrooms and died of cardiac arrest. The psilocin level was 0.03 μg/mL [14]. Though *Psilocybe* consumption appears more common, a fatal case from the consumption of *Russula subnigricans* mushroom has also been reported. In this case, a 51-year-old man suffered from rhabdomyolysis, acute kidney injury, severe hypocalcemia, respiratory failure, ventricular tachycardia, and cardiogenic shock before his death [15].

Most psilocybin fatalities reported have been reported as mixed drug toxicities. The drugs most often ingested with psilocybin include ethanol, cocaine, and cannabis. Interestingly, no toxic concentration levels of psilocybin have been published. Although we did not perform quantitative analysis, the history and symptoms observed support a conclusion that psilocin contributed to the subject's death. The cause of the death was determined to be a mixed drug toxicity resulting from the combination of ethanol, benzodiazepines, and psilocin. The manner of death was ruled as accidental.

References

[1] Baselt RC. Psilocybin. In: Baselt RC, editor. Disposition of toxic drugs and chemicals in man. Seal Beach, CA: Biomedical Publications; 2017. p. 1832—4.

[2] de Veen BT, Schellekens AF, Verheij MM, Homberg JR. Psilocybin for treating substance use disorders? Expert Rev Neurother 2017;17:203—12.

[3] Guzman G. New studies on hallucinogenic mushrooms: history, diversity, and applications in psychiatry. Int J Med Mushrooms 2015;17:1019—29.

[4] Johnson MW, Griffiths RR. Potential therapeutic effects of psilocybin. Neurotherapeutics 2017;14:734—40.

[5] Sommerkamp Y, Paz AM, Guzman G. Medicinal mushrooms in Guatemala. Int J Med Mushrooms 2016;18:9—12.

[6] Dinis-Oliveira RJ. Metabolism of psilocybin and psilocin: clinical and forensic toxicological relevance. Drug Metab Rev 2017;49:84—91.

[7] Albers C, Kohler H, Lehr M, Brinkmann B, Beike J. Development of a psilocin immunoassay for serum and blood samples. Int J Leg Med 2004;118:326—31.

[8] Borowiak KS, Ciechanowski K, Waloszczyk P. Psilocybin mushroom (*Psilocybe semilanceata*) intoxication with myocardial infarction. J Toxicol Clin Toxicol 1998;36:47—9.

[9] Sticht G, Kaferstein H. Detection of psilocin in body fluids. Forensic Sci Int 2000;113:403—7.

[10] Thornton SL, Bute S, Gerona RR. Magic BBQ? 10 year old with laboratory confirmed psilocin intoxication. Clin Toxicol 2014;52:764—5.

[11] Flogstad DL. Mushroom poisoning: three cases with one death. Clin Med (Northfield) 1962;69:2261—2.

[12] McCawley EL, Brummett RE, Dana GW. Convulsions from psilocybe mushroom poisoning. Proc West Pharmacol Soc 1962;5:27—33.

[13] Gerault A, Picart D. Fatal poisoning after a group of people voluntarily consumed hallucinogenic mushrooms. Bull Soc Mycol Franc 1996;112:1—14.

[14] Lim TH, Wasywich CA, Ruygrok PN. A fatal case of "magic mushroom" ingestion in a heart transplant recipient. Intern Med J 2012;42:1268—9.

[15] Cho JT, Han JH. A case of mushroom poisoning with *Russula subnigricans*: development of rhabdomyolysis, acute kidney injury, cardiogenic shock, and death. J Korean Med Sci 2016;31:1164—7.

Chapter 19.4

A US Army Captain faced discipline after cocaine positive test from coca tea consumption

Cecilia M. Rosales[1] and Uttam Garg[2,3]
[1]Souteastern Pathology Associates, Brunswick, GA, United States, [2]Department of Pathology and Laboratory Medicine, Children's Mercy Hospital, Kansas City, MO, United States, [3]University of Missouri School of Medicine, Kansas City, MO, United States

Case description

We report a case of a US Army Captain that tested positive for cocaine after a routine urine test for drugs of abuse at his workplace. He was accused of using cocaine and faced court proceedings for dishonorable discharge. Following an investigation, it was found that the captain was consuming a tea labeled as "Mate de Coca," a popular infusion used in South America that is made from the leaves of the coca plant.

The tea bags were submitted to our laboratory for cocaine analysis. The bags showed a writing that appeared to be in Spanish. On one side of the package was a label "Mate de Coca" with the image of a plant and the word "ZURIT."

To extract cocaine and its metabolites, benzoylecgonine (BE) and ecgonine methyl ester (EME), a single tea bag was placed in 100 mL of boiling deionized water for 10 minutes. The extract was analyzed by gas chromatography mass spectrometry (GC–MS) as described earlier by our laboratory [1]. Briefly, the extract was spiked with deuterated internal standards cocaine-d3, BE-d3, and EME-d3, and the mixture was buffered with phosphate buffer. The drugs were extracted using cation-exchange solid phase extraction. The drugs from the solid phase cartridges were eluted and the eluent was dried under the stream of nitrogen. The residue was incubated with pentafluoropropionic acid anhydride and pentafluoropropanol to form pentafluoropropionyl derivatives of BE and EME; cocaine is refractory to pentafluoropropionyl derivatization. The derivatized extract was dried, reconstituted in ethyl acetate, and injected into GC–MS. Quantitation of cocaine and its metabolites in the sample was performed using selected ion monitoring [1]. The analysis for cocaine and its metabolites showed the following results:

Cocaine:	5.35 mg/tea bag
BE:	0.26 mg/tea bag
EME:	0.64 mg/tea bag

After the trial, it was concluded that the positive cocaine result was due to coca tea consumption, and the captain was acquitted of the charges.

Discussion

The name "coca" comes from the Aymara word "khoka" that means "the tree." The Aymara is an indigenous population of Andes and Altiplano regions of South America [2].

Coca tea, with one of the trade names "Mate de Coca," is an herbal tea made using the raw or dried leaves of the coca plant which is native to South America. Many indigenous people from the Andes use the coca tea for medicinal purposes including its use as a fatigue reducer, stimulant, analgesic, and local anesthetic. Coca tea or chewing the raw coca leaves mixed with "Llipta" is often recommended to travelers to this area to prevent altitude sickness. Llipta, which improves the extraction of coca alkaloids, is a mixture of lime and/or ash typically from kiwicha or quinoa.

Coca tea is an infusion made by submerging the coca leaf or dipping a tea bag in hot water. It is a greenish-yellow liquid with a mild bitter flavor like green tea. The effects of drinking coca tea are a mild stimulation or mood lift with no significant numbing of the mouth. Unlike snorting cocaine, consumption of coca tea or chewing coca leaves do not cause the "rush" sensation. Cocaine is a principal alkaloid found in coca. It is a sympathomimetic stimulant that stimulates the release and blocks the reuptake of neurotransmitters dopamine and norepinephrine in the CNS. It is rapidly metabolized in the liver to major metabolites (BE and EME) and minor metabolites (norcocaine, p-hydroxycocaine, m-hydroxycocaine, p-hydroxybenzoylecgonine, and m-hydroxybenzoylecgonine) (Fig. 19.4.1) [3]. Cocaine and its major metabolites, BE and EME, are generally assayed in blood, plasma and/or urine to detect and monitor its abuse.

Coca tea is legal in Colombia, Peru, Bolivia, Argentina, Ecuador, and Chile; however, coca tea is illegal in the United States and the United Kingdom unless it is decocainized. Coca tea can be decocainized in a process similar to decaffeination but decocainized coca tea will still contain a small quantity of organic coca alkaloids. According to the US Customs and Border Protection, "It is illegal to bring coca leaves into the United States for any purpose, including the use for brewing tea or for chewing" [4]. A 2016 document by the US Food and Drug Administration states that coca leaves and any salt, compound, derivative or preparation of coca leaves is a controlled substance and is illegal in the United States [5].

Depending on the origin of the tea bags, an average of 4.86 and 5.11 mg of cocaine were found in tea bags from Bolivia and Peru, respectively [6]. The coca tea bag we tested had a comparable cocaine concentration of 5.35 mg/tea bag. An average "line" of cocaine bought on the street contains an estimated 20–50 mg of cocaine [7]. This amount approximates to the cocaine amount in 4–10 tea bags which can be consumed in a day by a moderate to heavy coca tea user. Several reports on cocaine and/or its metabolites' positive results after consuming coca tea have been published in the literature [6,8–10]. In one report, consumption of one cup of coca tea resulted in urine peak BE concentrations ranging from 3900 to 5000 ng/mL. In another report, five healthy adult volunteers consumed coca tea and underwent serial quantitative urine testing for cocaine metabolites. Using a BE cut-off of 300 ng/mL, all samples tested positive by 2 hours after coca tea ingestion. Results of three out of five participants remained positive at 36 hours. Mean urine BE concentrations in all postconsumption samples was 1777 ng/mL. The BE concentrations found in these studies were significantly higher than the typical SAMHSA cut-off of 300 ng/mL for screening and 150 ng/mL for confirmation. Furthermore, the Jockey Club in Great Britain commissioned a research study to find the effect of coca tea on cocaine positive testing as several jockeys tested positive for BE over a few years' time period. Urine samples collected at various time points within 24 hours after ingestion of a 250 ml infusion of "Mate de Coca" tea were analyzed. All samples tested positive for BE [9]. A report from Kippenberger discussed the detection of cocaine in oral fluid samples after the consumption of Bolivian coca tea [11]. In this study, oral fluid was taken from volunteers after consumption of approximately 200 mL of coca tea. In parallel, urine samples were collected for up to 42 hours postingestion and analyzed. Both cocaine and BE were detected in oral fluid. Cocaine was found in less than one hour and BE was found up to 27.5 hours after consumption. In all cases the urine screening showed longer positive results in comparison to oral fluid up to 42 hours postconsumption.

Similarly to our case, a publication in Drug Detection Report, discussed the case of a navy electrician who faced possible dishonorable discharge after a positive cocaine test was reported [12]. The subject claimed that the positive

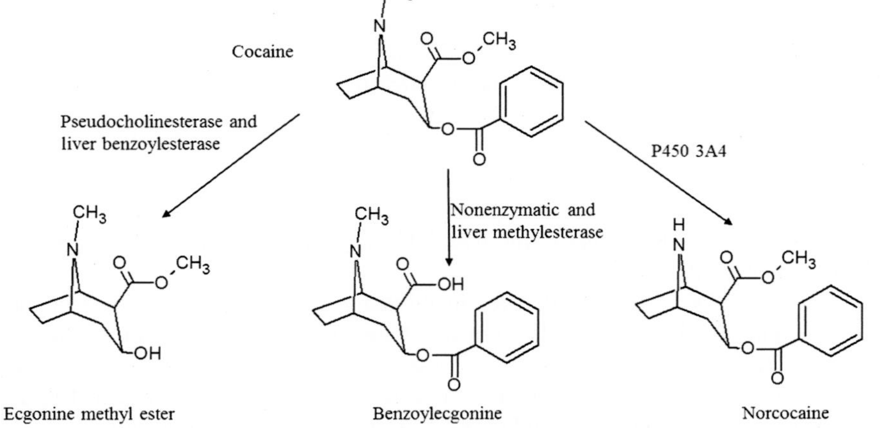

FIGURE 19.4.1 Pathways for cocaine metabolism.

test for cocaine was due to the consumption of coca tea. The report states that the San Diego Reference Laboratory tested two varieties of coca tea, "Mate de Coca," and found cocaine and its metabolites in the tea. The report also described that after drinking only one cup of the coca tea, the two laboratory employees tested positive for approximately one and one-half days.

In conclusion, it is apparent from these studies that certain coca teas such as "Mate de Coca" contain enough cocaine that their consumption can cause a positive urine test for cocaine and its metabolites. A thorough investigation of an individual's case history is recommended for appropriate interpretation of test results.

References

[1] Fleming SW, Dasgupta A, Garg U. Quantitation of cocaine, benzoylecgonine, ecgonine methyl ester, and cocaethylene in urine and blood using gas chromatography-mass spectrometry (GC-MS). Methods Mol Biol 2010;603:145–56.
[2] Biondich AS, Joslin JD. Coca: the history and medical significance of an ancient Andean tradition. Emerg Med Int 2016;2016:4048764.
[3] Inaba T, Stewart DJ, Kalow W. Metabolism of cocaine in man. Clin Pharmacol Ther 1978;23:547–52.
[4] US Customs and Border Protection. Can I bring in coca leaves for my personal use?. <https://help.cbp.gov/app/answers/detail/a_id/725/kw/coca> [accessed 21.01.19].
[5] US FDA. Code of Federal Regulations Title 21. <www.accessdata.fda.gov/scripts/cdrh/cfdocs/cfcfr/cfrsearch.cfm?fr = 1308.12> [accessed 21.01.19].
[6] Jenkins AJ, Llosa T, Montoya I, Cone EJ. Identification and quantitation of alkaloids in coca tea. Forensic Sci Int 1996;77:179–89.
[7] Heard K, Palmer R, Zahniser NR. Mechanisms of acute cocaine toxicity. Open Pharmacol J 2008;2:70–8.
[8] Mazor SS, Mycyk MB, Wills BK, Brace LD, Gussow L, Erickson T. Coca tea consumption causes positive urine cocaine assay. Eur J Emerg Med 2006;13:340–1.
[9] Turner M, McCrory P, Johnston A. Time for tea, anyone? Br J Sports Med 2005;39:e37.
[10] Cingolani M, Cippitelli M, Froldi R, Gambaro V, Tassoni G. Detection and quantitation analysis of cocaine and metabolites in fixed liver tissue and formalin solutions. J Anal Toxicol 2004;28:16–19.
[11] Kippenberger D. Detection of cocaine in oral fluid samples after the consumption of Bolivian coca tea and confirmation via GC-MS. ToxTalk 2003;27(3):9.
[12] Anon. Navy electrician faces dishonorable discharge after positive cocaine test. Drug Detection Report, 31 March; 2005, p. 51.

Chapter 20

Toxic gases

Saswati Das
Ram Manohar Lohia Hospital, New Delhi, India

Introduction

Gaseous poisons that kill either by displacing oxygen from the environment or by other complex mechanisms are known as asphyxiants or toxic gases.

Toxic gases or asphyxiants are classified as follows: (1) simple asphyxiants: these gases displace oxygen from ambient air and reduce the partial pressure of available oxygen. Simple asphyxiants cause death by excluding oxygen from the lung. Examples include carbon dioxide, nitrogen, aliphatic hydrocarbon gases (butane, ethane, methane, and propane), and noble gases (argon, helium, neon, radon, and xenon). (2) Respiratory irritants: these gases damage the respiratory tract by destroying the integrity of the mucosal barrier and produce inflammatory changes. Examples include ammonia, chlorine, formaldehyde, hydrogen sulfide, methyl bromide, methyl isocyanate, oxides of nitrogen, ozone, phosgene, and sulfur dioxide. Heavy metal−related gases also fall under this category (e.g., cadmium fumes, copper fumes, and mercury vapor). (3) Systemic asphyxiants/chemical asphyxiants: these are the inert gases and when these gases are inhaled in high concentration, they act by displacing or excluding oxygen. These gases produce significant systemic toxicity by specialized mechanisms such as combining with hemoglobin. Examples include carbon monoxide, carbon disulfide, and hydrogen cyanide (HCN). It must be noted that systemic toxicity may also be observed in the case of some simple asphyxiants and respiratory irritants, though it is not the principal feature. (4) Volatile compounds: volatile compounds have little or no irritant effect after absorption; they act as anesthetic agents or toxic to the liver, kidney, etc. Examples are as follows: aliphatic hydrocarbons, halogenated hydrocarbons, aromatic hydrocarbons, and anesthetic gases.

Carbon monoxide

Carbon monoxide is a colorless, odorless nonirritant gas that is lighter than air and insoluble in water. It is highly toxic to humans and animals in higher quantities. The normal concentration in atmosphere is less than 0.001% and a concentration of 0.1% can be lethal to humans. CO is formed normally in the body as a by-product of heme degradation; one molecule of CO is produced per molecule of heme degraded. This produces sufficient CO to result in low levels of carboxyhemoglobin (COHb) even in nonsmoking persons (up to 1%). In hemolytic anemia, COHb may raise up to 8%. CO like NO is a gaseous neurotransmitter in the central nervous system (CNS). It can diffuse and signal adjacent cells much like nitric oxide. Endogenously produced CO serves as a signaling molecule in multiple cellular functions such as inflammation, proliferation, and apoptosis. At present, CO is one of the most commonly encountered toxic gases in our environment and a leading cause of poisoning worldwide [1−3].

Exposure routes and pathways

Primary source of exposure to this colorless, odorless gas is via inhalation. CO is always produced whenever there is fire in confined spaces limiting the availability of oxygen. Some sources include [1,4−6] (1) coal gas, smoke from fires and fumes from defective heating appliances (gas and wood stoves, furnace, oil lamps, fire places, kerosene, and gas water heaters), leaking chimneys; (2) component of the fumes of coke kilns and lime kilns, heating unit used only occasionally and not well maintained and barbeques; (3) production due to explosion in mines and detonation of explosives; (4) tobacco smoke: average levels of COHb in an average smoker in about 4%; in a heavy smoker, it may be up to 20%. A smoker is estimated to be exposed to 400−5000 ppm of CO while actively smoking; and (5) dermal and inhalation exposures to the paint remover methylene chloride. Fumes of methylene chloride are metabolized within body to produce CO.

Mechanism of action

CO is readily absorbed after inhalation. Absorption of inhaled carbon monoxide occurs in the gas exchange region of the respiratory tract following inhalation. The affinity of CO to Hb is approximately 200−250 greater than that of oxygen. Carboxyhemoglobin is completely dissociable and carbon monoxide is liberated and eliminated through the lungs after exposure to CO ceases. Small amounts are oxidized to carbon dioxide. After binding to Hb to displace oxygen to form COHb, carbon monoxide is transferred rapidly throughout the body, where it produces asphyxia. The majority of body burden exists as COHb bound to hemoglobin of red cells, while approximately 10% is present in extravascular space [5]. A prominent biochemical feature of CO toxicity is a leftward shift of the oxyhemoglobin dissociation curve which decreases the release of oxygen from oxyhemoglobin to tissue [5]. Neuronal cell death and neuronal deficits produced by CO cannot be explained by simple mechanisms. This is because comparable levels of anemia fail to produce similar neuronal lesions. The following mechanism is thought to be an important pathway of CO-mediated neuronal cell death: CO displaces NO from platelets. This results in endothelial damage to brain microvasculature and activation of excitatory amino acids leading to neuronal cell loss and neuronal deficits. Further evidence supporting this mechanism is presented by the observation that NO synthase inhibitors have been shown to prevent CO toxicity via this mechanism [7].

Acute toxicity

The effects generally result from tissue hypoxia—headache, nausea, vomiting, dizziness, lethargy, and a feeling of weakness. Symptoms develop in severity as concentration of COHb increases. CO particularly affects organs that have high O_2 utilization cardio vascular system, central nervous system (CVS, CNS). There may be a cherry-red discoloration of skin. Neurological signs include poor concentration, confusion, disorientation, visual disturbance, syncope, and seizures. Small increments in carboxyhemoglobin levels may produce adverse cardiovascular effects [8−10]. Ophthalmoscopic examination may reveal retinal hemorrhages. Exposure during pregnancy may result in fetal loss. The absorption and elimination of carbon monoxide are slower in the fetal circulation than in the maternal circulation. Thus the fetus may experience toxicity when the mother is at low carbon monoxide level with no effects [11,12] (Table 20.1).

Chronic toxicity

Humans are exposed to low levels of carbon monoxide every day from automobile traffic, from smoking or being close to those who are cooking with natural gas or through other occupational means. At doses that produce carboxyhemoglobin concentration <10%, no appreciable symptoms are observed. Higher doses produce more pronounced toxic effects. Delayed neurological sequelae most likely involve white matter demyelination. Major manifestations—cognitive deficits, dementia, psychosis, mood disorders, personality changes. Humans exposed to moderate doses of CO during pregnancy have lower birth weight children [13] (Table 20.2).

Diagnosis and investigation

Carboxyhemoglobin (COHb) levels are the most useful test for diagnosing CO toxicity. COHb concentration is ∼1% in nonsmokers and ∼4% in smokers. A COHb concentration of more than 2% in a nonsmoker and >10% in a smoker confirms exposure to CO. Direct CO assays use spectrophotometry for quantitation of COHb. Cardiac monitoring with ECG for detecting ischemia and dysrhythmias and scintigraphy of heart are used for evaluation of heart injury in patients. Creatine phosphokinase is slightly raised as a result of rhabdomyolysis and troponin is elevated due to diffuse cardiac myonecrosis. Serum pH and lactic acid levels are monitored closely. Metabolic acidosis due to generation of lactic acid is commonly seen (Table 20.3).

TABLE 20.1 Symptoms of CO Poisoning

Level of Poisoning	COHb (%)	Symptoms
Mild poisoning	0–10	No appreciable symptoms
	10–20	Breathlessness on exertion, headache
	20–30	Emotional instability, exertional dyspnea, throbbing headache, irritability, disturbed judgment, and rapid fatigue
Moderate poisoning	30–40	Cherry-red coloration, chest pain, confusion, dizziness, impaired judgment, loss of muscular control, nausea and vomiting, perspiration, prolonged reaction time, impaired sensorimotor performance, severe headache, tinnitus, and decreased vigilance
Severe poisoning	40–50	Confusion, hallucinations, ataxia, collapse, and rapid respirations
	50–60	Deep pink or red coloration, weak pulse, rapid respiration, weak pulse, coma with intermittent convulsions, tachycardia, and trismus
	60–70	Hypotension, myocardial ischemia, incontinence of urine and feces, ventricular arrhythmias, and depth of coma increases
	70–80	Weak and thready pulse, depressed reflexes, profound coma, and irregular shallow respiration
	>80	Death from respiratory arrest

TABLE 20.2 Different Air Concentrations of CO and Its Effects

COHb (%)	Symptoms
0–10	No appreciable symptoms
10–20	Mild headache, breathless on moderate exertion
20–30	Nausea, vomiting, throbbing headache, drowsiness, and fainting

TABLE 20.3 Diagnosis of Carbon Monoxide Poisoning

Method	Advantages	Disadvantages
Spectrophotometric analysis	Quantitative assessment, rapid guide for therapy	Delay in analysis may lead to spuriously normal COHb levels. Costly and not readily available
Kunkel's test (5–6 drops of analysis 3% tannic acid added to patient's diluted blood gives a persistent crimson red coagulum)	Rapid field test	Only qualitative analysis
Pulse oximetry	Noninvasive bedside test	Unreliable because it grossly overestimates oxygen saturation in the presence of COHb
Arterial blood	Useful to confirm lactic acidosis, which is a marker of prolonged, serious exposure to carbon monoxide	Measures dissolved oxygen and thus overestimates the true oxygen saturation of Hb

Clinical management

Immediate removal from the source. Airway needs to be secured and adequate ventilation ensured. Initiate oxygen therapy with normobaric oxygen and hyperbaric oxygen (HBO). Benefits of HBO are as follows: (1) it decreases half-life of CO and (2) it increases amount of dissolved O_2 by 10 times [13,14].

Postmortem appearances of CO toxicity include cherry red coloration of skin, conjunctivae, mucus membranes, and all internal organs that are usually seen when concentration of CO is >40%. In dark-colored individuals, it can be seen more prominently in the inner aspects of lips, tongue, nail beds, palms, and soles. Other presentations of CO toxicity include serous effusions in body cavities and skin blisters in wrists, interdigital surfaces of fingers, calves, and knees. Trachea may contain soot and lungs may show congestion, pulmonary edema, pleural hemorrhage, and bronchopneumonic consolidation. In some cases, heart may have pericardial hemorrhages and focal areas of necrosis. Brain consistency is usually firmer than normal and meninges show hemorrhages. Basal ganglia necrosis and cavitation especially in globus pallidus and putamen are considered most characteristic lesions in delayed deaths due to CO toxicity. Punctiform and ring-shaped hemorrhages with edema in the white matter are also seen.

Chlorine

Chlorine is a commonly used highly toxic industrial gas having greenish-yellow color and an unpleasant irritating odor. It was discovered in 1772 by Carl Wilhelm Scheele. It is an oxidizing agent and can cause destruction of organic tissue.

Exposure routes and pathways

Exposure occurs primarily by inhalation and initial exposure may be asymptomatic. Chlorine is usually used for bleaching of wood pulp newsprint and for chlorination of water, disinfection, and bleaching. It has been used as a war gas. Exposure is likely in laboratories, bleaching powder factories, and other chemical works [15–17].

Mechanism of action

Chlorine has an irritating effect on all mucosal surfaces, because of formation of hydrochloric and hypochlorous acids ($Cl_2 + H_2O \rightarrow HCL + HOCl$). Although the exact mechanism of epithelial damage is unknown, oxidative injury is likely involved, as chlorine gas can combine with reactive oxygen species and other airway fluid constituents to form a variety of highly reactive oxidants [18]. Direct oxidative injury to the epithelium may occur immediately with exposure to chlorine, but further damage to the epithelium may occur with migration and activation of inflammatory cells such as neutrophils within the airway epithelium, with the subsequent release of oxidants and proteolytic enzymes. Through these mechanisms, chlorine exposure can result in injury not only to the lower airways but also to the eyes, skin, and upper airways. The airway is especially affected from the nose to the level of the bronchi [19].

Acute toxicity

In an event of acute chlorine exposure, an intense irritation to the eye tissue along with redness, lacrimation is observed. Gastroenteric symptoms include nausea, vomiting, and mild gastritis. Respiratory symptoms observed include spasm of glottis, extreme dyspnea, trachypnoea, oral mucositis, violent cough, hemoptysis, and increased respiratory tract sounds and wheezes [20].

Chronic toxicity

In chemical works, chronic poisoning may occur causing anemia, cachexia, emaciation, progressive wasting, gastritis, dental caries (because of acid production), bronchitis, and emphysema [21–23].

Diagnosis and investigation

In a case of chlorine intoxication, complete blood count exhibits erythrocytosis and lymphopenia. Relevant serum biochemistry includes hepatic enzymes and creatine kinase elevations. Arterial blood gas concentration is usually consistent with hypoxemia and hyperventilation. Chest X-ray shows widespread pulmonary alveolar infiltrates predominantly affecting the ventral portions of both lungs, consistent with noncardiogenic pulmonary edema. Chest X-ray taken in expiration shows an increase in lung volume, which indicates pulmonary edema. Specific biochemical markers to identify chlorine inhalation victims and to estimate inhaled dose would be useful for future studies of both accidental and occupational exposures. Some of them include surfactant proteins, Clara cell secretory proteins, and chlorotyrosine residues in protein, any of which might appear in various compartments, including blood plasma, sputum, bronchoalveolar lavage fluid, and/or lung tissue. Fatal dose in humans is an exposure to 1 part in 1000 that can cause death in about 5 minutes.

Clinical management

The first step in clinical management of chlorine toxicity is removal from toxic atmosphere, fresh air, steam inhalations, and supplemental oxygen. Nebulized sodium bicarbonate and IV electrolytes have also been used to treat chlorine toxicity [24]. Other supportive interventions include general treatment of shock, circulatory failure, pulmonary edema, and treatment with glucocorticoids [25,26].

Postmortem appearances

Cause of death from chlorine exposure is usually cardiac failure following pulmonary congestion and inflammatory edema of lungs. Common presenting features include signs of asphyxia, inflammation of the respiratory tract, congestion of lungs, lungs appear edematous, alveolar walls are usually ruptured, and hemorrhages and thrombosis are seen in lung beds. Gastrointestinal (GI) features include esophagitis and gastritis. Blood shows increased viscosity.

Hydrogen cyanide

Hydrogen cyanide, also known as prussic acid, is a chemical compound with the chemical formula HCN. It is a colorless, extremely poisonous, and flammable liquid that boils slightly above room temperature.

Exposure routes and pathways

The most common source of cyanide exposure is smoke inhalation from residential or industrial fires. HCN is used in dyeing, electroplating, fumigants, manufacturing processes, metal extraction, metal hardening, metal polishes, pesticides, photography, and printing.

Mechanism of action

Cyanide (CN) has a high affinity for Fe^{3+} ions and its physiological actions are brought forth by inhibition of cytochrome oxidase enzymes. In normal cells, cytochrome oxidase has a high concentration of Fe^{3+} ions. CN attaches to cytochrome-c oxidase and stops cellular respiration. Other enzymes that have high concentration of Fe^{3+} ions are carbonic anhydrase, decarboxylases, succinic dehydrogenase, and superoxide dismutase; these are inhibited as well. Their inhibition has no recognizable relation to the toxic action.

Toxicokinetics

Cyanide is rapidly absorbed from gastrointestinal tract, respiratory system, and via the skin. When orally consumed, absorption can be delayed particularly when consumed with food. Cyanide salts are acted upon by stomach HCl to generate HCN. Achlorhydric individuals may be difficult to be poisoned by cyanided salts. However, water in stomach still produces HCN. Cyanide precursors are present in many natural foods of CN detoxification. It converts cyanide to thiocyanate with enzyme rhodanese, a mitochondrial enzyme present in great amounts in liver and kidneys, acting as a catalyst. Glutathione acts as a sulfur donor. Humans can remove 1 mg of cyanide/kg/h by converting it to thiocyanate that is excreted mainly in urine as thiocyanate.

Acute toxicity

Cyanide is one of the most rapid of all poisons, which causes histotoxic anoxia. Cyanide toxicity should be considered in patients with sudden cardiovascular system collapse or in a fire victim with hemodynamic instability, increased lactic acid, or coma. When inhaled as a gas, action is instantaneous. An acute sense of constriction about the throat and chest, dizziness, vertigo, and insensibility can lead to death from respiratory failure. When orally ingested, symptoms may not appear immediately. With ingestion of small (less than fatal) doses symptoms include nausea, giddiness, head ache, confusion, and loss of muscular power. CNS symptoms include anxiety, confusion, drowsiness, excitement, presyncope, headache, perspiration and vertigo, stupor, coma, and death. Cardiovascular symptoms include hypertension, reflex bradycardia, sinus arrhythmia followed by hypotension, reflex tachycardia, and cardiovascular collapse.

Respiratory symptoms of cyanide poisoning include and odor bitter almonds and sometimes an ammoniacal smell. Initial respiratory response is tachypnea and dyspnea due to stimulation of chemoreceptors and respiratory system by CN. Later severe respiratory depression reduced respiratory rate termed as cyanosis.

Neuromuscular symptoms include convulsions (epileptiform or tonic; localized or generalized), cramps, opisthotonus and trismus, prostration, twitchings, and paralysis.

Gastrointestinal symptoms include nausea, vomiting, salivation bitter taste, and a feeling of throat constriction, numbness. KCN and NaCN cause corrosive burns on mouth, throat, and stomach, may cause epigastric pain. Other symptoms include bullous skin lesions, perspiration and hyperthermia, glassy eyes, prominent dilated, and unreactive pupils [27–29].

Chronic toxicity

Chronic cyanide toxicity is caused by continued inhalation of low concentration of CN over long periods, which occurs usually in smokers after a long period. Chronic cyanide toxicity commonly causes neuropathy.

Ingestion of 50–60 mg of HCN and 200–200 mg of NaCN or KCN-200 can be fatal. Air inhalation of HCN with a 1:50,000 concentration in inhaled air can be fatal in a few hours with 1:10,000 causing death in a few minutes and cyanide exposure at 1:2000 of room air can cause death almost immediately. HCN causes adverse effect almost immediately, whereas KCN or NaCN can take up to about 30 minutes to start acting, because of the delay in chemical conversion of salt to HCN by gastric juices.

Diagnosis and investigation

Relevant laboratory findings include an early decreased arteriovenous difference in the partial pressure of oxygen (PO_2) with progressive lactic acidosis. Timely measurements of blood and urine concentrations for suspected intoxicants are useful in guiding clinical therapy, especially when there is toxicity associated with the treatment agents. Analysis of cyanide in biological fluids is a difficult task for a variety of reasons [30]. Documentation of blood cyanide levels is useful in confirming the clinical diagnosis and in subsequent follow-up investigations. The red blood cells contain most of the cyanide in the blood, so an assay of whole blood is necessary. Cyanide concentrations in tissue, such as liver, lung, spleen, and heart, may be more accurate indicators of the blood cyanide intoxication levels. Estimates of tissue levels are necessary adjunctive studies in forensic cases [31].

Clinical management

Treatment should be started immediately because fatal period is very short. Restore the respiration by giving 100% oxygen. Remove unabsorbed CN by decontamination. Reverse cyanide-cytochrome combination by antidotes [32–36].

Postmortem appearances

Cause of death in the setting of cyanide poisoning is usually respiratory failure. The following features are characteristics of cyanide poisoning: (1) odor of bitter almonds from the body-odor is well marked in the cranial cavity; (2) because of oxyhemoglobin as oxygen could not be utilized by cells so the oxygen remains in the cells as oxyhemoglobin causing the color of cheeks to be pinkish in appearance; (3) eyes-bright, glistening, prominent, pupils dilated; (4) blood stained froth at mouth, in trachea and bronchi; (5) stomach wall may be eroded or blackened due to formation of alkaline hematin; (6) edematous lungs; (7) ecchymosed serous cavities; (8) congested viscera and may be discolored blue or dark green if methylene blue had been given as a treatment; (9) all blood vessels of the body, including veins contain oxygenated blood; (10) degenerative changes in the CNS; and (11) in fatal cases, blood can contain as high as 12 mg/L of cyanide.

Hydrogen sulfide

Hydrogen sulfide (H_2S) is readily water soluble, and, at physiological pH, about two-thirds exist as hydrogen sulfide ion (HS^-) and one-third as undissociated hydrogen sulfide [37]. Hydrogen sulfide is a colorless gas with a smell of rotten eggs. It is formed during the decomposition of organic substances containing sulfur; thus found in cesspools, privies, sewers, swamps, tannery vats, and volcanic gases. It is formed in human intestine by gut bacteria during life and in the body after death when putrefaction sets. It is also produced as a by-product during distillation of petroleum oil, in the manufacture of artificial silk, gas works, glue factories, tannery, vats and in other industries where sulfur compounds are used [38–40]. It is heavier than air; therefore it tends to collect at ground level.

Exposure routes and pathways

H_2S is a colorless gas and most human exposures occur via inhalation. Exposure may also occur via ingestion or skin contact. The general population may be exposed from industrial operations, from natural gas well during drilling operations. Exposure may be to low, medium, or high concentrations during an accidental release. Exposure to hydrogen sulfide may also occur when precursors enter the body and are changed to produce hydrogen sulfide.

Toxicokinetics and mechanism of action

The most prominent effect of hydrogen sulfide is the inhibition of cytochrome oxidase which is similar to that of cyanide and phosgene. Hydrogen sulfide adheres to cytochrome-c oxidase and inhibits its action. Cellular respiration comes to a halt. For this reason, its toxicity and rapidity of action are comparable with that of hydrogen cyanide and even treatment is essentially similar. Upon hydrogen sulfide exposure, lack of oxygen to brain causes sudden loss of consciousness. Hydrogen sulfide is often known as knockdown gas. Hydrogen sulfide interacts with a number of other enzymes and other macromolecules, including hemoglobin to formation of sulfhemoglobin, responsible for greenish discoloration of putrefaction, methemoglobin, and myoglobin. Most macromolecules are held together by disulfide bonds, which are easily disrupted by hydrogen sulfide. Olfactory paralysis occurs very quickly at higher levels. Although not a mechanism of death, it contributes to death, by removing warning signals [41].

Acute toxicity

Exposure to hydrogen sulfide may cause irritation, neurotoxicity, respiratory distress, pulmonary edema, headache, nausea, shortness of breath, dizziness, ataxia, chest pain, collapse, coma, and death. Certain effects have been observed after acute and/or short-term exposure at ranges of concentration. Upon exposure to low concentrations (<250 ppm) of hydrogen sulfide, cyanosis, irritation of all mucus membranes, dullness, and sleepiness can be observed. Death may result during sleep without the victim regaining consciousness. Discoloration of coins in the victims pocket due to oxidation may be clue to hydrogen sulfide exposure. Exposure to moderately high concentration of hydrogen sulfide (250–500 ppm) can result in nausea, vomiting, diarrhea, disorientation, dizziness, headache, and vertigo. Other symptoms include cough, labored breathing, apnea, palpitations, tachycardia, itching, severe pain, muscle cramps, prostration, and weakness. Metabolic acidosis may be observed secondary to anaerobic metabolism, which, in turn, results in CNS, respiratory, and myocardial depression. In addition to symptoms described previously, exposure to high concentration of hydrogen sulfide (750–1000 ppm) can lead to amnesia, confusion, delirium, hallucinations, convulsions, nystagmus, somnolence, respiratory paralysis, asphyxial seizures, delirium, and death [42–45].

Chronic toxicity

Long-term exposure to hydrogen sulfide has been reported to cause fatigue, headache, sleep disturbances, diminished memory, dizziness, nausea, vomiting, loss of appetite, weight loss, irregular heartbeat, lung congestion, and nerve damage [46–49].

Diagnosis and investigation

Hydrogen sulfide poisoning presents as a rotten egg odor from patient. Another characteristic feature of hydrogen sulfide poisoning is blackening of copper and silver coins in patients' pockets, darkening of jewelry. Laboratory investigation will show an increased blood sulfide. In addition, at a scene of poisoning a filter paper moistened with lead acetate will turn black.

Fatal dose can be an air concentration of >1000 ppm which can be instantly fatal. Depending on the exposure levels and durations, fatal levels can be reached in a few minutes to a few hours.

Clinical management

The first step in clinical management of toxicity from hydrogen sulfide is an immediate removal of the exposed from the source of hydrogen sulfide to fresh air. Oxygen therapy with HBO may be given. Nitrites are used as an antidote that induces methemoglobinemia. Hydrogen sulfide has a greater affinity for methemoglobin which releases its hold on cytochrome oxidase and attaches to methemoglobin leading to formation of sulfmethemoglobin that is spontaneously detoxified by body [50,51].

Postmortem appearances

Some characteristic postmortem features of hydrogen sulfide poisoning include odor of rotten eggs from the body, signs of asphyxia especially cyanosis, and demonstration of sulfmethemoglobin in the blood. In such cases, brain is edematous and there is greenish discoloration of gray matter due to sulfmethemoglobin. Congestion in liver and spleen is seen and blood and viscera are greenish.

Phosgene

Salient features: Phosgene is widely used as a chemical intermediate. It is used in metallurgy and in the production of pesticides, herbicides, and many other compounds. It is a by-product of chloroform biotransformation and can be generated from some chlorinated hydrocarbon solvents under intense heats. Phosgene has been used in chemical warfare.

Exposure routes and pathways

Inhalation and exposure to skin and mucous membranes are possible exposure routes. Potential for toxicity depends on concentration, route of exposure, and length of time exposed.

Toxicokinetics and mechanism of action

Phosgene is absorbed by the lungs and excreted via the liver and kidneys. Rapid onset ocular, nasal, and airway irritations from high levels of phosgene are caused by hydrochloric acid released during hydrolysis. The carbonyl group participates in acylation reactions with amino, hydroxyl, and sulfhydryl groups. These reactions may account for some toxic effects of phosgene. At alveolar capillary membranes, these reactions can cause fluid leakage into the interstitial lung. Leakage of fluid into the interstitium is opposed by lymphatic drainage, but as fluid accumulates, this drainage is overwhelmed. After a latent period, fluid reaches alveoli and peripheral airways, leading to increasingly severe dyspnea and pulmonary edema [3].

Acute toxicity

Exposure to high levels may cause death. Phosgene causes irritation to skin, eyes, nose, throat, and lungs. Phosgene exposure may be asymptomatic in the short term with effects delayed up to 48 hours. High concentrations may cause accumulation of fluids in the lungs or pneumonia and can produce choking, chest constriction, pain in breathing, coughing, blood in sputum, and heart failure. Exposure to eyes and mucous membrane can be very irritating. Buildup of phosgene in the liver and kidneys may produce damage [52,53].

Chronic toxicity

Chronic inhalation to low levels of phosgene not only can lead to some degree of tolerance to acute effects noted in humans but also can cause irreversible pulmonary changes, for example, emphysema and fibrosis [53,54].

Clinical management

The exposed individual should be removed from exposure. Clothing should be removed carefully avoiding further exposure. The body should be washed rapidly with soap and water and eyes flushed if needed. Individuals should be given immediate medical attention and monitored 48 hours for delayed effects.

References

[1] Cobb N, Etzl RA. Unintentional carbon monoxide related deaths in United States. JAMA 1991;266:659–63.
[2] National Center for Health Statistics. Vital statistics of the United States, 1988. Washington, DC: Government Printing Office; 1991. p. 89–1102.
[3] Raub JA, Mathieu-Nolf M, Hampson NB, Thom SR, et al. Carbon monoxide poisoning—a public health perspective. Toxicology 2000;145:1–14.
[4] Grace TW, Platt FW. Sub acute poisoning. JAMA 1981;246:1698–700.
[5] Meredith T, Vale A. Carbon monoxide poisoning. BMJ 1988;296:77–9.
[6] Mehta SR, Niyogi M, Kasthuri AS, et al. Carbon monoxide poisoning. J Assoc Physicians India 2001;49:622–5.
[7] Centers for Disease Control and Prevention (CDC). Deaths from motor-vehicle-related unintentional carbon monoxide poisoning—Colorado, 1996, New Mexico, 1980-1995, and United States, 1979-1992. MMWR Morb Mortal Wkly Rep 1996;45:1029–32.
[8] Burney RE. Mass carbon monoxide poisoning – 184 victims. Ann Emerg Med 1982;11:394–9.
[9] Cho IS. Delayed neurologic sequelae in carbon monoxide intoxication. Arch Neurol 1983;40:433–5.
[10] Hark IK, Kennedy PGE. Neurological manifestation of carbonmonoxide poisoning. Postgrad Med J 1998;64:213–16.
[11] Ginsberg MD, Myers RF. Fetal brain injury after maternal carbon monoxide intoxication: clinical and neuropathological aspects. Neurology 1976;26:15–23.
[12] Robkin MA. Carbon monoxide and the embryo. Int J Dev Biol 1997;11:283–9.
[13] Harper A, Croft Baker J. Carbon monoxide poisoning. Age Ageing 2004;33:105–9.
[14] Ernst A, Zibrak JD. Carbon monoxide poisoning. N Engl J Med 1998;339:1603–8.
[15] Martinez TT, Long C. Explosion risk from swimming pool chlorinators and review of chlorine toxicity. J Toxicol Clin Toxicol 1995;33:349–54.
[16] Babu RV, Cardenas V, Sharma G. Acute respiratory distress syndrome from chlorine inhalation during a swimming pool accident: a case report and review of the literature. J Intensive Care Med 2008;23:275–80.
[17] Gorguner M, Aslan S, Inandi T, Cakir Z. Reactive airways dysfunction syndrome in housewives due to a bleach-hydrochloric acid mixture. Inhal Toxicol 2004;16:87–91.
[18] van der Vliet A, Eiserich JP, Halliwell B, Cross CE. Formation of reactive nitrogen species during peroxidase-catalyzed oxidation of nitrite: a potential additional mechanism of nitric oxide-dependent toxicity. J Biol Chem 1997;272:7617–25.
[19] Winder C. The toxicology of chlorine. Environ Res 2001;85:105–14.
[20] Jones RN, Hughes JM, Glindmeyer H, Weill H. Lung function after acute chlorine exposure. Am Rev Respir Dis 1986;134:1190–5.
[21] Schwartz DA, Smith DD, Lakshminarayan S. The pulmonary sequelae associated with accidental inhalation of chlorine gas. Chest 1990;97:820–5.
[22] Charan NB, Lakshminarayan S, Myers GC, Smith DD. Effects of accidental chlorine inhalation on pulmonary function. West J Med 1985;143:333–6.
[23] Hasan FM, Gehshan A, Fuleihan FJ. Resolution of pulmonary dysfunction following acute chlorine exposure. Arch Environ Health 1983;38:76–80.
[24] Bosse GM. Nebulized sodium bicarbonate in the treatment of chlorine inhalation. J Toxicol 1994;32:233–41.
[25] Aslan S, Kandis H, Akgun M, Cakir A, Inandi T, Gorguner M. The effect of nebulized $NaHCO_3$ treatment on "RADS" due to chlorine gas inhalation. Inhal Toxicol 2000;18:895–900.
[26] Wang J, Winskog C, Edston E, Walther SM. Inhaled and intravenous corticosteroids both attenuate chlorine gas-induced lung injury in pigs. Acta Anaesthesiol Scand 2005;49:183–90.
[27] Hall AH, Rumack BH. Clinical toxicology of cyanide. Ann Emerg Med 1986;15:1067–74.
[28] Egekeze JO, Oehme FW. Cyanides and their toxicity: a literature review. Vet Q 1980;2:104–14.
[29] Izraeli S, Israeli A, Danon Y. Pharmacological treatment of cyanide poisoning. Harefuah 1988;114:338–42.
[30] Groff Sr WA, Stemler FW, Kaminskis A, Froehlich HL, Johnson RP. Plasma free cyanide and blood total cyanide: a rapid completely automated microdistillation assay. Clin Toxicol 1985;23:133–63.
[31] Sunshine I, Finkle B. The necessity for tissue studies in fatal cyanide poisoning. Int Archiv Gewerbepathol Gewerbehyg 1964;20:558–61.
[32] Lambert RJ, Kindler BL, Schaeffer DJ. The efficacy of superactivated charcoal in treating rats exposed to a lethal oral dose of potassium cyanide. Ann Emerg Med 1988;17:595–8.
[33] Graham DL, Laman D, Theodore J, Robin ED. Acute cyanide poisoning complicated by lactic acidosis and pulmonary edema. Arch Intern Med 1977;137:1051–5.
[34] Blake J. Observations and experiments on the mode in which various poisonous agents act on the animal body. Edinb Med Surg J 1840;53:35–49.

[35] Takano T, Miyazaki Y, Nashimoto I. Effect of hyperbaric oxygen on cyanide intoxication: in situ changes in intracellular oxidation reduction. Undersea Biomed Res 1980;7:191–7.
[36] Brivet F, Delfraissy JF, Bertrand P, Dormont J. Acute cyanide poisoning: recovery with non-specific supportive therapy. Intensive Care Med 1983;9:33–5.
[37] Reiffenstein RJ, Hulbert WC, Roth SH. Toxicology of hydrogen sulfide. Annu Rev Pharmacol Toxicol 1992;32:109–34.
[38] Hessel PA, Herbert FA, Melenka LS, Yoshida K, Nakaza M. Lung health in relation to hydrogen sulfide exposure in oil and gas workers in Alberta, Canada. Am J Ind Med 1997;31:554–7.
[39] Kilburn KH. Effects of hydrogen sulfide on neurobehavioral function. South Med J 2003;96:639–46.
[40] Kilburn KH, Thrasher JD, Gray MR. Low-level hydrogen sulfide and central nervous system dysfunction. Toxicol Ind Health 2010;26:387–405.
[41] Guidotti TL. Hydrogen sulfide: advances in understanding human toxicity. Int J Toxicol 2010;29(6):569–81.
[42] Osbern LN, Crapo RO. Dung lung: a report of toxic exposure to liquid manure. Ann Intern Med 1981;95:312–14.
[43] Smith RP, Kruszyna R, Kruszyna H. Management of acute sulfide poisoning: effects of thiosulfate, oxygen, and nitrite. Arch Environ Health 1976;85:756–8.
[44] Stine RJ, Slosberg B, Beacham BE. Hydrogen sulfide intoxication: a case report and discussion of treatment. Ann Intern Med 1976;85:756–8.
[45] Wang DX. A review of 152 cases of acute poisoning of hydrogen sulfide. Chin J Prev Med 1989;23:330–2.
[46] Kilburn K. Case report: profound neurobehavioural deficits in an oil field worker overcome by hydrogen sulfide. Am J Med Sci 1993;306:301–5.
[47] Snyder JW, Safir EF, Summerville GP, Middleberg RA. Occupational fatality and persistent neurological sequelae after mass exposure to hydrogen sulfide. Am J Emerg Med 1995;13:199–203.
[48] Wasch HH, Estrin WJ, Yip P, Bowler R, Cone JE. Prolongation of the P-300 latency associated with hydrogen sulfide exposure. Arch Neurol 1989;46:902–4.
[49] Tvedt B, Edland A, Skyberg K, Forberg O. Delayed neuropsychiatric sequelae after acute hydrogen sulfide poisoning: affection of motor function, memory, vision, and hearing. Acta Neurol Scand 1991;84:348–51.
[50] Guidotti TL. Hydrogen sulfide. Occup Med 1996;46(5):367–71 42 Smilkstein MJ, Bronstein AC, Pickett HM, Rumack BH. Hyperbaric oxygen therapy for severe hydrogen sulfide poisoning.
[51] Beck JF, Brandbury CM, Conors AJ, Donini JC. Nitrite as antidote for acute hydrogen sulfide intoxication? Am Ind Hyg Ass J 1981;42:805–9.
[52] The Merck Index. In: Budavari S, editor. An encyclopedia of chemicals, drugs, and biologicals. 11th ed. Rahway, NJ: Merckand Co. Inc. 1989.
[53] Sittig M. Handbook of toxic and hazardous chemicals and carcinogens. 2nd ed. Park Ridge, NJ: Noyes Publications; 1985.
[54] US Environmental Protection Agency. Integrated risk information system (IRIS) on phosgene. Washington, DC: National Center for Environmental Assessment, Office of Research and Development; 1999.

Further reading

Myers and Synder, 1985   Myers RAM, Synder SK. Subacute sequelae of carbon monoxide poisoning. Ann Emerg Med 1985;14:1163–7.

Chapter 20.1

Evaluation of toxicity following ammonia exposure: a case report

Andrew W. Lyon, Viktor A. Zherebitskiy and Fang Wu

Pathology & Laboratory Medicine, Saskatchewan Health Authority, Saskatoon, SK, Canada

Case description

A 48-year-old male donned personal protective equipment and respirator to repair a small leak from a valve on a large anhydrous ammonia tank truck. After the repair, he rinsed the equipment with water to remove traces of ammonia and to confirm the leaking had stopped. He removed the respirator and accidently slipped on the ammoniac water bucket and was splashed on the side of his face and one eye. He reported his irritated eye and throat following the ammonia exposure to his employer and continued to work and drive the truck for 3 days before seeking attention from an ophthalmologist. He was prescribed antiinflammatory and antibiotic eye creams (fluorometholone and erythromycin ointment). In the morning, 6 days after the exposure, he was found unconscious. Emergency paramedics resuscitated the patient. Laboratory investigations in hospital revealed elevation of liver enzymes (Table 20.1.1). During the next 24 hours the patient's condition deteriorated and he was pronounced dead.

Postmortem examination of the right facial area exposed to ammonia revealed slight yellow discoloration and no overt burn. The right eye was clear with no visual pathology. Both sclerae had yellow discoloration and the patient had mild jaundice. Traces of greasy residue were found at the side of the eye, consistent with use of eye ointment. The patient succumbed with multiorgan failure and had no evidence of significant airway inflammation, necrosis or fibrosis. Marked cerebral swelling was observed (bilateral mild hippocampal uncal herniation and prominent ventricular effacement). The liver had patches of centrilobular necrosis.

Discussion

Ammonia is a colorless gas at room temperature with a unique pungent odor. It is used in large quantities as an agriculture fertilizer in farming industries, refrigeration systems, and household cleaners. Anhydrous ammonia gas rapidly dissolves into moisture in the eyes, skin, mucous membranes of the nose, and upper respiratory system creating strong alkali ammonium hydroxide solutions and causing tissue irritation, inflammation, and necrosis [1]. Levels of exposure to gaseous ammonia are associated with different levels of adverse effects: (1) 25–50 ppm detectable odor—adverse effects unlikely; (2) 250–500 ppm moderate eye irritation, throat irritation, dangerous to life and health; and (3) 2500–5000 ppm necrosis of airway mucosa, acute lung injury and death [2], Table 20.1.2.

Plasma ammonia testing is not used to assess acute gaseous ammonia exposure because the rapid inflammatory response to gaseous ammonia in exposed tissue surfaces is a separate compartment from the internal tissues that are pH-buffered and perfused with blood. Plasma ammonia levels are dependent on the relative rates of endogenous ammonia production by amino acid catabolism and rate of ammonia removal by hepatic formation of urea. Hyperammonemic states are associated with loss of hepatic urea synthesis such as in acute liver failure, cirrhosis, and hepatic encephalopathy [3]. Treatment for ammonia toxic exposure is mainly supportive. Depending on the severity of exposure and respiratory tract injury, patients are treated with presumed respiratory distress or burns [1].

TABLE 20.1.1 Selected Clinical Chemistry Test Results at Hospital Admission of Day 6 After Ammonia Exposure

Tests	Result, S.I. Units (Reference Interval)	Result, Units (Reference Interval)
Bilirubin, direct	32 μmol/L (0–9)	1.8 μg/L (0–0.52)
Bilirubin, total	42 μmol/L (3–21)	2.5 μg/L (0.2–1.2)
Lactate dehydrogenase	1705 U/L (125–220)	1705 U/L (125–220)
Alkaline phosphatase	68 U/L (40–150)	68 U/L (40–150)
Aspartate transaminase	1120 U/L (5–34)	1120 U/L (5–34)
Alanine transaminase	587 U/L (0–55)	587 U/L (0–55)
Creatine kinase	1614 U/L (30–200)	1614 U/L (30–200)
Troponin I	0.288 μg/L (0.000–0.030)	0.288 μg/L (0.000–0.030)
Glucose	13.40 mmol/L (3.6–11.0)	241 mg/dL (65–198)
Urea	9.60 mmol/L (3.0–9.2)	26.9 mg/dL (8.4–25.8)
Creatinine	219 μmol/L (64–111)	2.5 mg/dL (0.7–1.2)

Test results are listed with both S.I. and conventional units. *S.I.*, System International.

TABLE 20.1.2 Attributes of Acute Gaseous Ammonia Toxicity [2]

- Burns and inflammation to eyes, nose, throat, and airways
- Burns and inflammation to skin
- Acute and severe hypoxia due to compromised lung function, loss of consciousness, and death
- Fluid and inflammatory infiltrates in lungs, chronic and mild hypoxia
- Note that laboratory analyses of plasma ammonia are not relevant to acute gaseous ammonia exposures

This patient had an occupational exposure to ammonia and it is relevant to consider if and how ammonia exposure and toxicity contributed to his death. Ammonia exposure to this patient was associated with an irritated eye that may have been painful but did not prevent the truck driver from working for several days. This ammonia exposure did not induce acute injury to airway tissues. Exogenous ammonia exposure is not expected to cause hepatic necrosis and the pattern of necrosis-suggested exposure to a toxin such as acetaminophen. This patient had a painful irritated eye and could have self-medicated with an over-the-counter antiinflammatory medication such as acetaminophen. When the potential for acetaminophen overdose was suggested by the pathologist, the coroner recalled an empty bottle of Tylenol tablets in the patient's room and personal effects. Postmortem toxicology studies subsequently detected 13 mg/L (84 μmol/L) acetaminophen in blood and the stomach contents had 570 mg/L (3800 μmol/L) acetaminophen.

Interpretation of the acetaminophen toxicity based on serum levels of that drug is highly dependent on the time interval between drug exposure and blood sample collection [4]. The levels of acetaminophen observed in this patient's blood days after exposure could not be determined with the Rumack–Matthew nomogram which is limited to 24 hours following ingestion [5]. It was determined that acetaminophen-mediated hepatotoxicity was a major contributor to this patient's mortality.

Academically, it is interesting to contrast the two ways that ammonia contributed morbidity and mortality in this case. First, acute exposure of ammonic water to the patient's face rapidly caused painful irritation and inflammation of an eye. The direct impact of ammonia was limited to the exposed tissues and the influence was immediate and painful. Alternatively, the patient apparently self-medicated with large doses of acetaminophen that resulted in hepatic toxicity. After several days of hepatic injury and necrosis, the patient likely became hyperammonemic (not confirmed by the laboratory in this case) and succumbed to hepatic encephalopathy. In hepatic encephalopathy elevated ammonia partitions into cerebral tissues resulting in swelling of the brain, uncal herniation, and ventricular effacement associated with altered levels of consciousness and death.

References

[1] Makarovsky I, Markel G, Dushnitsky T, Eisenkraft A. Ammonia — when something smells wrong. Isr Med Assoc J 2008;10:537—43.
[2] Roney N, Llados F, Little SS, Knaebel DB. ATSDR toxicological profile for ammonia. Atlanta, GA: Agency for Toxic Substances and Disease Registry; 2004. From: <https://www.atsdr.cdc.gov/toxprofiles/tp126.pdf> [accessed 31.12.18].
[3] Upadhyay R, Bleck TP, Busl KM. Hyperammonemia: what urea-lly need to know: case report of severe noncirrhotic hyperammonemic encephalopathy and review of the literature. Case Rep Med 2016;2016:8512721.
[4] Ghanem CI, Maria JP, Manautou JE, Mottino AD. Acetaminophen; from liver to brain: new insights into drug pharmacological action and toxicity. Pharmacol Res 2016;109:119—31.
[5] Smilkstein MJ, Knapp GL, Kulig KW, Rumack BH. Efficacy of oral N-acetylcysteine in the treatment of acetaminophen overdose. Analysis of the national multicenter study (1976 to 1985). N Engl J Med 1988;319:1557—62.

Chapter 20.2

Two fatalities involving 1,1-difluoroethane

C. Clinton Frazee[1,2], Lindsey J. Haldiman[3], Diane Peterson[3] and Uttam Garg[1,2]

[1]*Department of Pathology and Laboratory Medicine, Children's Mercy Hospital, Kansas City, MO, United States,* [2]*University of Missouri School of Medicine, Kansas City, MO, United States,* [3]*Office of the Jackson County Medical Examiner, Kansas City, MO, United States*

Case history 1

A 41-year-old white male with a history of alcoholism had not been seen in 3 days. A family member pursued a welfare check and found the subject unresponsive on the living room couch inside his residence. The subject was seated with one leg underneath him. There were numerous cans of compressed air visible throughout the living room and in an adjacent room. Vomit was present on both the floor and on the decedent's shirt, but there were no signs of trauma or foul play. Death was confirmed by EMS responders called to the scene.

Heart blood, subclavian blood, vitreous fluid, liver tissue, brain tissue, and gastric contents were submitted as per routine toxicological investigation. The heart blood was screened for volatiles (ethanol, methanol, isopropanol, and acetone) by headspace gas chromatography with flame ionization detector (GC−FID). The heart blood was also used for a comprehensive broad spectrum drug screening that utilized enzyme immunoassays (EIAs) for amphetamines, barbiturates, benzodiazepines, cannabinoids, cocaine metabolite, methadone, opiates, phencyclidine and propoxyphene, and drug-screen testing for >200 drugs by gas chromatography−mass spectrometry (GC−MS). Drug screen testing by GC−MS employed a liquid−liquid alkaline extraction using bicarbonate buffer (pH 11.0) and butyl acetate followed by mass spectrometer detection analysis in full-scan mode. Presumptive identification of analytes was made by spectral library match and relative retention time comparison with reference standards.

The nine panel EIA drug screen was negative for drugs of abuse. The GC−MS drug screen for >200 drugs indicated the presence of caffeine. The volatile screen was negative for ethanol, methanol, isopropanol, and acetone. Given the presence of canned air found at the scene, subclavian blood was submitted to a reference laboratory (NMS Labs, Willow Grove, PA) for 1,1-difluoroethane (difluoroethane) and 1,1,1,2-tetrafluoroethane (tetrafluoroethane) testing by GC−MS. Difluoroethane confirmed positive at 6.5 mg/L. No tetrafluoroethane was detected. All results are summarized in Table 20.2.1.

Case history 2

The subject is a 26-year-old white female who had a history of several police-related interactions involving huffing with compressed air cans. The subject used Uber services for transport to a local Walmart. She purchased four cans of compressed air from the Walmart and then ubered to a drop-off location near her home. She was found 3 days later between a building and some abandoned rail cars. She was still wearing the clothes observed on Walmart surveillance footage recorded at the time of purchase. The plastic bag containing the Walmart receipt was located near her person, as well as 12 additional cans of compressed air. The subject's rigor was easily broken and lividity was consistent with the position found. Insect activity was observed around the body, but no signs of injury or foul play were noted. Heart blood, subclavian blood, urine, liver tissue, brain tissue, and gastric contents were submitted for toxicological analyses. Toxicology testing was performed on heart blood as per aforementioned in the section "Case history 1."

The nine panel EIA drug screen was negative for drugs of abuse and no volatiles were detected by headspace GC−FID. The GC−MS drug screen for >200 drugs indicated the presence of diphenhydramine and diphenhydramine metabolite. The diphenhydramine and metabolite gas chromatographic peaks were relatively small in comparison to

TABLE 20.2.1 Postmortem Toxicology Results

	Results (Case 1)	Results (Case 2)
EIA (Heart Blood)		
Amphetamine	Negative	Negative
Barbiturates	Negative	Negative
Benzodiazepines	Negative	Negative
Cannabinoids	Negative	Negative
Cocaine metabolite	Negative	Negative
Methadone	Negative	Negative
Opiates	Negative	Negative
Phencyclidine	Negative	Negative
Propoxyphene	Negative	Negative
GC–FID Volatile Screening (Heart Blood)		
Ethanol	<10 mg/dL	<10 mg/dL
Acetone	<5 mg/dL	<5 mg/dL
Methanol	<5 mg/dL	<5 mg/dL
Isopropanol	<5 mg/dL	<5 mg/dL
GC–MS screen positive (heart blood)	Caffeine	Diphenhydramine
GC–MS Confirmation (Subclavian Blood)		
1,1-Difluoroethane	6.5 mg/L	140 mg/L

EIA, Enzyme immunoassay; *GC–FID*, gas chromatography with flame ionization detector; *GC–MS*, gas chromatography–mass spectrometry.

internal standard and no quantitation of diphenhydramine was pursued. Given the presence of compressed air at the scene, heart blood was submitted to a reference laboratory (NMS Labs, Willow Grove, PA) for inhalant panel testing. Difluoroethane was confirmed present at 140 mg/L. No other inhalants were detected.

Discussion

Difluoroethane ($C_2H_4F_2$) is a fluorocarbon gas that is colorless and odorless. It is primarily used as a refrigerant, Freon-152a, but it is also frequently found as a propellant in various gas duster products that are marketed as canned or compressed air. Though fluorocarbons were believed to be relatively nontoxic when they were first introduced into the consumer market, there are now numerous documented cases of tissue injury and rapid death following repeated use and/or abuse. Difluoroethane, an active ingredient of "dust propellants," is one of the most frequently abused fluorocarbons. Due to its easy access and availability, its abuse is increasing, particularly among teenagers. In fact, it has been estimated that approximately 11% of high-school children have used inhalants such as difluoroethane at least once. Its abuse is driven by its euphoric effect and addiction.

Like other volatile hydrocarbons, difluoroethane is lipophilic and quickly crosses the blood–brain barrier with immediate CNS effects. Peak blood concentrations occur 10–20 seconds after inhalation. In a study, healthy volunteers performing 50 W exercise were exposed to 0, 200, and 1000 ppm difluoroethane for 2 hours. Capillary blood, urine, and exhaled air were collected for 22 hours. Difluoroethane increased rapidly in blood and reached an average value of 0.5 and 2.3 mg/L in volunteers exposed to 200 and 1000 ppm, respectively [1]. Most of the drugs are exhaled unchanged with small amounts being metabolized or excreted through urine. Difluoroethane is metabolized to fluorocitrate which can both inhibit the citric acid cycle and lead to hypocalcemia [2].

The euphoric high that results from inhaling or "huffing" difluoroethane can last for 15–30 minutes. Clinical presentation varies and depends on dose and exposure time. Overdose symptoms include nausea, vomiting, abdominal pain, paresthesia, confusion, slurred speech, diaphoresis, pulmonary irritation, and altered mental status. Continued exposure results in the loss of consciousness, seizures, cardiac arrhythmias, cardiac arrest, and death [3]. A number of cases of clinical toxicity and fatality have been reported. A 34-year-old white male with a history of depression and seizure disorder was found unconscious in a parking lot with 15 empty cans of dust-offs [4]. He was arousable and was transported to an emergency department. ECG revealed sinus tachycardia and prolonged QTc of 472 ms. The patient had left and right ventricular dysfunction with an ejection fraction of 25%. The patient also had frostbite on his hand due to refrigerant exposure. Other cases involving cardiac abnormalities or toxicity have been reported [5−7]. Cates and Cook reported a case of severe cardiomyopathy following a history of recurrent and heavy abuse of difluoroethane [4]. Acute and chronic kidney injuries have also been reported after difluoroethane use. A 32-year-old Caucasian with a known history of depression and violent outburst was brought to emergency department following prolonged difluoroethane abuse during a suicide attempt. The patient had elevated creatinine and blood urea nitrogen (BUN) and decreased GFR of 54 mL/min/1.73 m^2. The urinalysis was significant for 3 + proteinuria, moderate blood, 14 RBCs, and hyaline cast [8]. Despite very little metabolism, an interesting case of skeletal fluorosis following chronic difluoroethane use has been described. A 28-year-old man with no history of health problems developed abnormal gait, left hip pain, loss of mobility in his right wrist and forearm, and progressive deformities following 2 years of difluoroethane use. Dual-energy X-ray absorptiometry of his lumbar spine, femoral neck, and total hip revealed bone mineral density Z-scores of +6.2, +4.8, and +3.0, respectively. Serum, urine, and bone fluoride levels were all elevated [9].

A number of fatal cases involving difluoroethane use have also been reported [3,5,10−12]. Difluoroethane abuse has been implicated in motor vehicle accidents with postmortem blood levels ranging from 35 to 86 mg/L [10,13]. Following acute inhalation of difluoroethane in 14 adults, postmortem difluoroethane peripheral levels averaged 124 mg/L and spanned a very large range of concentrations that are as low as 3 mg/L to as high as 380 mg/L [3]. In 12 adults, ages 18−44 years, who died of excessive abuse of difluoroethane had postmortem blood levels of 86−413 mg/L in heart blood, 20−133 mg/kg in brain, and 28−188 mg/kg in liver [14].

Difluoroethane is typically not included in a routine toxicology screen and could easily be overlooked without the appropriate case history information and/or anatomical pathology findings. We reviewed 27 difluoroethane findings in cases received by our laboratory from local police departments (12 cases) and medical examiners (15 cases) in an 8-year period ranging from 2010 to 2018. The median difluoroethane concentration in traffic stops was 13.5 mg/L with a median suspect age of 30.5 years old. The median difluoroethane concentration in fatalities was observed at 43 mg/L with a median decedent age of 39 years old. The user age in all 27 observed cases ranges from 17 to 56 years old. In both the sections "Case history 1" and "Case history 2," we present a decedent who had not been seen for 3 days prior to discovery. The difluoroethane concentration observed in the section "Case history 1" (6.1 mg/L) is not typical of what we have observed in our review of fatality submissions. In fact, it is among one of the lower levels that we have found reported for a difluoroethane-related fatality in literature. However, no other significant drugs or volatiles were detected by routine toxicological screening. The difluoroethane concentration observed in the section "Case history 2" is more consistent with our observations of difluoroethane-related fatalities to date. Because the subject bodies were in the beginning stages of decomposition in both cases, peripheral blood was not available and C/P comparisons that are helpful in determining time of death could not be made. Still, these two cases provide an interesting comparison of difluoroethane levels and illustrate the dangers related to canned air huffing. They also augment the importance of historical information and data collection from the scene when conducting routine toxicological investigations.

References

[1] Ernstgard L, Lind B, Andersen ME, Johanson G. Liquid-air partition coefficients of 1,1-difluoroethane (HFC152a), 1,1,1-trifluoroethane (HFC143a), 1,1,1,2-tetrafluoroethane (HFC134a), 1,1,1,2,2-pentafluoroethane (HFC125) and 1,1,1,3,3-pentafluoropropane (HFC245fa). J Appl Toxicol 2010;30:59−62.

[2] Arroyo JP, Johnson DC, Lewis JB, Al Sheyyab A, King A, Danter MR, et al. Treatment of acute intoxication from inhaled 1,2-difluoroethane. Ann Intern Med 2018;169:820−2.

[3] Vance C, Swalwell C, McIntyre IM. Deaths involving 1,1-difluoroethane at the San Diego County Medical Examiner's Office. J Anal Toxicol 2012;36:626−33.

[4] Cates AL, Cook MD. Severe cardiomyopathy after huffing dust-off. Case Rep Emerg Med 2016;2016:9204790.

[5] Avella J, Wilson JC, Lehrer M. Fatal cardiac arrhythmia after repeated exposure to 1,1-difluoroethane (DFE). Am J Forensic Med Pathol 2006;27:58−60.

[6] Joshi K, Barletta M, Wurpel J. Cardiotoxic (arrhythmogenic) effects of 1,1-difluoroethane due to electrolyte imbalance and cardiomyocyte damage. Am J Forensic Med Pathol 2017;38:115−25.

[7] Kumar S, Joginpally T, Kim D, Yadava M, Norgais K, Laird-Fick HS. Cardiomyopathy from 1,1-difluoroethane inhalation. Cardiovasc Toxicol 2016;16:370–3.
[8] Calhoun K, Wattenbarger L, Burns E, Hatcher C, Patel A, Badam M, et al. Inhaling difluoroethane computer cleaner resulting in acute kidney injury and chronic kidney disease. Case Rep Nephrol 2018;2018:4627890.
[9] Tucci JR, Whitford GM, McAlister WH, Novack DV, Mumm S, Keaveny TM, et al. Skeletal fluorosis due to inhalation abuse of a difluoroethane-containing computer cleaner. J Bone Mineral Res 2017;32:188–95.
[10] Broussard LA, Brustowicz T, Pittman T, Atkins KD, Presley L. Two traffic fatalities related to the use of difluoroethane. J Forensic Sci 1997;42:1186–7.
[11] Xiong Z, Avella J, Wetli CV. Sudden death caused by 1,1-difluoroethane inhalation. J Forensic Sci 2004;49:627–9.
[12] Yamada G, Takaso M, Kane M, Furukawa S, Hitosugi M. A fatality following difluoroethane exposure with blood and tissue concentrations. Clin Toxicol (Phila) 2018;56:1–2.
[13] Hahn T, Avella J, Lehrer M. A motor vehicle accident fatality involving the inhalation of 1,1-difluoroethane. J Anal Toxicol 2006;30:638–42.
[14] Baselt RC. Fluorocarbons. In: Baselt RC, editor. Disposition of toxic drugs and chemicals in man. Seal Beach, CA: Biomedical Publications; 2017. p. 916–8.

Chapter 20.3

A case of difluoroethane toxicity—sudden sniffing death syndrome

Milad Webb[1,2] and Jeffrey Jentzen[1,2]
[1]*Wayne County Medical Examiner's Office, Detroit, MI, United States,* [2]*Department of Pathology, University of Michigan Health System, Ann Arbor, MI, United States*

Case description

The decedent, a 39-year-old white female, was brought into the emergency department, allegedly suffering from a "huffing"-related overdose. The decedent presented pale and clammy, but in stable condition. An intravenous line and blood draw were attempted, when suddenly she appeared to have a seizure and became unresponsive. Despite resuscitation efforts in the emergency department, the decedent passed.

External examination

Initial examination of the body was done in the hospital emergency room. The body was supine on a hospital bed and covered with blankets to the chest. She was cool at the arms and warm at the legs. Rigor was not observed and lividity appeared consistent with body position and blanched easily. She was intubated, an IV to the back of her right hand and an attempt for an IV to the back of the left hand. There was fecal material between her legs. Her pupils were fixed and dilated. There were no rashes or skin lesions present on the face. There was no injury or trauma identified on the body.

Death investigation

Per investigation, and review of medical records, the decedent called emergency medical services at 0340 hours with concerns that she was overdosing while huffing air duster cans. While on the phone with emergency personnel, she stated that she had no suicidal ideations and she was using the air dusters to try to sleep. This was not her first time, and she claimed to have used four cans that day.

Upon the arrival of the emergency medical services at the home, the decedent was alert and oriented, laying on the floor of her bedroom. There was vomit in front of her. She complained of nausea but denied any pain. She was able to walk with assistance to the ambulance. Her vital signs were reported as stable. Upon arrival to the emergency department (0419 hours), she was able to walk into the hospital and into a bed.

She was evaluated by an emergency room physician (0425 hours). She reported that she had been "huffing" for 24–36 hours, continuously. She reported being nauseated and thirsty. The physician informed her that an intravenous line would be soon placed and they will be able to give her fluids. Vital signs at this time were reported to be stable (blood pressure 92/64; pulse 90; respirations 18; and SPO$_2$ 97%–99%).

An emergency room nurse entered into the decedent's room (0435 hours) and started to prepare for the blood draw and placement of the intravenous line. The vital signs are reported to be stable and unchanged at this point in time. As the nurse attempted to place the intravenous line, the decedent stated, "I don't feel good..." and trailed off. She simultaneously stiffened in the bed, writhed with extended neck (decerebrate posturing).

Suddenly, within seconds, all body movements stopped and the decedent was limp. Emergency staff called a "code" and noted that the decedent went into ventricular tachycardia followed by asystole. The time of arrest was 0443 hours, 57 minutes from the emergency call. The decedent was pronounced dead at 0503 hours, after 20 minutes of unsuccessful resuscitation.

Postmortem examination

The postmortem examination revealed no significant abnormality. Examination of the brain showed slight cerebellar tonsillar notching.

Toxicology

Postmortem toxicology was performed on the first draw blood requested from the emergency department. Along with the expanded panel for prescription and recreational drugs, the blood was tested for volatile compounds. The decedent's blood was positive for 1,1-difluoroethane (DFE) (52 mcg/mL) and mitragynine (18 ng/mL). Documented deaths associated with DFE abuses have reported in postmortem blood concentrations as low as 29 μg/mL.

DFE is an organofluorine compound that is a colorless gas. It is used as a refrigerant, where it is often listed as R-152a (refrigerant-152a) or HFC-152a (hydrofluorocarbon-152a). It is also used as a propellant for aerosol sprays and in gas duster products. In regards to environmental effects, DFE has an ozone depletion potential of zero, a lower global warming potential, and a shorter atmospheric lifetime (1.4 years) when compared to typical chlorofluorocarbons (CFCs) (55–140 years), making it an ideal commercial and industrial product [1]. DFE has been recognized as a substance of abuse that can lead to serious injury and death.

Inhalation of hydrocarbons can induce cardiac dysrhythmias. The mechanism that causes this disturbance in normal cardiac conduction is thought to be associated with increased sensitivity of the myocardium to endogenous catecholamines. Halogenated hydrocarbons are the most likely to cause cardiac sensitization. In addition, they are known to have negative inotropic, dromotropic, and chronotropic effects [1,2].

Various effects have been suggested as contributing to the cardiac sensitization. Slowing of repolarization and slowing of conduction create an opportunity for reentrant arrhythmias [3]. Several cardiac channels have been shown to alter their function after exposure to various hydrocarbons, to include inhibition calcium and potassium channels and alteration of sodium currents. It has been shown that inhibition of calcium and potassium channels by volatile anesthetics can facilitate afterdepolarizations, which is arrhythmogenic [1,3]. In addition, it is also believed that conduction through gap junctions is inhibited via dephosphorylation of connexin proteins [3–5].

Although the electrical function of the heart can be altered with acute exposure to hydrocarbons, prolonged use can cause structural damage that may also impede normal function. Samples of cardiac muscle taken from inhalant abusers have shown interstitial edema, intramyocardial hemorrhages, contraction band necrosis, swollen and ruptured myofibrils, and myocarditis and interstitial fibrosis [6,7].

The effects are often unpredictable in relation to concentration and exposure time and fatal effects can occur in first-time users. Toxicity is related to the dose, as well as the chemical characteristics of volatility, lipid solubility, viscosity, and surface tension. Those with higher solubility are better absorbed after inhalation [1].

Inhalants are a broad range of household and industrial chemicals whose volatile vapors or pressurized gases can be concentrated and breathed in via the nose or mouth to produce intoxication. The desired effects of inhalant abuse include euphoria, lightheadedness, and a general state of intoxication similar to that produced by alcohol or marijuana. The effects usually last for only 15–30 minutes but can be sustained by continuous or repeated use [8].

CFCs were developed in the 1930s, and heavy use continued until the 1970s when environmental impacts of these compounds became more readily known. Reports of human toxicity from inhalant abuse began in the 1950s as use of hydrocarbons became more widespread [1]. Freon (halogenated hydrocarbons used as a refrigerant) was abused at the beginning of the 1950s and was implicated in deaths in the 1970s [9]. Glue and other products containing hydrocarbons were initially abused in the 1960s [10].

Acute hydrocarbon exposure can result in a wide array of pathology, such as encephalopathy, pneumonitis, arrhythmia, acidosis, and dermatitis. Clinical effects can be somewhat predicted by substance, route of exposure, and dose. Intentional inhalational and accidental ingestion exposures with aspiration lead to the greatest morbidity and mortality.

Interestingly, once there is sufficient sensitization of the myocardium, there need to be a triggered surge to initiate the ultimate downward spiral to sudden cardiac death. In many cases, this trigger is thought to be masturbation. In this case, based on careful investigation, the trigger was the placement of a vascular line in the emergency room.

There is little in the medical literature on treatment and management for hydrocarbon inhalant exposure/abuse; however, what information is available indicates that patients who have ventricular arrhythmias following known inhalant abuse should receive amiodarone or lidocaine rather than epinephrine other catecholamines (e.g., norepinephrine) because these compounds can theoretically precipitate or worsen arrhythmias in the irritable myocardium [11].

In consideration of the investigation, postmortem examination, and postmortem toxicology, the cause of death was certified as DFE toxicity and the manner of death was certified as an accident.

References

[1] Tormoehlen LM, Tekulve KJ, Nañagas KA. Hydrocarbon toxicity: a review. Clin Toxicol 2014;52(5):479–89.
[2] Müller SP, Wolna P, Wünscher U, Pankow D. Cardiotoxicity of chlorodibromomethane and trichloromethane in rats and isolated rat cardiac myocytes. Arch Toxicol 1997;71(12):766–77.
[3] Himmel HM. Mechanisms involved in cardiac sensitization by volatile anesthetics: general applicability to halogenated hydrocarbons? Crit Rev Toxicol 2008;38(9):773–803.
[4] Zhou Y, Wu HJ, Zhang YH, Sun HY, Wong TM, Li GR. Ionic mechanisms underlying cardiac toxicity of the organochloride solvent trichloromethane. Toxicology 2011;290(2–3):295–304.
[5] Jiao Z, De Jesús VR, Iravanian S, Campbell DP, Xu J, et al. A possible mechanism of halocarbon-induced cardiac sensitization arrhythmias. J Mol Cell Cardiol 2006;41(4):698–705.
[6] Sakai K, Maruyama-Maebashi K, Takatsu A, Fukui K, Nagai T, et al. Sudden death involving inhalation of 1,1-difluoroethane (HFC-152a) with spray cleaner: three case reports. Forensic Sci Int 2011;206(1–3):e58–61.
[7] Kumar S, Joginpally T, Kim D, Yadava M, Norgais K, Laird-Fick HS. Cardiomyopathy from 1,1-difluoroethane inhalation. Cardiovasc Toxicol 2016;16(4):370–3.
[8] Ossiander EM. Volatile substance misuse deaths in Washington State, 2003-2012. Am J Drug Alcohol Abuse 2015;41(1):30–4.
[9] Caplan JP, Pope AE, Boric CA, Benford DA. Air conditioner refrigerant inhalation: a habit with chilling consequences. Psychosomatics 1970;53(3):273–6.
[10] Bass M. Sudden sniffing death. JAMA 1970;212(12):2075–9.
[11] Perry H. Inhalant abuse in children and adolescents. UpToDate Inc; 2019.

Chapter 20.4

Carbon monoxide poisoning

Julia E. Esswein[1] and D. Adam Algren[2,3,4]

[1]University of Missouri-Kansas City School of Medicine, Kansas City, MO, United States, [2]Division of Clinical Pharmacology, Toxicology, and Therapeutic Innovation, Children's Mercy Hospital, Kansas City, MO, United States, [3]Department of Emergency Medicine, Truman Medical Center, Kansas City, MO, United States, [4]University of Kansas Hospital Poison Control Center, Kansas City, MO, United States

Case history

A 34-year-old man with no known medical illness presented to the emergency department (ED) complaining of headache, malaise, and nausea from the past week. He was seen in the same ED approximately 3 days ago, concerned that he had the flu. A rapid flu test at that time was negative. He was diagnosed with a viral syndrome; however, his symptoms have not subsided, and his wife has also began to complain of headache and fatigue. Vitals for the patient are as follows: T 37°C, HR 104, RR 16, BP 131/86, O_2 sat 97%. When asked about sick contacts, the patient stated that his work was closed about a week ago due to a power outage and that he did not notice any sick colleagues when they returned to work. On further questioning, he revealed that his home had also lost power, and he had used a generator in his basement for several days.

Due to the history of recent indoor generator use, a venous blood gas was obtained and showed a carboxyhemoglobin level of 19%. He and his wife were placed on 100% oxygen nonrebreather masks and observed for 3 hours. Their levels improved, and they were discharged with instructions to move the generator outdoors and install carbon monoxide detectors in the home.

Case discussion

Carbon monoxide (CO) is a tasteless, colorless, and odorless gas that is formed by combustion reactions with limited access to oxygen, also known as an "incomplete combustion reaction" [1]. Sources of unintentional CO poisoning include coal-burning furnaces, portable generators, household fires, and vehicle and boat exhaust [2]. Because of its evasiveness of human senses, and its nonspecific symptoms, including headache, dizziness, nausea, and chest pain, coma and death may occur before patients seek medical care. Approximately 2244 deaths were due to unintentional CO poisoning between 2010 and 2015 [3]. According to a study performed by the CDC from 1999 to 2012, the home was the site exposure to CO in more than 60% of patients. Fortunately, the rate of intentional poisoning resulting in death has decreased by about half in the last two decades [4].

Carbon monoxide exerts its effects on the body by binding to hemoglobin, forming carboxyhemoglobin. Hemoglobin binds to CO with an affinity of 200 times that of oxygen [5]. Because of this high affinity, the oxygen dissociation curve is shifted to the left, impairing oxygen delivery to tissues, resulting in hypoxia [5]. CO also binds to cytochromes, myoglobin, and guanylyl cyclase, all of which are heme-containing compounds. Binding of CO to cytochromes impairs cellular respiration by interrupting the electron transport chain, leading to a lactic acidosis [5,6]. CO binding to guanylyl cyclase activates the C protein, which utilizes guanylyl triphosphate as an energy source, and thus the amount of cellular cyclic guanylyl monophosphate increases [6,7]. The most notable result of this signaling is a decrease in intracellular calcium, leading to smooth muscle relaxation, and vasodilation [7].

Carbon monoxide also binds to the heme-containing protein nitric oxide synthase on the surface of platelets and endothelial cells [8]. This damage to cell membranes causes neutrophil chemotaxis and initiates the inflammatory cascade [5], which leads to an increased production of nitric oxide (NO), which contributes to the pathogenesis of CO toxicity [8]. NO reacts with superoxide and produces peroxynitrite and causes DNA damage and ultimately cell death [9]. NO can also cause a cascade of oxidative stress, which can lead to brain lipid peroxidation, causing more long-term effects evident even after CO poisoning is treated [6]. CO-binding guanylyl cyclase and NO can lead to a profound hypotension due to vasodilation of blood vessels. Overwhelming of the compensatory tachycardia and increase in cardiac output make cardiovascular collapse likely [8,10,11].

CO toxicity has rather nonspecific symptoms, which can lead delayed or misdiagnoses. Suspicion should be high when a multiple family members present with such nonspecific symptoms, people who live in homes with poor ventilation, and people who use gas-burning appliances. Any victims of structure fires should also be tested [5]. An important distinguishing feature of the history is illicit when symptoms are worst, as they may improve as patients are removed from the exposure.

Acute toxicity symptoms include headache, nausea, dizziness, tachycardia, and syncope [6] but may be as severe as cognitive dysfunction, seizures, arrhythmias, and coma [10]. The exam may be significant for decreased Glascow Coma Scale or an abnormal mental status exam [6]. Underlying cardiac disease may be exacerbated by the lack of oxygen present in CO poisoning [6]. This hypoxia also predisposes patients to myocardial infarctions [11]. Physical exam of these patients is typically significant for a tachycardia [10]. It is important to note that finger pulse oximeter will be normal, as pulse oximetry cannot distinguish carboxyhemoglobin from oxyhemoglobin; thus a cooximeter must be used [6].

Hypoxia and hypotension also cause brain injury, via lipid peroxidation once reperfusion to the area occurs [5,6] due to free radical stress [6]. Although these insults may present during intoxication as syncope, confusion, or coma, these can have lasting effects. Persistent or delayed neurologic sequelae can result from CO poisoning and have a wide variety of neurologic and psychiatric manifestations, including confusion, ataxia, hallucinations, incontinence, and parkinsonism [6]. Patients may recover from these various symptoms; however, permanent neurologic sequelae may occur [10].

The distinguishing laboratory test to confirm a diagnosis of CO poisoning is a carboxyhemoglobin (CO-Hgb) level on blood gas [6]. It is important to note that the baseline CO-Hgb level in a nonsmoker is approximately 1%–3%, while in smokers, the baseline CO-Hgb level may be as high as 10% [6]. Another cause of an elevated baseline CO-Hgb is hemolytic disease, including spherocytosis, sickle cell anemia, thalassemia, and other hemolytic anemias [12]. In patients with CO exposure related to a fire, hydroxocobalamin is often administered as empiric treatment for possible cyanide toxicity [13]. However, there are reports that hydroxocobalamin can interfere with the light absorption used by cooximetry and may show falsely low levels of CO-Hgb [14]. This interference can last for several hours. Therefore ideally labs and the blood gas should be obtained prior to the administration of hydroxocobalamin [14].

For supportive care a complete blood count, electrolytes, and renal function and monitored as CO poisoning can cause an acute kidney injury as the result of rhabdomyolysis [6]. An ECG and troponin should be obtained in those that are symptomatic with chest pain [6]. A pregnancy test is important to obtain as carbon monoxide can affect the fetus. A CT scan of the brain may show damage related to hypoxia, including injury to the basal ganglia [6].

The goal of treatment for CO poisoning is to decrease the amount of CO-Hgb and return oxyhemoglobin to normal levels [15]. At this time, there is still debate over whether treatment with normobaric oxygen (NBO) or hyperbaric oxygen (HBO) is preferred. A majority of trials to date do not demonstrate benefit of HBO in preventing neurologic sequelae. Some studies suggest that HBO be used only if certain conditions are met, for example, if the patient is unconscious, if there is evidence of severe sequelae such as neurological or cardiovascular effects or acidosis, if the CO-Hgb level is >25%, or if the patient is pregnant [16–18]. In a study done by Weaver et al., HBO was associated with a decrease in neurologic sequelae by 46% at 6 weeks [19]. Of note, there were methodologic concerns with the study, including changing the primary outcome definition of midtrial, which could have biased the study in favor of HBO. Ultimately, there was no functional difference between the HBO and NBO groups [20]. In addition, in another study by Scheinkestel, no benefit to HBO was demonstrated and HBO complications occurred in 9%, compared to NBO complications which occurred at an incidence of 1% [21]. The most commonly noted complications were anxiety and aural barotrauma, though other, more serious complications are possible. Ultimately, the choice between NBO and HBO remains up to the treating physician at this time.

Carbon monoxide poisoning is the cause of a considerable number of ED visits due to accidental and intentional intoxication every year. The nonspecific symptoms can make it difficult to diagnose and can allow for continued exposure until more drastic symptoms present. Diagnosis must be made by blood-level measurement. Finally, treatment with HBO versus NBO remains controversial, as no definitive conclusions can be made as to whether the benefits of HBO outweigh the risks, but it may have more relevance if the poisoning is acute and clinically severe. Use of carbon monoxide detectors should be encouraged to prevent poisoning.

References

[1] Thompson M. Carbon monoxide − molecule of the month. UK: Winchester College. Available from: <http://www.chm.bris.ac.uk/motm/co/coh.htm> [accessed 27.04.19].

[2] Hall R, Earnest G, Hammond D, Dunn K, Garcia A. A summary of research and progress on carbon monoxide exposure control solutions on houseboats. J Occup Env Hyg 2014;11(7):D92−100.

[3] [No authors listed]. QuickStats: number of deaths resulting from unintentional carbon monoxide poisoning, by month and year—National Vital Statistics System, United States, 2010−2015. MMWR Morb Mortal Wkly Rep 2017;66:234.

[4] Sircar K, Clower J, Shin MK, Bailey C, King M, Yip F. Carbon monoxide poisoning deaths in the United States, 1999 to 2012. Am J Emerg Med 2015;33:1140−5.

[5] Maloney G. Carbon monoxide. In: Tintinalli J, Stapczynski J, Ma O, Yealy D, Meckler G, Cline D, editors. Tintinalli's emergency medicine: a comprehensive study guide, 8e. New York: McGraw-Hill; 2016.

[6] Kao L, Nañagas K. Carbon monoxide poisoning. Med Clin North Am 2005;89(6):1161−94.

[7] Denninger J, Marletta M. Guanylate cyclase and the NO/cGMP signaling pathway. Biochim Biophys Acta Bioenerg 1999;1411(2−3):334−50.

[8] Macnow T, Waltzman M. Carbon monoxide poisoning in children: diagnosis and management in the emergency department. Pediatr Emerg Med Pract 2016;13:1−24.

[9] Ramdial K, Franco M, Estevez A. Cellular mechanisms of peroxynitrite-induced neuronal death. Brain Res Bull 2017;133:4−11.

[10] Rose J, Wang L, Xu Q, McTiernan C, Shiva S, Tejero J, et al. Carbon monoxide poisoning: pathogenesis, management, and future directions of therapy. Am J Respir Crit Care Med 2017;195:596−606.

[11] Lippi G, Rastelli G, Meschi T, Borghi L, Cervellin G. Pathophysiology, clinics, diagnosis and treatment of heart involvement in carbon monoxide poisoning. Clin Biochem 2012;45:1278−85.

[12] Engel RR, Rodkey FL, Krill Jr CE. Carboxyhemoglobin levels as an index of hemolysis. Pediatrics 1971;47:723−30.

[13] Thompson J, Marrs T. Hydroxocobalamin in cyanide poisoning. Clin Toxicol 2012;50(10):875−85.

[14] Livshits Z, Lugassy D, Shawn L, et al. Falsely low carboxyhemoglobin level after hydroxocobalamin therapy. N Engl J Med 2012;367:1270−1.

[15] Weaver L. Carbon monoxide poisoning. N Engl J Med 2009;360(12):1217−25.

[16] Thom S. Hyperbaric-oxygen therapy for acute carbon monoxide poisoning. N Engl J Med 2002;347:1105−6.

[17] Brent J. What does the present state of knowledge tell us about the potential role of hyperbaric oxygen therapy for the treatment of carbon monoxide poisoning? Toxicol Rev 2005;24(3):145−7.

[18] Henry J. Hyperbaric therapy for carbon monoxide poisoning: to treat or not to treat, that is the question. Toxicol Rev 2005;24(3):149−50.

[19] Weaver L, Hopkins R, Chan K, et al. Hyperbaric oxygen for acute carbon monoxide poisoning. N Engl J Med 1999;347:1057−67.

[20] Buckley NA, Isbister GK, Stokes B, Juurlink DN. Hyperbaric oxygen for carbon monoxide poisoning. Toxicol Rev 2005;24(2):75−92.

[21] Scheinkestel C, Myles P, Cooper D, Millar I, Tuxen D, Bailey M, et al. Hyperbaric or normobaric oxygen for acute carbon monoxide poisoning: a randomised controlled clinical trial. Med J Aust 1999;1999(170):203−10.

Chapter 21

Toxic metals

Frederick G. Strathmann and Riley Murphy
NMS Labs, Horsham, PA, United States

Introduction

To study the elements is to study nearly every aspect of life itself. From history to politics, alchemy and mythology, elements have been a constant and influential presence. Elemental toxicity has been featured in numerous high profile cases in both the real and fictitious setting. Though not often at the top of a medical differential for disease or forensic investigation as cause of death, elemental toxicity remains a fascinating and sometimes underappreciated area of toxicology.

Few elements are more prevalent in the clinical and forensic toxicology setting than arsenic, mercury, and lead—a point emphasized by their ranking as the top three substances on the Agency for Toxic Substances and Disease Registry (ATSDR) 2017 Substance priority list of hazardous substance due to a combination of frequency, toxicity, and human exposure potential [1]. Arsenic and mercury have been known as both poison and medicine, with arsenic remaining an available treatment to this day for acute promyelocitic leukemia. Though widely studied and toxicity profiles known for decades, these three elements continue to dominate the headlines of academic and mainstream culture based on a continued need to address exposure occurrences in numerous contexts.

Mechanisms of action

Arsenic

Arsenic is listed as the No.1 toxicant on the ATSDR 2017 substance priority list of hazardous substances. Arsenic occurs in both organic and inorganic forms, the toxicity of each varying considerably. Ranging from most toxic to largely benign, inorganic, methylated, and organic arsenic are all encountered as analytical targets in the clinical and forensic setting. Inorganic arsenic has numerous proposed mechanisms of action underlying its toxicity. Arsenic avidly binds to a cofactor for pyruvate dehydrogenase via characteristic sulfhydryl group binding inhibiting the conversion of pyruvate to acetyl coenzyme A; however, generation of reactive oxygen species has also been proposed as the mechanism in the observed inhibition of pyruvate dehydrogenase activity [2]. With similar chemical properties to phosphorous, arsenic can substitute for phosphate in biochemical reactions [3] resulting in a phenomenon termed "arsenolysis" [4]. The most widely studied mechanism of action for arsenic is arguably the generation of reactive oxygen species such as superoxide anions, hydrogen peroxide, and hydroxyl radicals [5]. Carcinogenesis of arsenic is due to genetic and epigenetic alterations with arsenic being listed as a carcinogen since 1987 [6], as well as alterations in DNA repair, signal transduction, and cell proliferation [7]. More often than not, elevated total arsenic concentrations are attributed to organic arsenic, largely considered benign with the most prevalent source of exposure from recent seafood ingestion [8].

Lead

Lead is listed as the No.2 toxicant on the ATSDR 2017 substance priority list of hazardous substances. The toxicity of lead, well known today, was underappreciated as evidence by its widespread use in paints, ceramic products, gasoline, and solder. In 1978 the limit of lead in paint was reduced to <0.5% (from as much as 35% w/w) and exposure through inhalation was dramatically reduced with its removal from gasoline in approximately the same year after decades of struggle [9,10]. Despite the widely accepted toxicity of lead, exposure remains a constant concern for good reason, epitomized by the crisis in Flint Michigan regarding water-supply contamination [11]. The toxicity of lead includes several mechanisms. Inhibition of delta aminolevulinic acid dehydratase, delta aminolevulinic acid synthase, and ferrochelatase causes the accumulation of protoporphyrin in red blood cells and is still used as an indicator of lead toxicity [12]. The substitution of zinc for iron in the protoporphyrin IX structure results in the formation of zinc protoporphyrin, a relatively insensitive indicator of lead toxicity though widely used in occupational exposure assessment [13]. Like other toxic metals, lead forms covalent bonds with the sulfhydryl group of cysteines in proteins causing dysfunction and even structural changes [14]. The neurotoxicity of lead is exacerbated in children, as the developing nervous system is particularly susceptible to lead toxicity [12,14]. In addition, iron deficiency is highly correlated with and has been shown to increase the predisposition for lead toxicity, possibly through the upregulation of gastrointestinal iron transporters also capable of transporting lead [15].

Mercury

Mercury is listed as the No.3 toxicant on the ATSDR 2017 substance priority list of hazardous substances. Mercury has a long history with natural and industrial sources of exposure. In past years, mercury was used extensively in devices such as thermometers, barometers, manometers, and sphygmomanometers, in large part due to its high density and heat conduction characteristics. Concern over the toxicity of mercury has led to its replacement in numerous versions of these devices. Mercury continues to be used in dental amalgams and mercury vapor lamps, continuing the ongoing debate about the harm associated with low level mercury exposure [16,17]. Elemental mercury is relatively harmless, and if ingested, will pass through the body with exceedingly low absorption. Once used for therapeutic reasons, mercury-containing laxatives were commonly used for explorers to offset the gastrointestinal stress of encountering new dietary foliage [18]. Once ionized, inorganic mercury becomes toxic with alkylation producing species known to be lipophilic and exceedingly neurotoxic [19]. The toxicity of inorganic mercury occurs through several mechanisms. First, reaction with sulfhydryl groups of proteins causes a loss of protein structural integrity. Second, these structural changes can induce immunogenicity causing immune-mediate tissue damage. Third, organic mercury species bind lipophilic proteins causing disruption in cell types such as neurons and the myelin sheaths encapsulating them [20].

Pharmacokinetics and toxicokinetics

Arsenic

Arsenic exposure is complex due to the various forms in which arsenic exists. In humans, inorganic arsenic is metabolized to the methylated species primarily in the liver; however, further metabolism to the organic forms of arsenic does not occur [21]. The metabolism of arsenic into the less toxic methylated forms results in a urine excretion profile with inorganic forms seen shortly after ingestion, while methylated forms persist longer than 24 hours [22]. In blood, inorganic arsenic has relatively short half-life of 4–6 hours while methylated species have a half-life of 20–30 hours. The disappearance of arsenic into the phosphate pool of the body reduces the utility of arsenic in blood as a marker of exposure [23]. Arsenic excretion in individuals exposed to organic arsenic occurs over several days, with a reported half-life of organic arsenic of 4–6 hours [21]. Interestingly, arsenic may have been one of the original analytes to be detected in nails for forensic purposes [24], with Mees lines a hallmark characteristic of arsenic exposure though attributable to a disruption of nail growth rather than deposition of arsenic itself [21]. Brown rice has been continuously implicated in low level arsenic exposure [25,26], yet contamination of opium has also been reported as a source of exposure in various countries [27–29]. Urine is most often the specimen of choice for the assessment of arsenic toxicity, primarily due to the short half-life of arsenic in blood. Though often done initially as an assessment of total arsenic [30], fractionation or speciation of the specimen using chromatographic means or the use of a back extraction technique is required to distinguish toxic from nontoxic arsenic exposure.

Lead

The majority of ingested lead is excreted in stool. Absorbed lead is rapidly incorporated into bone and red blood cells with eventual distribution among all tissues [31]. Lipid-dense tissues such as the central nervous system are particularly sensitive to organic forms of lead [32]. Because of the sulfhydryl group binding of lead, hair is a particularly good specimen for detecting lead exposure due to the presence of cysteine-rich keratin [33]. Deposition of lead into bone occurs as lead and calcium is used interchangeably, resulting in both a long-term pool and a readily exchangeable pool contributing to endogenous lead exposure [34]. Lead is excreted from the body mainly through urine and feces, though lead may be excreted to a minor degree in sweat and saliva [35].

Toxic effects of lead can be acute after single ingestion of a large dose (e.g., ingestion of a lead salt [36]) or as is more common through long-term, chronic exposure [37]. However, in recent years several reports of acute lead toxicity in association with opium use have been seen. Between 2004 and 2016, 18 studies reported 324 cases of lead poisoning in opium abusers in Iran. Adulteration during preparation and distribution to increase the weight of opium were the leading explanation for the presence of lead [38]. Blood is the preferred specimen type for the assessment of lead toxicity, while urine may be useful in monitoring efficacy of chelation therapy.

Mercury

Elemental mercury has little to no toxicity due to poor absorption; however, absorption of elemental mercury is near 75% with inhalation exposures [39]. Once absorbed, elemental mercury is taken up by red blood cells, easily crosses the blood-brain barrier, and is widely distributed throughout the body, ultimately becoming trapped in tissues after oxidation [40]. In contrast, inorganic mercury does not easily cross the blood—brain barrier but instead is heavily bound to sulfhydryl groups in the plasma and in red blood cells [41]. The half-life of inorganic mercury ranges from approximately 20 to 60 days, with elimination predominantly through the urine and feces [42]. Organic mercury is readily absorbed, easily crosses the blood-brain barrier, and concentrates in several tissues including brain, liver, kidney, and hair [43]. Exposure to methyl mercury, with a half-life of approximately 50 days, occurs through ingestion of fish and sea mammals and is more rapidly metabolized to inorganic mercury underlying one reason for its observed damage to kidneys [20]. In comparison, ethyl mercury is less toxic, has a shorter half-life of approximately 7—10 days, and for many years was used as a preservative for vaccines in a sulfur-bound form known as thimerosal [44]. The use of thimerosal in vaccines remains a fiercely debated topic despite much of the original research indicating links to disorders such as autism having been largely discredited [45]. Analysis of blood, urine, and hair for mercury can be used to determine exposure.

Treatment

Arsenic

Like many metal toxicities, signs and symptoms of arsenic toxicity overlap with symptoms of other toxicants. Acute arsenic toxicity includes gastrointestinal stress, vomiting, diarrhea, and cardiac arrhythmias [46]. In chronic exposures, renal failure, cardiac arrhythmias, liver dysfunction, and peripheral neuropathy have all been reported [47,48]. Treatment for acute arsenic poisoning may involve decontamination though gastrointestinal upset leading to vomiting reduces the likely need in severe ingestions [21]. Two chelating agents are available in the United States for arsenic, dimercaprol [British anti-Lewisite (BAL)—originally developed during World War II in response to the use of lewisite [49]] and meso-2,3-dimercaptosuccinic acid (DMSA, succimer). Unithiol (sodium 2,3-dimercapto-1-propane sulfonate) may have the most favorable profile for the treatment of acute poisoning by inorganic arsenic though currently it has not been approved by the US Food and Drug Administration [50,51].

Lead

In adults, lead concentrations greater than 30 µg/dL as indicative of significant exposure [52] while concentrations greater than 60 µg/dL prompt strong consideration of chelation therapy. For pediatric cases, elimination of the phrase "lead level of concern," a shift to the use of 5 µg/dL as the reference value for lead-exposure hazard, and continued use of 45 µg/dL as the concentration prompting chelation therapy were all revisited by the Advisory Committee on Childhood Lead Poisoning Prevention and the Centers for Disease Control and Prevention [53]. Treatment for chronic lead exposure is to first identify the source of exposure, determine if others are at risk, and remove the source of exposure. In the case of elevated lead concentrations in blood, decontamination of a foreign body and/or chelation therapy may be required. Though several chelating agents are available, including BAL, DMSA, ethylenediaminetetraacetic acid (EDTA), penicillamine, and dimercaptopropane sulfonic acid, DMSA is currently the preferred treatment [54].

Mercury

For elemental mercury, inhalation is the route associated with toxicity and initial treatment is focused largely on management of any pulmonary effects [42]. BAL is considered a poor choice for chelation as it can redistribute mercury from peripheral to central sites, such as the central nervous system [55]. DMSA is administered in a manner similar to lead toxicity, with monitoring of the blood and urine to guide repeat dosage [51]. Treatment of inorganic mercury toxicity is similar to absorbed elemental mercury exposure; however, BAL may be substituted early in treatment when oral chelation is not tolerated or due to gastrointestinal complications [56]. Prevention of exposure to organic mercury is the only effective treatment for organic mercury exposure, as chelating agents for patients with toxic exposure to organic mercury has not been shown to be of benefit. Chelation with BAL is contraindicated because of the likelihood of increasing the redistribution of mercury to the brain. Treatment with oral DMSA has not be shown to be effective in reversing neurologic damage [57,58].

Analytical methods and clinical management implications

Elemental analytical techniques in the modern laboratory are similar in principle to the spectroscopic elemental analysis developed by Bunsen and Kirchhoff in the 1800s. In their revolutionary experiments, Bunsen and Kirchhoff discovered new elements by their atomic emission lines after introduction to a flame. Modern instruments in the laboratory still analyze elements based on the atomic emission or atomic absorption (AA) lines. In the 1980s the inductively coupled argon plasma developed for atomic emission measurements was coupled to a mass spectrometer to create the very first applications of inductively coupled plasma mass spectrometry (ICP-MS) [59]. There are elemental analysis techniques that do not rely upon the atomic emission or absorption lines associated with valence electron transitions. These include neutron activation analysis, X-ray fluorescence, ion selective electrodes, and wet chemical techniques. However, these are becoming less common in the elemental analysis laboratory in comparison to AA, atomic emission spectroscopy (AES)/optical emission spectroscopy, and ICP-MS. This section will focus primarily on ICP-MS techniques for one of the same reasons that ICP-MS is becoming increasingly popular in the elemental laboratory—the capability to analyze all elements of toxicological concern.

The founding principle of AA, AES, or ICP-MS is that an element in gaseous state generates a quantifiable signal that is proportional to its concentration. Some of the challenges to obtaining accurate results include inadequate sensitivity, interferences, nonlinearity, and matrix effects. One of the greatest strengths of ICP-MS is its excellent sensitivity, which allows for increasingly lower reporting limits for toxic elements. Interferences may be platform and analyte specific and have posed some of the greatest challenges to analysis by ICP-MS. While inadequate sensitivity and interferences challenge analysis at low concentrations, matrix effects and linearity can challenge accurate quantification at any concentration. Another advantage of the ICP-MS is its wide linear analytical measurement range. However, matrix interferences can pervade ICP-MS analysis as well as other techniques. Therefore particular effort must be made to during method development and validation to address interferences and matrix effects. Poor accuracy and precision in volumetric sampling can be problematic in any analytical technique. A challenge particular to analytical sample preparation for elemental analysis is contamination of reagents, more so for elements that are ubiquitous in nature. Spectroscopic challenges and challenges regarding sample preparation will be discussed.

Arsenic

The analysis of arsenic represents a technique that is challenging in terms of both matrix effects and interferences that are resolvable using modern techniques and instrumentation. The functioning principle of an ICP-MS is that elements are measured as a single atom most often a single charge. The principle begins to break down when ions are analyzed that contain multiple charges or multiple atoms, the latter referred to as polyatomic ions. While the high temperature of the plasma does dissociate most elemental bonds, some polyatomic ions can produce an interfering signal, especially for major constituents of biological fluids like Cl and O. Arsenic, which is monoisotopic at 75 amu, is most pervasively interfered by the polyatomic ions $^{40}Ar^{35}Cl^+$ and $^{40}Ca^{35}Cl^+$. This particular interference is the most problematic for a single quadrupole ICP-MS. Higher resolution mass analyzers such as the sector field can easily discriminate between subinteger differences in m/z that can be the case for single atom to polyatomic ions [60]. These sector field instruments are not as common due in part to cost considerations and lack of robust performance for high throughput and/or routine analyses. To circumvent the known interferences with arsenic measurement, many laboratories have chosen to measure arsenic with atomic absorption spectroscopy (AAS) or AES. Currently, however, most ICP-MS instruments come with additional components that aid in reducing polyatomic interferences.

One method of minimizing polyatomic interferences, which is becoming an increasingly standard component of ICP-MS instruments, is kinetic energy discrimination (KED). In this mode of operation, a collision cell (e.g., octopole and flatopole) is inserted before the quadrupole mass analyzer. By adding helium gas in the collision cell and applying an electric field as a potential energy barrier, the instrument is able to discriminate between the polyatomic ions and the monatomic ions. KED is an effective technique at reducing polyatomic interferences but does so at the cost of analytical sensitivity. The most recent advancement that is currently commercially available is the triple quadrupole ICP-MS or QQQ-ICP-MS. In these instruments, reaction gas such as oxygen or ammonia can selectively react with the analyte ion to intentionally form a polyatomic ion. This formation occurs in a reaction cell situated between two separate quadrupoles. Analyte that passes through the first quadrupole selectively reacts with the reaction mass and the new, shifted mass based on the polyatomic ion formed is selected for in the second quadrupole. Alternative scenarios exist, for example, an interfering ion may be selectively reacted resulting in its mass shift to differentiate it from the analyte. The introduction of KED and QQQ instruments has further enhanced the capabilities of ICP-MS analysis. The reduction of interferences is allowing laboratories to detect elements with high analytical sensitivity lower limits of detection. The QQQ in particular is making ICP-MS be an increasingly versatile technique.

An additional challenge to the analysis of arsenic is sensitivity to matrix effects. Strictly speaking, the signal generated by arsenic will vary depending on the matrix being measured (e.g., urine, serum, blood, or dilute acid). There are a number of variables that can affect the analyte signal, including variation in nebulization efficiency and ion throughput through the interface. The addition of a separate element to be used as an internal standard can help to correct for some changes in physical transport such as nebulization efficiency. Selection of an appropriate internal standard can be difficult and are typically chosen to match based on both mass and ionization potential. Perhaps the biggest constraint on an internal standard is that they must not be found endogenously in the sample being measured. However, it is not always possible to match the behavior of one element with another. In the case of arsenic the response will vary from matrix to matrix based on ionization efficiency in the plasma. Since the response of any element in an ICP-MS is proportional to the percentage of the element that is ionized, changes in the ionization efficiency will result poor compensation and inaccurate quantitation. In the case of arsenic, different sample matrices can have an effect on the ionization efficiency due to its high ionization energy. In particular, the introduction of carbon to the plasma has a significant enhancement on the ionization of certain elements with high ionization energies. The mechanism is attributed to a transfer of charge from the carbon ion to an element with similar ionization energy. A commonly used mechanism to obtain this result is to add 1%–2% of alcohol (e.g., isopropyl alcohol, methanol, and ethanol) to the diluent [61–63].

The chromatographic separation of arsenic species, often referred to as speciation or fractionation, is becoming increasingly popular despite the increase in instrument complexity. Improvements to the chromatographic technique including shorter run times pose some advantage over extraction techniques. Separation of species before analysis has another benefit relative to total arsenic analysis. During chromatography the chloride ion is separated from the arsenic species, the effect being the elimination of the polyatomic interferences of $^{40}Ar^{35}Cl^+$ and $^{40}Ca^{35}Cl^+$ from arsenic determination.

Lead

The analysis of lead is in many ways straightforward compared to other elements. Lead measurements by ICP-MS are not highly impacted from matrix effects or interferences. The typical sample preparation for ICP-MS analysis is a simple nitric acid dilution. A potential problem with acidification of whole blood, the primary matrix of choice for lead assessment is that it can lead to coagulation and protein precipitation. The challenge of blood dilution has historically been overcome with the brute force technique of acid digestion. These acid digestion techniques typically employ heat and an oxidizing agent such as nitric acid to oxidize blood proteins. While this technique is effective, it is labor-intensive and time-consuming and, therefore, expensive. More modern techniques use a simple dilution step for blood, wherein an alkaline diluent containing ammonia or other additives is used [64–67].

The other significant improvement alluded to earlier is the reduction of lead contamination within the laboratory. The challenge of lead contamination was most famously brought to light as a concern in environmental water testing in environmental protection agency (EPA) labs by Clair Patterson [68]. In the current day, problems of contamination are more generally acknowledged and most elemental labs employ at least some form of clean room protocol. For rare elements (e.g., thallium), clean room protocols are not necessary because there are no likely sources of contamination. However, some elements that can be tested for toxicity are also abundant (e.g., aluminum, manganese, nickel, and chromium) and require additional consideration, such as a clean environment and clean handling. Another source of contamination is the sample container itself. Ideally, the container will be screened for contamination since glass containers can leach elements such as barium into the sample, while some plastic containers may contain antimony.

Mercury

In its elemental form, mercury is difficult to analyze by ICP-MS for the same reason that it is poorly absorbed in the digestive tract. The uncharged form has properties similar to a noble gas and has very little affinity for the aqueous phase [69]. While nitric acid is a strong enough oxidizer to oxidize most elements, it is not strong enough to oxidize mercury. Some researchers have addressed this problem with the use of stronger oxidizing species to force the oxidation of mercury. Some oxidizing species found in the literature include gold [70] and chromate [71]. While these techniques have been effective, they have drawbacks such as contamination of the instrument with the oxidizing reagent. A separate approach to ICP-MS is the direct mercury analyzer, which takes advantage of the unique chemistry of mercury rather than attempting to treat it in a similar fashion to other elements. In this approach, a sample is heated under a stream of air, which transports volatilized mercury to a gold amalgamator that effectively concentrates the mercury. After the mercury is trapped in the amalgamator, the amalgam is heated and the mercury is detected by AA.

Forensic considerations

Arsenic

In a detailed case report by Tournel et al. of a successful suicide attempt by a 30-year-old woman that self-injected arsenate, the initial signs of arsenic toxicity (gastrointestinal distress) were initially overlooked by family members due to a history of ethanol intoxication and drug use [72]. Fourteen hours post ingestion, the woman died of sudden circulatory arrest. Postmortem fluid samples (blood, bile, gastric content, and urine) were collected and analyzed for arsenic, and all but urine was found to contain either arsenite (antemortem femoral blood) or a combination of arsenate and dimethylarsenic acid (DMA) (cardiac blood, bile, and gastric content). Of note, the lack of arsenic in urine and hair ruled out the potential for chronic arsenic exposure as a contributing factor. The source of arsenic used was found to be a fungicide taken from a brother that had recently traveled to Madagascar to grow spirulina.

Mercury

In the case of a 53 year-old female schoolteacher, evidence of mercury poisoning was collected 6 years after death [73]. At the time of her death, no toxicological testing was performed despite the woman being hospitalized twice in the course of several months with gastrointestinal issues, including abdominal pain, nausea, vomiting, and diarrhea. During the last hospitalization where kidney failure and circulatory—respiratory insufficiency was noted, myocardial infarction was reported as the cause of death. Several years later, new circumstances led police to the woman's son-in-law where invoices for several mercury compounds, arsenic and related chelators was found in proximity to a textbook entitled *Handbook on Toxicology*. After exhumation of the body, inner organs and hair were collected for analysis with little other tissue being available for testing because of advanced putrefaction. Numerous heavy metals were found in concentrations typical for the general population, including lead, cadmium, and chromium. Arsenic was not detected. All inner organ samples (intestine, liver, and heart) were found to contain high concentrations of mercury in addition to a minor elevation in tested hair. Of note, the claim by the suspect that a nearby trash dump was the source of mercury exposure was refuted by the assessment of air, soil, and other environmental samples from the surrounding area.

References

[1] Centers for disease control and prevention. Substance priority list I ATSDR, <https://www.atsdr.cdc.gov/spl/>; [accessed 25.10.18].
[2] Samikkannu T, Chen CH, Yih LH, Wang AS, Lin SY, Chen TC, et al. Reactive oxygen species are involved in arsenic trioxide inhibition of pyruvate dehydrogenase activity. Chem Res Toxicol 2003;16:409—14.
[3] Roggenbeck BA, Banerjee M, Leslie EM. Cellular arsenic transport pathways in mammals. J Environ Sci (China) 2016;49:38—58.
[4] Nemeti B, Anderson ME, Gregus Z. Glutathione synthetase promotes the reduction of arsenate via arsenolysis of glutathione. Biochimie 2012;94:1327—33.
[5] Sun HJ, Rathinasabapathi B, Wu B, Luo J, Pu LP, Ma LQ. Arsenic and selenium toxicity and their interactive effects in humans. Environ Int 2014;69:148—58.

[6] Galanis A, Karapetsas A, Sandaltzopoulos R. Metal-induced carcinogenesis, oxidative stress and hypoxia signalling. Mutat Res 2009;674:31–5.
[7] Hughes MF, Beck BD, Chen Y, Lewis AS, Thomas DJ. Arsenic exposure and toxicology: a historical perspective. Toxicol Sci 2011;123:305–32.
[8] Heinrich-Ramm R, Mindt-Prufert S, Szadkowski D. Arsenic species excretion after controlled seafood consumption. J Chromatogr B Anal Technol Biomed Life Sci 2002;778:263–73.
[9] Richmond-Bryant J, Meng Q, Davis JA, Cohen J, Svendsgaard D, Brown JS, et al. A multi-level model of blood lead as a function of air lead. Sci Total Environ 2013;461-462:207–13.
[10] Needleman H. Lead poisoning. Annu Rev Med 2004;55:209–22.
[11] Zahran S, McElmurry SP, Sadler RC. Four phases of the flint water crisis: evidence from blood lead levels in children. Environ Res 2017;157:160–72.
[12] Papanikolaou NC, Hatzidaki EG, Belivanis S, Tzanakakis GN, Tsatsakis AM. Lead toxicity update. A brief review. Med Sci Monit: Int Med J Exp Clin Res 2005;11:RA329–36.
[13] Wildt K, Berlin M, Isberg PE. Monitoring of zinc protoporphyrin levels in blood following occupational lead exposure. Am J Ind Med 1987;12:385–98.
[14] de Burbure C, Buchet JP, Leroyer A, Nisse C, Haguenoer JM, Mutti A, et al. Renal and neurologic effects of cadmium, lead, mercury, and arsenic in children: evidence of early effects and multiple interactions at environmental exposure levels. Environ Health Perspect 2006;114:584–90.
[15] Kwong WT, Friello P, Semba RD. Interactions between iron deficiency and lead poisoning: epidemiology and pathogenesis. Sci Total Environ 2004;330:21–37.
[16] Roberts HW, Charlton DG. The release of mercury from amalgam restorations and its health effects: a review. Oper Dent 2009;34:605–14.
[17] Kern JK, Geier DA, Bjorklund G, King PG, Homme KG, Haley BE, et al. Evidence supporting a link between dental amalgams and chronic illness, fatigue, depression, anxiety, and suicide. Neuro Endocrinol Lett 2014;35:537–52.
[18] Kean S. 1st Back Bay pbk. ed The disappearing spoon: and other true tales of madness, love, and the history of the world from the periodic table of the elements, 391. New York: Back Bay Books; 2011. p. 9.
[19] Costa LG, Aschner M, Vitalone A, Syversen T, Soldin OP. Developmental neuropathology of environmental agents. Annu Rev Pharmacol Toxicol 2004;44:87–8110.
[20] Clarkson TW, Magos L, Myers GJ. The toxicology of mercury—current exposures and clinical manifestations. N Engl J Med 2003;349:1731–7.
[21] Caravati EM. Arsenic and arsine gas Vol In: Dart RC, editor. Medical toxicology. 3rd ed. Philadelphia, PA: Lippincott, Williams & Wilkins; 2004. p. 1393–401.
[22] Vahter M. What are the chemical forms of arsenic in urine, and what can they tell us about exposure? Clin Chem 1994;40:679–80.
[23] Hall M, Chen Y, Ahsan H, Slavkovich V, van Geen A, Parvez F, et al. Blood arsenic as a biomarker of arsenic exposure: results from a prospective study. Toxicology 2006;225:225–33.
[24] Lappas NT, Lappas CM. Analytical samples Vol Forensic toxicology: principles and concepts. Waltham, MA: Elsevier; 2016. p. 123.
[25] Zavala YJ, Duxbury JM. Arsenic in rice: I. Estimating normal levels of total arsenic in rice grain. Environ Sci Technol 2008;42:3856–60.
[26] Davis MA, Mackenzie TA, Cottingham KL, Gilbert-Diamond D, Punshon T, Karagas MR. Rice consumption and urinary arsenic concentrations in U.S. Children. Environ Health Perspect 2012;120:1418–24.
[27] Wijesekera AR, Henry KD, Ranasinghe P. The detection and estimation of (a) arsenic in opium, and (b) strychnine in opium and heroin, as a means of identification of their respective sources. Forensic Sci Int 1988;36:193–209.
[28] Narang AP, Chawla LS, Khurana SB. Levels of arsenic in Indian opium eaters. Drug Alcohol Depend 1987;20:149–53.
[29] Balachandra AT, Balasooriya BA, Athukorale DN, Perera CS, Henry KD. Chronic arsenic poisoning in opium addicts in Sri Lanka. Ceylon Med J 1983;28:29–34.
[30] Hackenmueller SA, Strathmann FG. Total arsenic screening prior to fractionation enhances clinical utility and test utilization in the assessment of arsenic toxicity. Am J Clin Pathol 2014;142:184–9.
[31] Christensen JM, Seiler HG, Sigel A, Sigel H, Kristiansen J, Seiler HG. Vol. 1994 ed Lead. Handbook on metals in clinical and analytical chemistry. CRC; 1994. p. 753.
[32] Verity MA. Comparative observations on inorganic and organic lead neurotoxicity. Environ Health Perspect 1990;89:43–8.
[33] Gil F, Hernández AF, Márquez C, Femia P, Olmedo P, López-Guarnido O, et al. Biomonitorization of cadmium, chromium, manganese, nickel and lead in whole blood, urine, axillary hair and saliva in an occupationally exposed population. Sci Total Environ 2011;409:1172–80.
[34] Jacobs JJ, Skipor AK, Patterson LM, Hallab NJ, Paprosky WG, Black J, et al. Metal release in patients who have had a primary total hip arthroplasty. A prospective, controlled, longitudinal study. J Bone Jt Surg Am 1998;80:1447–58.
[35] Skerfving S, Bergdahl IA. Lead Vol In: Nordberg G, Fowler B, Nordberg M, Friberg L, editors. Handbook on the toxicology of metals. 3rd ed. Amsterdam; Boston: Academic Press; 2007. p. 599–643.
[36] Baselt RC. Vol Lead. Disposition of toxic drugs and chemicals in man. 9th ed. Seal Beach, CA: Biomedical Publications; 2011. p. 894–6.
[37] Dart RC, Hurlbut KM, Boyer-Hassen LV. Lead Vol In: Dart RC, editor. Medical toxicology. 3rd ed. Philadelphia, PA: Lippincott, Williams & Wilkins; 2004. p. 1423–31.
[38] Soltaninejad K, Shadnia S. Lead poisoning in opium abuser in Iran: a systematic review. Int J Prev Med 2018;9:3.
[39] Cherian MG, Hursh JB, Clarkson TW, Allen J. Radioactive mercury distribution in biological fluids and excretion in human subjects after inhalation of mercury vapor. Arch Environ Health 1978;33:109–14.

[40] Hursh JB, Sichak SP, Clarkson TW. In vitro oxidation of mercury by the blood. Pharmacol Toxicol 1988;63:266–73.
[41] Smith JC, Allen PV, Turner MD, Most B, Fisher HL, Hall LL. The kinetics of intravenously administered methyl mercury in man. Toxicol Appl Pharmacol 1994;128:251–6.
[42] Dart RC, Sullivan JB. Mercury Vol In: Dart RC, editor. Medical toxicology. 3rd ed. Philadelphia, PA: Lippincott, Williams & Wilkins; 2004. p. 1437–48.
[43] Bjorklund G, Dadar M, Mutter J, Aaseth J. The toxicology of mercury: current research and emerging trends. Environ Res 2017;159:545–54.
[44] Pichichero ME, Cernichiari E, Lopreiato J, Treanor J. Mercury concentrations and metabolism in infants receiving vaccines containing thiomersal: a descriptive study. Lancet 2002;360:1737–41.
[45] Eggertson L. Lancet retracts 12-year-old article linking autism to MMR vaccines. CMAJ 2010;182:E199–200 Canada.
[46] Szymanska-Chabowska A, Antonowicz-Juchniewicz J, Andrzejak R. Some aspects of arsenic toxicity and carcinogenicity in living organisms with special regard to its influence on cardiovascular system, blood and bone marrow. Int J Occup Med Environ Health 2002;15:101–16.
[47] Platanias LC. Biological responses to arsenic compounds. J Biol Chem 2009;284:18583–7.
[48] Jomova K, Jenisova Z, Feszterova M, Baros S, Liska J, Hudecova D, et al. Arsenic: toxicity, oxidative stress and human disease. J Appl Toxicol 2011;31:95–107.
[49] Howles JK. The treatment of heavy metal poisoning with British anti-Lewisite. N Orleans Med Surg J 1949;101:435–40.
[50] Wax PM, Thornton CA. Recovery from severe arsenic-induced peripheral neuropathy with 2,3-dimercapto-1-propanesulphonic acid. J Toxicol Clin Toxicol 2000;38:777–80.
[51] Blanusa M, Varnai VM, Piasek M, Kostial K. Chelators as antidotes of metal toxicity: therapeutic and experimental aspects. Curr Med Chem 2005;12:2771–94.
[52] Fewtrell LJ, Pruss-Ustun A, Landrigan P, Ayuso-Mateos JL. Estimating the global burden of disease of mild mental retardation and cardiovascular diseases from environmental lead exposure. Environ Res 2004;94:120–33.
[53] Centers for Disease Control and Prevention. CDC response to advisory committee on childhood lead poisoning prevention recommendations in "low level lead exposure harms children: a renewed call of primary prevention", <http://www.cdc.gov/nceh/lead/acclpp/cdc_response_lead_exposure_recs.pdf> 2012; [accessed 15.04.15].
[54] Kim HC, Jang TW, Chae HJ, Choi WJ, Ha MN, Ye BJ, et al. Evaluation and management of lead exposure. Ann Occup Environ Med 2015;27:30.
[55] Kosnett MJ. The role of chelation in the treatment of arsenic and mercury poisoning. J Med Toxicol 2013;9:347–54.
[56] Jones AL, Flanagan RJ. Dimercaprol Vol In: Dart RC, editor. Medical toxicology. 3rd ed. Philadelphia, PA: Lippincott, Williams & Wilkins; 2004. p. 185–6.
[57] Baum CR. Treatment of mercury intoxication. Curr Opin Pediatr 1999;11:265–8.
[58] Sears ME. Chelation: harnessing and enhancing heavy metal detoxification—a review. Sci World J 2013;2013:219840.
[59] Houk RS, Fassel VA, Flesch GD, Svec HJ, Gray AL, Taylor CE. Inductively coupled argon plasma as an ion-source for mass-spectrometric determination of trace-elements. Anal Chem 1980;52:2283–9.
[60] Fiket Z, Mikac N, Kniewald G. Arsenic and other trace elements in wines of eastern Croatia. Food Chem 2011;126:941–7.
[61] McShane WJ, Pappas RS, Wilson-McElprang V, Paschal D. A rugged and transferable method for determining blood cadmium, mercury, and lead with inductively coupled plasma-mass spectrometry. Spectrochim Acta, B—Atomic Spectrosc 2008;63:638–44.
[62] Dressler VL, Pozebon D, Curtius AJ. Introduction of alcohols in inductively coupled plasma mass spectrometry by a flow injection system. Anal Chim Acta 1999;379:175–83.
[63] Hu ZC, Hu SH, Gao S, Liu YS, Lin SL. Volatile organic solvent-induced signal enhancements in inductively coupled plasma-mass spectrometry: a case study of methanol and acetone. Spectrochim Acta, B—Atomic Spectrosc 2004;59:1463–70.
[64] Lu Y, Kippler M, Harari F, Grander M, Palm B, Nordqvist H, et al. Alkali dilution of blood samples for high throughput ICP-MS analysis-comparison with acid digestion. Clin Biochem 2015;48:140–7.
[65] Barany E, Bergdahl IA, Schutz A, Skerfving S, Oskarsson A. Inductively coupled plasma mass spectrometry for direct multi-element analysis of diluted human blood and serum. J Anal At Spectrom 1997;12:1005–9.
[66] Delves HT, Campbell MJ. Measurements of total lead concentrations and of lead isotope ratios in whole-blood by use of inductively coupled plasma source-mass spectrometry. J Anal At Spectrom 1988;3:343–8.
[67] Schutz A, Bergdahl IA, Ekholm A, Skerfving S. Measurement by ICP-MS of lead in plasma and whole blood of lead workers and controls. Occup Environ Med 1996;53:736–40.
[68] Kehoe RA. Contaminated and natural lead environments of man. Arch Environ Health 1965;11:736–9.
[69] Norrby LJ. Why is mercury liquid - or, why do relativistic effects not get into chemistry textbooks. J Chem Educ 1991;68:110–13.
[70] Chan MHM, Chan IHS, Kong APS, Osaki R, Cheung RCK, Ho CS, et al. Cold-vapour atomic absorption spectrometry underestimates total mercury in blood and urine compared to inductively-coupled plasma mass spectrometry: an important factor for determining mercury reference intervals. Pathology 2009;41:467–72.
[71] Nixon DE, Burritt MF, Moyer TP. The determination of mercury in whole blood and urine by inductively coupled plasma mass spectrometry. Spectrochim Acta, B—Atomic Spectrosc 1999;54:1141–53.
[72] Tournel G, Houssaye C, Humbert L, Dhorne C, Gnemmi V, Becart-Robert A, et al. Acute arsenic poisoning: clinical, toxicological, histopathological, and forensic features. J Forensic Sci 2011;56(Suppl. 1):S275–9.
[73] Lech T. Detection of mercury in human organs and hair in a case of a homicidal poisoning of a woman autopsied 6 years after death. Am J Forensic Med Pathol 2015;36:227–31.

Chapter 21.1

Toxicity of heavy metals—case study

Frederick G. Strathmann
NMS Labs, Horsham, PA, United States

Case description

A 50-year-old female was found in her home disoriented, vomiting who eventually became unresponsive before being taken to the emergency department (ED). At the time of arrival to the ED the patient was incoherent and not responding appropriately to prompting. The patient progressed to severe muscle rigidity with violent thrashing causing self-injury. Other signs included nystagmus and foaming at the mouth. The patient had no history of a seizure disorder, but the medical staff were informed that numerous herbal supplements were found in the home. Organophosphates were initially considered; however, the classic SLUDGE toxidrome (salivation, lacrimation, urination, diarrhea, GI upset, and emesis) was not noted. Over the course of several hours, bleeding gums and dark brown urine were observed. The patient was provided Ativan, fluids, and palliative care while laboratory testing ensued. Approximately 12 hours after presentation to the ED, the patient was pronounced dead.

Extensive laboratory testing was conducted including immunoassay and mass spectrometry-based screening of urine and blood. In addition, several powders and solutions found in the decedents home were submitted for testing. Excluding medications given during treatment and what was ultimately determined to be a false-positive immunoassay for amphetamine in urine, all toxicology findings were unremarkable except an elevated arsenic urine test.

Laboratory report 1—total arsenic

The laboratory report indicated a total urine arsenic concentration of 91 mcg/L. Of note, the interpretive data provided indicated that the concentration of arsenic was consistent with unexposed individuals; however, in the "reference comments" section it was noted, "urinary arsenic concentrations are typically <50 mcg/L in unexposed individuals. With occasional seafood consumption the normal urinary arsenic concentration may be increased to between 200 and 1700 mcg/L."

Laboratory report 2—total inorganic arsenic

The laboratory report indicated a total *inorganic* urine arsenic concentration of 48 mcg/L and 99 mcg/g creatinine after normalization to creatinine to account for concentration of the urine (481.3 mg/L creatinine reported). The interpretive data indicated that the "diagnosis of arsenic exposure is based on 24-hour urinary arsenic and creatinine concentrations of equal to or greater than 50 mcg/L, 100 mcg/g creatinine or 100 mcg total arsenic; inorganic arsenic concentrations are typically less than 20 mcg/L."

Discussion and follow-up

Arsenic exerts its toxic affects by inhibiting cellular respiration through the targeting of mitochondrial ATP production. Symptoms of acute arsenic ingestion typically appear within 30 minutes but may be delayed if arsenic is consumed with food. The most commonly involved organ symptoms include the central and peripheral nervous systems, cardiovascular system, gastrointestinal system, genitourinary system, hematopoietic system, and the dermatologic system. Due to the acute nature of the included presentation, only those relevant to acute ingestion are highlighted. The reader is referred to Chapter 22, Venoms, for more details on arsenic toxicity and detailed analytical approaches to testing.

Central nervous system

Acute arsenic exposure has been associated with seizures, delirium, and encephalopathy progressing to coma and death.

Cardiovascular system

Numerous abnormalities have been reported including conduction blocks, QT interval prolongation and T-wave changes. Ventricular tachycardia and fibrillation have been reported after acute ingestion. Arsenic is known to cause dilatation of blood vessels and may progress to severe shock.

Gastrointestinal system

Metallic taste and garlic odor to the breath have been associated with arsenic ingestion. Difficulty swallowing, nausea, vomiting, abdominal pain, and watery/bloody diarrhea have been reported after acute arsenic ingestion. As a result, electrolyte and fluid imbalances will typically be present with hypovolemic shock occurring. Liver damage may be indicated by elevated liver enzymes such as aspartate aminotransferase (AST) and alanine aminotransferase (ALT).

Hematopoietic system

Hemolysis (lysing of red blood cells) has been reported in conjunction with arsenic exposure. Intravascular coagulation has also been reported.

Miscellaneous symptoms

Excessive salivation, darkened urine, and skin rash have all been reported.

Findings from therapeutic exposures

Arsenic trioxide is indicated for induction of remission and consolidating in adult patients with relapsed/refractory acute promyelocytic leukemia. Symptoms associated with arsenic trioxide administration include the following:

1. QT interval prolongation and complete atrioventricular block
2. ventricular arrhythmia
3. hyperglycemia
4. hypokalemia
5. neutropenia
6. increased ALT
7. hemorrhage
8. diarrhea
9. nausea
10. mortality from disseminated intravascular coagulation (DIC)

Background on the clinical assessment of arsenic

Urine is the preferred specimen for assessing arsenic exposure in large part due to the short half-life of arsenic in blood ($t_{1/2}$ of 12−36 hours). The current American Conference of Governmental Industrial Hygienists biological exposure index for inorganic arsenic and its methylated metabolites is 35 mcg/L in urine at the end of the workweek.

One critical aspect of arsenic testing interpretation was the speciation or fractionation of total arsenic into inorganic, methylated, and organic arsenic in report 2. Inorganic arsenic is considered the most clinically significant form with the highest degree of toxicity. Methylated arsenic is of moderate toxicity. Organic arsenic is considered relatively benign and its presence in the urine is typically a result of seafood exposure. Minor amounts of methylated species, in the presence of organic arsenic, are also consistent with seafood exposure.

Inorganic arsenic exists in a trivalent or pentavalent form (the sum of both comprising an inorganic laboratory value) and are metabolized in humans to both methyl and dimethyl forms of arsenic (the sum of both comprising a methylated laboratory value). Humans are incapable of producing organic arsenic from inorganic or methylated forms and there is no evidence for organic arsenic breakdown in humans into methylated or inorganic arsenic.

Specific interpretation

The initial total arsenic level of 91 mcg/L was of concern as the majority of clinical laboratories use 35 mcg/L as the upper limit for reflex to fractionation to determine which arsenic species are present. Due to the method used by the performing laboratory for the speciation, the secondary results included only inorganic arsenic at a concentration of 48 mcg/L. A concentration of 43 mcg/L (91 total − 48 inorganic) remain unclassified and cannot be conclusively determined based upon the provided information. However, the provided concentration of 48 mcg/L of inorganic arsenic is above the 35 mcg/L threshold for clinically significant inorganic and/or methylated arsenic.

Timing between potential exposure and testing is a key component to appropriate case interpretation. In similarly reported, nonfatal cases of acute arsenic ingestion, urine specimens collected at various time points post ingestion ranged from 40 to 970 mcg/L, with the range of concentrations highlighting the time-dependent excretion of arsenic. As it is uncertain when the arsenic ingestion occurred in this case, it is not possible to surmise if the measured concentrations of arsenic are before, equal to, or after the peak of arsenic excretion into the urine. Of note, the majority of excreted arsenic is in the methylated forms with less than 20% present in the inorganic forms. A ratio of 1:1 (methylated: inorganic), as is possible but currently remains unproven in this case, is consistent with a proposed though heavily debated "methylation threshold hypothesis" where higher concentrations of inorganic arsenic are detected in urine due to an inability to methylate at high concentrations.

The reported signs, symptoms, and the indicated cause of death, taken together with the presence of inorganic arsenic in the urine, are consistent with inorganic arsenic. Although the concentrations of arsenic in the urine in this case are less than those reported in known, acute ingestions of arsenic, uncertain timing of collection postingestion is a complication in providing adequate context to the observed arsenic concentrations.

The remaining 43 mcg/L of unclassified arsenic could be determined using a method capable of speciating arsenic into inorganic, methylated, and organic forms (such as liquid chromatography−inductively coupled plasma mass spectrometry). This would provide further confirmation that the source of arsenic was inorganic if no organic arsenic is found. Identification of the residual 43 mcg/L as methylated arsenic species would be consistent with metabolism of inorganic arsenic.

Arsenic is rarely measured in blood due to the short half-life noted. However, because of the acute nature and severity of symptoms at the time of presentation in this case, measurement of blood arsenic may have been of utility in correlating exposure to inorganic arsenic with the clinical presentation. Blood arsenic concentration would have also provided a mechanism for comparison to lethal concentrations of arsenic reported in the literature and help to classify the observed urine concentrations as a chronic or acute exposure. Of importance, blood collected prior to or shortly after clinical intervention would be of most value in the measurement of arsenic.

Further reading

Apostoli P, Bartoli D, Alessio L, Buchet JP. Biological monitoring of occupational exposure to inorganic arsenic. Occup Environ Med 1999;56:825−32.

Barbey JT, Pezzullo JC, Soignet SL. Effect of arsenic trioxide on QT interval in patients with advanced malignancies. J Clin Oncol 2003;21:3609−15.

Correia N, Carvalho C, Frioes F, Araujo JP, Almeida J, Azevedo A. Haemolytic anaemia secondary to arsenic poisoning: a case report. Cases J 2009;2:7768.

Davis MA, Mackenzie TA, Cottingham KL, Gilbert-Diamond D, Punshon T, Karagas MR. Rice consumption and urinary arsenic concentrations in U.S. children. Environ Health Perspect 2012;120:1418−24.

Ficker E, Kuryshev YA, Dennis AT, et al. Mechanisms of arsenic-induced prolongation of cardiac repolarization. Mol Pharmacol 2004;66:33−44.

Goldsmith S, From AHL. Arsenic-induced atypical ventricular tachycardia. N Engl J Med 1980;303:1096−8.

Graeme KA, Pollack Jr CV. Heavy metal toxicity, Part I: arsenic and mercury. J Emerg Med 1998;16:45−56.

Hughes MF. Biomarkers of exposure: a case study with inorganic arsenic. Environ Health Perspect 2006;114:1790−6.

Hughes MF, Beck BD, Chen Y, Lewis AS, Thomas DJ. Arsenic exposure and toxicology: a historical perspective. Toxicol Sci 2011;123:305−32.

Ibrahim D, Froberg B, Wolf A, Rusyniak DE. Heavy metal poisoning: clinical presentations and pathophysiology. Collaboratory Med 2006;26:67−97.

Marchiset-Ferlay N, Savanovitch C, Sauvant-Rochat M-P. What is the best biomarker to assess arsenic exposure via drinking water? Environ Int 2012;39:150−71.

Mertz W, Underwood EJ, Arsenic. Trace elements in human and animal nutrition: Academic Pr, 1986: 499.

Raber G, Raml R, Goessler W, Francesconi KA. Quantitative speciation of arsenic compounds when using organic solvent gradients in HPLC-ICPMS. J Anal At Spectrom 2010;25:570−6.

Ratnaike RN. Acute and chronic arsenic toxicity. Postgrad Med J 2003;79:391−6.

Sears ME. Chelation: harnessing and enhancing heavy metal detoxification—a review. Sci World J 2013;2013:219840.

Tchounwou PB, Centeno JA, Patlolla AK. Arsenic toxicity, mutagenesis, and carcinogenesis—a health risk assessment and management approach. Mol Cell Biochem 2004;255:47−55.

Tyrrell J, Melzer D, Henley W, Galloway TS, Osborne NJ. Associations between socioeconomic status and environmental toxicant concentrations in adults in the USA: NHANES 2001−2010. Environ Int 2013;59:328−35.

Yoshida T, Yamauchi H, Fan Sun G. Chronic health effects in people exposed to arsenic via the drinking water: dose-response relationships in review. Toxicol Appl Pharmacol 2004;198:243−52.

Chapter 21.2

Lithium toxicity—case study

Prashant Nasa[1] and Deven Juneja[2]
[1]*Critical Care Medicine, NMC Specialty Hospital, Dubai, UAE,* [2]*Institute of Critical Care Medicine, Max Superspeciality Hospital, Saket, India*

Case description

This 65-year-old gentleman presented in our hospital emergency with an altered mental status and decreased oral intake since last two days. On neurological assessment, patient was anxious, disoriented to time and place, resting tremors, tone and deep reflexes were increased. His initial laboratory investigations showed increased total leucocyte count, renal dysfunction and nonanion gap metabolic acidosis. The provisional diagnosis of sepsis was made started on IV fluids and antibiotics. On detailed evaluation, patient's previous medication history includes lithium for manic depressive psychosis. The patient lithium levels were checked and found to be high. The other investigations (serum lithium levels, thyroid profile, parathyroid levels, urine specific gravity, and urinary electrolytes) were also performed (Table 21.2.1).

In view of severe lithium intoxication with renal dysfunction, the patient was immediately started on intermittent hemodialysis (HD). The patients serial (12 hourly) lithium levels were monitored and dialysis sessions were stopped after clinical improvement and two consecutive lithium levels less than 1 mEq/L, 12 hours apart. The patient neurological status started showing improvement and subsequent evaluated by psychiatrist. The patient was discontinued from lithium treatment and was started on alternative antipsychotics as per the psychiatrist with regular follow-up.

The final diagnosis was lithium overdose, acute kidney injury, nephrogenic diabetes insipidus (NDI), nonanion gap metabolic acidosis, hyperparathyroidism and hypothyroidism.

Discussion and follow-up

Lithium is a commonly prescribed medication in psychiatry as its salt lithium carbonate for the treatment of bipolar disorder, major depression, and mania. Lithium is minimal protein bound ('10%) and is predominantly excreted through kidney [1]. Lithium is freely filtered at a rate dependent on glomerular filtration rate (GFR) majority of which is reabsorbed at proximal tubules ($\sim$60%) and ascending limb of loop of Henle ($\sim$20%) [1]. The therapeutic dosing is thus subjected to renal function, volume status which may affect GFR and drug-interactions. The drugs that can increase lithium toxicity either decrease GFR (nonsteroidal antiinflammatory drugs, renin-angiotensin system inhibitors) or may increase proximal tubular reabsorption (carbonic anhydrase inhibitors and spironolactone) [2]. The drugs that may decrease volume status (thiazide diuretics) or some unknown mechanism (calcium channel blockers) can also cause lithium toxicity [2]. The unintentional overdosing is commonest mechanism of poisoning because of either narrow therapeutic index or many interactions (disease or drug) [3].

The signs and symptoms of lithium intoxication mainly depend on the duration of exposure and amount of ingestion which are broadly classified into acute, acute on chronic or chronic intoxication [4]. The neurological toxicity is most severe and generally responsible for mortality, but other organ such as kidney, endocrine, and gastrointestinal contributes to morbidity [1,3]. The gastrointestinal symptoms such as nausea and vomiting are common in acute intoxication. The other organs such as cardiac (dysrhythmias) and neurological symptoms are rare and seen in severe toxicity. In chronic toxicity the renal involvement in the form of NDI is most common [4]. The severity of intoxication closely approximates clinical symptoms, while serum levels may be falsely low in chronic toxicity due to slow diffusion of lithium to intracellular compartment [4]. The neurological symptoms are common with acute on chronic and chronic toxicity. The neurotoxicity includes neuromuscular excitability (coarse unintentional tremors, hyperreflexia, myoclonic jerks,

TABLE 21.2.1 Significant Laboratory Investigations and Sessions of Hemodialysis

	Day 1	Day 2	Day 3	Day 4	Day 5	Day 6	Day 7	Day 8
Hemoglobin (mg/dL)	12.2	10.9		10.8		10.7		
TLC (mm^3)	27.6	22.3	16.7	15.3	14.8	12.6		9.3
Urea (mg/dL)		46	74	58	70	66		
Creatinine (mg/dL)	2.3	2.2	2.4	1.57	1.55	1.42		1.69
Sodium (mEq/L)	152.8	143	148	152	149	156	144	142
Potassium (mEq/L)	5.9	4.5	5.7	6.1		5.6	4.1	4.2
Lithium levels (mmol/L)	2.5 and 1.0	1.4 and 1.0	0.6 and 0.8	0.35	0.3		0.25	
Hemodialysis	S1	S2 and S3						

S, Session; *TLC*, total leucocyte count.

TABLE 21.2.2 Indications for Extracorporeal Therapy for Lithium Elimination

Life threatening neurotoxicity (unconsciousness and seizures), dysrhythmias, or hemodynamic instability irrespective of lithium levels (recommendation strong)
Renal impairment (GFR ≤45 mL/min/1.73 m^2, serum creatinine ≥ 176 mmol/L) with lithium levels >4.0 mEq/L (recommendation strong)
Lithium levels are >5.0 mEq/L or severe confusion (recommendation weak)
Drop in lithium levels of <1.0 mEq/L from initial level in more than 36 h of fluid resuscitation and other supportive management (recommendation weak)

GFR, Glomerular filtration rate.

and fasciculations), higher cortical symptoms (confusion and agitation) and cerebellar signs (nystagmus and ataxia), rarely seizures, and coma [1,4]. The risk factors for neurotoxicity include coexisting NDI, renal impairment, age more than 50 years, and hypothyroidism [4]. The other renal manifestations include renal tubular acidosis, tubulointerstitial nephritis, and even nephrotic syndrome. The endocrine features are also seen with chronic toxicity, and hypothyroidism is most common because of interference in thyroid hormone release by its preferential uptake of lithium in thyroid gland [4].

The mortality with lithium overdose has reduced significantly in recent years from 9% in 1975 to less than 1% [4,5]. This may be because of routine therapeutic drug levels monitoring and/or early identification and aggressive management of lithium intoxication, including early HD [5].

The initial management of lithium toxicity is similar to general management of any poisoning and includes stabilization of airway, breathing, and circulation. Special consideration should be given to the assessment of neurotoxicity (primary organ involved) and renal function. The activated charcoal has no effect as lithium does not bind to charcoal [4]. At least one retrospective study showed some benefit of sodium polystyrene sulfonate (SPS) and whole bowel irrigation either alone or in combination in early decontamination and reduced incidence of severe poisoning [6]. There is a risk of hypokalemia with use of SPS and serum potassium should be checked frequently if it is being used [7].

Fluid resuscitation is an effective strategy in lithium clearance as it restores the volume status and thereby improves renal perfusion. A percentage of 0.9 saline may also reduce renal tubular reabsorption and hence the preferred choice of fluid in initial resuscitation despite preexisting hypernatremia [6]. Lithium has many characteristics like low molecular weight (74 Da), water solubility, small volume of distribution (0.6–1 L/kg), and insignificant protein binding, which makes it one of the most readily dialyzable toxins [1]. Extracorporeal therapy (ECT) is useful in removing lithium from body, but due to lack of randomized controlled trials is not indicated as first line treatment [8]. ECT can be used only in patients with either renal failure (renal elimination of lithium is not possible) or hydration is not safe (congestive heart failure and liver failure) and/ or in emergency (neurotoxicity and life-threatening arrhythmias) [9] (Table 21.2.2).

Our patient has features of acute on chronic intoxication with neurotoxicity, renal dysfunction, NDI with hypothyroidism and hyperparathyroidism which indicate for urgent ECT. There is another important consideration in patients with renal failure, the plasma elimination half-life of lithium is prolonged from 12−27 to 36 hours or more [9]. This is also important in elderly patients and with chronic toxicity. HD is most effective ECT modality, while continuous renal replacement therapy (CRRT) is acceptable alternative though clearance is slow and less efficient than HD [9]. CRRT can be useful to prevent rebound for preventing rebound of lithium and can be used after initial HD session in the case of rebound [10].

ECT clear only extracellular (serum) stores of lithium from the body; rebound of lithium is thus seen after HD because of its intracellular stores and/or continued absorption from gastrointestinal tract. The rebound is maximal 6−12 hours after HD and hence serial levels 12 hourly should be done after HD unless there is no evidence of rebound [9]. The HD can be stopped once levels are less than 1 mEq/L or clinical resolution [9].

In summary, the lithium is an effective drug in many psychiatric disorders but because of narrow therapeutic margin with potential serious side effects, proper patient's selection, indication, regular clinical, organ (especially renal), and drug monitoring is advocated. HD is an efficient tool for lithium elimination and randomized studies are required to establish its clear role and appropriate indications.

References

[1] Greller HA. Lithium. In: Hoffman RS, Howland MA, Lewin NA, Nelson LS, Goldfrank LR, editors. Goldfrank's toxicologic emergencies. 10th ed. New York: McGraw-Hill Education; 2015.

[2] Finley PR. Drug interactions with lithium: an update. Clin Pharmacokinet 2016;55(8):925−41.

[3] Gummin DD, Mowry JB, Spyker DA, Brooks DE, Fraser MO, Banner W. Annual report of the American association of poison control centers' national poison data system (NPDS): 34th annual report. Clin Toxicol (Phila) 2016;55(10):1072−252.

[4] McKnight RF, Adida M, Budge K, Stockton S, Goodwin GM, Geddes JR. Lithium toxicity profile: a systematic review and meta-analysis. Lancet. 2012;379:721−8.

[5] Baird-Gunning J, Lea-Henry T, Hoegberg LCG, Gosselin S, Roberts DM. Lithium poisoning. J Intensive Care Med 2017;32(4):249−63.

[6] Bretaudeau Deguigne M, Hamel JF, Boels D, Harry P. Lithium poisoning: the value of early digestive tract decontamination. Clin Toxicol (Phila) 2013;51(4):243−8.

[7] Roberts D, Gosselin S. Variability in the management of lithium poisoning. SemDialysis 2014;27(4):390−4.

[8] Lavonas EJ. Hemodialysis for lithium poisoning. Cochrane Database Syst Rev 2015. Sep 16;(9):CD007951.

[9] Decker BS, Goldfarb DS, Dargan PI, Friesen M, Gosselin S, Hoffman RS, et al. Extracorporeal treatment for lithium poisoning: systematic review and recommendations from the EXTRIP workgroup. Clin J Am Soc Nephrol 2015;10(5):875−87.

[10] Meertens JH, Jagernath DR, Eleveld DJ, et al. Hemodialysis followed by continuous venovenous hemofiltration in lithium intoxication, a model and a case. Eur J Intern Med 2009;20:70−3.

Chapter 21.3

Mercury poisoning from a high seafood diet: a case report

Vijayalakshmi Nandakumar, Sarah Delaney and Paul J. Jannetto
Department of Laboratory Medicine & Pathology, Mayo Clinic, Rochester, MN, United States

Case report

Patient was a 75-year-old nonhispanic male with a history of elevated glucose, hyperlipidemia and coronary heart disease, who presented to the doctor for a routine visit. Approximately a year ago, fairly suddenly he had developed visual changes and double vision. He did not have any eye swelling, dryness, tearing, nasal or sinus congestion or drainage, or other constitutional discomforts. It was also noted that he did not have any preceding head or eye trauma, history of thyroid disease, difficulty chewing or swallowing, shortness of breath, generalized weakness, gait disturbance, or any other neurological difficulties. As part of his evaluation for double vision, he had brain magnetic resonance imaging studies that did not show any abnormalities. He was considered a candidate for the right fourth nerve palsy and weakness of the extraocular muscles, for which he was treated with prisms in his lenses. For the next 2 months, he went through multiple changes in his lenses, prisms, and other adjustments. Eventually, his lenses did begin to correct and his diplopia had stabilized.

At the time of his current visit, he reported frequent episodes of fatigue despite his sound night sleep and otherwise excellent health. During his visit, he also reported a very high ocean fish and seafood intake in pursuit of a healthy food regimen. Given his symptoms of diplopia, fatigue, and a possibility of dietary exposure to mercury, blood heavy metal screen was ordered. The results showed a mild elevation of mercury 21 ng/mL (reference range: 0–9 ng/mL) with normal levels of arsenic, cadmium, and lead. He was counseled to keep a watch on his seafood intake since he had been consuming fish up to four times per week that was the most likely source of the elevated mercury.

On his follow-up appointment, 1 year later, his blood mercury levels were increased (43 ng/mL) to a concerning level. Although watchful, he admitted he was still consuming fish. At this time, he also complained of a sudden redness and swelling of his eyes and rash on his forehead. It was diagnosed as shingles by the local emergency department, but given the characteristics of his rash, associated neuropathic pain, and persistent fatigue, it remained unclear if this was secondary to his mercury toxicity. Appropriate medications were prescribed to relieve his pain, and he was advised to stop his consumption of fish/seafood until his mercury levels returned to normal. His mercury levels were evaluated again a month later after no longer eating fish where it had decreased to 30 ng/mL. Additional blood mercury levels at 2 months were reported as 18 ng/mL. Six months later, it had reached normal levels and continued to remain in the normal range with his controlled seafood diet.

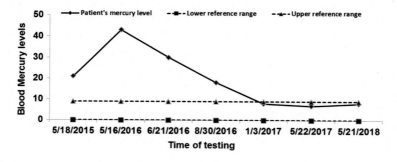

Discussion

Mercury is a toxic element present in the earth's crust and is released into the environment in different forms due to both natural events and human activities. The three common forms of its occurrence are metallic (also known as elemental), inorganic, and organic, and all forms are toxic. Elemental mercury (Hg^0) is volatile and can be readily absorbed by lungs; however, it is poorly absorbed through skin or gastrointestinal (GI) tract. Inorganic mercury compounds occur when mercury combines with chlorine, sulfur, or oxygen, and these compounds vary widely in their solubility and absorptive properties [1]. Organic mercury, particularly methylmercury and dimethylmercury, is found most widely in nature and is readily absorbed through lungs, GI tract, and sometimes even through skin [2]. Once released into the atmosphere, organic mercury dissolves in freshwater and seawater and bioaccumulates in the food chain [3]. Due to its high assimilation and slow elimination in the living organisms, methylmercury displays properties of biomagnification, where concentrations are higher in the predator tissue than in prey tissue. Consuming foods high in mercury such as fish and shellfish contributes to about 80%–90% of organic mercury in human body and about 75%–90% of it is methylmercury [4]. Although seafood is an excellent source of protein with low saturated fat content, there is an accrual of tightly bound methylmercury in the fish tissue proteins as a result of absorption from water passing over the gills. Methylmercury has an exceptionally long biological half-life, is barely eliminated by cooking, and is the form to which most human exposure occurs. Consequently, mercury levels in the blood are highly dependent on the species and the quantity of fish consumed [5].

Humans are exposed to elemental mercury primarily by inhalation and to methylmercury primarily by ingestion. The frequency of exposure to elemental mercury is low in the general population and is more often due to deliberate uses from occupational, dental, and cultural practices. The most common form of mercury exposure comes from seafood consumption. The toxicity from organic mercury exposure is reliant on the amount of fish consumed, the concentrations of mercury in different kinds of fish, as well as, in the physiological conditions of the individual [6]. For example, the allowable seafood intake in a pregnant woman is lower because mercury can easily be transferred to a fetus resulting in developmental deformities and fetal impairment of nervous system. According to FDA, fish that are highest in mercury include tilefish from the Gulf of Mexico, shark, swordfish, king mackerel, as well as, marlin, bluefish, grouper, Spanish Gulf mackerel, and Chilean sea bass [7]. The commercial fish considered safe for consumption contain <0.3 mcg/g of mercury, but some game fish, containing >2.0 mcg/g and, if consumed on a regular basis can contribute to significant body burdens [8].

Mechanisms for the toxic effects of mercury are believed to be similar, and the relative toxicities of the different forms of mercury (e.g., metallic, monovalent, and divalent cations and methyl- and phenylmercury compounds) are related, in part, to its differential accumulation in sensitive tissues. This theory is supported by the observation that mercury rapidly accumulates in the kidneys and specific areas of the central nervous system [9,10]. Mechanistically, mercury toxicity is dependent on the ability of an organism to extract mercury from the environmental matrix (external bioavailability) and the ability of the organs (GI tract or lung) to absorb and liberate mercury from the matrix into bloodstream to gain access to target organs (internal bioavailability) [11]. Once in the bloodstream, mercury is subjected to very complex and inadequately understood processes involving binding, redox cycling, methylation, and demethylation, which once again influences the internal bioavailability of mercury [12]. Ultimately, there are multiple mechanisms by which mercury causes toxicity. One of the major mechanisms is due to the high-affinity binding of divalent mercuric ion to the thiol or sulfhydryl groups of proteins that occur in both the extracellular and intracellular membranes and organelles [13,14]. In general, sulfhydryl groups play an integral role in the structure or function of most proteins that is disrupted when mercury binds to them and alters the protein structure/function [8].

The initial distribution phase of methylmercury from the blood to tissues takes about 3 days. It readily crosses the blood–brain and placental barriers and accumulates there [15]. There is a uniform tissue distribution of methylmercury unlike inorganic mercury, except in red blood cells where the concentration is 10–20 times greater than in the plasma. The brain:blood concentration ratio is about 5:1. Methylmercury also accumulates in hair during the process of formation of hair strands. The hair:blood concentration ratio is approximately 250:1 in humans at the time of incorporation into hair [16]. Methylmercury undergoes biotransformation to inorganic mercury by demethylation, particularly in the gut. Excretion of methylmercury is predominantly via the feces. Methylmercury is also slowly broken down in the gut to inorganic mercury, and most, if not all, of the mercury excreted is in the inorganic form [17].

Mercury initially affects the nervous system, leading to neurological symptoms such as anxiety, depression, and memory problems, but as the levels of mercury increase in the body, more complex symptoms such as muscle weakness, metallic taste in the mouth, changes in vision, hearing, or speech, lack of motor skills, and gait can develop [18]. The symptoms associated with mercury exposure are ambiguous and nonspecific and, if overlooked, can result in long-term complications such as neurological damage, impotency, and cardiovascular risks [19]. In this case report the symptoms associated with the patient included a sudden onset of visual changes, double vision, and persistent fatigue presented in a very nebulous fashion.

During clinical testing, it may be desirable to have insights into the specific species of mercury present in the sample due to different toxicity potentials and biological and preventive implications associated with it. However, speciation requires addition testing (more laborious and costly) and since most of the mercury in seafood is methylmercury, it is customary to just analyze total mercury and assume that it is in the methyl form [20]. Some other matrices that can be used for biomonitoring mercury include hair, blood, urine, and cord blood [21]. The presence of mercury in blood indicates recent or current exposure, and there is a direct relationship between mercury concentrations in human blood and consumption of fish contaminated with methylmercury. Urine is the main excretory pathway of elemental and inorganic mercury in humans, and hence urinary mercury is the best estimate of exposure to inorganic and/or elemental mercury. Hair integrates methylmercury during its formation and shows a relatively direct relationship with blood mercury levels, providing an accurate, reliable, and long-term marker of exposure to methylmercury intake levels [16]. In this patient, hair mercury levels were not measured since his blood mercury levels were well within the intermediate range (10−50 ng/mL) of control and management [22]. In addition, his urine mercury levels were negative, which confirmed nonexposure to elemental and inorganic mercury.

It is highly crucial to identify and eliminate the source of mercury, which can be a strenuous process due to the nebulous presentation of the clinical symptoms. The patient described in this case report started having symptoms approximately a year before he was tested. He spends his winters in Naples, Florida, where he eats a lot of seafood. So, he was given a brochure from Florida state department of health with information on the list of fish, their mercury levels, and exposure to humans. Even though the levels were not remarkably high, it was steadily rising until the patient was placed on a strict low/no seafood diet. Noticeably, the patient's mercury levels did not reach the normal range for approximately 12 months from the time he was seen. Mercury has a half-life of 70 days and at least five half-lives are required for the steady-state levels to drop down to normal value and reflect on the affected's well-being. In this patient, chelation therapy was not needed since the blood mercury levels were well below 40 µg/L and reducing the source of exposure proved sufficient to decrease his body burden of mercury [23].

Conclusion

This case demonstrates the importance of the assessment of food and dietary history during the patient's routine clinical visits, which prompted the evaluation of blood mercury levels in this individual. Awareness about mercury toxicity in different fish and seafood should be made available to the people by both the local and state public health authorities. The FDA currently advises that pregnant women and women of childbearing age who may become pregnant limit their consumption of shark and swordfish to no more than one meal per month. This advice is given because methylmercury levels are relatively high in these fish species. Health-care professionals should be judicious in recognizing and managing mercury toxicity from seafood intake despite its nonspecific presentation of clinical symptoms.

References

[1] Park J-D, Zheng W. Human exposure and health effects of inorganic and elemental mercury. J Prev Med Public Health = Yebang Uihakhoe chi 2012;45(6):344−52.

[2] Bernhoft RA. Mercury toxicity and treatment: a review of the literature. J Environ Public Health 2012;2012:460508.

[3] Knobeloch LM, et al. Imported seabass as a source of mercury exposure: a Wisconsin case study. Environ Health Perspect 1995;103(6):604−6.

[4] Björnberg KA, et al. Methyl mercury and inorganic mercury in Swedish pregnant women and in cord blood: influence of fish consumption. Environ Health Perspect 2003;111(4):637−41.

[5] Mahaffey KR, Clickner RP, Bodurow CC. Blood organic mercury and dietary mercury intake: National Health and Nutrition Examination Survey, 1999 and 2000. Environ Health Perspect 2004;112(5):562−70.

[6] Stern S, et al. Perinatal and lifetime exposure to methylmercury in the mouse: blood and brain concentrations of mercury to 26 months of age. Neuro Toxicol 2001;22(4):467−77.

[7] https://www.fda.gov/food/metals/mercury-concentrations-fish-fda-monitoring-program-1990-2010.

[8] de Burbure C, et al. Renal and neurologic effects of cadmium, lead, mercury, and arsenic in children: evidence of early effects and multiple interactions at environmental exposure levels. Environ Health Perspect 2006;114(4):584−90.

[9] Passow H, Rothstein A. The binding of mercury by the yeast cell in relation to changes in permeability. J Gen Physiol 1960;43(3):621−33.

[10] Somjen GG, et al. The uptake of methyl mercury (203Hg) in different tissues related to its neurotoxic effects. J Pharmacol Exp Ther 1973;187(3):602.

[11] Rice KM, et al. Environmental mercury and its toxic effects. J Prev Med Public Health = Yebang Uihakhoe Chi 2014;47(2):74–83.
[12] Compeau G, Bartha R. Methylation and demethylation of mercury under controlled redox, pH and salinity conditions. Appl Environ Microbiol 1984;48(6):1203–7.
[13] Clarkson TW. The toxicology of mercury. Crit Rev Clin Lab Sci 1997;34(4):369–403.
[14] Passow H, Rothstein A, Clarkson TW. The general pharmacology of the heavy metals. Pharmacol Rev 1961;13(2):185–224.
[15] Jan AT, et al. Heavy metals and human health: mechanistic insight into toxicity and counter defense system of antioxidants. Int J Mol Sci 2015;16(12):29592–630.
[16] George GN, et al. The chemical forms of mercury in human hair: a study using X-ray absorption spectroscopy. J Biol Inorg Chem: JBIC 2010;15(5):709–15.
[17] Bose-O'Reilly S, et al. Mercury exposure and children's health. Curr Probl Pediatric Adolesc Health Care 2010;40(8):186–215.
[18] Brodkin E, et al. Lead and mercury exposures: interpretation and action. CMAJ 2007;176(1):59–63.
[19] Fernandes Azevedo B, et al. Toxic effects of mercury on the cardiovascular and central nervous systems. J Biomed Biotechnol 2012;2012:949048 p.
[20] Reis A, et al. Overview and challenges of mercury fractionation and speciation in soils. TrAC Trends Anal Chem 2016;82:109–17.
[21] Keil DE, Berger-Ritchie J, McMillin GA. Testing for toxic elements: a focus on arsenic, cadmium, lead, and mercury. Lab Med 2011;42(12):735–42.
[22] Ye B-J, et al. Evaluation of mercury exposure level, clinical diagnosis and treatment for mercury intoxication. Ann Occup Environ Med 2016;28:5 p.
[23] Clarkson TW, et al. Tests of efficacy of antidotes for removal of methylmercury in human poisoning during the Iraq outbreak. J Pharmacol Exp Ther 1981;218(1):74.

Chapter 21.4

Trust your gut: a case of persistent gastrointestinal disturbances

Sarah R. Delaney, Vijayalakshmi Nandakumar and Paul J. Jannetto
Department of Laboratory Medicine & Pathology, Mayo Clinic, Rochester, MN, United States

Case report

A 22-year-old female presented with severe reflux-type dyspepsia that had been progressively worsening for approximately 4 years. She reported severe symptoms of postprandial acid regurgitation, nausea, abdominal bloating, abdominal wall pain, and loud eructation. The patient had multiple food intolerances and followed a diet limited to mainly rice-based foods.

Upon evaluation, her weight was 53.4 kg with a normal body mass index (BMI) of 19.9 kg/m^2. Her physical exam revealed a soft abdomen, with severe left-sided abdominal tenderness. She was unable to perform any abdominal wall assessment due to worsening of pain. Abdominal imaging and colonoscopy were unremarkable, with the exception of mild (grade 2) esophagitis as shown on her esophagogastroduodenoscopy. Due to these unremarkable findings, she was diagnosed with irritable bowel syndrome and underwent further investigation.

Laboratory results showed normal chemistries, thyroid-stimulating hormone (TSH) and morning cortisol concentrations. Further tests indicated that she had mild normocytic anemia, with equivocal iron findings (Table 21.4.1). Despite following a restricted diet, all of her nutrient markers, vitamin B6, vitamin B12, vitamin D, and folate, were all within normal range. However, the urine heavy metal screen was remarkable—reporting a total arsenic concentration of 82 mcg/L (<18 mcg/L). As a result, arsenic speciation was performed to determine which species were present and the inorganic arsenic was elevated at 66 mcg/mL (<20 mcg/L) with an organic arsenic concentration of 16 mcg/L (Table 21.4.1).

Discussion

Arsenic is a naturally occurring metalloid found abundantly throughout the environment primarily existing in two forms: organic and inorganic [1]. The organic forms (arsenobetaine and arsenocholine) are generally considered to be nontoxic, while the inorganic species (trivalent and pentavalent arsenic) are toxic and pose the greatest risk to human health [1]. Sources of human exposure can include contaminated ground water, industrial manufacturing, wood preservatives, cigarette smoke, cosmetics, and various dietary items [2]. Dietary intake, specifically seafood—including fish, shellfish, and algae—are considered to be the greatest contributors to arsenic exposure. The majority of marine-derived foods contain the organic, nontoxic forms of arsenic (arsenobetaine and arsenocholine), or arsenic-containing ribose derivatives (arsenosugars) which are unlikely to have clinically significant effects following exposure [3]. However, some commercially available seaweeds, especially brown algae varieties, may have high percentages of the total arsenic present as inorganic arsenic (>50%) [4,5]. Rice products can also contain high concentrations of inorganic arsenic. The relative amount of inorganic arsenic varies depending on the source; however, concentrations can reach 90% of the total arsenic species in rice [6]. Clinically significant effects can result from acute exposure greater than 2 mg/kg/day, or from repeated lower exposure ranging from 0.03 to 0.1 mg/kg/day [2]. It is estimated that the daily dietary intake of inorganic arsenic can range from 1 to 20 mcg/day in the United States through various sources including diet and drinking water [2].

TABLE 21.4.1 Laboratory Data

Analyte	Result	Reference Range
Blood Tests		
Hemoglobin	11.9 g/dL	12.0–15.5 g/dL
Serum iron	112 mcg/dL	35–145 mcg/dL
Ferritin	9 mcg/dL	11–307 mcg/L
Total iron binding	409 mcg/dL	250–400 mcg/dL
Arsenic Speciation, Urine		
Arsenic, total	82 mcg/L	<18 mcg/L
Inorganic arsenic	66 mcg/L	<20 mcg/L
Organic arsenic	16 mcg/L	–

Inorganic arsenic species are readily absorbed following inhalation or ingestion, taken up by red blood cells, and distributed throughout the body [7]. Serum concentrations peak 30–60 minutes following ingestion, metabolized via methylation to monomethylarsonic acid and dimethylarsinic acid, and excreted through the urine [8]. Trivalent arsenic is the most toxic form of inorganic arsenic, which avidly binds to sulfhydryl groups, inactivating nearly 200 enzymes primarily involved in cellular energy processes and DNA synthesis and repair [3]. Specifically, trivalent arsenic inhibits the cofactor for pyruvate dehydrogenase, and various other enzyme systems required for gluconeogenesis, glucose uptake, and glutathione metabolism [7].

The primary targets of arsenic toxicity are the gastrointestinal (GI) tract (site of highest exposure), skin, kidneys, and peripheral nervous system [3]. Patients with clinically significant acute exposure to arsenic may develop a wide range of GI symptoms including nausea, vomiting, severe abdominal pain, esophagitis, gastritis and watery diarrhea [3]. The most commonly recognized symptom of arsenic toxicity is peripheral neuropathy; however, exposure can also result in seizures, cardiac arrhythmias, acute nephropathy, and shock [3].

This case presents a 22-year-old female with multiple GI complaints and mild anemia. The most notable result was an elevated inorganic arsenic concentration. Ingestion of inorganic arsenic commonly produces symptoms of stomach ache, nausea, vomiting, and diarrhea. Other effects include decreased production of red and white blood cells and impaired nerve function causing a "pins and needles" sensation in the extremities (hands/ft) [1]. The patient consumed a diet higher in rice and rice-based products than typical, due to her inability to tolerate many other foods. As mentioned previously, rice can contain higher amounts of inorganic arsenic, but in a typical North American diet these levels are negligible. In this case, it was thought that the higher than normal consumption of rice and rice-based foods could be causing the elevated inorganic arsenic concentrations to a level where symptoms may present.

Following this finding, the patient consulted with a dietitian and determined a new dietary regime, consisting of mainly soft mechanical foods, while minimizing rice intake. Subsequently, a total arsenic urine test was repeated, and the total concentration was <18 mcg/L. As a result of the normal total arsenic value, speciation was not performed.

Ultimately, the patient was diagnosed with Central Sensitization Pain Syndrome and was enrolled in a comprehensive pain rehabilitation program. To manage her dyspepsia, she initiated acid suppression therapy (pantoprazole 40 mg bid and famotidine 40 mg qd) and functional dyspepsia capsules, which ultimately improved some of her acid-related symptoms. To continue to manage her other symptoms of irritable bowel syndrome, the patient underwent hypnosis therapy and initiated amitriptyline (10 mg qd), which she tolerated well.

While the patient had another underlying condition, the presence of her severe GI-related symptoms was partially explained by her persistent exposure to inorganic arsenic through her high intake of rice and rice-based foods. Although the patient's symptoms of abdominal pain and nausea are a common hallmark of arsenic exposure, they are also not specific to arsenic toxicity and could have multiple etiologies. Nevertheless, removal of the patient's main source of exposure to arsenic, along with aggressive acid suppressive therapy, led to the overall improvement of her symptoms and shows the importance of screening for heavy metals in complicated cases.

References

[1] Agency for Toxic Substances and Disease Registry. Environmental health and medicine education, arsenic toxicity. U.S. Department of Health & Human Services, Public Health Service; 2018.
[2] Chung JY, Yu SD, Hong YS. Environmental sources of arsenic exposure. J Prev Med Public Health 2014;47(5):253–7.
[3] Ratnaike RN. Acute and chronic arsenic toxicity. Postgrad Med J 2003;79:391–6.
[4] Almela S, Algora V, Benito MJ, et al. Heavy metal, total arsenic, and inorganic arsenic contents of algal food products. J Agric Food Chem 2002;50(4):918–23.
[5] Laparra JM, Vélez D, Montoro R, et al. Estimation of arsenic bioaccessibility in edible seaweed by an in vitro digestion method. J Agric Food Chem 2003;51(20):6080–5.
[6] Hojsak I, Braegger C, Bronsky J, et al. Arsenic in rice: a cause for concern. J Pediatr Gastroenterol Nutr 2015;60(1):142–5.
[7] Mundy SW. Arsenic. In: Hoffman RS, Lewin NA, Howland MA, et al., editors. Goldfrank's toxicologic emergencies. 10th ed. New York: McGraw-Hill Education; 2015. p. 1169.
[8] Lauwerys RR, Hoet P. Industrial chemical exposure – guidelines for biological monitoring. Boca Raton, FL: Lewis Publishers; 2000.

Chapter 22

Venoms

Jennifer A. Lowry[1,2]

[1]Department of Pediatrics, Children's Mercy Hospital, Kansas City, MO, United States, [2]University of Missouri School of Medicine, Kansas City, MO, United States

Introduction

Bites and envenomations to humans occur worldwide. Of the creatures that bite and sting, venomous snakes account for more cases of severe morbidity and mortality [1,2]. The World Health Organization (WHO) estimates that more than 5 million people are bitten each year and account for more than 100,000 deaths and significant more morbidity from amputations and permanent disabilities (Fig. 22.1).

The majority of snake envenomations occur in Africa, Asia, and Latin America. However, in 2017, United States Poison Control Centers (USPCC) were involved in more than 46,000 bites and envenomations to humans, including 6700 cases related to snake envenomations [3]. Due to underreporting, it is believed that these numbers greatly underestimate the true prevalence of the problem. The majority of bites by venomous snakes affect women, children, and agricultural workers in poor rural communities in low- and middle-income countries. More than 146 million people live in areas without quality healthcare and disproportionate access to antivenom. An additional 600 million live more than 1 hour from population centers where adequate treatment could occur [4]. WHO characterized snakebite envenoming as a "highest priority neglected tropical disease" in June 2017 for these reasons. As a result of these efforts, the Venomous Snakes Distribution Database was developed to aid health-care providers and other users to identify snakes in their area and information on antivenom available in the area.

Snakes are not the only creatures that bite and sting. Snakes are members of the reptile class, which have other venomous creatures, including amphibians (e.g., toads, frogs, salamanders, and newts), lizards (e.g., Gila monster and beaded lizard), and sea snakes, all of which produce venom. Arthropods such as arachnids (e.g., spiders and scorpions), insects (e.g., bees, wasps, and mosquitoes), and marine animals may also inject venom into their prey as a defense mechanism or for sustenance. While decreased mortality rates occur compared to snake envenomations, the associated morbidity and health-care utilization for these exposures are significant. More than 5000 spider cases were called to USPCCs in 2017 with the majority of cases having minor outcomes [3]. Despite the title of this chapter, neither all venoms can be discussed nor all of the creatures who produce them. Thus the focus will be on more common envenomations that occur throughout the world with emphasis placed on the toxicology of the venom and how to treat it.

Terrestrial snakes

All snakes belong to the order *Squamata* and suborder *Serpentes*, that is, characterized by scaly skin, absence of movable eyes, and elongated bodies without appendages. Venomous snakes inhabit all continents except Antarctica and some islands (Ireland and New Zealand). Approximately 15% of snake species are venomous. The three primary venomous snake families are the Atractaspididae, Elapidae (elapids), and Viperidae (vipers). Viperidae can be divided into two subfamilies, Viperinae (true vipers) and Crotalinae (pit vipers).

Mechanism of action and toxicokinetics

Venom is used by the snake to facilitate the capture of prey by immobilizing or killing, assist in digestion, and as a defense mechanism when threatened by apparent predators. Snake venom is a mixture of biologically active proteins

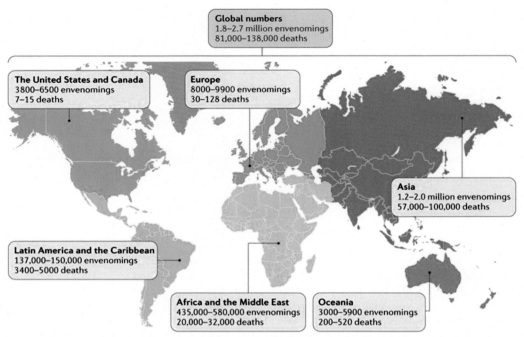

FIGURE 22.1 Geographical distribution of the estimated number of snakebite envenomations and death. *Reproduced with permission from Gutierrez JM, Calvete JJ, Habib AG, et al. Snakebite envenoming. Nature Reviews Disease Primers.*

and polypeptides, which comprises more than 95% of the venom inoculation. Other components include carbohydrates, lipids, and amines [5]. Most of the potent snake venoms target specific targets mostly affecting the neuromuscular junction and/or hematologic systems. In the wild, this would result in immobilization of their prey, aid in digestion, and a mechanism of defense [6].

Neurotoxins, clinically, result in paralytic effects at the neuromuscular junction which, if left untreated, could lead to death from respiratory failure [7]. Neurotoxins may have effect at pre- and/or postsynaptic junctions. Presynaptic toxins result in neurotransmitter release and, ultimately, complete cessation of neurotransmitter release from the significant damage to the axonal structure. This results in progressive flaccid paralysis that can be complete in 24 hours. The time to recovery is dependent on how soon antivenom is given, as once paralysis is complete, recovery rate is dependent on axonal repair. Postsynaptic toxins bind to acetylcholine receptors but do not damage the cells. Thus flaccid paralysis may occur but is reversible with antivenom therapy.

Myotoxins may also be present in elapid and vipers. These toxins cause myolysis of skeletal muscle with little to no effect on cardiac or smooth muscle. The damage is to the cell but not the basement membrane, which allows muscle regeneration within days of the bite. However, significant muscle damage can result in increased release of myoglobin, creatine kinase, and potassium into systemic circulation leading to renal damage in some patients. Antivenom has little effect on myolysis once the process is established. Myotoxins may also work with necrotoxins to result in the local injuries and necrosis that is seen most commonly after a snake envenomation.

The hematologic system is a common target for venom from the majority of venomous snake families. Toxins include those that act as procoagulants, anticoagulants, platelet aggregators and promoters, and direct hemorrhagins. The net effect of these toxins is to induce bleeding, but some components result in thrombus formation and consumption of fibrinogen and/or disseminated intravascular (IV) coagulation.

Envenomation from snakes involves the injection of venom through the skin of the prey that enters the systemic circulation. Depending on the snake and venom components, the location on the body of the envenomation, and the amount of venom injected, local and systemic effects may occur. The rate of absorption and development of systemic effects can be altered depending on the route of inoculation [IV vs intramuscular (IM) vs subcutaneous (SC)]. When venom enters through the IM or SC route, absorption may be slower with decreased bioavailability compared to the IV route [8–11]. Similarly, the terminal half-life was longer after IM injection compared to IV administration, although this is not consistent between studies [12,13].

The majority of venom is not absorbed into the systemic circulation and will result in local tissue injury. Venom, when transported, enters the lymphatic system and enters the circulation through the thoracic duct. Once it reaches the systemic circulation, the hematologic toxins will have reached their site and exert their effect. Neurotoxins and myotoxins may have a delayed effect, as time is needed to reach their targets.

Analytical methods and clinical management implications

Diagnosing a snake envenomation is easy if the patient witnesses the envenomation and is awake/alert when presenting to a health-care facility. However, not all envenomations are known or witnessed. Thus the health-care provider must consider that an envenomation has occurred even if it is not in the presenting history. For example, a child presenting to the emergency department when a hand becomes increasingly swollen and painful out of character for the mechanism described (e.g., hand scratched when retrieving a ball from under a bush).

Qualitative or quantitative levels of venom cannot be found in clinical laboratories with few exceptions. The only commercial Snake Venom Detection Kit (SVDK) is used in Australia and Papua New Guinea [14]. SVDK is an enzyme-linked immunosorbent assay (ELISA) applicable to common envenomous snakes in that region of the World, including Tiger snake, Brown snake, Black snake, Death adder, and Taipan snake venoms. This test is not intended to be used to diagnose snake envenomations but largely to determine the appropriate antivenom therapy for the patient.

Thus the diagnosis must be made considering the history and physical symptoms. The extent and severity of the envenomation can be determined by laboratory analysis (Table 22.1).

Urinalysis, coagulation studies, complete blood count, creatine kinase, electrolytes, blood urea nitrogen (BUN), and creatinine are the analyses helpful in following the progression of systemic effects. Coagulation studies can include a whole blood clotting time in rural hospitals without laboratory support to a full laboratory work up (e.g., prothrombin time, activated partial thromboplastin time, fibrinogen level, and cross-linked or D dimer degradation products of fibrin/fibrinogen).

Laboratory assessment should occur as soon as the patient is stabilized upon entering the health-care facility. If the patient's initial tests are normal, they should be repeated in 2 and 3 hours later (5 hours after initial assessment) to ensure that no delayed effects from the envenomation have occurred. With some snake envenomations (e.g., Malayan pit viper bite), effects may be delayed or recur after several days.

TABLE 22.1 Local, Systemic, Hematologic, and Neurologic Signs and Symptoms of Snakebite Envenomation.

Local	Systemic	Hematologic	Neurologic
Pain	Tachycardia[a]	Anemia	Diplopia
Localized bleeding	Dyspnea[a]	Thrombocytopenia	Perioral paresthesias or metallic taste
Erythema	Chest pain	Petechiae	Numbness/tingling (widespread)
Edema	Nausea or vomiting[a]	Gingival bleeding	Fasciculations (widespread)
Ecchymosis	Hypotension	Epistaxis	Altered mental status
Blistering	Angioedema	Retinal hemorrhage	Cranial nerve dysfunction, especially ptosis (Mohave toxin)
Joint stiffness	Myalgia/cramps	Internal bleeding	
Numbness/tingling (localized)	Rhabdomyolysis	Coagulopathies	
Cramps/fasciculations (localized)		Disseminated intravascular coagulation	

[a] Can be from envenomation or autonomic responses to pain and anxiety, therefore not used as a sole indicator of systemic signs of envenomation.

Source: Reproduced with permission from Kanaan NC, Ray J, Stewart M, et al. Wilderness Medical Society Practice Guidelines for the treatment of pit viper envenomations in the United States and Canada. Wilderness Environ Med 2015;26:472–87.

TABLE 22.2 Emergency Medicine Care of Crotaline Envenomations.

Envenomation	Observation	Laboratory Studies	Treatment
Dry/no bite	≥8 h	Initial laboratory studies[a]	No antivenom
Minor: nonprogressive symptoms without systemic signs	12–24 h	Initial laboratory studies; repeat laboratory studies[b] every 4–6 h and before discharge	Consider antivenom only if high-risk areas affected (e.g., hand or face)
Moderate: progressive symptoms and/or systemic signs	Admit	Initial laboratory studies; repeat every 1 h after antivenom until initial control	Antivenom administration, supportive care
Severe: progressive symptoms with systemic signs and/or end-organ damage	Admit	Initial laboratory studies; repeat every 1 h after antivenom until initial control	Antivenom administration, supportive care

[a] Initial laboratory studies include complete blood count with platelets, basic metabolic panel, liver function tests, prothrombin time/international normalized ratio, partial thromboplastin time, total creatine kinase, fibrinogen, urinalysis.
[b] Repeat labs include complete blood count with platelets, prothrombin time/international normalized ratio, and fibrinogen.

Source: Reproduced with permission from Kanaan NC, Ray J, Stewart M, et al. Wilderness Medical Society Practice Guidelines for the treatment of pit viper envenomations in the United States and Canada. Wilderness Environ Med 2015;26:472–87.

Treatment

Treatment of snake envenomations begins at the time of the envenomation in the field, but nothing will significantly alter the course of the envenomation unless the wrong thing is done. Without delaying transport the wound should be cleaned (soap and water) and dressed. Procedures such as oral suctioning, mechanical suction, bleeding of the bite site, electrotherapy, cryotherapy, and tourniquet placement or pressure bandaging are not recommended, as all of these may result in increased harm to the patient [15].

Antivenom therapy is the only specific therapy for snake envenomations. Its use is dependent on specificity of the antivenom for the snake and the ability to obtain the product in developing countries. Antivenom is antibody to venoms from specific snakes. Older formulations contain the whole immunoglobulin G (IgG) antibody to the venom; however, newer products are Fab or F(ab′)$_2$ fragments of the IgG [16]. Antivenom is produced by inoculating mammals (e.g., horses or sheep) with snake venom and collecting the formed antibody. In most countries, horses continue to be used as the host animal to supply whole IgG. While adverse effects can occur with all formulations, immediate (anaphylaxis) and delayed (serum-sickness) is more likely to occur with horse serum. Nevertheless, antivenom should never be withheld in persons with symptoms from snake envenomations due to the high morbidity and mortality.

Antivenom should be used if there is a indication of snake envenomation and as soon as possible (Table 22.2).

The choice of antivenom is specific to the snake involved. Often times, the snake is not visualized and the choice for antivenom is based on those endemic to the region. However, snakes, not endemic to the area, may be commonly maintained as pets and result in envenomations. In the United States, zoos with reptile exhibits maintain stockpiles of antivenom for exotic snakes. In addition, the venomous snakes distribution database may be utilized worldwide.

Other than antivenom, symptomatic and supportive care should be provided to the patient. Antibiotics are not recommended unless a secondary infection occurs. Tetanus immunizations should be verified and/or updated.

Arthropods

Arthropod is an invertebrate animal phylum that includes arachnids, insects, myriapods, and crustaceans. Common to all of them is the presence of an exoskeleton, a segmented body, and paired jointed appendages. Some species have wings. Over a million species have been described accounting for more than 80% of the living animal species. Despite this relatively few are venomous with even less resulting in direct harm to humans.

Arachnids and ticks

Arachnids, including spiders, scorpions, and ticks, are found in all human habitats globally. Indirectly, they are primarily beneficial to humans as a result of decreasing pest (e.g., insect) populations. While rare direct interaction may result in harm to humans.

Spiders (order Araneae) are different from other Arachnids in which they have a slender waist that separates the cephalothorax from the abdomen. There is estimated to be approximately 38,000 species of spiders with very few being venomous to humans [17]. In 2017 USPCCs were notified of more than 5000 cases of spider bites or envenomations to humans. The majority occurred in adults over 20 years of age and 34% requiring treatment in a medical facility. Minor outcomes predominated with death occurring in one patient and attributed to *Loxosceles reclusa* [3]. No accurate numbers are available to estimate the global accounts of spider envenomations.

The WHO considers only four major genera of medically significant envenomations to humans. These include *Atrax, Latrodectus, Loxosceles*, and *Phoneutria* (Table 22.3).

While many other spiders have venom, the toxicity from these mentioned have resulted in significant harm to humans. Bites from these spiders not only result in local reactions but may also cause life-threatening systemic symptoms. Scorpions (orders Scorpionida, Uropygi, Amblypygi, and Palpigradi) are a diverse group of arachnids that are found throughout the world, predominantly in tropical and subtropical regions. Globally, approximately 800 species exist; however, only a limited number are toxic to humans [18]. Their anatomy is different than spiders as they have a hard exoskeleton, anterior pinching claws, and a tail with a bulbous enlargement where the poison gland and stinger are located. In 2017 USPCCs documented more than 12,000 bites from scorpions with the majority of patients having minor to moderate outcomes, although only 1500 having presented to a health-care facility [3]. In the United States the only clinically significant scorpion is the Bark Scorpion, which is found in the SW region. More commonly, medically important scorpions can be found in India, Middle East, Africa, and South America.

Ticks (subclass Acarida and superorder Parasitiformes) are vectors of human disease and are more associated with infectious diseases than with a toxin. However, more than 40 species of ticks can produce a neuro toxidrome resulting in paralysis. Most commonly occurring in the spring—summer months, tick paralysis can occur due to prolonged tick attachment on mammals by particular species of ticks, including *Dermacentor andersoni* (United States Pacific Northwest, United States West, and Southwestern Canada), *Dermacentor variabilis* (United States Southeast), and *Ixodes holocyclus* (Eastern Australia) [19]. Treatment is removal of the offending tick with resolution of symptoms within 12 hours.

TABLE 22.3 Medically Significant Spiders.

Spider Species	Common Name	Distribution	Clinical Effects
Phoneutria spp.	Wandering spider	Brazil	Excitatory neurotoxic effects, pain, malaise. Rarely lethal
Atrax spp.	Funnel-web spider	Australia	Excitatory neurotoxic effects, catecholamine storm. Potentially lethal
Lactrodectus spp.	Black widow spider	Global	Excitatory neurotoxic effects, pain, malaise. Rarely lethal
Loxosceles spp.	Recluse spider	Global	Local tissue necrosis, systemic effects such as hemolysis, shock and DIC. Occasionally lethal
Hexophthalma spp.	Six-eyed sand spider	Africa	Local tissue necrosis, systemic effects such as hemolysis, shock and DIC. Rarely lethal
Missulena spp.	Mouse spider	Australia	Excitatory neurotoxic effects, local pain, bleeding. Rarely lethal
Theraphosidae spp.	Tarantula	Global	Urticarial hairs. Old-world tarantulas with localized pain, swelling and muscle cramping after envenomation. Rarely lethal

Source: Adapted from Culin J, Goodnight ML. Arachnid. Encyclopedia Britannica. <https://www.britannica.com/animal/arachnid> [accessed 25.03.19].

Mechanism of action and toxicokinetics

Similar to snakes, spider venoms are a complex mixture of substances and peptides that aid the spider in food accessibility and predator responses. The toxicity of the peptides tends to fall in two categories: neurotoxins or necrotoxins. However, hematologic effects also occur, rarely, with spiders but fairly commonly with the *Loxosceles* spp.

The medically important *Latrodectus* sp. (e.g., Black Widow Spider) produces potent neurotoxins. A number of proteins are present that may participate in the clinical presentation, but the most likely toxin is a high-molecular weight, presynaptic neurotoxin called alpha-latrotoxin. Variability in the potency occurs between species and seasons [20]. Latrotoxins act by stimulating the release of neurotransmitters (e.g., norepinephrine, gamma-aminobutyric acid, and acetylcholine) from presynaptic nerve terminals [21]. The toxin may also cause degeneration of the motor end plates, which can result in denervation at the site. Lastly, the venom may result in a large influx of calcium into the cell and depletion of acetylcholine due to destabilization of nerve cell membranes. The result is development of irritation at the site (e.g., erythema, urticaria, or characteristic halo-shaped target lesion) followed by generalized pain and muscle spasms. Diaphoresis is common and, in fact, the rapid development of irritability, hypertension, and sweating in a child is suggestive of *Latrodectus* envenomation. Most symptoms develop in 3–4 hours and resolve within 2–3 days; however, residual effects are known to occur. Other medically important spiders with neurotoxin effects include *Phoneutria* spp. (PhTx3 causing calcium channel blockade in neural synapses) and *Atrax* spp. (alpha-atracotoxin causing inhibition of sodium channels in peripheral nerves).

Loxosceles spp. envenomations may result in local (dermonecrotic) and systemic effects (hemolytic). The most toxic species is *L. reclusa* which is found in Mid-South Central United States. The knowledge of the venom from *Loxosceles* has evolved over the past century with still much unknown (Fig. 22.2).

The venom consists of phospholipases (e.g., hyaluronidase, collagenase, esterase, phospholipase, and sphingomyelinase) that result in dermonecrosis at the site of the bite. Spreading factors are also present that result in dependent spreading increasing the degree of necrosis. While hemolysis from *L. reclusa* may be secondary to sphingomyelinase, the venom is a mixture of proteins acting synergistically causing disruption of inflammatory cascades and resulting in hematologic effects [22]. The venom, per volume, is more potent than that of the rattlesnake resulting classic necrotic

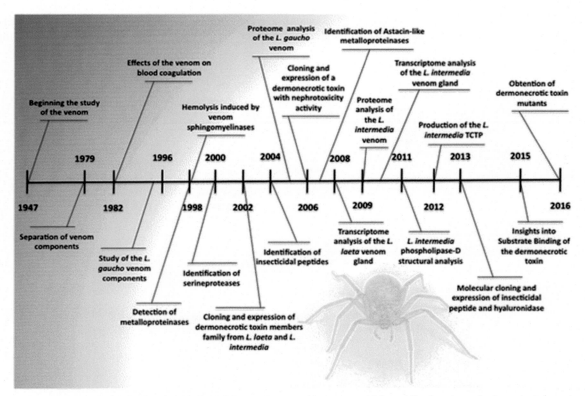

FIGURE 22.2 Major historical evolution on the knowledge on brown spider venom. Main publications in toxinology on *Loxosceles* spiders. *Reproduced with permission from Chaves-Moreira D, Senff-Ribeiro A, Wille ACM, et al. Highlights in the knowledge of brown spider toxins. J Venomous Animals Toxins Trop Dis 2017;23:6.*

lesion and, less commonly, in systemic effects, including coagulopathies, hemolysis, and, possibly, death. Jaundice may develop due to hemolysis, which may be the first sign of a systemic response; however, most patients develop an early rash which may confuse the diagnosis [23]. Hemolysis may occur at any time that other systemic symptoms (e.g., fever and malaise) are present and can be delayed. Early recognition is important as death may occur. While case reports have blamed other spider envenomations for dermonecrosis, these are often misdiagnoses. Examples of wrongly convicted spiders include wolf spiders (South America), white-tailed spiders (Australia), hobo spiders (NW, United States), and yellow sac spiders (North America and South Africa). Diagnosis of spider envenomations is one of the exclusions if the spider is not seen on the patient.

Of the over 800 species of scorpions, only 25 have venom that is considered fatal to humans; most belong to the family Buthidae (e.g., *Leiurus*, *Hottentotta*, *Centruroides*, and *Androctonus*) and found worldwide. The neurotoxin is a small protein containing hyaluronidase components that blocks the inactivation of sodium channels, decreasing the duration and amplitude of neuron action potentials, and increasing the release of acetylcholine. Parasympathetic and sympathetic results may occur. This results in pain, muscle fasciculations, increased secretions (e.g., respiratory, gastric, and pancreatic), and cardiac dysrhythmias [24].

Envenomation by ticks is most commonly seen in animals; however, humans, especially girls under the age of 10, have been affected rarely. The venom resulting in the paralysis is thought to be found in the tick's salivary gland and enters the body during feeding. The greatest amount of venom produced between the 5th and 7th day of tick attachment. Weakness in the lower extremities progresses quickly to an ascending paralysis and may result in respiratory failure and death [21].

Analytical methods and clinical management implications

Most arachnid envenomations are diagnosed clinically with a few exceptions. Unfortunately, misdiagnoses occur frequently largely due to the misconception of the geographical range of spiders and the broad differential diagnosis. Spider envenomations are a diagnosis of exclusion unless the spider is seen at the time of envenomation. It is imperative that symptomatic and supportive care be given while other diagnoses are considered and ruled out.

Researchers have developed an ELISA swab to detect surface *Loxosceles* venom at the envenomation site. In some instances the venom has been detected as long as 2 weeks after envenomation [25]. This test has been used in clinical situations, as well, to accurately determine the underlying cause of a dermonecrotic wound [26]. However, this test is not widely or commercially available.

Some patients who have been envenomated by a *Loxosceles* spp. do not develop the classic wound presentation but can develop systemic effects. These patients are more likely to be misdiagnosed. Due to the lack of venom at the wound site, the swab ELISA test may have a false-negative result. While not available clinically, efforts have been made for immunodetection of venom components in biological fluids with some success [27].

Laboratory assessment is otherwise not available for the diagnosis of arachnid envenomations. However, laboratory assessment is necessary to determine the extent of systemic effects from envenomations resulting in hematologic changes. For those patients with *Loxosceles* envenomations, obtaining a complete blood count (anemia and thrombocytopenia), liver function tests (liver enzymes and bilirubin), lactate dehydrogenase LDH (extravascular hemolysis), plasma free hemoglobin/haptoglobin (intravascular hemolysis, and urinalysis (blood, red blood cells, and urobilinogen) should be obtained regularly while the patient is febrile and/or in the presence of other systemic symptoms. Hemolysis may occur at any time when other systemic symptoms (e.g., fever and malaise) are present and can be delayed. Early recognition is important as death may occur.

Treatment

In contrast to snake envenomations, antivenom is not widely available for most arachnid envenomations. Spider antivenom is available and effective for neurotoxic envenomations from *Atrax* spp. (Funnel web spiders—Australia), *Phoneutria* spp. (Banana spiders—Brazil), and *Latrodectus* spp. (Widow spiders—worldwide). Funnel web spider envenomations can be life-threatening with frequent deaths noted for those who do not receive antivenom. Antivenom is commonly used for moderate to severe envenomations from the Brazilian banana spider especially in the very young and elderly. Widow antivenom is recommended for moderate to severe envenomations worldwide but is less likely to be given in the United States. In fact, a shortage exists for antivenom in the United States. This may be, in part, due to the original product as a horse serum—derived product which can result in significant adverse reactions, including anaphylaxis and serum sickness. However, a more highly purified equine $F(ab)_2$ antibody is being studied for use and is proposed to result in less adverse reactions.

Historically, scorpion envenomations were treated with antivenom therapy consisting of whole IgG derived from goat serum. As this was highly immunogenic, anaphylaxis and serum sickness were common. After its discontinuation in 2001, antivenom was not available until 2011 when the US Food and Drug Administration approved a more highly purified equine F(ab)$_2$ antibody to *Centruroides* sp. [21]. In a recent metaanalysis, antivenom against *Centruroides* sp. was found to be effective in reversing the clinical syndrome faster than no antivenom; however, it is usually reserved for patients with significant toxicity [18,28,29]. Globally, clinical improvement has been seen with the use of antivenom for scorpion envenomations from *Mesobuthus tamulus*, *Leiurus quinquestriatus*, and *Androctonus* spp., but these findings are not consistent across all studies. Of the three species, envenomations from *M. tamulus* has been more likely to respond to antivenom [18,29].

Insects

The order *Hymenoptera* contain three medically important families: *Apidae* (honeybees and bumblebees), *Vespidae* (yellow jackets, wasps, and hornets), and *Formicidae* (fire ants). In the United States, *Hymenoptera* account for the majority of fatal envenomations [30]. Despite this, few notify the USPCCs as only 4100 cases, including two deaths, were documented in 2017 [3].

Hymenoptera tend to avoid humans unless they are disturbed. Most tend to live in colonies or hives, which, if threatened, can result in massive envenomations [24,31]. While the methods of envenomation are different, a few types are worth mentioning here. Yellow jackets can reside in colonies of several hundred to thousand insects. Honeybees have a barbed stinger that will remain in the skin resulting in the ability to only sting once. The stingers of other vespids (e.g., yellow jackets, wasps, and hornets) and bumblebees are not barbed and can retain the stinger if it can pull itself out. The Africanized honeybee has migrated north resulting in envenomations in populations that had not previously experienced them. The African honeybee tends to swarm in large populations and can travel far distances. While their venom is similar to other *Hymenoptera*, the more number of envenomations at one time are potentially fatal. Human toxicity occurs from massive envenomation resulting in anaphylactoid reactions or from immunoglobulin E-mediated anaphylaxis. Death from anaphylaxis is more common.

Fire ants are native to the United States; however, significant toxicity occurs after envenomation from the red fire ant (*Solenopsis invicta*). The red fire ants were introduced to Southern United States in the early 1900s and have spread their population northward. Similar to other *Hymenoptera*, death can occur from anaphylaxis or massive envenomation. Research shows that 90% of venom is injected within the first 20 seconds after a sting with 100% injected at 1 minute [32].

Mechanism of action and toxicokinetics

The venom from the *Hymenoptera* is similar with only slight variations. More than 75 venom components have been identified across 31 species. The primary component is melittin which, after envenomation, enters the phospholipid bilayer of the cell membrane and destroys red blood cells, platelets, and the vascular endothelium. Phospholipase A1 is present in wasps and fire ants. Phospholipase A2 is present in honeybees, bumblebees, and some wasps. It is considered the primary allergen causing histamine release and inflammatory reactions [33]. Hyaluronidase is believed to facilitate the movement of venom through tissues surrounding the sting site [32].

Fire ants have water-insoluble alkaloid venom. As previously mentioned, the primary component is phospholipase A1, Hyaluronidase, and the enzyme *N*-acetyl-beta-glucosaminidase. This venom has hemolytic, cytotoxic, and neurotoxic effects as it inhibits Na-K ATPase, reducing cellular respiration and uncoupling oxidative phosphorylation.

Analytical methods and clinical management implications

The diagnosis of *Hymenoptera* envenomations is a clinical diagnosis, and no clinical laboratory test can aid in the diagnosis. The majority of deaths occur as a result of anaphylaxis and rapid, aggressive therapy is required at the scene to aid in maintaining an airway.

Anaphylactoid reactions can result in vomiting, diarrhea, hemoglobinuria, acute renal failure, thrombocytopenia, and rhabdomyolysis. Thus laboratory tests should include a complete blood count, urinalysis, creatine, BUN, and creatine phosphokinase.

Treatment

Treatment is symptomatic and supportive as antivenom is not available clinically. Antihistamines, topical corticosteroids, and cool compresses can aid in the cutaneous effects. Patient with anaphylaxis and/or systemic toxicity should receive aggressive supportive care at the scene, emergency department, and intensive care. Epinephrine is most commonly used to treat anaphylaxis as well as antihistamines and corticosteroids. Any person found to have an anaphylactic reaction after envenomation should be prescribed epinephrine upon discharge to carry with them at all times.

Marine species

The number of venomous marine species is vast, despite this, few reports are made annually regarding envenomations. In 2017 USPCCs had 1020 contacts regarding "aquatic" bites and envenomations from fish, jellyfish, and other coelenterate stings, and other or unknown marine animal bites and/or envenomations [3]. Less than half were treated in a health-care facility with all patients having no effect to moderate symptoms. No major symptoms or deaths were reported. However, venom from marine life has resulted in severe morbidity and mortality. Sponges, jellyfish, vertebrate fish species, coral, stingrays, and sea snakes are venomous and can result in harm to humans [34,35]. In fact, marine envenomations are usually the result of jellyfish stings or penetrating injuries from venomous creatures [36]. Having basic knowledge in the signs, symptoms, and immediate treatment can prevent significant morbidity and mortality.

Mechanism of action and toxicokinetics

About 5000 species of sponges exist in the ocean. Envenomations may occur from direct contact with the organism or its cohabitation with other venomous creatures. The surface toxins (e.g., crinotoxin) are secreted from the collagenous skeleton which is made of spongin. These toxins may result in skin irritation to swimmers and divers.

Jellyfish (Cnidaria) consists of more than 10,000 species that are further characterized into four classes. In the majority of cases, envenomation occurs through the release of nematocysts which contain multiple enzymes, including phospholipase A2. *Chironex fleckeri* (box jellyfish) is one of the most venomous species with tentacles that secrete venom with the ability to result in death.

Mollusks include the cone snail and octopus. The cone snail has a hollow probiscus with a radicular tooth that may inject venom when provoked. Envenomation results in tetany and blockade of sodium, potassium, and calcium channels. The blue-ringed octopus, another extremely venomous marine species, secretes tetrodotoxin, a potent sodium channel blocker.

Marine vertebrates, when provoked, may also envenomate their prey through punctures. Stingrays are found throughout the world and can envenomate attackers or prey by mechanical injury, and envenomation via a gland at the base of the tail that secretes venom. Scorpionfish, lionfish, and stonefish secrete venom from their dorsal pelvic and anal spines. The venom from the stonefish causes increased capillary permeability that can result in significant morbidity. Similarities may exist between the scorpionfish and stonefish, which may be beneficial in treatment [37].

Analytical methods and clinical management implications

The diagnosis is generally made considering the history and physical symptoms. Most marine envenomations result in severe intense pain. Systemic effects such as anaphylactoid reactions and penetrating trauma should be treated immediately. Patients with jellyfish envenomations may present with a delayed (20–30 minutes) systemic pain and cardiac effects (Irukandji-like syndrome). Clinical laboratory analysis is dependent on the presenting symptoms, including a troponin for those with a delayed presentation. However, a recent prospective study of lionfish envenomations in 117 patients found hypophosphatemia (12%), elevated aspartate aminotransferase (10%), and thrombocytopenia (3%) suggesting that electrolytes, liver function tests, and a complete blood count be assessed in patients [38].

Treatment

Treatment is dependent on the characteristics of the venom (e.g., heat-labile and acid labile). For example, symptoms from a sponge or fire coral envenomations can be relieved by acetic acid (vinegar) soak. Comparatively, symptoms from stingrays and other vertebrate envenomations can be relieved from hot water emersion. Generally, pain returns with removal from the hot water. The temperature should be carefully monitored to prevent burn to the patient.

Depending on the type of jellyfish, acetic acid may worsen the pain. For minor jellyfish envenomations, rinsing with seawater may be enough to alleviate the pain. However, for more toxic envenomations, acetic acid and removal of the tentacles may be needed. In the more severe cases, patients should be transported to the hospital for ongoing care.

Antivenom is available for jellyfish and stonefish envenomations. This should be considered for severe jellyfish envenomations especially those with cardiovascular collapse. However, early cardiopulmonary resuscitation (CPR) is more likely the key intervention to reduce fatalities. Antivenom may be given for stonefish envenomations if the pain is severe. As previously mentioned, antivenom may be used for severe pain due to similarities in the venom between stonefish and other vertebrate fish.

Mammals

Venomous mammals are rare but were more prominent in history. Venomous mammals that are currently in existence include insectivores (e.g., solenodons, shrews, and vampire bats), the male platypus, primates (e.g., slow loris), hedgehogs, and African crested rat. However, the hedgehog and African crested rat are not venomous themselves but use toxin from others (Bufo toad and Poison Arrow Tree, respectively) on their outer coating to poison predators. These mammals use their venom to catch prey or fend of predators; thus humans are rarely impacted. If a human is envenomed, pain is most likely to be the only symptom with the exception of infections [39,40].

References

[1] World Health Organization. Addressing the burden of snakebite envenoming. Seventy-first World Health Assembly. WHA71.5. May 26, 2018. <https://www.who.int/neglected_diseases/mediacentre/WHA_71.5_Eng.pdf?ua=1> [accessed 13.02.19].
[2] Gutierrez JM, Calvete JJ, Habib AG, et al. Snakebite envenoming. Nature Reviews Disease Primers, 2017. <https://doi.org/10.1038/nrdp.2017.63>.
[3] Gummin DD, Mowry JB, Spyker DA, et al. Annual report of the American Association of Poison Control Centers' National Poison Data System (NPDS): 35th Report. Clin Tox 2017;2018(56):1213–415.
[4] Longbottom J, Shearer FM, Devine M, et al. Vulnerability to snakebite envenoming: a global mapping of hotspots. Lancet 2018;392:673–84.
[5] Sanhajariya S, Duffull SB, Isbister GK. Pharmacokinetics of snake venom. Toxins (Basel) 2018;10:73. Available from: https://doi.org/10.3390/toxins10020073.
[6] White J. Overview of snake envenoming. Critical care toxicolog: diagnosis and management of the critically poisoned patient. 2nd ed. Springer International Publishing; 2017. p. 2279–318.
[7] Rolan TD. Neurotoxic snakes of the Americas. Neurol Clin Pract 2015;5:383–8.
[8] Tan CH, Sim SM, Gnanathasan CA, et al. Pharmacokinetics of the Sri Lankan hump-nosed pit viper (hypnale hypnale) venom following intravenous and intramuscular injections of the venom into rabbits. Toxicon 2014;79:37–44.
[9] Bryvar M, Kurtovic T, Grenc D, et al. Vipera ammodytes bites treated with antivenom viperatab: a case series with pharmacokinetic evaluation. Clin Toxicol 2017;55:241–8.
[10] Hart AJ, Hodgxon WC, O'Leary M, Isbister GK. Pharmacokinetics and pharmacodynamics of the myotoxic venom of *Pseudechis australis* (mulga snake) in the anesthetized rat. Clin Toxicol 2014;52:604–10.
[11] Paniagua D, Jimenez L, Romero C, et al. Lymphatic route of transport and pharmacokinetics of *Micrurus fulvius* (coral snake) venom in sheep. Lymphology 2012;45:114–53.
[12] Yap MK, Tan NH, Sim SM, et al. Pharmacokinetics of *Naja sumatrana* (equatorial spitting cobra) venom and its major toxins in experimentally envenomed rabbits. PLoS Negl Trop Dis 2014;8:e2890.
[13] Yap MK, Tan NH, Sim SM, Fung SY. Toxicokinetics of *Naja sputatrix* (javan spitting cobra) venom following intramuscular and intravenous administrations of the venom into rabbits. Toxicon 2013;68:18–23.
[14] Snake Venom Detection Kit. Package Insert. Seqirus Pty Ltd, 2017. <https://www.seqirus.com.au/docs/743/718/03100000HFeb17.pdf> [accessed 13.03.19].
[15] Kanaan NC, Ray J, Stewart M, et al. Wilderness Medical Society Practice Guidelines for the treatment of pit viper envenomations in the United States and Canada. Wilderness Environ Med 2015;26:472–87.
[16] Alirol E, Lechevalier P, Zamatto F, et al. Antivenoms for snakebite envenoming: what is in the research pipeline? PLoS Negl Trop Dis 2015;9: e0003896.
[17] Culin J, Goodnight ML. Arachnid. Encyclopedia britannica. <https://www.britannica.com/animal/arachnid> [accessed 25.03.19].
[18] White J. Overview of scorpion envenoming. Critical care toxicolog: diagnosis and management of the critically poisoned patient. 2nd ed. Springer International Publishing; 2017. p. 2239–53.
[19] Diaz JH. A 60-year meta-analysis of tick paralysis in the United States: a predictable, preventable, and often misdiagnosed poisoning. JMT 2010;6:15–21.
[20] Barceloux DG. Spiders. Medical toxicology of natural substances. 1st ed John Wiley and Sons, Inc; 2008. p. 925–49.
[21] Erickson TB, Cheema N. Arthropod envenomation in North America. Emerg Med Clin N Am 2017;35:355–75.

[22] Chaves-Moreira D, Senff-Ribeiro A, Wille ACM, et al. Highlights in the knowledge of brown spider toxins. J Venom Anim Toxins Trop Dis 2017;23:6.
[23] Vetter RS. Clinical consequences of toxic envenomation by spiders. Toxicon 2018;152:65–70.
[24] Hodgson E. Toxins and venoms. Prog Mol Biol Transl Sci 2012;112:373–416.
[25] McGlasson DL, Green JA, Stoecker WV, Babcock JL, Calcara DA. Duration of *Loxosceles reclusa* venom detection by ELISA from swabs. Clin Lab Sci 2009;22:216–22.
[26] Akdeniz S, Green JA, Stoecker WV, Gomez HF, Keklikci SU. Diagnosis of loxoscelism in two Turkish patients confirmed with an enzyme linked immunosorbent assay (ELISA) and non-invasive tissue sampling. Dermatol Online J 2007;13:11.
[27] Jiacomini I, Silva SK, Aubrey N, et al. Immunodetection of the "brown" spider (*Loxosceles intermedia*) dermonecrotoxin with an scFv-alkaline phosphatase fusion protein. Immunol Lett 2016;173:1–6.
[28] Rodrigo C, Gnanathasan A. Management of scorpion envenoming: a systemic review and meta-analysis of controlled clinical trials. Syst Rev 2017;6:74.
[29] Coorg V, levatan RD, Gerkin RD, Muenzer J, Ruha AM. Clinical presentation and outcomes associated with different treatment modalities for pediatric bark scorpion envenomation. JMT 2017;13:66–70.
[30] Forrester JA, Holstege CP, Forrester JD. Fatalities from venomous and nonvenomous animals in the United States (1999-2007). Wilderness Environ Med 2012;23:146–52.
[31] Levine M. Pediatric envenomations: don't get bitten by an unclear plan of care. Pediatric Emerg Med Pract 2014;11:1–16.
[32] West FL, McKeown NJ, Hendrickson RG. Massive hymenoptera envenomation in a 3 year old. Pediatric Emerg Care 2011;27:46–8.
[33] Perez-Riverol A, Lasa AM, dos Santos-Pinto JRA, Palma MS. Insect venom phospholipases A1 and A2: roles in the envenoming process and allergy. Insect Biochem Mol Biol 2019;105:10–24.
[34] Barker G, Montrief T. Marine Envenomations. Modern Resident Blog. November 29, 2018. <https://aaemrsa.blogspot.com/2018/11/marine-envenomations.html> [accessed 11.10.19].
[35] Reese E, Depenbrock P. Water envenomations and stings. Curr Sports Med Rep 2014;13:126–31.
[36] Berling I, Isbister G. Marine envenomations. Austr Fam Phys 2015;44:28–32.
[37] Gomes HI, Menezes TN, Carnielli JB, et al. Stonefish antivenom neutralizes the inflammatory and cardiovascular effects induced by scorpionfish *Scorpaena plumieri* venom. Toxicon 2011;57:992–9.
[38] Resiere D, Cerland L, De Haro L, et al. Envenomation by the invasive *Pterois volitans* species (lionfish) in the French West Indies — a two-year prospective study in Martinique. Clin Toxicol 2016;54:313–18.
[39] Sen Nag O. Venomous mammals living in the world today. World Atlas, 2017. <https://www.worldatlas.com/articles/venomous-mammals-living-in-the-world-today.html> [accessed 13.09.19].
[40] Kowalski K, Marciniak P, Rosiński G, Rychlik L. Evaluation of the physiological activity of venom from the Eurasian water shrew *Neomys fodiens*. Front Zool 2017;14:46.

Chapter 23

Case Studies on Other Drugs

Chapter 23.1

Naloxone-responsive respiratory depression in a patient with a negative urine drug screen

Nicola J. Rutherford-Parker and Jennifer M. Colby
Department of Pathology, Microbiology and Immunology, Vanderbilt University Medical Center, Nashville, TN, United States

Case description

A 24-year-old male presented to the emergency department via ambulance for the fourth time in 36 hours following syncopal episodes. He had a history of frequent emergency visits with complaints of dizziness, syncope, chest pain/pressure, and shortness of breath. His medical history was also significant for postural orthostatic tachycardia syndrome and psychiatric conditions including factitious disorder. Previous episodes of hyperkalemia and bradycardia were attributed to misuse of his prescribed medications.

Upon examination in the emergency department, the patient appeared somnolent and displayed shallow, short breaths with occasional apnea. There was a concern for possible opioid overdose prompting admission to the medical intensive care unit. Laboratory testing demonstrated hypercapnic respiratory failure (Table 23.1.1) and a negative urine drug screen (Table 23.1.2). The patient's symptoms resolved upon administration of naloxone, but he denied ingestion of any substances that would contribute to his condition. Following discharge from this visit, the patient continued to have frequent emergency department visits for syncope, chest pain and shortness of breath. In all cases, his urine drug screen was negative but his symptoms resolved with naloxone therapy. During one admission, a search of the patient's belongings was performed, due to his history of misusing medications, and loperamide was discovered.

Discussion and follow-up

Loperamide (brand name Imodium), a piperidine derivative, is a readily available over-the-counter medication used for the treatment of diarrhea. Loperamide is a μ-opioid receptor agonist that primarily acts on the receptors located in the gastrointestinal tract. When used at dosages recommended by the manufacturer, it has a limited ability to cross the blood–brain barrier, which is necessary to produce the euphoria associated with opioids. As a result, loperamide is historically considered to have a low abuse potential. Recent reports demonstrate that supratherapeutic doses of loperamide, up to 200 mg/day, can be used to alleviate opioid withdrawal symptoms and may even produce euphoria [1]. For the treatment of diarrhea the manufacturer recommends a maximum dose of 8 mg/day of over-the-counter products and 16 mg/day by prescription. The majority of symptoms reported in a retrospective study of loperamide misuse were respiratory depression, cardiac dysrhythmias and central nervous system depression [2]. Nine out of the 224 cases investigated displayed cardiotoxicity; however, most of these individuals were chronic users. Only rarely did loperamide abuse result in death [2].

Following the discovery of loperamide in his belongings, the patient was treated and discharged but continued to visit the emergency department for recurrent syncopal episodes as well as seizure-like activity and tachycardia. Testing for loperamide was performed after the patient visited the emergency department multiple times over a 2-day period. During the final admission in this period, his serum loperamide concentration was 9.7 ng/mL (expected concentrations

TABLE 23.1.1 Notable Laboratory Test Results Upon Presentation.

Analyte	Result (Reference Interval)
Arterial pH	6.94 (7.35–7.45)
Arterial pCO_2	134 (35–45 mm Hg)
Arterial pO_2	143 (80–100 mm Hg)
Arterial HCO_3	29 (21–28 mmol/L)

TABLE 23.1.2 Urine Drug Screen Panel With Results at Admission.

Test	Cutoff (ng/mL)	Result
Amphetamines	500	Negative
Barbiturates	200	Negative
Benzodiazepines	200	Negative
Buprenorphine	5	Negative
Cannabinoids	50	Negative
Cocaine metabolite	200	Negative
Methadone	300	Negative
Opiates	300	Negative
Oxycodone	300	Negative
Tricyclic antidepressants	300	Negative

following usual dosing are less than 3 ng/mL according to the reporting laboratory). The patient continued to have episodes of naloxone-responsive respiratory depression, and during one particularly severe episode, he displayed additional symptoms consistent with opioid overdose (e.g., miosis) and required multiple resuscitations. A measurement taken on a sample from this admission showed that his serum loperamide concentration was 47 ng/mL.

Loperamide misuse is growing, as demonstrated by recent literature [1,2]. Treatment for loperamide overdose includes naloxone, to reverse the opioid effects of the drug, and supportive measures. Along with fentanyl, loperamide should be considered as a cause of naloxone responsive somnolence and respiratory depression in the presence of a negative urine drug screen. This is of particular importance during the current efforts to restrict the prescription of opioids, as long-time users may look elsewhere to ward off the symptoms of withdrawal, and loperamide is particularly appealing due to its availability over-the-counter.

National Institute of Drug Abuse (NIDA) has issued an alert about loperamide abuse. It stated that loperamide abuse was uncommon as of this alert being issued (September 2016). The alert mentioned that loperamide abuse had been reported all over the United States. The calls to poison centers were reported to have increased exponentially in the year 2014. The American Association of Poison Control Centers reported that calls involving the intentional abuse or misuse of loperamide increased from 87 in 2010 to 190 in 2014 nationwide. Users' reports of effects of loperamide abuse are conflicting. Physiological physical consequences of loperamide abuse can be severe. Loperamide abuse can cause fainting, abdominal pain, constipation, cardiovascular toxicity (including racing heart and even cardiac arrest), pupil dilation, and kidney failure from urinary retention. Based on NIDA's report, the risk of physical harm following loperamide abuse is high. Of note opioid withdrawal symptoms such as severe anxiety, vomiting, and diarrhea when loperamide abuse is stopped are also reported [3].

Overall, we report a case of recurrent loperamide abuse with a negative drug screen in a 24-year-old male whose symptoms were responsive to naloxone treatment.

References

[1] Daniulaityte R, Carlson R, Falck R, Cameron D, Perera S, Chen L, et al. 'I just wanted to tell you that loperamide WILL WORK': a web-based study of extra-medical use of loperamide. Drug Alcohol Depend 2013;130:241–4.
[2] Lasoff DR, Koh CH, Corbett B, Minns AB, Cantrell FL. Loperamide trends in abuse and misuse over 13 years: 2002-2015. Pharmacotherapy 2017;37:249–53.
[3] <https://archives.drugabuse.gov/emerging-trends/loperamide-misuseabuse>.

Chapter 23.10

Cocaine, is it really there? Differing sensitivities of immunoassay drug screen and mass spectrometry

Hana Vakili[1], Khushbu Patel[1,2] and Patricia M Jones[1,2]
[1]Department of Pathology, University of Texas Southwestern Medical Center, Dallas, TX, United States, [2]Children's Medical Center, Dallas, TX, United States

Case description

A 23-month-old girl with no significant medical history other than an upper respiratory tract infection 3 weeks prior to presentation was admitted to the emergency department with lethargy and seizure activity. At presentation the patient tested positive for cocaine by a drug of abuse immunoassay screen, which prompted a comprehensive drug profile testing by quadrupole time-of-flight mass spectrometry (MS-TOF). Results of both tests are seen in Table 23.10.1, with no cocaine found in the MS-TOF assay. The laboratory was contacted regarding the discrepancy between the positive drugs of abuse screen and negative comprehensive drug profile. Upon investigation, it was found that the results were from two different urine samples. The emergency room staff had sent a second urine sample an hour and twenty minutes after the first sample that was much more dilute (as noted by urine creatinine measurements) compared to the original sample. It was the second sample that tested negative for benzoylecgonine on the comprehensive drug profile testing.

As part of the investigation, both samples were reanalyzed on the drugs of abuse screen immunoassay and the original sample was run on the comprehensive drug profile testing. The results of these tests are shown in Table 23.10.2. Both samples repeated as negative by the immunoassay drugs of abuse screen and further investigation of the original presumptive positive result determined that the value was 158 ng/mL with the threshold cut-off of 150 ng/mL. On repeat, the value fell below the cut-off. The result of the MS-TOF assay on the original sample showed that it was indeed positive for benzoylecognine-the primary cocaine metabolite. This assay has a limit of detection (LOD) of 10 ng/mL, although a positive cut-off of 50 ng/mL is used for reporting. Results of this test indicate that the patient was exposed to cocaine.

Discussion

Most of the reported pediatric cocaine exposures in the medical literature are intrauterine exposures from maternal abuse or adolescent recreational abuse cases. However, the past couple of decades have seen an increasing number of unintentional cocaine exposures in pediatric patients under the age of 6 years from passive ingestion or inhalation [1–3]. The first challenge in treating cocaine-associated emergencies is to correctly identify this group of patients. A broad range of conditions may mimic the classic symptomatic of acute cocaine intoxication. Observational data suggests that passive exposure to cocaine may manifest as more frequent respiratory symptoms in infants, with and without fever [2], and more frequent generalized and focal seizures in children below 8 years [4]. Acute toxicity from cocaine is known to cause seizures in adults as well as children [4]. Cocaine exposure associated febrile seizures are commonly seen in the emergency department with an incidence of 2%–5%, with most cases occurring between 6 months and 5 years of age [1–3].

TABLE 23.10.1 Initial Lab Test Results.

Sample Date	Sample Time	Test	Drug Class	Result
01/12	14:00	Drugs of abuse screen	PCP screen	Negative
			Benzodiazepam	Negative
			Cocaine metabolites	Presumptive positive[a]
			Amphetamines	Negative
			Cannabinoids	Negative
			Opiates	Negative
			Barbiturates	Negative
			Methylenedioxymethamphetamine	Negative
		Test	**Reported positive for:**	
01/12	15:20	Comprehensive urine toxicology profile	Acetaminophen	
			Levetiracetam	
			Lorazepam	
			Phenytoin	

PCP, Phencyclidine.
[a]The positive cut-off for the drug of abuse screen was 150 ng/mL. The sample collected at 14:00 had an estimated concentration of 158 ng/mL (positive), but on repeat was 139 ng/mL (negative).

TABLE 23.10.2 Rerun Test Results.

Sample Date	Sample Time	Test	Drug Class	Result
01/12	14:00	Drugs of abuse screen	PCP screen	Negative
			Benzodiazepam	Negative
			Cocaine metabolites	Negative[a]
			Amphetamines	Negative
			Cannabinoids	Negative
			Opiates	Negative
			Barbiturates	Negative
			Methylenedioxymethamphetamine	Negative
01/12	15:20	Drugs of abuse screen	PCP screen	Negative
			Benzodiazepam	Negative
			Cocaine metabolites	Negative
			Amphetamines	Negative
			Cannabinoids	Negative
			Opiates	Negative
			Barbiturates	Negative
			Methylenedioxymethamphetamine	Negative
		Test	**Reported positive for:**	
01/12	14:00	Comprehensive urine toxicology profile	Cocaine	
			Levetiracetam	
			Trimethoprim	

PCP, Phencyclidine.
[a]The positive cut-off for the drug of abuse screen was 150 ng/mL. The sample collected at 14:00 had an estimated concentration of 158 ng/mL (positive), but on repeat was 139 ng/mL (negative).

The patient described in this case, presented with tonic—clonic and intermittent decerebrate posturing, seizure, and slight fever (temperature on presentation was 38.3 °C). In the emergency department, the patient was given Ativan 1.3 mg × 3 followed by 60 mg/kg levetiracetam during which seizures did stop briefly but then resumed. She was then given a fourth dose of Ativan and then 20 mg/kg fosphenytoin after which clinical seizures stopped. Seizure time was estimated at 75 minutes. From the toxicology standpoint, it was reasonable to suspect cocaine toxicity as a cause of the patient's seizures given that the MRI and lumbar puncture results were unremarkable. Thus based on the fact that the screen and comprehensive tests on the initial urine sample were both positive for the benzoylecgonine, this does raise the concern that the child had been exposed to cocaine with the low levels suggesting either a low dose exposure or exposure that may have occurred a few days prior to her presentation.

Cocaine is well absorbed by all routes of exposure and is extensively and rapidly metabolized in the liver [5]. There are two main types of urine drug tests—screening and confirmatory tests. Initial drug tests or screens are usually performed using immunoassay based methods [6,7]. Because cocaine is metabolized so rapidly, the metabolite benzoylecgonine is the compound detected by essentially all drug tests, including standard urine drug-of-abuse screening immunoassays for cocaine testing [8]. Utilizing immunoassays for toxicology provides relatively rapid results; however, they have many weaknesses that can result in false-positive and false-negative results. It is vital to understand how to interpret urine immunoassays by considering the cut-off values, detection times, and false-positive results [9]. Even though the cocaine immunoassay is considered one of the most specific amongst toxicology screening immunoassays, there are rare reported cases of false-positive urine cocaine screenings results that were not confirmed by mass spectrometry [10]. In all the reported cases of false-positive cocaine by immunoassay, a definitive cause for the interference was undetermined [11,12]. Positive results on immunoassays should be confirmed by mass spectrometry on the same specimen as tested for the screening test. It is also crucial to take the time factor into consideration as the amount of time a drug can be detected in the urine, its detection window, determines whether the test will be positive [13]. To evaluate detection times of a drug or substance, both drug characteristics and patient factors need to be considered. The half-life estimates for cocaine metabolites detection in urine range from 14.6 to 52.4 hours depending on an individual's metabolic rate [5,14]. As seen in this case the second urine specimen collected less than two hours later from this patient was negative on both immunoassay and confirmatory tests. Upon reviewing the patient record, it was found that the second sample was obtained after administration of intravenous fluid. Because benzoylecgonine was actively being excreted and urine was much more dilute due to fluid administration, the concentration of benzoylecgonine in the second sample was below 10 ng/mL (LOD).

In general, there are not enough studies to establish an age-specific cut-off value for cocaine testing in the pediatric population. 50 ng/mL is a commonly used cut-off for benzoylecgonine in the pediatric population [1–3]. Furthermore, drugs of abuse cut-offs for both immunoassays and confirmatory methods are fairly high to keep false-positive rates low for their most common use, workplace drug testing. Positive urine drug-of-abuse tests for cocaine are not typically confirmed by further testing by hospital laboratories because the assays have very high sensitivity and specificity (99%) for benzoylecgonine [15]. However, the unique aspect of pediatric toxicology testing is that both providers and laboratories have a definitive obligation to confirm drugs of abuse positive screening results particularly when a child is too young to use drugs willingly and appears to have symptoms consistent with exposure to a drug of abuse. In addition, there are both legal and social implications associated with positive drug testing in the pediatric population.

References

[1] Rosenberg NM, Meert KL, Knazik SR, Yee H, Kauffman RE. Occult cocaine exposure in children. Am J Dis Child 1991;145:1430–2.
[2] Lustbader AS, Mayes LC, McGee BA, Jatlow P, Roberts WL. Incidence of passive exposure to crack/cocaine and clinical findings in infants seen in an outpatient service. Pediatrics 1998;102:e5.
[3] Mott SH, Packer RJ, Soldin SJ. Neurologic manifestations of cocaine exposure in childhood. Pediatrics 1994;93:557–60.
[4] Armenian P, Fleurat M, Mittendorf G, Olson KR. Unintentional pediatric cocaine exposures result in worse outcomes than other unintentional pediatric poisonings. J Emerg Med 2017;52:825–32.
[5] Jufer RA, Wstadik A, Walsh SL, Levine BS, Cone EJ. Elimination of cocaine and metabolites in plasma, saliva, and urine following repeated oral administration to human volunteers. J Anal Toxicol 2000;24:467–77.
[6] Armbruster DA, Schwarzhoff RH, Hubster EC, Liserio MK. Enzyme immunoassay, kinetic microparticle immunoassay, radioimmunoassay, and fluorescence polarization immunoassay compared for drugs-of-abuse screening. Clin Chem 1993;39:2137–46.
[7] Jaffee WB, Trucco E, Levy S, Weiss RD. Is this urine really negative? A systematic review of tampering methods in urine drug screening and testing. J Subst Abuse Treat 2007;33:33–42.
[8] Moore FM, Simpson D. Detection of benzoylecgonine (cocaine metabolite) in urine: a cost-effective low risk immunoassay procedure. Med Lab Sci 1990;47:85–9.

[9] Moeller KE, Kissack JC, Atayee RS, Lee KC. Clinical interpretation of urine drug tests: what clinicians need to know about urine drug screens. Mayo Clin Proc 2017;92:774–96.

[10] Kim JA, Ptolemy AS, Melanson SE, Janfaza DR, Ross EL. The clinical impact of a false-positive urine cocaine screening result on a patient's pain management. Pain Med 2015;16:1073–6.

[11] De Giovanni N, Fucci N. Hypothesis on interferences in kinetic interaction of microparticles in solution (KIMS) technology. Clin Chem Lab Med 2006;44:894–7.

[12] Reisfield GM, Haddad J, Wilson GR, Johannsen LM, Voorhees KL, Chronister CW, et al. Failure of amoxicillin to produce false-positive urine screens for cocaine metabolite. J Anal Toxicol 2008;32:315–18.

[13] Moeller KE, Lee KC, Kissack JC. Urine drug screening: practical guide for clinicians. Mayo Clin Proc 2008;83:66–76.

[14] Nickley J, Pesce AJ, Krock K. A sensitive assay for urinary cocaine metabolite benzoylecgonine shows more positive results and longer half-lives than those using traditional cut-offs. Drug Test Anal 2017;9:1214–16.

[15] Eskridge KD, Guthrie SK. Clinical issues associated with urine testing of substances of abuse. Pharmacotherapy 1997;17:497–510.

Chapter 23.11

False-positive levorphanol (opioid) due to dextromethorphan on urine drug screen by time-of-flight MS

Patricia M Jones and Khushbu Patel
Department of Pathology, University of Texas Southwestern Medical Center and Children's Medical Center, Dallas, TX, United States

Case description

A 15-year-old overweight male presented to the psychiatric clinic with a past medical history of substance abuse disorder, attention deficit disorder, and major depressive disorder. At the time of his clinic visit, he was also having suicidal thoughts and had a chest cold. The patient's history of substance abuse was primarily for marijuana. In the past the patient had been caught with an illicit hydrocodone prescription and had reported taking hydrocodone. At the time of the current visit the patient had a point-of-care (POC) urine drug screen in the clinic, as well as having a urine sample collected for drugs of abuse screen and a comprehensive urine drug profile to be assayed in the laboratory by time-of-flight mass spectrometry (LC–QTOF). Results of all three tests are shown in Table 23.11.1.

The POC and drugs of abuse screen results were negative; however, the comprehensive LC–QTOF-based drug screen was positive for six compounds. The patient was prescribed methylphenidate, venlafaxine, and hydroxyzine for his ADHD, depression, and anxiety, respectively. Though smoking status was not documented in the patient's chart, nicotine and cotinine were detected in this urine sample. Dextromethorphan is a cough suppressant found in over-the-counter cough medicine. It was the presence of levorphanol, a synthetic opioid analgesic, that caught the caregiver's attention. The physician called the laboratory to inquire whether this could be a mistake in either analysis or patient sample identity, or considering the patient's history, whether the patient was taking a non-prescribed opioid.

Discussion and follow-up

Opioids are a large class of compounds that bind to the opioid rector to produce analgesic, sedative, and euphoric effects. They are primarily prescribed as analgesics for pain management. The use of opioids at high doses produces a feeling of euphoria and has sedative effects. There is a high dependence liability associated with chronic opioid use, which leads to impaired coordination and decision making. Depressed respiration, hypothermia, and coma are associated with toxic overdose and can be lethal. In this case the patient had first used hydrocodone at the age of 14 following a dental procedure. He reported abuse at that time taking more than the prescribed dose. The patient had stated that he likes the feeling of being high from opiates.

Alere *i*Cup Dx Drug Screen Cup, a POC test was used for drug screening. This assay is based on competitive immunochemical reactions, in which drug or drug metabolites present in the patient's urine sample compete with the chemically labeled drug (drug–protein conjugate) in the assay for the limited antibody binding sites present. Interferences in the form of false-positive and false-negative results are not uncommon and include immunoassay interferences cause by cross-reactivity and colorimetric interferences caused by colored or adulterated urine samples. The manufacturer's package insert did not state levorphanol as a reactive compound.

The urine drug screen was performed in the laboratory on the Dimension Vista platform. This immunoassay is based on competitive binding of the drug in the sample and drug labeled with glucose-6-phosphate dehydrogenase. The concentration of the drug in the sample is proportional to the unbound drug conjugate, which is available for catalyzing the

TABLE 23.11.1 Drug tests performed and their reported results.

Test	Drug Class	Result
DAU-POC	Amphetamine	Negative
	Barbiturates	Negative
	Benzodiazepines	Negative
	Cocaine	Negative
	Ecstasy	Negative
	Methylenedioxymethamphetamine	Negative
	Methadone	Negative
	Opiates	Negative
	Oxycodone	Negative
	Phencyclidine	Negative
	Tricyclic antidepressants	Negative
	Cannabinoids	Negative
Test	**Drug Class**	**Result**
Drugs of abuse screen	PCP screen	Negative
	Benzodiazepam	Negative
	Cocaine metabolites	Negative
	Amphetamines	Negative
	Cannabinoids	Negative
	Opiates	Negative
	Barbiturates	Negative
	Methylenedioxymethamphetamine	Negative
Test	**Reported positive for:**	
Comprehensive urine toxicology profile	Cotinine, nicotine	
	Dextromethorphan	
	Hydroxyzine	
	Levorphanol	
	Methylphenidate	
	Venlafaxine	

PCP, Phencyclidine.

oxidation of glucose-6-phosphate to produce NADH. The readout of this reaction is detected at 340 nm and is proportional to the concentration of the drug in the sample. The opiate screen uses an antibody that is reactive to morphine. However, the antibody cross-reacts with other opiates including codeine, hydrocodone, oxycodone, and levorphanol. According to the manufacturer's package insert, the limit of detection for levorphanol is 7680 ng/mL.

The assay used for the comprehensive drug profile is a laboratory-developed test that analyzes prescreened urine samples on a hybrid quadrupole-time of flight mass spectrometer. The assay was developed with 138 targeted drug compounds for which standards were purchased and validations were performed for retention time and spectral data. In the assay, high mass accuracy parent and product ion information are acquired in a non-targeted fashion. Analysis is based on a combination of mass accuracy, retention time and library matching. The 138 validated drugs make up the targeted compounds reported by the assay, and they are reported when the instrument gives an acceptable match on all

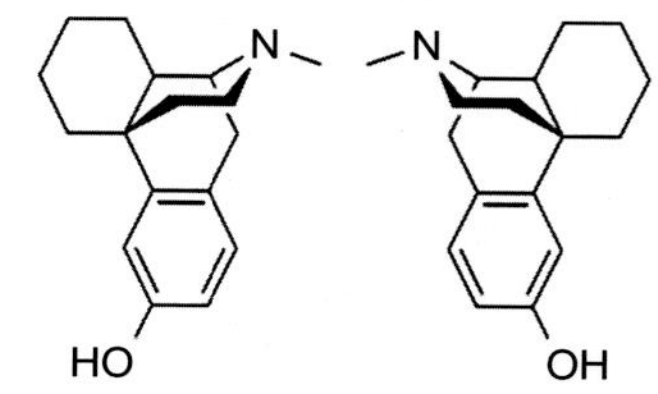

FIGURE 23.11.1 Stereoisomers dextrorphan and levorphanol.

established parameters for that compound in the sample. During assay validation, the limit of detection for levorphanol was determined to be 10 ng/mL.

In this case the compounds listed in Table 23.11.1 are the targeted compounds found in this patient's sample. Despite the negative POC and drugs of abuse screen, the comprehensive profile detected three therapeutic drugs the patient was taking (methylphenidate, venlafaxine, and hydroxyzine), as well as nicotine and over the counter drug, dextromethorphan. In addition, the assay detected levorphanol. Given the patients' inclination for drug use, the physician contacted the lab to inquire about the presence of levorphanol in the urine sample given that the opiate screen on the POC and drug of abuse screen was negative for opioids.

The discrepant results between the comprehensive LC−QTOF drug screen and immunoassay screen are possible due to the differing analytical sensitives of the two methods. However, upon further investigation by the laboratory, it was noticed that levorphanol is a stereoisomer of dextrorphan (Fig. 23.11.1), the main metabolite of dextromethorphan. Dextromethorphan does not produce the same analgesic and sedative effects as opioids, which exhibit nervous system and respiratory depression. LC−MS assays are not able to distinguish between stereoisomers, and thus dextrorphan in the sample will be recognized as levorphanol. Only enantiomeric chiral LC−MS/MS analysis can help differentiate the two compounds.

The primary caregiver was notified that with dextromethorphan present, the detected levorphanol was most likely the metabolite dextrorphan. The lab has since instituted a rule that if levorphanol is detected in a patient sample that also contains dextromethorphan, a comment will be appended to the results stating that the detection of Levorphanol is likely due to the presence of Dextromethorphan.

Conclusion

This case illustrates that it is paramount that the laboratory and clinical staff be aware of potential interferences and limitations of the various testing modalities used for drug screening. An open channel of communication between the laboratory and care team is crucial when a result does not fit the clinical picture.

Further reading

<https://www.ncbi.nlm.nih.gov/pubmed/29565959>.
<https://academic.oup.com/labmed/article/41/8/457/2657548>.

Chapter 23.12

Comparison of methamphetamine detection in urine and oral fluid

Sarah Smiley and Amadeo Pesce
Precision Diagnostics, LLC, San Diego, CA, United States

Case description

A 38-year-old female enlisted in an outpatient rehabilitation center for substance abuse disorder was being treated for methamphetamine dependence. Upon initial entrance into treatment the patient continued use of methamphetamine confirmed with high concentrations in urine. The patient, however, remained abstinent the past few months of treatment, with negative results for all illicit substances tested. Due to recent suspicious behavior exhibited by the patient during previous visits, the center collected both urine and oral fluid specimens minutes apart.

The urine and oral fluid specimens were sent to the laboratory (Precision Diagnostics, San Diego, CA) where liquid chromatography–tandem mass spectrometry (LC–MS/MS) protocol was validated for the quantification of methamphetamine in urine and oral fluid using AB Sciex 6500 LC–MS systems. The subsequent cutoffs for urine and oral fluid were set to 50 and 1 ng/mL, respectively.

The patient's test results for methamphetamine were concluded as negative in urine, but positive at 6 ng/mL in oral fluid. Specimen validity, including creatinine, oxidant, pH, and specific gravity, were all in range per SAMHSA guidelines. It should be noted amphetamine was not present as a metabolite in either the urine or oral fluid samples. Upon questioning the rehabilitation center, it was confirmed the patient was in good health and observed collections were completed for the urine and oral fluid samples; therefore we can rule out the likelihood of dilution, adulteration, or substitution of the urine sample.

Discussion

Traditional drug testing has typically considered urine as the standard; however, drug abusers are frequently motivated to "tamper" with their urine specimens by means of dilution, substitution, and/or adulteration [1]. This is particularly the case if specimen collection is unobserved, a main driver in considering oral fluid as a suitable alternative matrix to test for drugs of abuse [2,3]. Oral fluid can offer a noninvasive collection that can be easily observed to lessen issues with sample integrity [1–5].

Methamphetamine is a CNS stimulant commonly abused illicitly, and may also aid in the treatment of obesity, hyperactivity disorders, Parkinson's disease, and nasal congestion, and regularly monitored among drug abuse patients. Many factors play a role in detection of methamphetamine between urine and oral fluid testing matrices including detection windows, route of administration, drug metabolism, pH, and cutoff concentrations.

Detection windows exhibited between urine and oral fluid can be heavily influenced by acute versus chronic use, time of dose, and route of administration [6]. An analyte present in urine but negative in oral fluid is likely due to end elimination of a particular drug while detection in oral fluid typically requires a free form of the drug [4]. Oral fluid exhibits shorter detection windows with an average of 1 or 2 days while urine detection windows are approximately 1.5–4 days [8]. Methamphetamine detection windows have been reported in urine, to be 24–60 hours after a single dose, and up to 6 days with chronic use. In oral fluid, methamphetamine can be detected for at least 24 hours after a single use and 3 days or more with chronic use [8]. On the other hand, the route of administration of a drug, for instance, inhalation, a common route of use among methamphetamine users, could results in residual contamination of the oral fluid cavity due to recent use [7].

In addition, drugs and metabolites present in oral fluid via passive diffusion from the blood to saliva from the salivary glands offering a window into recent drug use by detection of predominantly the parent drug(s) [2,4,5]. This explains the reason we do not see amphetamine in conjunction with methamphetamine in the oral fluid.

The metabolic pathways of many drugs also affect detection between urine and oral fluid. Many drugs that are administered orally go through first-pass metabolism via major enzyme groups, for example, the CYP enzymes and further into phase II metabolism rendering drugs more polar via glucuronidation [5,9]. Because oral fluid collects the free drug(s) through passive diffusion from systemic circulation, metabolites of drugs are not readily detected, while still detected in urine [5].

Adversely, basic drugs such as the amphetamine drug class, opiates, and many illicit drugs exhibit high affinity for the oral fluid cavity [2]. The transfer across membranes and excretion in saliva is highly influenced by methamphetamine's lipophilic nature, low plasma-membrane protein binding; therefore many illicit drugs will be ionized due to the lower pH in saliva resulting in ion trapping [5]. Ion trapping will help aid in detection of methamphetamine in the oral fluid sample we see in this patient [10, 11]. Drugs that are heavily protein bound, for example, benzodiazepines, may only be detected at low concentrations or not detectable at all in oral fluid [2,4,5].

Urinary excretion of drugs can be heavily influenced by urinary pH. In acidic urine results the basic methamphetamine remains in an ionized state thus reducing reabsorption by the kidneys shortening the half-life in the body. In contrast, more alkaline urine slows excretion and prolongs the detection time [6,12]. However, in the case of this patient, the validity results indicated a normal urine pH of 6.1; therefore we can conclude this would not typically affect the methamphetamine results in urine.

Lastly, the cutoff differences between urine and oral fluid play a role in the detection of analytes between the two matrices. Due to the lower volume collected in oral fluid, it is anticipated to result in lower concentrations of a drug(s) when detected. This requires a more sensitive solid-phase extraction technique in order to detect analytes in oral fluid before entering into LC−MS system. The methamphetamine cutoff in urine was validated to 50 ng/mL while oral fluid cut off was validated at 1 ng/mL.

Due to the high affinity for methamphetamine in oral fluid [5] the low cutoff of 1 ng/mL, and history of patient use, we concluded the oral fluid results confirm methamphetamine use. The concentration of methamphetamine in urine likely fell below the cutoff of 50 ng/mL, while capable of detection in oral fluid. Oral fluid was the choice of specimen by the provider to circumvent sample tampering when collecting the urine. Oral fluid can be an alternative when adulteration, dilution, or substation is in question.

The negative results in urine indicate urine may be an inadequate matrix when monitoring patients for methamphetamine where oral fluid can be considered a suitable alternative for methamphetamine and other drugs of abuse. Another possible cause of lack of amphetamine in the urine could be a relatively recent exposure to a very small quantity of methamphetamine. Nevertheless, this case exemplifies the scenario where oral fluid can be used to monitor compliance and abstinence in high-risk populations. In particular, oral fluid can be beneficial when monitoring for long term use of illicit drugs and weak bases, for example, amphetamines, cocaine, opiates, and opioids [2]. The limitations of oral fluid as a suitable matrix include shorter detection windows compared to urine, effects of pH variations, drugs present at lower concentrations than urine due to smaller volume, the potential for oral contaminations, and the need for sensitive and specific laboratory LC−MS/MS methodologies [8,13].

References

[1] Fritch D, Blum K, Nonnemacher S, Haggerty BJ, Sullivan MP, Cone EJ. Identification and quantification of amphetamines, cocaine, opiated, and phencyclidine in oral fluid by liquid chromatography-tandem mass spectrometry. J Anal Toxicol 2009;33:569−77.
[2] Bosker WM, Heustis MA. Oral fluid testing for drugs of abuse. Clin Chem 2009;55(11):1910−31.
[3] Miller KL, Puet BL, Roberts A, Hild C, Carter J, Black DL. Urine drug testing results and paired oral fluid comparison from patients enrolled in long term medication-assisted treatment in Tennessee. J Subst Abuse Treat 2017;76:36−42.
[4] Dummer OH. Review: pharmacokinetics of illicit drugs in oral fluid. Forensic Sci Int 2005;133−42.
[5] Allen KR. Screening for drugs of abuse: which matrix, oral fluid or urine? Ann Clin Biochem 2011;48:531−41.
[6] Baselt RC. In disposition of toxic drugs and chemicals in man. 10th ed. Seal Beach, CA: Biomedical Publications; 2014. p. xxiii−xxxiii.
[7] Vindenes V, Yttredal B, Oiestad EL, Waal H, Bernard JP, et al. Oral fluid is a viable alternative for monitoring drug abuse: detection of drugs in oral fluid by liquid chromatography-tandem mass spectrometry and comparison to the results from urine samples from patients treated with methadone or buprenorphine. J Anal Toxicol 2011;35:32−9.
[8] Verstraete AG. Detection times of drugs of abuse in blood, urine, and oral fluid. Ther Drug Monit 2004;26(2):200−5.

[9] Christo PJ, Manchikanti L, Ruan X, Bottro M, Hansen H, et al. Urine drug testing in chronic pain. Pain Physician 2011;14:132−43.
[10] Huestis MA, Cone EJ. Methamphetamine disposition in oral fluid, plasma, and urine. NY Acad Sci 2007;1098:104−21.
[11] Bahmandabadi L, Akhgari M, Jokar F, Sadeghi HB. Quantification determination of methamphetamine in oral fluid by liquid-liquid extraction and gas chromatograph/mass spectrometry. Hum Exp Toxicol 2016;36(2):195−202.
[12] Andas HT, Enger A, Oiestad AL, Vindenes V, Christophersen AS, Huestis MA, et al. Extended detection of amphetamine and methamphetamine in oral fluid. Drug Monit 2016;38:114−19.
[13] Liu H, Lee H, Hsu Y, Huan M, Liu RH, et al. Direct injection LC-MS-MS analysis of opiates, methamphetamine, buprenorphine, methadone and their metabolites in oral fluid from substitution therapy patients. J Anal Toxicol 2015;39:472−80.

Chapter 23.2

Assessing medication compliance in palliative care: what methodology should be utilized?

Stacy E.F. Melanson
Department of Pathology, Division of Clinical Laboratories, Brigham and Women's Hospital, Harvard Medical School, Boston, MA, United States

Case description

A 50-year-old male presented to the palliative care clinic for follow up of his advanced metastatic testicular cancer. He recently underwent surgery to remove a left scapular mass and concluded a prolonged course of radiation to the resection site. Since his last visit, he reported increased fatigue and loss of appetite accompanied by a 5-lb weight loss. At presentation the patient exhibited the following: temperature, 98.2°F; blood pressure, 115/60 mmHg; heart rate, 75 beats/min. Physical exam was remarkable for an improvement in left arm abduction to 45 degrees. His cancer-related pain was well-controlled on his chronic opioid regimen—a combination of morphine and oxycodone.

Over the past few months, toxicology screens by immunoassay in this patient were consistently positive for opiates and oxycodone and negative for amphetamines, benzodiazepines, buprenorphine, cocaine metabolite, and methadone. The positive opiate and oxycodone screens suggested compliance with morphine and oxycodone, respectively. Therefore at previous visits, the provider continued to refill the patient's morphine and oxycodone prescriptions. At this visit the provider used a validated opioid risk stratification tool which suggested the patient was at risk for medication noncompliance due to his remote history of substance abuse. To better assess compliance, the provider requested that the urine toxicology screen results be confirmed by liquid chromatography–tandem mass spectrometry (LC–MS/MS). After receiving the results (Table 23.2.1) the provider contacted the laboratory director to assist with result interpretation. The clinical team requested technical consultation from the clinical toxicologist to confirm if the patient was compliant with his medications.

Discussion

The recent dramatic increase in opioid misuse has made the skilled and informed management of patients with both chronic cancer and noncancer pain imperative. Urine drug testing (UDT) is considered standard of care when managing patients taking chronic opioid medications. It is one important tool, along with others such as pill counting and opioid risk stratification, to assess medication adherence and detect aberrant behavior [1–6]. Furthermore, recent guidelines and research recommend replacing traditional UDT by immunoassay screens with front-line definitive testing such as LC–MS/MS as this testing is more sensitive and specific than immunoassay drug screens [5–11].

This case illustrates the importance of confirmatory testing to assess compliance in the palliative care setting. The previous positive immunoassay screens for oxycodone and opiates were misleading and the provider erroneously concluded that the patient was compliant with his medications. By requesting more sensitive and specific testing, the provider had definitive results for individual drugs and metabolites as well as quantitative concentrations for opioids. As shown in Table 23.2.1, the parent compounds, oxycodone and morphine, were detected in the patients' urine explaining

TABLE 23.2.1 Results of Immunoassay Screens and Confirmation Testing by Liquid Chromatography–Tandem Mass Spectrometry (LC–MS/MS). *EDDP*, 2-ethylidene-1,5-dimethyl,3-3-diphenylpyrrolidine; *MDA*, 3,4-methylenedioxyamphetamine; *MDMA*, 3,4-methylenedioxymethamphetamine.

Drug Class	Drug	Cutoff (Immunoassay) (ng/mL)	Immunoassay Screen Results	Cutoff (LC–MS/MS) (ng/mL)	LC–MS/MS Results (ng/mL)
Amphetamines	Amphetamine	1000	Negative	25	*Not detected*
	MDA			25	*Not detected*
	MDMA			25	*Not detected*
	Methamphetamine			25	*Not detected*
Benzodiazepines	7-Aminoclonazepam	100	Negative	25	*Not detected*
	Alpha-hydroxy-alprazolam			25	*Not detected*
	Clonazepam			5	*Not detected*
	Diazepam			5	*Not detected*
	Lorazepam			25	*Not detected*
	Lorazepam-glucuronide			25	*Not detected*
	Nordiazepam			25	*Not detected*
	Oxazepam			25	*Not detected*
	Oxazepam-glucuronide			25	*Not detected*
	Temazepam			25	*Not detected*
Buprenorphine	Buprenorphine	5	Negative	5	*Not detected*
	Norbuprenorphine			5	*Not detected*
	Buprenorphine–glucuronide			5	*Not detected*
	Norbuprenorphine–glucuronide			5	*Not detected*
	Naloxone			100	*Not detected*
Cocaine metabolite	Benzoylecgonine	150	Negative	25	**Detected**
Fentanyl	Fentanyl	Not performed	Not performed	2	**Detected**
	Norfentanyl			2	**Detected**
Methadone	Methadone	300	Negative	5	*Not detected*
	Methadone metabolite (EDDP)			5	*Not detected*
Opiates/opioids	6-Acetylmorphine (heroin metabolite)	Not performed	Not performed	5	*Not detected*
	Codeine	300	**Positive**	25	*Not detected*
	Hydrocodone			25	*Not detected*
	Hydromorphone			25	*Not detected*
	Hydromorphone-glucuronide			25	*Not detected*
	Morphine			25	**>2000 ng/mL**
	Morphine-3-beta-glucuronide			25	*Not detected*
	Morphine-6-beta-glucuronide			25	*Not detected*
	Noroxycodone	300	**Positive**	25	*Not detected*
	Oxycodone			25	**>2000 ng/mL**
	Oxymorphone			25	*Not detected*
Tramadol	O-Desmethyltramadol	200	Negative	25	*Not detected*
	Tramadol			5	*Not detected*

Bold indicates the table headers and positive LC-MS/MS results.
Italic indicates the not detected LC-MS/MS results.

the positive results by immunoassay. However, none of the metabolites of oxycodone or morphine were detected [i.e., absence of noroxycodone or oxymorphone (the oxycodone metabolites) and absence of morphine-3-beta-glucuronide or morphine-6-beta-glucuronide (the morphine metabolites)]. The definitive testing results suggested that the patient was dropping or shaving the pill directly into their urine to simulate compliance.

Simulated compliance or pill shaving can occur, particularly with buprenorphine, and providers should understand how to detect it [12]. Buprenorphine is a synthetic partial opioid agonist. Suboxone, a combination of buprenorphine and the opioid antagonist naloxone, is commonly prescribed to treat opioid use disorder. The naloxone was included in the formulation to deter abuse; if patient snorts or injects dissolved Suboxone, the effects of naloxone will predominant preventing the high and possibly precipitating withdrawal. Similar to simulating compliance with morphine or oxycodone, as was seen in this patient, the addition of Suboxone directly into a patients' urine leads to high concentrations of the parent compound, buprenorphine, and low or undetectable concentrations of the metabolites, norbuprenorphine, buprenorphine−glucuronide and norbuprenorphine−glucuronide. Naloxone concentrations will also be high in simulated compliance. Hull et al. [12] described how the norbuprenorphine−buprenorphine ratio (in hydrolyzed specimens) can be utilized to assess compliance in patients prescribed buprenorphine, with a ratio of <0.02 indicating simulated compliance. To this authors' knowledge, there are no published articles on the utility of ratios to determine simulated compliance with other medications, but ratios may be helpful if low levels of metabolite are present. Regardless, definitive testing is necessary to ensure patients have not dropped or shaved their medication directly into their urine as immunoassay screens can be misleading.

In this case, when confirmatory testing was performed on nonhydrolyzed specimens, glucuronide metabolites were reported. This is contrasted by the cases described by Hull et al. [12] in which specimens were pretreated with β-glucuronidase prior to analysis and total drug concentrations were reported. In this authors' opinion, analysis of nonhydrolyzed specimens has several advantages including faster specimen preparation and better assessment of compliance, particularly for morphine [13]. Providers should understand which methodology is provided by their laboratory and how to use the results to assess compliance.

Definitive testing by LC−MS/MS also has the advantages of lower cutoffs, for many drugs, and a broader test menu. In this patient, cocaine use was missed by the immunoassay screen which had a cutoff of 150 ng/mL. The LC−MS/MS method was 6-fold more sensitive (cutoff of 25 ng/mL) and revealed illicit cocaine use. Furthermore, the immunoassay panel did not include a screen for fentanyl and/or metabolites. Both fentanyl and its primary metabolite, norfentanyl, were detected by LC−MS/MS in this patient. The LC−MS/MS findings suggested that patient recently used both cocaine and fentanyl. Cocaine is often laced with fentanyl which may have explained the presence of fentanyl and/or fentanyl metabolites.

Although there are evidence and recommendations that definitive testing should replace immunoassays to assess medication compliance, the design of the definitive testing panel is a matter of debate [14]. As discussed above, some laboratories hydrolyze specimens prior to analysis and others do not. The decision on whether to provide quantitative or qualitative results is also controversial. In this case, qualitative results were sufficient to detect illicit drug use, whereas quantitative results may have assisted with detecting simulated compliance especially if metabolites were present but at very low concentrations relative to the parent compound. Therefore a laboratory may choose to provide a combination of quantitative and qualitative results based on their patient population.

The laboratory director and provider reviewed the results and discussed how to detect simulated compliance. The laboratory director also discussed the benefits of definitive testing and suggested that the provider orders definitive testing up front for all patients in which an assessment of medication compliance is necessary. The aberrant findings were documented in the medical record and discussed with the patient. The patient was switched to methadone and referred to psychiatry for management of his substance abuse disorder.

References

[1] Manchikanti L, Atluri S, Trescot AM, Giordano J. Monitoring opioid adherence in chronic pain patients: tools, techniques, and utility. Pain Physician 2008;11(2 Suppl.):S155−80.
[2] Trescot AM, Helm S, Hansen H, Benyamin R, Glaser SE, Adlaka R, et al. Opioids in the management of chronic non-cancer pain: an update of American Society of the Interventional Pain Physicians' (ASIPP) Guidelines. Pain Physician 2008;11(2 Suppl.):S5−62.
[3] Melanson SEF, Kredlow MI, Jarolim P. Analysis and interpretation of drug testing results from patients on chronic pain therapy: a clinical laboratory perspective. Clin Chem Lab Med 2009;47(8):971−6.
[4] Michna E, Jamison RN, Pham L-D, Ross EL, Janfaza D, Nedeljkovic SS, et al. Urine toxicology screening among chronic pain patients on opioid therapy: frequency and predictability of abnormal findings. Clin J Pain 2007;23(2):173−9.

[5] Argoff CE, Alford DP, Fudin J, Adler JA, Bair MJ, Dart RC, et al. Rational urine drug monitoring in patients receiving opioids for chronic pain: consensus recommendations. Pain Med 2018;19(1):97–117.

[6] Jannetto PJ, Bratanow NC, Clark WA, Hamill-Ruth RJ, Hammett-Stabler CA, Huestis MA, et al. Executive summary: American Association of Clinical Chemistry Laboratory Medicine Practice Guideline—using clinical laboratory tests to monitor drug therapy in pain management patients. J Appl Lab Med An AACC Publ 2018;2(4):489–526.

[7] Cross TG, Hornshaw MP. Can LC and LC-MS ever replace immunoassays? J Appl Bioanal 2016;2(4):108–16.

[8] Pesce A, Rosenthal M, West R, West C, Crews B, Mikel C, et al. An evaluation of the diagnostic accuracy of liquid chromatography-tandem mass spectrometry versus immunoassay drug testing in pain patients. Pain Physician 2010;13(3):273–81.

[9] Petrides AK, Melanson SE. LC-MS/MS testing for monitoring compliance in pain management: the cost and technical benefits of a phased approach. Clinical Lab News 2016;42:22–3.

[10] Snyder ML, Fantz CR, Melanson S. Immunoassay-based drug tests are inadequately sensitive for medication compliance monitoring in patients treated for chronic pain. Pain Physician 2017;20(2S):SE1–9.

[11] Darragh A, Snyder ML, Ptolemy AS, Melanson S. KIMS, CEDIA, and HS-CEDIA immunoassays are inadequately sensitive for detection of benzodiazepines in urine from patients treated for chronic pain. Pain Physician 2008;17(4):359–66.

[12] Hull MJ, Bierer MF, Griggs DA, Long WH, Nixon AL, Flood JG. Urinary buprenorphine concentrations in patients treated with suboxone as determined by liquid chromatography-mass spectrometry and CEDIA immunoassay. J Anal Toxicol 2014;32:516–21.

[13] Genecheva R, Petrides A, Kantartjis M, Tanasijevic M, Dahlin JL, Melanson S. Clinical benefits of direct-to-definitive testing for monitoring compliance in pain management. Pain Physician 2018;21:E583–92.

[14] Petrides AK, Melanson SEF. Urine drug testing: debates over best practices to assess compliance and manage the opioid crisis. What are the critical components of a definitive testing panel?. Clinical Lab News, 2019.

Chapter 23.3

A case of suicide involving diphenhydramine

Lindsey J. Haldiman[1], Andrea Ho[2], Diane C. Peterson[1], C. Clinton Frazee III[3] and Uttam Garg[3,4]

[1]Office of the Jackson County Medical Examiner, Kansas City, MO, United States, [2]Department of Pathology and Laboratory Medicine, Truman Medical Center, Kansas City, MO, United States, [3]Department of Pathology and Laboratory Medicine, Children's Mercy Hospital, Kansas City, MO, United States, [4]University of Missouri School of Medicine, Kansas City, MO, United States

Case history

A 22-year-old male with a history of anxiety, depression, suicidal ideation, and drug abuse was found deceased in his apartment by his friends. He was last known alive approximately 2 days prior to his being found when he was seen by a neighbor. He was found unclothed lying on his bedroom floor. Early signs of decomposition were noted including marbling, skin slippage, foul odor, and multifocal green discoloration of the skin. EMS was called and confirmed death at the scene. Foul play was not suspected. A scene investigation revealed multiple empty bottles of diphenhydramine (DPH) and dextromethorphan throughout the apartment. Numerous notes (some incoherent) to family and friends were written on various objects including the refrigerator, bathroom mirror, and on multiple pink balloons scattered throughout the apartment (Figs. 23.3.1–23.3.4). His family reported that he had recently lost his job, had no money, and was struggling with depression. Autopsy revealed cerebral edema, marked pulmonary edema, and a bolus mass of blue pills within the stomach.

Heart blood, chest fluid, gastric contents, liver and brain tissue, urine, and vitreous humor were submitted for toxicological investigation. The heart blood and vitreous humor were screened for volatiles (ethanol, methanol, isopropanol, and acetone) by headspace gas chromatography (GC) with flame ionization detector. The heart blood was also used for a comprehensive broad-spectrum drug-screening that utilized enzyme immunoassays (EIAs) for amphetamines, barbiturates, benzodiazepines, cannabinoids, cocaine metabolite, methadone, opiates, phencyclidine, and propoxyphene; and drug-screen testing for >200 drugs by GC mass spectrometry (GC–MS). Drug screen testing by GC–MS employed a liquid–liquid alkaline extraction using bicarbonate buffer (pH 11.0) and butyl acetate followed by mass spectrometer detection analysis in full scan mode. Presumptive identification of analytes was made by spectral library match and relative retention time comparison with reference standards.

The nine-panel EIA drug screen was positive for opiates. GC/MS screen identified the presence of DPH, DPH metabolite, dextromethorphan, dextromethorphan/levomethorphan metabolite, nicotine, and cotinine. The heart blood volatiles panel revealed ethanol at a concentration of 54 mg/dL. Methanol, acetone, and isopropanol were negative. A volatiles panel was performed on vitreous humor and was negative for methanol, ethanol, acetone, and isopropanol. An opiate confirmation screen was negative for codeine, hydrocodone, and morphine. Quantitation of DPH in the heart blood revealed a concentration of 110 mg/L. Quantitation of dextromethorphan/levomethorphan in the heart blood revealed a concentration of 0.27 mg/L. The cause of death was found to be DPH overdose. The manner of death was classified as suicide.

Discussion

DPH is a histamine H1-receptor antagonist and is widely used as an over-the-counter allergy, cold, and sleep aid. When taken according to the recommended prescribed dosage, serum concentrations rarely exceed 0.1 mg/L. Although considered relatively safe, side effects of DPH include anticholinergic effects with inhibition of the salivary and sweat glands

FIGURE 23.3.1 Notes and Objects found throughout apartment of the deceased.

FIGURE 23.3.2 Notes and Objects found throughout apartment of the deceased.

leading to dry mucous membranes and hot dry skin; redness of the face resulting from peripheral vasodilation; impaired thermoregulation and, thus, increased body temperature; agitation due to altered central nervous system regulation; and visual problems due to dilation of the pupils [1–4]. This toxidrome is classically remembered by the mnemonic "red as a beet, dry as a bone, blind as a bat, mad as a hatter, and hotter than Hades." Further, DPH penetrates the blood–brain barrier, exhibiting muscarinic receptor blocking properties and subsequently producing sedation, antiemesis, and hypnotic effects [2]. Antihistamines can also cause cardiac conduction abnormalities such as sinus tachycardia, ventricular tachycardia, and prolonged QT interval [2,5–6].

Antihistamines such as DPH are implicated in thousands of overdoses each year, mostly classified as mixed drug overdoses. DPH-related deaths in adults are relatively uncommon. Suicidal deaths due to antihistamine overdose are rare. In 1977 Backer et al. described the first case of a suicidal adult death due to DPH where toxicological analysis was available. The case involved a 37-year-old man found dead in bed with empty pill bottles next to his body.

FIGURE 23.3.3 Notes and Objects found throughout apartment of the deceased.

FIGURE 23.3.4 Notes and Objects found throughout apartment of the deceased.

His DPH level was 31 mg/L [7]. We reviewed all cases of suicide involving DPH intoxication from the years 1997 to 2016 at the Jackson County Medical Examiner's Office in Kansas City, MO. A total of 18 cases were identified (Table 23.3.1). Of those, 14 were mixed drug intoxications and only 4 cases reported DPH by itself. One case involved DPH intoxication and subsequent suffocation using a plastic bag. The blood concentrations of DPH found in the 14 Jackson County Medical Examiner cases reviewed have ranged from 0.248 to 110 mg/L with our describes case having the highest level. A review of literature reporting DPH overdose reveals only one other case with a blood concentration higher than that seen in this presented case and that level was reported at 119 mg/L [8].

Case reports involving individuals who have survived DPH overdose report hyperthermia, tachycardia, delirium, seizures, and psychosis [4,8]. Mild symptoms of antihistamine effects including dry mouth, fever, and mydriasis are noted after ingesting doses above 300 mg. Moderate symptoms such as agitation, hallucinations, and cardiac rhythm abnormalities are seen after ingesting above 500 mg. Ingestion of 1000 mg can cause psychosis, seizures, coma, and death [8]. A blood concentration of 5.0 mg/L is considered to be the minimum lethal concentration for DPH monointoxication [1,8]. A review of the scene of the current presented case indicates that the individual may have been experiencing some level of psychosis. Specifically, there were multiple pink balloons scattered throughout the apartment and numerous handwritten, incoherent messages were found on various objects throughout the apartment. In addition, he was found unclothed, a finding which could indicate that he was experiencing hyperthermia (i.e., impaired thermoregulation).

In summary, DPH is a widely used, over-the-counter medication. DPH overdose is not common, but when it occurs, DPH overdose results in a wide variety of symptoms which can range from anticholinergic effects, blockage of fast calcium channels and muscarinic effects. Suicide deaths due solely to DPH ingestion are relatively rare. However, DPH is readily obtainable and can effectively be used as a suicidal drug agent. It is important for clinicians to recognize the symptoms of DPH overdose before death occurs. It is also important for those individuals investigating deaths to understand the lethal potential of DPH when conducting a thorough scene investigation and to proceed with comprehensive toxicological analysis when DPH ingestion is suspected.

TABLE 23.3.1 Suicides Involving Diphenhydramine Intoxication From 1997 to 2016 at the JCMEO in Kansas City, MO.

Date	Age	Gender/Race	Cause of Death	Concentrations (Source of Blood)
5/2/1997	28	M/W	DPH and zolpidem toxicity	DPH 1.8 mg/L (heart)
				Zolpidem 0.03 mg/L (heart)
12/13/1997	35	F/W	Drug toxicity (DPH and acetaminophen)	DPH 23.6 mg/L (femoral)
				Acetaminophen 1046 mg/L (femoral)
7/15/2000	20	F/W	DPH toxicity	DPH 8.6 mg/L (heart)
				DPH 9.6 mg/L (femoral)
7/7/2004	46	M/W	Digoxin and DPH toxicity	DPH 0.248 mg/L (femoral)
				Digoxin 0.0061 mg/L (femoral)
7/7/2004	52	M/W	DPH toxicity	DPH 2.46 mg/L (heart)
3/1/2005	35	M/W	DPH toxicity and suffocation by plastic bag	DPH 18.9 mg/L (heart)
3/11/2005	28	F/B	DPH and mirtazapine toxicity	DPH 10.8 mg/L (heart)
				Mirtazapine 0.618 mg/L (heart)
5/3/2006	73	F/W	DPH and ethylene glycol toxicity	DPH 1.08 mg/L (femoral)
				Ethylene glycol 164 mg/dL (heart)
1/29/2007	62	M/W	Mixed drug (acetaminophen and DPH intoxication)	DPH 0.72 mg/L (femoral)
				Acetaminophen 98 mg/L (femoral)
9/23/2008	19	F/W	Mixed drug (oxycodone, DPH, and tramadol) intoxication	DPH 12.0 mg/L (femoral)
				Oxycodone 1.3 mg/L (femoral)
				Tramadol 1.6 mg/L (femoral)
10/6/2009	44	F/B	Mixed drug (zolpidem, DPH, and quetiapine) intoxication	DPH 10.0 mg/L (heart)
				Zolpidem 1.4 mg/L (heart)
				Quetiapine 0.67 mg/L (heart)
11/6/2010	39	M/W	Ethanol, acetaminophen, and DPH toxicity	DPH 2.90 mg/L (femoral)
				Acetaminophen 85 mg/L (femoral)
				Ethanol 379 mg/dL (femoral)
9/8/2012	26	F/W	Combined ethanol and acute drug (DPH) intoxication	DPH 0.88 mg/L (Chest)
				Ethanol 57 mg/dL (Chest)
2/24/2015	17	F/H	DPH intoxication	DPH 33 mg/L (heart)
6/12/2015	64	M/W	Combined ethanol and multidrug intoxication (diphenhydramine and quetiapine)	9.9 mg/L (femoral blood)
6/22/2015	56	F/W	DPH, hydrocodone, atomoxetine, and alprazolam intoxication	DPH 4.2 mg/L (femoral)
				Hydrocodone 2.014 mg/L (femoral)
				Atomoxetine 2.8 mg/L (femoral)
				Alprazolam 0.108 mg/L (femoral)
7/1/2015	46	F/W	Acute methamphetamine and DPH intoxication	DPH 2.20 mg/L (femoral)
				Methamphetamine 1.177 mg/L (femoral)
2/1/2016	22	M/NA	DPH intoxication	DPH 110 mg/L (heart)

DPH, Diphenhydramine.

References

[1] Eckes L, Tsokos M, Herre S, Gapert R, Hartwig S. Toxicological identification of diphenhydramine (DPH) in suicide. Forensic Sci Med Pathol 2013;9(2):145−53.
[2] Jeffrey AD, Lytle-Saddler T. Diphenhydramine overdose in a 26-year-old woman. J Emerg Nurs 2008;34(6):543−4.
[3] Karch SB. Diphenhydramine toxicity: comparisons of postmortem findings in diphenhydramine-, cocaine-, and heroin-related deaths. Am J Forensic Med Pathol 1998;19(2):143−7.
[4] Tanaka T, Takasu A, Yoshino A, Terazumi K, Ide M, Nomura S, et al. Diphenhydramine overdose mimicking serotonin syndrome. Psychiatry Clin Neurosci 2011;65(5):534.
[5] Radovanovic D, Meier PJ, Guirguis M, Lorent JP, Kupferschmidt H. Dose-dependent toxicity of diphenhydramine overdose. Hum Exp Toxicol 2000;19(9):489−95.
[6] Sype J, Khan IA. Prolonged QT interval with markedly abnormal ventricular repolarization in diphenhydramine overdose. Int J Cardiol 2005;99(2):333−5.
[7] Backer R, Pisano R, Sopher J. Diphenhydramine suicide—case report. J Anal Toxicol 1977;1:227−8.
[8] Pragst F, Herre S, Bakdash A. Poisonings with diphenhydramine—a survey of 68 clinical and 55 death cases. Forensic Sci Int 2006;161(2−3):189−97.

Chapter 23.4

Tizanidine intoxication in a postmortem case

Ross J. Miller[1], Brehon Davis[2,3], C. Clinton Frazee, III[2,3], Marius C. Tarau[4], Mary H. Dudley[4] and Uttam Garg[2,3]

[1]Office of the Chief Medical Examiner, Tulsa, OK, United States, [2]Department of Pathology and Laboratory Medicine, Children's Mercy Hospital, Kansas City, MO, United States, [3]University of Missouri School of Medicine, Kansas City, MO, United States, [4]Office of the Jackson County Medical Examiner, Kansas City, MO, United States

Case description

The decedent was a morbidly obese (body mass index = 42.8 kg/m^2) 55-year-old white female who was found unresponsive by her husband approximately 3 hours after she had told him she was going to take a nap. Emergency personnel arrived, began resuscitation efforts that continued for approximately 40 minutes, and transported her to a hospital where she was pronounced deceased. Emergency personnel recovered an empty bottle of tizanidine (90 pills to be taken once a day, prescribed 15 days prior) from the scene and brought it to the hospital. Antemortem history included significant depression in relation to the death of a family member; she had missed several appointments with her primary care provider. Additional history included hypertension, diabetes mellitus, prior gastric bypass surgery, bipolar disorder, fibromyalgia, and previous suicide attempts.

The subject's body was transported to Jackson County Medical Examiner's Office and was examined in the morning after she was pronounced deceased. The external examination showed head and neck cyanosis/congestion. The sclerae and the conjunctivae show multiple petechiae. Medical interventions, including an intubation tube, defibrillator and electrocardiogram pads, a cervical stabilization collar, and an intravenous catheter, were present. An ecchymotic contusion associated with needle puncture sites is present on the right anterior neck. Nonlethal blunt force injuries including two contusions of the right abdomen and an abraded contusion of the left elbow are also present. The internal examination revealed mildly to moderately congested lungs (right, 830 g; left, 510 g), hepatosplenomegaly (liver, 2540 g; spleen, 470 g), a stomach that is status post remote appearing gastric bypass surgery, and granular kidney cortices. The anterior left second and third ribs are fractured and are consistent with having occurred during cardiopulmonary resuscitation efforts.

Postmortem heart blood, femoral blood, vitreous fluid, urine, gastric contents, brain tissue, and liver tissue were submitted for toxicological analyses. An enzyme multiplied immunoassay technique (EMIT) and liquid−liquid alkaline extraction followed by gas chromatography/mass spectrometry (GC−MS) utilized femoral blood for a broad spectrum drug screen. Due to its inherent importance to this publication, it should be noted that tizanidine is not included in EMIT or GC/MS testing. GC/flame ionization detection and a Conway reagent were used to screen femoral blood for volatile compounds (ethanol, acetone, isopropanol, methanol).

While no volatile substances were detected, EMIT and GC−MS testing revealed the presence of doxepin, diphenhydramine, and diphenhydramine metabolite. Diphenhydramine and metabolite presented as small peaks relative to internal standard with GC−MS analysis and, thus, were not quantified. A sample of femoral blood was sent to an external laboratory for doxepin and tizanidine quantitation. Using liquid chromatography/tandem MS (LC−MS/MS), tizanidine was measured at a concentration of 1.0 mg/L. Doxepin was determined to be present at 0.054 mg/L (reference range 0.05−0.150 mg/L) using GC.

Apart from obesity and evidence of circulatory and respiratory failure (head and neck cyanosis/congestion, congested organs), there was no gross or microscopic evidence of significant disease or trauma that would explain her death. The autopsy findings in combination with the toxicological profile make our case unique and suggest that tizanidine was the major contributor of death. The medical examiner in this case certified the death certificate with a probable cause of death as acute tizanidine intoxication and the probable manner of death as suicide.

Discussion

Tizanidine (Sirdalud, Ternelin, Zanaflex) is a synthetic, alpha-2 adrenergic antagonist, imidazoline derivative that is used for short-term treatment of muscle spasticity [1, 2]. It is structurally similar to clonidine, but functions as a central nervous system (CNS) muscle relaxant. Although the exact mechanism is not well established, it appears to have antagonistic effects on the alpha-2 adrenergic receptors at the level of the spinal nerves [1, 2]. This drug–receptor interaction inhibits the activity of ventral (motor) neurons involved in the pain reflex. It is commonly prescribed for treating conditions associated with multiple sclerosis, amyotrophic lateral sclerosis, spastic diplegia, CNS and spinal trauma, and/or generalized back pain. Less commonly, it may also be prescribed as an anticonvulsant, as a sleep aide, and for treatment of fibromyalgia and headaches (migraine).

Tizanidine is supplied as an oral medication and formulated as a hydrochloride salt in tablet and capsule forms at doses of 2, 4, and 8 mg [3]. The recommended dosage ranges between 2 and 8 mg three to four times a day. After ingestion the entire drug is essentially metabolized in the liver via enzymes of the cytochrome P450 superfamily (CYP1A2) with an oral bioavailability of 40% and a half-life of 2–4 hours [3–6]. The parent drug is broken into at least seven inactive metabolites via oxidation; only 2% of the dose is eliminated unchanged in the urine and/or feces. Blood concentrations of tizanidine following therapeutic use generally do not exceed 0.025 mg/L.

Common adverse effects associated with the use of tizanidine include dizziness, sedation, hallucinations, asthenia, dry mouth, blurred vision, bradycardia, and hypotension [2,3,7–11]. Abrupt discontinuation of tizanidine can result in a withdrawal syndrome with symptoms that include nausea, vomiting, dysthermia, hypertension, reflex tachycardia, hypertonicity, and anxiety [12]. Approximately 5% of those that use tizanidine will present with transient elevations in liver enzymes aspartate, aminotransferase, and alanine aminotransferase without significant complications. However, rare reports of induced severe hepatotoxicity rarely with subsequent acute liver failure and death have been reported [13]. Investigation of the decedent's medical history did not indicate symptoms of hepatotoxicity and gross and microscopic examination of the liver was unremarkable.

Fatal cases involving nonmedical use of psychoactive drugs including tizanidine have been described [14]. Deaths involving intentional overdose with tizanidine seem extremely rare. A previously published case report described a situation in which death resulted from an intentional intoxication with this medication [15]. This case described a 57-year-old woman who was discovered unresponsive approximately 6 hours after being seen alive. At that time, she was noted to be intoxicated at a bar. She had a history of ethanol and prescription drug abuse. She also had a history of suicide attempts and a scene investigation found a suicide note near the body. Toxicological analysis of heart blood indicated toxic levels of tizanidine (2.3 mg/L) in combination with toxic levels of ethanol (0.16 g/dL) and diazepam (1.1 mg/L). The medical examiner ruled the cause of death as ethanol and combined drug intoxication and the manner of death as suicide.

The case we present is similar to the previously published postmortem case in many ways. Both cases involved middle-aged females with a substance abuse history. Both women had a history of suicide attempts and both investigations found evidence that supported the manner of death as a suicide. However, the toxicological profiles of both decedents differed. The level of tizanidine was higher in the published case (2.3 mg/L). Toxic levels of ethanol and diazepam were also present. The synergistic effect of these drugs, when taken in combination, would increase the decedent's risk for CNS depression, or decreased brain function, and the central regulation of breathing. The only toxic drug level in our case is the tizanidine, present at a concentration of 1.0 mg/L. Although not as high as the published case, it is still at a level well above the expected therapeutic maximum. Also, given that the source of blood was femoral (peripheral) and less likely to be affected by postmortem redistribution, the level measured here is likely more representative of the actual level at the time of death. The previously reported case measured tizanidine in heart blood. Doxepin was also quantified, but at a level that was determined to not be a toxic level by itself.

In conclusion, fatalities resulting from intentional tizanidine overdose are very rare. Blood concentrations of tizanidine following therapeutic use generally do not exceed 0.025 mg/L. The decedent presented here has a tizanidine femoral blood level of 1.0 mg/L. Tizanidine was the only drug present at a toxic level in her tested samples and, apart from morbid obesity, there were no other contributing gross or microscopic factors leading to her death. This suggests that an overdose death from tizanidine can happen at a much lower concentration than that which has been previously published. Tizanidine is not detected by our GC/MS postmortem drug screen, and thus the finding of an empty tizanidine prescription bottle at the decedent's house was critical in our death investigation. In its totality, this case indicates a death resulting from tizanidine toxicity and emphasizes the importance of a broad-scoped death investigation when determining causes and manners of death.

References

[1] Henney III HR, Runyan JD. A clinically relevant review of tizanidine hydrochloride dose relationships to pharmacokinetics, drug safety and effectiveness in healthy subjects and patients. Int J Clin Pract 2008;62:314—24.
[2] Wagstaff AJ, Bryson HM. Tizanidine. A review of its pharmacology, clinical efficacy and tolerability in the management of spasticity associated with cerebral and spinal disorders. Drugs 1997;53:435—52.
[3] Baselt RC. Tizanidine. In: Baselt RC, editor. Disposition of toxic drugs and chemicals in man. Seal Beach, CA: Biomedical Publications; 2017. p. 2121—2.
[4] Emre M, Leslie GC, Muir C, Part NJ, Pokorny R, Roberts RC. Correlations between dose, plasma concentrations, and antispastic action of tizanidine (Sirdalud). J Neurol Neurosurg Psychiatry 1994;57:1355—9.
[5] Granfors MT, Backman JT, Laitila J, Neuvonen PJ. Tizanidine is mainly metabolized by cytochrome p450 1A2 in vitro. Br J Clin Pharmacol 2004;57:349—53.
[6] Tse FL, Jaffe JM, Bhuta S. Pharmacokinetics of orally administered tizanidine in healthy volunteers. Fundam Clin Pharmacol 1987;1:479—88.
[7] Amino M, Yoshioka K, Ikari Y, Inokuchi S. Long-term myocardial toxicity in a patient with tizanidine and etizolam overdose. J Cardiol Cases 2016;13:78—81.
[8] Cortes J, Hall B, Redden D. Profound symptomatic bradycardia requiring transvenous pacing after a single dose of tizanidine. J Emerg Med 2015;48:458—60.
[9] Kaddar N, Vigneault P, Pilote S, Patoine D, Simard C, Drolet B. Tizanidine (Zanaflex): a muscle relaxant that may prolong the QT interval by blocking IKr. J Cardiovasc Pharmacol Ther 2012;17:102—9.
[10] Luciani A, Brugioni L, Serra L, Graziina A. Sino-atrial and atrio-ventricular node dysfunction in a case of tizanidine overdose. Vet Hum Toxicol 1995;37:556—7.
[11] Spiller HA, Bosse GM, Adamson LA. Retrospective review of tizanidine (Zanaflex) overdose. J Toxicol Clin Toxicol 2004;42:593—6.
[12] Suarez-Lledo A, Padulles A, Lozano T, Cobo-Sacristan S, Colls M, Jodar R. Management of tizanidine withdrawal syndrome: a case report. Clin Med Insights Case Rep 2018;11: 1179547618758022.
[13] de Graaf EM, Oosterveld M, Tjabbes T, Stricker BH. A case of tizanidine-induced hepatic injury. J Hepatol 1996;25:772—3.
[14] Haukka J, Kriikku P, Mariottini C, Partonen T, Ojanpera I. Non-medical use of psychoactive prescription drugs is associated with fatal poisoning. Addiction 2018;113:464—72.
[15] Sklerov JH, Cox DE, Moore KA, Levine B, Fowler D. Tizanidine distribution in a postmortem case. J Anal Toxicol 2006;30:331—4.

Chapter 23.5

Methemoglobinemia due to dietary nitrate

Devin L. Shrock and Matthew D. Krasowski
Department of Pathology, University of Iowa Hospitals and Clinics, Iowa City, IA, United States

Case histories

Six-week-old, previously healthy female twin infants were brought to a local hospital by their parents for gray skin discoloration which developed over the course of the afternoon. The infants' birth history was significant for late preterm birth at 36 weeks' gestation via Cesarean section. One twin developed perinatal respiratory distress which resolved after intubation for less than 1 day; the other infant had an unremarkable perinatal course. The parents reported no signs or symptoms of illness prior to presentation and denied sick contacts at home. There was no known exposure to chemicals, pesticides, or other toxic substances. The infants resided in an Amish community. Their diet comprised breast milk supplemented by bottles containing a mixture of goat milk, well water, and sorghum syrup.

Physical examination of the infants was significant for generalized cyanosis with mottled, gray skin discoloration, cool extremities, and tachycardia. Point of care pulse oximetry showed both infants had decreased oxygen saturations (91%, 94%). Laboratory examination also revealed anemia in both infants (8.5, 8.7 g/dL; reference range: 12–16 g/dL). Additional testing with venous co-oximetry demonstrating significantly elevated methemoglobin concentrations in both infants (63%, 49%; reference range: <2%). Venous blood gas analysis showed severe metabolic acidosis in the infant with the higher methemoglobin concentration (Table 23.5.1). Following transfer of the infants to a tertiary care medical center, venous co-oximetry again demonstrated markedly elevated methemoglobin with the blood gas analyzer reporting >30% methemoglobin in both infants (exact quantitation not displayed above the upper limit of quantification). This interinstitutional difference in upper limit of quantification of methemoglobin prompted a call from clinicians to the clinical chemistry laboratory to request more specific of quantification of methemoglobin for response to treatment. However, the blood gas analyzers used in the hospital does not provide a specific quantification of methemoglobin concentration above 30%.

The infants were intubated and treated emergently with methylene blue. They each received a single transfusion of packed red blood cells for methemoglobin-associated hemolytic anemia. Following treatment, both infants had resolution of methemoglobinemia within 24 hours of methylene blue administration and were discharged home a few days later in stable condition.

Discussion

Methemoglobin is formed endogenously when one or more of the ferrous (Fe^{2+}) irons of the hemoglobin complex are oxidized to a ferric state (Fe^{3+}), rendering it unable to carry oxygen. Erythrocytes possess a mechanism for reduction of methemoglobin via a nicotinamide adenine dinucleotide (NAD)–dependent reaction catalyzed by cytochrome b5 methemoglobin reductase. Under normal circumstances, this system maintains a methemoglobin concentration of ~1% [1]. Methemoglobinemia occurs when the rate of methemoglobin production exceeds endogenous reduction capacity.

There are two forms of methemoglobinemia—congenital and acquired. Causes of congenital methemoglobinemia include inherited defects in the hemoglobin molecule (hemoglobin M disease) and type 1 (erythrocytic, partial) or type 2 (generalized) deficiencies of cytochrome b5 or cytochrome b5 reductase. Acquired methemoglobinemia develops as a

TABLE 23.5.1 Venous Blood Gas Analysis and Co-oximeter Panel Results at Initial Presentation.

	Infant A	Infant B	Ref. Range
pH	6.91	7.36	7.30–7.40
pCO_2 (Torr)	33	42	32–45
pO_2 (Torr)	24	33	50–65
Base excess (mEq/L)	−25	−2	−2–2
HCO_3 (calculated) (mEq/L)	7	24	22–26
Total CO_2 (mEq/L)	8	25	24–32
Total hemoglobin (g/dL)	8.5	8.7	12–16
Methemoglobin (%)	63	48	0.4–1.5
Oxyhemoglobin (%)	39.5	52.6	40–70
Carboxyhemoglobin (%)	4.7	2.5	1–3

TABLE 23.5.2 Signs and Symptoms of Methemoglobinemia [2,3].

Methemoglobin Concentration (% of Hemoglobin)	Clinical Signs and Symptoms
0–10	None
10–20	Generalized cyanosis, blue-gray skin discoloration
20–50	Headache, fatigue, anxiety, confusion, dizziness, syncope, tachycardia, dyspnea, tachypnea, weakness
50–70	Coma, seizures, dysrhythmias, metabolic acidosis
>70	Potentially fatal

result of ingestion or exposure to certain drugs, chemicals, dietary, or other environmental sources with direct oxidant properties or oxidant metabolites. Substances known to induce methemoglobinemia include topical anesthetics (benzocaine, lidocaine, prilocaine), dapsone, phenazopyridine, aniline, nitrites, nitrates, solvents, pesticides, and other chemicals [2].

Methemoglobinemia results in a functional anemia, the severity of which is dependent on both methemoglobin concentration as well as the baseline functional reserve of the patient to compensate for the abrupt reduction in oxygen-carrying capacity. Toxic effects of methemoglobinemia increase with methemoglobin concentration. The most common presenting sign is generalized cyanosis with gray-blue skin discoloration. Additional signs and symptoms of hypoxia develop at increasing methemoglobin concentrations with concentrations of greater than 70% often fatal [3] (Table 23.5.2). Individuals with baseline disease resulting in impaired oxygen transport (anemia, respiratory disease) or oxygen delivery (cardiac or significant peripheral vascular disease) may show more severe cyanosis and cardiovascular impairment at lower methemoglobin levels.

The clinical laboratory plays a vital role in evaluation of the acutely cyanotic patient. Arterial blood gas analysis and pulse oximetry are commonly used diagnostic tools, and it is important to understand potential interpretive pitfalls in patients with dyshemoglobins in circulation.

Blood gas analysis in patients with methemoglobinemia may show falsely normal partial pressure of oxygen (pO_2) and oxygen saturation (sO_2). pO_2 is a measurement of free unbound oxygen. In patients with methemoglobinemia, oxygen administration causes a rise in pO_2 which is unaffected by the presence of methemoglobin, without improvement in clinical cyanosis. This pO_2 dissociation is a helpful clue in distinguishing methemoglobinemia from other causes of cyanosis. Oxygen saturation results may also appear falsely normal, as calculation performed by the analyzer assumes a normal oxygen dissociation curve and the absence of a significant proportion of abnormal hemoglobin, neither of which is true in cases of methemoglobinemia [1].

Pulse oximeters use a spectrophotometric method to measure arterial oxygen saturation based on the normal peak absorption frequencies of hemoglobin and oxyhemoglobin (660 and 940 nm). The ratio between the two values is then used to determine percentage oxygen saturation. This calculation assumes the absence of abnormal hemoglobin species. When dysfunctional hemoglobin species are present, including methemoglobin and carboxyhemoglobin, oxygen saturation may be overestimated [4]. Pulse oximetry is also susceptible to interference by intravenous dyes, including methylene blue, the primary treatment for methemoglobinemia. Following administration of methylene blue, pulse oximetry may show a transient false depression of oxygen saturation which resolves within minutes [5].

Laboratory co-oximeters use multiwavelength spectrophotometric analysis to directly measure carboxyhemoglobin and methemoglobin in addition to oxyhemoglobin and reduced hemoglobin. This approach, available on some blood gas analyzers, represents a way to accurately determine levels of methemoglobin. Results are then expressed as both functional and fractional oxygen saturation [4]. The characteristic absorption band of methemoglobin is 635 nm. Fixed wavelength co-oximeters interpret readings in the 630 nm range as methemoglobin and are susceptible to interference by methylene blue due to overlap in absorbance frequencies. Blood gas analyzers that use the Evelyn—Malloy method for methemoglobin detection (often considered a gold standard method) are not susceptible to this interference, as they measure the change in optical density as methemoglobin is converted to cyanmethemoglobin in the presence of sodium cyanide [6].

Aside from avoidance of the inciting agent, asymptomatic patients with methemoglobin concentrations of less than 20% may not require additional treatment except for close observation. For symptomatic patients and those with methemoglobin concentrations greater than 20%, the most common treatment is intravenous administration of methylene blue. Treatment takes advantage of an alternative method of methemoglobin reduction via NADH phosphate (NADPH)-methemoglobin reductase. Under normal conditions, this enzyme reduces a negligible amount of methemoglobin; however, it has an affinity for dyes including methylene blue. The enzyme reduces methylene blue to leukomethylene blue, which in turn reduces methemoglobin to functional hemoglobin [2]. Treatment with methylene blue is not without risk. When administered in excess or to patients with glucose-6-phosphate dehydrogenase deficiency, methylene blue may cause methemoglobinemia. Additional supportive care measures should be tailored to patient symptoms but may include transfusion and oxygen supplementation.

Case follow-up

While there was no known family history of methemoglobinemia, the initial clinical differential diagnosis included both acquired and congenital forms, as both environmental and genetic causes of methemoglobinemia could produce the observed symptoms in twins. Testing of well water and other household items performed by the state public health authorities identified an extremely elevated nitrate level (9350 μg/g) in the bottle of sorghum syrup used to fortify the infants' bottles.

Sodium nitrate and sodium nitrite are used for preservative properties in the food processing industry. They are most commonly used in cured meats, fish, and some cheeses. Nitrates and nitrites are also common well water contaminants due to nitrogenous waste in fertilizer runoff [2,7]. Ingested nitrates are not toxic; however, following ingestion, intestinal bacterial flora convert nitrates to toxic nitrites which are capable of inducing methemoglobin formation. Symptoms of nitrite toxicity develop rapidly, with hypoxic symptoms of methemoglobinemia presenting within minutes to hours of ingestion. Symptoms of acute nitrate/nitrite toxicity are attributable to the degree of methemoglobinemia.

At the recommendation of the state public health department, a bulletin was issued to encourage consumers to dispose of any unused amounts of the implicated sorghum syrup brand, with a product recall subsequently undertaken. In addition to avoidance of contaminated sorghum syrup, the parents were instructed to supplement breast milk with formula prepared with sterile water.

References

[1] Haymond S, et al. Laboratory assessment of oxygenation in methemoglobinemia. Clin Chem 2005;51(2):434—44.
[2] Wright RO, Lewander WJ, Woolf AD. Methemoglobinemia: etiology, pharmacology, and clinical management. Ann Emerg Med 1999;34 (5):646—56.
[3] Cvetkovic D, et al. Sodium nitrite food poisoning in one family. Forensic Sci Med Pathol 2019;15:102—5.
[4] Wagner JL, Ruskin KJ. Pulse oximetry: basic principles and applications in aerospace medicine. Aviat Space Environ Med 2007;78(10):973—8.

[5] Ralston AC, Webb RK, Runciman WB. Potential errors in pulse oximetry. III: Effects of interferences, dyes, dyshaemoglobins and other pigments. Anaesthesia 1991;46(4):291–5.
[6] Evelyn KA, Malloy HT. Microdetermination of oxyhemoglobin, methemoglobin, and sulfhemoglobin in a single sample of blood. J Biol Chem 1938;126(2):655–62.
[7] Puckett LJ, Tesoriero AJ, Dubrovsky NM. Nitrogen contamination of surficial aquifers—a growing legacy. Environ Sci Technol 2011;45(3):839–44.

Chapter 23.6

Cyanide toxicity—a case study

Kamisha L. Johnson-Davis[1,2]

[1]*Department of Pathology, University of Utah, Salt Lake City, UT, United States,* [2]*Clinical Toxicology, ARUP Laboratories, Salt Lake City, UT, United States*

Case description

A 45-year-old male was brought to the emergency department due to smoke inhalation from a domestic fire. The patient was experiencing nausea, vomiting, and rapid breathing. Upon medical evaluation, the patient had muscle weakness and slow reflexes. The patient's blood pressure was 75/45 mmHg and heart rate was 42 beats per minute. The patient had a seizure, 30 minutes after arrival to the emergency department, then became unresponsive. Lab results, in the table below, revealed that the patient had metabolic acidosis from elevated lactate concentrations. Resuscitation and support care therapies were administered. The patient was given sodium bicarbonate to manage the lactic acidosis, 100% oxygen and the cyanide kit antidote was administered to treat acute cyanide poisoning. The antidote consisted of amyl nitrite, sodium nitrite, and sodium thiosulfate. The patient was able to regain consciousness after 2 hours and was discharged a couple of days after routine monitoring to determine that the patient had normal vital signs, and the abnormal test results, upon presentation to the emergency department, were within the normal reference range.

Laboratory Test	Patient Result	Normal Reference Range
Hemoglobin (g/dL)	14.2	12–17.5
Blood urea nitrogen (BUN; mg/dL)	7	7–18
Creatinine (mg/dL)	0.8	0.6–1.2
Sodium (mmol/L)	**147**	135–145
Potassium (mmol/L)	4.2	3.5–5.1
Chloride (mmol/L)	106	98–106
Glucose (mg/dL)	**350**	70–115
Lactate (mmol/L)	**15**	0.5–2
Arterial pH	**6.75**	7.35–7.45
Bicarbonate (mmol/L)	**10**	22–29
Osmolality (mOsm/kg)	**305**	275–295
Anion gap	**35.2**	7–16

The tests that are in bold font have patient results that are outside of the normal reference range.

Introduction

Hydrogen cyanide was first extracted in 1782 by a chemist from Sweden [1]. It is a lethal toxin that can cause death within a short period of time after exposure. Cyanide was used in World War II as a weapon for mass genocide in concentration camps [2]. In 1978 cyanide was added to Flavor-aid and was used to kill over 900 people, under the direction of Jim Jones in Guyana, for mass suicide/murder. In 1982 Tylenol bottles were tampered with cyanide which led to the deaths of seven people in Chicago [3,4]. In addition, cyanide is also used in the United States for court-ordered executions by gas chamber [5].

Cyanide (CN) group contains a carbon atom that is triple bonded to a nitrogen atom, with a molecular weight of 26.02 g/mol and 27.02 g/mol for hydrogen cyanide. It can be produced in the form of a solid, liquid or gas. It is a weak acid and it has a pKa of 9.2 [6]. Hydrogen cyanide is a colorless gas that has an odor similar to bitter almonds. The liquid form of hydrogen cyanide can have a color that ranges from colorless to light blue and the cyanide salts are a white power. Cyanide is soluble in alcohol and water [6]. Common cyanide containing compounds include inorganic

TABLE 23.6.1 Cyanide Species.

Cyanide Species	Molecular Formula	Molecular Mass
Hydrogen cyanide	HCN	27.03
Sodium cyanide	NaCN	49.02
Potassium cyanide	KCN	65.11
Calcium cyanide	$Ca(CN)_2$	92.12
Copper cyanide	CuCN	89.56
Gold cyanide	AuCN	223
Mercury cyanide	$Hg(CN)_2$	252.6
Lead cyanide	$Pb(CN)_2$	117.4
Zinc cyanide	$Zn(CN)_2$	259.2
Potassium silver cyanide	$KAg(CN)_2$	198.01
Sodium ferrocyanide	$Na_4Fe(CN)_6$	303.91
Potassium ferrocyanide	$K_4Fe(CN)_6$	368.35
Potassium ferricyanide	$K_3Fe(CN)_6$	329.95
Cyanogen	$(CN)_2$	52.04
Cyanogen chloride	CNCl	61.47
Cyanogen bromide	CBrN	105.92
Acetonitrile	C_2H_3N	41.05
Acrylonitrile	C_3H_3N	53.06
Butyronitrile	C_4H_7N	69.11
Sodium nitroprusside	$Na_2[Fe(CN)_5NO]$	261.97
Amygdalin	$C_{20}H_{27}NO_{11}$	457.4
Linamarin	$C_{10}H_{17}NO_6$	247.24
Acetone cyanohydrin	$(CH_3)_2C(OH)CN$	85.1

hydrogen cyanide (HCN), cyanide salts, such as potassium cyanide (KCN), metal cyanides, such as zinc cyanide ($Zn(CN)_2$), cyanogen, cyanogen halides (cyanogen chloride) and aliphatic nitriles (acetonitrile) (Table 23.6.1). In acidic solutions, cyanide salts can release hydrogen cyanide gas; therefore cyanide solutions are typically kept at basic pH [7].

Cyanide can be produced in bacteria and fungi [8,9] and it naturally occurs in the seeds of apples, cherries, peaches, apricots, and plums. It is also found in several plant species, such as cassava root, lima beans, and almonds, which contain cyanogenic glycosides such as amygdalin (laetrile) and linamarin that release cyanide upon digestion. [10,11]. Laetrile is also known as vitamin B17 and is used as an antitumor treatment to block cell growth and proliferation. It can be produced in a pill form of 500 mg and contain 30 mg of cyanide [12,13], which has been shown to cause cyanide poisoning [12–14]. However, laetrile is not recommended for use as an antitumor agent to treat cancer. Cassava roots are harvested in several African countries to produce flour and other starches. Cassava roots must be processed in a manner to reduce the risk of cyanide exposure from ingestion. The root is typically soaked for about 4 days, dried or heated to start the process of removing linamarin [11]. Linamarin is then hydrolyzed by β-glycosidase linamarase to form glucose and cyanohydrins. The cyanohydrins are then metabolized to ketone and hydrogen cyanide, which is released as gas [11]. Symptoms of cyanide toxicity from cassava root include headaches, nausea, vomiting, dizziness, and seizures [11].

Hydrogen cyanide is used and produced in industrial manufacturing, such as electroplating for the production of jewelry, rubber, and plastic manufacturing, metal extracting processes from ore, photography, and paper manufacturing.

It is also available as a rodenticide and pesticide in some fumigating processes [5]. Domestic fires are the most common cause of cyanide exposure in industrialized countries [15–19]. Hydrogen cyanide production from fires occurs when materials containing either natural or synthetic nitrogen–containing polymers such as wool, silk, plastics (polyacrylonitriles), and rubber undergo incomplete combustion [20]. Cyanide gas is then released when the temperature reaches 600 °F [20]. In the United States, there are over 1.3 million fires that are reported annually, and in 2017 there were about 3400 fire-related deaths [21]. There have been over 3000 human cyanide exposures, according to the Toxic Exposure Surveillance System [1]. In addition, about 35% of all fire victims will have toxic concentrations of cyanide in their blood [1,2]. Cyanide can be detected in about 60% of individuals who had fire-related deaths [22]. In 2007 it was reported that there were 247 reported cases of chemical exposures to cyanide in the United States, according to the National Poison Data System of a Poison Control Centers annual report, and five of the cases were fatal [23].

Cyanide can also be released from cigarette smoke, emitted from fossil fuel combustion, as well as waste and biomass incinerators [9]. Cigarettes that do not contain a filter can release 500 μg hydrogen cyanide; however, cigarettes with filters generate 100 μg in mainstream smoke [9]. Hydrogen cyanide concentrations can differ between mainstream smoke and sidestream smoke. Mainstream smoke concentrations can range from 280 to 550 μg/cigarette and from 53 to 111 μg/cigarette in sidestream smoke [9]. The hydrogen–cyanide ratio between sidestream:mainstream smoke concentrations can range from 0.06 to 0.50 [9]. Cyanide exposures can also occur from plane and automobile accidents. Airplanes and cars tend to contain large amounts of plastic materials that generate cyanide as a product of pyrolysis in fires [15].

Another potential source of cyanide toxicity is the administration of sodium nitroprusside (SNP) in the treatment of hypertension, for the reduction of blood pressure during certain surgical procedures, and for the treatment of severe heart failure [24]. SNP contains five molecules cyanide, and the drug is metabolized in the body to yield cyanide, which is bound to protein and/or metabolized to thiocyanate. Thiocyanate toxicity also may occur in patients with renal or hepatic impairment [25]. Of note, thiocyanate levels do not reflect the concentration of cyanide.

Mechanism of action

Cyanide exposure can occur through ingestion, inhalation, or transdermal. At physiological pH, hydrogen cyanide is unionized and can rapidly distribute throughout the body and enter cellular membranes and mitochondria. In the mitochondria, cyanide produces its toxic effects by binding, reversibly, to the ferric ions in the cytochrome A_3 enzyme and inhibit its activity [26–30]. When cytochrome A_3 activity is blocked, it will prevent oxidative phosphorylation, which functions to generate adenosine triphosphate (ATP) for aerobic organisms [26–30]. When cellular respiration is blocked, cells will try to generate ATP through anaerobic process to convert pyruvate to lactate. Metabolic acidosis will occur due to elevated concentrations of lactic acid. Cyanide can also bind to methemoglobin, to form cyanomethemoglobin; the formation of cyanomethemoglobin will reverse the inhibition of cytochrome oxidase by the removal of cyanide. Cyanide exposure can cause respiratory depression, asphyxia, hypoxia, and hypotension [5,31]. It also functions to inhibit up to 40 enzymes that are metalloproteins, such as superoxide dismutase and carbonic anhydrase [19,32]. Cyanide exposure can also cause cellular necrosis, demyelination of neurons, CNS toxicity, and seizures [5,33].

Pharmacokinetics and toxicokinetics

Cyanide quickly absorbs through mucous membranes of the respiratory, gastrointestinal tract, and it can be absorbed through the skin. Once absorbed in the bloodstream, cyanide will distribute throughout the body, and it can be found in areas of the body that have increased blood flow, such as the lungs, liver, brain, and blood [9]. Cyanide has a half-life of 0.7–2.1 hours in whole blood, a volume of distribution estimated to be up to 0.5 L/kg and it has an elimination half-life of 6–66 hours [6,34,35]. In chronic exposures, cyanide does not accumulate in tissues or blood [9]. The body has several mechanisms to detoxify small concentrations of cyanide. Rhodanese is a mitochondrial enzyme that can be found in the liver and functions as a trans-sulfurase to transfer the sulfur from thiosulfate to cyanide to form thiocyanate [5]. Thiocyanate is a less toxic form that is water soluble and eliminated by the kidneys. Thiocyanate has a volume of distribution of 0.25 L/kg and has an elimination half-life of three days [5]. About 60%–80% of cyanide is metabolized by the rhodanese enzyme for thiocyanate. The body stores of thiosulfate are the rate-limiting step for the rhodanese enzyme [5]. The enzyme β-mercaptopyruvate-cyanide sulfurtransferase is another enzyme that converts cyanide to thiocyanate by transferring a sulfur group from mercaptopyruvate to cyanide [5]. Cyanide can also bind with cysteine to form 2-aminothizoline-4-carboxylic acid [36] and cobalt to form cyanocobalamin (Vitamin B_{12}) [37].

Signs and symptoms of toxicity

The Occupational Safety and Health Administration standards state that air cyanide should not exceed 5 mg/m^3 and hydrogen cyanide exposure should not exceed 10 parts per million (ppm) or mg/L [38]. Moreover, health dangers can occur at 50 ppm and concentrations of 150 ppm and higher have the potential to be fatal [6,38]. Studies have been reported that cyanide deaths have occurred when concentrations were greater than 100 ppm [6]. An adult lethal dose is about 200 mg of potassium cyanide and 100 mg of hydrocyanic acid [6]. Cyanide exposed individuals may experience symptoms of toxicity when blood concentrations are 40 mol/L and higher [1]. Blood cyanide concentrations of 0.006–0.041 mg/L can be detected in individuals who smoke, and nonsmokers can have a concentration of 0.004 mg/L from normal biological metabolism [6,39]. The cyanide metabolite, thiocyanate, has a blood concentration range of 3–12 mg/L in smokers and 1–4 mg/L in nonsmokers [40]. Symptoms of toxicity can occur within minutes of exposure from inhalation and ingestion. When hydrogen cyanide is inhaled, individuals who are exposed may smell a bitter, almond odor. Signs and symptoms of toxicity include headache, dizziness, confusion, and mydriasis, which are due to tissue hypoxia. Exposed individuals may experience nausea and vomiting, and symptoms can progress to seizures, unconsciousness, paralysis, and coma [31]. Respiratory and cardiovascular symptoms include tachypnea, tachycardia, apnea, hypotension, pulmonary edema, and lactic acidosis [6,31]. Patients with cyanide poisoning will have a cherry red color skin, due to excess oxygen in the bloodstream. Chronic dermal exposure can cause dermatitis, rash, and pruritus [6]. Blood concentrations of cyanide can range from 0.4 to 230 mg/L in exposures that ended in fatality [6,41]. Severe toxicity can cause persistent neurologic consequences, such as tremor, slowed speech, and neuronal damage. Of note, the fetus in pregnant patients is at risk for cyanide exposure and toxicity [23].

Treatment

One of the first steps to managing patients with cyanide poisoning is to remove the patient from the source of exposure, whether it be from inhalation, dermal, or ingestion exposure [5,31,42]. It is important to establish and stabilize the patient's airway, breathing, and circulation. If exposure has occurred through the skin, it is important to remove the clothing as soon as possible [5,31,42]. Cardiac and respiratory monitoring should be performed and the patient should be administered the cyanide antidote kit and 100% oxygen. Hyperbaric oxygen is a treatment that is still controversial but it can also be administered to inhibit cyanide from binding to cytochrome oxidase A$_3$ [5,31,42]. Activated charcoal can be given to the exposed individual if they are conscious and the oral exposure occurred within 1 hour [5,31]. Gastric lavage may be another option for decontamination; however, ipecac is contraindicated because it may cause seizures and coma [5,31].

There are four different mechanisms of action for cyanide antidotes. There are antidotes which function to produce methemoglobin, bind to cyanide to produce a nontoxic form, antidotes that enhance the metabolism of cyanide to a less toxic form and function to reduce cyanide absorption [42]. The cyanide kit is approved by the US Food and Drug Administration and consists of three antidotes: amyl nitrite, sodium nitrite, and sodium thiosulfate. Amyl nitrite ($C_5H_{11}NO_2$) is administered by inhalation from a 0.3 mL ampule for 15–30 seconds [31,42]. It is discontinued once sodium nitrite ($NaNO_2$) is administered intravenously at a dose of 10 mg/kg over 3–5 minutes. In children the dose of sodium nitrite is 0.2 mL/kg [31]. Nitrites will function to oxidize the iron in hemoglobin to the ferric form to generate methemoglobin at concentrations of 10%–18% [42,43]. It will take 20%–30% methemoglobin to bind to cyanide for optimal treatment [42,44]. The limitations of using nitrite antidotes include hypotension through the mechanism of vasodilation and the reduction of oxygen capacity from the formation of methemoglobin [5,42,45]. Cyanide has a preference for binding to the ferric form of iron in methemoglobin than the ferric ion in cytochrome oxidase A$_3$ in mitochondria. Cyanomethemoglobin is then detoxified by the rhodanese enzyme for thiocyanate for renal elimination [5,42]. Sodium thiosulfate ($Na_2S_2O_3$) is the third antidote in the kit. It is administered intravenously at a dose of 0.5 g/kg for 30 minutes in adults and in children, the dose is 7 g/m^2 [31,42]. Sodium thiosulfate functions as a sulfate donor for the rhodanese and mercaptopyruvate sulfurtransferase enzymes to form thiocyanate [5]. Thiocyanate is a less toxic form of cyanide that can be eliminated by the kidneys [46]. Thiosulfate and oxygen are also recommended for treatment in pregnant patients that are exposed to cyanide [23,47].

Hydroxycobalamin is a safer antidote for acute cyanide poisoning because it can rapidly bind to cyanide without impacting oxygen binding in hemoglobin [42,48,49]. It is a precursor of Vitamin B$_{12}$ and it contains cobalt, which can bind to cyanide to form cyanocobalamin (Vitamin B$_{12}$) [48,49]. The standard dose is 70 mg/kg in children, and 5 g are administered to adults. Hydroxycobalamin is given intravenously over 15 minutes, and the cobalt in hydroxycobalamin binds to cyanide to form cyanocobalamin, which is vitamin B$_{12}$. Cyanide has more affinity for hydroxycobalamin than the cytochrome oxidase A$_3$ and therefore will remove the cyanide inhibition of cellular respiration [49,50].

Dicobalt edetate is another antidote that functions as a chelator for cobalt to bind to cyanide to form a complex for renal elimination [42]. Each mole of cobalt can bind 6 moles of cyanide [51]. Dicobalt edetate treatment is limited, due to its adverse side effects such as vomiting risk of anaphylactic shock, ventricular arrhythmias and hypotension [42]. Lastly, dimethylaminophenol (4-DMAP) is an antidote that produces methemoglobin. Although it is more effective at producing methemoglobin than amyl and sodium nitrite, it can cause nephrotoxicity, tissue necrosis and hemolysis [52,53].

Laboratory evaluation

The following are laboratory tests that are used to evaluate cyanide poisoning: chemistry tests for sodium, chloride, potassium, and bicarbonate to evaluate the anion gap to assess metabolic acidosis [31]. A serum lactate concentration should be measured to identify lactic acidosis. Lactate concentrations greater than or equal to 10 mmol/L have high sensitivity and specificity for cyanide toxicity in patients exposed from smoke inhalation [54]. A complete blood count, arterial blood gas, carboxyhemoglobin, and methemoglobin concentrations are also used in the assessment of cyanide toxicity [31]. Hydroxycobalamin is known to cause interference with colorimetic assays, such as carboxyhemoglobin and methemoglobin, due to its red color [31]. For general testing, a glucose point of care test can evaluate if the patient has hypoglycemia, a drug screen can be used to rule out acetaminophen or salicylate exposure and a pregnancy test for women who are in the age range for childbearing [31]. An electrocardiogram can be used to rule out poisonings from drugs that cause torsades de pointes. Blood cyanide concentration can be useful to monitor decontamination efforts if results are not received in time to confirm the diagnosis [31]. Blood thiocyanate concentrations should not be used to assess cyanide exposure [5]. Nonsmokers generally have cyanide concentrations of less than 0.5 mg/L [55]. Cyanide concentrations greater than 0.5 are associated with adverse effects and concentration greater than 2.5 mg/L are associated with coma and death, if a treatment options are unsuccessful [55]. Cyanide concentrations are not stable in serum, since 70% of cyanide in blood is bound to hemoglobin. In addition, cyanide stability is also affected by metabolism to thiocyanate. Consequently, whole blood is the preferred specimen for analysis [56,57]. Cyanide stability is impacted by temperature and the recommended storage conditions for specimens to preserve the cyanide concentrations are refrigerated and frozen temperatures [56–58]. Since cyanide is unstable, it is important for specimen collection, and transport and analysis need to be performed as soon as possible.

Analytical methods and clinical management implications

Cyanide concentrations in blood specimens or environmental sample collections can be analyzed by spectrophotometry with microdiffusion. Whole blood cyanide analysis is performed using a Conway microdiffusion cell for the production of hydrogen cyanide and the trapping of this gas in a basic solution prior to colorimetric analysis. Hydrogen cyanide gas is formed by acid lyse of red blood cells and absorbed into a dilute alkaline solution. Cyanide is quantitatively determined by producing a cyanogen chloride with the addition of Chloramine-T. A pyridine color reagent containing barbituric acid and HCl is added to produce a chromogen, which absorbs at 580 nm [59,60].

Cyanide spectrophotometric assays can also be coupled to capillary electrophoresis (CE) for enzymatic conversion of cyanide to thiocyanate using the rhodanese enzyme and thiosulfate [61]. CE affords high resolution, sensitivity, specificity and short analysis time. It can also be coupled to fluorescence detection [62]. Cyanide can be analyzed through derivatization with 2,3-naphthalenedialdehyde and taurine to form 1-cyanobenz[f]isoindole, which is a fluorescent compound, and analyzed by ultraviolet detection with a detection limit was 0.1 ng/mL [62].

Cyanide colorimetric test strips can be used to determine cyanide concentrations in solutions and food. Some test strips employ a chemical reaction of pyridine–pyrazolone reagent with cyanide to form a blue complex in a buffer solution at pH 7 [63]. Another test strip kit may utilize the reaction between cyanide and chloride to form cyanogen chloride, then glutaconic dialdehyde. The compound in the test strip will react with glutaconic dialdehyde to form a red dye [64]. Cyanide is then measured semiquantitatively and results can be reported within minutes. The limitations for using test strips are that oxidizing reagent can cause interference, temperature and the analysis of dark-colored solutions may interfere with the accuracy of the results.

Potentiometric methods with ion-specific electrode can be used to measure cyanide in samples. The samples are drawn, through a cassette that contains a cellulose ester membrane filter, into the impinger which contains 10 mL of 0.1 N NaOH. The filter is then placed into a vial, desorbed with 25 mL of 0.1 N NaOH and analyzed by a cyanide ion-specific electrode and a potentiometer [38]. Free cyanide can also be analyzed by a cyanide ion electrode to determine cyanide concentrations in drinking water [9]. The advantages of the analytical method are that it is simple and has an analytical measurement range is 0.25–25 ppm. The disadvantages are that the impinger solution can be lost during collection, cyanide particulates on the filter can be converted to HCN when exposure to air and the electrode can be damaged by strong reducing solutions [38]. The antidote, sodium thiosulfate can interfere with potentiometric and colorimetric analysis, due to its red color [9].

Gas chromatography mass spectrometry can be used to measure cyanide in various matrices. Mass spectrometry affords improved specificity and sensitivity in comparison to other analytical methods. Gas chromatography can also be coupled to a nitrogen-specific detector [65,66] or electron capture detector [67].

References

[1] Graham, J., & Traylor, J. (2018). Cyanide toxicity. StatPearls Publishing, Treasure Island (FL), 25 Jun 2018, PMID: 29939573. <https://europepmc.org/article/NBK/NBK507796>.

[2] Hanley ME, Murphy-Lavoie HM. Hyperbaric, cyanide toxicity. StatPearls Publishing, Treasure Island (FL), 28 Feb 2018 PMID: 29489235. <https://europepmc.org/article/NBK/NBK482265>.

[3] Beck M, Monroe S. The Tylenol scare: the death of seven people who took the drug triggers a nationwide alert—and a hunt for a madman. Newsweek 1982;100(15):32–6.

[4] Dunea G. Death over the counter. British Med J (Clinical research ed) 1983;286(6360):211–12.

[5] Erdman, W. A. (2003). Cyanide. Ellenhorn's medical toxicology: diagnosis and treatment of human poisoning, 3rd ed. Lippincott Williams & Wilkins, Baltimore, 1155-1168.

[6] Baselt, Randall C., and R. H. Cravey. "Disposition of toxic drugs and chemicals in man. Foster City. (2014): 76.

[7] Singh BM, Coles N, Lewis P, Braithwaite RA, Nattrass M, FitzGerald MG. The metabolic effects of fatal cyanide poisoning. Postgrad Med J 1989;65(770):923–5.

[8] Short SM, van Tol S, MacLeod HJ, Dimopoulos G. Hydrogen cyanide produced by the soil bacterium *Chromobacterium* sp. Panama contributes to mortality in *Anopheles gambiae* mosquito larvae. Sci Rep 2018;8(1):8358.

[9] WHO, (2004). Hydrogen cyanide and cyanides: human health aspects. <https://www.who.int/ipcs/publications/cicad/en/cicad61.pdf>.

[10] Ferraro V, Piccirillo C, Tomlins K, Pintado ME. Cassava (*Manihot esculenta* Crantz) and Yam (*Dioscorea* spp.) crops and their derived foodstuffs: safety, security and nutritional value. Crit Rev Food Sci Nutr 2016;56(16):2714–27.

[11] Tshala-Katumbay DD, Ngombe NN, Okitundu D, David L, Westaway SK, Boivin MJ, et al. Cyanide and the human brain: perspectives from a model of food (cassava) poisoning. Ann NY Acad Sci 2016;1378(1):50–7.

[12] Dang T, Nguyen C, Tran PN. Physician beware: severe cyanide toxicity from amygdalin tablets ingestion. Case Rep Emerg Med 2017;2017:4289527.

[13] Juengel E, Thomas A, Rutz J, Makarevic J, Tsaur I, Nelson K, et al. Amygdalin inhibits the growth of renal cell carcinoma cells in vitro. Int J Mol Med 2016;37(2):526–32.

[14] Litovitz TL, Larkin RF, Myers RA. Cyanide poisoning treated with hyperbaric oxygen. Am J Emerg Med 1983;1(1):94–101.

[15] Walsh DW, Eckstein M. Hydrogen cyanide in fire smoke: an underappreciated threat. Emerg Med Serv 2004;33(10):160–3.

[16] Antonio AC, Castro PS, Freire LO. Smoke inhalation injury during enclosed-space fires: an update. J Bras Pneumol 2013;39(3):373–81.

[17] Eckstein M, Maniscalco PM. Focus on smoke inhalation—the most common cause of acute cyanide poisoning. Prehospital Disaster Med 2006;21(2):s49–55.

[18] Parker-Cote JL, Rizer J, Vakkalanka JP, Rege SV, Holstege CP. Challenges in the diagnosis of acute cyanide poisoning. Clin Toxicol (Phila) 2018;56(7):609–17.

[19] Anseeuw K, Delvau N, Burillo-Putze G, De Iaco F, Geldner G, Holmstrom P, et al. Cyanide poisoning by fire smoke inhalation: a European expert consensus. Eur J Emerg Med: Off J Eur Soc Emerg Med 2013;20(1):2–9.

[20] Alarie Y. Toxicity of fire smoke. Crit Rev Toxicol 2002;32(4):259–89.

[21] U.S. Fire Administration (2019). <https://www.usfa.fema.gov/data/statistics/> [accessed January 2019].

[22] Grabowska T, Skowronek R, Nowicka J, Sybirska H. Prevalence of hydrogen cyanide and carboxyhaemoglobin in victims of smoke inhalation during enclosed-space fires: a combined toxicological risk. Clin Toxicol (Phila) 2012;50(8):759–63.

[23] Hamad E, Babu K, Bebarta VS. Case files of the university of Massachusetts toxicology fellowship: does this smoke inhalation victim require treatment with cyanide antidote? J Med Toxicol 2016;12(2):192–8.

[24] Udeh CI, Ting M, Arango M, Mick S. Delayed presentation of nitroprusside-induced cyanide toxicity. Ann Thorac Surg 2015;99(4):1432–4.

[25] Hottinger DG, Beebe DS, Kozhimannil T, Prielipp RC, Belani KG. Sodium nitroprusside in 2014: a clinical concepts review. J Anaesthesiol Clin Pharmacol 2014;30(4):462–71.

[26] Albaum HG, Tepperman J, Bodansky O. The in vivo inactivation by cyanide of brain cytochrome oxidase and its effect on glycolysis and on the high energy phosphorus compounds in brain. J Biol Chem 1946;164:45–51.

[27] Piantadosi CA, Sylvia AL, Jobsis FF. Cyanide-induced cytochrome a, a3 oxidation-reduction responses in rat brain in vivo. J Clin Investig 1983;72(4):1224–33.

[28] Nicholls P, van Buuren KJ, van Gelder BF. Biochemical and biophysical studies on cytochrome aa 3. 8. Effect of cyanide on the catalytic activity. Biochim Biophys Acta 1972;275(3):279–87.

[29] van Buuren KJ, Nicholis P, van Gelder BF. Biochemical and biophysical studies on cytochrome aa 3. VI. Reaction of cyanide with oxidized and reduced enzyme. Biochim Biophys Acta 1972;256(2):258–76.

[30] van Buuren KJ, Zuurendonk PF, van Gelder BF, Muijsers AO. Biochemical and biophysical studies on cytochrome aa 3. V. Binding of cyanide to cytochrome aa 3. Biochim Biophys Acta 1972;256(2):243−57.
[31] Hamel J. A review of acute cyanide poisoning with a treatment update. Crit Care Nurse 2011;31(1):72−81.
[32] Ardelt BK, Borowitz JL, Isom GE. Brain lipid peroxidation and antioxidant protectant mechanisms following acute cyanide intoxication. Toxicology 1989;56(2):147−54.
[33] Lessell S. Experimental cyanide optic neuropathy. Arch Ophthalmol (Chicago, IL: 1960) 1971;86(2):194−204.
[34] Bright JE, Marrs TC. Pharmacokinetics of intravenous potassium cyanide. Hum Toxicol 1988;7(2):183−6.
[35] Hall AH, Doutre WH, Ludden T, Kulig KW, Rumack BH. Nitrite/thiosulfate treated acute cyanide poisoning: estimated kinetics after antidote. J Toxicol Clin Toxicol 1987;25(1-2):121−33.
[36] Wood JL, Cooley SL. Detoxication of cyanide by cystine. J Biol Chem 1956;218(1):449−57.
[37] Hall AH, Rumack BH. Hydroxycobalamin/sodium thiosulfate as a cyanide antidote. J Emerg Med 1987;5(2):115−21.
[38] United States Department of Labor, Occupational Safety and Health Administration. Cyanide. OSHA Method ID-120, December 1988. <https://www.osha.gov/dts/sltc/methods/validated/id120/id120.html> [accessed January 2019].
[39] Wilson J, Matthews DM. Metabolic inter-relationships between cyanide, thiocyanate and vitamin B 12 in smokers and non-smokers. Clin Sci 1966;31(1):1−7.
[40] Pettigrew AR, Fell GS. Simplified colorimetric determination of thiocyanate in biological fluids, and its application to investigation of the toxic amblyopias. Clin Chem 1972;18(9):996−1000.
[41] Rehling CJ. Poison residues in human tissues. Prog Chem Toxicol 1967;3:363−86.
[42] Reade MC, Davies SR, Morley PT, Dennett J, Jacobs IC. Review article: management of cyanide poisoning. Emerg Med Australas: EMA 2012;24(3):225−38.
[43] Chen KK, Rose CL. Nitrite and thiosulfate therapy in cyanide poisoning. J Am Med Assoc 1952;149(2):113−19.
[44] Barillo DJ. Diagnosis and treatment of cyanide toxicity. J Burn Care Res 2009;30(1):148−52.
[45] Bebarta VS, Brittain M, Chan A, Garrett N, Yoon D, Burney T, et al. Sodium nitrite and sodium thiosulfate are effective against acute cyanide poisoning when administered by intramuscular injection. Ann Emerg Med 2017;69(6):718−725. e714.
[46] Chen KK, Rose CL, Clowes G. Amyl nitrite and cyanide poisoning. JAMA 1933;100:1921−2.
[47] Culnan DM, Craft-Coffman B, Bitz GH, Capek KD, Tu Y, Lineaweaver WC, et al. Carbon monoxide and cyanide poisoning in the burned pregnant patient: an indication for hyperbaric oxygen therapy. Ann Plast Surg 2018;80(3 Suppl. 2):S106−s112.
[48] Thompson JP, Marrs TC. Hydroxocobalamin in cyanide poisoning. Clin Toxicol (Phila) 2012;50(10):875−85.
[49] Borron SW, Baud FJ, Megarbane B, Bismuth C. Hydroxocobalamin for severe acute cyanide poisoning by ingestion or inhalation. Am J Emerg Med 2007;25(5):551−8.
[50] Shepherd G, Velez LI. Role of hydroxocobalamin in acute cyanide poisoning. Ann Pharmacother 2008;42(5):661−9.
[51] Pickering WG. Cyanide toxicity and the hazards of dicobalt edetate. Br Med J (Clinical Research Ed) 1985;291(6509):1644.
[52] Beasley DM, Glass WI. Cyanide poisoning: pathophysiology and treatment recommendations. Occup Med (Oxford, Engl) 1998;48(7):427−31.
[53] Megarbane B, Delahaye A, Goldgran-Toledano D, Baud FJ. Antidotal treatment of cyanide poisoning. J Chin Med Assoc: JCMA 2003;66(4):193−203.
[54] Baud FJ, Barriot P, Toffis V, Riou B, Vicaut E, Lecarpentier Y, et al. Elevated blood cyanide concentrations in victims of smoke inhalation. N Engl J Med 1991;325(25):1761−6.
[55] Dries DJ, Endorf FW. Inhalation injury: epidemiology, pathology, treatment strategies. Scand J Trauma Resusc Emerg Med 2013;21:31.
[56] McAllister JL, Roby RJ, Levine B, Purser D. Stability of cyanide in cadavers and in postmortem stored tissue specimens: a review. J Anal Toxicol 2008;32(8):612−20.
[57] Ballantyne B, Bright J, Williams P. An experimental assessment of decreases in measurable cyanide levels in biological fluids. J Forensic Sci Soc 1973;13(2):111−17.
[58] Ballantyne B, Bright JE, Williams P. The post-mortem rate of transformation of cyanide. Forensic Sci 1974;3(1):71−6.
[59] Feldstein M, Klendshoj NC. The determination of cyanide in biologic fluids by microdiffusion analysis. J Lab Clin Med 1954;44(1):166−70.
[60] Feldstein M, Klendshoj NC. The determination of cyanide in biological fluids by microdiffusion analysis. Can J Med Technol 1955;17(1):29−32.
[61] Papezova K, Glatz Z. Determination of cyanide in microliter samples by capillary electrophoresis and in-capillary enzymatic reaction with rhodanese. J Chromatogr A 2006;1120(1-2):268−72.
[62] Chinaka S, Tanaka S, Takayama N, Tsuji N, Takou S, Ueda K. High-sensitivity analysis of cyanide by capillary electrophoresis with fluorescence detection. Anal Sci: Int J Jpn Soc Anal Chem 2001;17(5):649−52.
[63] Hanna Instruments, HI 3855 Cyanide Test Kit, Instruction manual. <http://www.hannacan.com/PDF/manHI3855.pdf> [accessed January 2019].
[64] Millipore Sigma, Cyanide Test. <https://www.emdmillipore.com/US/en/product/Cyanide-Test,MDA_CHEM-110044#anchor_PI> [accessed January 2019].
[65] Sun S, Li Y, Lv P, Punamiya P, Sarkar D, Dan Y, et al. Determination of prometryn in vetiver grass and water using gas chromatography-nitrogen chemiluminescence detection. J Chromatogr Sci 2016;54(2):97−102.
[66] Roda G, Arnoldi S, Dei Cas M, Ottaviano V, Casagni E, Tregambe F, et al. Determination of cyanide by microdiffusion technique coupled to spectrophotometry and GC/NPD and propofol by fast GC/MS-TOF in a case of poisoning. J Anal Toxicol 2018;42(6):e51−7.
[67] Maseda C, Matsubara K, Hasegawa M, Akane A, Shiono H. [Determination of blood cyanide using head-space gas chromatography with electron capture detection]. Nihon Hoigaku Zasshi 1990;44(2):131−6.

Chapter 23.7

A case of unknown pill ingestion—bupropion toxicity

Heather A. Paul[1,2] and S.M. Hossein Sadrzadeh[1,2]
[1]*Alberta Precision Laboratories, Calgary, AB, Canada,* [2]*Department of Pathology and Laboratory Medicine, Cumming School of Medicine, University of Calgary, Calgary, AB, Canada*

Case description

A 15-year-old previously healthy female was brought in by ambulance following episodes of general tonic−clonic seizures. The previous night, she had stayed up all night talking with family members following a recent family stressor. That morning, she complained of a headache and was given an unknown purple-colored pill by a relative to help with the pain. Within 1 hour of ingesting the pill, she began to feel drowsy and lightheaded and went to sleep for a few hours. Throughout the day, the patient continued to be drowsy and exhibit abnormal behavior such as "staring off into space." By early evening, the patient exhibited her first witnessed seizure activity, during which she abruptly brought her arms up to her chest, fell to the ground, with stiff and shaking limbs. Overall, the episode lasted between 1 and 5 minutes, although the patient was noted to be grinding her teeth and was confused for approximately 10 minutes afterward; it was at this point that the ambulance was called. While in the ambulance, the patient exhibited a second episode of seizure activity which lasted around 45 seconds and was resolved with intranasal midazolam. Upon arrival to the emergency department (ED), the patient's level of consciousness increased, although her dizziness remained.

Family medical history included febrile seizures in her younger sister and epilepsy in her younger brother. The patient denied any alcohol use and concerns of self-harm. She was on no medications and had no allergies. Laboratory results included several abnormal values consistent with the postictal phase of grand-mal seizures including high glucose, critically high lactate, critically low pH, and low carbon dioxide [1] (consistent with a low calculated HCO_3^-); the patient also had low hemoglobin and low hematocrit which were also seen on her complete blood count (Table 23.7.1). Of note, her anion gap and serum osmolality were within the reference interval (Table 23.7.1). Serum toxicology testing was negative for acetaminophen, ethanol, and salicylates.

Given the ingestion of an unknown pill, a urine sample was collected for toxicology testing. Notably, the urine drug screen by immunoassay was presumptive positive for both benzodiazepines and amphetamines; however, the patient denied any other drug use except for the "purple pill." At our laboratory, if samples from inpatients are presumptive positive for amphetamines by immunoassay, drug confirmation testing by liquid chromatography tandem mass spectrometry (LC−MS/MS) is performed. In this case, no amphetamine or methamphetamine was detected, raising the question of the cause of the patient's presumptive positive results. Importantly, the ED had also contacted the local poison information service in regard to the unknown purple-colored pill the patient had ingested. Based on the description of the pill, together with the delayed seizure activity, bupropion was suggested as a possibility for the pill's identity. Bupropion was subsequently detected in the patient's urine sample by gas-chromatography mass spectrometry, as was midazolam. The lack of detection of amphetamine or methamphetamine by LC−MS/MS indicated that the "presumptive positive" results were false-positive results. Indeed, there are reports in the literature indicating cross-reactivity of bupropion with the antibodies in several amphetamines immunoassays [2]. The patient's condition improved while in hospital, and she had no recurrence of seizures.

TABLE 23.7.1 Selected Laboratory Values.

	Analyte	Result (Reference Interval)
Venous blood gas	pH	7.23[a] (7.30–7.40)
	pCO$_2$	38 (36–46 mmHg)
	Base excess	−11[a] (−5–1 mmol/L)
	HCO$_3^-$ (calculated)	16[a] (20–24 mmol/L)
	Total hemoglobin	111[a] (120–160 g/L)
	Hematocrit	0.330[a] (0.360–0.480 L/L)
	Carboxyhemoglobin	0.8 (0.0%–3.0%)
	Methemoglobin	1.1 (0.0%–1.5%)
	Sodium	143 (133–145 mmol/L)
	Potassium	4.1 (3.5–5.0 mmol/L)
	Chloride	110 (98–111 mmol/L)
	Lactate	6.3[a] (≤2.0 mmol/L)
	Glucose	6.5[a] (3.9–6.1 mmol/L)
	Ionized calcium	1.19 (1.15–1.35 mmol/L)
	Anion gap	15 (4–16 mmol/L)
	Carbon dioxide content	15[a] (21–31 mmol/L)
	Osmolality, serum	295 (280–300 mmol/kg)
	Alanine transaminase	7 (1–35 U/L)
	Lipase	26 (0–60 U/L)
	Creatinine, serum	59 (40–100 μmol/L)
	Urea	4.3 (2.0–7.0 mmol/L)

[a]Abnormal results.

Discussion

Bupropion, sold under the brand names Wellbutrin and Zyban, is an aminoketone belonging to the class of substituted cathinones that is prescribed for depression and smoking cessation, acting as a norepinephrine–dopamine reuptake inhibitor and noncompetitive nicotinic acetylcholine receptor agonist [3]. It is also used off-label as treatment for attention deficit hyperactivity disorder in adults [3], children, and adolescents [4]. In the adolescent population, bupropion may be used as a preferred antidepressant due to its ability to increase energy and promote weight loss and to avoid the side effects associated with tricyclic antidepressants (TCAs) [5]. Bupropion is currently available in immediate, sustained, and extended-release formulations. Notably, the major side effect related to bupropion use is grand mal seizures [6], which is attributed to the ability of bupropion to lower the seizure threshold [7]. While delayed seizures occurring up to 8 hours postingestion in overdoses of immediate-release and up to 14 hours in overdoses of sustained-release formulations have been described [8], bupropion-associated seizures can also occur with treatment doses [9]. Indeed, a retrospective study of drug-related new-onset generalized seizures found that all cases of seizures attributed to bupropion occurred in the context of therapeutic doses (<450 mg/day) [6]. In particular, coexisting factors, such as a previous history of seizures or the use of other drugs that can lower the seizure threshold, may predispose individuals to bupropion-related seizures [6]. Finally, adolescents who intentionally overdose on bupropion have a higher incidence of seizures and other serious outcomes compared to those seen in TCAs overdoses [5]. In fact, a recent study showed that almost one-third of bupropion overdoses in adolescents results in seizures [5]. The final diagnosis in the case described above was seizures that were provoked by the ingestion of an unknown pill, which was most likely bupropion, in a patient with an unrecognized reduced threshold for seizures. Her low hemoglobin and hematocrit results were believed to be due to iron deficiency associated with heavy menstruation and not related to the seizure activity.

Here, a description of the unknown pill in combination with the expertise of the clinical toxicologist at the poison and drug information service allowed the identification of the potential cause of seizure activity in this patient. A recent review of resource utilization in the Mayo Clinic Arizona Emergency Department over 4 years in cases of suspected poisonings and toxicities found that compared to time-matched controls (i.e., similar ED arrival time), patients with suspected poisonings had an increased length of stay in the ED and a higher admission rate to an advanced care bed (e.g., the intensive care unit) [10]. While they did not examine the use of poison center consultation, they raised the possibility that the increased use of resources in suspected poisonings may be due to an underutilization of this service, and that early involvement of clinical toxicologists and other specialists with expertise in poisonings may lead to faster triage and discharge [10]. The consultation of poison information specialists played an important role in the early identification of the most likely cause for the patient's seizures in the case described above.

Notably, here, the patient's urine sample was presumptive positive for both amphetamines and benzodiazepines. However, positive results from laboratory evaluations for substance abuse using urine drug screening by immunoassays must be interpreted with caution [11]. While immunoassays are rapid and inexpensive method for determining whether the patient has been exposed to drugs of abuse, low specificity due to cross-reactivities between structurally similar molecules with the same antibody is known to cause false-positive results [12]. Knowledge of this limitation is paramount to proper interpretation of both positive and negative results. In the context of positive results, best practice involves confirmatory testing using mass spectrometry [12]. Unfortunately, due to the highly specialized nature of mass spectrometry, this type of analysis is not available at all clinical laboratories. Further, even if mass spectrometry is available, it generally takes a day or two (or even longer) to receive the results [12]. Thus familiarity with the most common drugs that can generate false-positive results on an immunoassay system is beneficial.

Although the above case is not a typical substance abuse case, it is important to note that adolescents will typically abuse what is both cheap and readily available [11]. In this case the patient was given the unknown pill by a relative to help with her headache. In suspected acute poisonings and/or toxicities in adolescent population, it is helpful to look for the source of the drug within the close circle of family members or friends, for example from family members' prescription medications. Involving the clinical toxicologists from poison centers is also of great importance to identify the abused substance(s). Finally, while the misuse of bupropion is uncommon, it does occur [13], and its use as an alternative to TCAs is increasing in the adolescent population [5]. As such, health care providers should be familiar with its potential side effects, in particular seizures, as well as the possibility false-positive results on amphetamine immunoassays.

References

[1] Orringer CE, Eustace JC, Wunsch CD, Gardner LB. Natural history of lactic acidosis after grand-mal seizures. N Engl J Med 2010;297:796—9.

[2] Casey ER, Scott MG, Tang S, Mullins ME. Frequency of false positive amphetamine screens due to bupropion using the Syva EMIT II immunoassay. J Med Toxicol 2011. Available from: https://doi.org/10.1007/s13181-010-0131-5.

[3] Verbeeck W, Bekkering GE, Van den Noortgate W. Bupropion for attention deficit hyperactivity disorder (ADHD) in adults. Cochrane Database Syst Rev 2017;10:CD009504.

[4] Stuhec M, Munda B, Svab V, Locatelli I. Comparative efficacy and acceptability of atomoxetine, lisdexamfetamine, bupropion and methylphenidate in treatment of attention deficit hyperactivity disorder in children and adolescents: a meta-analysis with focus on bupropion. J Affect Disord 2015;178:149—59.

[5] Sheridan DC, Lin A, Zane Horowitz B. Suicidal bupropion ingestions in adolescents: increased morbidity compared with other antidepressants. Clin Toxicol 2018;56:360—4.

[6] Pesola GR, Avasarala J. Bupropion seizure proportion among new-onset generalized seizures and drug related seizures presenting to an emergency department. J Emerg Med 2002;22:235—9.

[7] Hill S, Sikand H, Lee J. A case report of seizure induced by bupropion nasal insufflation. Prim Care Companion J Clin Psychiatry 2009;09:67—9.

[8] Starr P, et al. Incidence and onset of delayed seizures after overdoses of extended-release bupropion. Am J Emerg Med 2009;27:911—15.

[9] Bayir A, et al. Seizures after overdoses of bupropion intake. Balk Med J 2012;30:248—9.

[10] Traub SJ, Saghafian S, Buras MR, Temkit MH. Resource utilization in emergency department patients with known or suspected poisoning. J Med Toxicol 2017;13:238—44.

[11] Wang GS, Hoyte C. Common substances of abuse. Pediatr Rev 2018;39:403—14.

[12] Saitman A, Park HD, Fitzgerald RL. False-positive interferences of common urine drug screen immunoassays: a review. J Anal Toxicol 2014;38:387—96.

[13] Naglich AC, Brown ES, Adinoff B. Systematic review of preclinical, clinical, and post-marketing evidence of bupropion misuse potential. Am J Drug Alcohol Abuse 2019;1—14. Available from: https://doi.org/10.1080/00952990.2018.1545023.

Chapter 23.8

Laboratory confirmed massive donepezil ingestion

Stephen Thornton and Todd Crane
Department of Emergency Medicine, University of Kansas Health System, Kansas City, KS, United States

Case description

A 67-year-old male with a history of hypertension and dementia presented to the emergency department 1 hour after reportedly ingesting 290 mg of donepezil in a suicide attempt. On presentation the patient was alert and oriented but complaining of a headache and nausea. He had access to lisinopril, hydrochlorothiazide, oxycodone, and alendronate but denied taking any other medications except the donepezil. He had prescribed 10 mg of donepezil once a day approximately 1 week prior to his overdose. His initial vital signs were a heart rate of 82 beats per minute (bpm), respiratory rate of 16 with an oxygen saturation of 99% on room air and a temperature of 36.7°C. His physical exam was notable for 3 mm pupils bilaterally with diaphoresis and vomiting. His cardiovascular and pulmonary exams were normal. No lacrimation, salivation, or diarrhea was noted. He was given 4 mg of intravenous ondansetron and 50 g of activated charcoal. His laboratory evaluation consisting of a complete blood count and metabolic panel was normal. Serum acetaminophen, salicylate and ethanol concentrations were undetectable. A seven-panel urine drug immunoassay screen was negative. His EKG demonstrated a sinus rhythm with a QTc of 441 ms. The longest document QTc duration during his hospitalization was 450 ms.

Within 90 minutes of his arrival, his heart rate decreased to 59 bpm. His blood pressure was 168/77 mmHg and on reexamination his mental status remained unchanged. The lowest heart rate recorded was 50 bpm approximately 4 hours from presentation. Atropine was considered but never administered. His blood pressure remained elevated throughout his hospitalization. He never developed bronchorrhea, fasciculations, muscle weakness, or seizures. On the second hospital day, he became acutely agitated and required physical restraints and was sedated with haloperidol and quetiapine. Due to his acute change in mental status, a head CT, lumbar puncture, and EEG were performed all which were normal. He remained agitated and trialed on multiple medications including olanzapine and valproic acid without substantial improvement. On approximate hospital day 14, he developed an ileus requiring a feeding tube placement. He was finally discharged to a long-term care facility after a 40-day hospital admission.

Serial donepezil levels were obtained from NMS Labs (Willow Grove, PA) using high-performance liquid chromatography/ tandem mass spectrometry. Levels at presentation and 3.5 hours after presentation were 240 and 130 ng/mL, respectively. The patient was admitted for observation. He had no further episodes of bradycardia. A serum donepezil level 10 days from presentation was still detectable at 8.9 ng/mL. He was discharged after a 40-day hospital course which was complicated by ileus, delirium, and placement issues.

Discussion

Donepezil is a centrally acting, reversible inhibitor of acetylcholinesterase commonly used in the treatment of dementia, specifically the Alzheimer's type [1]. It was first marketed in 1996 and is now one of the most common antidementia drugs prescribed [2]. It is taken orally and has close to 100% bioavailability [1]. The peak onset of donepezil is 3–5 hours after ingestion and therapeutic levels are reported to be 35–75 ng/mL [3,4]. It is metabolized by metabolized by CYP2D6 and CYP3A4 and then renally eliminated [3]. Donepezil has been found to have an extremely long half-life of approximately of up to 100 hours in elderly patients [5]. This explains why donepezil could still be detected in this patient's serum even 10 days after the overdose.

Considering the widespread use of donepezil, it is prudent to understand the effects of supratherapeutic ingestions. As donepezil inhibits acetylcholinesterase activity there is concern that a donepezil overdose could lead to cholinergic excess. Cholinergic excess can manifest with relatively mild symptoms such as miosis, diaphoresis, salivation, vomiting, and diarrhea, and/or it could potentially cause bronchorrhea, bradycardia, seizures, neuromuscular weakness, and even paralysis [6]. There is a paucity of published literature on intentional donepezil overdoses. The reported cases of donepezil overdoses that do exist involve doses of less than 50 mg resulting from therapeutic errors or unintentional ingestions [7–11]. In several of these cases, patients did require atropine, but only for mild bradycardia. No seizures or fasciculations were reported in any prior case.

The amount of donepezil reportedly ingested in this case is the largest in the medical literature and is supported by the fact that we report serum donepezil level in the medical literature, almost four times the upper limits of normal therapeutic range. Despite this, our patient developed only mild bradycardia which responded to supportive care. QT prolongation has also been reported with donepezil but even with a donepezil concentration of 240 ng/mL, our patient's QTc was only mildly prolonged [12]. This patient did develop significant agitation and delirium, which could be attributed to his donepezil ingestion. While the prolonged duration of the delirium is consistent with donepezil's long half-life, the onset of symptoms was greater than 24 hours from the exposure which is inconsistent with a medication that is rapidly absorbed. Similarly, it is difficult to attribute this patient's ileus to the donepezil ingestion as this is a complication more commonly expected with anticholinergic drug overdoses not a cholinergic drug such as donepezil [13].

In conclusion, this case suggests that even large ingestions of donepezil with significantly elevated serum levels can be expected to cause only minimal cholinergic symptoms which can successfully be managed with observation and supportive care.

References

[1] Wilkinson DG. The pharmacology of donepezil: a new treatment of Alzheimer's disease. Expert Opin Pharmacother 1999;1(1):121–35.
[2] Rodrigues Simões MC, et al. Donepezil: an important prototype to the design of new drug candidates for Alzheimer's disease. Mini Rev Med Chem 2014;14(1):2–19.
[3] Shigeta M, Homma A. Donepezil for Alzheimer's disease: pharmacodynamic, pharmacokinetic, and clinical profiles. CNS Drug Rev 2001;7(4):353–68.
[4] Hiemke C, Baumann P, Bergemann N, et al. AGNP consensus guidelines for therapeutic drug monitoring in psychiatry: update 2011. Pharmacopsychiatry 2011;44:195–235.
[5] Ohnishi A, Mihara M, Kamakura H, et al. Comparison of the pharmacokinetics of E2020, a new compound for Alzheimer's disease, in healthy young and elderly subjects. J Clin Pharmacol 1993;33:1086–91.
[6] Adeyinka A, Kondamudi NP. Cholinergic crisis. StatPearls [Internet]. Treasure Island, FL: StatPearls Publishing; 2018.
[7] Shepherd G. Donepezil overdose: a tenfold dosing error. Ann Pharmacother 1999;33(7):812–15.
[8] Pourmand A, Shay C, Redha W, Aalam A, Mazer-Amirshahi M. Cholinergic symptoms and QTc prolongation following donepezil overdose. Am J Emerg Med 2017;35(9):1386.e1–3.
[9] Yano H, Fukuhara Y, Wada K, Kowa H, Nakashima K. A case of acute cholinergic adverse effects induced by donepezil overdose: a follow-up of clinical course and plasma concentration of donepezil. Rinsho Shinkeigaku 2003;43(8):482–6.
[10] Greene YM, Noviasky J, Tariot PN. Donepezil overdose. J Clin Psychiatry 1999;60(1):56–7.
[11] Garlich FM, Balakrishnan K, Shah SK, Howland MA, Fong J, Nelson LS. Prolonged altered mental status and bradycardia following pediatric donepezil ingestion. Clin Toxicol (Phila) 2014;52(4):291–4.
[12] Kitt J, Irons R, Al-Obaidi M, Missouris C. A case of donepezil-related torsades de pointes. BMJ Case Rep 2015;2015.
[13] Isbister GK, Oakley P, Whyte I, Dawson A. Treatment of anticholinergic-induced ileus with neostigmine. Ann Emerg Med 2001;38(6):689–93.

Further reading

Atri A, Molinuevo J, Lemming O, Wirth Y, Pulte I, Wilkinson D. Memantine in patients with Alzheimer's disease receiving donepezil: new analysis of efficacy and safety for combination therapy. Alzheimer's Res Ther 2013;5:6.
Geldmacher DS. Donepezil (Aricept®) for treatment of Alzheimer's disease and other dementing conditions. Expert Rev Neurother 2004;4(1):5–16.
Indira M, Andrews M, Rakesh T. Incidence, predictors, and outcome of intermediate syndrome in cholinergic insecticide poisoning: a prospective observational cohort study. Clin Toxicol 2013;51(9):838–45.
Johansson P, Almqvist E, Johansson J, Mattsson N, Andreasson U, Hansson O, et al. Cerebrospinal fluid (CSF) 25-hydroxyvitamin D concentration and CSF acetylcholinesterase activity are reduced in patients with Alzheimer's disease. PLoS One 2013;8(11):e81989.

Mayeux R, Stern Y. Epidemiology of Alzheimer disease. Cold Spring Harbor Perspect Med 2012;2(8):a006239.

Meguro K, Akanuma K, Meguro M, Kasai M, Ishii H, Yamaguchi S. Lifetime expectancy and quality-adjusted life-year in Alzheimer's disease with and without cerebrovascular disease: effects of nursing home replacement and donepezil administration — a retrospective analysis in the Tajiri Project. BMC Neurol 2015;15:227.

O'Brien J, Burns A, BAP Dementia Consensus Group. Clinical practice with anti-dementia drugs: a revised (second) consensus statement from the British Association for Psychopharmacology. J Psychopharmacol 2011;(8):997—1019.

Tariot P, Cummings J, Katz I, Mintzer J, Perdomo C, Schwam E, et al. A randomized, double-blind, placebo-controlled study of the efficacy and safety of donepezil in patients with Alzheimer's disease in the nursing home setting. J Am Geriatrics Soc 2001;49(12):1590—9.

Chapter 23.9

Missing oxycodone metabolites confirm suspected drug diversion

Geza S. Bodor
Department of Pathology, University of Colorado School of Medicine, UC Health Laboratories, Aurora, CO, United States

Case presentation

The 36-year-old male patient had been on 30 mg per day oxycodone prescription for chronic, nonmalignant pain. He regularly requested refills before his next scheduled prescription was due, sometimes 2 weeks early. He claimed that his pain was very poorly controlled therefore he needed to take more medication than prescribed. He participated in random urine drug testing but the results of the Syva EMIT II Plus opiate immunoassay screen were repeatedly negative at the 300 ng/mL morphine cut-off level. The prescribing provider suspected drug diversion and called the laboratory for an explanation. She was informed that the negative results could be due to the low cross-reactivity of oxycodone in the urine opiate immunoassay or it could be due to drug diversion, and she was instructed to order oxycodone confirmation testing using the residual sample from the screening assay. The confirmation testing was negative for oxycodone by LC–MS/MS method and the clinician was notified that diversion was a real possibility. The patient claimed he may have forgotten to take his medicine the previous day but he was advised by his clinician that he would be excluded from the pain treatment program if his urine test was found negative again.

On the next random urine screen the patient was positive by the opiate immunoassay and by an oxycodone immunoassay as well but the provider wanted additional confirmation to prove that he was taking his oxycodone and he did not spike his urine; therefore confirmation testing was performed by an LC–MS/MS method specific for oxycodone and its metabolites. The oxycodone confirmation test by the LC–MS/MS method confirmed the presence of oxycodone at >2000 ng/mL (the upper limit of AMR of the LC–MS/MS assay) but was negative (<20 ng/mL) for noroxycodone and oxymorphone. The clinical nurse practitioner treating the patient requested further explanation for these findings before deciding if the patient can be kept in the program.

Discussion

Ordering the appropriate follow-up test after an unexpected opiate result and interpreting the result are very common questions to the laboratory. Many times the unexpected result is a negative finding on the urine opiate drug screen in patients receiving semisynthetic opiates or synthetic opioids, although clinicians who prescribe narcotic analgesics often question positive findings as well when they expect a negative result. Result interpretation is another area where laboratory consultation commonly sought because of the presence of opiate metabolites that are prescription drugs by themselves. The reason for these calls is inadequate knowledge of opioid testing and metabolism by prescribing clinicians who often request help from the laboratory after ordering the wrong test for the drug they are prescribing. This gap in knowledge has been documented in a group of medicine residents [1], family physicians and physicians who practice in other fields, too [2,3]. The knowledge gap appears to be independent of whether the practitioners routinely prescribe opioids or not.

Our case is a typical example of how many of these laboratory consultations go. The clinician treating our patient has called the laboratory three times requesting help interpreting the screening test result, asking for recommendation on appropriate confirmation test ordering and, eventually, to discuss interpretation of the confirmation results. While the first two calls to the laboratory were common topics, interpretation of the second confirmation result is less frequently encountered in our practice. The clinician's suspicion of diversion of the prescribed opioid did not fit the result of positive oxycodone confirmation by LC—MS/MS and the highly elevated concentration of oxycodone (>2000 ng/mL). Additional questions were raised about the meaning of the less than 20 ng/mL noroxycodone and oxymorphone results. The clinician was not sure if this was expected or if they were relevant to her suspicion of drug diversion. Explanation of the various findings required detailed discussion of oxycodone metabolism and opiate and opioid testing methods.

Oxycodone is a semisynthetic opiate, a substituted phenanthrene, and is structurally similar to codeine and morphine. It is a narcotic analgesic with high addictive potential. Its wide spread use is a major culprit in the recent opioid epidemic. It is available under various brand names from several manufacturers in immediate or extended release oral formulations and as injectable preparation. Bioavailability of the oral preparations is 60%—90%. It is an agonist of the mu-, kappa-, and delta-opioid receptors and it is commonly used for the treatment of acute and chronic pain. It is metabolized extensively [4] by various enzymes such as CYP3A5 (N-demethylation), CYP2D6 (O-demethylation), 6-keto-reductase (reduction), and UGT2B7 (glucuronide conjugation, Fig. 23.9.1). Polymorphism of the CYP enzyme system contributes to between-individual variation in oxycodone metabolism. A small fraction of the metabolites is excreted as sulfate conjugates, too. Oxycodone metabolism, along with metabolism of the other prescription opioids and opiates, had been reviewed in detail by DePriest et al. [4].

After undergoing metabolism in the liver, oxycodone and its metabolites are primarily excreted in the urine. The half-life of orally administered oxycodone is approximately 3 hours (immediate release) and 4.5 hours (extended release), and those of noroxycodone and oxymorphone are approximately 6 and 9 hours, respectively. Oxycodone (free and conjugated) constitutes approximately 9% of the dose in urine and its two major metabolites by quantity, oxymorphone (free and conjugated) and noroxymorphone (free and conjugated), constitute 10% and 37% of the dose, respectively [4]. Both of these metabolites are clinically active and contribute to the analgesic effects of oxycodone. One of the metabolites, oxymorphone, is also available as a prescription opiate analgesic often initiating a call to the laboratory when its presence is reported in a patient who is in oxycodone treatment program.

Oxycodone testing can be performed by immunoassay or by chromatography—mass spectrometry—based methods. Immunoassays are most commonly utilized in the clinical laboratory for urine drug screening, using immunoassays originally developed for workplace drug screening. Oxycodone was not included on the list of mandated analytes for workplace drug testing until recently therefore many laboratories only use opiate assays that were standardized to detect morphine or codeine at the regulated concentration of 2000 ng/mL, although many assay manufacturers offer a lower calibrator, usually of 300 ng/mL concentration, that could be utilized for nonworkplace testing. Unfortunately, assay cross-reactivities with other opiates than morphine or codeine are extremely variable. For example, our screening assay, the Syva EMIT II. Plus, calibrated to 300 ng/mL morphine as cut off, gives positive result with oxycodone or oxymorphone if they are present at 1500 or 9300 ng/mL, respectively, corresponding to 20% and 3% cross-reactivity. Immunoassays from other manufacturers demonstrate similar performance characteristics, though they may show actual cross-reactivities numerically different from ours. Even the 1500 ng/mL oxycodone cutoff concentration of an opiate immunoassay may be too high to detect usual concentrations of the drug in patients taking low doses of oxycodone. If a laboratory used the same brand of screen but calibrated to the SHAMSA mandated cutoff of 2000 ng/mL morphine, oxycodone, at 20% cross-reactivity, would not even be detected even in severe overdose cases. Those laboratories that use opiate immunoassay screen do not routinely report their expected concentration for positive result for the other opiates leaving the clinician in the dark regarding what to expect and what negative results may mean. The clinician must call the laboratory and hope to receive accurate answer if he or she needs specific information.

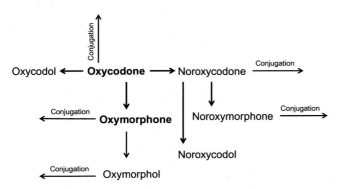

FIGURE 23.9.1 **Oxycodone and its metabolites.** Oxymorphone (bold) is a prescription drug and an oxycodone metabolite as well. The enzymes CYP2D6 and CYP3A5 are responsible for demethylation, 6-keto reductase for reduction and UGT2B7 for glucuronide conjugation.

Oxycodone immunoassays are available in the US market, usually calibrated to either 100 or 300 ng/mL cutoff concentrations. A few of those immunoassays have been evaluated by independent research but not all of them have known and well-documented cross-reactivities. For example, one publication investigated the cross-reactivity of the HEIA (Immunalysis Corporation) and DRI (Microgenics/Thermo Fisher) brand oxycodone immunoassays [5]. Both were confirmed to show 100% cross-reactivity with oxymorphone but the DRI assay did not show any reactivity with noroxycodone or noroxymorphone, two of oxycodone's metabolites. Because oxymorphone is also a prescription drug by itself and, therefore, could be available for abuse, it could cause positive result in the DRI assay even when oxycodone is not present. Even more important, both assays showed significant cross reactivity with naloxone, another opiate that is included in Suboxone, a prescription buprenorphine combination, and is also used for resuscitation in cases of opioid overdose. Not all assay formulations, used for oxycodone testing on automated instruments, have been evaluated extensively by independent investigators. In addition, many providers and pain clinics use point-of-care devices for opiate or oxycodone testing, but these devices are even more of an enigma when it comes to opioid drug testing. Lack of sufficiently low limit of detection, specificity, and missing objective performance information render immunoassay testing marginally acceptable for pain management testing.

The recommended testing method for opiates, including oxycodone, or opioids is chromatography—mass spectrometry—based analysis with sufficiently low cutoff concentration to catch even low dose administration. Traditionally, this specificity was achieved by gas chromatography—mass spectrometry (GC—MS) but more recently liquid chromatography—tandem mass spectrometry (LC—MS/MS) based methods are becoming dominant. GC-MS methods require derivatization [6] and may miss low-concentration metabolites or require digestion of conjugated metabolites to remove glucuronide. LC—MS/MS methods, on the other hand, can detect unexpected compounds, such as designer opioids, or previously unreported metabolites [7] even when present at very low concentrations. Another advantage of LC—MS/MS methods is exceptional selectivity; the capability of identifying and quantitating opiates, synthetic opioids, and their metabolites simultaneously, even when several compounds are present in the same sample [8]. Multiple methods have been described in the literature and private and commercial laboratories may develop their methods, but if good laboratory practices are followed and CAP rules and FDA guidelines for LDT testing are observed, LC—MS/MS analytical methods are superior to immunoassays.

Patients who are addicted to narcotic analgesics are often used other than prescribed drugs, including prescription opioids, illicit drugs such as heroin, fentanyl, or buprenorphine analogs, but abuse of benzodiazepines, marijuana, or stimulants is not uncommon either. Opiate immunoassay cross-reactivity with illicit opiates or designer opioids can mask the abuse of these compounds and create the impression of compliance when it does not exist. On the other hand, the use of immunoassays with insufficiently low cutoffs or testing with an immunoassay that does not detect the prescribed compound can produce false-negative results leading to erroneous conclusion of diversion. Finally, detection of the parent compound without knowledge of the presence of metabolites can hide diversion that is concealed by sample spiking. The only way to prevent all these mistakes occurring is by the use of LC—MS/MS methods for pain management testing.

Conclusion

Based on the above, the initial negative immunoassay results could be explained by insufficient cross-reactivity of the opiate immunoassay with oxycodone. However, the LC—MS/MS confirmation of the leftover sample showing no detectable amount of oxycodone in that sample created a strong foundation for suspicion of diversion.

Our patient had very high parent drug (oxycodone) concentration in his urine, indicating significant absorption if its origin was the ingested drug. In such case, detectable concentrations of oxycodone's metabolites would be also expected because of oxycodone's metabolism during first pass through the liver and the high proportional excretion of oxymorphone and noroxycodone. We did not detect any of these metabolites although they are excreted at a higher proportion than the parent drug itself. This finding can only be explained by introduction of oxycodone into the urine sample outside of the body, as it is seen during spiking of the sample with the intention of hiding proof of diversion. After this explanation was discussed with the clinician she decided to discontinue the patient's prescription oxycodone.

References

[1] Starrels JL, et al. They don't know what they don't know: internal medicine residents' knowledge and confidence in urine drug test interpretation for patients with chronic pain. J Gen Intern Med 2012;27(11):1521—7.
[2] Reisfield GM, et al. Family physicians' proficiency in urine drug test interpretation. J Opioid Manag 2007;3(6):333—7.

[3] Reisfield GM, et al. Urine drug test interpretation: what do physicians know? J Opioid Manag 2007;3(2):80–6.
[4] DePriest AZ, et al. Metabolism and disposition of prescription opioids: a review. Forensic Sci Rev 2015;27(2):115–45.
[5] Jenkins AJ, Poirier 3rd JG, Juhascik MP. Cross-reactivity of naloxone with oxycodone immunoassays: implications for individuals taking Suboxone. Clin Chem 2009;55(7):1434–6.
[6] Goldberger BA, Chronister CW, Merves ML. Quantitation of oxycodone in blood and urine using gas chromatography-mass spectrometry (GC-MS). Methods Mol Biol 2010;603:453–60.
[7] Baldacci A, et al. Identification of new oxycodone metabolites in human urine by capillary electrophoresis-multiple-stage ion-trap mass spectrometry. J Chromatogr A 2004;1051(1–2):273–82.
[8] Bodor GS. Pain management testing by liquid chromatography tandem mass spectrometry. Clin Lab Med 2018;38(3):455–70.

Index

Note: Page numbers followed by "*f*" and "*t*" refer to figures and tables, respectively.

A

AA. *See* Atomic absorption (AA)
AB Sciex 6500 LC–MS systems, 501
AB Sciex iMethod Forensic Toxicology Library, 23
Abalone (mollusc), 23–24
ABC. *See* Airway, breathing, and circulation control (ABC)
Abdominal imaging, 433
Abdominal radiographs, 29
Abstral, 347–348
ACC/AHA. *See* American College of Cardiology/American Heart Association (ACC/AHA)
Accidental death involving psilocin
  case history, 379
  discussion, 379–382
Acepromazine, 224
Acetaldehyde-protein adducts, 41–42
Acetaminophen, 67–68, 69*f*, 70, 72, 343
  toxicity
    case discussion, 75–77
    case history, 75
    laboratory studies, 76*t*
Acetanilide, 68, 68*f*
Acetone, 41
  cyanohydrins, 474*t*
Acetonitrile (C_2H_3N), 473–474, 474*t*
Acetyl fentanyl, 340
Acetylcholine, 442
Acetylcysteine, 72
6-Acetylmorphine (6-AM), 20, 335
Acetylsalicylates, 70, 365
Acetylsalicylic acid, 67–68, 68*f*, 72
ACMIA. *See* Antibody conjugated magnetic immunoassay (ACMIA)
AcMPAG. *See* Acylglucuronide (AcMPAG)
Aconite, 364
Aconitum napellus (Monkhood), 364
Acrylonitrile (C_3H_3N), 474*t*
Actiq, 347–348
Acute ethanol, 38–39
Acute ketamine toxicity, 305
Acute kidney injury (AKI), 151, 161
Acute lymphocytic leukemia (ALL), 141
Acute nicotine intoxication, death due to
  case description, 261
  postmortem toxicology results on heart blood, 262*t*
Acute overdose, 75–76

Acute renal failure (ARF), 253
Acute toxicity, 388
Acylglucuronide (AcMPAG), 323
Adenosine, 10
Adenosine triphosphate (ATP), 475
ADEs. *See* Adverse drug events (ADEs)
ADH. *See* Alcohol dehydrogenase (ADH)
Adverse drug events (ADEs), 99
AEDs. *See* Antiepileptic drugs (AEDs)
AES. *See* Atomic emission spectroscopy (AES)
Affinity, 8–9, 14
Agency for Toxic Substances and Disease Registry (ATSDR), 413
Airway, breathing, and circulation control (ABC), 335
AKI. *See* Acute kidney injury (AKI)
Alanine aminotransferase (ALT), 422
Albuterol, 223
Alcohol by volume (alc/vol), 37
Alcohol dehydrogenase (ADH), 38–40, 60
Alcohols, 37, 196, 199, 254, 371
  ethanol, 37–39
  ethylene glycol, 45–46
  isopropanol, 45
  methanol, 44–45
Aldehyde dehydrogenase, 60
Alendronate, 485
Alere *i*Cup Dx Drug Screen Cup, 497
Aliphatic nitriles, 473–474
Alkylating agents, 142, 198
ALL. *See* Acute lymphocytic leukemia (ALL)
Alpha-amino-3-hydroxy-5-methyl-4-isoxazolepropionic acid receptor (AMPA receptor), 123
Alpha-phosphoribosylpyrophosphoric acid (PRPP), 147–148
Alprazolam, 253, 300
Altered mental status, 311
6-AM. *See* 6-Acetylmorphine (6-AM)
American Association of Poison Control Centers, 209–210, 452
American College of Cardiology/American Heart Association (ACC/AHA), 196
Amikacin, 93
4-Amino-4-deoxy-*N*10-methylpteroic acid (DAMPA), 147, 158–159
Aminoglycoside overdoses, 93
Amiodarone, 197
Ammonia, 397–398, 398*t*
  acute gaseous ammonia toxicity, 398*t*

  case description, 397
AMPA receptor. *See* Alpha-amino-3-hydroxy-5-methyl-4-isoxazolepropionic acid receptor (AMPA receptor)
Amphetamine, 228–230, 243–244, 253–255, 257–258, 311, 379, 456*t*
  case description, 243–244
  class drugs, 267
  metabolism, 245*f*
  pharmacokinetics, 229–230
L-Amphetamine, 257
Amphibians, 437
Amphotericin B, 90
AMR. *See* Analytical measuring ranges (AMR)
Amygdalin ($C_{20}H_{27}NO_{11}$), 474*t*
Amyl nitrite ($C_5H_{11}NO_2$), 476
Analgesics, 67
  analytical methods, 72
  discovery and history, 67–68
  mechanism of action, 70–71
  pharmacokinetics, toxicokinetics, and treatment, 71–72
  toxicity, 69–70
Analogue Enforcement Act, 287–289
Analytical measuring ranges (AMR), 166–167
Anaphylactoid reactions, 76–77, 444
Anaphylaxis, 440
ANCA-associated vasculitis, 252
Androctonus, 443–444
Angel dust. *See* Phencyclidine (PCP)
Angiogenesis inhibitors, 198
Angiotensin-converting enzyme (ACE) inhibitors, 197
Anhydrous ammonia gas, 397
Anion gap, 44, 64
4-ANPP, 340
Antemortem history, 465
Anthracyclines, 197
Antibiotics, 83, 86
  analytical methods and clinical management implications, 93–95
  antibiotic classes, 84*t*
  azole antifungals, 89*f*
  β-lactam ring structure, 87*f*
  cardiac toxicity, 85
  gram stain and susceptibility testing, 83–84
  graphical representation, 90*f*
  inhibition
    of bacterial cell wall synthesis, 86–88

505

Antibiotics (*Continued*)
    of fungal cell wall synthesis or cell wall disruption, 89
    of nucleic acid synthesis, 89
    of protein synthesis, 88
  mechanism of action, 86–89
  nephrotoxicity and hepatotoxicity, 86
  neurotoxicity, 85–86
  pharmacokinetics, toxicokinetics, and treatment, 89–93
  serum–drug concentration, 94f
  structures
    of common cephalosporins, 87f
    of linezolid, 88f
    of vancomycin, 88f
  therapeutic drug monitoring, 86
  and toxicity, 84–85
Antibody conjugated magnetic immunoassay (ACMIA), 319, 321
Anticonvulsants, 121
  analytical considerations, 127–128
  clinical management, 125–127
  by generation, 121t
  mechanism of action, 122–123
  pharmacokinetics and toxicokinetics, 123–124, 124t
Antidepressants, 99–100
  analytical methods for, 105–107
  antipsychotics and, 100t
Antidotes, 473, 476–477
Antiepileptic drugs (AEDs), 121, 123
Antihistamines, 38–39
Anti-inflammatory drugs, 67
  analytical methods, 72
  discovery and history, 67–68
  mechanism of action, 70–71
  pharmacokinetics, toxicokinetics, and treatment, 71–72
  toxicity, 69–70
Antimalarial drugs, 196
Antimetabolites, 142, 198
Antimicrobials, 83
Antineoplastic drugs, 141, 143t
  analytical methods and clinical management implications, 148
  childhood cancer, 141
  cytotoxic agents, 142–144
  hormones and hormone regulators, 145
  mechanisms of action, 145
  methotrexate, 145
  pathway-targeted therapies, 144–145
  pharmacokinetics, 146–147
  thiopurine analogs, 145
  treatment, 147–148
Antipsychotics, 100–102, 100t
  analytical methods, 105–107
  antidepressant drugs/drug classes associated with ED visits, 101t
  data on adverse drug events—associated with ED visits, 101t
  management, 104–105
  pharmacokinetics and toxicokinetics and treatment, 102–105

  monoamine oxidase inhibitor overdose, 102–104
  SSRI, SNRI, and atypical antidepressant overdose, 102, 103t
  tricyclic antidepressant overdose, 102
Antivenom therapy, 440
Antizol, 44–45
Apidae, 444
Apparent, 8–9
Aprobarbital, 211
Ara-C. *See* Cytosine arabinoside (Ara-C)
Arachidonic acid, 70–71
Arachnids, 437, 440–444
  analytical methods and clinical management implications, 443
  mechanism of action and toxicokinetics, 442–443
  medically significant spiders, 441t
  treatment, 443–444
ARF. *See* Acute renal failure (ARF)
Arg181Cys, 334
L-Arginine, 126
Arrhythmias, 266
Arsenic, 3, 413, 433
  analytical methods and clinical management implications, 416–417
  cardiovascular system, 422
  central nervous system, 422
  discussion and follow-up, 421–422
  findings from therapeutic exposures, 422–423
  background on clinical assessment of arsenic, 422–423
  forensic considerations, 418
  gastrointestinal system, 422
  hematopoietic system, 422
  laboratory report
    for total arsenic, 421
    for total inorganic arsenic, 421
  mechanisms of action, 413
  miscellaneous symptoms, 422
  pharmacokinetics and toxicokinetics, 414
  specific interpretation, 423
  treatment, 415
Arsenobetaine, 433
Artemisia absinthium L., 365
Arthralgias, 252
Arthropods, 437, 440–445
  arachnids and ticks, 441–444
  analytical methods and clinical management implications, 443
  mechanism of action and toxicokinetics, 442–443
  medically significant spiders, 441t
  treatment, 443–444
ARUP Laboratories confirmatory drug-screen test, 247
Ascomycin, 320
Asialotransferrin, 41
L-Asparaginase, 142
Aspartate aminotransferase (AST), 422
Aspirin, 67–71, 261, 365
Asthma, 223
Ativan. *See* Lorazepam

Atomic absorption (AA), 416
Atomic absorption spectroscopy (AAS), 416
Atomic emission, 416
Atomic emission spectroscopy (AES), 416
Atorvastatin, 223
ATP. *See* Adenosine triphosphate (ATP)
Atractaspididae, 437
Atrax, 441, 443
Atrial fibrillation, 219
Atropine, 366
ATSDR. *See* Agency for Toxic Substances and Disease Registry (ATSDR)
Attention deficit hyperactivity disorder (ADHD), 311
Atypical antidepressant overdose, 102
Ayahuasca, 297
Aymara, 383
Azathioprine, 145–147
Azole antifungal drug overdose, 92, 92t
Azole antifungals, 85

B
Bacterial cell wall synthesis inhibition, 86–88
BAL. *See* British anti-Lewisite (BAL)
Barbiturates, 38–39, 209, 379
  analytical methods and clinical-management implications, 213–215
  drug class, 210f
  mechanism of action, 210–211
  pharmacokinetics and toxicokinetics, 211–212
  treatment, 212–213
Bark Scorpion, 441
Baseline effect, 15
Bath salts, 230, 261
BCR-ABL protein kinase, 144
BE. *See* Benzoylecgonine (BE)
Benzodiazepines, 6, 38–39, 209–210, 219, 300, 311–312, 379, 456t
  analytical methods and clinical-management implications, 213–215
  drug class, 210f
  immunoassays, 214
  mechanism of action, 210–211
  metabolism, 212f
  pharmacokinetics and toxicokinetics, 211–212
  treatment, 212–213
Benzoylecgonine (BE), 228, 232, 251, 383, 493
β-blockers, 197
β-glucuronidase, 220–221
β-mercaptopyruvate-cyanide sulfurtransferase, 475
Bifunctional alkylating agents, 142
Bilirubin, 75–76
Bioavailability, 11
Biological fluids, 7
Biomarkers of ethanol use, 41–42
Bipolar disorder, 243
Bites, 437, 441
Blood ethanol analysis, 39–40
Blood gas analysis, 470–471
Blood urea nitrogen (BUN), 402–403, 439

Blood/breath ethanol ratio, 40
Blood−brain barrier, 451
Body mass index (BMI), 433
Box jellyfish (*Chironex fleckeri*), 445
Bradykinins, 70
Breath ethanol, 40−41
Breathalyzer, 39
British anti-Lewisite (BAL), 415−416
Brivaracetam, 121
Broad-spectrum drugs, 83
Bromocriptine, 39
Bromo-DragonFLY, 297
Brown rice, 414
Bufalin, 364
BUN. *See* Blood urea nitrogen (BUN)
Buprenorphine, 341, 456t, 457
Buprenorphine/norbuprenorphine, 20
Bupropion, 482−483
  toxicity
    case description, 481
    discussion, 481−483
    laboratory values, 482t
Busulfan, 142
Butalbital, 209
2-Butanone, 40
Butterbur, common (*Petasites hybridus*), 362
Butyronitrile (C_4H_7N), 474t
Butyrylfentanyl, 340

C

$C_2H_4F_2$. *See* Difluoroethane (DFE)
Caffeine
  blood results on admission, 240t
  case description, 239
  concentrations, 240f
  discussion and follow-up, 239−241
Calcineurin inhibitor (CNI), 317−319
Calcium channel blockers, 197
Calcium cyanide ($Ca(CN)_2$), 474t
Calcium oxalate crystals, 47−48, 52f
Calcium supplements, 359−360
Camptothecins, 142
Cancer treatment, 196−197
Cannabidiol (CBD), 127, 169−170, 183−184, 187
  case description, 183
Cannabinoids (CB), 169, 184, 187−188, 379
  analytical methods for, 172−173
  compounds, 170
  pharmacokinetics and toxicokinetics, 170−172
  receptor, 170
Cannabis, 295, 382
Cannabis sativa, 183
Capillary electrophoresis (CE), 477
Carbamazepine, 121−122, 125, 127, 140f
Carbamazepine-10,11-epoxide (CBZE), 122
Carbohydrate deficient transferrin (CDT), 41
Carbon monoxide (CO), 387−390, 409−410
  poisoning, 410
    case history, 409
Carboxyhemoglobin (COHb), 387−388, 410
Carboxylesterase, 265

Carcinogenesis of arsenic, 413
Cardiac arrest, 205
Cardiac drugs
  cardiotoxicity
    associated with drugs of abuse, 199−200
    associated with noncancer therapeutic drugs, 199
  chemotherapy associated myocardial toxicity, 197−199
  drug-induced cardiotoxicity, 196−197
  drug−drug interaction, 197, 198t
  endocrine therapy, 201
  lifestyle and genetic triggers, 196
  therapeutic drug monitoring of cardiac drugs, 201
Cardiac monitoring, 198−199
Cardiac toxicity, 85
Cardiomyopathy, 219
Cardio-oncology, 196−197
Cardiopulmonary resuscitation (CPR), 203, 284, 446
Cardiotoxic herbs, 363−365
Cardiotoxicity, 195−197
  associated with drugs of abuse, 199−200
  associated with noncancer therapeutic drugs, 199
Cardiovascular diseases (CVDs), 195
Cardiovascular risk, 195
Cardiovascular system (CVS), 195, 388
  arsenic impact in, 422
Carfentanil, 357
Carisoprodol, 224
Catha edulis, 230
Cathinones, 230−231
  pharmacokinetics, 231
Cathinones, 273−274
CB. *See* Cannabinoids (CB)
CBC. *See* Complete blood counts (CBC)
CBD. *See* Cannabidiol (CBD)
CBZE. *See* Carbamazepine-10,11-epoxide (CBZE)
CDT. *See* Carbohydrate deficient transferrin (CDT)
CE. *See* Capillary electrophoresis (CE)
CEDIA. *See* Cloned enzyme donor immunoassay (CEDIA)
Celecoxib, 70
Celiac disease, 267
Cell wall disruption, 89
Centers for Disease Control and Prevention (CDC), 273
Central nervous system (CNS), 28, 38−39, 75, 85−86, 209, 244, 387−388, 466
  arsenic impact in, 422
  depressants, 38−39
  depression, 81
  stimulants, 227
    amphetamines, 228−230
    analytical considerations, 232−234
    cathinones, 230−231
    cocaine, 227−228
    designer drugs, 232−233
    oral fluid testing, 233−234

  therapeutic and toxic concentrations for CNS depressants, 213t
Central Sensitization Pain Syndrome, 434
Centrifugal apheresis technique, 203−204
Centruroides, 443−444
Cerebral spinal fluid (CSF), 151
Cerebral vasculitis, 251−252
Cetuximab, 144−145
CFCs. *See* Chlorofluorocarbons (CFCs)
Chan Su (Chinese medicine), 364
Chaparral, 360, 362
"Chasing the dragon", 339
Checkpoint inhibitors, 198
Chelating agents, 415
Chemiluminescent microparticle immunoassay (CMIA), 319, 321
Chemotherapy associated myocardial toxicity, 197−199
Chest pain, 199
Chewing tobacco, 261−263
CHF. *See* Congestive heart failure (CHF)
Childhood cancer, 141
Children's Oncology Group (COG), 161
Chironex fleckeri. *See* Box jellyfish (*Chironex fleckeri*)
Chlordiazepoxide, 209
Chlorine, 390−391
  acute toxicity, 390
  chronic toxicity, 390
  clinical management, 391
  diagnosis and investigation, 390−391
  exposure routes and pathways, 390
  mechanism of action, 390
  postmortem appearances, 391
Chlorofluorocarbons (CFCs), 406
Cholinergic toxidrome, 28
Chromatographic analysis, specimen preparation for, 23−24
Chromatographic methods, 20, 47
Chronic cyanide toxicity, 392
Chronic ethanol, 38−39
Chronic ketamine toxicity, 305−306
Chronic kidney disease, 267
Chronic low-dose aspirin, 69−70
Chronic myelogenous leukemia (CML), 144
Chronic pain, 351−352
Chronic toxicity, 388
Cigarettes, 261−263, 475
Cinchona bark, 68
Citrobacter freundii, 87
CK. *See* Creatine kinase (CK)
Classical anticonvulsants. *See* First-generation drugs
Classical benzodiazepines, 211
Classical hallucinogens. *See* Psychedelics
Claviceps purpurea, 298
Clearance (Cl), 8, 11
Clinical depression, 99
Clinical laboratories, 19−20
Clinical management implications, 93−95
Clinical toxicology, 3
  analytical workflows in toxicology laboratory, 4−6
  scope of, 4

Clobazam, 121, 123, 127, 188
  intoxication
    case description, 187
    laboratory values for various analytes in our patient, 189t
    metabolism of clobazam, 188f
Clomipramine, 369
Clonazepam, 118f, 121, 211–212, 215, 369
Clonazolam, 292
Clonazolam abuse, 292f
  case descriptions, 291
Cloned enzyme donor immunoassay (CEDIA), 4–5, 94–95, 320
  CEDIA Plus, 321
  method, 375
Clozapine, 114, 115t
CMIA. See Chemiluminescent microparticle immunoassay (CMIA)
CML. See Chronic myelogenous leukemia (CML)
CMP. See Comprehensive metabolic panel (CMP)
CN. See Cyanide (CN)
CNI. See Calcineurin inhibitor (CNI)
CNS. See Central nervous system (CNS)
Coagulation studies, 439
Cobalt, 476
Coca tea consumption, 383–385
  case description, 383
  pathways for cocaine metabolism, 384f
Cocaethylene, 228, 254, 265–266
Cocaine, 196, 199, 227–228, 251–254, 265–266, 382–384, 457, 495
  case description, 493
  discussion, 493–495
  lab test results, 494t
  metabolite, 379, 456t
  pathways for cocaine metabolism, 384f
  pharmacokinetics, 227–228
  rerun test results, 494t
Cocaine-induced vasculitis
  case description, 251
  discussion, 251–252
Codeine, 343
COG. See Children's Oncology Group (COG)
COHb. See Carboxyhemoglobin (COHb)
Colchicine, 196
College of American Pathologists (CAP), 127
Colonoscopy, 433
Colorimetric assays, 72
Coltsfoot (*Tussilago farfara*), 362
Combination immunosuppressive therapy, 317
Comfrey (*Symphytum officinale*), 360, 362
Commuted tomography scan (CT scan), 243–244
Competitive inhibitors, 12–14
Complete blood counts (CBC), 91, 172, 439
Compliance with pharmacotherapy, 5–6
Comprehensive metabolic panel (CMP), 151
Comprehensive urine toxicology profile, 498t
Concentration–effect curve, 14–15
Confirmation methods, 173, 292
Confirmatory testing, 232
Congenital mitral stenosis, 219

Congestion of kidneys and liver, 261
Congestive heart failure (CHF), 265
Conocybe, 297, 379–380
Conservative management, 243–245
Continuous renal replacement therapy (CRRT), 80–82, 125–126, 161–162, 427
Continuous veno-venous hemodiafiltration (CVVHDF), 239
Convallatoxin, 363
Copelandia, 297
Copper cyanide (CuCN), 474t
Cord tissue, 215
Coronary artery disease, 219
Corticosteroids, 317
Cotinine, 261
COX. See Cyclooxygenase (COX)
COX-2. See Cyclooxygenase-2 (COX-2)
CPK. See Creatine phosphokinase (CPK)
CPR. See Cardiopulmonary resuscitation (CPR)
Crack cocaine, 227
Creatine kinase (CK), 253, 281, 311, 439
Creatine phosphokinase (CPK), 243, 388
Creatinine, 439
  normalization, 251
Cross-reactivity profiles and clinical management of MTX
  analytical precision, 167t
  case description, 165–166
  methotrexate concentrations, 166t
CRRT. See Continuous renal replacement therapy (CRRT)
Crustaceans, 440
CsA. See Cyclosporin A (CsA)
CSF. See Cerebral spinal fluid (CSF)
CT scan. See Commuted tomography scan (CT scan)
CVDs. See Cardiovascular diseases (CVDs)
CVVHDF. See Continuous veno-venous hemodiafiltration (CVVHDF)
Cyanide (CN), 391–392, 473–474
  colorimetric test strips, 477
  exposure, 475
  gas, 474–475
  salts, 391, 473–474
  species, 474t
  spectrophotometric assays, 477
  toxicity, 473–475
    analytical methods and clinical management implications, 477–478
    case description, 473
    laboratory evaluation, 477–478
    mechanism of action, 475
    pharmacokinetics, 475
    signs and symptoms, 476–477
    toxicokinetics, 475
    treatment, 476–477
Cyanocobalamin, 476
Cyanogen ($(CN)_2$), 473–474, 474t
Cyanogen bromide (CBrN), 474t
Cyanogen chloride (CNCl), 473–474, 474t
Cyanogen halides, 473–474
Cyanomethemoglobin, 476
Cyclooxygenase (COX), 70, 80–81
Cyclooxygenase-2 (COX-2), 68

inhibitors, 70–71
Cyclophosphamide, 142
Cyclopropylfentanyl, 340
Cyclosporin A (CsA), 317–321
  analytical issues, 320–321
  immunoassays, 319
  LC–MS/MS methods for, 319–320
Cyclosporin D, 320
CYP isoforms. See Cytochrome P450 isoforms (CYP isoforms)
CYP1A2. See Cytochrome P450 superfamily (CYP1A2)
CYP2D6. See Cytochrome P450 2D6 (CYP2D6)
CYP2E1, 38
CYP3A4. See Cytochrome P450 3A4 (CYP3A4)
CYP3A5, 490
CYP450 2E1, 75
Cysteine, 72
Cytarabine, 142
Cytochrome P450 2D6 (CYP2D6), 145, 352, 374, 485–486, 490
Cytochrome P450 3A4 (CYP3A4), 211–212, 352, 374, 485–486
Cytochrome P450 isoforms (CYP isoforms), 199, 211
Cytochrome P450 superfamily (CYP1A2), 466
Cytosine arabinoside (Ara-C), 142
Cytotoxic agents, 142–144

D

d-Ala, N-Me-Phe, Gly-ol-enkephalin (DAMGO-enkephalin), 373–374
DART-MS. See Direct analysis in real time-mass spectrometry (DART-MS)
Datura, 366
Datura stramonium. See Jimson weed (*Datura stramonium*)
DAU-POC, 498t
DAWN. See Drug Abuse Warning Network (DAWN)
DEA. See Drug Enforcement Administration (DEA)
Delta OPioid receptor (DOP), 333–334
Delta receptor (δ receptor), 333
Demethylxyyangonin, 370
L-Deprenyl. See Selegiline
Depression, 243
Dermacentor andersoni, 441
Dermacentor variabilis, 441
Dermal absorption, 334–335
Designer drugs, 214, 232–233, 269, 274–275, 287–289
  cathinones, 273–274
  designer opiates, 272–273
  fentanyl and three challenging to distinguish analogs, 274f
  isometric pairs of challenging cathinone derivatives, 274f
  mescaline and synthetic analogs, 269f
  PCP and dissociative analogs, 275f

potent hallucinogens 25I-NBOMe and Bromo-Dragonfly, 275f
synthetic cannabinoids, 271–272
L-Desmethylselegiline, 257
Deuterium-labeled (d12), 320
Dexedrine, 258
Dextroamphetamine, 243
case description, 243–244
Dextromethorphan (DXM), 299, 497
Dextrorphan, 499f
DFC drug. See Drug-facilitated crime drug (DFC drug)
DFE. See Difluoroethane (DFE)
DHFR. See Dihydrofolate reductase (DHFR)
Diabetic ketoacidosis, 56
falsely elevated EG results in patient with, 55–56
Diacetylmorphine. See Heroin
Diagnostic and Statistical Manual of Mental Disorders (DSM-5), 99
2,4-Diamino-N (10)-methylpteroic acid (DAMPA), 152
Diamorphine, 339–340
Diaphoresis, 442
Diazepam, 121, 209, 211, 466
DIC. See Disseminated intravascular coagulation (DIC)
DiCBZ. See 10,11-Dihydro-10-hydroxycarbamazepine (DiCBZ)
3,4-Dichloro-N-[2-(dimethylamino) cyclohexyl]-N-methylbenzamide). See U-47700
Dicobalt edetate, 476–477
Dietary intake, 433
Dietary Supplement Health and Education Act (1994), 359
Diethylene glycol, 46
Difluoroethane (DFE), 402–403, 406
case histories, 401–402
postmortem toxicology results, 402t
toxicity
case description, 405–407
death investigation, 405–406
external examination, 405
postmortem examination, 406
toxicology, 406–407
Digibind, 363
Digitalis purpurea. See Foxglove (*Digitalis purpurea*)
Digoxin, 197, 223
10,11-Dihydro-10-hydroxycarbamazepine (DiCBZ), 139
Dihydrofolate reductase (DHFR), 152–153, 166
Dihydrokavain, 370
1-Dihydromephedrone, 231
Dihydromethysticin, 370
Dihydropyrimidine dehydrogenase (DPD), 142
Dimension Vista platform, 497–498
Dimercaprol, 415
Dimercaptopropane sulfonic acid, 415
2,5-Dimethoxy-4-methylamphetamine (DOM), 297
Dimethylaminophenol (4-DMAP), 477

Dimethylarsenic acid (DMA), 418
Diphenhydramine (DPH), 459–462, 462t
case history, 459, 460f, 461f
Diphenhydramine peak, 287
Diplopterys cabrerana, 297
Direct analysis in real time-mass spectrometry (DART-MS), 375
Direct injection analysis, 40
Disialotransferrin, 41
Disseminated intravascular coagulation (DIC), 284
Dissociative drugs, 298–299
analysis, 300–301
Dissociatives, 295
Disulfiram, 39, 197
4-DMAP. See Dimethylaminophenol (4-DMAP)
DMEs. See Drug metabolizing enzymes (DMEs)
DMSA. See Meso-2,3-dimercaptosuccinic acid (DMSA)
DMT. See N-dimethyltryptamine (DMT)
DOM. See 2,5-Dimethoxy-4-methylamphetamine (DOM)
Domestic fires, 474–475
Donepezil, 485–486
Donepezil ingestion
case description, 485
discussion, 485–486
Doxorubicin (DOX), 196
DPD. See Dihydropyrimidine dehydrogenase (DPD)
DPH. See Diphenhydramine (DPH)
DPT. See N,N-Dipropyltryptamine (DPT)
DRE. See Drug recognition expert (DRE)
DRI (Microgenics/Thermo Fisher), 491
Drivers under influence of 5F-ADB
case description, 177–178
cases with field sobriety test, 179t
SCs, 178–181, 178f
Driving under the influence (DUI), 177
Driving under the influence of drugs (DUID), 210
Dronabinol, 170
Drug abuse, 267
Drug Abuse Warning Network (DAWN), 227
Drug Enforcement Administration (DEA), 375
Drug metabolizing enzymes (DMEs), 11
Drug recognition expert (DRE), 178–180
Drug screening, 261, 379
Drug-facilitated crime drug (DFC drug), 210
Drug-induced cardiotoxicity, 196–197
Drug-induced liver injury, 374
Drug-screening, 379
Drug–drug interaction, 197, 198t, 353
Drugs of abuse screen, 498t
DSM-5. See Diagnostic and Statistical Manual of Mental Disorders (DSM-5)
DUI. See Driving under the influence (DUI)
DUID. See Driving under the influence of drugs (DUID)
Duragesic, 347–348
DXM. See Dextromethorphan (DXM)
Dynorphins, 333

Dysrhythmias, 425–426
Dystonic reactions, 139–140

E

Eastern groundsel (*Senecio vernalis*), 362
Ecgonine methyl ester (EME), 383
E-cigarettes, 261–263, 263f
ECMO. See Extracorporeal membrane oxygenation (ECMO)
Ecstasy, 247
ECT. See Extracorporeal therapy (ECT)
ED. See Emergency department (ED)
EDDP. See 2-Ethylidene-1,5-dimethyl-3,3-diphenylpyrrolidine (EDDP)
e-deoxy-4-amino-N10-methylpeteroic acid (DAMPA), 161
EEG. See Electroencephalogram (EEG)
EG. See Ethylene glycol (EG)
EGCG. See Epigallocatechin 3-gallate (EGCG)
EGFR. See Epidermal growth factor receptor (EGFR)
eGFR. See Estimated glomerular filtration rate (eGFR)
EI ionization. See Electron-impact ionization (EI ionization)
EIAs. See Enzyme immunoassays (EIAs)
Elapidae, 437
Electrocardiogram, 29, 305
Electrocardiography, 239
Electrochemical fuel-cell sensors, 41
Electroencephalogram (EEG), 131
Electrolytes, 439
Electron-impact ionization (EI ionization), 21
Elemental mercury (Hg^0), 414–415, 430
inhalation, 416
Elemental toxicity, 413
ELISA. See Enzyme-linked immunosorbent assay (ELISA)
EMCDDA. See European Monitoring Center for Drugs and Drug Addiction (EMCDDA)
EME. See Ecgonine methyl ester (EME)
Emergency department (ED), 99, 101t, 227, 291, 409, 421, 481
Emergency medical services (EMS), 203, 291
Emergency room (ER), 187
EMIT. See Enzyme-multiplied immunoassay technique (EMIT)
Emit II Plus Urine Drug Screen Assay, 243–244
EMS. See Emergency medical services (EMS)
Endocrine therapy, 201
Endomyocardial biopsy, 197
Endorphins, 333
Enkephalins, 333
Entactogens, 295
Enterobacter aerogenes, 47
Enterobacter spp., 87
Envenomations, 437
from snakes, 438
local, systemic, hematologic, and neurologic signs and symptoms, 439t
treatment, 440

Envenomations (*Continued*)
  by ticks, 443
Environmental protection agency (EPA), 417—418
Enzymatic assay, 56—57, 323
Enzymatic methods, 39—40, 47
Enzyme immunoassays (EIAs), 117, 135, 223, 261, 277, 287, 347, 379, 401, 459
Enzyme-linked immunosorbent assay (ELISA), 19—20, 321, 375, 439
Enzyme-multiplied immunoassay technique (EMIT), 4—5, 94—95, 148, 154, 267, 320, 465
Ephedra, 364
Ephedra equisetina, 364
Ephedra sinica, 364
Ephedrine, 247, 364
  urine drug screen results after spiking with known, 248t
Epidermal growth factor receptor (EGFR), 144
Epigallocatechin 3-gallate (EGCG), 362—363
Epinephrine infusion, 79, 81
ER. See Emergency room (ER)
ER capsules. See Extended-release capsules (ER capsules)
ErbB, 144—145
Ergot, 298
Ergotism, 298
Erythroxylum coca, 265
Escherichia coli, 23—24, 158—159, 173
Escitalopram, 79, 369
Eslicarbazepine acetate, 121—122
Esmolol, 241
Esomeprazole, 223
Estimated glomerular filtration rate (eGFR), 151, 253
EtG. See Ethyl glucuronide (EtG)
Ethanol, 28, 37—39, 254—255, 379, 382
  analytical considerations, 43
  biomarkers of ethanol use, 41—42, 42t
  blood ethanol analysis, 39—40
  breath ethanol, 40—41
  interpretation issues, 42
  metabolism, 38f
  oral fluid, 41
  pharmacology, 38—39
    pharmacodynamics, 38—39
    pharmacokinetics, 38
    treatment of ethanol abuse, 39
  postanalytical considerations, 43
  preanalytical considerations, 42—43
  specimen collection, 39
  specimen types, 39, 42t
  vitreous fluid, 41
Ethosuximide, 121—122
Ethylenediaminetetraacetic acid (EDTA), 415
Ethyl glucuronide (EtG), 41
Ethyl sulfate (EtS), 41
Ethylene glycol (EG), 37, 45—46, 51, 63—64
  chromatogram for, 47f, 53f
  chromatograph, 64f
  falsely elevated EG results in patient with diabetic ketoacidosis, 55—56
  initial and posthemodialysis tests results for patient, 55t
  ingestion, 51
  pharmacology, 45—46
  poisoning, 51, 64—65
  toxicity, 51—52
  treatment and management of ethylene glycol toxicity, 46
2-Ethylidene-1,5-dimethyl-3,3-diphenylpyrrolidine (EDDP), 117
Etizolam, 210
EtS. See Ethyl sulfate (EtS)
Euphoria, 191—192
European Monitoring Center for Drugs and Drug Addiction (EMCDDA), 231
Everolimus, 317, 325—326
  immunoassays, 326
  LC—MS/MS methods for, 326
Extended-release capsules (ER capsules), 203
Extracorporeal membrane oxygenation (ECMO), 31
Extracorporeal therapy (ECT), 426—427

F

FAEEs. See Fatty acid ethyl esters (FAEEs)
False-positive levorphanol
  case description, 497
  dextrorphan and levorphanol, 499f
  discussion and follow-up, 497—499
  drug tests, 498t
FAS. See Fetal alcohol syndrome (FAS)
FASD. See Fetal alcohol spectrum disorders (FASD)
Fatal hepatitis, 362
Fatty acid ethyl esters (FAEEs), 41
FDA. See Food and Drug Administration (FDA)
Federal Analogue Act, 270
Felbamate, 121, 123
Fentanyl, 20, 333—335, 340, 347—348, 456t
Fentanyl-laced pills, death involving, 349f
  case history, 347
  discussion, 347—349
Fentora, 347—348
Fetal alcohol spectrum disorders (FASD), 38—39
Fetal alcohol syndrome (FAS), 38—39
FIDs. See Flame ionization detectors (FIDs)
First-generation drugs, 121
"First-pass" hepatic metabolism, 204—205
Flame ionization detectors (FIDs), 37, 337—338
Flashback, 296
Flaxseed oil, 359—360
Flubromazepam, 117—119, 118f
  case description, 117
  cross-reactivity as change in absorbance, 118t
Fluconazole, 91
Fluid resuscitation, 426
Flumazenil, 127, 212—213
Flunitrazepam, 210
Fluorescence polarization immunoassay (FPIA)
Fluorescent Polarization ImmunoAssay (FPIA), 4—5, 94—95, 148, 154, 320
Fluoropyrimidines, 142
Fluoroquinolone antibiotics, 85
5-Fluorouracil (5-FU), 142
Foam cone, 334
Fomepizole, 44—45, 52, 60—61
Food and Drug Administration (FDA), 122, 153, 359
Forensic laboratories, 19—20
Forensic toxicology, 3
  analytical workflows in toxicology laboratory, 4—6
  scope of, 4
Formaldehyde, 44
Formate, 60
Formic acid, 44, 60
  methanol and, 44
Formicidae, 444
Foxglove (*Digitalis purpurea*), 363
FPIA. See Fluorescent Polarization ImmunoAssay (FPIA)
Fractionation. See Speciation
Fragmentation energy, 21
Free base, 227
Free cyanide, 477—478
Free phenytoin measurements
  case description, 131
  laboratory results, 132t
Free test, 337, 339
5-FU. See 5-Fluorouracil (5-FU)
FUB-AMB, 281
Fungal cell wall synthesis inhibition, 89
Fused silica columns, 20—21

G

G protein—coupled receptors (GPCR), 333
G6PDH. See Glucose-6-phosphate dehydrogenase (G6PDH)
GA. See Gestational age (GA)
GABA. See Gamma-aminobutyric acid (GABA)
Gabapentin, 121—123, 125, 135—137
  fatality involving massive overdose
    case history, 135
    postmortem toxicology, 136t
Gamma-aminobutyric acid (GABA), 122, 135—136, 136f, 442
  $GABA_A$ receptors, 211
  subtype A receptors, 211
  uptake transporter, 122
γ-glutamyltransferase (GGT), 41
Gamma-interferon, 266
Gas chromatography (GC), 4—5, 20—21, 37, 44, 56, 63—64, 105—107, 172, 337, 459, 478
Gas chromatography with flame ionization detector (GC—FID), 40, 277, 347, 401
Gas chromatography with nitrogen—phosphorus detection (GC—NPD detection), 21
Gas chromatography—mass spectrometry (GC—MS), 19, 21—22, 47, 109, 117,

135, 139, 184, 214, 223, 232, 261, 267, 277, 287, 300–301, 338, 379, 383, 401, 459, 465, 478, 491
Gastroenteric symptoms, 390
Gastroesophageal reflux disease, 223
Gastrointestinal (GI) discomfort, 68
Gastrointestinal disturbances
  case report, 433
  discussion, 433–434
  laboratory data, 434t
Gastrointestinal system, arsenic impact in, 422
Gastrointestinal tract (GI tract), 434
GC. See Gas chromatography (GC)
GC–FID. See Gas chromatography with flame ionization detector (GC–FID)
GC–MS. See Gas chromatography–mass spectrometry (GC–MS)
GC–NPD detection. See Gas chromatography with nitrogen–phosphorus detection (GC–NPD detection)
General screening methods, 19–24
  chromatographic methods, 20
  GC, 20–21
  GC–MS, 21–22
  GC–NPD detection, 21
  immunoassay screening, 19–20
  LC–MS, 22–23
  specimen preparation for chromatographic analysis, 23–24
Generalized seizures, 121
Generally recognized as safe (GRAS), 359–360
Genetic polymorphisms, 145, 353
  case description, 113–116
  clinical history, 113
  2D6 Variants, 116t
  decedent found unresponsive, 114f
  external examination, 113
  genetic testing, 115–116
  postmortem examination, 113, 114t
  toxicology, 114–115
  tramadol, 115
Genetic testing, 115–116, 116t
Gentamicin, 11, 93
Germander, 360
Gestational age (GA), 59
GFR. See Glomerular filtration rate (GFR)
GGT. See γ-glutamyltransferase (GGT)
GI tract. See Gastrointestinal tract (GI tract)
Glasgow Coma Scale (GCS), 410
Glitazones, 199
Glomerular filtration rate (GFR), 425
Glucarpidase, 147, 158–159
Glucarpidase cleaves, 153
Glucarpidase rescue MTX concentration
  case report, 161–163
  comparison of blood, 162t
Glucocorticoids, 196
Glucose-6-phosphate dehydrogenase (G6PDH), 162–163
Glucuronide (MPAG), 323
  conjugates, 211–212
  metabolites, 214
Glutamate, 298–299

Glutathione, 72
Glycolic acid, 64
Glycols, 37, 46–48
  analysis and related analytes, 47–48
Glyoxylic acid, 64
Gold cyanide (AuCN), 474t
GPCR. See G protein–coupled receptors (GPCR)
Gram stain, 83–84
Gram-negative bacteria, 83
Gram-positive bacteria, 83
GRAS. See Generally recognized as safe (GRAS)
Guanosine diphosphate (GDP), 333
Guanosine triphosphate (GTP), 333
Gymnopilus mushroom, 379–380

H

Hair analysis, 128
Hair testing, 215
Half-life, 7
  plasma concentration *vs.* time curves, 8f
Hallucinogenic designer drugs, 274–275
Hallucinogens, 295, 314
  analysis of psychedelics, dissociatives, and other hallucinogens, 300–301
  dissociative drugs, 298–299
  psychedelics, 295–298
  toxicokinetics and clinical management of overdose, 299–300
Halogen, 210
Halogenated hydrocarbons, 406
Haloperidol, 295–296
HCN. See Hydrogen cyanide (HCN)
HD. See Hemodialysis (HD)
HDMTX. See High-dose methotrexate (HDMTX)
Headspace analysis, 40
Health-care practitioners, 282
Heart blood alprazolam level, 277
Heart disease, 195, 223
Heavy metals toxicity
  cardiovascular system, 422
  case description, 421
  central nervous system, 422
  discussion and follow-up, 421–422
  findings from therapeutic exposures, 422–423
    background on clinical assessment of arsenic, 422–423
    gastrointestinal system, 422
    hematopoietic system, 422
    laboratory report
      for total arsenic, 421
      for total inorganic arsenic, 421
    miscellaneous symptoms, 422
    specific interpretation, 423
HEIA (Immunalysis Corporation), 491
Helix pomatia. See Snail (*Helix pomatia*)
Hematologic system, 438
Hematopoietic system, arsenic impact in, 422
Hemodialysis (HD), 59, 81–82, 241, 425

laboratory investigations and sessions of, 426t
  after methanol inhalation, 61
Hemolysis, 442–443
Henry's law, 40
Hepatitis C, 243
Hepatotoxicity, 86, 161–162
HER2. See Human epidermal growth factor receptor 2 (HER2)
Herbal supplements, popularity and safety of, 359–360, 360t
Herbs and liver toxicity, 360–363
Herbs with nephrotoxicity, 365
Heroin, 254, 333, 335–337, 339–340, 349f
  case description, 343
  discussion, 343–345
Heterogeneous immunoassays, 19–20
HGN. See Horizontal gaze nystagmus (HGN)
High-dose methotrexate (HDMTX), 157, 161
High-performance liquid chromatography. See High-pressure liquid chromatography (HPLC)
High-pressure liquid chromatography (HPLC), 4–5, 95, 105–107, 139, 148, 154, 272
High-resolution mass spectrometry (HRMS), 22–23, 172, 214, 337–338
Histological stains, 83
HMG-CoA reductase inhibitors, 199
Homogenous immunoassay, 72
Horizontal gaze nystagmus (HGN), 177
Hottentotta, 443
HPLC. See High-pressure liquid chromatography (HPLC)
HPRT. See Hypoxanthine-guanine phosphoribosyltransferase (HPRT)
HRMS. See High-resolution mass spectrometry (HRMS)
5-HT. See 5-Hydroxytryptamine (5-HT)
Human epidermal growth factor receptor 2 (HER2), 197
Hydrochloride salt, 227
Hydrochlorothiazide, 485
Hydrocodone, 356
Hydrogen cyanide (HCN), 391–392, 473–475, 474t
  acute toxicity, 391–392
  chronic toxicity, 392
  clinical management, 392
  diagnosis and investigation, 392
  exposure routes and pathways, 391
  mechanism of action, 391
  postmortem appearances, 392
  toxicokinetics, 391
Hydrogen sulfide (H_2S), 393–394
  acute toxicity, 393
  chronic toxicity, 393
  clinical management, 394
  diagnosis and investigation, 393–394
  exposure routes and pathways, 393
  postmortem appearances, 394
  toxicokinetics and mechanism of action, 393
Hydrolysis, 221
Hydromorphone, 379
10-Hydroxy-10,11-dihydrocarbamazepine, 122

5-Hydroxy-2-pentylpiperidine, 366
4-Hydroxy-3-methoxymethcathinone, 231
Hydroxycobalamin, 476–477
17-Hydroxymitragynine, 375
7-Hydroxymitragynine, 373–375
5-Hydroxypropafenone, 204–205
4-Hydroxytamoxifen, 145
5-Hydroxytryptamine (5-HT), 295–296
    5-HT$_{2a}$ receptors, 295–296
    5-HT$_{2c}$ receptors, 295–296
Hydroxyzine, 499
Hymenoptera, 444
Hyoscyamine, 366
Hyperalgesia, 334
Hyperammonemic states, 397
Hyperbaric oxygen (HBO). *See* Normobaric oxygen (NBO)
Hyperglycemia, 199
Hyperlipidemia, 199
Hypertension, 223, 265
Hypnotic toxidrome, 28
Hypocalcemia, 51–52
Hypoglycemia, 27
Hypokalemia, 241
Hyponatremia, 139–140
Hypoperfusion, 81
Hypoxanthine-guanine phosphoribosyltransferase (HPRT), 146–147
25I-NBOMe, 297
IA injection. *See* Intraarterial injection (IA injection)
Ibuprofen, 67
Ibuprofen ingestion
    case description, 79–80
    laboratory values at indicated time after patient's initial presentation, 80t

I

iCa. *See* Ionized calcium (iCa)
ICP. *See* Intracranial pressure (ICP)
ICP-MS. *See* Inductively coupled plasma mass spectrometry (ICP-MS)
IgG. *See* Immunoglobulin G (IgG)
IL-2. *See* Interleukin-2 (IL-2)
ILE. *See* Intravenous lipid emulsion (ILE)
Illicit drug abuse, during pregnancy, 191–192
Imatinib, 144
Immediate release tablets (IR tablets), 204–205
Immunoassay, 219, 220t
    CsA, 319
    MPA, 323
    screening, 19–20
    sirolimus, 324
    tacrolimus, 321
    techniques, 337
Immunochromatographic techniques, 4–5, 375
Immunoglobulin G (IgG), 440
Immunosuppressants, therapeutic drug monitoring of, 317
    CNI, 318–319
    creatinine levels, 318f
    CsA, 319–321
    drugs used in solid organ transplantation, 318t
    immunoassays, 319t
    MPA, 322–324
    mTOR inhibitors, 324–326
    tacrolimus, 321–322
Immunosuppressive drugs, 317
IMPDH. *See* Inosine monophosphate dehydrogenase (IMPDH)
Indomethacin, 70
Inductively coupled plasma mass spectrometry (ICP-MS), 416, 418
Infrared spectrometry (IR spectrometry), 40–41
Inhalants, 406
Inorganic arsenic, 422–423
    species, 434
Inorganic mercury, 414–415, 430
Inosine monophosphate dehydrogenase (IMPDH), 322
Insectivores, 446
Insects, 437, 440, 444–445
    analytical methods and clinical management implications, 444
    mechanism of action and toxicokinetics, 444
    treatment, 445
Insomnia, 360
Intentional ingestion, 205
Interferences, 56
Interleukin-2 (IL-2), 266, 318
Intermediate-acting barbiturates, 209
Intraarterial injection (IA injection), 243, 245
Intracranial pressure (ICP), 308
Intramuscular (IM) inoculation, 438
Intravascular coagulation (IV coagulation), 438
Intravenous drug abuse (IV drug abuse), 243
Intravenous lipid emulsion (ILE), 30
    therapy, 205
Intrinsic activity, 14
Ion trap detectors, 22
Ionized calcium (iCa), 206
Ionsys, 347–348
IR spectrometry. *See* Infrared spectrometry (IR spectrometry)
IR tablets. *See* Immediate release tablets (IR tablets)
Isopropanol, 37, 45
Isotopic-labeled drug analogs, 23
Itraconazole, 91
IV coagulation. *See* Intravascular coagulation (IV coagulation)
IV drug abuse. *See* Intravenous drug abuse (IV drug abuse)
Ixodes holocyclus, 441

J

Jaundice, 442–443
Jeavons syndrome, 187
Jellyfish, 445
Jimson weed (*Datura stramonium*), 366

K

Kappa OPioid receptor (KOP), 333–334
Kappa receptor (κ receptor), 333
Kava, 360
Kava tea consumption, 369
    case presentation, 369
    discussion, 369–371
    laboratory results, 370t
Kavain, 370
    toxicity, 370t
Kavalactones, 360–361, 369–370
KED. *See* Kinetic energy discrimination (KED)
Ketamine, 295, 299
    acute ketamine toxicity, 305
    as anesthetic and use in pain management, 307–308
    chronic ketamine toxicity, 305–306
    mechanism of action, 307
    pediatric ketamine toxicity, 306
    pharmacodynamic drug interactions, 307
    pharmacokinetics and pharmacodynamics, 307
    pharmacologic management of complex regional pain syndrome, 308
    pharmacology and use in outpatient anesthesia, 308
K-hole, 299
KIMS. *See* Kinetic Interaction of Microparticles in Solution (KIMS)
Kinetic energy discrimination (KED), 417
Kinetic Interaction of Microparticles in Solution (KIMS), 4–5
Kombucha tea, 362
Kommon, 306
Kratom (*Mitragyna speciosa*), 373–374
    analysis in toxicology laboratory, 375
    case presentation, 373
    legality, 375
    pharmacology, 374
    toxicity, 374
Krebs cycle, 64

L

Laboratory testing, 172
Laboratory-based screening methods, 270
Laboratory-developed test, 498–499
Lacosamide, 121, 123, 126
Lactate. *See* Lactic acid
Lactate dehydrogenase (LDH), 39–40
Lactic acid, 57, 388
Lactic acidosis, 241
Lactose, 227
Laetrile, 474
Lamotrigine, 121, 123, 125
Latrodectus sp., 441–443
Lazanda, 347–348
LC. *See* Liquid chromatography (LC); Locus coeruleus (LC)
LC-TOF/MS. *See* Liquid chromatography time of flight mass spectrometry (LC-TOF/MS)
LC–MS. *See* Liquid chromatography–mass spectrometry (LC–MS)

LC–MS/MS. *See* Liquid chromatography–tandem mass spectrometry (LC–MS/MS)
LC–QTOF. *See* Liquid chromatography quadrupole time-of-flight mass spectrometry (LC–QTOF)
LC–UV methods for MPA, 323
LDH. *See* Lactate dehydrogenase (LDH)
Lead
 analytical methods and clinical management implications, 417–418
 mechanisms of action, 413–414
 pharmacokinetics and toxicokinetics, 414–415
 treatment, 415
Lead cyanide (Pb(CN)$_2$), 474t
Left ventricular dysfunction (LV dysfunction), 197
Legal marijuana, 180
Leiurus, 443
Leiurus quinquestriatus, 444
Lennox–Gastaut syndrome, 122, 127, 170
Leucovorin, 147, 165
Leukocytosis, 311
Levamisole, 252, 277
Levetiracetam, 121, 123
Levorphanol, 499f
Lewisite, 415
Lidocaine, 227
Lily of valley plant, 359
Limit of detection (LOD), 493
Limit of quantitation (LOQ), 184
Linamarin (C$_{10}$H$_{17}$NO$_6$), 474, 474t
Linezolid, 91
Lipid lowering drugs, 196
Lipid-dense tissues, 414–415
Lipo Kinetix, 362
Lipodystrophy, 199
Liquid chromatography (LC), 191, 214, 337, 360–361
Liquid chromatography quadrupole time-of-flight mass spectrometry (LC–QTOF), 291, 497
 LC–QTOF-based drug screen, 497
Liquid chromatography time of flight mass spectrometry (LC-TOF/MS), 373
Liquid chromatography–mass spectrometry (LC–MS), 22–23, 41, 232, 239, 241, 318
Liquid chromatography–tandem mass spectrometry (LC–MS/MS), 4–5, 19, 95, 105–107, 127, 135, 151, 162–163, 173, 184, 219–220, 220t, 253, 277, 292, 343–344, 351, 455, 456t, 465, 481, 491, 501
 for cyclosporin A, 319–320
 for everolimus, 326
 for MPA, 323
 for sirolimus, 324–325
 for tacrolimus, 321
Liquid chromatography–tandem mass spectrometry (LC–MS/MS), 501
Liquid–liquid alkaline extraction, 465
Liquid–liquid extraction (LLE), 21, 337

Lisdexamfetamine, 311
Lisinopril, 485
Lithium toxicity
 case description, 425
 discussion and follow-up, 425–427
 indications for extracorporeal therapy for lithium elimination, 426t
 laboratory investigations and sessions of hemodialysis, 426t
Liver
 failure, 75–77
 tissue, 135
 toxicity, 360
 transplantation, 77
Lizards, 437
LLE. *See* Liquid–liquid extraction (LLE)
Llipta, 383
Locus coeruleus (LC), 296
LOD. *See* Limit of detection (LOD)
Long-acting barbiturates, 209
Long-acting benzodiazepines, 209
Loperamide, 451–452
LOQ. *See* Limit of quantitation (LOQ)
Lorazepam, 203, 211–212, 215, 219
Lormetazepam, 220
Losartan potassium, 267
Loxosceles reclusa, 441
Loxosceles spp., 441–442
 envenomations, 442
LSD. *See* Lysergic acid diethylamide (LSD)
LV dysfunction. *See* Left ventricular dysfunction (LV dysfunction)
Lysergic acid diethylamide (LSD), 16, 274–275, 284, 295, 298, 313–314
 case descriptions, 313–314

M

mAbs. *See* Monoclonal antibodies (mAbs)
"Magic mushroom", 380–382
Major depressive disorder, 257
6-MAM. *See* 6-Monoacetylmorphine (6-MAM)
Mammalian target of rapamycin inhibitors (mTOR inhibitors), 324–326
 everolimus, 325–326
 sirolimus, 324–325
Mammals, 446
Manchurian Mushroom or Manchurian Fungus tea. *See* Kombucha tea
Mannitol, 227
MAO enzyme. *See* Monoamine oxidase enzyme (MAO enzyme)
MAO-B enzyme, 257
MAOIs. *See* Monoamine oxidase inhibitors (MAOIs)
Marijuana, 184, 191–192, 253–254
Marine species, 445–446
 analytical methods and clinical management implications, 445
 mechanism of action and toxicokinetics, 445
 treatment, 445–446
Mass spectrometry (MS), 4, 37, 172, 214, 337–338, 360–361, 478
Mate de Coca, 383–385

Matrifen, 347–348
Mayo Clinic Arizona Emergency Department, 483
MCV. *See* Mean corpuscular volume (MCV)
MDA. *See* 3,4-Methylenedioxyamphetamine (MDA)
MDMA. *See* 3,4-Methylenedioxymethamphetamine (MDMA)
MDME. *See* Miami-Dade Medical Examiner Department (MDME)
MDPV. *See* Methylenedioxypyrovalerone (MDPV)
ME. *See* Medical examiner (ME)
Meadow (*Spirea* sp.), 67–68
Mean corpuscular volume (MCV), 41
Meconium, 191–192, 215
Medical examiner (ME), 4
Medication compliance in palliative care
 case description, 455
 discussion, 455–457
 results of immunoassay screens and confirmation testing, 456t
Mephedrone, 230–231
Mephobarbital, 210
Meprobamate, 224
6-Mercaptopurine (6-MP), 145
Mercury, 413
 analytical methods and clinical management implications, 418
 forensic considerations, 418
 mechanisms of action, 414
 pharmacokinetics and toxicokinetics, 415
 poisoning from high seafood diet
  case report, 429
  discussion, 430–431
 treatment, 415
Mercury cyanide (Hg(CN)$_2$), 474t
Mescal button, 296–297
Mescaline, 296–297
Meso-2,3-dimercaptosuccinic acid (DMSA), 415–416
Mesobuthus tamulus, 444
Messenger RNA (mRNA), 88
MET. *See* Methamphetamine (MET)
Metabolic acidosis, 44, 59, 81, 388
Metabolites, 335
 cross reactivity, 323
Metal cyanides, 473–474
Metallic mercury. *See* Elemental mercury (Hg^0)
Methadone, 118f
 case description, 117
 cross-reactivity as change in absorbance, 118t
Methadone, 379, 456t
Methamphetamine (MET), 228–229, 247, 253, 257–258, 267–268, 501
 detection comparison in urine and oral fluid
  case description, 501
  discussion, 501–502
D-Methamphetamine, 267–268
L-Methamphetamine, 257
Methanol, 37, 44–45, 60

Methanol (*Continued*)
  and formic acid assay, 44
  inhalation during pregnancy
    case description, 59–61
    initial laboratory findings and treatment, 59t
  pharmacology, 44
  treatment and management of methanol toxicity, 44–45
Methcathinone, 230
Methemoglobin, 469–471, 470t
Methemoglobinemia, 469
  due to dietary nitrate
    blood gas analysis and co-oximeter panel results, 470t
    case follow-up, 471
    case histories, 469
    discussion, 469–471
    signs and symptoms, 470t
Methicillin-resistant Staphylococcus aureus (MRSA), 85
Methionine, 72
Methohexital, 210
Methotrexate (MTX), 141, 145, 151–154, 152t, 153f, 158t, 161, 162t, 166–167
  analytical methods and clinical management implications, 148
  case description, 151, 157
  mechanisms of action, 145
  pharmacokinetics, 146
  treatment, 147
Methoxyacetylfentanyl, 340
Methyl benzoylecgonine. *See* Cocaine
Methyl parathion, 15
Methyl salicylate, 365
1-Methyl-4-phenyl-1,2,3,6-tetrahydropyridine (MPTP), 269
1-Methyl-4-phenyl-4-propionoxypiperidine (MPPP), 269
  synthetic analog, 270f
Methylated arsenic, 422
Methylated barbiturates, 210
Methylation threshold hypothesis, 423
3,4-Methylenedioxyamphetamine (MDA), 277–279
3,4-Methylenedioxymethamphetamine (MDMA), 20, 227–229, 247, 277, 278f, 284, 295
  case description, 277
  postmortem toxicology, 278t
Methylenedioxypyrovalerone (MDPV), 230, 287, 288f
  case description, 287
  postmortem toxicology results, 288t
6-Methylmercaptopurine (6-MMP), 146–147
Methylmercury, 430
Methylone, 230
  case descriptions, 283–284
Methylphenidate, 499
4-Methylpyrazole (4-MP), 44–45, 56–57
6-S-Methylthioinosine-5′-monophosphate (6-MTIMP), 146–147
Methysticin, 370
Metoprolol, 223
Metronidazole, 267
MI. *See* Myocardial infraction (MI)
Miami-Dade Medical Examiner Department (MDME), 284
MIC. *See* Minimum inhibitory concentration (MIC)
Michaelis–Menten equation, 12, 14–15
Microdiffusion method, 39
Microgenics CEDIA for tacrolimus, 321
Midazolam, 209, 307
Mild cerebral edema, 379
MILIS. *See* Multicenter investigation of infarct size (MILIS)
Minimum inhibitory concentration (MIC), 83–84
Mirtazapine, 369
Missing oxycodone metabolites confirm suspected drug diversion
  case presentation, 489
  discussion, 489–491
Mistletoe, 360
Mitogen-activated protein kinases, 144–145
Mitotane, 142
Mitragyna speciosa. *See* Kratom (*Mitragyna speciosa*)
Mitragynine, 373–375
6-MMP. *See* 6-Methylmercaptopurine (6-MMP)
Mollusks, 445
Monkhood. *See Aconitum napellus* (Monkhood)
6-Monoacetylmorphine (6-MAM), 343–344
Monoamine oxidase enzyme (MAO enzyme), 257
Monoamine oxidase inhibitors (MAOIs), 99–100, 103, 105t
Monoclonal antibodies (mAbs), 144, 198
Monohydroxycarbamazepine, 122
Morphine, 335, 339, 343, 455–457
6-MP. *See* 6-Mercaptopurine (6-MP)
4-MP. *See* 4-Methylpyrazole (4-MP)
MPA. *See* Mycophenolic acid (MPA)
MPAG. *See* Glucuronide (MPAG)
MPPP. *See* 1-Methyl-4-phenyl-4-propionoxypiperidine (MPPP)
MPTP. *See* 1-Methyl-4-phenyl-1,2,3,6-tetrahydropyridine (MPTP)
mRNA. *See* Messenger RNA (mRNA)
MRSA. *See* Methicillin-resistant Staphylococcus aureus (MRSA)
MS. *See* Mass spectrometry (MS)
MS-TOF. *See* Quadrupole time-of-flight mass spectrometry (MS-TOF); Time-of-flight mass spectrometry (MS-TOF)
6-MTIMP. *See* 6-S-Methylthioinosine-5′-monophosphate (6-MTIMP)
mTOR inhibitors. *See* Mammalian target of rapamycin inhibitors (mTOR inhibitors)
MTX. *See* Methotrexate (MTX)
Mu OPioid receptor (MOP), 333–334
Mu receptor (μ receptor), 333
Multicenter investigation of infarct size (MILIS), 196
Multidose-activated charcoal, 30
Multiple sclerosis, 466
Mushrooms, 379–380
Mycobacteria, 83
Mycobacterium tuberculosis, 85
Mycophenolate sodium, 322
Mycophenolic acid (MPA), 317, 322–324
  analytical issues, 323–324
  immunoassays, 323
  LC–UV and LC–MS/MS methods for, 323
Myocardial infarction, 418
Myocardial infarction (MI), 196
Myoglobin, 254
Myotoxins, 438
Myriapods, 440
mzCloud database, 23

N

N,N-Dipropyltryptamine (DPT), 311–312
N-acetyl-p-aminophenol, 68
N-acetyl-p-benzoquinone imine (NAPQI), 72, 75
N-acetylcysteine (NAC), 30, 72, 75–77
N-bomb, 297
N-depropylpropafenone, 204–205
N-desmethyl-4-hydroxytamoxifen, 145
N-desmethyl-LSD, 299
N-dimethyltryptamine (DMT), 297, 311
  DMT-associated trauma, 311–312
N-methyl-D-aspartic acid receptor (NMDA receptor), 298–299, 307
N/OFQ. *See* Nociceptin/Orphanin FQ (N/OFQ)
N/OFQ receptor (NOP), 333
Nabiximols, 170
NAC. *See* N-acetylcysteine (NAC)
NAD. *See* Nicotinamide adenine dinucleotide (NAD)
NADH phosphate (NADPH), 471
Naloxone, 55, 340–341, 457
Naloxone-responsive respiratory depression
  case description, 451
  discussion and follow-up, 451–452
  urine drug screen panel, 452t
Naltrexone, 39, 369
NAPQI. *See* N-acetyl-p-benzoquinone imine (NAPQI)
NARCAN formulation, 340–341
Narcotics, 333
National Forensic Laboratory Information System (NFLIS), 348
National Institute of Drug Abuse (NIDA), 452
National Medical Services (NMS), 261
National Poison Data System, 209–210
NBO. *See* Normobaric oxygen (NBO)
NDI. *See* Nephrogenic diabetes insipidus (NDI)
Nebulized sodium bicarbonate, 391
Necrotoxins, 442
Neonatal abstinence syndrome, 374
Neoplastic disease, 144
Neoral, 319
Nephrogenic diabetes insipidus (NDI), 425
Nephrotoxicity, 86
Neprilysin inhibitors, 197
Neurological signs, 388

Neuromuscular symptoms, 392
Neurotoxicity, 85–86, 425–426
Neurotoxins, 438, 442
Neurotransmitters, 442
Neutropenic enterocolitis, 142–144
NF-AT. See Nuclear factor of activated T lymphocytes (NF-AT)
NFLIS. See National Forensic Laboratory Information System (NFLIS)
Nicotinamide adenine dinucleotide (NAD), 39–40, 56, 162–163, 469
Nicotine, 261–263
NIDA. See National Institute of Drug Abuse (NIDA)
Nitric acid, 418
Nitric oxide (NO), 409–410
Nitrite toxicity, 471
Nitrogen mustards, 142
Nitrogen–phosphorus detectors (NPD), 21, 337–338
Nitrous oxide, 295
NMDA receptor. See N-methyl-D-aspartic acid receptor (NMDA receptor)
NMS. See National Medical Services (NMS)
Nociceptin/Orphanin FQ (N/OFQ), 333
Nomogram, 71–72
Nonclinical toxicology, 4
Nonlinear pharmacokinetics, 12–14
Nonsteroidal anti-inflammatory drugs (NSAIDs), 67, 69–70
NOP. See N/OFQ receptor (NOP)
Nordiazepam, 212
Norepinephrine, 370, 442
Norketamine, 305
Normobaric oxygen (NBO), 390, 410, 476
Noroxycodone, 489–491
Novel benzodiazepines, 210
NPD. See Nitrogen–phosphorus detectors (NPD)
NSAIDs. See Nonsteroidal anti-inflammatory drugs (NSAIDs)
Nuclear factor of activated T lymphocytes (NF-AT), 318
Nucleic acid synthesis, inhibition of, 89
NUDT15, 147

O

O-phosphoryl-4-hydroxy-N, N-dimethyltryptamine. See Psilocybin
Occupational Safety and Health Administration, 476
Olanzapine, 115
  case description, 109–111
  toxicity in infant, 109, 110t
Oleander, 359, 363
Oleandrin, 363
OLS. See One leg stand (OLS)
One leg stand (OLS), 177
Opiates, 333, 343, 379, 456t, 502
Opioid use disorder (OUD), 334
Opioids, 38–39, 333, 373, 456t, 497, 499
  absorption, 334–335
  abuse, 311
  and dependence, 358
  analytical methods
    confirmation, 338
    interpretation, 338–339
    sample preparation, 337
    screening, 337–338
  buprenorphine, 341
  fentanyl and fentanyl analogs, 340
  heroin, 339–340
  mechanism of action, 333–334
  metabolism, 352–353
    case description, 351
  morphine, 339
  naloxone, 340–341
  opioid use disorder and treatment, 335–337
  pharmacodynamics, 334–335, 336t
  pharmacokinetics, 334–335, 336t
  withdrawal syndrome, 334
Opium poppy (Papaver somniferum L.), 333
Optical emission spectroscopy, 416
Oral bioavailability, 339
Oral fluid, 173, 215
  ethanol, 39, 41
  methamphetamine detection in
    case description, 501
    discussion, 501–502
    testing, 233–234
Organic arsenic, 422
Organic mercury, 414–415, 430
ORL-1. See Nociceptin/Orphanin FQ (N/OFQ)
Osmolal gap, 44–45, 64
Osteosarcoma, 157–159, 159t
OUD. See Opioid use disorder (OUD)
Over-the-counter (OTC), 21–22, 67
Overdose, 80–81, 241
  patient management, 27
    evaluation, 27–29
    treatment, 29–31
  toxicokinetics and clinical management of, 299–300
Oxalic acid, 64
Oxazepam, 212
Oxcarbazepine, 79, 121–123
  antibiotics, 85
  overdose, 139–140
    case description, 139
  structures of, 140f
2-OXO-3-OH-LSD, 299
2-OXO-LSD, 299
Oxycodone, 20, 356, 455–457, 485, 490, 490f
  immunoassays, 491
  metabolites
    case presentation, 489
    discussion, 489–491
Oxygen saturation (sO_2), 470–471
Oxymorphone, 352, 489, 491

P

Palliative care, 455–457
Panaeolina mushroom, 379–380
Panaeolus, 297, 379–380
Papaver somniferum L. See Opium poppy (Papaver somniferum L.)
Paroxysmal atrial fibrillation, 204–205
Partial pressure of oxygen (pO_2), 470–471
Partial seizures, 121
Particle-enhanced turbidimetric inhibition immunoassay, 94–95
Patella vulgata, 23–24
Pathway-targeted therapies, 144–145
Pavilion for Women at Texas Children's Hospital (PFWTCH), 247
PCP. See Phencyclidine (PCP)
PD. See Pharmacodynamics (PD)
PecFent, 347–348
Pediatric intensive care unit (PICU), 203
Pediatric ketamine toxicity, 306
PEG. See Polyethylene glycol (PEG)
Penicillamine, 415
Penicillium species, 322
Pennyroyal, 360, 362
Pentobarbital, 209
2-Pentyl-3,4,5,6-tetrahydropyridine, 366
2-Pentylpiperidine, 366
Perampanel, 121, 123
Perinatal respiratory distress, 469
Petasites hybridus. See Butterbur, common (Petasites hybridus)
Peyote buttons, 296–297
Peyote cactus, 296–297
PFWTCH. See Pavilion for Women at Texas Children's Hospital (PFWTCH)
PG. See Propylene glycol (PG)
P-glycoprotein, 374
Pharmacodynamics (PD), 7, 14–15
  of opioids, 334–335, 336t
Pharmacokinetics (PK), 7–12, 89–93, 102–105, 170–172
  of cyanide toxicity, 475
  of opioids, 334–335, 336t
Phenacetin, 68
Phenazepam, 118f
Phencyclidine (PCP), 117, 275, 295, 299, 379
  PCP in utero, 191–192
Phenethylamines, 295–296
Phenobarbital, 121–122, 126, 209, 211, 213
Phenothiazines, 224
Phentolamine, 243, 245
Phenylacetone, 229–230
Phenylboronic acid, 47
Phenylpropanolamine, 362
Phenytoin, 13, 65, 121–122, 131–132
Phoneutria, 441, 443
Phosgene, 394–395
  acute toxicity, 394
  chronic toxicity, 394
  clinical management, 395
  exposure routes and pathways, 394
  toxicokinetics and mechanism of action, 394
Phosphatidylethanol, 41–42
Phosphatidylinositol-3-kinases/protein kinase B (PI3K/AKT), 144–145
Phospholipases, 442–443
Physostigmine, 366
PI3K/AKT. See Phosphatidylinositol-3-kinases/protein kinase B (PI3K/AKT)
PICU. See Pediatric intensive care unit (PICU)

Pill shaving, 457
Pink. *See* U-47700
"Pins and needles" sensation, 434
"Pinzor", 291, 292f
Piper methysticum, 369–370
PK. *See* Pharmacokinetics (PK)
Plasma ammonia levels, 397
Plasma exchange, 80
Pluteus, 379–380
Pluteus mushroom, 379–380
Point-of-care (POC), 497, 499
Poisonous plants, 365–366
Polyangiitis, 252
Polyatomic ions, 416
Polyethylene glycol (PEG), 30
Polypharmacy, 352
  ingestion, 203–204
Polypropylene glycols, 46
Polysubstance abuse, 254
  case description, 253
  patient follow-up, 253–254
Posaconazole, 91–92
Positron emission tomography (PET), 151
Postmortem
  chest blood, 135
  examination, 113, 114t
  toxicology, 114, 406
    testing, 283
Postsynaptic stimulation, 265
Postsynaptic toxins, 438
Posttraumatic stress disorder (PTSD), 253
Potassium chloride, 223
Potassium cyanide (KCN), 473–474, 474t
Potassium ferricyanide ($K_3Fe(CN)_6$), 474t
Potassium ferrocyanide ($K_4Fe(CN)_6$), 474t
Potassium oxalate, 39
Potassium silver cyanide ($KAg(CN)_2$), 474t
Potency, 14
Potentiometric methods, 477–478
Preclinical toxicology, 4
Prednisone, 145
Pregabalin, 121–123
Pressure of oxygen (pO_2), 470–471
Primidone, 121–122, 213
Propafenone, 203–204
  elimination, 206
  postmortem blood level, 205
  toxicity, 205
1,2-Propanediol, 46
1,3-Propanediol, 46, 63–64
  chromatograph, 64f
1-Propanol, 40
Propoxyphene, 379
Propranolol, 369
Propylene glycol (PG), 46, 63–64
  chromatograph, 64f
  poisoning, 65
Prostaglandin-G2, 70–71
Prostaglandins, 70
Protein synthesis inhibition, 88
PRPP. *See* Alpha-phosphoribosylpyrophosphoric acid (PRPP)
Pseudomonas aeruginosa, 87
Psilocin, 297, 311–312, 379–382
  case history, 379
  postmortem toxicology results, 381t
Psilocybe, 297, 379–380
Psilocybe semilanceata, 380
Psilocybin, 295–297, 379–380, 381f, 382
  fatalities, 382
  metabolism, 381f
Psychedelics, 295–298
  analysis, 300–301
Psychosis, 100–101
Psychotria viridis, 297, 311–312
PTSD. *See* Posttraumatic stress disorder (PTSD)
Pulmonary congestion, 379
Pulmonary edema, 261
Pulse oximeters, 471
Purple pill, 481
Pyrrolizidine alkaloids, 362

Q

QMS. *See* Quantitative microsphere system (QMS)
QQQ-ICP-MS. *See* Triple quadrupole ICP-MS (QQQ-ICP-MS)
QT interval prolongation, 197
QTOF. *See* Quadrupole-TOF (QTOF)
Quadrupole, 22
Quadrupole time-of-flight mass spectrometry (MS-TOF), 493
Quadrupole-TOF (QTOF), 21–22
Quantitative microsphere system (QMS), 326
Quetiapine, 79, 369
Quinine, 68

R

R-MVP. *See* Rituximab, MTX, Procarbazine (R-MVP)
Radiation therapy, 198
Radiative Energy Attenuation technology (REA technology), 39–40
Rage, erythema, dilated pupils, delusions, amnesia, nystagmus, excitation, and skin drying (RED DANES), 300
REA technology. *See* Radiative Energy Attenuation technology (REA technology)
Recreational drug users, 210
Red blood cells (RBCs), 443
Red fire ant (*Solenopsis invicta*), 444
Red flower oil, 365
Reference interval (RI), 253
Respiratory depression, 334
Reverse transport, 229
Rhabdomyolysis, 80, 254–255, 311–312
  associated with laboratory-confirmed FUB-AMB use, 281–282
  case description, 253
  case description, 281
  patient follow-up, 253–254
Rhodanese, 475
RI. *See* Reference interval (RI)
Ribonucleic acid (RNA), 88
Ricin, 366
Ricinus communis, 366
Right upper extremity (RUE), 243
Right upper quadrant (RUQ), 75
Risperidone, 224, 295–296
Rituximab, MTX, Procarbazine (R-MVP), 151
RNA. *See* Ribonucleic acid (RNA)
RUE. *See* Right upper extremity (RUE)
Rufinamide, 121–122
Rumack–Matthew nomogram, 72, 75–76, 76f
Russula subnigricans, 380–382

S

Salicin, 67–68, 67f
Salicylates, 67–69, 71–72
Salicylic acid, 67–68, 68f, 70, 365
Saliva collection, 233
Salivation, lacrimation, urination, diarrhea, GI upset, and emesis toxidrome (SLUDGE toxidrome), 421
Salix alba. *See* White willow (*Salix alba*)
SAMHSA. *See* Substance Abuse Mental Health Service Administration (SAMHSA)
Sativex, 170
Schizophrenia, 100–101
SCN1A (voltage-gated sodium channel), 15
Scopolamine, 366
Scorpions, 441
SCr. *See* Serum creatinine (SCr)
SCs. *See* Synthetic cannabinoids (SCs)
Second-generation drugs, 121
Sedation, 230
Seizures, 121
Selective ion monitoring mode (SIM mode), 21–22
Selective reaction monitoring (SRM), 105–107
Selective serotonin reuptake inhibitors (SSRIs), 99–100, 102, 352
Selegiline, 257, 258f
Senecio vernalis. *See* Eastern groundsel (*Senecio vernalis*)
Serological testing, 251–252
Serotonergic psychedelics, 295–296
Serotonin modulators (SRMs), 99–100
Serotonin receptors, 295–296
Serotonin toxicity, 28–29
Serotonin–norepinephrine reuptake inhibitors (SNRIs), 99–100, 102
Serratia marcescens, 87
Sertraline, 224
Serum creatinine (SCr), 151, 152t
Serum osmoles, 60
Serum pH levels, 388
Serum-sickness, 440
Sesquiterpenoids, 362
SFSTs. *See* Standard field sobriety tests (SFSTs)
Sheiner–Tozer equation, 132–133
Short-acting barbiturates, 209
Short-acting benzodiazepines, 209
Short-acting beta-blockers, 241
Sigmoid E_{max} model, 14–15

Signal transducer and activator of transcription (STAT), 144–145
Silymarin, 360
SIM mode. *See* Selective ion monitoring mode (SIM mode)
Simulated compliance, 457
Single quadrupole detector, 21–22
Sirolimus, 317, 324–325
　analytical issues, 325
　immunoassays, 324
　LC–MS/MS methods for, 324–325
Sleeping pills, 261
SLUDGE toxidrome. *See* Salivation, lacrimation, urination, diarrhea, GI upset, and emesis toxidrome (SLUDGE toxidrome)
Small molecules, 144
Smoking cannabis leaves, 170–171
Snail (*Helix pomatia*), 23–24, 173
Snake venom, 437–438
Snake Venom Detection Kit (SVDK), 439
Snakebite envenoming, 437
Snakes, 437
SNP. *See* Sodium nitroprusside (SNP)
SNRIs. *See* Serotonin–norepinephrine reuptake inhibitors (SNRIs)
Sodium 2,3-dimercapto-1-propane sulfonate, 415
Sodium cyanide (NaCN), 474t
Sodium ferrocyanide ($Na_4Fe(CN)_6$), 474t
Sodium fluoride, 39
Sodium nitrate, 471
Sodium nitrite ($NaNO_2$), 471, 476
Sodium nitroprusside (SNP), 475
Sodium nitroprusside ($Na_2(Fe(CN)_5NO)$), 474t
Sodium thiosulfate ($Na_2S_2O_3$), 476
Solenopsis invicta. *See* Red fire ant (*Solenopsis invicta*)
Solid phase extraction (SPE), 21, 337
SPE. *See* Solid phase extraction (SPE)
Speciation, 417
Specimen preparation for chromatographic analysis, 23–24
Spectrophotometric method, 39
"Speedball" (Heroin and cocaine combination), 339–340
Spider venoms, 442
　historical evolution on knowledge, 442f
Spiders, 441
　medically significant, 441t
Spirea sp. *See* Meadow (*Spirea* sp.)
SRM. *See* Selective reaction monitoring (SRM)
SRMs. *See* Serotonin modulators (SRMs)
S-shaped lens, 22
SSRIs. *See* Selective serotonin reuptake inhibitors (SSRIs)
Standard field sobriety tests (SFSTs), 177
STAT. *See* Signal transducer and activator of transcription (STAT)
Statins, 199
Stereoisomers, 499f
Streptococcus pneumoniae, 85
Streptomyces tsukubaensis, 321
Stropharia, 297, 379–380
Subcutaneous (SC) inoculation, 438

Sublimaze, 347–348
Suboxone, 457
Substance Abuse Mental Health Service Administration (SAMHSA), 19–20, 227
Subsys, 347–348
Succimer. *See* Meso-2,3-dimercaptosuccinic acid (DMSA)
SVDK. *See* Snake Venom Detection Kit (SVDK)
Sympathomimetic toxidrome, 28
Sympathomimetics, 243
Symphytum officinale. *See* Comfrey (*Symphytum officinale*)
Synergistic toxicity, 28
Synthetic cannabinoids (SCs), 178–181, 271–272, 281
　THC and select group of, 271f
Synthetic cathinones, 230–231, 287–289
Syva EMIT 2000, 321
Syva EMIT II, 258
Syva EMIT II Plus, 490
　opiate immunoassay screen, 489

T

Tacrolimus, 317–318, 321–322
　analytical issues, 321–322
　immunoassays, 321
　LC–MS/MS methods for, 321
Tamoxifen, 145
Tamsulosin, 223
Tandem mass spectrometry, 191, 214, 338
Targeted analysis methods, 24
TCAs. *See* Tricyclic antidepressants (TCAs)
TDM. *See* Therapeutic drug monitoring (TDM)
TDP. *See* Torsades de pointes (TDP)
Temazepam, 212
Terrestrial snakes, 437–440
　analytical methods and clinical management implications, 439
　emergency medicine care of crotaline envenomations, 440t
　local, systemic, hematologic, and neurologic signs and symptoms of snakebite envenomation, 439t
　mechanism of action and toxicokinetics, 437–439
　treatment, 440
Tetrahydrocannabinol (THC), 169, 171, 178, 183, 187–188, 191–192, 253
THC-COOH, 184
Tetrahydrofolate (THF), 166
Tetrahydrofolic acid (FH_4), 145
Tetrasialotransferrin, 41
6-TG. *See* 6-Thioguanine (6-TG)
6-TGMP. *See* 6-Thioguanosine monophosphate (6-TGMP)
TGNs. *See* Thioguanine nucleotides (TGNs)
6-TGTP. *See* 6-Thioguanosine triphosphate (6-TGTP)
THC. *See* Tetrahydrocannabinol (THC)
Therapeutic drug monitoring (TDM), 86, 93–94, 121–122, 131–133, 141, 201, 317, 323

　of cardiac drugs, 201
Therapeutic plasma exchange (TPE), 81–82, 203–206
THF. *See* Tetrahydrofolate (THF)
Thiazolidinediones (TZDs), 199
Thin layer chromatography, 20
Thiobarbiturates, 210
Thiocyanate, 475–476
Thioguanine nucleotides (TGNs), 145
6-Thioguanine (6-TG), 145–147
6-Thioguanosine monophosphate (6-TGMP), 146–147
6-Thioguanosine triphosphate (6-TGTP), 146–147
6-Thioinosine 5′-monophosphate (6-TIMP), 146–147
Thiopental, 209–210
Thiopurine analogs, 141–142, 145
　analytical methods and clinical management implications, 148
　mechanisms of action, 145
　pharmacokinetics, 146–147
　treatment, 147–148
Thiopurine S-methyltransferase (TPMT), 146–147
Third-generation drugs, 121
Thunder God Vine, 364–365
Thyroid stimulating hormone (TSH), 291, 433
Tiagabine, 121–122, 126
Ticks, 441–444
　analytical methods and clinical management implications, 443
　mechanism of action and toxicokinetics, 442–443
　medically Significant Spiders, 441t
　treatment, 443–444
Time-of-flight (TOF), 21–22, 270
Time-of-flight mass spectrometry (MS-TOF), 493
Time-of-light mass spectrometry (TOFMS), 127
6-TIMP. *See* 6-Thioinosine 5′-monophosphate (6-TIMP)
Tizanidine, 466
Tizanidine intoxication
　case description, 465–466
　discussion, 466–467
TNF, 266
Tobacco, 254
Tobramycin, 93
TOF. *See* Time-of-flight (TOF)
TOFMS. *See* Time-of-light mass spectrometry (TOFMS)
Tolerance, 334
Topiramate, 121, 123
Topoisomerases, 142
Torsades de pointes (TDP), 85
Torsemide, 223
Toss-and-wash method, 373
Total phenytoin measurements
　case description, 131
　laboratory results, 132t
Total test, 337, 339
TOX/SEE urine drug screen cartridge, 247–248

Toxic gases, 387, 389t
　acute toxicity, 388
　air concentrations of, 389t
　carbon monoxide (CO), 387–390
　chlorine, 390–391
　chronic toxicity, 388
　clinical management, 390
　diagnosis and investigation, 388–389
　exposure routes and pathways, 387–388
　hydrogen cyanide, 391–392
　hydrogen sulfide (H_2S), 393–394
　mechanism of action, 388
　phosgene, 394–395
Toxic herbals and plants in United States, 359, 365
　active ingredients, 361t
　cardiotoxic herbs, 363–365
　herbs and liver toxicity, 360–363
　herbs with nephrotoxicity, 365
　poisonous plants, 365–366
　popularity and safety of herbal supplements, 359–360
　supplements, 360
Toxic metals
　arsenic, 413
　　analytical methods and clinical management implications, 416–417
　　forensic considerations, 418
　　mechanisms of action, 413
　　pharmacokinetics and toxicokinetics, 414
　　treatment, 415
　lead
　　analytical methods and clinical management implications, 417–418
　　mechanisms of action, 413–414
　　pharmacokinetics and toxicokinetics, 414–415
　　treatment, 415
　mercury, 413
　　analytical methods and clinical management implications, 418
　　forensic considerations, 418
　　mechanisms of action, 414
　　pharmacokinetics and toxicokinetics, 415
　　treatment, 415
Toxicity, antibiotics and, 84–85
Toxicodynamics, 15–16
Toxicokinetics, 12–14, 89–93, 102–105, 170–172
　of cyanide toxicity, 475
Toxicology, 114–115, 406–407
　laboratories, 19
　　alternate specimen types, 24–25
　　methods for general screening, 19–24
　　methods for targeted analysis, 24
　　use of clinical information, 25
　laboratory testing, 180
　reports, 177
　testing, 401
Toxi-Lab system, 20
Toxin-induced cardiogenic shock, 27
TPE. See Therapeutic plasma exchange (TPE)
TPMT. See Thiopurine S-methyltransferase (TPMT)
Tramadol, 115, 456t
Transdermal patches, 261–263

Transesterification, 265
Transfer RNA (tRNA), 88
Transferrin, 41
Trastuzumab, 144, 198
Trauma, 254
Triazolam, 209
Tricyclic antidepressants (TCAs), 99–100, 104t, 482–483
　overdose, 102
Triglyceride gap, 47
1,3,7-Trimethylxanthine. See Caffeine
Triple quadrupole ICP-MS (QQQ-ICP-MS), 417
Tripterygium wilfordii, 364–365
Trivalent arsenic, 434
tRNA. See Transfer RNA (tRNA)
Tryptamines, 295–297, 311–312
　trauma, 311–312
Tryptophan, 295–296
Tussilago farfara. See Coltsfoot (*Tussilago farfara*)
Tussilagone, 362
Tyrosine kinase inhibitor (TKI) therapy, 144, 198
TZDs. See Thiazolidinediones (TZDs)

U

U-47700
　case description, 355, 356f
　discussion, 355–358
UDS. See Urine immunoassay drug screen (UDS)
UDT. See Urine drug testing (UDT)
UGT2B7, 490–491
Ultrashort-acting barbiturates, 209
Ultraviolet (UV) spectrophotometry, 105–107
Uncompetitive inhibitors, 12–14
Uncontrolled type 2 diabetes, 219
United States Poison Control Centers (USPCC), 437, 441
Unithiol. See Sodium 2,3-dimercapto-1-propane sulfonate
Unknown pill ingestion, 482–483
Urinalysis, 439
Urinary excretion of drugs, 502
Urine, 422, 431
　methamphetamine detection in
　　case description, 501
　　discussion, 501–502
　test, 383
Urine benzodiazepine immunoassay screening, false-negative results in
　case description, 219
　discussion and follow-up, 219–221
Urine drug screen, 219, 248
　results after spiking with known ephedrine, 248t
　test, 257
Urine drug testing (UDT), 343, 344t, 351, 369, 455–457
Urine immunoassay drug screen (UDS), 29
US Food and Drug Administration, 476
Usnic acid, 362
USPCC. See United States Poison Control Centers (USPCC)

V

Valproic acid, 121–122, 126
Vancomycin, 88, 92–93
Vancomycin-resistant enterococci, 85
Vasculitic syndromes, 252
Venlafaxine, 499
Venoms, 437–438
　arthropods, 440–445
　　insects, 444–445
　mammals, 446
　marine species, 445–446
　of snakes, 3
　terrestrial snakes, 437–440
Veno-occlusive disease, 362
Venous blood gas, 409, 469, 470t, 482t
Vespidae, 444
Vigabatrin, 121, 123, 126
Viperidae, 437
Vitamin B_{12}. See Cyanocobalamin
Vitamin B_{17}. See Laetrile
Vitamin supplements, 359–360
Vitreous fluid ethanol, 41
Vitros 5600 analyzer, 253
Volatile alcohols, 37
Volatile screening, 261
Volume of distribution, 8
Voriconazole, 91

W

Walk and turn (WAT), 177
Warfarin, 223
Wellbutrin, 482–483
White flower oil, 365
White willow (*Salix alba*), 67–68
WHO. See World Health Organization (WHO)
Whole bowel irrigation technique, 30
Widow antivenom, 443
Winter–Tozer equation (WT equation), 132–133
World Health Organization (WHO), 359, 437
WT equation. See Winter–Tozer equation (WT equation)

X

Xanax, 253

Y

Yangonin, 370
Yeast fermentation, 37
Yohimbine, 365

Z

Zidovudine, 196
Zinc cyanide ($Zn(CN)_2$), 473–474, 474t
Zolpidem, 223–224
　case history, 223
　metabolism, 224f
Zonisamide, 121, 123, 126
ZURIT, 383
Zyban, 482–483

Printed in the United States
By Bookmasters